PRECALCULUS MATHEMATICS

Flanders/Price Series 2nd Edition

Harley Flanders
Florida Atlantic University
and
Justin J. Price
Purdue University

SAUNDERS COLLEGE PUBLISHING
Philadelphia

SAUNDERS COLLEGE PUBLISHING/CBS
Educational and Professional
Publishing, A Division of CBS Inc.
Address Orders to: 383 Madison Ave.,
New York, NY 10017
Address Editorial correspondence to:
West Washington Square
Philadelphia, PA 19105

LIBRARY OF CONGRESS
CATALOG CARD NO.: 80-53934

Flanders, Harley, & Justin J. Price
Precalculus.
Philadelphia, Pa. : Saunders College
576 p.
8101 801010

PRECALCULUS MATHEMATICS ISBN 03 057723 3

1234 147 98765432

CBS COLLEGE PUBLISHING

PREFACE

This book is a natural development from its first edition (titled *Elementary Functions and Analytic Geometry*) and from our subsequent Algebra and Trigonometry series. Our objective is to teach the essentials of algebra, trigonometry, and analytic geometry, with applications. We hope to provide students with a solid working knowledge, which they will be able to apply in other courses and in their occupations. To this end we emphasize practical skills, problem solving, and computational techniques. We always try to justify theory by down-to-earth applications.

FEATURES OF THIS EDITION

We have rewritten about 85% of the text, mostly to improve its teachability, and have added many realistic applications.

Examples

The examples form the most important part of the text. Through them the student learns how to do mathematics himself rather than be told about it. There are many more informal examples (set off with the heading *Example*) than in the first edition, and the number of formal (numbered) worked examples has been increased 233%, from 105 to 350. A particular effort has been made to illustrate every type of exercise.

Exercises and Answers

The number of exercises has been increased 107%, from 2005 to 4152. Almost all of this increase has been at the easy and middle levels of difficulty. As always we try to make the exercises interesting, and to make the applications realistic and relevant.

While the production of the book was in the galley proof stage, we ourselves wrote the Solutions Manual (available to instructors), which includes a solution for every exercise in the text. In this way the exercise sections themselves were thoroughly shaken out, and should be error-free. An accuracy reviewer independently checked all solutions, so the *Solutions Manual* is guaranteed reliable, as is the *Answer Section* for odd-numbered exercises (printed in the back of the book and including all graphs).

Each chapter of the book closes with a section of about 20 review exercises.

Figures

We have taken great pains to have the art work simple, uncluttered, and clear. Our goal was figures that students can read easily and reproduce themselves, and that illustrate what they are supposed to. In the text alone there are 38% more illustrations than in the first edition.

Computations

The revolution in semiconductor technology has made inexpensive scientific hand-held calculators of immense power available to everyone. This is necessarily changing the teaching of school and college mathematics at every level and is causing a rethinking of many traditional topics. Obviously hand calculations using log and trig tables are obsolete; it is a waste of a student's time to drill him on skills that he will never use in life.

Our 1975 algebra and trigonometry series were the very first precalculus books to introduce calculators. The material was appreciated, and followed more or less by subsequent authors. We expect our matter-of-fact use of calculators in this edition to set the standard.

Since students often do not know the capabilities of their calculators, we instruct as we go along, trying to be fair to left-to-right versus algebraic logic. We slightly favor left-to-right because we like it better and because most four-function calculators have this logic, so the student is more likely to be accustomed to chain calculations than to full algebraic hierarchy.

For students who have only a four-function calculator, we include primitive log and trig tables.

This is not a "calculator precalculus" any more than a book written 15 years ago was a "log table precalculus". Calculations then were done by log tables, and calculations now are done by calculator; only the tool has changed. However—and this is important—calculations are now so easy that we can use concrete numerical evidence and experimentation freely in a way no one imagined possible 15 years ago. We believe numerical evidence should be used to reinforce theoretical discussions when possible. For instance the theoretical discussion on the existence of a root of a real cubic is certainly helped by a few keystrokes leading to 1.324718 as a 6-place approximation to the root of $x^3 - x - 1 = 0$. On the other hand, we do not force the calculator into every situation imaginable. This sort of thing was tried with "computerized calculus" a dozen years ago and was a disaster. As we have said, this is definitely not a "calculator precalculus course", nor is it a "baby programming" course.

Design

The design of this book, evolved from four previous designs of our textbooks, has two main purposes. First, the student always knows precisely where he is. For instance, important statements are boxed. Worked examples begin with a clear statement of the *problem*. Then the *solution* is set off. Finally, the *answer* is stated. The student knows the example has ended because there is a line down the left side from its beginning to its end. It is important for the student to do his own work this way: state the problem clearly, solve it, state the answer that is called for in the problem.

In the text, *italics* are used for stress. Defined terms are in **boldface,** so the student can spot these terms easily. In general, the design is open while not wasteful of space.

The other main purpose of the design is to make mathematical expressions and formulas easy to read. This involves much detail on the horizontal and vertical spacing of the math. We have replaced most of the English punctuation marks (which often obscure math) by large spaces. Whenever in the body of a paragraph a math symbol is next to a word, the space separation is 50% more than the separation between words. We hope this quality feature (done over the printer's dead body!) will set a new standard. Careful scrutiny will show other details, such as extra spaces between math symbols and punctuation marks.

CONTENTS

The book splits naturally into five main parts.

Fundamentals of Algebra

In Chapter 1, the standard topics from first year algebra are reviewed: real numbers, operations, exponents and radicals, polynomials, factoring, and rational expressions. The emphasis is on drill in the use of the rules of elementary algebra. There is no set theory (which we feel is inappropriate and distracts from developing skill and accuracy). There is little work on foundations and the names of the rules. *Features:* (1) A discussion of parentheses, priorities of operations, and calculator priorities. (2) A section on common algebra errors.

Chapter 2 covers linear and quadratic equations, pairs of linear equations, radical and other types of equations, applications, inequalities, and absolute value. Much of this will be review to most students, but the material on order may be new to many, and the chapter should be appreciated by students who may be shaky in their previous mastery of the equation material. *Features:* (1) A section on how to set up word problems. There are no solutions in this section. The idea is to isolate and deal systematically with something most students dread. Why? Because usually no one shows them how to do it. (2) A careful discussion of equations versus identities. (3) An unusual selection of applications—carefully distinguished from the usual algebra puzzles (How old is John?) that are sometimes passed off as "applications". (4) A calculator routine for solving a quadratic, just in case it comes from real life and the coefficients are not small integers.

Functions

This is the core of the volume. The main tool for understanding a function is its graph, and there is great emphasis on graphing.

Chapter 3 covers functions from a real interval to the reals. (Almost nothing is said about abstract functions from sets to sets—a topic for another course.) The material includes plane coordinates, domains, graphs, composite functions, linear and quadratic functions, and graphs of relations—equations and inequalities. *Features:* (1) Much practical work on domains and graphs. (2) Black box interpretation of functions. (3) Much use of translating graphs. (4) Graphing inequalities and systems of inequalities. (5) Some unusual exercises on control systems with feedback.

Chapter 4 covers polynomial and rational functions. There is much emphasis on key features that fix the graph: symmetries, zeros, asymptotes. Features: (1) A section on "tips" in graphing. (2) Material on graphing polynomials and rational functions that can be completely factored into linear factors.

Chapter 5 includes exponential functions, inverse functions, logarithm functions, applications, and how to calculate. Calculation is included here because, traditionally, computation was done via log tables. Our emphasis of course is on computing on a calculator. However, there is material on reading tables and on interpolation. The main emphasis is on the functions, their graphs, and their applications. *Features:* (1) An unusually rich selection of real life applications involving exponential growth and decay, and exponential and logarithm functions in other ways. (2) A careful discussion of inverse functions and their graphs. (2) Casual use of calculators throughout for function values, making otherwise abstract notions very concrete. (4) Optional material on rates of growth of exponential and logarithm functions versus each other and other functions. (5) Considerable material on general calculator "know how" that is useful on all calculators, such as use of memory, hidden digits, constant factors, and calculator accuracy. (6) General numerical "know how" on accuracy, round-off, significant digits, relative error, etc. (7) Optional calculator material on polynomial evaluation and root finding by an efficient combination of 10-section and interpolation—very efficient on equations like $e^x = 2 - x^3$. (8) An optional section on business calculations that is essentially a mini-course on the math of finance: compound interest, annuities, sinking funds, payment of debts. Some tandem problems are included that fully exploit the calculator.

Chapter 6 introduces the trigonometric functions. We begin with angles in the domain $0° \leq \theta \leq 360°$ in standard position and define $\sin\theta = y/r$, etc. As quickly as possible we compute the functions of 45°, 30°, etc. and plot rough graphs. This occupies the first two sections; in the third we introduce radian measure and $\sin\theta$, etc. for any real θ, essentially by periodicity. Thus quickly, without fuss and philosophy, we have the functions and are able to work with them. *Features:* (1) One can go directly from Section 2 to Chapter 8 (triangle solving) if practical work is desirable before the bulk of this chapter and all of Chapter 7. (2) Approximations like $\cos\theta \approx 1 - \frac{1}{2}\theta^2$ near 0 are discussed with calculator experiments. (3) There is a discussion of periodic functions in general with some non-trigonometric examples. (4) No universal covering spaces (wrapping function).

Chapter 7 covers trigonometric identities, the inverse trigonometric functions, and trigonometric equations. The trig identities material is standard—it is certainly hard to say anything new here. We do claim some restraint—there are hundreds, not thousands of exercises—consistent with the relative importance of the topic. (It is possible to go overboard.) The inverse trigonometric functions are introduced and studied via their graphs. There are two sections on trigonometric equations, one standard, the other (optional) on equations like $\sin\theta = \theta^2$ that require approximation methods. *Feature:* An optional purely descriptive section on how trigonometric identities are important in studying vibrations. Beats and audio modulation are discussed.

Triangles

This part consists of Chapter 8 only. It contains the usual material on solving right triangles, the law of sines, and the law of cosines. *Features:* (1) Discussion of degrees, minutes and seconds, but almost all work in decimal degrees (which fit a calculator). (2) Numerous interesting and practical applications, far more variety than usual, and far more realism. (3) Flowcharts for the law of sines. (4) A neat proof of the law of sines via area. (5) An optional section on the tools of practical distance and angle measurement, and their accuracy. Problems on sensitivity. (6) The chapter can be taken immediately after Chapter 6, Section 2.

Further Algebra Topics

Chapter 9 contains two independent topics. The first four sections consist of introductory linear algebra. The main goal is to teach quickly certain basic skills that are needed in many subsequent courses: (1) Solving linear systems by elimination, including recognition of inconsistent and underdetermined systems. (2) Working systems in matrix form, multiplying and inverting matrices. (3) Working with determinants and Cramer's rule.

The rest of the chapter is a unit on polynomials, their zeros, long division, and partial fractions.

Chapter 10 covers sequences, summation, induction, permutations, combinations, and the binomial theorem. *Features:* (1) "Telescoping" as a method for finding sums. (2) An unusual and interesting set of induction exercises, most of which are *not* proving summation formulas. (3) Inductive definitions.

Chapter 14 covers complex numbers: complex arithmetic, the complex plane, polar form, zeros of polynomials, and De Moivre's theorem.

Analytic Geometry

Chapter 11 contains the linear part of plane analytic geometry. The topics are vectors, addition and multiplication by scalars, lines, parametric form, inner product, and normal form. We believe that vectors are the natural tool for linear geometry and that every calculus student will learn vectors eventually; it is a waste of his time to do this material twice, first coordinatewise, then vectorially.

Chapter 12 contains circles, tangents, loci, polar coordinates, conics, and rotation of axes. Features: (1) The equation of the ellipse is first derived in polar form, which is easily changed to rectangular form. This avoids the usual mess with radicals. (2) Rotation of axes is first given in polar form, from which rectangular form easily follows.

Chapter 13 is a brief introduction to coordinates and vectors in space, lines and planes, and the quadric surfaces in standard position. There is much to be said for a precalculus introduction to space geometry, particularly since there is usually none in high school mathematics. Early in calculus, before any formal work in space geometry, the student is expected to draw and visualize space figures (e.g., volumes of revolution, forces on dams). This is where he often has great difficulty because of his lack of any previous work in space geometry.

ACKNOWLEDGMENTS

We are grateful to Stanley Lukawecki (Clemson University), Ken Seydel (Skyline College), and the late Darla Stewart (Catonsville Community College), who read the first edition critically and made many suggestions. We are also grateful to Thomas Butts (Ohio State University), Mark Hale (University of Florida), and Douglas Hall (Michigan State University), who reviewed the manuscript of this edition and supplied much useful criticism. Philip Gillett (University of Wisconsin) checked the galley proofs and the solution manuscript for mathematical accuracy, and we owe him a great debt for doing a most careful and painstaking task.

Mark Glucksman (El Camino College) has written an excellent study guide to accompany this text, including lists of concepts and procedures, additional examples and exercises, and chapter review tests. He has also, in the process, contributed to the accuracy of this book.

The staff at Saunders College Publishing has been most helpful and professional, particularly editors William Karjane and Jay Freedman and artist Linda Savalli. Finally, we thank our typist, Kathleen Meyer, for her fine and speedy work.

REWARD

In spite of all efforts to root out errors, some may still remain. But we believe these are so few that we safely offer for its first detection in a book of this series \$5 per typo and \$10 per serious error to students and staff at institutions using this book.

Harley Flanders
Justin J. Price

TABLE OF CONTENTS

1 BASIC ALGEBRA

2 EQUATIONS AND INEQUALITIES

3 FUNCTIONS AND GRAPHS

4 POLYNOMIAL AND RATIONAL FUNCTIONS

5 EXPONENTIAL AND LOGARITHM FUNCTIONS

6 TRIGONOMETRIC FUNCTIONS

7 IDENTITIES AND INVERSE FUNCTIONS

8 SOLVING TRIANGLES

9 LINEAR AND POLYNOMIAL ALGEBRA

10 DISCRETE ALGEBRA

11 VECTORS IN THE PLANE

12 PLANE ANALYTIC GEOMETRY

13 SOLID ANALYTIC GEOMETRY

14 COMPLEX NUMBERS

1 BASIC ALGEBRA

1 REVIEW OF THE REAL NUMBERS

The real numbers are the common numbers of everyday life. We are all familiar with their arithmetic. For our study of algebra, let us review briefly their basic properties.

The real number system is equipped with two operations, addition and multiplication. We add two real numbers a and b to form their **sum** $a + b$; we multiply them to form their **product** ab (also written $a \cdot b$, or $a \times b$). These operations satisfy certain basic identities.

Associative Laws

$$a + (b + c) = (a + b) + c \qquad a(bc) = (ab)c$$

Commutative Laws

$$a + b = b + a \qquad ab = ba$$

Distributive Laws

$$a(b + c) = ab + ac \qquad (a + b)c = ac + bc.$$

The first associative law guarantees that $a + b + c$ has a unique meaning whether you interpret this sum as $a + (b + c)$ or as $(a + b) + c$. Similarly, abc has a unique meaning.

The distributive laws connect addition and multiplication. We use them all the time without realizing it, for example in ordinary multiplication:

Long multiplication	Explanation
$\begin{array}{r} 14 \\ \times 23 \\ \hline 42 \\ 280 \\ \hline 322 \end{array}$	$\begin{aligned} 14 \times 23 &= 14 \times (3 + 20) \\ &= 14 \times 3 + 14 \times 20 \\ &= 42 + 280 \\ &= 322 \end{aligned}$ [The distributive law was used to pass from the first line to the second.]

The real number system contains certain special elements that have important algebraic properties.

Zero and Unity Laws There is a real number 0 (**zero**), the **additive identity,** and a real number 1 (**one**), the **multiplicative identity,** such that

$$0 + a = a + 0 = a \qquad 1 \cdot a = a \cdot 1 = a$$

Inverse Laws

For each real number a there exists a unique real number $-a$, its **additive inverse,** such that

$$a + (-a) = (-a) + a = 0$$

For each real number a different from 0, there exists a unique real number a^{-1}, its **multiplicative inverse,** such that

$$aa^{-1} = a^{-1}a = 1$$

The additive inverse $-a$ is also called the **negative** of a. The multiplicative inverse a^{-1} is also called the **reciprocal** of a and is written

$$\frac{1}{a} \quad \text{or} \quad 1/a \qquad (a \neq 0)$$

Rules for Inverses

$$-(-a) = a \qquad -(a + b) = (-a) + (-b)$$

$$(-a)b = a(-b) = -(ab) \qquad (-a)(-b) = ab$$

If a and b are different from 0, then

$$(a^{-1})^{-1} = a \qquad (-a)^{-1} = -(a^{-1}) \qquad (ab)^{-1} = a^{-1}b^{-1}.$$

Examples

$$-(-4) = 4 \qquad -(6 + 2) = (-6) + (-2)$$

$$(-5) \cdot 3 = 5 \cdot (-3) = -(5 \cdot 3) \qquad (-7)(-4) = 7 \cdot 4$$

$$\frac{1}{\frac{1}{2}} = 2 \qquad \frac{1}{-3} = -\tfrac{1}{3} \qquad \frac{1}{4 \cdot 5} = \tfrac{1}{4} \cdot \tfrac{1}{5}$$

The following two rules are important when solving equations; they can be derived from the existence of inverses.

Cancellation Laws

If $a + b = a + c$, then $b = c$.

If $ab = ac$ and $a \neq 0$, then $b = c$.

Next, some important properties of the number zero:

Properties of Zero

$0 \cdot a = a \cdot 0 = 0.$

The number 0 has no multiplicative inverse.

If $ab = 0$, then either $a = 0$ or $b = 0$ (or both). Similarly, if $abc = 0$, then $a = 0, b = 0$, or $c = 0$, etc.

Differences and Quotients

Subtraction is defined in terms of addition. Division (by a non-zero number) is defined in terms of multiplication.

Differences and Quotients

$a - b$ means $a + (-b)$

$\frac{a}{b}$ means $a \cdot b^{-1} = a \cdot \frac{1}{b} \quad b \neq 0$

Note that division by 0 is undefined because 0 has no reciprocal.

Rules for Differences

$a - (b + c) = (a - b) - c \qquad a - (b - c) = (a - b) + c$

Examples

$12 - (5 + 3) = 12 - 8 = 4 \qquad (12 - 5) - 3 = 7 - 3 = 4$

$12 - (5 - 3) = 12 - 2 = 10 \qquad (12 - 5) + 3 = 7 + 3 = 10$

Warning The expression $a - b - c$ is *always* interpreted as $(a - b) - c$, *never* as $a - (b - c)$. For instance, $4 - 6 - 1 = (4 - 6) - 1 = -2 - 1 = -3$.

The rules for quotients look familiar from the arithmetic of fractions. But they apply to all real numbers, not just fractions.

Rules for Quotients (all denominators non-zero)

$\frac{a}{b} = \frac{c}{d}$ if and only if $ad = bc$

$\frac{a}{b} + \frac{c}{d} = \frac{ad + bc}{bd} \qquad \frac{a}{b} \cdot \frac{c}{d} = \frac{ac}{bd}$

$\frac{a}{b} = \frac{ka}{kb} \; (k \neq 0) \qquad \left(\frac{c}{d}\right)^{-1} = \frac{d}{c} \qquad \frac{\frac{a}{b}}{\frac{c}{d}} = \frac{ad}{bc}$

Examples $\frac{2}{3} + \frac{6}{5} = \frac{2 \cdot 5 + 3 \cdot 6}{3 \cdot 5} = \frac{10 + 18}{15} = \frac{28}{15}$

$$\frac{7}{9} \cdot \frac{\pi}{4} = \frac{7 \cdot \pi}{9 \cdot 4} = \frac{7\pi}{36} \qquad \frac{\frac{11}{8}}{\frac{2}{0.05}} = \frac{(11)(0.05)}{8 \cdot 2} = \frac{0.55}{16}$$

Order

In addition to its arithmetic, the real number system has an **order** relation. Any pair of different real numbers can be compared; one is smaller than the other. When a is smaller than b, we write

$$a < b$$

and read "a is less than b." We also write

$$b > a$$

and read "b is greater than a." If a is less than or equal to b, we write

$$a \leq b \quad \text{equivalently} \quad b \geq a.$$

Examples $5.18 < 7.62 \quad -5.18 < 7.62 \quad -7.62 < -5.18 \quad -7.62 < 5.18$

We often think of the real number system as a number line (Fig. 1) like the scale on a thermometer, except that it extends *endlessly* in both directions. Geometrically, $a < b$ means that a is to the left of b on the number line. What is more, the *distance* from a to b then equals $b - a$.

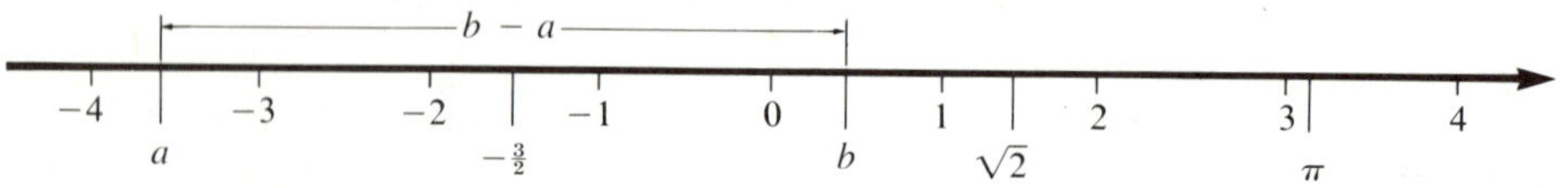

Fig. 1 The number line

If $a > 0$, then a is **positive;** if $a < 0$, then a is **negative.** If $a \geq 0$, then a is **non-negative.** Each real number a has a **sign,** written sgn a. If $a > 0$, then sgn $a = 1$; if $a < 0$, then sgn $a = -1$; finally, sgn $0 = 0$. (The operation sgn is standard in programming languages and occurs on some programmable calculators.)

Examples sgn $3.14159 = 1$ sgn $(-8.017) = -1$

All numbers to the right of zero on the number line have sign 1; all numbers to the left of zero have sign -1.

Rules of Order

If $a > 0$ and $b > 0$, then $a + b > 0$.

If $a < 0$ and $b < 0$, then $a + b < 0$.

If a and b are non-zero and have the same sign, then $ab > 0$; if they have opposite signs, then $ab < 0$.

An important consequence of the third law in the box is this: if $a \neq 0$, then $a \cdot a > 0$. Also $0 \cdot 0 = 0$, so in all cases $a \cdot a \geq 0$. We abbreviate $a \cdot a$ by a^2.

If a is any real number, then $a^2 \geq 0$.

Equality holds only if $a = 0$.

We'll have a lot more to say about order and the number line in Chapter 2.

Integers and Rational Numbers

These are especially important classes of real numbers. The **positive integers** (**whole numbers**) are 1, 2, 3, 4, · · · . The **negative integers are** -1, -2, -3, -4, · · · . The collection of positive integers and negative integers, together with 0, is called the **integers.**

A real number that can be expressed as the quotient of two integers is a **rational number;** for example,

$$\frac{9}{4} \qquad \frac{-3}{10} \qquad 5 = \frac{5}{1} \qquad \frac{1}{4} \qquad \frac{107}{-62}.$$

The rational numbers are precisely those real numbers whose decimal representations repeat eventually; for example,

$$\tfrac{1}{3} = 0.3333\cdots \qquad -\tfrac{2}{3} = -0.6666\cdots$$
$$\tfrac{3}{11} = 0.27272727\cdots \qquad 14 = 14.0000\cdots$$
$$\tfrac{241}{300} = 0.80333333\cdots \qquad \tfrac{2}{5} = 0.40000\cdots$$

Not all real numbers are rational. For instance, it is known that π and $\sqrt{2}$ are irrational (not rational).

Exercises

Use the distributive law to compute

1 11×1492 **2** 16×1003 **3** 19×211
4 18×407 **5** 383×22 **6** $579 \times (-11)$.

Derive the relation; state which law of algebra is used at each step

7 $(5x)(6y) - 10u = 10(3xy - u)$
8 $x + 2y + (x + 2y)^2 = (1 + x + 2y)(2y + x)$
9 $6xyz + x^2y^2z = xyz(yx + 6)$
10 $(x + 1)(x + 2)(x + 3) - x(x + 1)(x + 2) = 3(x + 2)(x + 1)$
11 $(a + b + c + 1)^2 - (a + b + c + 1) = (a + b + c)(a + b + c + 1)$
12 $abc(a + b + c) - abc^2 = abc(a + b)$
13 $(x + y)(x - y) = x^2 - y^2$
14 $(x + y + z)^2 = x^2 + y^2 + z^2 + 2(xy + yz + zx)$.
15 Is division commutative? Why?
16 Is subtraction commutative? Why?
17 Is division associative? Why?
18 Is subtraction associative? Why?

Derive from the rules for inverses and differences

19 $b - a = -(a - b)$ **20** $(a + c) - (b + c) = a - b$
21 $(a - b) - (a - c) = c - b$
22 $(a - b)(c - d) = (ac + bd) - (ad + bc)$.

Derive from the rules for inverses and quotients (assuming x and y non-zero)

23 $xy(x^{-1} - y^{-1}) = y - x$ **24** $(x + y)(x^{-1} - y^{-1}) = yx^{-1} - xy^{-1}$
25 $x^{-1}(x + y)y^{-1} = x^{-1} + y^{-1}$ **26** $x^{-1}y^{-1}(x^2 + y^2) = xy^{-1} + yx^{-1}$
27 Prove $(1 + a)^{-1} = 1 + b$ if and only if $a + b + ab = 0$
28 Prove $(1 - a)^{-1} = 1 - b$ if and only if $a + b = ab$

Which is greater

29 4^3 or 3^4 **30** 3^5 or 4^4 **31** $(2^2)^2$ or $2^{(2^2)}$
32 $(3^3)^3$ or $3^{(3^3)}$ **33** $(-3)^3$ or 3^2 **34** $(-3)^5$ or $(-5)^3$?

Which lies to the left and which to the right on the number line

35 3.572 3.752 **36** -15.72 -16.31

37 0.131 -0.019 **38** -0.131 -0.018?

Carry out the operations and express the answer as a fraction in lowest terms

39 $\frac{14}{5} + \frac{1}{3}$ **40** $\frac{1}{15} + \frac{1}{6}$ **41** $(\frac{2}{3})(\frac{30}{50})$

42 $\frac{3}{2}(\frac{4}{9} - \frac{2}{21})$ **43** $(\frac{1}{2} + \frac{1}{3})(\frac{1}{2} - \frac{1}{3})$ **44** $\frac{102}{35}(2 - \frac{3}{17})$

45 $\frac{1}{2} + \frac{1}{3} - \frac{1}{4}$ **46** $\frac{2}{3} - \frac{5}{6} + 1$ **47** $\frac{24}{77}(\frac{4}{15} - \frac{1}{2} - \frac{21}{20})$

48 $\dfrac{\frac{2}{3} - \frac{1}{4}}{1 + \frac{1}{12}}$ **49** $\frac{3}{4}(2 - \frac{16}{21})$ **50** $\dfrac{\frac{1}{6} + \frac{1}{9} - \frac{1}{18}}{\frac{1}{6} - \frac{1}{9} + \frac{1}{18}}$

51 $2\frac{2}{3} \div 1\frac{7}{9}$ **52** $5\frac{3}{5} \div (3\frac{7}{8} - 1\frac{1}{4})$.

2 ALGEBRAIC NOTATION

The four operations of arithmetic are the basis of all computations. Let x and y be real numbers. Their sum, difference, product, and quotient are denoted by

$$x + y \qquad x - y \qquad xy \qquad \frac{x}{y}.$$

The product is also denoted by

$$x \cdot y \quad \text{and} \quad x \times y.$$

The quotient is also denoted by

$$x/y \quad \text{and} \quad x \div y.$$

Mixtures of numbers and letters are allowed, as in

$$x + 4 \qquad 5 - xy \qquad \pi x \qquad \frac{y}{10}.$$

Note that $y/10$ and $\frac{1}{10}y$ are the same.

EXAMPLE 1 Write an expression for
(a) one more than twice a number
(b) a number decreased by one-half of another number
(c) the average of two numbers.

Solution (a) Call the number x. Then twice x is $2x$, and one more is $2x + 1$.
(b) Call the first number x and the second y. From x subtract $\frac{1}{2}y$, obtaining $x - \frac{1}{2}y$.
(c) The average of x and y is their sum divided by 2. Their sum is $x + y$, so the average is

$$\frac{x + y}{2}, \quad \text{that is,} \quad \tfrac{1}{2}(x + y).$$

Answer (a) $2x + 1$ (b) $x - \frac{1}{2}y$ (c) $\frac{1}{2}(x + y)$

Priority of Operations

Algebraic notation is meant to be absolutely precise. We must always avoid any possible misunderstanding. For example, the expression $3x + y$ might be interpreted in two ways: as "y added to 3 times x" or as "3 times the sum of x and y." Which is it?

Such cases are covered by a convention: multiplications and divisions precede additions and subtractions. So to compute $3x + y$, we first multiply x by 3, then add y. For insurance, we could write

$$3x + y = (3x) + y \quad \text{and} \quad x + yz = x + (yz),$$

but we won't because everybody follows the convention.

Be careful with expressions like $x/y + 1$. By convention, division precedes addition so $x/y + 1$ means $(x/y) + 1$, not $x/(y + 1)$. These two expressions are not the same. For example, if $x = 8$ and $y = 2$, then

$$x/y + 1 = \frac{x}{y} + 1 = 5 \quad \text{but} \quad x/(y + 1) = \frac{x}{y + 1} = \frac{8}{3}.$$

Parentheses

Sometimes parentheses are essential to avoid ambiguity. For example, if you want to subtract from x the number that is 1 more than y, you must write

$$x - (y + 1) \quad \text{not} \quad x - y + 1.$$

These expressions are different. In fact, from the rules for differences (p. 3),

$$x - (y + 1) = x - y - 1.$$

If you want the average of x and y, write $(x + y)/2$. You cannot omit the parentheses because $x + y/2$ means $x + \frac{1}{2}y$, which is different. You could also write $\frac{1}{2}(x + y)$, but not $\frac{1}{2}x + y$, which again is different. For example, if $x = 6$ and $y = -6$, then

$$\tfrac{1}{2}(x + y) = \frac{x + y}{2} = 0 \qquad x + \tfrac{1}{2}y = 3 \qquad \tfrac{1}{2}x + y = -3,$$

all different!

It is good policy to insert parentheses in case of doubt. Usually no harm can be done, but often lots of harm can be done by omitting them.

There are two rules for evaluating expressions with parentheses. The first says that you must complete all calculations inside of parentheses before doing anything outside. Thus in

$$(3 + 7)\cdot(5 - 2)$$

you must complete the addition $3 + 7$ and the subtraction $5 - 2$ before doing the multiplication.

The second rule says that in case of nested parentheses (parentheses within parentheses (like this)), you must complete the innermost one (or ones) first. Then if more nested parentheses remain, repeat the process. For instance,

$$\begin{aligned} &14 - ((25 - 2) - (7 + (9 - 1))) \\ &\quad = 14 - (23 - (7 + 8)) \\ &\quad = 14 - (23 - 15) = 14 - 8 = 6. \end{aligned}$$

All the parentheses in this calculation make the first line hard to read. For greater ease of reading we often substitute brackets [ ] and braces { } for some of the parentheses. For example

$$14 - ((25 - 2) - (7 + (9 - 1))) = 14 - \{(25 - 2) - [7 + (9 - 1)]\}.$$

Calculator Priorities

A left-to-right calculator has its own priorities. It performs operations as they are keyed in, a procedure which sometimes agrees with the conventions of algebra and sometimes does not. For example, to compute $3 \cdot 9 + 1$, you key in

3 [×] 9 [+] 1 [=] .

The calculator does the computations from left to right, that is, it computes

$$(3 \times 9) + 1 = 28,$$

which agrees with algebra. But suppose you want $21 + \frac{5}{13}$ and key in

2 1 [+] 5 [÷] 1 3 [=] .

The left-to-right calculator computes

$$(21 + 5) \div 13 = 2,$$

which is wrong; according to the conventions of algebra,

$$21 + 5/13 = 21 + \tfrac{5}{13} = 21.384615 \cdots .$$

Wait, don't throw away your left-to-right calculator! It gives you the correct answer this way:

5 [÷] 1 3 [+] 2 1 [=] .

Many calculators use *algebraic* logic with priorities for the arithmetic operations as in algebra. On such a calculator the sequence

2 1 [+] 5 [÷] 1 3 [=]

will give $21.3846 \cdots$ because the division is done before the addition. As this example shows, you may have to arrange computations according to the logic of your particular model. For a further discussion of calculators, see p. 208.

Algebraic Abbreviations

The expression $x + x$ makes sense, but it's simpler to write $2x$. Similarly, we abbreviate

$$x + x + x \text{ by } 3x, \qquad x + x + x + x \text{ by } 4x, \qquad \text{etc.}$$

There are corresponding abbreviations for products. Even though xx makes sense, we write x^2. Similarly, we abbreviate

$$xxx \text{ by } x^3, \qquad xxxx \text{ by } x^4, \qquad \text{etc.}$$

In this notation, $x = x^1$.

These abbreviations apply to repeated additions or multiplication of any algebraic expression. For example,

$$(7x - 5y + 1) + (7x - 5y + 1) + (7x - 5y + 1) = 3(7x - 5y + 1),$$

$$(x - \tfrac{1}{2})(x - \tfrac{1}{2})(x - \tfrac{1}{2})(x - \tfrac{1}{2}) = (x - \tfrac{1}{2})^4.$$

EXAMPLE 2 The side of a cube has length s. Write formulas for

(a) the volume (b) the surface area.

Solution (a) The volume of any rectangular box is the product of its length, width, and height. For the cube, each of these dimensions is s. Hence

$$\text{Volume} = s \cdot s \cdot s \qquad \text{briefly} \qquad V = s^3.$$

(b) The cube has 6 faces, each a square of side s, hence area $s \cdot s = s^2$. Therefore

$$\text{Surface area} = s^2 + s^2 + s^2 + s^2 + s^2 + s^2 \quad \text{briefly} \quad A = 6s^2.$$

Answer (a) $V = s^3$ (b) $A = 6s^2$

Exercises

Express in algebraic notation

1 one number added to one-half of another number
2 half of one number subtracted from twice another number
3 double the sum of a and b
4 the average of x and 1
5 three percent of a number
6 5% of the amount by which a number exceeds 100
7 the time it takes to go 350 miles at r miles per hour
8 the area of a triangle of base b and height h
9 the area of a triangle whose height h is $\frac{1}{3}$ the base b
10 the area of a circle of radius r
11 the area of a circle of radius $r + 1$
12 the area of a circular ring bounded by two concentric circles
13 the sum of two numbers multiplied by their difference
14 the surface area of a rectangular box.
15 It is said that for a perfect wedding the age of the groom should be 17 years more than half the age of the bride. Express this as an algebraic formula.

Compute the value for (a) $x = 4, y = 3$ and (b) $x = -4, y = 3$

16 $5x - 2y$ **17** $(x + y)/7$ **18** $x + y/7$
19 $x^2/4 + 3$ **20** $(x - 3y)^2$ **21** $(x + 1)(y - 1)$
22 $x/(y + 3)$ **23** $x/y + 3$ **24** $(x^2 - 1)(x + y)$.

Substitute numbers to show that the expressions are different

25 $(x + 2)y$ and $x + 2y$ **26** $3x(y + 1)$ and $3xy + 1$
27 $(x + y)^2$ and $x + y^2$ **28** $1/(x + y)$ and $1/x + 1/y$
29 $4 - 2(x - y)$ and $2x - 2y$ **30** $5 - (3x - 5)$ and $-3x$.

Find the result of the given sequence of calculator steps assuming (a) left-to-right logic (b) algebraic logic

31 6 [−] 2 [×] 3 [=] **32** 6 [×] 2 [−] 3 [=]
33 8 [÷] 4 [+] 6 [÷] 2 [=] **34** 8 [+] 4 [÷] 6 [+] 2 [=]
35 a [×] b [−] c [÷] d [+] e [=] **36** a [−] b [×] c [+] d [÷] e [=]

Using only [+], [−], [×], [÷], [=] find a sequence of calculator steps to compute the given quantity assuming (a) left-to-right logic (b) algebraic logic

37 $\dfrac{x + 3}{y}$ **38** $5(x - y)$ **39** $\dfrac{x}{yz}$

40 $\dfrac{a+b}{c} - d$ **41** $\frac{1}{2}\left(\dfrac{a}{b} + c\right)$ **42** $a\left(b + \dfrac{c}{d} + e\right)$.

Compute

43 $1 - (2 - (3 - 4))$ **44** $6 + 5(4 + 3(2 + 1))$

45 $(2 + 3(7 + 4))((17 - 12)(13 - 8) + 2)$

46 $((6 - 4)(7 - 2) + (1 + 2)(6 - 3))(20 - 5(10 - 7))$

47 $((18 - 3)/5 + 5)/4 + ((18 - 3)/5 - 5)/2$

48 $1 + (1 \div (1 + (1 \div (1 + (1 \div (1 + 1))))))$.

3 INTEGRAL EXPONENTS

In Section 2 we introduced the abbreviations

$$a^1 = a \quad a^2 = a \cdot a \quad a^3 = a \cdot a \cdot a \quad \cdots \quad a^n = \underbrace{a \cdot a \cdot a \cdot a \cdot a \cdots a}_{n \text{ factors}}.$$

Read a^n as "a to the n-th power" or "a to the n". The number n is an **exponent.** The number a^n is called the n-th **power** of a.

Examples

$$(-2)^3 = (-2)(-2)(-2) = -8 \qquad 10^4 = 10 \cdot 10 \cdot 10 \cdot 10 = 10{,}000$$

Look what happens when you multiply a^5 by a^3:

$$a^5a^3 = (\underbrace{aaaaa}_{5})(\underbrace{aaa}_{3}) = \underbrace{aaaaaaaa}_{8 \text{ factors}} = a^8.$$

Thus $a^5a^3 = a^8$. Similarly, if you multiply m factors a by n factors a, you get $m + n$ factors a. Therefore

$$a^m a^n = a^{m+n}$$

To multiply, add exponents.

Now look what happens when you divide a^5 by a^3:

$$\frac{a^5}{a^3} = \frac{aaaaa}{aaa} = \frac{aaa}{aaa}\,\frac{aa}{1} = aa = a^2.$$

Thus $a^5/a^3 = a^{5-3}$. You subtract exponents because the three factors in the denominator cancel three of the factors in the numerator. By the same reasoning

$$\frac{a^m}{a^n} = a^{m-n} \quad \text{when} \quad m > n.$$

Examples

(1) $\dfrac{2^5}{2^3} = 2^{5-3} = 2^2$ that is $\dfrac{32}{8} = 4$

(2) $\dfrac{10^6}{10^4} = 10^2$ that is $\dfrac{1{,}000{,}000}{10{,}000} = 100$

It would be nice if we could also subtract exponents when the denominator contains as many or more factors than the numerator. That would require zero and negative exponents. Since $a^n/a^n = 1$, we should have $a^{n-n} = a^0 = 1$. So we define

$$a^0 = 1 \qquad (a \neq 0).$$

We do this only for $a \neq 0$. The symbol 0^0 is undefined in mathematics.

Next we observe that

$$\frac{a^3}{a^5} = \frac{a \cdot a \cdot a}{a \cdot a \cdot a \cdot a \cdot a} = \frac{1}{a \cdot a} = \frac{1}{a^2}$$

so we should have $a^{3-5} = a^{-2} = 1/a^2$. In general, if n is a positive integer, we define

$$a^{-n} = \frac{1}{a^n} \qquad (a \neq 0).$$

Examples $\quad 10^{-3} = \dfrac{1}{10^3} = \dfrac{1}{1000} \qquad 2^{-6} = \dfrac{1}{2^6} = \dfrac{1}{64}$

Note that a^{-1} means $1/a^1 = 1/a$. Thus $a^{-1} \cdot a = 1$. This is consistent with Section 1, where a^{-1} meant the multiplicative inverse of a. Also $a^{-n} \cdot a^n = 1$, so a^{-n} is the multiplicative inverse of a^n, that is, $(a^n)^{-1}$. Therefore a^{-n} can be interpreted in several ways, all equivalent:

$$a^{-n} = (a^n)^{-1} = (a^{-1})^n = \frac{1}{a^n}.$$

We have now defined a^n for every integer exponent n. Here a is any real number except that $a \neq 0$ when $n \leq 0$. Now let us state (without proofs) the basic rules for working with exponents.

Rules of Exponents If a and b are non-zero real numbers and if m and n are integers, then

(1) $a^m a^n = a^{m+n}$ (2) $\dfrac{a^m}{a^n} = a^{m-n}$ (3) $(a^m)^n = a^{mn}$

(4) $(ab)^n = a^n b^n$ (5) $\left(\dfrac{a}{b}\right)^n = \dfrac{a^n}{b^n}$.

We emphasize that these rules are valid for *all* integers m and n, not just positive ones.

Particular Cases

Rule (3) $\quad (a^4)^3 = a^4 \cdot a^4 \cdot a^4 = a^{4+4+4} = a^{12} = a^{4 \cdot 3}$

$$(a^4)^{-3} = \frac{1}{(a^4)^3} = \frac{1}{a^{12}} = a^{-12} = a^{4(-3)}$$

Rule (4) $(ab)^3 = (ab)(ab)(ab) = (aaa)(bbb) = a^3b^3$

$$(ab)^{-3} = \frac{1}{(ab)^3} = \frac{1}{a^3b^3} = \frac{1}{a^3}\cdot\frac{1}{b^3} = a^{-3}b^{-3}$$

The rules can be extended in various ways. For instance, by the associative law and repeated applications of (3) and (4) we prove

$$(a^m b^n c^p)^r = a^{mr} b^{nr} c^{pr}.$$

EXAMPLE 1 Use the rules of exponents to simplify

(a) $(xy)^2(x^2y^3)^{-1}$ (b) $(x^2y^{-3})^{-5}$.

Solution (a) $(xy)^2(x^2y^3)^{-1} = (x^2y^2)\dfrac{1}{x^2y^3} = \dfrac{x^2}{x^2}\dfrac{y^2}{y^3} = \dfrac{1}{y} = y^{-1}$.

Alternatively, $(xy)^2(x^2y^3)^{-1} = (x^2y^2)(x^{-2}y^{-3}) = x^{2-2}y^{2-3} = x^0y^{-1} = y^{-1}$.

(b) $(x^2y^{-3})^{-5} = (x^2)^{-5}(y^{-3})^{-5} = x^{2(-5)}y^{(-3)(-5)} = x^{-10}y^{15}$.

Answer (a) $y^{-1} = 1/y$ (b) $x^{-10}y^{15} = y^{15}/x^{10}$

EXAMPLE 2 Express $\dfrac{2^{-3}\cdot 8^5}{4^3\cdot 16}$ as a power of 2.

Solution $\dfrac{2^{-3}\cdot 8^5}{4^3\cdot 16} = \dfrac{2^{-3}(2^3)^5}{(2^2)^3(2^4)} = \dfrac{2^{-3}\cdot 2^{15}}{2^6\cdot 2^4} = \dfrac{2^{12}}{2^{10}} = 2^2$.

Answer 2^2

EXAMPLE 3 Use the rules of exponents to compute $\dfrac{2^6\cdot 5^7}{25\cdot 10^4}$.

Solution $\dfrac{2^6\cdot 5^7}{25\cdot 10^4} = \dfrac{2^6\cdot 5^7}{5^2\cdot(2\cdot 5)^4} = \dfrac{2^6\cdot 5^7}{5^2\cdot 2^4\cdot 5^4} = \dfrac{2^6\cdot 5^7}{2^4\cdot 5^6} = 2^2\cdot 5 = 20$.

Answer 20

Scientific Notation

One important practical application of exponents is in computations. For scientific work, we need an efficient way of writing and computing with very large or very small numbers, such as

32,000,000,000   1,876,000   0.00000 00000 006

Imagine multiplying such numbers as they are written!

The idea of scientific notation is to express each positive number as a small number times a power of 10, more precisely, in the form $c \times 10^n$, where $1 \le c < 10$ and n is an appropriate exponent.

Examples

$$140 = 1.4 \times 10^2 \qquad 0.05 = 5 \times 10^{-2}$$
$$2550 = 2.55 \times 10^3 \qquad 0.0031 = 3.1 \times 10^{-3}$$
$$1{,}876{,}000 = 1.876 \times 10^6 \qquad 0.000988 = 9.88 \times 10^{-4}$$
$$32{,}000{,}000{,}000 = 3.2 \times 10^{10} \qquad 0.00000\,00000\,006 = 6 \times 10^{-13}$$

The mass of a neutron is approximately

$$0.00000\ 00000\ 00000\ 00000\ 00016 = 1.6 \times 10^{-24} \text{ gram.}$$

EXAMPLE 4 Multiply $(140)(32{,}000{,}000{,}000)(0.00000\ 00000\ 006)$.

Solution $(1.4 \times 10^2)(3.2 \times 10^{10})(6 \times 10^{-13})$

$$= (1.4)(3.2)(6) \times 10^{2+10-13} = 26.88 \times 10^{-1}.$$

Answer 2.688

EXAMPLE 5 Compute $\dfrac{(14000)(0.00003)(8800000)}{(1100)(0.000002)}$.

Solution

$$\frac{(1.4 \times 10^4)(3 \times 10^{-5})(8.8 \times 10^6)}{(1.1 \times 10^3)(2 \times 10^{-6})} = \frac{(1.4)(3)(8.8)}{(1.1)(2)} \times 10^{4-5+6-3+6}$$

$$= 16.8 \times 10^8.$$

Answer 1.68×10^9

Very large and very small numbers are displayed in scientific calculators by means of scientific notation. For example, the display

3.94016 67

means 3.94016×10^{67}. Your calculator has a key such as EXP or EE that allows you to enter a suitable power of 10 (usually up to 99 or down to -99). For example, to key in the number shown you would press

3 . 9 4 0 1 6 EXP 6 7.

To key in 1.6×10^{-24}, you would press

1.6 EXP 2 4 +/− .

The key +/− (or a similar key) changes the exponent from 24 to -24.

Scientific notation on a calculator is also called **floating point** notation.

Calculating Powers

Note the key y^x on your calculator. It is used for the (approximate) calculation of powers. To calculate y^x, you first key in y, then y^x, then the exponent x, and finally =.

Examples

(1) $(3.52)^7$ 3 . 5 2 y^x 7 = 6695.7410

(2) $(145.3)^{-12}$ 1 4 5 . 3 y^x 1 2 +/− = $1.1293086 \times 10^{-26}$

Exercises

Compute

1 3^3 **2** $(-3)^3$ **3** 6^{-2} **4** $(\frac{1}{5})^2$

5 $(\frac{1}{2})^{-2}$ **6** $(\frac{1}{3})^0$ **7** $(-\frac{2}{3})^5$ **8** $(-\frac{3}{2})^{-1}$

9 $2^{-3}\left(\frac{1}{3}\right)^2$ **10** $\dfrac{2^5\cdot 5^3}{2^6\cdot 5^{-2}}$ **11** $2^4 3^2 6^{-2} 8^{-1}$ **12** $10^4 25^{-3} 16^{-1}$.

Express as a power of 2

13 4^2 **14** 8^{-3} **15** $\left(\frac{1}{2}\right)^7$ **16** $(2^4\cdot 16^{-2})^3$

17 $2^4\left(\frac{1}{2}\right)^3 8^2 16^{-1}$ **18** $2\cdot 4^2\cdot 8^3$ **19** $\left(\frac{1}{2}\right)^{-8}\left(\frac{1}{4}\right)^2$ **20** $\frac{1}{2}\cdot\frac{1}{4}\cdot\frac{1}{8}\cdot\frac{1}{16}$.

Express as simply as possible, with no negative exponents

21 $\dfrac{(xy)^6}{xy^2}$ **22** $\dfrac{1}{x^3}(x^2)^3x^{-4}$ **23** $a^2(a^{-1}+a^{-3})$

24 $(aba^{-4})(a^3b^{-2})^0$ **25** $(8a^3b)^{-4}(2a/b)^{12}$ **26** $(-5x^2y^{-3})^{-20}$

27 $(-xy^2)^3(-2x^2y^2)^{-4}$ **28** $(xy)^{-5}(2xy^2)(3xy)^3$ **29** $(4x^3y^2z)^2(4x^3y^2z)^{-7}$

30 $(2x^2y^2)^3(-3x^4y)^{-2}$ **31** $\dfrac{x^{-2}}{y^{-2}}+\dfrac{y^2}{x^2}$ **32** $\dfrac{(2pqr)^2}{(p^3q)^2(pr)^{-1}}$

33 $\left(\dfrac{a}{b}\right)^{-2}\left(\dfrac{b}{a}\right)^0$ **34** $\left(\dfrac{ab^2c^3d^4}{a^4b^3c^2d}\right)^2$.

Express in scientific notation

35 0.4 **36** 0.0081 **37** 2.37 **38** 0.6008

39 0.000076 **40** 0.00000 00059 **41** 17 **42** 5280

43 12400 **44** 3.005 **45** 3,832,000 **46** 452,000,000,000,000.

Compute and express the result in scientific notation

47 $(180)(30{,}000{,}000)(0.00012)$ **48** $\dfrac{(20{,}100)(0.006)}{(0.00000\,02)(402{,}000)}$

49 $(0.002)^3(0.00004)(0.00000\,5)(6{,}000{,}000{,}000)$ **50** $\dfrac{1}{(800)(200{,}000)^2(0.00001)^4}$

51 $(200{,}000)^6$ **52** $(10)(200)(3000)(40{,}000)(500{,}000)$

53 $\dfrac{(0.00000\,02)^2}{(5000)^4}$ **54** $(8000)(30{,}000)^5(0.00000\,1)^6$

55 the number of inches in 100 miles

56 the number of cubic centimeters in a cubic kilometer

57 the total mass in grams of 3,000,000 neutrons

58 the length in kilometers of a light-year, the distance light travels in a year. Take the speed of light to be 300,000 km/sec and a year to be 365 days.

Estimate on a calculator

59 $(3.5)^{10}$ **60** $(1.723)^{35}$ **61** $(1.01)^{96}$

62 $(1.01)^{-120}$ **63** $(1.5708\times 10^7)^{-9}$ **64** $(2.718\times 10^{-12})^5$.

4 ROOTS AND RADICALS

Square Roots

Given a real number a, a number x is a **square root** of a if $x^2 = a$. Each positive number has two square roots, one positive and one negative. We denote the *positive* square root of a by the symbol $\sqrt{a}$. For example, $\sqrt{9} = 3$ because $3 > 0$ and $3^2 = 9$. Similarly,

$$\sqrt{25} = 5 \qquad \sqrt{\tfrac{4}{49}} = \tfrac{2}{7} \qquad \sqrt{0.0016} = 0.04.$$

The negative square root of a is $-\sqrt{a}$ since $(-\sqrt{a})^2 = (\sqrt{a})^2 = a$. The number 0 has only one square root and we write $\sqrt{0} = 0$.

Square roots obey the following algebraic rules.

Rules for square roots Let a and b be positive real numbers. Then

(1) $(\sqrt{a})^2 = a$ (2) $\sqrt{a^2} = a$

(3) $\sqrt{ab} = \sqrt{a}\sqrt{b}$ (4) $\sqrt{\dfrac{a}{b}} = \dfrac{\sqrt{a}}{\sqrt{b}}$.

Rules (1) and (2) come directly from the definition of square roots. Let us check rule (3):

$$\sqrt{a}\sqrt{b} > 0 \quad \text{and} \quad (\sqrt{a}\sqrt{b})^2 = (\sqrt{a})^2(\sqrt{b})^2 = ab.$$

Thus $\sqrt{a}\sqrt{b} = \sqrt{ab}$. Rule (4) is checked in a similar way.

Square roots of complicated quantities can often be simplified by rule (3). For instance,

$$\sqrt{48} = \sqrt{16\cdot 3} = \sqrt{16}\sqrt{3} = 4\sqrt{3}.$$

In general, by rule (3),

$$\sqrt{r^2b} = \sqrt{r^2}\sqrt{b} = r\sqrt{b} \qquad (r > 0, \quad b > 0).$$

Thus a squared factor under the square root sign can come out in front, but it loses its square on the way out.

EXAMPLE 1 Simplify (a) $\sqrt{36\cdot 49}$ (b) $\sqrt{500}$ (c) $\sqrt{\dfrac{75}{16}}$.

Solution (a) By rule (3),

$$\sqrt{36\cdot 49} = \sqrt{36}\sqrt{49} = 6\cdot 7 = 42.$$

(b) By rule (3),

$$\sqrt{500} = \sqrt{100\cdot 5} = \sqrt{100}\sqrt{5} = 10\sqrt{5}.$$

(c) Use rule (4), then rule (3):

$$\sqrt{\frac{75}{16}} = \frac{\sqrt{75}}{\sqrt{16}} = \frac{\sqrt{25\cdot 3}}{4} = \frac{\sqrt{25}\sqrt{3}}{4} = \frac{5\sqrt{3}}{4}.$$

Answer (a) 42 (b) $10\sqrt{5}$ (c) $\frac{5}{4}\sqrt{3}$

Note In the following examples, and in the exercises, all letters represent positive numbers.

EXAMPLE 2 Simplify (a) $\sqrt{25x^6}$ (b) $\sqrt{225p^4q^7}$ (c) $\sqrt{uv}\sqrt{\dfrac{8}{u^3v}}$.

Solution (a) Use rule (3):

$$\sqrt{25x^6} = \sqrt{25}\sqrt{x^6} = 5\sqrt{(x^3)^2} = 5x^3.$$

(b) Use rule (3) several times:

$$\sqrt{225p^4q^7} = \sqrt{225}\sqrt{p^4}\sqrt{q^7} = 15p^2\sqrt{q^6q}$$
$$= 15p^2\sqrt{q^6}\sqrt{q} = 15p^2q^3\sqrt{q}.$$

(c) Use rule (3), then rule (4):

$$\sqrt{uv}\sqrt{\frac{8}{u^3v}} = \sqrt{uv\left(\frac{8}{u^3v}\right)} = \sqrt{\frac{8}{u^2}} = \frac{\sqrt{8}}{\sqrt{u^2}} = \frac{\sqrt{4\cdot 2}}{u} = \frac{2\sqrt{2}}{u}.$$

Answer (a) $5x^3$ (b) $15p^2q^3\sqrt{q}$ (c) $2\sqrt{2}/u$

EXAMPLE 3 Compute

(a) $(4+\sqrt{7})(5-2\sqrt{7})$ (b) $(\sqrt{x}+\sqrt{y})(\sqrt{x}-\sqrt{y})$.

Solution (a) Use the distributive law and $(\sqrt{7})^2 = 7$:

$$\begin{aligned}(4+\sqrt{7})(5-2\sqrt{7}) &= (4)(5) + (\sqrt{7})(5) - (4)(2\sqrt{7}) - (\sqrt{7})(2\sqrt{7})\\ &= 20 + 5\sqrt{7} - 8\sqrt{7} - 2(\sqrt{7})^2\\ &= 20 - 3\sqrt{7} - (2)(7) = 6 - 3\sqrt{7}.\end{aligned}$$

(b) $(\sqrt{x}+\sqrt{y})(\sqrt{x}-\sqrt{y}) = (\sqrt{x})^2 - (\sqrt{y})^2 = x - y$.

Answer (a) $6 - 3\sqrt{7}$ (b) $x - y$

Warning The rules for square roots apply only to multiplication and division. Square roots have no simple behavior with respect to addition and subtraction. In particular,

$$\sqrt{a+b} \text{ IS NOT } \sqrt{a} + \sqrt{b}\ !$$

For example, $\sqrt{25} = \sqrt{9+16} \neq \sqrt{9} + \sqrt{16}$ since $5 \neq 3 + 4$.

Rationalizing the Denominator

Sometimes it is convenient to eliminate square roots from the denominator of a quotient. If the denominator is $r\sqrt{a}$, where r is a rational number, we just multiply numerator and denominator by $\sqrt{a}$. For example,

$$\frac{1}{\sqrt{3}} = \frac{1}{\sqrt{3}}\cdot\frac{\sqrt{3}}{\sqrt{3}} = \frac{\sqrt{3}}{3} \qquad \frac{7}{2\sqrt{5}} = \frac{7}{2\sqrt{5}}\cdot\frac{\sqrt{5}}{\sqrt{5}} = \frac{7\sqrt{5}}{10}.$$

If the denominator is $\sqrt{a}+\sqrt{b}$, we multiply numerator and denominator by $\sqrt{a}-\sqrt{b}$, changing the denominator into $a - b$. Similarly, if the denominator is $\sqrt{a}-\sqrt{b}$, we multiply by $\sqrt{a}+\sqrt{b}$. For example,

$$\frac{1}{\sqrt{5}+\sqrt{2}} = \frac{1}{\sqrt{5}+\sqrt{2}}\,\frac{\sqrt{5}-\sqrt{2}}{\sqrt{5}-\sqrt{2}} = \frac{\sqrt{5}-\sqrt{2}}{(\sqrt{5})^2-(\sqrt{2})^2} = \frac{\sqrt{5}-\sqrt{2}}{3},$$

$$\frac{1}{\sqrt{3}-1} = \frac{1}{\sqrt{3}-1}\,\frac{\sqrt{3}+1}{\sqrt{3}+1} = \frac{\sqrt{3}+1}{(\sqrt{3})^2-1^2} = \frac{\sqrt{3}+1}{3-1} = \frac{\sqrt{3}+1}{2}.$$

We can use the same method also to eliminate square roots from the numerator.

EXAMPLE 4 Express

(a) $\dfrac{\sqrt{2}}{7+3\sqrt{2}}$ without square roots in the denominator

(b) $\dfrac{\sqrt{x+h}-\sqrt{x}}{h}$ without square roots in the numerator.

Solution (a) Multiply numerator and denominator by $7 - 3\sqrt{2}$:

$$\frac{\sqrt{2}}{7 + 3\sqrt{2}} = \frac{\sqrt{2}}{7 + 3\sqrt{2}}\,\frac{7 - 3\sqrt{2}}{7 - 3\sqrt{2}} = \frac{7\sqrt{2} - 3\cdot 2}{7^2 - (3\sqrt{2})^2}$$

$$= \frac{7\sqrt{2} - 6}{49 - 9\cdot 2} = \frac{7\sqrt{2} - 6}{31} = \frac{-6 + 7\sqrt{2}}{31}.$$

(b) Multiply numerator and denominator by $\sqrt{x + h} + \sqrt{x}$:

$$\frac{\sqrt{x + h} - \sqrt{x}}{h} = \frac{\sqrt{x + h} - \sqrt{x}}{h}\,\frac{\sqrt{x + h} + \sqrt{x}}{\sqrt{x + h} + \sqrt{x}} = \frac{(x + h) - x}{h(\sqrt{x + h} + \sqrt{x})}$$

$$= \frac{h}{h(\sqrt{x + h} + \sqrt{x})} = \frac{1}{\sqrt{x + h} + \sqrt{x}}.$$

Answer (a) $\dfrac{-6 + 7\sqrt{2}}{31}$ (b) $\dfrac{1}{\sqrt{x + h} + \sqrt{x}}$

Higher Order Roots

Given a real number a, a number x is an n-th **root** of a if $x^n = a$. Here n is an integer, $n \geq 2$. Each positive real number a has a unique positive n-th root, which we denote by the symbol $\sqrt[n]{a}$. For example,

$$\sqrt[3]{125} = 5 \quad \text{because} \quad 5 > 0 \quad \text{and} \quad 5^3 = 125,$$

$$\sqrt[4]{81} = 3 \quad \text{because} \quad 3 > 0 \quad \text{and} \quad 3^4 = 81.$$

We also write $\sqrt[n]{0} = 0$ since $0^n = 0$. The symbol $\sqrt[n]{}$ is called a **radical.** Note that $\sqrt[2]{}$ is the same as $\sqrt{}$.

Remark If n is even, then $-\sqrt[n]{a}$ is also an n-th root of the positive number a.

If n is odd, then negative numbers also have n-th roots. For instance -2 is a 5-th root of -32. Nevertheless, we shall stick to n-th roots of positive numbers only. The following rules for n-th roots are not 100% correct without qualifications if roots of negative numbers are allowed.

> **Rules for *n*-th roots** Let a and b be positive real numbers. Then
>
> (1) $(\sqrt[n]{a})^n = a$ (2) $\sqrt[n]{a^n} = a$ (3) $\sqrt[n]{ab} = \sqrt[n]{a}\sqrt[n]{b}$
>
> (4) $\sqrt[n]{\dfrac{a}{b}} = \dfrac{\sqrt[n]{a}}{\sqrt[n]{b}}$ (5) $\sqrt[m]{\sqrt[n]{a}} = \sqrt[mn]{a}$.

Rules (1)–(4) are just like the corresponding rules for square roots. Rule (5) is new; it corresponds to the rule of exponents $(r^m)^n = r^{mn}$. In fact, if we set

$$r = \sqrt[m]{\sqrt[n]{a}},$$

then r is the positive number whose m-th power is $\sqrt[n]{a}$, that is, $r^m = \sqrt[n]{a}$. Thus r^m is the n-th root of a, that is, $(r^m)^n = a$. Therefore, $r^{mn} = a$, so r is the mn-th root of a, which proves (5). For example,

$$\sqrt{\sqrt{x}} = \sqrt[4]{x} \qquad \sqrt[4]{\sqrt[3]{x}} = \sqrt[12]{x}.$$

EXAMPLE 5 Simplify (a) $\sqrt[3]{16x^3y^6}$ (b) $\sqrt[4]{\dfrac{48}{x^{12}y^4}}$ (c) $\sqrt[3]{\sqrt{27}}$.

Solution (a) By rule (3),

$$\sqrt[3]{16x^3y^6} = \sqrt[3]{16}\sqrt[3]{x^3}\sqrt[3]{(y^2)^3} = \sqrt[3]{16}\,xy^2.$$

But $16 = 2^4 = 2^3 \cdot 2$, so $\sqrt[3]{16} = \sqrt[3]{2^3}\sqrt[3]{2} = 2\sqrt[3]{2}$. Hence the answer is $2xy^2\sqrt[3]{2}$.

(b) By rules (3) and (4),

$$\sqrt[4]{\frac{48}{x^{12}y^4}} = \frac{\sqrt[4]{48}}{\sqrt[4]{x^{12}y^4}} = \frac{\sqrt[4]{16\cdot 3}}{\sqrt[4]{x^{12}}\sqrt[4]{y^4}} = \frac{\sqrt[4]{16}\sqrt[4]{3}}{x^3y} = \frac{2\sqrt[4]{3}}{x^3y}.$$

(c) $\sqrt[3]{\sqrt{27}} = \sqrt[6]{27} = \sqrt{\sqrt[3]{27}} = \sqrt{3}$.

Answer (a) $2xy^2\sqrt[3]{2}$ (b) $\dfrac{2\sqrt[4]{3}}{x^3y}$ (c) $\sqrt{3}$

Remark Note that we write the answer to (a) in the form $2xy^2\sqrt[3]{2}$ rather than $2\sqrt[3]{2}\,xy^2$. In general we prefer to write radical factors on the right because it is so easy to confuse symbols like $\sqrt[3]{2}\,x$ and $\sqrt[3]{2x}$. Just draw the bar a bit too long and you aren't sure which one is meant.

Calculating Roots

Note the key [√] on your calculator. It calculates (approximately) the square root of the displayed number.

Examples

(1) $\sqrt{1276.5}$ 1 2 7 6 . 5 [√] 35.72814

(2) $\sqrt{1.4\times 10^{77}}$ 1 . 4 [EXP] 7 7 [√] 3.7416574 ×10 38

If your calculator has a key [$y^{1/x}$], you can use it to estimate $\sqrt[x]{y}$ directly. Why this method works will become clear in the next section.

Example

$\sqrt[7]{45}$ 4 5 [$y^{1/x}$] 7 [=] 1.7225555

Without the key [$y^{1/x}$] you can work indirectly, using both [y^x] and [$1/x$].

Example

$\sqrt[7]{45}$ 4 5 [y^x] 7 [$1/x$] [=] 1.7225555

Check 1 . 7 2 2 5 5 5 5 [y^x] 7 [=] 45.000005

Exercises

[All letters in the exercises represent positive numbers.]

Simplify

1 $\sqrt{81}$ **2** $\sqrt{144}$ **3** $\sqrt{\dfrac{1}{9}}$ **4** $\sqrt{\dfrac{49}{25}}$

5 $\sqrt{9a^6}$ **6** $\sqrt{\frac{1}{4}y^2}$ **7** $\sqrt{50}$ **8** $\sqrt{128}$

9 $\sqrt{\dfrac{1}{12}}$ **10** $\sqrt{ab^2c^5}$ **11** $\sqrt{\dfrac{18x^3}{(y+z)^4}}$ **12** $\dfrac{\sqrt{6}}{\sqrt{3}}$

13 $\sqrt[3]{1000}$ **14** $\sqrt[3]{\frac{1}{8}}$ **15** $\sqrt[3]{\frac{27}{64}}$ **16** $\sqrt[3]{27 \cdot 64 \cdot 125}$

17 $\sqrt[4]{\frac{1}{16}}$ **18** $\sqrt[4]{81}$ **19** $\sqrt[4]{32}$ **20** $\sqrt[4]{162}$

21 $\sqrt[3]{8a^3b^9}$ **22** $\sqrt[4]{10{,}000u^2}$ **23** $\sqrt[5]{64u^5v^6}$ **24** $\sqrt[3]{\frac{a^7}{24b^9}}$.

Find an equivalent expression without radicals in the denominator

25 $\frac{1}{\sqrt{6}}$ **26** $\frac{\sqrt{2}}{\sqrt{6} + \sqrt{2}}$ **27** $\frac{\sqrt{x} + \sqrt{y}}{\sqrt{x} - \sqrt{y}}$ **28** $\frac{1}{\sqrt{x} - 2\sqrt{y}}$

29 $\frac{4 - \sqrt{3}}{4 + \sqrt{3}}$ **30** $\frac{2}{8 - 3\sqrt{5}}$ **31** $\frac{1}{7\sqrt{2} - 2\sqrt{3}}$ **32** $\frac{1}{\sqrt[3]{4}}$.

Find an equivalent expression without radicals in the numerator

33 $\frac{\sqrt{2} + \sqrt{3}}{4}$ **34** $\frac{2 + \sqrt{3}}{2 - \sqrt{3}}$

35 $\frac{\sqrt{x + h} - \sqrt{x - h}}{2h}$ **36** $\frac{3\sqrt{2} - 2\sqrt{3}}{6}$.

Compute and simplify

37 $(a + b\sqrt{2})(a - b\sqrt{2})$ **38** $(\sqrt{x} + 3\sqrt{y})^2$ **39** $(\sqrt{x})^4$

40 $\left(\sqrt{x} - \frac{1}{\sqrt{x}}\right)^2$ **41** $\sqrt{\frac{5x}{3y}}\sqrt{27x^3y^2}$ **42** $\sqrt{xy}\left(\sqrt{\frac{x}{y}} + \sqrt{\frac{y}{x}}\right)$.

Find an equivalent expression involving at most one radical sign

43 $\sqrt{x\sqrt{x^6}}$ **44** $\sqrt{3}(1 + \sqrt{3})$ **45** $\sqrt{2}\sqrt{xy}\sqrt{10yz}$ **46** $\sqrt[3]{\frac{16xy}{z}}\sqrt[3]{4y^2z^7}$.

Verify the formulas

47 $\sqrt{a^2 + x^2} = a\sqrt{1 + \left(\frac{x}{a}\right)^2}$ **48** $\sqrt{x^2 + 1} - \frac{1}{\sqrt{x^2 + 1}} = \frac{x^2}{\sqrt{x^2 + 1}}$

49 $\frac{2}{\sqrt{3} + 1} - \frac{1}{\sqrt{3} - 1} = \frac{\sqrt{3} - 3}{2}$ **50** $\frac{1}{3y}\sqrt{9x^2 - y^2} = \sqrt{(x/y)^2 - \frac{1}{9}}$.

Prove for n-th roots

51 rule (1) **52** rule (2) **53** rule (3) **54** rule (4).

Use the approximations $\sqrt{2} \approx 1.4$, $\sqrt{3} \approx 1.7$, $\sqrt{5} \approx 2.2$ to estimate

55 $\sqrt{6}$ **56** $\sqrt{10}$ **57** $\sqrt{3/5}$ **58** $\sqrt{2/3}$

59 $\sqrt{30}$ **60** $\sqrt{75}$ **61** $\sqrt{40}$ **62** $\sqrt{216}$.

Estimate on a calculator

63 $\sqrt{2991}$ **64** $\sqrt[3]{17}$ **65** $\sqrt[10]{10}$

66 $\sqrt[5]{371293}$ **67** $\sqrt[9]{1.4 \times 10^{30}}$ **68** $\sqrt{8.039 \times 10^{91}}$

69 $\sqrt[12]{5.62 \times 10^{-17}}$ **70** $\sqrt[14]{6.77 \times 10^{-551}}$ **71** $\sqrt[3]{8 + 17\sqrt{5}}$

72 $\sqrt[5]{16 + \sqrt[4]{78}}$ **73** $\sqrt{2 + \sqrt{2 + \sqrt{2}}}$ **74** $\sqrt[3]{4^4 + 5^5 + 6^6}$.

Verify on a calculator

75 $\sqrt{5 + 2\sqrt{6}} = \sqrt{2} + \sqrt{3}$ **76** $\sqrt{5 - 2\sqrt{6}} = \sqrt{3} - \sqrt{2}$

77 $\sqrt{9 - 2\sqrt{14}} = \sqrt{7} - \sqrt{2}$ **78** $\sqrt{8 + 2\sqrt{15}} = \sqrt{3} + \sqrt{5}$.

79 Prove the relation in Ex. 75. **80** Prove the relation in Ex. 77.

5 RATIONAL EXPONENTS

What meaning can we give to $a^{m/n}$, where m/n is a rational number not necessarily an integer? If the rules of exponents are to hold, we should have

$$(a^{m/n})^n = a^{(m/n)n} = a^m.$$

This means that $a^{m/n}$ must be the n-th root of a^m. Therefore we *define*

$$a^{m/n} = \sqrt[n]{a^m},$$

where m/n is a rational number with $n > 0$.

Examples

$$8^{2/3} = \sqrt[3]{8^2} = \sqrt[3]{64} = 4 \qquad 25^{3/2} = \sqrt{(25)^3} = \sqrt{(5^2)^3} = \sqrt{5^6} = 5^3 = 125$$

$$9^{-1/2} = \sqrt{9^{-1}} = \sqrt{\tfrac{1}{9}} = \tfrac{1}{3}.$$

According to the definition of $a^{m/n}$, if $m = 1$,

$$a^{1/n} = \sqrt[n]{a},$$

so $a^{1/n}$ is just the n-th root of a in new clothes.

Examples $36^{1/2} = 6 \qquad 1000^{1/3} = 10 \qquad (\tfrac{1}{32})^{1/5} = \tfrac{1}{2}.$

Remark There is a subtle point in the definition of $a^{m/n}$. The same rational number might be expressed in two ways. For instance, suppose $m/n = p/q$. The definition then gives two possibilities,

$$a^{m/n} = \sqrt[n]{a^m} \qquad \text{and} \qquad a^{p/q} = \sqrt[q]{a^p}.$$

These had better be the same, or the definition is plain nonsense. They are the same, as is verified by showing that their qn-th powers are equal.

Example $\tfrac{15}{6} = \tfrac{10}{4}$ $(=\tfrac{5}{2})$ so we claim that $7^{15/6} = 7^{10/4}$.

Indeed, $7^{15/6} = \sqrt[6]{7^{15}} = \sqrt[6]{7^{6\cdot 2}\cdot 7^3} = 7^2\cdot\sqrt[6]{7^3} = 7^2\sqrt{7}$

and $7^{10/4} = \sqrt[4]{7^{10}} = \sqrt[4]{7^{4\cdot 2}\cdot 7^2} = 7^2\cdot\sqrt[4]{7^2} = 7^2\sqrt{7}.$

Rules of Exponents

The definition of $a^{m/n}$ is a useful one because all the rules of exponents for integer exponents carry over to fractional exponents.

Rules of exponents If m/n is rational with $n > 0$ and if $a > 0$, then

(1) $a^{m/n} = \sqrt[n]{a^m} = (\sqrt[n]{a})^m$.

If s and t are rational and if a and b are positive, then

(2) $a^s a^t = a^{s+t}$ (3) $\dfrac{a^s}{a^t} = a^{s-t}$

(4) $(a^s)^t = a^{st}$ (5) $a^s b^s = (ab)^s$.

The first equality in (1) is the definition of $a^{m/n}$: take the m-th power of a and then extract its n-th root. The second equality

$$\sqrt[n]{a^m} = (\sqrt[n]{a})^m$$

says something new, that you can extract the n-th root first, then take the m-th power, and you get the same thing. To prove it we show that $(\sqrt[n]{a})^m$ has the property that defines $a^{m/n}$; its n-th power is a^m:

$$[(\sqrt[n]{a})^m]^n = (\sqrt[n]{a})^{mn} = [(\sqrt[n]{a})^n]^m = a^m.$$

We prove Rule (2) by writing s and t with a common denominator and using rules for *integer* exponents and radicals. Thus we write

$$s = \frac{m}{n} \qquad t = \frac{p}{n} \qquad n > 0.$$

Then

$$a^s a^t = a^{m/n} a^{p/n} = \sqrt[n]{a^m}\sqrt[n]{a^p} = \sqrt[n]{a^m a^p} = \sqrt[n]{a^{m+p}}$$
$$= a^{(m+p)/n} = a^{(m/n)+(p/n)} = a^{s+t}.$$

Rule (3) is proved similarly. Rules (4) and (5) are proved by careful use of the rules for integer exponents and radicals.

We remark that the rules for rational exponents include as special cases the rules for integer exponents. The rule $\sqrt[n]{a}\sqrt[n]{b} = \sqrt[n]{ab}$ is just the special case $s = 1/n$ of Rule (5).

EXAMPLE 1 Simplify using rules for exponents

(a) $(9u^4)^{-3/2}$ (b) $\left(\dfrac{x^3}{8y^{-6}}\right)^{1/3}$ (c) $(16x^4y^8z^{13})^{1/4}$.

Solution

(a) $(9u^4)^{-3/2} = 9^{-3/2}(u^4)^{-3/2} = (3^2)^{-3/2}u^{(4)(-3/2)} = 3^{-3}u^{-6} = \dfrac{1}{27u^6}$.

(b) $\left(\dfrac{x^3}{8y^{-6}}\right)^{1/3} = \dfrac{(x^3)^{1/3}}{(8y^{-6})^{1/3}} = \dfrac{x}{8^{1/3}y^{-6/3}} = \dfrac{x}{2y^{-2}} = \dfrac{xy^2}{2}$.

(c) $(16x^4y^8z^{13})^{1/4} = (2^4x^4y^8z^{12}z)^{1/4}$

$$= (2^4)^{1/4}(x^4)^{1/4}(y^8)^{1/4}(z^{12})^{1/4}z^{1/4} = 2xy^2z^3\sqrt[4]{z}.$$

Answer (a) $\dfrac{1}{27u^6}$ (b) $\dfrac{xy^2}{2}$ (c) $2xy^2z^3\sqrt[4]{z}$

EXAMPLE 2 Express using only one radical

(a) $\sqrt[3]{9}\sqrt{\frac{1}{3}}$ (b) $\dfrac{\sqrt{r^3s^5}}{\sqrt[4]{r^2s}}$.

Solution (a) $\sqrt[3]{9}\sqrt{\frac{1}{3}} = \sqrt[3]{3^2}\sqrt{3^{-1}} = \sqrt[6]{3^4}\sqrt[6]{3^{-3}} = \sqrt[6]{3^4 \cdot 3^{-3}} = \sqrt[6]{3}$

(b) $\dfrac{\sqrt{r^3s^5}}{\sqrt[4]{r^2s}} = (r^3s^5)^{1/2}(r^2s)^{-1/4} = r^{3/2}s^{5/2}r^{-1/2}s^{-1/4}$

$$= r^{3/2-1/2}s^{5/2-1/4} = rs^{9/4} = rs^{2+1/4} = rs^2s^{1/4} = rs^2\sqrt[4]{s}.$$

Answer (a) $\sqrt[6]{3}$ (b) $rs^2\sqrt[4]{s}$

Calculating Powers

To estimate y^x on your calculator, first calculate (or approximate) x in decimal form, then use the [y^x] key as usual.

Examples

(1) $(81)^{1/4}$ 8 1 [y^x] . 2 5 [=] **3**

(2) $(27)^{2/3}$ 2 7 [y^x] . 6 6 6 6 6 6 7 [=] **9.0000010**

Note that the sequence

2 7 [y^x] 2 [÷] 3 [=]

does not calculate $(27)^{2/3}$, it calculates $27^2/3 = 243$. Thus to estimate something like $5^{3/17}$ you must proceed in two steps:

3 [÷] 1 7 [=] **0.1764706**

5 [y^x] . 1 7 6 4 7 0 6 [=] **1.3284575**

The work can be shortened by using the memory feature of your calculator. See p. 211, where this is discussed. The key sequence

3 [÷] 1 7 [=] [STO] 5 [y^x] [RCL] [=]

will compute $5^{3/17}$ without your having to key in the decimal approximation to 3/17. (The work can also be shortened if your calculator has parenthesis keys.)

Exercises

[All letters in these exercises represent positive numbers.]

Compute

1 $27^{2/3}$ **2** $64^{3/2}$ **3** $25^{-1/2}$ **4** $16^{-5/4}$

5 $(1000)^{5/3}$ **6** $(1{,}000{,}000)^{5/6}$ **7** $(\frac{4}{49})^{-3/2}$ **8** $(0.001)^{-5/3}$.

Express as a power of 2

9 $4\sqrt[3]{2}$ **10** $(\sqrt{2})^{2/3}$ **11** $8(16 \cdot 2^{-7/2})$ **12** $\left(\dfrac{32}{\sqrt[3]{2}}\right)^{1/6}$

13 $\sqrt{2} + \sqrt{2}$ **14** $\sqrt[3]{16}\sqrt[4]{\frac{1}{4}}$ **15** $\left(\dfrac{1}{2\sqrt{2}}\right)^{3/2}$ **16** $\sqrt{2^n}\sqrt[3]{2^k}$.

Simplify

17 $(25x^4)^{-3/2}$ **18** $\left(\dfrac{8}{u^6}\right)^{2/3}$ **19** $\left(\dfrac{u^4}{v^{12}}\right)^{3/4}$

20 $\left(\dfrac{27a^3}{8b^3c^6}\right)^{-4/3}$ **21** $(x^4y^6z^{-8})^{5/2}$ **22** $(x^{-3/2}\sqrt{y})^{-2}$

23 $(x^{4/3}y^{-2/3})^3$ **24** $(xy^2)^{1/3}(x^2y)^{-2/3}$ **25** $(8x\sqrt{x})^{5/3}$

26 $(x^{1/2} + y^{3/2})^2$ **27** $u^{1/3}(2u^{2/3} - u^{-1/6})$ **28** $v^{1/3}(8v^6)^{-2/3}$.

Express in terms of integral exponents and at most a single radical

29 $\sqrt{2} \cdot \sqrt[3]{2}$ **30** $\sqrt[3]{\sqrt{x^{1/4}}}$ **31** $\sqrt[3]{4}/\sqrt[6]{16}$

32 $\dfrac{\sqrt{xy}}{\sqrt[4]{x^2y}}$ **33** $\left(\dfrac{\sqrt{3a}}{\sqrt[3]{6a^2}}\right)^4$ **34** $\sqrt{b\sqrt{b}}$

35 $\sqrt{x}\cdot\sqrt{x^2}\cdot\sqrt{x^3}$ **36** $\sqrt{x}\cdot\sqrt[3]{x}\cdot\sqrt[4]{x}$ **37** $\sqrt{\sqrt{\sqrt{a}}}$

38 $\sqrt{a\sqrt{a\sqrt{a}}}$ **39** $\sqrt{3}\sqrt[5]{2}$ **40** $\sqrt[3]{27a^4b^2\sqrt{64ab}}$.

Express without radicals, using only positive exponents; simplify as much as possible

41 $(\sqrt[3]{xy^2})^{-3/5}$ **42** $\sqrt[5]{(xy^2)^{-10/3}/(x^2y)^{-15/7}}$

43 $(\sqrt[4]{x^{14}y^{-21/5}})^{-3/7}$ **44** $\sqrt[3]{x^{5/6}/x^{-5/6}}$

45 $\dfrac{\sqrt[3]{x}}{(x^{5/6})^{42}(x^{51})^{-2/3}}$ **46** $\left(\dfrac{x}{y}\right)^{1/5}\left(\dfrac{y}{z}\right)^{2/5}\left(\dfrac{z}{w}\right)^{3/5}\left(\dfrac{w}{x}\right)^{4/5}$.

Estimate on a calculator

47 $8^{3/4}$ **48** $10^{6/5}$ **49** $(2.79)^{4/7}$

50 $(6.207)^{13/21}$ **51** $(7.434\times10^{13})^{11/19}$ **52** $(1.043\times10^{-40})^{7/6}$.

6 POLYNOMIALS

In this section we study properties of the most common algebraic expressions, polynomials. A **polynomial** (in one variable x) is an algebraic expression of the form

$$a_nx^n + a_{n-1}x^{n-1} + \cdots + a_2x^2 + a_1x + a_0,$$

where $a_n, a_{n-1}, \cdots, a_2, a_1, a_0$ are real numbers called the **coefficients** of the polynomial.

Examples $x^2 + 6x + 5$ $\quad 2x^3 + \frac{1}{10}x - \frac{1}{4}\pi$ $\quad -8x^5 + x^3 + 17x^2 - 5$

The **degree** of a polynomial is the highest exponent occurring with non-zero coefficient. For example, the degrees of the polynomials above are 2, 3, and 5 respectively. Polynomials have special names according to degree:

Constant polynomial (zero degree) $a_0, \quad a_0 \neq 0$
Linear polynomial (first degree) $a_1x + a_0, \quad a_1 \neq 0$
Quadratic polynomial (second degree) $a_2x^2 + a_1x + a_0, \quad a_2 \neq 0$
Cubic polynomial (third degree) $a_3x^3 + a_2x^2 + a_1x + a_0, \quad a_3 \neq 0$
***n*-th degree** polynomial $a_nx^n + a_{n-1}x^{n-1} + \cdots + a_2x^2 + a_1x + a_0, \quad a_n \neq 0.$

Technically, the constant 0 is a polynomial, but it is not assigned a degree. In an n-th degree polynomial, the coefficient a_n is called the **leading coefficient,** and the term a_nx^n is called the **leading term.**

Remark We can write

$$a_0 + a_1x + \cdots + a_{n-1}x^{n-1} + a_nx^n$$

just as well as $a_nx^n + \cdots + a_1x + a_0$, and we shall use either form as is convenient. (The leading term is still a_nx^n.)

Polynomial Algebra

To add or subtract polynomials, add or subtract coefficients of like powers. This is not an arbitrary rule, but a consequence of the laws of algebra. Polynomials are sums; add them like any other sums.

EXAMPLE 1 Compute

(a) $(2x^2 + 6x + 5) + (3x^2 - 8x - 1)$

(b) $(x^4 + 4x^2 + x + 12) - (3x^2 + 7x)$.

Solution Use the commutative and associative laws to rearrange and regroup the terms. Then use the distributive law to combine like powers of x:

(a)
$$\begin{aligned}(2x^2 + 6x + 5) &+ (3x^2 - 8x - 1)\\ &= (2x^2 + 3x^2) + (6x - 8x) + (5 - 1)\\ &= (2 + 3)x^2 + (6 - 8)x + (5 - 1)\\ &= 5x^2 - 2x + 4.\end{aligned}$$

(b)
$$\begin{aligned}(x^4 + 4x^2 + x + 12) &- (3x^2 + 7x)\\ &= x^4 + (4x^2 - 3x^2) + (x - 7x) + 12\\ &= x^4 + (4 - 3)x^2 + (1 - 7)x + 12\\ &= x^4 + x^2 - 6x + 12.\end{aligned}$$

Answer (a) $5x^2 - 2x + 4$ (b) $x^4 + x^2 - 6x + 12$

Remark In practice, nobody adds polynomials with so much fuss. You just combine like powers by inspection.

To multiply polynomials, treat them like any other sums: multiply each term of the first by each term of the second using the rule for exponents.

$$x^m \cdot x^n = x^{m+n}.$$

Then collect all terms in the same power of x.

EXAMPLE 2 Compute

(a) $(3x + 2)(5x - 1)$ (b) $(x - 2)(4x^2 + 5x + 6)$.

Solution (a) Multiply each term of $3x + 2$ by each term of $5x - 1$:

$$\begin{aligned}(3x + 2)(5x - 1) &= (3x)(5x) + (2)(5x) + (3x)(-1) + (2)(-1)\\ &= 15x^2 + 10x - 3x - 2 = 15x^2 + 7x - 2.\end{aligned}$$

(b) Same technique, only you get six terms instead of four:

$$\begin{aligned}(x - 2)(4x^2 + 5x + 6) &= (x)(4x^2) - (2)(4x^2)\\ &\quad + (x)(5x) - (2)(5x) + (x)(6) - (2)(6)\\ &= 4x^3 - 8x^2 + 5x^2 - 10x + 6x - 12\\ &= 4x^3 - 3x^2 - 4x - 12.\end{aligned}$$

Answer (a) $15x^2 + 7x - 2$ (b) $4x^3 - 3x^2 - 4x - 12$

EXAMPLE 3 Express as $a_nx^n + \cdots + a_1x + a_0$

(a) $(x - 4)(x^2 + 1) + (2x + 1)(3x + 1)$

(b) $3(x - 1)(x + 4) + 2(x - 1)(5x - 7)$.

Solution (a) Compute each product, then add:

$$(x-4)(x^2+1)+(2x+1)(3x+1)$$
$$=(x^3-4x^2+x-4)+(6x^2+5x+1)$$
$$=x^3+2x^2+6x-3.$$

(b) You could multiply out each product as in (a), but it saves work to remove the common factor $x-1$ first:

$$3(x-1)(x+4)+2(x-1)(5x-7)$$
$$=(x-1)[3(x+4)+2(5x-7)]$$
$$=(x-1)(13x-2)=13x^2-15x+2.$$

Answer (a) x^3+2x^2+6x-3 (b) $13x^2-15x+2$

Example 4 Find the coefficient of x^2 in the product

$$(5x^3+2x^2+7x-3)(7x^4-12x^3+x^2+2x+10).$$

Solution No need to multiply it all out. Just collect all products of terms that produce x^2:

$$(2x^2)(10)+(7x)(2x)+(-3)(x^2)=(20+14-3)x^2=31x^2.$$

Answer 31

Because of the rules of exponents, the degree of the product of non-zero polynomials is the sum of their degrees. For if we multiply $a_mx^m+\cdots+a_0$ by $b_nx^n+\cdots+b_0$, we obtain

$$a_mb_nx^{m+n}+\cdots+a_0b_0.$$

Thus the degree of the product is $m+n$ since $a_mb_n\neq 0$, being the product of the non-zero leading coefficients of the factors.

> The degree of the product of non-zero polynomials is the sum of their degrees.

Exercises

Compute

1 $(x+4)+(5x-3)$

2 $(3x-2)+(x+6)$

3 $(8x^2+5x+6)+(3x-4)$

4 $(x^2+x+7)-(4x+1)$

5 $(3x^2+7x-2)-(9x^2+x+1)$

6 $(15x^2-8x-11)-(-10x^2+4x+5)$

7 $(\frac{1}{3}x^2-2x-6)+(\frac{4}{3}x^2-\frac{1}{2})$

8 $(4.1x^2+7.05x)-(2x^2-1.34x-0.02)$

9 $(x^3+2x^2-4x-5)+(6x^2-12x+1)$

10 $(x^4+7x^3+8x^2+2x+1)-(3x^3+4x^2+2x-1)$

11 $(-x^5+x+9)-2(x^4+x^3+x^2+x+1)$

12 $3(x^5-4x^4-6x^3-7)-(x^5+x^4+7x^2+x)$

13 $5(x^8+6x^6+4x^4+2x^2)+6(x^6-5x^5-4x^3)$

14 $3(-x^{10}+x+1)+4(x^3-4x^2+2x+2).$

Multiply

15 $(x+2)(x-3)$

16 $(x-4)(x-7)$

17 $(2x+5)(3x+1)$

18 $(8x-3)(x+2)$

19 $(x+6)(x^2+x-1)$

20 $(2x-1)(x^2+3x+2)$

21 $(x^3+x)(x^2+1)$

22 $(2x^3-1)(-x^2+x)$

23 $(x^5+\frac{1}{2}x^2)(x^2-x-\frac{1}{2})$

24 $(x^4-3x)(2x^2+5x+1)$

25 $x(x+1)(x^2+x+1)$

26 $(x+1)(x+2)(x+3)$

27 $(x-1)(x^4+x^3+x^2+x+1)$

28 $(x+1)(x^4-x^3+x^2-x+1)$

29 $(1+x)(1+x^2)$

30 (cont.) $(1+x)(1+x^2)(1+x^4)$

31 (cont.) Can you see a pattern?

32 $(1-x)(1+x)$

33 (cont.) $(1-x)(1+x)(1+x^2)$

34 (cont.) $(1-x)(1+x)(1+x^2)(1+x^4)$

35 (cont.) Can you see a pattern?

36 $(1-x^2)(1+x+x^2+x^3+x^4)$

37 $(x^2+x+1)(x^2-x+1)$

38 $(x^2+2x+2)(x^2-2x+2)$.

Express as $a_nx^n+\cdots+a_1x+a_0$

39 $(2x+3)(x^2+1)+(x-1)(x-2)$

40 $x(x+1)(x^2+1)-(2x+1)(2x-1)$

41 $(x+a)(x+b)$

42 $(x+a)(x+b)-(x+a)(x+c)$

43 $(x^2+1)(2x-1)+(x^2+1)(x^2+2)$

44 $x(3x^2+x-1)+(x+1)(x^2+2x)$

45 $(ax+b)(bx+a)$

46 $(x^2+c)(x^2-d)$

47 $(x+2)(4x+5)+3(2x-1)(3x-1)-4(x+1)(x-3)$

48 $(x+1)^2(x+4)-x(x^2+4x+6)+5(x-1)(2x-1)$

49 $(x+2)^2(x^2+1)+x(x+2)(x+3)+(x+2)(3x^2-5x-5)$

50 $(x+1)(x+2)(x+3)(x+4)-x(x+1)(x+2)(x+3)$.

Find the coefficient of x^3 without computing the whole product

51 $(x^2+3x+1)(2x-1)$

52 $(x^3-7x+6)(x^2+1)$

53 $x^2(2x-5)(x+6)$

54 $(x+1)(2x-1)(4x-1)$

55 $(x^4-6x^3+2x^2+5x+2)(x^3+x+4)$

56 $(x^3+2x^2+3x+4)(6x^3+7x^2-x-5)$

57 $(1+x^3)(1+x^4)(1+x^5)(1+x^6)$

58 $(1+2x)(1+3x^2)(1+4x^3)(1+5x^4)$

59 $x(2x+1)(3x+1)-(x^2+2)(3x-1)$

60 $3(x+5)(x-7)(x-9)-2(x^2-3)(x^4-1)+(x^2+x+7)(2x^2-x-1)$.

7 POLYNOMIALS IN SEVERAL VARIABLES

A **polynomial** in two variables x and y is a sum of terms of the form

$$1 \quad x \quad y \quad x^2 \quad xy \quad y^2 \quad x^3 \quad x^2y \quad xy^2 \quad y^3 \quad \cdots$$

with various coefficients. A term such as ax^ky^m is called a **monomial.** Its **degree** is the sum $k+m$ of its exponents. Thus the degree of $5x^2y^4$ is 6; the degree of $-8xy^3$ is 4. The **degree** of a *polynomial* is the highest degree of the monomials it contains. For example,

$$x - 5y + 2 \qquad x^2 + xy + 4y^2 + 2x \qquad \tfrac{1}{2}x^4y^2 - x^3y - 6x^2y^2 + 3y + 4$$

are polynomials in two variables of degree 1, 2, and 6, respectively.

A **polynomial** in three variables x, y, and z is a sum of **monomials** of the form $ax^ky^mz^n$, of **degree** $k + m + n$. The **degree** of a polynomial is the highest degree of the monomials it contains. For example,

$$x + 3y + 2z - 7 \qquad x^2 + y^2 + z^2 - 6yz + 8x - y + 2$$

$$y^2z^2 - xyz - 4x - 1$$

are polynomials in three variables of degrees 1, 2, and 4, respectively.

The algebra of polynomials in several variables is just like that of polynomials in one variable. Polynomials in several variables are sums. They are added, subtracted, and multiplied like any other sums.

EXAMPLE 1 Compute

(a) $(x^2 + 3y^2 + xy + 2x) + (4x^2 - y^2 + 2xy + 1)$

(b) $(4x^3 + x^2y + xy^2 + z^2 + xyz) - (4x^3 - 2x^2y - xyz + 5xz)$.

Solution (a) Rearrange and combine similar terms:

$$\begin{aligned}(x^2 + 3y^2 + xy + 2x) &+ (4x^2 - y^2 + 2xy + 1)\\ &= (x^2 + 4x^2) + (3y^2 - y^2) + (xy + 2xy) + 2x + 1\\ &= 5x^2 + 2y^2 + 3xy + 2x + 1.\end{aligned}$$

(b) $(4x^3 + x^2y + xy^2 + z^2 + xyz) - (4x^3 - 2x^2y - xyz + 5xz)$

$$\begin{aligned}&= (4x^3 - 4x^3) + (x^2y + 2x^2y) + xy^2 + z^2 + (xyz + xyz) - 5xz\\ &= 3x^2y + xy^2 + z^2 + 2xyz - 5xz.\end{aligned}$$

Answer (a) $5x^2 + 2y^2 + 3xy + 2x + 1$

(b) $3x^2y + xy^2 + z^2 + 2xyz - 5xz$

EXAMPLE 2 Multiply

(a) $(2x - y)(x^2 + xy + 3y^2)$ (b) $(xy + z)(xy^2 + y^2z^2 + y)$.

Solution (a) Multiply each term in the first parentheses by each term in the second, use rules of exponents, then collect similar terms:

$$\begin{aligned}(2x &- y)(x^2 + xy + 3y^2)\\ &= 2x\cdot x^2 - y\cdot x^2 + 2x\cdot xy - y\cdot xy + 2x\cdot 3y^2 - y\cdot 3y^2\\ &= 2x^3 + (-x^2y + 2x^2y) + (-xy^2 + 6xy^2) - 3y^3\\ &= 2x^3 + x^2y + 5xy^2 - 3y^3.\end{aligned}$$

(b) $(xy + z)(xy^2 + y^2z^2 + y)$

$$\begin{aligned}&= xy\cdot xy^2 + z\cdot xy^2 + xy\cdot y^2z^2 + z\cdot y^2z^2 + xy\cdot y + z\cdot y\\ &= x^2y^3 + xy^2z + xy^3z^2 + y^2z^3 + xy^2 + yz.\end{aligned}$$

Answer (a) $2x^3 + x^2y + 5xy^2 - 3y^3$

(b) $x^2y^3 + xy^2z + xy^3z^2 + y^2z^3 + xy^2 + yz$

Perfect Squares

If we square $x + y$, we obtain

$$(x + y)^2 = (x + y)(x + y) = x^2 + xy + yx + y^2 = x^2 + 2xy + y^2.$$

$$\boxed{(x + y)^2 = x^2 + 2xy + y^2.}$$

This formula holds for all real numbers x and y. Therefore, it holds just as well if x and y are replaced by any expressions that represent real numbers.

Examples $(a - b)^2 = a^2 - 2ab + b^2$ $\quad (x + \frac{1}{2})^2 = x^2 + x + \frac{1}{4}$

$$(2a + 3b)^2 = (2a)^2 + 2(2a)(3b) + (3b)^2 = 4a^2 + 12ab + 9b^2.$$

Recognizing squares is an important skill. For example, given the expression $4a^2 + 12ab + 9b^2$, would you spot it as $(2a + 3b)^2$ in disguise?

When you suspect a perfect square you should look for three terms: two squares plus twice the product of the unsquared quantities. In other words, look for a combination like this:

$$(\text{first})^2 + 2(\text{first})(\text{second}) + (\text{second})^2.$$

EXAMPLE 3 Express as a perfect square if possible

(a) $x^2 - 10x + 25$ (b) $x^2y^2 + 6xyz^2 + 9z^4$

(c) $36x^2 + 6xy + \frac{1}{4}y^2$ (d) $9x^2 + 8xy + 16y^2$.

Solution (a) Does $x^2 - 10x + 25$ fall into the pattern of $a^2 + 2ab + b^2$? Yes, if $a = x$ and $b = -5$:

$$(x - 5)^2 = x^2 + 2x(-5) + (-5)^2 = x^2 - 10x + 25.$$

(b) The logical choice is $a^2 = x^2y^2$ and $b^2 = 9z^4$, that is, $a = xy$ and $b = 3z^2$:

$$(xy + 3z^2)^2 = (xy)^2 + 2(xy)(3z^2) + (3z^2)^2 = x^2y^2 + 6xyz^2 + 9z^4.$$

(c) Try $a = 6x$ and $b = \frac{1}{2}y$:

$$(6x + \tfrac{1}{2}y)^2 = (6x)^2 + 2(6x)(\tfrac{1}{2}y) + (\tfrac{1}{2}y)^2 = 36x^2 + 6xy + \tfrac{1}{4}y^2.$$

(d) The only logical choices are $a = 3x$ and $b = 4y$:

$$(3x + 4y)^2 = (3x)^2 + 2(3x)(4y) + (4y)^2 = 9x^2 + 24xy + 16y^2.$$

We obtain $24xy$, not $8xy$; therefore the given expression (d) is not a square.

Answer (a) $(x - 5)^2$ (b) $(xy + 3z^2)^2$

(c) $(6x + \frac{1}{2}y)^2$ (d) not a square

Exercises

Compute

1 $(x + 2y + 3) + (4x - y - 1)$ **2** $(2x - y + 8) + (3x + 2y - 5)$

3 $(-4x + 5y + 6z) - (x - y + z + 2)$ **4** $(y + 2z) - (3x + y + z + 1)$

5 $(x^2 + 3y^2) + (x^2 + y^2 - 3y)$

6 $(x^2 + 6xy + 2y^2) - (x^2 + 4xy - y^2 - x - 2)$

7 $(x^2 + y^2 + z^2 + xy + yz + xz) - (3x^2 + y^2 + z^2 + xy + x + y)$

8 $x(3x^2 + xy^3 + yz^4) - y(xz^4 + xy + x^2y^2)$.

Multiply

9 $(x + y)(2x + 3y)$

10 $(2x - y)(3x - y)$

11 $(5x + 2y)(5x - 2y)$

12 $(\frac{1}{2}x - \frac{1}{3}y)(2x - 3y)$

13 $(4u - 3v)^2$

14 $(uv + 2w)^2$

15 $(x + y)^3$

16 $(x - y)(x + y)(x^2 + y^2)$

17 $(x + y + z)(2x - y)$

18 $(y - z)(3x + 4y + z + 1)$

19 $(x + y - 1)(x^2 - y^2 + x)$

20 $(x + y + z)(x + y - z)$

21 $(x^2 + y^2)(x^2 - 3y^2)$

22 $(x^4 - y^4)(x^4 + y^4)$

23 $(x + y + z)^2$

24 $(ax + by + cz)(cx + by + az)$

25 $(r + 1)(s + 1)(t + 1)$

26 $(r + s)(r - s)^2$

27 $(a + b)(a + 2b)(a + 3b)$

28 $(a + b)(b + c)(c + a)$

29 $(x^2 + 2xy - 2y^2)(x^2 - 2xy + 2y^2)$

30 $(x + y + z)(x^2 + y^2 + z^2 - xy - yz - zx)$.

Write out

31 $(x + 2y)^2$

32 $(x - 3y)^2$

33 $(x - 4y)^2$

34 $(x + 5y)^2$

35 $(3a + 2b)^2$

36 $(t^2 + 5w)^2$

37 $(u^2 + v^2)^2$

38 $(4w - 5z^3)^2$.

Mental arithmetic: compute

39 102^2

40 999^2.

Is the expression a perfect square? If so, of what?

41 $x^2 + 16x + 64$

42 $y^4 + 2y^2 + 1$

43 $z^6 + 4z^3 + 4$

44 $16a^2 + 8a + 1$

45 $x^2y^2 + 3xy + 9$

46 $9c^2 - 30cd + 25d^2$

47 $(x^2 + 1)^2 + 2(x^2 + 1) + 1$

48 $9u^2 + 9u + 1$

49 $a^2b^8 - 4ab^4c^2 + 4c^4$

50* $x^2 + 2xy + y^2 + 4xz + 4yz + 4z^2$.

Explain these party games

51 Take a number from 1 to 10. Square the number one larger and square the number one smaller. Subtract. Divide by the number you started with. Now you have 4.

52 Multiply your number by the number 4 larger. Add 4. Take the square root. Subtract the number you started with. Now you have 2.

53 If n is a positive integer, show that $3n^2 + 6n + 5$ is the sum of the squares of three consecutive integers.

54 Consider two consecutive positive integers. Which is larger, the average of their squares or the square of their average?

Find the coefficient of x^2y in

55 $(2x^2 + 3x - 4)(y^2 + 8y + 1)$

56 $(x + y)(x^2 + y^2)(1 + x)$

57 $(x + 4y)^3$

58 $(1 + x + y + xy + x^2 + y^2)(1 + 2x + 3y + 2xy + x^2)$.

8 FACTORING

Factoring is the process of expressing a polynomial as a product of polynomials of lower degree (factors). It is useful for simplifying algebraic expressions, for solving certain types of equations, and for numerical work. When we write

$$x^2 + 5x + 6 = (x + 2)(x + 3),$$

we are factoring the left-hand side. If a given polynomial has integer coefficients, we generally look for factors with integer coefficients.

A familiar kind of factoring is removing a common polynomial factor.

Examples

(1) $x^4 - 6x^3 - 2x^2 = x^2(x^2 - 6x - 2)$

(2) $(x - 3)(x^2 + 4) + 2(x - 3)(5x - 1)$

$$= (x - 3)[(x^2 + 4) + 2(5x - 1)]$$
$$= (x - 3)(x^2 + 10x + 2).$$

Most factoring depends on recognizing certain polynomial products. In Section 7, we had some practice in recognizing perfect squares. Here are some other useful factored forms:

Factoring Formulas

(1) $x^2 + (c + d)x + cd = (x + c)(x + d)$

(2) $acx^2 + (ad + bc)x + bd = (ax + b)(cx + d)$

(3) $x^2 - y^2 = (x + y)(x - y)$

(4) $x^3 + y^3 = (x + y)(x^2 - xy + y^2)$

(5) $x^3 - y^3 = (x - y)(x^2 + xy + y^2)$

(6) $x^n - y^n = (x - y)(x^{n-1} + x^{n-2}y + x^{n-3}y^2 + \cdots + xy^{n-2} + y^{n-1})$.

These formulas are checked by multiplication. Each holds for all real values of x and y. Therefore, x and y may be replaced by any expressions that represent real numbers. For example, (3) is a formula for the difference of any two squares. From it follow such formulas as

$$4a^2 - 9b^2 = (2a)^2 - (3b)^2 = (2a + 3b)(2a - 3b)$$
$$36 - t^4 = 6^2 - (t^2)^2 = (6 + t^2)(6 - t^2)$$
$$(u + v)^2 - 49 = (u + v)^2 - 7^2 = (u + v + 7)(u + v - 7).$$

Let us look at some examples of factoring. In these examples, and in the exercises, the instruction "factor" means factor completely into factors with integer coefficients.

EXAMPLE 1 Factor (a) $x^2 + 10x + 9$ (b) $2x^2 - 7x + 3$.

Solution (a) Try $(x + c)(x + d)$. By formula (1), we need numbers c and d for which $c + d = 10$ and $cd = 9$. Such a pair is $c = 1$, $d = 9$. So

$$x^2 + 10x + 9 = (x + 1)(x + 9)$$

as is checked by multiplying.

(b) Try $(ax + b)(cx + d)$. By formula (2), we need $ac = 2$. So try $a = 1$ and $c = 2$, that is, $(x + b)(2x + d)$. Now we need $bd = 3$ and $2b + d = -7$. A little trial and error shows that $b = -3$ and $d = -1$ work, so

$$2x^2 - 7x + 3 = (x - 3)(2x - 1).$$

Answer (a) $(x + 1)(x + 9)$ (b) $(x - 3)(2x - 1)$

Example 2 Factor $x^5 - 8x^3 + 16x$.

Solution First remove the common factor x:

$$x^5 - 8x^3 + 16x = x(x^4 - 8x^2 + 16)$$

The polynomial in parentheses is a perfect square, $(x^2 - 4)^2$. Also, by formula (3), we have $x^2 - 4 = (x + 2)(x - 2)$. Therefore

$$x^5 - 8x^3 + 16x = x(x^2 - 4)^2 = x[(x + 2)(x - 2)]^2.$$

Answer $x(x + 2)^2(x - 2)^2$

Example 3 Factor (a) $x^3 + 8$ (b) $27x^3 - 125y^3$.

Solution (a) Use formula (4) for the sum of x^3 and 2^3:

$$\begin{aligned} x^3 + 8 = x^3 + 2^3 &= (x + 2)(x^2 - x \cdot 2 + 2^2) \\ &= (x + 2)(x^2 - 2x + 4). \end{aligned}$$

(b) Use formula (5) for the difference of cubes:

$$\begin{aligned} 27x^3 - 125y^3 &= (3x)^3 - (5y)^3 \\ &= (3x - 5y)[(3x)^2 + (3x)(5y) + (5y)^2] \\ &= (3x - 5y)(9x^2 + 15xy + 25y^2). \end{aligned}$$

Answer (a) $(x + 2)(x^2 - 2x + 4)$
(b) $(3x - 5y)(9x^2 + 15xy + 25y^2)$

Remark It's not a bad idea to check your factoring by multiplying out.

Example 4 Without long division or a calculator, show that 999,992 is divisible by 98.

Solution 999,992 is the difference of cubes:

$$\begin{aligned} 999{,}992 &= 1{,}000{,}000 - 8 = (100)^3 - 2^3 \\ &= (100 - 2)(100^2 + 100 \cdot 2 + 2^2) \\ &= 98(10000 + 200 + 4) = (98)(10204). \end{aligned}$$

Example 5 Factor $x^4 - 1$.

Solution This is a difference of squares:

$$x^4 - 1 = (x^2)^2 - 1^2 = (x^2 + 1)(x^2 - 1).$$

The second factor is also a difference of squares, hence

$$x^4 - 1 = (x^2 + 1)(x + 1)(x - 1).$$

Answer $(x^2 + 1)(x + 1)(x - 1)$

Sometimes an expression can be factored by grouping terms suitably. The next example illustrates the idea.

EXAMPLE 6 Factor (a) $xy - y + 2x - 2$

(b) $ab^2 + bc - c^2 - abc$.

Solution (a) We note that

$$xy - y = y(x - 1) \quad \text{and} \quad 2x - 2 = 2(x - 1).$$

Hence

$$xy - y + 2x - 2 = y(x - 1) + 2(x - 1) = (y + 2)(x - 1).$$

(b) We group the first term with the fourth, and the second term with the third, then factor each group:

$$(ab^2 - abc) + (bc - c^2) = ab(b - c) + c(b - c)$$
$$= (ab + c)(b - c).$$

Answer (a) $(y + 2)(x - 1)$ (b) $(ab + c)(b - c)$

Remark Alternatively in (a),

$$xy - y + 2x - 2 = (xy + 2x) + (-y - 2) = x(y + 2) - (y + 2)$$
$$= (x - 1)(y + 2)$$

and a similar alternative for (b) is possible.

What is Factoring Really?

All examples in this section have involved integer coefficients, both in the given polynomials and in their factors. However, there is a big difference between seeking factors with *integer* coefficients and factors with *real* coefficients. For example, if we do not require integer coefficients, it is easy to factor $x^2 - 5$:

$$x^2 - 5 = x^2 - (\sqrt{5})^2 = (x - \sqrt{5})(x + \sqrt{5}).$$

But if we insist on integer coefficients, it can't be done, as some trial-and-error will show.

Not every polynomial can be factored, even into factors with real coefficients. Consider the polynomial $x^2 + 1$. Suppose it had factors with real coefficients,

$$x^2 + 1 = (ax + b)(cx + d) \qquad a \neq 0, \quad c \neq 0.$$

Set $x = -b/a$:

$$\left(-\frac{b}{a}\right)^2 + 1 = \left[a\left(\frac{-b}{a}\right) + b\right]\left[c\left(\frac{-b}{a}\right) + d\right]$$
$$= (-b + b)\left(-\frac{cb}{a} + d\right) = 0.$$

Hence $(-b/a)^2 = -1$. But this is impossible because the square of a real number is non-negative. Therefore $x^2 + 1$ cannot be factored.

We have seen that not every polynomial can be factored, and whether a given polynomial can might depend on what coefficients are allowed. There is no sure-fire method for deciding whether a given polynomial can be factored; this may be a hard problem. Certainly it is not obvious that $x^3 + y^5$ *cannot* be factored, no matter what coefficients you allow. Nor is it obvious that $x^4 + 4$ *can* be factored into two quadratic factors with integer coefficients, $x^2 + 2x + 2$ and $x^2 - 2x + 2$.

Exercises

Factor

1 $x^2 + 4x + 3$
2 $x^2 + 11x + 30$
3 $x^2 + x - 6$
4 $x^2 - 8x + 15$
5 $x^2 - 9x + 14$
6 $x^2 + 5x - 36$
7 $x^2 - 6x + 9$
8 $x^3 - 4x^2 + 3x$
9 $x^4 + x^3 - 20x^2$
10 $3x^2 + 4x + 1$
11 $2x^2 + 11x + 12$
12 $10x^2 - 17x + 3$
13 $4x^2 + 4x + 1$
14 $x^6 + 7x^3 + 12$
15 $x^4 - 5x^2 + 6$
16 $x^4 - x^2$
17 $4x^2 - 81$
18 $9y^2 - 100$
19 $(x + y)^2 - z^2$
20 $9a^2 - 4 + (3a - 2)(a + 1)$
21 $a^4 - 16b^2$
22 $(u^2 - 1) + 2(u + 1)$
23 $(v^4 - 81) + v^3 + 9v$
24 $b^3 - 27$
25 $8c^3 + 27$
26 $x^4 - 8x$
27 $8(x + y)^3 + 27x^3$
28 $(2a + 3b)^2 - (a - b)^2$
29 $x^5 - 1$
30 $x^5 - 32$
31 $(x^5 - 1) - (x^2 - 1)$
32 $x^6 - 64$
33 $x^8 - 1$
34 $(x^2 - 1)^2 - 6(x^2 - 1) + 9$
35 $8x^3y^6 + 1000$
36 $(16x^4 - 1) + (4x^4 - 3x^2 - 1)$
37 $x^2 - y^4z^6$
38 $(x + y + z)^2 - (x + y)^2$
39 $a^6 - 18a^4b^2 + 81a^2b^4$
40 $(a + b)^3 + (a - b)^3$
41 $r^4 - 4(s + t)^2$
42 $(r + s)^3(r - s) - (r + s)(r - s)^3$
43 $25 - (a + b + c + 5)^2$
44 $107^2 - 93^2$.

Factor by grouping

45 $4x + 3y + 6xy + 2$
46 $ax + 3axy + 3by^2 + by$
47 $2pq - 8q + 5p - 20$
48 $x^6 - 2x^4 + 5x^2 - 10$
49 $u^4 + u^2 - v^2 - v^4$
50 $4rt - 10qs - 5st + 8qr$.

51 Verify formula (4) by multiplication.
52 Verify formula (5) by multiplication.
53 Write out formula (6) for $n = 5$ and verify it by multiplication.
54 Work Example 5 using formula (6) and grouping.
55 Without dividing, show that 1027 is divisible by 13.
56 Factor $x^2 + 2xy - 8y^2$ by writing $-8y^2$ in the form $y^2 - 9y^2$.

9 RATIONAL EXPRESSIONS

Quotients of polynomials are called **rational expressions.**

Examples

$$\frac{2x + 5}{x^2 - 9} \qquad \frac{x^3 - 7x^2 + x + 4}{5x^4 + 9x + 1} \qquad \frac{y}{x} \qquad \frac{x + 4}{2y - 7} \qquad \frac{x^2z}{y^2 + z^2 + 1}.$$

Let us agree that all denominators are understood to be non-zero. Thus, in the first example above, $x \neq \pm 3$, even though this is not stated explicitly.

A rational expression is in **lowest terms** if numerator and denominator have no common polynomial factors.

EXAMPLE 1 Express in lowest terms

(a) $\dfrac{x+y}{(x+y)(2x+y)}$ (b) $\dfrac{3x-6}{x^2-4x+4}$ (c) $\dfrac{x^2-1}{x^3-1}$.

Solution (a) Simply cancel the common factor $x+y$:

$$\frac{x+y}{(x+y)(2x+y)} = \frac{1}{2x+y}.$$

(b) Factor the numerator and denominator, then cancel the common factor:

$$\frac{3x-6}{x^2-4x+4} = \frac{3(x-2)}{(x-2)^2} = \frac{3}{x-2}.$$

(c) $$\frac{x^2-1}{x^3-1} = \frac{(x-1)(x+1)}{(x-1)(x^2+x+1)} = \frac{x+1}{x^2+x+1}.$$

Answer (a) $\dfrac{1}{2x+y}$ (b) $\dfrac{3}{x-2}$ (c) $\dfrac{x+1}{x^2+x+1}$

Since rational expressions represent quotients of real numbers, they satisfy the rules of algebra for quotients (p. 3). To add, express the summands as quotients with a *common denominator,* then add the numerators. To multiply, multiply the numerators and multiply the denominators. To divide, invert the divisor, then multiply.

EXAMPLE 2 Compute

(a) $\dfrac{x}{3x-y} + \dfrac{y}{3x+y}$ (b) $\dfrac{2x+1}{x^2-4} + \dfrac{2}{x+2} - \dfrac{1}{x-2}$.

Solution (a) A common denominator is $(3x-y)(3x+y)$:

$$\frac{x}{3x-y} + \frac{y}{3x+y} = \frac{x(3x+y)}{(3x-y)(3x+y)} + \frac{(3x-y)y}{(3x-y)(3x+y)}$$
$$= \frac{x(3x+y)+(3x-y)y}{(3x-y)(3x+y)} = \frac{3x^2+4xy-y^2}{9x^2-y^2}.$$

(b) A common denominator is $x^2-4 = (x+2)(x-2)$:

$$\frac{2x+1}{x^2-4} + \frac{2}{x+2} - \frac{1}{x-2} = \frac{2x+1}{x^2-4} + \frac{2(x-2)}{x^2-4} - \frac{x+2}{x^2-4}$$
$$= \frac{(2x+1)+2(x-2)-(x+2)}{x^2-4} = \frac{3x-5}{x^2-4}.$$

Answer (a) $\dfrac{3x^2+4xy-y^2}{9x^2-y^2}$ (b) $\dfrac{3x-5}{x^2-4}$

EXAMPLE 3 Compute and express in lowest terms

(a) $\dfrac{xy}{(x-1)^2} \cdot \dfrac{x^2-1}{y^2+y}$ (b) $\dfrac{\dfrac{x+1}{x^2+1}}{\dfrac{x}{x+3}}$

Solution (a) Multiply numerators and denominators, factor, then cancel common factors:

$$\frac{xy}{(x-1)^2}\cdot\frac{x^2-1}{y^2+y}=\frac{xy(x^2-1)}{(x-1)^2(y^2+y)}=\frac{xy(x-1)(x+1)}{(x-1)^2y(y+1)}$$
$$=\frac{x(x+1)}{(x-1)(y+1)}.$$

(b) Invert the divisor and multiply:

$$\frac{\dfrac{x+1}{x^2+1}}{\dfrac{x}{x+3}}=\frac{x+1}{x^2+1}\cdot\frac{x+3}{x}=\frac{(x+1)(x+3)}{x(x^2+1)}.$$

There are no common factors; the product is already in lowest terms.

Answer (a) $\dfrac{x(x+1)}{(x-1)(y+1)}$ (b) $\dfrac{(x+1)(x+3)}{x(x^2+1)}$

Example 4 Simplify $\dfrac{\dfrac{1}{x}-\dfrac{1}{x+1}}{1+\dfrac{1}{x}}$.

Solution One way to handle a 4-story fraction like this is to simplify the numerator and denominator separately, then to divide the two fractions. A second way, sometimes shorter, is to multiply numerator and denominator by a common denominator. In this case multiply by $x(x+1)$:

$$\frac{\dfrac{1}{x}-\dfrac{1}{x+1}}{1+\dfrac{1}{x}}=\frac{x(x+1)\left(\dfrac{1}{x}-\dfrac{1}{x+1}\right)}{x(x+1)\left(1+\dfrac{1}{x}\right)}$$
$$=\frac{(x+1)-x}{x(x+1)+(x+1)}=\frac{1}{(x+1)^2}.$$

Answer $\dfrac{1}{(x+1)^2}$

Exercises

Express in lowest terms

1 $\dfrac{x-4}{x^2-16}$ **2** $\dfrac{x^4+3x^2}{x}$ **3** $\dfrac{x+1}{x^3+1}$

4 $\dfrac{x^2-x-30}{x^2-25}$ **5** $\dfrac{x^2-y^2}{x^3-y^3}$ **6** $\dfrac{(x+y)(x^2-xy+y^2)}{x^4+xy^3}$

7 $\dfrac{x+y-2z}{(x+y)^2-4z^2}$ **8** $\dfrac{2xyz+3xz^2}{4xy^2+12xyz+9xz^2}$.

Compute

9 $\dfrac{1}{x-1}-\dfrac{1}{x}$ **10** $\dfrac{3}{x-1}+\dfrac{4}{x+1}$

11 $\dfrac{1}{2x+3}-\dfrac{1}{2x+5}$ **12** $\dfrac{1}{x}-\dfrac{1}{2x+1}$

13 $\dfrac{x}{x^2-4}+\dfrac{1}{x+2}+\dfrac{3}{x-2}$

14 $\dfrac{3x}{x^2-4}+\dfrac{1}{(x-2)^2}$

15 $\dfrac{x}{y}+\dfrac{y}{x}$

16 $\dfrac{1}{x}+\dfrac{1}{xy}+\dfrac{1}{xy^2}$

17 $\dfrac{x+y}{x+2y}-\dfrac{y}{x+3y}$

18 $1-\dfrac{y^2}{x^2+y^2}$.

Multiply and express in lowest terms

19 $\dfrac{x}{1-x}\cdot\dfrac{x^2-1}{x^2}$

20 $\dfrac{2x+3}{x^2+1}\cdot\dfrac{x}{x^2-1}$

21 $\dfrac{x}{y}\cdot\dfrac{x+1}{y+1}$

22 $\dfrac{x^2+xy}{x+3y}\cdot\dfrac{2y}{x^2+y^2}$

23 $\dfrac{1}{x+y}\cdot\left(\dfrac{1}{x}+\dfrac{1}{y}\right)$

24 $\dfrac{1}{x-y}\cdot\left(\dfrac{y}{x}-\dfrac{x}{y}\right)$

25 $xyz\left(\dfrac{1}{x}+\dfrac{2}{y}-\dfrac{3}{z}\right)$

26 $\left(\dfrac{x+y}{x-2y}\right)^2\cdot\dfrac{x^2-3xy+2y^2}{(x^2-y^2)^2}$.

Simplify the quotient

27 $\dfrac{\dfrac{x+3}{x+4}}{\dfrac{2x}{x+4}}$

28 $\dfrac{\dfrac{x+y}{x+2y}}{\left(\dfrac{x+y}{xy}\right)^2}$

29 $\dfrac{\dfrac{a^2-1}{a^3+a}}{\dfrac{a+1}{a^2+1}}$

30 $\dfrac{(a+b)^2}{\dfrac{a^2-b^2}{ab}}$

31 $\dfrac{\dfrac{u^2}{v^2}-\dfrac{v^2}{u^2}}{\dfrac{u}{v}+\dfrac{v}{u}}$

32 $\dfrac{\dfrac{1}{t-3}+\dfrac{4}{t^2-9}}{2+\dfrac{1}{t+3}}$

33 $\left(\dfrac{u^2}{v^2}-\dfrac{v^2}{u^2}\right)\Big/\left(\dfrac{u^4}{v^4}-2+\dfrac{v^4}{u^4}\right)$

34 $\dfrac{x^4-\dfrac{1}{y^4}}{1-x^2y^2}$

35 $\dfrac{(xu-yv)^2+(xv+yu)^2}{x^2+y^2}$

36 $\dfrac{x^2+\dfrac{x}{y}-\dfrac{6}{y^2}}{x^2+4\dfrac{x}{y}+\dfrac{3}{y^2}}$

37 $\dfrac{\dfrac{3}{r+1}+\dfrac{1}{3r+1}}{\dfrac{1}{r+1}-\dfrac{1}{3r+1}}$

38 $\dfrac{\dfrac{s^2}{t}-\dfrac{t^2}{s}}{s^2+st+t^2}$.

Simplify, and express as a quotient of polynomials

39 $\dfrac{1}{2x-1}-\dfrac{2}{(2x-1)^2}+\dfrac{3}{(2x-1)^3}$

40 $(x^2-3x+2)\left(\dfrac{1}{x-1}+\dfrac{1}{x-2}+\dfrac{1}{x-3}\right)$

41 $\dfrac{1}{(x+1)^2}-\dfrac{1}{(2x+1)^2}$

42 $\dfrac{1}{x(x+1)}+\dfrac{1}{(x+1)(x+2)}+\dfrac{1}{(x+2)(x+3)}$

10 COMMON ERRORS IN ALGEBRA

The basic rules of algebra discussed in this chapter are indispensable in this course and all other mathematics courses you will take. It is worth the effort to master them now so you will avoid silly mistakes in the future. Let us list some of the most common types of errors in algebra.

1 Incorrect use of rules of real numbers

WRONG	RIGHT
$x/(y+2) = x/y + 2$	Keep the parentheses: $x/(y+2)$
$(x+3)(x+4) = x + 3\cdot x + 4$	$(x+3)(x+4) = x^2 + 7x + 12$
$(x+y)^2 = x + y^2$	Keep the parentheses: $(x+y)^2$
$\dfrac{1}{x+y} = \dfrac{1}{x} + \dfrac{1}{y}$	Leave as $\dfrac{1}{x+y}$.

2 Poor use of parentheses

WRONG	RIGHT
$3(x+1) = 3x + 1$	$3(x+1) = 3x + 3$
$(\frac{1}{2}x)(\frac{1}{2}y) = \frac{1}{2}xy$	$(\frac{1}{2}x)(\frac{1}{2}y) = \frac{1}{4}xy$
$x - (2y+1) = x - 2y + 1$	$x - (2y+1) = x - 2y - 1$
$-3x^2 + 15x - 6 = -3(x^2 + 5x + 2)$	$-3x^2 + 15x - 6 = -3(x^2 - 5x + 2)$

3 Mistakes with exponents and radicals

WRONG	RIGHT
$(2x)^3 = 2x^3$	$(2x)^3 = 8x^3$
$(x^3)^4 = x^7$	$(x^3)^4 = x^{12}$
$x^3x^4 = x^{12}$	$x^3x^4 = x^7$
$\sqrt{3x} = 3\sqrt{x}$	$\sqrt{3x} = \sqrt{3}\sqrt{x}$
$\sqrt{x+y} = \sqrt{x} + \sqrt{y}$	Leave as $\sqrt{x+y}$.

4 Incorrect use of formulas

WRONG	RIGHT
$\dfrac{1}{x} + \dfrac{1}{y} = \dfrac{1}{x+y}$	$\dfrac{1}{x} + \dfrac{1}{y} = \dfrac{x+y}{xy}$
$(x+y)^2 = x^2 + y^2$	$(x+y)^2 = x^2 + 2xy + y^2$
$x^3 + y^3 = (x+y)(x^2 + xy + y^2)$	$x^3 + y^3 = (x+y)(x^2 - xy + y^2)$.

5 Sloppy writing

$\sqrt{x + 1}$ (radical sign extending ambiguously)	Is it $\sqrt{x+1}$ or $\sqrt{x} + 1$?
$\dfrac{1}{x + y}$+3 (ambiguous fraction bar)	Is it $\dfrac{1}{x+y+3}$ or $\dfrac{1}{x+y} + 3$?
x−5 (ambiguous)	Is it x^{-5} or $x - 5$?
$\dfrac{1}{2}$ x (ambiguous)	Is it $\frac{1}{2}x$ or $\dfrac{1}{2x}$ or $\dfrac{1}{2^x}$?

These examples illustrate some, but not all, of the common mistakes made in algebra. Get into the habit of checking your work. Read it over and see if it says what you mean. Are your formulas ambiguous? Are they legible?

Exercises

Find the mistake in each formula and correct it.

1 $\left(\dfrac{x}{5}\right)^2 = \dfrac{x^2}{5}$

2 $(3x + 1)^2 = 9x^2 + 1$

3 $\sqrt{4x + 4} = 4\sqrt{x + 1}$

4 $(-x)^4 = -x^4$

5 $\sqrt{2^9} = 2^3$

6 $\dfrac{x^4}{x^2 + x^4} = x^2 + 1$

7 $\sqrt[3]{\sqrt[3]{x}} = \sqrt[6]{x}$

8 $(x - y)(x + y) = x^2 + y^2$

9 $\sqrt{x^2 + 1} = x + 1$

10 $\left(\dfrac{1}{\sqrt{x} + \sqrt{y}}\right)^2 = \dfrac{1}{x + y}$

11 $2a(a^2 + 4a + 1) = 2a^3 + 8a^2 + 1$

12 $\sqrt{3x + 5} + \sqrt{2x + 1} = \sqrt{5x + 6}$

13 $a^3 + b^3 = (a + b)^3$

14 $\sqrt[3]{a^2} = a^{-2/3}$

15 $a^{-2}a^{-2} = a^4$

16 $(x/y^2)^2 = x/y^4$

17 $\sqrt{x^3 + 2x^2 + x} = x(x + 1)$

18 $x + x + x + x = x^4$

19 $(25x)(4x) = 100x$

20 $\sqrt{x} + \sqrt{2x} = \sqrt{3x}$

21 $\dfrac{x + y}{x + z} = 1 + \dfrac{y}{z}$

22 $x^{-2} = \dfrac{-2}{x}$

23 $(x + 1)(y + 1)(z + 1)$
$= xyz + x + y + z + 1$

24 $1 + \dfrac{1}{x} = \dfrac{2}{1 + x}$

25 $\dfrac{4x^3 - 7x^2 + 1}{x} = 4x^2 - 7x + 1$

26 $\dfrac{1/x}{1/y} = \dfrac{1}{xy}$

27 $(x + y)^3 = x^3 + 3xy + y^3$

28 $x^{p/q} = x^p/x^q$

29 $\dfrac{x^2}{x + y} = \dfrac{x}{1 + y}$

30 $\sqrt[3]{8 + 27} = 2 + 3 = 5$

31 $-2^4 = 2^{-4}$

32 $(-2)^{-4} = -2^{-4}$

33 $\dfrac{3 + x}{x + 2} = \dfrac{3}{2}$

34 $\dfrac{1}{x} - \dfrac{1}{y} = \dfrac{1}{x - y}.$

REVIEW EXERCISES

1 Derive the relation

$$3x + 3y + 4x^2 - 4y^2 = (3 + 4x - 4y)(x + y).$$

State which law of algebra is used at each step.

2 Which is greater: $3 \cdot 5 \cdot 7 \cdot 9$ or 8^4?

3 Express as a fraction in lowest terms $\dfrac{\frac{1}{2} - \frac{1}{3} + \frac{1}{4}}{\frac{1}{3} - \frac{1}{4} + \frac{1}{5}}$.

4 Express in algebraic notation the difference of two numbers divided by the sum of their squares.

5 Compute $(x + y/x)/(x^2 - y + 1)$ for $x = \frac{2}{3}$, $y = \frac{1}{2}$. Express the answer as a fraction in lowest terms.

6 Compute 1 [×] 2 [+] 3 [×] 4 [−] 5 [×] 6 [=]

(a) in left-to-right logic (b) in algebraic logic.

7 Express $[3^5(\frac{1}{3})^2]^2/[27\cdot(\frac{1}{9})^2]$ as a power of 3.

8 Express $(50{,}000{,}000{,}000)(0.00000\ 00000\ 002)$ in scientific notation.

9 Estimate $(356.187\times 10^5)^{11}$ on a calculator.

10 Simplify $\dfrac{2+\sqrt{5}}{5-2\sqrt{5}}$ to an expression without radicals in the denominator.

11 Estimate $\sqrt[4]{1+\sqrt[3]{2^4+3^5}}$ on a calculator.

12 Express $3^{4/5}\sqrt[7]{9}/\sqrt[4]{(27)^6}$ as a power of 3.

13 Simplify $(x^{-1/3}y^{4/5})^{5/6}/x^2y^3$.

14 Compute $(1-x)(1+x^2)(1-x^3)$.

15 Find the coefficient of x^4 in

$$(1-x+x^2-x^3+x^4-x^5)(2-3x+4x^2-5x^3).$$

16 Find the coefficient of x^2y^2 in $(x^2-3xy+4y^2)(2x^2+xy-5y^2)$.

Factor

17 $32y^5-1$ **18** $2x-6y^2-4xy+3y$.

19 Multiply and express in lowest terms $\dfrac{x-y}{x+y}\left(\dfrac{x^2}{y^2}-\dfrac{y^2}{x^2}\right)$.

20 Simplify $\dfrac{\dfrac{x^2-y^2}{3xy}}{\dfrac{x+y}{12y}}$.

2 EQUATIONS AND INEQUALITIES

1 EQUATIONS AND IDENTITIES

Can you see a fundamental difference between these equations?

(i) $x^2 - 1 = (x + 1)(x - 1)$ (ii) $x^2 - 1 = 99$

Equation (i) is a true statement for every real number x, but equation (ii) is true only if $x = 10$ or $x = -10$.

Since (i) is always true, $x^2 - 1$ and $(x + 1)(x - 1)$ are exactly the same thing, only expressed in different ways. We call such an equation an **identity.** Another identity is

$$(x + y)^2 = x^2 + 2xy + y^2,$$

a true statement no matter what real values are substituted for x and y.

Two further examples of identities are

(iii) $1 + \sqrt{x} = \sqrt{x} + 1$ (iv) $\dfrac{1}{x} - \dfrac{1}{x + 2} = \dfrac{2}{x(x + 2)}$.

In (iii), each side is defined only for non-negative values of x, but for all those values the two sides are equal. In (iv) each side is defined for all real numbers except 0 and -2, but where defined they agree.

In general an identity is a statement that two expressions (1) are defined for the same set of values of the letters involved, and (2) are equal no matter which of those values are substituted for the letters.

Sometimes we relax requirement (1) a little bit. For instance

$$\frac{x - 1}{x^2 - 1} = \frac{1}{x + 1}$$

is called an identity even though the two sides are not defined for exactly the same values of x. The left-hand side is defined for all real x except $x = 1$ and $x = -1$; the right-hand side is defined for all real x except $x = -1$. But apart from $x = 1$ and $x = -1$, the numbers

$$\frac{x-1}{x^2-1} \quad \text{and} \quad \frac{1}{x+1}$$

are equal, so the equation is still considered an identity.

An equation that is not an identity is called a **conditional equation.** This kind of equation is a question, asking what values of x make it a true statement. Each such value is called a **solution** or **root** of the equation. To solve a conditional equation is to find its set of solutions.

Remark In practice, conditional equations are loosely referred to as equations.

Sometimes identities are called formulas. But usually the word **formula** is used for an equation in which the letters involved denote physical or geometrical quantities. For instance

$d = rt$	distance = rate × time
$A = \frac{1}{2}bh$	area = $\frac{1}{2}$ × base × height
$E = mc^2$	energy = mass × (speed of light)^2

are formulas expressing laws of nature when the letters have the indicated meanings.

Solutions of Equations

The sentence

$$2x + 1 = 7$$

is certainly an equation, not an identity. To solve it, we first subtract 1 from both sides of the equation:

$$(2x + 1) - 1 = 7 - 1 \qquad 2x = 6.$$

Next, we divide both sides by 2:

$$\frac{2x}{2} = \frac{6}{2} \qquad x = 3.$$

Therefore $x = 3$ is the only possible solution of the equation $2x + 1 = 7$. But this does not say that $x = 3$ *is* a solution. To show that $x = 3$ is a solution, we must show that $2x + 1 = 7$ is true when $x = 3$. It is:

$$2 \cdot 3 + 1 = 6 + 1 = 7.$$

Therefore $x = 3$ is *the* solution of $2x + 1 = 7$.

Now let us consider the equation

$$x^2 - x = 25 - x.$$

Our first step is to add x to both sides of the equation:

$$(x^2 - x) + x = (25 - x) + x$$
$$x^2 = 25$$

Now we compare the equations

$$x^2 - x = 25 - x \quad \text{and} \quad x^2 = 25.$$

Any solution x of the first is a solution of the second. Conversely, any solution x of the second is a solution of the first, since we can subtract x from both sides of

$x^2 = 25$ to obtain $x^2 - x = 25 - x$. But if $x^2 = 25$, then x is a square root of 25, that is, $x = 5$ or $x = -5$. Therefore $x = 5$ and $x = -5$ are the solutions of $x^2 - x = 25 - x$.

Two equations are **equivalent** if their solution sets are the same. In the first example above we showed that

$$2x + 1 = 7 \qquad \text{and} \qquad x = 3$$

are equivalent equations. In the second example we showed that

$$x^2 - x = 25 - x \qquad \text{and} \qquad x^2 = 25$$

are equivalent equations. The methods used in these examples illustrate some important procedures in action.

> Adding (subtracting) the same quantity to (from) both sides of an equation produces an equivalent equation.
>
> Multiplying (dividing) both sides of an equation by the same non-zero quantity produces an equivalent equation.

Using these procedures, you can often change an equation into one with the same solutions, but easier to solve.

Note that multiplication by 0 is not allowed. Take the equation $x = 3$ for example. It holds only for one value of x. But if you multiply both sides by 0, you get the equation $0 \cdot x = 0$, which holds for all values of x.

EXAMPLE 1 Solve $4x - 3 = 9$.

Solution Add 3 to both sides:

$$(4x - 3) + 3 = 9 + 3 \qquad 4x = 12.$$

Divide both sides by 4:

$$\frac{4x}{4} = \frac{12}{4} \qquad x = 3.$$

Check $4 \cdot 3 - 3 = 12 - 3 = 9$.

Answer 3

EXAMPLE 2 Solve $x^2 + x = 24 - x$.

Solution Add x to both sides:

$$x^2 + 2x = 24.$$

Now add 1:

$$x^2 + 2x + 1 = 25.$$

The left side is $(x + 1)^2$, so the equation is

$$(x + 1)^2 = 25.$$

By inspection, $x + 1 = 5$ or $x + 1 = -5$. The first possibility yields $x = 4$, the second, $x = -6$.

Check $4^2 + 4 = 20 = 24 - 4 \qquad (-6)^2 + (-6) = 30 = 24 - (-6)$.

Answer 4 $\quad -6$

Remark 1 In the solution we replaced $x^2 + 2x + 1$ by $(x + 1)^2$. You are always allowed to replace any expression by an identical expression. If, for example, $x^2 - 1$ appears somewhere, you may replace it by $(x + 1)(x - 1)$ without having to apologize.

Remark 2 It makes no sense to solve an *identity*, since all meaningful values are solutions. For instance, nobody would want to *solve* the identity $x^2 - 4 = (x + 2)(x - 2)$. However, sometimes a complicated identity may be disguised and look like a conditional equation. If you try to solve, you will probably end up with something like $0 = 0$ or $x = x$. So if this ever happens, look carefully at the equation you started with. Probably it is an identity.

We have had examples of equations with one solution and with two solutions. Other possibilities exist as well. For instance, consider a *factored* equation such as

$$x(x + 3)(2x - 1) = 0.$$

Now a product ABC is 0 if and only if $A = 0$ or $B = 0$ or $C = 0$. (See p. 3.) Therefore the left-hand side of *this* equation is 0 if and only if $x = 0$ or $x + 3 = 0$ or $2x - 1 = 0$. Hence there are *three* solutions:

$$x = 0 \qquad x = -3 \qquad x = \tfrac{1}{2}.$$

Some equations have no real solutions, for instance,

$$x^2 + 3 = 0.$$

The left side can never be 0 since x^2 is non-negative for every real number x. Learn to be suspicious; just because an equation is written down does not guarantee that it has any solutions.

Exercises

State whether the equation is an identity or a conditional equation

1 $5x + 4 = 2x$

2 $\dfrac{1}{x + 1} = 2$

3 $4(2x - 3) + 1 = 6(x - 1) + 2x - 5$

4 $x^2 + 2x = x(x + 1) + x$

5 $x^3 + 2x = 1$

6 $x^2(x - 3) = 0$

7 $1 + \dfrac{1}{x} = \dfrac{x + 1}{x}$

8 $\sqrt{1 + x^2} = x + 3$

9 $4x^2 - 7x = 2x + 1$

10 $(x + 1)^2 - x^2 = 2(x + 1) - 1.$

Find all solutions

11 $4x = 24$

12 $\frac{1}{3}x = 2$

13 $x^2 + 4x + 4 = 9$

14 $3x + 7 = 4(x + 1) - x$

15 $x(x - 2)(x - 7) = 0$

16 $2x^2 + 1 = 3x^2 - 8$

17 $\sqrt{x - 2} = 1$

18 $1/x^2 = 9$

19 $(x^2 - 4)(x^2 - 9) = 0$

20 $x^2 = x$

21 $(x - 1)^2 = x - 1$

22 $(x + 1)^2 = x^2$

23 $(x - 2)^2 = x^2$

24 $(x - 1)^2 = (x + 1)^2$

25 $x^4 + x^2 + 1 = 0$

26 $(x - 1)^2 = -(x + 1)^2$

27 $\dfrac{x^2 + 3}{x^2 + 4} = 1$

28 $\dfrac{x^2 + 3}{x^2 + 4} = 0$

29 $(3x - 2)(x + 6) = 0$

30 $(2x + 1)(2x + 3) = 0$

31 $x(x^2 - 1)(x^2 + 1) = 0$

32 $(x + 1)(x^2 + 4)(5x - 2) = 0$

33 $(x^2 - 9)(2x - 15) = 0$

34 $(3x + 3)(x^2 - 16) = 0$

35 $(x^2 + 9)(2x^2 + 1) = 0$

36 $(x^4 + 2)(x^4 - 1) = 0.$

2 SETTING UP EQUATIONS

In working specific problems, we exploit the brevity and generality of algebra. Here is a simple example. Suppose we wanted to know whether there are any temperatures where Fahrenheit and Celsius readings are the same. Now the formula

$$F = \frac{9}{5}C + 32$$

expresses the relation between the two scales in all cases. We are asking if ever $F = C$. Therefore in algebraic language our question is whether the equation

$$C = \frac{9}{5}C + 32$$

can ever hold.

The battle is now 90% won; we can solve this simple equation by standard algebraic techniques. It has exactly one solution, $C = F = -40$. By using the generality of algebra, we are sure to have considered all possibilities. Conclusion: -40 °F and -40 °C represent the same temperature, and there is no other temperature where this happens.

Algebra is a powerful tool for solving a variety of problems in many fields, such as the natural sciences, the social sciences, business, and technology. Let us analyze how an algebra problem is solved. Generally the process has two parts.

(*1*) *Set-up* Translating the problem "into math," that is, expressing it as an equation or system of equations.

(*2*) *Solution* Solving the equation(s).

Often the harder part is the set-up. That's why this entire section is devoted to setting up equations. The actual solution of the equations is left for later sections.

It requires practice to translate word problems into equations. No single method works in all cases. Still there are some guidelines that can help you sharpen this important skill.

(a) Take your time in setting up a problem. Read it several times to make sure you know what is given and what is to be found.

(b) Assign letters to various quantities in the statement of the problem.

(c) Find an equation involving the unknown quantity. Sometimes the desired equation is disguised in the wording of the problem. Be suspicious of expressions like "the same as" or "as much as," or even "is"; they may indicate an equation lurking somewhere.

(d) If you get stuck, check whether you have used *all* of the data.

Let us look at some examples of setting up equations, starting with very easy ones.

EXAMPLE 1 17% of what number is 42?

Solution Call the unknown number x. The expression "is 42" signals an equation

$$(17\% \text{ of } x) = 42.$$

Set-up $\frac{17}{100}x = 42$

Example 2 My friend Steve, a slow driver, sets out on a trip at a constant speed of 40 mph. I follow one hour later, driving at a steady 55 mph. How long will it be until I catch up?

Solution The unknown is the time required to catch up. Call it t (hours). Now where is an equation? What quantities are equal? Well, "catch up" means that the distances we travel are equal. I'll drive t hours and Steve will drive one hour longer, $t + 1$ hours. So

$$(\text{my distance in } t \text{ hours}) = (\text{Steve's distance in } t + 1 \text{ hours}).$$

At 55 mph, I cover $55t$ miles in t hours. At 40 mph, he covers $40(t + 1)$ miles in $t + 1$ hours.

Set-up $55t = 40(t + 1)$

Remark Be consistent with units. Since Example 2 is stated in miles and hours, keep all distances in miles and all times in hours. If I leave 30 *minutes* after Steve, the correct equation is

$$55t = 40(t + \tfrac{1}{2}) \qquad \text{not} \qquad 55t = 40(t + 30).$$

Example 3 One angle of a triangle is 20°. The second angle is three times as big as the third. Find the second and third angles.

Solution Call the angles A, B, C. We are given that $A = 20°$ and $B = 3C$. But that's not enough information to compute B and C. Have we used *all* the data? No, we haven't used the fact that A, B, and C are angles of a *triangle*. So far A, B, and C can be any old angles. But in a *triangle*

$$A + B + C = 180°.$$

Now we substitute $A = 20$ and $B = 3C$.

Set-up $20 + 3C + C = 180$

Example 4 Find the size of a square photograph if a square photograph 3 cm longer on a side has 50% greater area.

Solution The larger photo *is* 50% bigger. The "is" indicates an equal sign:

$$(\text{area of large photo}) = 1.5 \times (\text{area of small photo}).$$

Let s denote the length in cm of the side of the small photo. Then its area is s^2. The side of the large photo is $s + 3$, hence its area is $(s + 3)^2$.

Set-up $(s + 3)^2 = 1.5\, s^2$

The next example illustrates another guideline: in physical and geometrical problems, try to draw a figure.

Example 5 A tinsmith wants to cut a square from each corner of a 3-ft by 6-ft rectangular sheet and fold up the edges to make an open rectangular box of volume 4.5 ft³. Find the dimensions of the box.

Solution In this problem a drawing helps enormously. First we show the rectangle with the squares cut out (Fig. 1a), then the folding-up (Fig. 1b, page 46).

We let x be the side of the square and label all dimensions accordingly. The sides of the box are x, $3 - 2x$, and $6 - 2x$. The desired equation must follow from

$$\text{Volume} = 4.5.$$

Since the volume of the box is the product of its three dimensions, the set-up is

$$x(3 - 2x)(6 - 2x) = 4.5$$

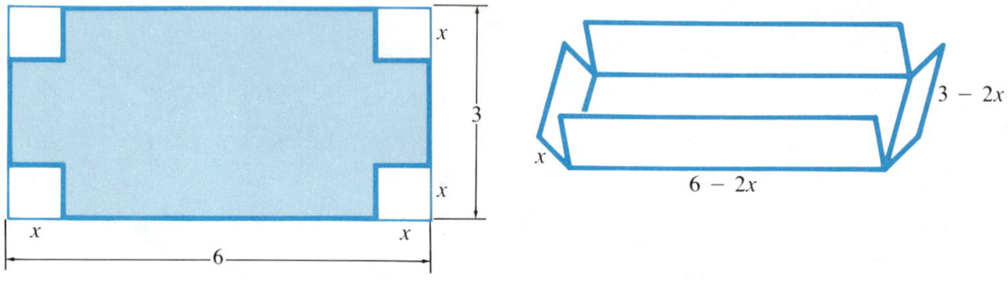

(a) sheet with corners cut out (b) open rectangular box

Fig. 1

Alternative solution Often, explicit numbers just get in the way and confuse the issue. Substituting letters for the data can provide both clarity and generality. In this example, for instance, let

L = length of sheet = 6 ft

W = width of sheet = 3 ft

V = required volume = 4.5 ft^3

x = side of cut-out square.

Then the box has dimensions

length $= L - 2x$ width $= W - 2x$ height $= x$

so $V = x(L - 2x)(W - 2x)$.

Now replace L, W, and V by their numerical values.

Set-up $x(3 - 2x)(6 - 2x) = 4.5$

EXAMPLE 6 I want to invest \$10,000 so it will grow to \$14,000 in 4 years. At what annual interest rate must I invest?

Solution One way to get started is by taking a guess, say 10%. Let's test this guess. In one year at 10%, the \$10,000 grows to

$$10{,}000 + \tfrac{1}{10}(10{,}000) = 11{,}000.$$

In the second year this grows to

$$11{,}000 + \tfrac{1}{10}(11{,}000) = 12{,}100.$$

In the third and fourth years the growth is

$$12{,}100 + \tfrac{1}{10}(12{,}100) = 13{,}310 \qquad 13{,}310 + \tfrac{1}{10}(13{,}310) = 14{,}641.$$

Hence 10% was too high. We could continue guessing, trying to home in on the right answer, but that involves a lot of computation (9% is also too high, 8% is too low, $8\frac{1}{2}$% is too low, etc.) There must be a better way.

Let's start again, using the generality of algebra. Let

A = amount invested = \$10,000

F = final amount after 4 years = \$14,000

r = unknown annual interest rate, in percent

$i = r/100$.

After one year my money has grown to

$$A + Ai = A(1 + i).$$

After two years, to

$$[A(1 + i)] + [A(1 + i)]i = [A(1 + i)](1 + i) = A(1 + i)^2.$$

Similarly after three and four years, to

$$A(1 + i)^3 \quad \text{and} \quad A(1 + i)^4.$$

Thus we want

$$A(1 + i)^4 = F \qquad A\left(1 + \frac{r}{100}\right)^4 = F.$$

Now substitute the data of the problem.

Set-up The solution r in % of $(10{,}000)\left(1 + \dfrac{r}{100}\right)^4 = 14{,}000$

Remark We'll learn how to solve such equations in Chapter 5. To 3 places, $r \approx 8.776\%$, where $\approx$ denotes "approximately equal."

In general, if a problem requires one unknown quantity, the set-up will lead to one equation. However, if there are two unknowns, you will need a system of two equations, so enough information must be given to set up a pair of equations.

Example 7 Find two numbers whose product is 5 and the sum of whose cubes is 30.

Solution There are two unknowns. Call one of them x and the other y. The set-up is easy because the problem is stated in algebraic terms. "Whose product is 5" translates to $xy = 5$, and "the sum of whose cubes is 30" translates to $x^3 + y^3 = 30$.

Set-up

$$\begin{cases} xy = 5 \\ x^3 + y^3 = 30 \end{cases}$$

Remark For the solution, see Example 8 on page 77.

Example 8 Find the sides of a right triangle if its perimeter is 90 cm and its area is 180 cm².

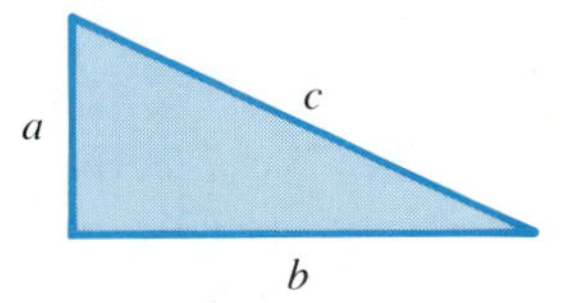

Solution Call the legs of the triangle a and b, and the hypotenuse c. The perimeter is $a + b + c$, and the area is $\frac{1}{2}$(base)(height), that is, $\frac{1}{2}ab$. Hence

$$\begin{cases} a + b + c = 90 \\ \frac{1}{2}ab = 180 \end{cases}$$

This is a system of two equations in *three* unknowns, not enough information to determine all three. But we have not used all the data! This is a *right* triangle; by the Pythagorean Theorem, $c^2 = a^2 + b^2$. So we replace c by $\sqrt{a^2 + b^2}$ and obtain a system of two equations in two unknowns.

Set-up
$$\begin{cases} a + b + \sqrt{a^2 + b^2} = 90 \\ \frac{1}{2}ab = 180 \end{cases}$$

Remark The solution is $(9, 40, 41)$ and can be found by the methods of Section 7.

Algebra Puzzles

These amuse (sometimes) and provide muscle-developing exercises. The wording is often deliberately confusing, so the problem may be actually easier than it looks. The following example is a typical puzzle. It is completely impractical; no one would determine ages this way!

EXAMPLE 9 Jack is twice as old as Sue was when Jack was as old as Sue is now. Double Jack's age increased by triple Sue's age is 68. How old are Jack and Sue now?

Solution Set

$$J = \text{Jack's age now},$$
$$S = \text{Sue's age now}.$$

The second statement in the problem is easy; it says

$$2J + 3S = 68.$$

The first statement is terribly confusing. Setting up a little translation table helps to decipher it.

	Now	When Jack was as old as Sue is now
Jack's age	J	S
Sue's age	S	?

To fill in the question mark, ask yourself, "How long ago was Jack's age S?" Answer: $J - S$. So you must subtract $J - S$ from Sue's age now to get Sue's age then. Therefore the question mark should be replaced by

$$S - (J - S) = 2S - J.$$

Hence the first statement in the problem says

$$J = 2(2S - J).$$

Set-up
$$\begin{cases} 2J + 3S = 68 \\ J = 2(2S - J) \end{cases}$$

Remark We'll learn how to handle such systems in Section 3. The answer is $J = 16$, $S = 12$.

Exercises

Set up the problem as an equation or a system of equations

1 At what temperature is the Fahrenheit reading 20 higher than the Celsius reading?

2 Is there a number that exceeds its reciprocal by 4?

3 Find three consecutive integers whose product is 15600.

4 Joe rents a car from agency A at \$80 for the first week, then \$20 per day. Bill rents a car from agency H at a straight \$15 per day. They keep their cars the same length of time and find their rental bills are the same. For how many days did they have the cars?

5 The Post Office (1979) book rate was 48¢ for the first pound, 18¢ per pound for the next 6 pounds, then 11¢ per pound after that. A certain shipment of books went for an average of 14.95¢ per pound. How much did it weigh?

6 I am playing red and black at the roulette table, betting one chip at a time. After 50 plays, I have lost 6 chips. How many times have I won?

7 A \$32 calculator is on sale at 10% off. However, because sales tax is added, the actual cash saving is \$3.36. What is the sales tax rate?

8 A tank contains 1000 gallons of gasoline. More gasoline begins flowing in at a constant rate of 20 gal/min. At the same time alcohol begins flowing in at 10 gal/min. When will the concentration of alcohol reach 10%?

9 A library discarded 20% of its old books to make space for new purchases. It then acquired 3500 new books, bringing its total colllection to 500 more books than it had originally. How many was that?

10 The first angle of a triangle is double the second, which in turn is triple the third. What are the angles of the triangle?

11 Is there a number such that when you add 14 to the number, divide the result by 3, then add 10, you get the original number back again?

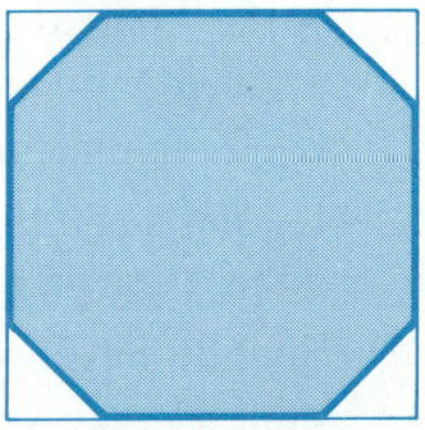

12 A tinsmith wants to lighten a square of sheet metal 10% by snipping off four small equal isosceles right triangles, one from each corner. If the square is 50 cm on each side, what are the legs of the triangles?

13 How many minutes after 4:00 will the minute hand of a clock first lie directly over the hour hand?

14 The number of bacteria in a certain colony doubles every three days. In 15 days there are 128,000 bacteria in the colony. How many were there originally?

15 A process consists of multiplying a number by itself and adding 3. If the process is applied three times, starting with a certain number, and the result is 619,372, what was the number?

16 If I spend \$20 cash at a good restaurant, then I will have left in my pocket 50% of what I would have had if I had gone to a modest restaurant and paid only \$10. How much did I start with?

17 I pay out \$50, then deposit half of what's left. Then I repeat this process twice more. I am left with $2\frac{1}{2}$% of what I started with. How much was that?

18 John, who has just turned 16, will receive a \$5000 bequest on his 21st birthday. But inflation is lowering the real value of money by 6% per year. What will the real value of his bequest be when he receives it?

19 A group of people charter a plane for \$30,000. At the last minute, 70 more people join the group, lowering the cost per person by \$26.25. How many people are now in the group?

20 Near the surface of a planet without atmosphere, a body falls $\frac{1}{2}gt^2$ meters in t seconds, starting from rest. (Here g is a constant.) Suppose two bodies are released from rest, the second d_0 meters below the first and t_0 seconds later. When will the first body pass the second?

21 After lengthening my stride (pace) by 2 inches, I can now walk a mile in 132 fewer paces. What is the length of my new stride?

22 Two joggers start simultaneously from opposite ends of the mile-long Golden Gate Bridge. The first jogs 5 mi/hr, the second 7 mi/hr. They pass each other, continue to

the ends of the bridge, turn and jog back. When do they pass again, and how far has each jogged at that instant?

23 How can you make a sum of one dollar with pennies and nickels using exactly 56 coins?

24 Find a two-digit whole number such that if you insert 0 between the digits, the resulting three-digit number is 9 times the original number.

25 The denominator of a certain fraction exceeds the numerator by 3. If both numerator and denominator are increased by 1, the value of the fraction increases by $\frac{1}{10}$. Find the fraction.

26 A certain roll of single ply bathroom tissue has 1000 sheets, each $4\frac{1}{2} \times 4\frac{1}{2}$ in. If the inner and outer radii of the roll are $\frac{3}{4}$ in. and $2\frac{3}{8}$ in, respectively, how thick is the paper?

27 Find the base of an isosceles triangle of height 3 and perimeter 12.

28 From here to the bus stop is 200 meters if I follow the sidewalk to the corner, turn a right angle, and continue along the sidewalk. But if I cut straight across the grass, the distance is 170 m. How far is it from here to the corner?

29 One leg of a right triangle is $\frac{1}{4}$ the length of the hypotenuse, and the area of the triangle is 100 cm^2. How long is the other leg?

30 Two partners, A and B, hold all the stock in a company. If A can buy $\frac{2}{9}$ of B's holdings he will have a controlling 51% interest. What percentage of the stock does A hold now?

31 A disc jockey fills 11 minutes of his half hour show with blab and the remaining time with rock selections of lengths $2\frac{1}{2}$, 3, and $3\frac{1}{2}$ minutes. If he plays at least one record of each length, find how many of each he plays.

32 If John's age three years ago is doubled, the result is the same as 3 plus the square of John's age 12 years ago. How old is John?

33 Split 30 into two parts, so that double one part is triple the other.

34 Split 30 into two parts, the difference of whose squares is 120.

35 During the night, each wolf are two sheep, reducing the flock of sheep by 20%. If there were 99 more sheep than wolves in the field the previous evening, how many sheep were there left in the morning?

36 A length L of rope is cut into two pieces, each of which is folded into a square. If one square has triple the area of the other, how long are the pieces?

37 When Jim was half as old as Sue is now, Sue was two years older than Jim is now. When Jim was one third as old as Sue is now, Sue was two years younger than Jim is now. How old are Jim and Sue?

38 Bill is twice as old as Ruth was 4 years ago. When Ruth reaches Bill's present age, the sum of their ages will be 76. How old are they now?

3 LINEAR EQUATIONS

A linear equation is one of the form

$$ax + b = 0 \qquad (a \neq 0)$$

or an equation that can be brought into that form by **transposing** terms (another expression for adding the same quantity to both sides). For example

$$2x + 1 = -7 \qquad x - \tfrac{1}{2} = 2x + 1 \qquad x^2 + 2x = x^2 + 3$$

are linear equations. They can be brought into the forms

$$2x + 8 = 0 \qquad x + \tfrac{3}{2} = 0 \qquad 2x - 3 = 0$$

by transposing terms. Other examples of linear equations are

$$\frac{x - 1}{3} = 4x + 5 \qquad 8(4x + 1) + 7(3x - 5) = 4(2x + 1).$$

Solution of Linear Equations

To solve the equation $ax + b = 0$, we subtract b from both sides, then divide both sides by a:

$$ax + b = 0 \qquad ax = -b \qquad x = -b/a.$$

This short calculation shows that the equation has only one possible solution, $x = -b/a$. But $-b/a$ is a solution, since if $x = -b/a$, then

$$ax + b = a(-b/a) + b = -b + b = 0.$$

> A linear equation $ax + b = 0 \quad (a \neq 0)$ has exactly one solution,
>
> $$x = -\frac{b}{a}.$$

EXAMPLE 1 Solve (a) $3x + 1 = 8$ (b) $4(x - 2) = 2x + 5$.

Solution (a) Subtract 1 from both sides:

$$3x = 7.$$

Now multiply by $\tfrac{1}{3}$:

$$\tfrac{1}{3}(3x) = \tfrac{1}{3} \cdot 7 \qquad x = \tfrac{7}{3}.$$

(b) Expand the left side:

$$4x - 8 = 2x + 5.$$

Add 8 to both sides, then subtract $2x$:

$$4x = 2x + 13 \qquad 2x = 13.$$

Hence $x = \tfrac{13}{2}$.

Answer (a) $\tfrac{7}{3}$ (b) $\tfrac{13}{2}$

Some equations can be brought into linear form by various algebraic manipulations.

EXAMPLE 2 Solve $\dfrac{3}{x - 6} = \dfrac{1}{2x - 4}$.

Solution The equation is undefined for $x = 2$ and $x = 6$. So just by writing it down, we are agreeing to exclude these values. To eliminate fractions, we multiply both sides by $(x - 6)(2x - 4)$. This quantity is non-zero since $x \neq 2$ and $x \neq 6$. Therefore the multiplication will yield a new equation with the same solutions as the given one, unless $x = 2$ or $x = 6$ happens to be a solution of the new equation:

$$\left(\frac{3}{x-6}\right)(x-6)(2x-4) = \left(\frac{1}{2x-4}\right)(x-6)(2x-4)$$

$$3(2x-4) = x-6.$$

This equation is linear. We solve it in the usual way:

$$6x - 12 = x - 6 \qquad 5x = 12 - 6 = 6 \qquad x = \tfrac{6}{5}.$$

Check Substitute $x = \frac{6}{5}$:

$$\frac{3}{x-6} = \frac{3}{\frac{6}{5}-6} = \frac{3\cdot 5}{6-6\cdot 5} = \frac{15}{6-30} = -\tfrac{15}{24} = -\tfrac{5}{8}$$

$$\frac{1}{2x-4} = \frac{1}{2(\frac{6}{5})-4} = \frac{5}{2\cdot 6 - 4\cdot 5} = \frac{5}{12-20} = -\tfrac{5}{8}.$$

Answer $\frac{6}{5}$

Some equations involve several unknowns. Take

$$x^2y + 3y = 4x^3 - xy - 5$$

for example. This equation looks complicated, but is actually a linear equation in y. With a little rearrangement of terms, it can be written as

$$(x^2 + x + 3)y = 4x^3 - 5,$$

which is of the form $ay = b$. It follows that

$$y = \frac{4x^3 - 5}{x^2 + x + 3}.$$

We say that we have solved for y **in terms of** x, or expressed y in terms of x.

Example 3 The relation between Fahrenheit and Celsius temperatures is $F = \frac{9}{5}C + 32$. Express C in terms of F.

Solution This is a linear equation for the unknown C. Solve:

$$F - 32 = \tfrac{9}{5}C \qquad \tfrac{5}{9}(F - 32) = C.$$

Answer $C = \frac{5}{9}(F - 32)$

Example 4 Solve $\dfrac{1}{x} - \dfrac{1}{y} = 10$ for x in terms of y.

Solution Since $xy \neq 0$ we may multiply both sides by xy:

$$y - x = 10xy$$

$$y = 10xy + x$$

$$y = (10y + 1)x$$

$$\frac{y}{10y+1} = x.$$

Answer $x = \dfrac{y}{10y+1}$ $(y \neq 0 \text{ and } y \neq -\frac{1}{10})$

Systems of Linear Equations

So far we have been solving linear equations in one unknown. Now let us look at equations in two unknowns. A **linear equation** in two unknowns x and y is an equation of the form

$$ax + by = c \qquad (a \text{ and } b \text{ not both } 0).$$

A (two-by-two) **system** of linear equations in x and y is a pair of equations

$$\begin{cases} a_1x + b_1y = c_1 & (a_1 \text{ and } b_1 \text{ not both } 0) \\ a_2x + b_2y = c_2 & (a_2 \text{ and } b_2 \text{ not both } 0). \end{cases}$$

A **solution** of the system is a pair (x, y) of real numbers that makes both equations correct statements at the same time, that is, **simultaneously.**

The strategy for solving a system is to eliminate one variable.

EXAMPLE 5 Solve the system $\begin{cases} x + y = 3 \\ x - y = 1. \end{cases}$

Solution 1 Suppose (x, y) is any solution. Then both equations are true statements. Adding them will produce another true statement, *and* will eliminate y at the same time:

$$\begin{array}{r} x + y = 3 \\ \underline{x - y = 1} \\ 2x \qquad = 4 \end{array}$$

The y drops out; from $2x = 4$ follows $x = 2$. Hence if the pair (x, y) is a solution, then $x = 2$. But if $x = 2$, then from $x + y = 3$ follows $y = 1$. Therefore the only possible solution is $(2, 1)$. But $(2, 1)$ *is* a solution. For if $x = 2$ and $y = 1$, then

$$\begin{cases} x + y = 2 + 1 = 3 \\ x - y = 2 - 1 = 1. \end{cases}$$

Solution 2 This time eliminate x by subtracting the two equations:

$$\begin{array}{r} x + y = 3 \\ \underline{x - y = 1} \\ 2y = 2 \end{array}$$

Hence $y = 1$, and from $x + y = 3$ follows $x = 2$.

Answer $x = 2 \qquad y = 1$

EXAMPLE 6 Solve $\begin{cases} 5x + 2y = 5 \\ 4x + 3y = 6. \end{cases}$

Solution 1 Adding or subtracting the equations will not eliminate either variable as in Example 5. But this technique will work if we first multiply the equations by suitable nonzero constants. To eliminate y, multiply the first equation by 3 and the second by -2, then add:

$$\begin{array}{r|l} 3 & 5x + 2y = 5 \\ -2 & 4x + 3y = 6 \end{array} \qquad \begin{array}{rcr} 15x + 6y &=& 15 \\ -8x - 6y &=& -12 \\ \hline 7x \quad &=& 3. \end{array}$$

Hence $x = \frac{3}{7}$. From $5x + 2y = 5$ we have

$$5(\tfrac{3}{7}) + 2y = 5 \qquad 2y = 5 - \tfrac{15}{7} = \tfrac{20}{7} \qquad y = \tfrac{10}{7}.$$

Solution 2 Another way of eliminating y is to solve one equation for y in terms of x, then substitute the result into the other equation. From

$$5x + 2y = 5 \quad \text{follows} \quad y = \tfrac{1}{2}(5 - 5x) = \tfrac{5}{2}(1 - x).$$

Substitute into $4x + 3y = 6$:

$$4x + 3 \cdot \tfrac{5}{2}(1 - x) = 6$$

$$8x + 15(1 - x) = 12$$

$$-7x + 15 = 12$$

$$-7x = -3.$$

Hence $x = \frac{3}{7}$. To find y, use the expression for y in terms of x:

$$y = \tfrac{5}{2}(1 - x) = \tfrac{5}{2}(1 - \tfrac{3}{7}) = \tfrac{5}{2} \cdot \tfrac{4}{7} = \tfrac{10}{7}.$$

Answer $x = \frac{3}{7} \qquad y = \frac{10}{7}$

Not every system has a solution. For example,

$$\begin{cases} x + 2y = 1 \\ x + 2y = 3 \end{cases}$$

has no solution. For if it did, we would be forced to the false conclusion that $1 = 3$. A system with no solution is called **inconsistent.**

At the opposite extreme are systems that have infinitely many solutions, for example

$$\begin{cases} 3x + 6y = 12 \\ \ \ x + 2y = 4. \end{cases}$$

This system is really a single equation in disguise, since the first equation is three times the second. Any attempt to eliminate one unknown will simultaneously eliminate the other, so the result will be $0 = 0$. Such a system is called **underdetermined.** It has infinitely many solutions since there are infinitely many pairs (x, y) that satisfy a single linear equation such as $x + 2y = 4$, for example $(0, 2)$, $(-2, 3)$, $(\frac{13}{2}, -\frac{5}{4})$, etc.

Exercises

Solve

1 $4x + 1 = 25$

2 $\frac{1}{2}x - 4 = 3$

3 $2x + 3 = 6x - 5$

4 $-3(2x + 1) = 4(4x - 3)$

5 $2(\frac{1}{3}x + 1) + 3(\frac{2}{3}x + 8) = -4$

6 $5x + 2 = \frac{3}{2}x - 11$

7 $\dfrac{2x - 1}{2x + 1} = \dfrac{1}{4}$

8 $\dfrac{2x - 1}{2x + 1} = \dfrac{x + 2}{x + 4}$

9 $\dfrac{1}{x} = \dfrac{4}{3x+1}$

10 $2x - \dfrac{x+5}{4} = \dfrac{3x-1}{2} + 1$

11 $(x+1)^2 = (x+2)^2 + 5$

12 $x^3 + 5x + 3 = x^2(x-2) + 2(x^2+4x+1)$

13 $ax + b = cx + d \quad (a \neq c)$

14 $(ax+1)(bx+1) = (ax+3)(bx-2) \quad (3a \neq 2b)$

15 $\dfrac{6x}{x+5} = 1 + \dfrac{2}{x+5}$

16 $2x + \dfrac{1}{3x-1} = \dfrac{6x^2+7x-4}{3x-1} + 2$

17 $\dfrac{1}{x^2-9} + \dfrac{4}{x-3} = \dfrac{2}{x+3}$

18 $\dfrac{1}{x^2-4} - \dfrac{3}{x-2} = \dfrac{8}{x^2-4}$

19 $a^2(x+1) + b^2(x+2) + c^2(x+3) = 1 - x$

20 $(x+a)^2 = (x+b)^2 \quad (a \neq b)$

21 $\dfrac{x(x+2)}{x^2+4} = 1$

22 $\dfrac{2x^2-9x-4}{x^2+11x+16} = 2.$

Solve for x in terms of y

23 $2x + 3y - 4 = 0$

24 $x(y+3) = 4x + 5y + 1$

25 $\dfrac{1}{x} + \dfrac{2}{y+1} = 3$

26 $\dfrac{x+y}{x-y} = 4$

27 $y = \dfrac{2x+1}{3x+5}$

28 $\dfrac{x}{y} = \dfrac{x+y+1}{2y-3}$

29 $\dfrac{x}{a} + \dfrac{y}{b} = 1$

30 $\dfrac{a}{x} + \dfrac{b}{y} = 1$

31 $y^3 + 7xy^2 + xy + x - 3y = 1$

32 $(2x-y)(y^2+1) = xy.$

Solve for h

33 $\pi r^2 h + 2\pi rh = 50$

34 $h = r^2(\pi - h).$

Solve for r^2

35 $\dfrac{r^2 s}{r+s^2} = r - s^2$

36 $r^2 s^2 + 1 = \dfrac{r^2}{s^2}.$

Find a value of a for which the equation

37 $3x + 1 = 2x - 4 + a(5x+2)$ has no solution

38 $\dfrac{2x+a}{3x+a} = \dfrac{1}{9}$ has solution $x = 5$.

Solve the system

39 $\begin{cases} x + 3y = 4 \\ 2x - y = 1 \end{cases}$

40 $\begin{cases} x - y = 4 \\ x + 2y = 1 \end{cases}$

41 $\begin{cases} x - 2y = -2 \\ 2x - y = 5 \end{cases}$

42 $\begin{cases} 2x + y = 2 \\ -3x + y = -3 \end{cases}$

43 $\begin{cases} x - 6y = -17 \\ x + 2y = 7 \end{cases}$

44 $\begin{cases} x - \frac{1}{2}y = 4 \\ 2x + y = 6 \end{cases}$

45 $\begin{cases} 2x - 7y = -1 \\ 3x + 4y = 2 \end{cases}$

46 $\begin{cases} 6x + 5y = -3 \\ 5x - 6y = \frac{23}{3} \end{cases}$

47 $\begin{cases} 2x - 3y = 1 \\ 6x - 9y = 2 \end{cases}$

48 $\begin{cases} x + 4y = 7 \\ 2x + 8y = 10 \end{cases}$

49 $\begin{cases} x^2 + y^2 = 5 \\ x^2 - y^2 = 3 \end{cases}$

50 $\begin{cases} x + y^2 = 10 \\ x + 3y^2 = 28. \end{cases}$

4 APPLICATIONS OF LINEAR EQUATIONS

In this section we consider some problems that lead to linear equations and linear systems.

EXAMPLE 1 The Audio Shop advertises a clearance sale on stereo equipment: 30% off after the first \$50. If you have \$275 to spend, what is the pre-sale price of a system that you can now afford?

Solution If the regular price is x dollars, then the sale price is $50 + 0.7(x - 50)$. Hence, for \$275 you can buy a system whose regular price x satisfies

$$50 + 0.7(x - 50) = 275.$$

Solve for x:

$$50 + 0.7x - 35 = 275$$

$$0.7x = 260$$

$$x = \frac{260}{0.7} \approx 371.42$$

Answer \$371.42

EXAMPLE 2 Express the repeating decimal as a fraction

(a) $0.585858 \cdots$ (b) $0.4761761761 \cdots$

Solution (a) Let $x = 0.585858 \cdots$. If we multiply x by 100, we get $58.585858 \cdots$, a number whose decimal part is x again. In other words,

$$100x = 58 + x$$

hence $99x = 58 \qquad x = \frac{58}{99}.$

(b) Let $x = 0.4761761761 \cdots$. The digits do not repeat from the beginning as in part (a), so we first multiply by 10 and split off the repeating part:

$$10x = 4.761761761 \cdots = 4 + 0.761761761 \cdots$$

$$10x - 4 = 0.761761761 \cdots.$$

Now this looks like part (a). We set $y = 10x - 4$ and multiply by 1000:

$$y = 0.761761761 \cdots$$

$$1000y = 761 + 0.761761 \cdots = 761 + y$$

$$999y = 761$$

$$y = \tfrac{761}{999}.$$

But $y = 10x - 4$, therefore $10x = y + 4$, so

$$x = \frac{y + 4}{10} = \frac{\frac{761}{999} + 4}{10} = \frac{761 + 4 \cdot 999}{10 \cdot 999} = \tfrac{4757}{9990}.$$

Answer (a) $\frac{58}{99}$ (b) $\frac{4757}{9990}$

EXAMPLE 3 (A true story) I send a shipment of household goods by sea to an Asian port. Insurance on the shipment is charged at the rate of 4% of the *total value* of the shipment. This total value is the sum of the value of goods, the cost of sea freight, and *the cost of the insurance.* Suppose my goods are worth \$30,000 and the sea freight cost \$4500. How much does the insurance cost?

Solution At first glance this problem seems impossible. You have to know how much the insurance costs in order to compute how much the insurance costs! However, before giving up, let's see what the algebra set-up looks like. Set

$$G = \text{value of goods} = \$30{,}000$$

$$F = \text{freight cost} = \$4500$$

$$I = \text{unknown insurance}.$$

Then the recipe is

$$I = (0.04)(G + F + I).$$

This is not impossible at all; it is just a linear equation for I. Solving, we have

$$I - (0.04)I = (0.04)(G + F) \qquad (0.96)I = (0.04)(G + F)$$

$$I = \frac{0.04}{0.96}(G + F) = \frac{0.04}{0.96}(34500) = 1437.50.$$

Answer \$1437.50

EXAMPLE 4 This headline appeared in the Chicago Tribune on November 22, 1978:

U.S. SWALLOWS HERSHEY BAR PRICE BOOST

The Hershey Company raised the price of its 20-cent chocolate bar to 25 cents. At that time, the President's Council on Wage and Price Stability was imposing a 9.5% limit on price increases. To comply, Hershey also increased the weight of its bar, which previously weighed 1.05 ounces. If the actual price increase was $9\frac{3}{8}$% *per ounce,* how much did the new 25-cent bar weigh?

Solution Let x be the new weight. The price per ounce was

$$\frac{20}{1.05} \text{ (old bar)} \qquad \frac{25}{x} \text{ (new bar)}.$$

The second ratio is $9\frac{3}{8}$% greater than the first:

$$\frac{25}{x} = (1.09375)\frac{20}{1.05}$$

Solving for x we find

$$x = \frac{25(1.05)}{20(1.09375)} = 1.2$$

Answer 1.2 ounces

Rate Problems

A quantity that changes with time, changes at a **constant** rate if the amount of change in any time interval of duration D is kD, where k is a fixed number. This k is the amount of change in one unit of time and is called the **rate of change,** sometimes just the **rate.**

This important concept applies to all sorts of quantities. If the quantity is money in a savings account, the rate is called the **interest rate.** If it is distance traveled by a vehicle, the rate is called the **speed.**

Constant rate is often assumed in the statement of a problem. When you say that a typist types 50 words a minute, you mean that he types 250 words in five minutes no matter when the five minute period starts, before lunch, after lunch, tomorrow, anytime. If a machine turns out 17 parts in 3 hours, its production rate is $\frac{17}{3}$ parts/hr.

EXAMPLE 5 Suppose A can paint a barn in 8 hours and B can do it in 5 hours. How long will it take them working together to do the job?

Solution A works at the rate of $\frac{1}{8}$ barns/hour, hence paints $\frac{1}{8}t$ barns in t hours. Similarly, B paints $\frac{1}{5}t$ barns in t hours. Working together they paint

$$\tfrac{1}{8}t + \tfrac{1}{5}t = \tfrac{13}{40}t \quad \text{barns in } t \text{ hours.}$$

The time t required to paint one barn satisfies

$$\tfrac{13}{40}t = 1 \quad \text{hence} \quad t = \tfrac{40}{13} = 3\tfrac{1}{13}.$$

Answer $3\frac{1}{13}$ hr

EXAMPLE 6 I borrow \$200 at 12% simple interest and another \$300 at 14%. What is the effective interest rate on my \$500 loan?

Solution If I borrow the principal amount P at the simple interest rate r, then the amount of interest I due after time t is

$$I = Prt.$$

The interest due on the first loan is

$$I_1 = (200)(0.12)t = 24t,$$

and that due on the second is

$$I_2 = (300)(0.14)t = 42t.$$

The total interest due is

$$I = I_1 + I_2 = 24t + 42t = 66t.$$

Suppose these two loans are equivalent to a single loan of \$500 at interest rate r. Then $I = 500rt$, hence

$$500rt = 66t \qquad r = \tfrac{66}{500} = 0.132.$$

Answer 13.2%

In problems involving distance traveled by a vehicle at constant speed, the basic relation is

$$\text{distance} = \text{rate} \times \text{time}, \quad d = rt.$$

There are all sorts of problems built upon this relation. In some way or other, you are given two of the three quantities and you must solve for the third. So it boils down to applying the formula $d = rt$ or one of the equivalent forms,

$$r = d/t \quad \text{or} \quad t = d/r.$$

EXAMPLE 7 Two river boats have a top speed of 10 mph in still water. They depart at the same time and go at full speed, one upstream and one downstream. When the first has gone 40 miles, the second has gone 60 miles. How fast is the current?

Solution What quantities are equal? The word "when" means "at the same time that." Therefore the *times* traveled by the two boats are equal. Call their common time t. Now compute t by the formula $t = d/r$ applied to each boat:

$$t = \frac{(\text{distance traveled by boat 1})}{(\text{speed of boat 1})} = \frac{d_1}{r_1}$$

$$t = \frac{(\text{distance traveled by boat 2})}{(\text{speed of boat 2})} = \frac{d_2}{r_2}$$

Let c be the (constant) current speed. Then

$$r_1 = 10 - c \qquad r_2 = 10 + c.$$

From the two expressions for t follows

$$\frac{d_1}{r_1} = \frac{d_2}{r_2}, \quad \text{that is,} \quad \frac{40}{10 - c} = \frac{60}{10 + c}.$$

Solve for c:

$$40(10 + c) = 60(10 - c) \qquad 400 + 40c = 600 - 60c$$

$$100c = 200 \qquad c = 2.$$

Answer 2 mph

Mixture Problems

Certain practical problems involve mixtures of various substances. Usually solving them requires more common sense than algebra. A typical example: Suppose 50 grams (g) of a 12% salt solution is mixed with 30 g of an 18% salt solution. What is the strength of the resulting mixture?

The mixture contains $50 + 30 = 80$ g total solution, of which

$$(0.12)(50) + (0.18)(30) = 6.0 + 5.4 = 11.4 \text{ g}$$

is salt. Consequently the proportion of salt in the mixture is

$$\frac{11.4}{80} = 0.1425$$

so the answer is 14.25%.

EXAMPLE 8 How many quarts of 75% alcohol must be mixed with 15 quarts of 45% alcohol to produce 70% alcohol?

Solution Let x denote the number of quarts of 75% alcohol. Then the mixture contains $15 + x$ quarts of solution including $0.75x + (0.45)(15)$ quarts of pure alcohol. We want this number to be 70% of $15 + x$, that is

$$0.75x + (0.45)(15) = (0.70)(15 + x).$$

Solve:

$$0.05x = (0.70)(15) - (0.45)(15) = (0.25)(15)$$

$$x = \frac{(0.25)(15)}{0.05} = 75.$$

Answer 75 qts

Problems Leading to Linear Systems

EXAMPLE 9 I want to invest \$5000, part in a high risk mining stock that grows at the rate of 8% annually and part in a relatively safe steel stock that grows at 3% annually. How much of each should I buy to achieve a 5% growth rate?

Solution Let x be the amount in the mining stock and y the amount in the steel stock. Then $x + y = 5000$.

Investment x grows by $0.08x$ and investment y by $0.03y$ annually, and these must total $(0.05)(5000)$. Thus we have the system

$$\begin{cases} x + y = 5000 \\ 0.08x + 0.03y = (0.05)(5000). \end{cases}$$

Multiply the first equation by -0.03 and add to the second. The result is

$$0.05x = (0.02)(5000) \qquad x = 2000.$$

Answer \$2000 in mining, \$3000 in steel

The economics of supply and demand provides some interesting examples. Generally, demand for a product is high when its price is low, and decreases as the price increases. Generally, the supply of a product increases as the price increases. The **equilibrium price** is the price at which supply equals demand.

EXAMPLE 10 For a certain cassette player, the quantity demanded (Q_d units) is related to the unit price (P dollars) by

$$Q_d = 10^5(100 - P).$$

The quantity supplied (Q_s units) is related to the price by

$$Q_s = 10^5(P - 10).$$

Find the equilibrium price and the quantity sold at that price.

Solution We must find the value of P for which $Q_s = Q_d$ and the common value $Q = Q_s = Q_d$. Therefore we must solve the system

$$\begin{cases} Q = 10^5(100 - P) \\ Q = 10^5(P - 10) \end{cases} \quad \text{that is} \quad \begin{cases} Q + 10^5 P = 10^7 \\ Q - 10^5 P = -10^6. \end{cases}$$

Add these equations, then subtract them:

$$2Q = 10^7 - 10^6 \qquad 2 \times 10^5 P = 10^7 + 10^6$$

$$2Q = 9 \times 10^6 \qquad 2P = 110$$

$$Q = 4.5 \times 10^6 \qquad P = 55.$$

Answer $P = \$55$, $Q = 4\frac{1}{2}$ million units

Remark Note the meaning of the given equations. At price 0, the demand is for 10^7 units, and at price \$10 the supply is 0. At price \$100, the demand is 0 and the supply is 9 million units.

Exercises

1 A bookstore sells a card for \$5 that entitles you to a 15% discount on all books. If you have \$80 to spend, what is the pre-discount price of books you can now buy?

2 A car rental agency rents cars with no mileage fees at \$30/day for the first four days, then \$12/day. For how many days will the rental be \$300?

3 How many minutes after 4:00 will the minute hand of a clock first lie directly over the hour hand?

4 At what time between 2:00 and 3:00 will the hands of a clock point in exactly opposite directions?

5 A stock broker sets her commission at \$25 per transaction plus 1.5% of the amount of the transaction. A second broker sets his commission at \$10 plus 1.8% of the amount. Find the amount of a transaction for which the two charge equal commissions.

6 One plumber charges \$19 per call plus \$9.60/hour. A second plumber charges a flat \$13.60/hour. For a job of how many hours are their charges equal?

Express the repeating decimal as a fraction

7 $0.1111 \cdots$

8 $0.232323 \cdots$

9 $0.123123123 \cdots$

10 $0.321321321 \cdots$

11 $0.45616161 \cdots$

12 $0.08989898 \cdots$.

13 Lloyd's of London insures a house (value \$90,000) and also the premium payment. If the insurance rate is 1% per year, how much is the annual premium?

14 U.S. government regulations require an average efficiency of at least 27.5 miles per gallon for all cars sold by a manufacturer in 1985. A certain manufacturer projects 1985 sales of 150,000 standard sized cars averaging 25.6 mpg. How many compacts averaging 29.0 mpg must it sell in 1985 to average exactly 27.5 mpg on all models sold?

15 If your midterm grades are 85, 62, 73, what do you need on the final to average 75, assuming the final counts the same as a midterm? What if the final carries double weight?

16 Suppose 10 men can do 20 jobs in 15 days. How long does it take one man to do one job?

17 Suppose an average professor can do a certain job in 9 hours and an average student can do the same job in 6 hours. If 2 professors and 5 students work together, how long will the job take?

18 If A can plow a certain field in 5 hours, B in 4 hours, and C in 3 hours, then working together, how long will it take them to plow the field?

19 Two large computers working in parallel can complete a calculation in 20 minutes. The smaller one alone takes 50 minutes. How much time does it take the larger one alone?

20 Suppose one pump can fill a certain tank in h_1 hours and a second pump can fill the tank in h_2 hours. How long will it take them together to fill 2 such tanks?

21 Suppose one pump can fill a tank in 8 hours and a second pump can empty the (full) tank in 12 hours. Suppose the tank is half full and both pumps are started. When will the tank be full?

22 I borrow \$100 from a friend, no interest, and I borrow \$200 from the bank at 8% interest. What is the effective interest rate on the total?

23 I borrow \$340 at 4% and \$150 at $7\frac{1}{2}$%. What is the effective interest rate on the total?

24 I borrow \$500 on my insurance policy at 4% and lend a friend \$200 at 6%. What is the effective interest rate on the \$300 I am actually holding?

25 Through my broker I buy \$10,000 worth of XYZ bond, which pays 14% interest. However, I buy "on margin"; I only put up 60% of the cost and borrow the rest from the brokerage house at $12\frac{1}{4}$% interest. Find the effective rate of return on my investment.

26 If the point is 4 in a "craps" game, the probability P of winning is the sum of the probability $\frac{3}{36}$ of hitting 4 on the next roll plus P times the probability $\frac{27}{36}$ of hitting neither 4 nor 7 on the next roll. Find P.

27 How much pure water must be added to 100 g of 30% salt solution to produce an 18% solution?

28 How much pure salt must be added to 200 g of 5% salt solution to triple its strength?

29 How many grams of 60-40 solder (60% tin, 40% lead) must be mixed with 40-60 solder to produce 500 g of 55-45 solder?

30 How many gallons of 100 proof whiskey (50% alcohol) must be mixed with 80 proof whiskey to produce 50 gal of 95 proof whiskey?

31 The current in a river flows at 2 mph. A boat can go 60 miles downstream in the same time as it can go 36 miles upstream. What is the boat's speed in still water?

32 An eastbound 747 aided by a tailwind of 60 mph flies 1000 miles while a westbound 747 flying against the same wind flies 800 miles. What is the common speed of these planes in still air?

33 A team has 11 wins and 9 losses. How many consecutive games must it win to have won $\frac{2}{3}$ of its games?

34* A long convoy of heavy trucks is moving at 20 mph. A messenger on motorcycle at 60 mph can go from the end of the convoy to the front and back again in 6 minutes. How long is the convoy?

35 Risky stock X pays 10% dividend and safe bond B pays 6% dividend. How should I invest \$10,000 to have an income of \$850 per year?

36 How should \$20,000 be split between a 4% savings account and a 7% investment fund to yield a 6% return?

37 Suppose demand and supply in units and price in dollars of a certain toy are related by

$$Q_d = 2 \times 10^6(1 - \tfrac{1}{15}P) \qquad Q_s = 5 \times 10^4(P - 5).$$

Find the equilibrium price and quantity sold.

38 Suppose demand and supply per week and price per egg in cents are related by

$$\frac{Q_d}{10^9} + \frac{P}{10} = 1 \qquad Q_s = \tfrac{1}{3} \times 10^9(P - 4).$$

Find the equilibrium of eggs and production per week.

39 A bar of precious metal found by a treasure hunter is a mixture of gold of density 18.8 g/cm^3 and silver of density 10.6 g/cm^3. The bar weighs 15810 g and has volume 950 cm^3. How many cm^3 of each metal does the bar contain?

40 Tankers A and B discharge oil at the same rate; tanker C at a different rate. Pumping together, all three fill a storage tank in 40 hours. On one delivery, A arrives first and starts pumping. Five hours later, B and C arrive and start pumping. After an additional 38 hours, the tank is filled. How long would it take each tanker alone to fill the tank?

5 QUADRATIC EQUATIONS

A **quadratic equation** is an equation of the form

$$ax^2 + bx + c = 0 \qquad (a \neq 0),$$

or one that can be transformed into this type by transposing and collecting terms. For example

$$3x^2 - 5x = -x^2 + 6x + 1.$$

If all terms on the right are transposed to the left, we get

$$(3x^2 - 5x) - (-x^2 + 6x + 1) = 0$$

$$4x^2 - 11x - 1 = 0.$$

Another example is

$$(x + 3)(2x - 9) = x^2 - 4x.$$

We expand the left side, transpose the terms on the right, then combine

$$2x^2 - 3x - 27 = x^2 - 4x$$

$$(2x^2 - 3x - 27) - (x^2 - 4x) = 0$$

$$x^2 + x - 27 = 0.$$

Simple quadratic equations such as

$$x^2 + 1 = 0 \qquad x^2 = 0 \qquad x^2 - 1 = 0$$

show that there may be no real solutions, or one real solution, or two real solutions. We shall see that there are no other possibilities.

Solution by Factoring

This method depends on a basic property of real numbers. (See properties of zero, p. 3.)

> If r and s are real numbers such that $rs = 0$, then either $r = 0$, or $s = 0$, or both.

In words, if a product of numbers is zero, then one of the factors must be zero. We exploit this property to solve quadratic equations $ax^2 + bx + c = 0$ when we can factor the left-hand side.

EXAMPLE 1 Solve

(a) $x^2 - 2x = 0$ (b) $x^2 - 3x - 4 = 0$.

Solution (a) Factor the left side:

$$x(x - 2) = 0.$$

The product is 0 only if either

$$x = 0 \quad \text{or} \quad x - 2 = 0, \quad \text{that is,} \quad x = 2.$$

Thus 0 and 2 are the only possible solutions. However, they *are* solutions since each value makes the left side 0.

(b) Factor the left side:

$$(x - 4)(x + 1) = 0.$$

The product is 0 only if

$$x - 4 = 0 \quad \text{or} \quad x + 1 = 0, \qquad \text{that is, only if} \quad x = 4 \quad \text{or} \quad x = -1.$$

Answer (a) 0 2 (*b*) 4 −1

EXAMPLE 2 Solve

(a) $3x^2 - 7x + 2 = 0$ (b) $15x^2 - 14x - 8 = 0$.

Solution (a) Test the various possible factors $ax \pm b$, where a is a divisor of 3 and b is a divisor of 2. Thus $a = 1$ or 3, and $b = 1$ or 2. After testing, we find

$$(3x - 1)(x - 2) = 0.$$

Hence either

$$3x - 1 = 0 \text{ so } x = \tfrac{1}{3}, \quad \text{or} \quad x - 2 = 0 \text{ so } x = 2.$$

(b) There are many possible factors:

$$\begin{Bmatrix} x \\ 3x \\ 5x \\ 15x \end{Bmatrix} \pm \begin{Bmatrix} 1 \\ 2 \\ 4 \\ 8 \end{Bmatrix}$$

After testing a number of combinations, we hit on

$$(5x + 2)(3x - 4) = 0.$$

Hence either

$$5x + 2 = 0 \text{ so } x = -\tfrac{2}{5}, \quad \text{or} \quad 3x - 4 = 0 \text{ so } x = \tfrac{4}{3}.$$

Answer (a) $\frac{1}{3}$ 2 (b) $-\frac{2}{5}$ $\frac{4}{3}$

This method works only if you can find a factorization. It is practical only when the coefficients are integers and fairly small ones at that. Let's describe a more general method.

Completing the Square

Consider the quadratic equation

$$(x - 2)^2 = 9$$

It is easy to solve because the left-hand side is a perfect square. By inspection, either $x - 2 = 3$ or $x - 2 = -3$. Hence $x = 5$ and $x = -1$ are solutions.

Now look at the equation

$$x^2 + 6x + 4 = 0.$$

The left side is not a perfect square, but something similar is: $x^2 + 6x + 9$. So add and subtract 9 on the left-hand side:

$$(x^2 + 6x + 9) - 9 + 4 = 0$$

$$(x + 3)^2 = 5.$$

Therefore $x + 3$ must be one of the two square roots of 5, that is, $x + 3 = \pm\sqrt{5}$. Hence

$$x = -3 + \sqrt{5} \quad \text{or} \quad x = -3 - \sqrt{5}.$$

The technique we have used is called **completing the square.** It provides a general method for solving quadratic equations.

Completing the Square Given any expression

$$x^2 + bx,$$

add and subtract $(\frac{1}{2}b)^2$:

$$\begin{aligned} x^2 + bx &= x^2 + 2(\tfrac{1}{2}b)x \\ &= x^2 + 2(\tfrac{1}{2}b)x + (\tfrac{1}{2}b)^2 - (\tfrac{1}{2}b)^2 \\ &= (x + \tfrac{1}{2}b)^2 - \tfrac{1}{4}b^2. \end{aligned}$$

Examples

$$x^2 - 10x = x^2 - 2\cdot 5x = x^2 - 2\cdot 5x + 5^2 - 5^2 = (x - 5)^2 - 25$$

$$x^2 + 3x = x^2 + 2\cdot\tfrac{3}{2}x = x^2 + 2\cdot\tfrac{3}{2}x + (\tfrac{3}{2})^2 - (\tfrac{3}{2})^2 = (x + \tfrac{3}{2})^2 - \tfrac{9}{4}.$$

Let us test completing the square on some equations.

EXAMPLE 3 Solve by completing the square

(a) $x^2 - 10x + 17 = 0$ (b) $x^2 - 3x + 8 = 0$.

Solution (a) Since $-10 = 2\cdot(-5)$, add and subtract $(-5)^2 = 25$:

$$x^2 - 10x + 17 = x^2 - 10x + 25 - 25 + 17 = (x - 5)^2 - 8.$$

The given equation is equivalent to

$$(x - 5)^2 - 8 = 0 \qquad (x - 5)^2 = 8.$$

Hence $x - 5 = \pm\sqrt{8}$,

$$x = 5 + \sqrt{8} \quad \text{or} \quad x = 5 - \sqrt{8}.$$

(b) Since $-3 = 2\cdot(-\frac{3}{2})$, add and subtract $(-\frac{3}{2})^2 = \frac{9}{4}$:

$$x^2 - 3x + 8 = x^2 - 3x + (-\tfrac{3}{2})^2 - \tfrac{9}{4} + 8 = (x - \tfrac{3}{2})^2 + \tfrac{23}{4}.$$

The given equation is equivalent to

$$(x - \tfrac{3}{2})^2 + \tfrac{23}{4} = 0, \quad \text{that is,} \quad (x - \tfrac{3}{2})^2 = -\tfrac{23}{4}.$$

But this is impossible since the square of a real number cannot be negative. Therefore there are no real solutions.

Answer (a) $5 \pm \sqrt{8}$ (b) no real solutions

We have completed the square to solve quadratic equations of the form $x^2 + bx + c = 0$. The method applies also to quadratic equations $ax^2 + bx + c = 0$, where $a \neq 1$. We merely divide the equation by a to obtain an equivalent equation

$$x^2 + \frac{b}{a}x + \frac{c}{a} = 0.$$

EXAMPLE 4 Solve by completing the square

$$3x^2 - 5x - 1 = 0.$$

Solution Divide by 3 to obtain the equivalent equation:

$$x^2 - \tfrac{5}{3}x - \tfrac{1}{3} = 0.$$

Now complete the square on the left-hand side:

$$\begin{aligned} x^2 - \tfrac{5}{3}x - \tfrac{1}{3} &= x^2 + 2\cdot(-\tfrac{5}{6})x - \tfrac{1}{3} \\ &= x^2 + 2\cdot(-\tfrac{5}{6})x + (-\tfrac{5}{6})^2 - (-\tfrac{5}{6})^2 - \tfrac{1}{3} \\ &= (x - \tfrac{5}{6})^2 - \tfrac{25}{36} - \tfrac{1}{3} \\ &= (x - \tfrac{5}{6})^2 - \tfrac{37}{36}. \end{aligned}$$

The equation becomes

$$(x - \tfrac{5}{6})^2 - \tfrac{37}{36} = 0, \quad \text{that is,} \quad (x - \tfrac{5}{6})^2 = \tfrac{37}{36}.$$

Hence

$$x - \tfrac{5}{6} = \pm\sqrt{\tfrac{37}{36}} = \pm\tfrac{1}{6}\sqrt{37}, \quad \text{that is,} \quad x = \tfrac{5}{6} \pm \tfrac{1}{6}\sqrt{37}.$$

Answer $\tfrac{5}{6} \pm \tfrac{1}{6}\sqrt{37}$

Warning The most frequent mistake in completing the square is forgetting to divide the middle coefficient by 2. Even an experienced teacher can make a mistake like the following:

$$\begin{aligned} 5x^2 + 2x + 1 &= 5(x^2 + \tfrac{2}{5}x) + 1 \\ &= 5[x^2 + \tfrac{2}{5}x + (\tfrac{2}{5})^2 - (\tfrac{2}{5})^2] + 1 \\ &= 5(x + \tfrac{2}{5})^2 - \tfrac{4}{5} + 1 = 5(x + \tfrac{2}{5})^2 + \tfrac{1}{5}. \end{aligned}$$

Of course $(\tfrac{1}{5})^2$ should have been added and subtracted, not $(\tfrac{2}{5})^2$, and the correct result is

$$5x^2 + 2x + 1 = 5(x + \tfrac{1}{5})^2 + \tfrac{4}{5}.$$

Good practice for avoiding this common error is *always* to write $bx = 2\cdot(\tfrac{1}{2}b)x$. This adds an extra step to be sure, but it is well worth the trouble. For instance, after the first line in the calculation above we should add the line

$$= 5[x^2 + 2\cdot(\tfrac{1}{5})x] + 1,$$

which leads naturally to the (correct!) next line

$$= 5[x^2 + 2\cdot(\tfrac{1}{5})x + (\tfrac{1}{5})^2 - (\tfrac{1}{5})^2] + 1.$$

We close the section with an example showing that completing the square has uses other than solving equations.

EXAMPLE 5 While fooling around with a new calculator, I notice that whenever I multiply a number between 1 and 100 by 100 minus that number, the product never exceeds 2500. For instance $37 \times (100 - 37) = 37 \times 63 = 2331 \le 2500$. Explain.

Solution Let N be the number. Then the product is

$$N(100 - N) = -N^2 + 100N.$$

Complete the square:

$$-N^2 + 100N = -(N^2 - 2\cdot 50N)$$
$$= -[N^2 - 2\cdot 50N + 50^2 - 50^2]$$
$$= -(N - 50)^2 + 2500.$$

Therefore

$$N(100 - N) = 2500 - (N - 50)^2 \le 2500.$$

Note that the inequality holds for any real number N, not just numbers between 1 and 100. (This doesn't say much for $N < 0$ or $N > 100$; why?)

Hindsight Set $N = 50 + x$. Then $100 - N = 50 - x$ and

$$(50 + x)(50 - x) = 2500 - x^2 \le 2500.$$

Exercises

Solve by factoring

1 $3x^2 - 4x = 0$ **2** $x(x + 1) + 2x = 0$ **3** $x^2 - 5x - 14 = 0$
4 $x^2 + 7x + 6 = 0$ **5** $6x^2 - 5x + 1 = 0$ **6** $4x^2 - 12x + 9 = 0$
7 $5x^2 + 17x - 12 = 0$ **8** $12x^2 + 17x + 6 = 0$.

Express each polynomial in the form $a(x + p)^2 + q$

9 $x^2 - 8x$ **10** $x^2 + 4x$ **11** $x^2 + 6x + 1$
12 $x^2 - 3x - 6$ **13** $x^2 + 2kx$ **14** $x^2 + 2bx + c$
15 $4x^2 + 12x$ **16** $4x^2 - 10x$ **17** $2x^2 - \frac{1}{2}x$
18 $2x^2 + 3x$ **19** $-x^2 + 4x$ **20** $-3x^2 + 4x$
21 $3x^2 - 5x + 1$ **22** $-5x^2 + 2x$ **23** $-3x^2 + x + 5$
24 $-6x^2 - 5x + 1$.

Solve by completing the square

25 $x^2 + 4x + 2 = 0$ **26** $x^2 + 2x - 5 = 0$
27 $2x^2 + 4x + 1 = 0$ **28** $9x^2 - 6x + 1 = 0$
29 $x^2 + 6x + 11 = 0$ **30** $3x^2 + x + 4 = 0$
31 $6(x^2 + x) + 7x = 3x - 1$ **32** $3x^2 + x = 5x + 1$
33 $(2x + 3)^2 = x - 8$ **34** $(x + 1)(2x - 3) = x^2 + x + 7$
35 $\frac{1}{2}x^2 - 4x - \frac{9}{2} = 0$ **36** $x + 3 = -\frac{1}{2}x^2$.

6 THE QUADRATIC FORMULA

The method of completing the square will work for every quadratic equation. But instead of completing the square each time, let us do it once and for all. We take the most general quadratic equation:

$$ax^2 + bx + c = 0 \qquad (a \ne 0).$$

First we divide by a:

$$x^2 + \frac{b}{a}x + \frac{c}{a} = 0, \quad \text{that is,} \quad x^2 + 2\left(\frac{b}{2a}\right)x + \frac{c}{a} = 0,$$

then complete the square, adding $(b/2a)^2$ to each side of the equation:

$$x^2 + 2\left(\frac{b}{2a}\right)x + \left(\frac{b}{2a}\right)^2 + \frac{c}{a} = \left(\frac{b}{2a}\right)^2$$

$$\left(x + \frac{b}{2a}\right)^2 + \frac{c}{a} = \left(\frac{b}{2a}\right)^2$$

We subtract c/a from both sides, then put the right-hand side over a common denominator:

$$\left(x + \frac{b}{2a}\right)^2 = \left(\frac{b}{2a}\right)^2 - \frac{c}{a} = \frac{b^2}{4a^2} - \frac{c}{a} = \frac{b^2 - 4ac}{4a^2},$$

$$\left(x + \frac{b}{2a}\right)^2 = \frac{b^2 - 4ac}{4a^2}.$$

The result is

$$\left(x + \frac{b}{2a}\right)^2 = \frac{D}{4a^2}, \quad \text{where} \quad D = b^2 - 4ac.$$

The sign of D is crucial. If $D < 0$, then the right side is negative with no real square roots; hence *there are no real solutions.*

If $D = 0$, the equation reduces to

$$\left(x + \frac{b}{2a}\right)^2 = 0, \quad \text{that is,} \quad x + \frac{b}{2a} = 0.$$

Hence *there is exactly one real solution,* $x = -b/2a$.

If $D > 0$, then $D/4a^2$ is positive with two distinct square roots,

$$x + \frac{b}{2a} = \frac{\sqrt{b^2 - 4ac}}{2a} \quad \text{or} \quad x + \frac{b}{2a} = -\frac{\sqrt{b^2 - 4ac}}{2a}$$

hence *there are two real solutions,*

$$x = \frac{-b + \sqrt{b^2 - 4ac}}{2a} \quad \text{and} \quad x = \frac{-b - \sqrt{b^2 - 4ac}}{2a}.$$

We abbreviate the two possibilities by use of the plus-minus sign:

$$x = \frac{-b \pm \sqrt{b^2 - 4ac}}{2a}.$$

Solutions of Quadratic Equations

The solutions of the quadratic equation

$$ax^2 + bx + c = 0 \quad a \neq 0,$$

depend on the sign of its **discriminant**

$$D = b^2 - 4ac.$$

(1) If $D < 0$, there are no real solutions.

(2) If $D = 0$, there is one real solution, $x = \dfrac{-b}{2a}$.

(3) If $D > 0$, there are two real solutions given by the **quadratic formula**

$$x_1 = \frac{-b + \sqrt{D}}{2a} \quad \text{and} \quad x_2 = \frac{-b - \sqrt{D}}{2a}.$$

Although a quadratic equation may have no real solutions, it always has solutions in a larger number system called the *complex number system*. This will be discussed in Chapter 14.

Suggestion You will probably memorize the quadratic formula,

$$x = \frac{-b \pm \sqrt{b^2 - 4ac}}{2a}.$$

However, there is some danger of becoming too dependent on the letters a, b, c. For instance, could you solve $bx^2 + cx + a = 0$ with the formula? To avoid trouble, identify the roles of a, b, c. They are the coefficient of x^2, the coefficient of x, and the constant term, or if you prefer, the first, middle, and last coefficients. Then you can remember the quadratic formula as

$$x = \frac{-(\text{middle}) \pm \sqrt{(\text{middle})^2 - 4(\text{first})(\text{last})}}{2(\text{first})}.$$

EXAMPLE 1 Solve for x by the quadratic formula

(a) $3x^2 - 10x + 4 = 0$ (b) $x^2y^3 + 5xy - 50 = 0$.

Solution (a) Use the quadratic formula with $a = 3$, $b = -10$, and $c = 4$:

$$x = \frac{-(-10) \pm \sqrt{(-10)^2 - 4\cdot 3\cdot 4}}{2\cdot 3} = \frac{10 \pm \sqrt{52}}{2\cdot 3} = \frac{5 \pm \sqrt{13}}{3}.$$

(b) Treat this equation as a quadratic in x:

$$(y^3)x^2 + (5y)x - 50 = 0.$$

By the quadratic formula with $a = y^3$, $b = 5y$, and $c = -50$,

$$x = \frac{-5y \pm \sqrt{(5y)^2 - 4y^3(-50)}}{2y^3} = \frac{-5y \pm \sqrt{25y^2 + 25(8y^3)}}{2y^3}$$

$$= \frac{-5y \pm 5y\sqrt{1 + 8y}}{2y^3} = \frac{-5 \pm 5\sqrt{1 + 8y}}{2y^2}.$$

Answer (a) $\dfrac{5 \pm \sqrt{13}}{3}$ (b) $\dfrac{-5 \pm 5\sqrt{1 + 8y}}{2y^2}$ $(y \geq -\frac{1}{8}, y \neq 0)$

EXAMPLE 2 Solve for x: $\dfrac{7}{x} = 3 - \dfrac{1}{x - 1}$.

Solution It is understood by the very statement of the problem that $x \neq 0$ and $x \neq 1$. Multiply both sides by $x(x - 1)$ to clear of fractions:

$$7(x - 1) = 3x(x - 1) - x$$

$$7x - 7 = 3x^2 - 4x$$

$$0 = 3x^2 - 11x + 7.$$

By the quadratic formula,

$$x = \frac{11 \pm \sqrt{11^2 - 4\cdot 3\cdot 7}}{2\cdot 3} = \frac{11 \pm \sqrt{37}}{6}.$$

Answer $\dfrac{11 \pm \sqrt{37}}{6}$

EXAMPLE 3 Without solving the equation, find how many real solutions it has:

(a) $4x^2 + 20x + 15 = 0$ (b) $2x^2 - x + \frac{3}{8} = 0$.

Solution (a) The discriminant is

$$D = 20^2 - 4(4)(15) = 400 - 240 = 160 > 0.$$

Hence there are two real solutions.

(b) $D^2 = (-1)^2 - 4(2)(\frac{3}{8}) = 1 - 3 = -2 < 0$.

Hence there are no real solutions.

Answer (a) two (b) none

Warning There are two frequent mistakes in the use of the quadratic formula that occur when $a \neq 1$. Whenever you see $a \neq 1$, your warning bell should ring. The first mistake is forgetting a in the denominator:

$$3x^2 + 7x + 4 = 0 \qquad x = \frac{-7 \pm \sqrt{49 - 4 \cdot 3 \cdot 4}}{2} = -4, -3.$$

These mistakes occur because 9 out of 10 of the quadratics you solve in real life have $a = 1$, so you just forget part of the formula.

The second mistake is forgetting 2 in the denominator:

$$3x^2 + 7x + 4 = 0 \qquad x = \frac{-7 \pm \sqrt{49 - 4 \cdot 3 \cdot 4}}{3} = -\tfrac{8}{3}, -2.$$

If $a = 2$, double trouble, and all the bells should peal!

$$2x^2 + 7x + 3 = 0 \qquad x = \frac{-7 \pm \sqrt{49 - 4 \cdot 2 \cdot 3}}{2} = -6, -1.$$

Not so!

Solution on a Calculator

Computing the solutions of a quadratic equation requires taking a square root and doing some unpleasant arithmetic if the coefficients are not nice. Just imagine solving

$$24.996x^2 - 837.09x - 6.0047 = 0$$

by hand! Yet with an inexpensive scientific calculator, the computation can be done fairly easily.

Note The following discussion requires some familiarity with a calculator having a memory. You might wish to look ahead at Section 5.8.

We assume that our calculator has a single memory into which we can store the displayed number by a key we shall call [STO], and from which we can recall a stored number to the display by a key [RCL]. We assume that our calculator works on left to right logic, and that each push of an arithmetic operation key ([+], [−], [×], [÷]) completes the previous operation, but [x^2] and [$\sqrt{\ }$] do not. For example, keying the sequence 3 [−] 1 [+] 25 [$\sqrt{\ }$] [=] produces $3 - 1 + \sqrt{25} = 7$. If your calculator has different rules, you may have to modify the program we suggest.

Now let's solve

$$ax^2 + bx + c = 0$$

on our calculator. As usual, we let $D = b^2 - 4ac$. Assuming $D > 0$, the solutions are given by the quadratic formula:

$$r_1 = \frac{-b - \sqrt{D}}{2a} = \frac{-b}{2a} - \frac{\sqrt{D}}{2a} \quad \text{and} \quad r_2 = \frac{-b + \sqrt{D}}{2a} = \frac{-b}{2a} + \frac{\sqrt{D}}{2a}.$$

Note that

$$r_2 = r_1 + 2\left(\frac{\sqrt{D}}{2a}\right).$$

This relation suggests a program, which we sketch in a "flowchart".

Compute $\sqrt{D}/2a$ and store it.

↓

Compute $-b/2a$.

↓

Subtract the stored number:

$$r_1 = -\frac{b}{2a} - \frac{\sqrt{D}}{2a}.$$

↓

Add the stored number twice:

$$r_2 = r_1 + \frac{\sqrt{D}}{2a} + \frac{\sqrt{D}}{2a} = -\frac{b}{2a} + \frac{\sqrt{D}}{2a}.$$

We list in Table 1 the steps of a possible program, together with the display and memory contents at each step.

Table 1

Key	Display	Memory
4	4	
×	4	
a	a	
×	$4a$	
c	c	
=	$4ac$	
STO	$4ac$	$4ac$
b	b	$4ac$
x^2	b^2	$4ac$
−	b^2	$4ac$
RCL	$4ac$	$4ac$
=	D	$4ac$
√	$\sqrt{D}$	$4ac$
÷	$\sqrt{D}$	$4ac$
2	2	$4ac$
[continue next column]		
÷	$\sqrt{D}/2$	$4ac$
a	a	$4ac$
=	$\sqrt{D}/2a$	$4ac$
STO	$\sqrt{D}/2a$	$\sqrt{D}/2a$
$-b$	$-b$	$\sqrt{D}/2a$
÷	$-b$	$\sqrt{D}/2a$
2	2	$\sqrt{D}/2a$
÷	$-b/2$	$\sqrt{D}/2a$
a	a	$\sqrt{D}/2a$
−	$-b/2a$	$\sqrt{D}/2a$
RCL	$\sqrt{D}/2a$	$\sqrt{D}/2a$
+	r_1	$\sqrt{D}/2a$
RCL	$\sqrt{D}/2a$	$\sqrt{D}/2a$
+	$-b/2a$	$\sqrt{D}/2a$
RCL	$\sqrt{D}/2a$	$\sqrt{D}/2a$
=	r_2	$\sqrt{D}/2a$

Now let us test this program on the equation

$$3.42x^2 - 2.57x + 0.23 = 0$$

Along the way we find

$$D = 3.4585 \qquad \sqrt{D} \approx 1.8597043 \qquad \sqrt{D}/2a \approx 0.2718866.$$

At the end,

$$r_1 \approx 0.1038444 \qquad r_2 \approx 0.6476176.$$

Here is a checking routine for r_2. (If r_2 is correct, it is pretty certain that r_1 is also.) After the last step in the solution program, continue with Table 2.

Table 2

Key	Display	Memory
STO	r_2	r_2
×	r_2	r_2
a	a	r_2
+	ar_2	r_2
b	b	r_2
×	$ar_2 + b$	r_2
RCL	r_2	r_2
+	$ar_2^2 + br_2$	r_2
c	c	r_2
=	$ar_2^2 + br_2 + c$	r_2

[continue next column]

If r_2 is indeed a solution, the final display should be zero. For our example, the final display is

$$-4.\times 10^{-10} = -0.00000\ 00004,$$

zero for all practical purposes. (The minute discrepancy is a machine error and may be different on different calculators.)

Of course, you can also check r_1 by keying it into the memory and applying the same sequence of steps.

A few comments about our program. Storing numbers that are used several times in a program is good practice; it cuts the risk of entry errors. With just one memory available, we chose to store $\sqrt{D}/2a$, generally a number with many digits, rather than a and b, which may be simpler numbers. If we had four memories we would store a, b, and c also.

If you have to solve a batch of quadratic equations, the ideal tool is a programmable calculator. It can remember the whole program, so for each equation you only have to key in a, b, and c.

Exercises

Without solving the equation, find how many real solutions it has

1 $9x^2 - 13x + 5 = 0$

2 $x^2 + 14x + 31 = 0$

3 $ax^2 - 2x + \frac{1}{a} = 0$

4 $2x^2 - 8x - 3 = 0$

5 $41x^2 + 10{,}000x + 57 = 0$

6 $3x^2 + 13x + 10{,}000 = 0.$

Solve by the quadratic formula

7 $x^2 - 5x + 6 = 0$

8 $x^2 + 5x + 6 = 0$

9 $x^2 - 2x - 4 = 0$

10 $x^2 + 8x + 4 = 0$

11 $x^2 + 10x + 2 = 0$

12 $x^2 + 5x - 3 = 0$

13 $2x^2 + x - 4 = 0$

14 $3x^2 - 6x - 1 = 0$

15 $x^2 + 2x - 99 = 0$

16 $x^2 - 16x + 60 = 0$

17 $x^2 - 5x + 7 = 0$

18 $3x^2 - x + 4 = 0$

19 $4x^2 + 5x - 6 = 0$

20 $12x^2 + 17x + 6 = 0$

21 $5x^2 - 13x - 6 = 0$

22 $6x^2 + 11x + 5 = 0$

23 $10x^2 - 8x + 1 = 0$

24 $4x^2 + 8x - 1 = 0$

25 $2x^2 - 5x + 5 = 0$

26 $7x^2 - 17x + 1 = 0$

27 $9x^2 - 6x - 4 = 0$

28 $36x^2 + 24x + 1 = 0$

29 $25x^2 - 10x - 1 = 0$

30 $3x^2 - 2x + 1 = 0.$

If $ax^2 + bx + c = 0$ and $D > 0$, find

31 the sum of the roots

32 the product of the roots.

Solve for x

33 $x + \dfrac{1}{x} = 2$

34 $\dfrac{x}{x^2 + 1} = \frac{1}{3}$

35 $\dfrac{1}{x + 1} + \dfrac{3}{x - 1} = 5$

36 $\dfrac{x - 2}{3x - 8} = \dfrac{x + 5}{5x - 2}$

37 $x^2 + 6xy^2 = y$

38 $x(x + y + 2) = y(2x + 1)$

39 $\dfrac{x + y}{x - y} = \dfrac{x}{3x + y}$

40 $\dfrac{x}{y} + \dfrac{y}{x} = 2y$

41 $6x^2 - 5xy + y^2 = 0$

42 $x^2 + 2xy - 5y^2 = 0.$

Solve with a calculator

43 $3.41x^2 - 2.57x - 0.23 = 0$

44 $1.01x^2 - 4.30x + 1.22 = 0$

45 $2.05x^2 + 3.56x + 0.94 = 0$

46 $0.47x^2 - 1.87x + 0.53 = 0$

47 $0.092x^2 - 5.77x + 2.91 = 0$

48 $0.0032x^2 + 10.75x - 35.22 = 0$

49 $1073x^2 + 547x - 380 = 0$

50 $425x^2 - 1093x + 136 = 0$

51 $0.35x^2 + 0.12x + 0.011 = 0$

52 $201x^2 - 327x + 133 = 0.$

7 OTHER TYPES OF EQUATIONS

Equations of Quadratic Type

An equation of the form

$$ax^4 + bx^2 + c = 0$$

is really a disguised quadratic since it can be expressed as

$$ay^2 + by + c = 0 \quad \text{where} \quad y = x^2.$$

We find y by the quadratic formula, then find $x = \pm\sqrt{y}$. Similarly the equations

$$ax^6 + bx^3 + c = 0 \quad \text{and} \quad d(x^2 - 3x)^2 + e(x^2 - 3x) + f = 0$$

are quadratic relative to the quantities $y = x^3$ and $y = x^2 - 3x$.

EXAMPLE 1 Solve $x^4 - 2x^2 - 2 = 0$.

Solution Write the equation as

$$y^2 - 2y - 2 = 0 \quad \text{where} \quad y = x^2.$$

Solve for y by the quadratic formula:

$$y = \frac{2 \pm \sqrt{4 + 8}}{2} = \frac{2 \pm \sqrt{12}}{2} = \frac{2 \pm 2\sqrt{3}}{2} = 1 \pm \sqrt{3}.$$

Hence

$$x^2 = 1 + \sqrt{3} \quad \text{or} \quad x^2 = 1 - \sqrt{3}.$$

The first possibility for x^2 leads to the solutions

$$x = \pm\sqrt{1 + \sqrt{3}}.$$

The second does not lead to real solutions because $1 - \sqrt{3} < 0$.

Answer $\pm\sqrt{1 + \sqrt{3}}$

EXAMPLE 2 Solve $x - 10\sqrt{x} + 23 = 0$, where $\sqrt{x}$ denotes (as always) the non-negative square root of x.

Solution The equation is quadratic in $y = \sqrt{x}$:

$$y^2 - 10y + 23 = 0.$$

From the quadratic formula,

$$y = 5 + \sqrt{2} \quad \text{or} \quad y = 5 - \sqrt{2}.$$

Both $5 + \sqrt{2}$ and $5 - \sqrt{2}$ are positive, hence acceptable as the square root of x. Now $x = y^2$, so

$$x = (5 + \sqrt{2})^2 = 27 + 10\sqrt{2} \quad \text{or} \quad x = (5 - \sqrt{2})^2 = 27 - 10\sqrt{2}.$$

Check If $x = 27 \pm 10\sqrt{2}$, then

$$\begin{aligned} x - 10\sqrt{x} + 23 &= (27 \pm 10\sqrt{2}) - 10(5 \pm \sqrt{2}) + 23 \\ &= (27 - 50 + 23) \pm 10\sqrt{2} \mp 10\sqrt{2} = 0. \end{aligned}$$

Answer $27 \pm 10\sqrt{2}$

Remark When the dual signs $\pm$ and $\mp$ appear in an expression, you either take both top signs or both bottom signs. For example,

$$\begin{aligned} (\pm a) + (\pm b) &= a + b \quad \text{or} \quad -a - b \\ (\pm a) + (\mp b) &= a - b \quad \text{or} \quad -a + b \\ (\pm a)(\pm b) &= ab. \end{aligned}$$

Polynomial Equations

Solving a polynomial equation

$$a_n x^n + a_{n-1}x^{n-1} + \cdots + a_1 x + a_0 = 0$$

is usually difficult when its degree n is greater than 2. However, under certain favorable circumstances the equation can be solved, for instance, when you are lucky or skillful enough to factor the left-hand side.

EXAMPLE 3 Solve $x^5 - 16x = 0$.

Solution Factor the left side:

$$\begin{aligned} x^5 - 16x &= x(x^4 - 16) = x(x^2 + 4)(x^2 - 4) \\ &= x(x^2 + 4)(x + 2)(x - 2) = 0. \end{aligned}$$

A product is 0 precisely when one of its factors is 0. The factor $x^2 + 4$ is never

0, but each of the other three factors can be 0: for the values 0, -2, and 2, respectively.

Answer $0 \quad -2 \quad 2$

Example 4 Solve $(x^3 + 1)(2x^2 + 7x + 1) = 0$.

Solution The left side is already a product; it can be 0 only if one of the factors is 0. Therefore, solving the given equation amounts to solving both

$$x^3 + 1 = 0 \quad \text{and} \quad 2x^2 + 7x + 1 = 0.$$

Solve the first equation by factoring:

$$x^3 + 1 = (x + 1)(x^2 - x + 1) = 0.$$

Set both factors equal to 0. The first yields $x = -1$; the second yields the equation

$$x^2 - x + 1 = 0.$$

This quadratic has no real solution because its discriminant is negative:

$$b^2 - 4ac = 1 - 4 = -3 < 0.$$

Solve the second equation, $2x^2 + 7x + 1 = 0$, by the quadratic formula:

$$x = \frac{-7 \pm \sqrt{49 - 8}}{4} = \frac{-7 \pm \sqrt{41}}{4}.$$

Answer $-1 \quad \frac{1}{4}(-7 \pm \sqrt{41})$

Equations Involving Radicals

Equations involving square roots, cube roots, etc. require care.

Example 5 Solve $\sqrt{x + 3} = x - 9$.

Solution It is natural to square both sides since that gets rid of the square root:

$$\begin{aligned} \sqrt{x + 3} &= x - 9 \\ x + 3 &= (x - 9)^2 \\ x + 3 &= x^2 - 18x + 81 \\ x^2 - 19x + 78 &= 0. \end{aligned}$$

Any solution of the given equation is a solution of this quadratic. By the quadratic formula,

$$x = \frac{19 \pm \sqrt{19^2 - 4 \cdot 78}}{2} = \frac{19 \pm 7}{2} = 13, 6.$$

Check the given equation:

$$\sqrt{13 + 3} = \sqrt{16} = 4 = 13 - 9 \qquad \sqrt{6 + 3} = \sqrt{9} = 3 \neq 6 - 9.$$

Thus 13 is a solution, but 6 is not! Since there are no other possibilities, 13 is the only solution.

Answer 13

Example 5 shows the danger of squaring both sides of an equation. Although each solution of a given equation $A = B$ is also a solution of $A^2 = B^2$, the latter equation may have other solutions also. These we call **extraneous solutions** since they do not satisfy the given equation. As a very simple example, take

$$x = 2.$$

Squaring both sides gives

$$x^2 = 4 \qquad x = 2 \quad \text{or} \quad -2.$$

Obviously -2 is an extraneous solution.

> The operation of raising both sides of an equation to the same power does not necessarily lead to an equivalent equation. Whenever this technique is used, all solutions must be checked.

We note that in solving $A = B$, any extraneous solutions found from $A^2 = B^2$ come from the equation $A = -B$. For if $A^2 = B^2$, then

$$A^2 - B^2 = 0$$

$$(A + B)(A - B) = 0$$

so either $A = B$ or $A = -B$. For example, the extraneous solution $x = 6$ found in Example 5 is actually the solution of

$$\sqrt{x + 3} = -(x - 9).$$

Example 6 Solve $\sqrt{2x + 4} - \sqrt{x + 3} = 1$.

Solution We could square both sides, but it's simpler to isolate one of the radicals first (usually the more complicated one):

$$\sqrt{2x + 4} = 1 + \sqrt{x + 3}.$$

Now we square:

$$2x + 4 = 1 + 2\sqrt{x + 3} + (x + 3).$$

We'll have to square again, but first we simplify:

$$x = 2\sqrt{x + 3}$$

$$x^2 = 4(x + 3)$$

$$x^2 - 4x - 12 = 0$$

$$(x - 6)(x + 2) = 0.$$

Hence $x = 6$ or $x = -2$. Let us check the given equation:

$$\sqrt{2 \cdot 6 + 4} - \sqrt{6 + 3} = \sqrt{16} - \sqrt{9} = 4 - 3 = 1$$

$$\sqrt{2(-2) + 4} - \sqrt{-2 + 3} = \sqrt{0} - \sqrt{1} = 0 - 1 \neq 1.$$

Therefore 6 is a solution, but -2 is extraneous.

Answer 6

Systems

Let us consider systems of two equations in two unknowns, where one or both of the equations are non-linear.

EXAMPLE 7 Solve the system $\begin{cases} x^2 + y^2 = 1 \\ x + 2y = 1. \end{cases}$

Solution As with systems of linear equations, we try to eliminate one of the variables. From the linear equation, we have

$$x = 1 - 2y.$$

We substitute this expression for x into $x^2 + y^2 = 1$ and obtain a quadratic in y:

$$(1 - 2y)^2 + y^2 = 1 \qquad 1 - 4y + 4y^2 + y^2 = 1$$

$$5y^2 - 4y = 0 \qquad y(5y - 4) = 0.$$

Hence $y = 0$ or $y = \frac{4}{5}$. The corresponding values of x come from the relation $x = 1 - 2y$:

$$x = 1 - 2\cdot 0 = 1 \qquad x = 1 - 2\cdot\tfrac{4}{5} = -\tfrac{3}{5}.$$

Answer $x = 1 \quad y = 0 \qquad x = -\frac{3}{5} \quad y = \frac{4}{5}$

EXAMPLE 8 Solve the system $\begin{cases} xy = 5 \\ x^3 + y^3 = 30. \end{cases}$

Solution Solve the first equation for y in terms of x and substitute the result into the second equation:

$$y = \frac{5}{x} \qquad x^3 + \left(\frac{5}{x}\right)^3 = 30.$$

Multiply both sides by x^3 and transpose all terms to the left:

$$x^6 - 30x^3 + 125 = 0.$$

This is an equation of quadratic type. Substitute $z = x^3$ and solve for z:

$$z^2 - 30z + 125 = 0$$

$$(z - 5)(z - 25) = 0.$$

Hence $z = 5$ or $z = 25$. But $x = \sqrt[3]{z}$, so $x = \sqrt[3]{5}$ or $\sqrt[3]{25}$. Since $y = 5/x$, the corresponding values of y are

$$\frac{5}{\sqrt[3]{5}} = \frac{\sqrt[3]{125}}{\sqrt[3]{5}} = \sqrt[3]{25} \quad \text{and} \quad \frac{5}{\sqrt[3]{25}} = \frac{\sqrt[3]{125}}{\sqrt[3]{25}} = \sqrt[3]{5}.$$

Answer $(x, y) = (\sqrt[3]{5}, \sqrt[3]{25})$ and $(x, y) = (\sqrt[3]{25}, \sqrt[3]{5})$

Remark Both of the given equations are symmetric in x and y; that is, they are unchanged if x and y are interchanged. This explains the symmetry in the answers, $(\sqrt[3]{5}, \sqrt[3]{25})$ and $(\sqrt[3]{25}, \sqrt[3]{5})$.

EXAMPLE 9 Solve the system $\begin{cases} \sqrt{x} - \sqrt{y} = 1 \\ x + y = 13. \end{cases}$

Solution Square the first equation:

$$x - 2\sqrt{xy} + y = 1.$$

But $x + y = 13$ by the second equation, hence

$$13 - 2\sqrt{xy} = 1 \qquad \sqrt{xy} = 6.$$

Squaring this last equation, we arrive at a system without radicals:

$$\begin{cases} xy = 36 \\ x + y = 13 \end{cases}$$

(This system is symmetric in x and y; we can expect symmetric solutions.) From the second equation, $y = 13 - x$, so

$$x(13 - x) = 36 \qquad x^2 - 13x + 36 = 0$$

$$(x - 4)(x - 9) = 0.$$

Hence $x = 4$ or 9. The corresponding values of $y = 13 - x$ are 9 and 4. Thus the only possible solutions of the original equations are $(x, y) = (4, 9)$ and $(9, 4)$. In either case $x + y = 13$. But

$$\sqrt{4} - \sqrt{9} = 2 - 3 = -1 \neq 1 \qquad \sqrt{9} - \sqrt{4} = 3 - 2 = 1,$$

so $(4, 9)$ is an extraneous solution.

Answer $(x, y) = (9, 4)$

Remark It is not surprising that one of the solutions $(9, 4)$ and $(4, 9)$ was extraneous. The original system was not symmetric in x and y. In hindsight, when it was replaced by a symmetric system, we should have smelled an extraneous solution.

Exercises

Solve

1 $x^4 - 7x^2 + 6 = 0$

2 $x^4 - 5x^2 - 14 = 0$

3 $10x^4 - 7x^2 + 1 = 0$

4 $\frac{1}{2}x^4 = 3x^2 - 1$

5 $x - 5\sqrt{x} + 6 = 0$

6 $x - 6\sqrt{x} + 9 = 0$

7 $x^6 - 2x^3 - 80 = 0$

8 $x^8 - 15x^4 - 16 = 0$

9 $\dfrac{1}{x^4} = \dfrac{6}{x^2} - 8$

10 $\left(\dfrac{x}{4x + 1}\right)^2 = \dfrac{x}{4x + 1} + 2$

11 $\dfrac{x^2 + 2}{x^2 + 5} = \dfrac{1}{x^2 - 2}$

12 $1 - x^2 = 1 + \dfrac{1}{x^2}$

13 $x^4 + 5x^3 + 6x^2 = 0$

14 $x^6 = x^3$

15 $(2x + 1)^3 = (2x + 1)(x^2 + 9x + 3)$

16 $x^2(x - 1) = 4(x - 1)$

17 $(x^2 - 4x + 4)(x^2 + 3x + 1) = 0$

18 $(x - 3)(2x - 3)(5x + 3) = 0$

19 $(x^2 - 9)(4x^2 - 1) = 0$

20 $(x^2 - 16)(25x^2 - 9) = 0$

21 $x^7 + 5x^6 + x + 5 = 0$

22 $x^4 - (2x - 1)^2 = 0$

23 $x^8 - 1 = 0$

24 $x^8 - x^5 = 0.$

Solve

25 $\sqrt{4 - x} = x - 2$

26 $\sqrt{x + 3} = 9 - x$

27 $2\sqrt{x + 5} = x - 3$

28 $\sqrt{18 + x} = x - 2$

29 $\sqrt{x} + \sqrt{x+3} = 3$

30 $\sqrt{2x+1} + \sqrt{5-x} = 4$

31 $2\sqrt{x} - \sqrt{x-5} = \sqrt{x+7}$

32 $2\sqrt{x+4} + \sqrt{x-5} = \sqrt{9x+7}$

33 $\sqrt[4]{8x^2+1} = \sqrt{3x-1}$

34 $\sqrt{x} = \sqrt{10 + 3\sqrt{x}}$.

Solve the system

35 $\begin{cases} x + y = 16 \\ xy = 63 \end{cases}$

36 $\begin{cases} x^2 + y^2 = 25 \\ -x - 2y = 11 \end{cases}$

37 $\begin{cases} x^2 + xy + y^2 = 3 \\ x + y = 1 \end{cases}$

38 $\begin{cases} x^2 + 2xy + y^2 = 9 \\ x - y = 1 \end{cases}$

39 $\begin{cases} 2x^2 + xy = 1 \\ 3x - \frac{1}{2}y = 1 \end{cases}$

40 $\begin{cases} \frac{1}{9}x^2 + \frac{1}{4}y^2 = 1 \\ -2x + y = 6 \end{cases}$

41 $\begin{cases} \frac{1}{9}x^2 - \frac{1}{25}y^2 = 1 \\ \frac{1}{3}x + \frac{2}{5}y = 1 \end{cases}$

42 $\begin{cases} x^4 + y^4 = 1 \\ x^2 + 2y^2 = 1 \end{cases}$

43 $\begin{cases} xy = 30 \\ x^3 + y^3 = 341 \end{cases}$

44 $\begin{cases} xy = 2 \\ x^3 + y^3 = 5 \end{cases}$

45 $\begin{cases} xy = 6 \\ x^3 - y^3 = 19 \end{cases}$

46 $\begin{cases} xy = 1 \\ x^4 - 32y^4 = 14 \end{cases}$

47 $\begin{cases} x + y = 25 \\ \sqrt{x} - \sqrt{y} = 1 \end{cases}$

48 $\begin{cases} x - y = 7 \\ \sqrt{x} - \sqrt{y} = 1. \end{cases}$

8 APPLICATIONS OF NON-LINEAR EQUATIONS

Example 1 Find all right triangles whose hypotenuse is two units longer than one leg and one unit longer than the other leg.

Solution Let x be the hypotenuse. Then the two legs are $x - 2$ and $x - 1$. By the Pythagorean Theorem,

$$x^2 = (x-2)^2 + (x-1)^2 = (x^2 - 4x + 4) + (x^2 - 2x + 1).$$

Rearranging and combining terms, we have

$$x^2 - 6x + 5 = 0 \qquad (x-5)(x-1) = 0.$$

The only feasible solution is $x = 5$.

Answer There is only one such right triangle; its sides are 3, 4, 5.

The next example illustrates an important point.

Example 2 Find two numbers whose sum is 19 and whose product is 100.

Solution Call the numbers x and y. Then

$$\begin{cases} x + y = 19 \\ xy = 100. \end{cases}$$

From the first equation $y = 19 - x$. Substitute into the second:

$$x(19 - x) = 100 \qquad x^2 - 19x + 100 = 0.$$

The discriminant is $D = 19^2 - 400 = -39 < 0$, hence there is no solution.

Answer No solution

The important point is that just because you state a problem, it doesn't necessarily have to have an answer.

Let's try to salvage something from the wreckage of Example 2. What went wrong? If we look at products giving 100 such as $5 \cdot 20$, $4 \cdot 25$, $2 \cdot 50$, we see that the sum of the factors exceeds 19. Apparently we did not allow $x + y$ to be big enough. But how big is big enough?

EXAMPLE 3 If the product of two positive numbers is 100, what is the least their sum can be?

Solution Let the sum be s. Then

$$\begin{cases} x + y = s \\ xy = 100. \end{cases}$$

We want the smallest value of s for which this system has solutions. We *try* to solve by the method of Example 2 and end up with the quadratic equation

$$x^2 - sx + 100 = 0.$$

The discriminant is $s^2 - 400$. Hence solutions exist provided $s^2 - 400 \geq 0$. The smallest positive value of s that does the trick is $s = 20$. (The corresponding solution is $x = y = 10$.)

Answer 20

EXAMPLE 4 A carpenter wishes to saw out a brace in the shape of a right triangle with hypotenuse 16 in. and one leg twice as long as the other. He will do it by trial-and-error with a framing square (Fig. 1). Use algebra to find to the nearest sixteenth of an inch the lengths of the legs.

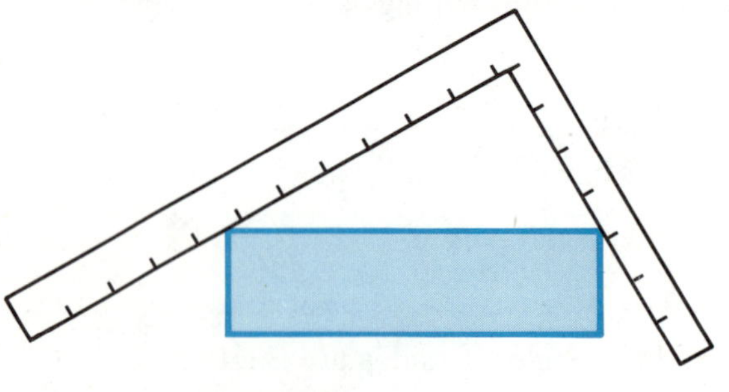

Fig. 1

Solution Call the legs x and $2x$. By the Pythagorean theorem

$$x^2 + (2x)^2 = 16^2 \qquad 5x^2 = 256$$

$$x^2 = \frac{256}{5} = 51.2 \qquad x = \sqrt{51.2} \approx 7.15542.$$

The nearest sixteenth to 0.15542 is $2/16 = 1/8$ (because the nearest integer to $(0.15542)(16) = 2.48672$ is 2). Hence $x \approx 7\frac{1}{8}$.

Similarly, $2x \approx 14.31084 \approx 14\frac{5}{16}$.

Answer $7\frac{1}{8}$ $14\frac{5}{16}$

EXAMPLE 5 A rectangle is divided into a square and a smaller rectangle, as shown in Figure 2 (next page). Suppose the ratio of length to width is the same for both rectangles. Compute that ratio.

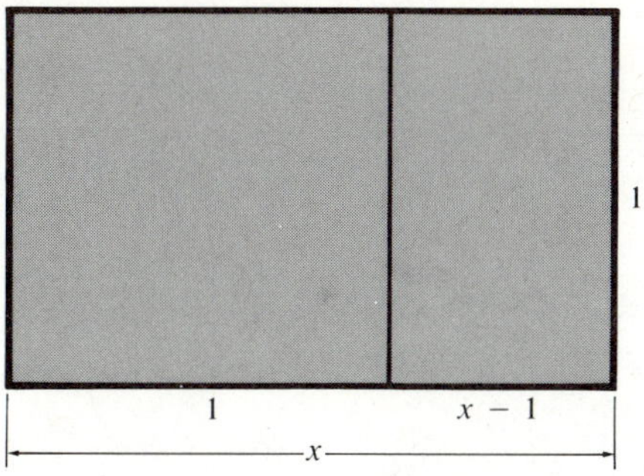

Fig. 2

Solution Let the given rectangle have length x and width 1. That way the desired ratio is just x. From the figure, the smaller rectangle has length 1 and width $x - 1$. Since the ratios of length to width are equal,

$$\frac{x}{1} = \frac{1}{x-1} \qquad x(x-1) = 1 \qquad x^2 - x - 1 = 0.$$

By the quadratic formula, the positive solution is

$$x = \frac{1+\sqrt{5}}{2} \approx 1.618034 \cdots$$

Answer $\frac{1}{2}(1+\sqrt{5})$

Remark The ratio in Example 5 was known to the ancient geometers, who called it the **Golden Section.** This number has many fascinating properties. For instance, it satisfies $x(x-1) = 1$, hence

$$x - 1 = \frac{1}{x} \qquad x = 1 + \frac{1}{x}.$$

We can replace the x on the right by $1 + 1/x$:

$$x = 1 + \cfrac{1}{1+\cfrac{1}{x}} = 1 + \cfrac{1}{1+\cfrac{1}{1+\cfrac{1}{x}}}, \text{ etc.}$$

Example 6 From a cliff above a lake, I drop a stone into the water. Exactly 3 seconds later, I hear the splash. Assuming that the stone falls $16t^2$ feet in t seconds, and that the speed of sound in air is 1100 ft/sec, how far down was the lake?

Solution Let t_1 be the time it took for the stone to fall, and t_2 the time for the sound of the splash to reach me. We are given that

$$t_1 + t_2 = 3.$$

Let d be the distance from the cliff to the water. Express t_1 and t_2 in terms of d. First,

$$16t_1^2 = d \quad \text{so} \quad t_1 = \sqrt{\frac{d}{16}} = \frac{\sqrt{d}}{4}.$$

Next,

$$1100t_2 = d \quad \text{so} \quad t_2 = \frac{d}{1100}.$$

Substitute into $t_1 + t_2 = 3$:

$$\frac{\sqrt{d}}{4} + \frac{d}{1100} = 3 \qquad 275\sqrt{d} + d = 3 \cdot 1100 = 3300$$

$$d + 275\sqrt{d} - 3300 = 0.$$

This is an equation of quadratic type. Let $x = \sqrt{d}$:

$$x^2 + 275x - 3300 = 0.$$

By the quadratic formula,

$$x = \frac{-275 \pm \sqrt{275^2 + 4 \cdot 3300}}{2} = \frac{-275 \pm \sqrt{88825}}{2}$$

$$\approx \frac{-275 \pm 298.035}{2}.$$

The positive solution is $x \approx 11.5175$. Hence $d = x^2 \approx 132.65$.

Answer 132.65 ft

The next example cannot be solved entirely by algebra, but algebra can carry you far enough that you can guess the answer.

Example 7 Find all positive fractions in lowest terms such that if both numerator and denominator are increased by 1, the fraction is increased by 10%.

Solution Let x/y be a fraction of the desired type. Then x and y are positive integers with no common factor. The problem requires

$$\frac{x+1}{y+1} = \frac{x}{y} + \frac{1}{10}\cdot\frac{x}{y}, \qquad \text{that is,} \qquad \frac{x+1}{y+1} = \frac{11}{10}\cdot\frac{x}{y}.$$

Clear of fractions:

$$10xy + 10y = 11xy + 11x \qquad 10y = xy + 11x.$$

This single equation is not enough to determine both x and y. However, we have more information: x and y are positive integers, which limits the possibilities. Let us solve for y:

$$(10 - x)y = 11x \qquad y = \frac{11x}{10 - x}.$$

The numerator is $11x > 0$; since $y > 0$, the denominator $10 - x$ must also be positive. The only possibilities for x, therefore, are $1, 2, 3, \cdots, 9$. Let us list the corresponding values of y:

x	1	2	3	4	5	6	7	8	9
y	$\frac{11}{9}$	$\frac{11}{4}$	$\frac{33}{7}$	$\frac{22}{3}$	11	$\frac{33}{2}$	$\frac{77}{3}$	44	99

Only three y-values are integers, 11, 44, and 99. The corresponding fractions are

$$\frac{x}{y} = \frac{5}{11}, \frac{8}{44}, \frac{9}{99},$$

and of these, only $\frac{5}{11}$ is in lowest terms.

Answer $\frac{5}{11}$

Exercises

1 Find the sides of an isosceles triangle of area 1 whose slant side is twice its base.

2 One side of a rectangle is $\sqrt{2}$ units longer than the other side, and the area is 12 sq units. Find the sides of the rectangle.

3 Is there a number that exceeds its reciprocal by 4?

4 Is there a number that exceeds its square root by 6?

5 Find the base of an isosceles triangle of height 3 and perimeter 12.

6 Express 30 as the sum of two numbers the difference of whose squares is 120.

7 A group of people charter a plane for \$30,000. At the last minute, 70 more people join the group, lowering the cost per person by \$26.25. How many people are in the group?

8 After lengthening my stride (pace) by 2 inches, I can now walk a mile in 132 fewer paces. What is the length of my new stride?

9 How long until I hear the splash if I drop a stone into the water from a height of 100 ft? (Refer to Example 6, page 81.)

10 Near the surface of the moon, a body falls $1.6t^2$ meters in t seconds, starting from rest. Suppose two bodies are released from rest, the second body 35.2 m below the first and 2 sec later. When will the first pass the second?

11 Show that the sum of a positive number and its reciprocal is always at least 2. [*Hint* Complete the square.]

12 The denominator of a certain fraction in lowest terms exceeds the numerator by 3. If both numerator and denominator are increased by 1, the value of the fraction increases by $\frac{1}{10}$. Find the fraction.

13 Find two consecutive positive integers whose cubes differ by 919. [*Hint* Factor.]

14 Find three consecutive positive odd integers the sum of whose squares is 1883.

15 A tinsmith wants to lighten a square of sheet metal 10% by snipping off four small equal isosceles right triangles, one from each corner. If the square is 50 cm on each side, what are the legs of the triangles?

16 From here to the bus stop is 200 meters if I follow the sidewalk to the corner, turn a right angle, and continue along the sidewalk. But if I cut straight across the grass, the distance is 170 m. How far is it from here to the corner?

17 Suppose r is a solution of $ax^2 + bx + c = 0$, where $ac \neq 0$. Show that $1/r$ is a solution of $cx^2 + bx + a = 0$.

18 Suppose r is a solution of $x^2 + x - 1 = 0$. Show that r is also a solution of $x^3 - 2x + 1 = 0$ and of $x^4 + 3x - 2 = 0$.

19 Find two numbers whose product is twice their sum, and whose difference is 2.

20 Find two numbers whose sum is 48 and whose product is 10 times their difference.

21 Take a two-digit integer. Multiply it by the integer you get when you interchange its digits, then subtract 10 times the sum of the squares of the original digits. The result is always divisible by 101. Why?

22 Find all positive fractions F in lowest terms such that if both numerator and denominator are increased by 1, then the fraction is increased to $\frac{16}{15}F$.

23 A length L of rope is cut into two pieces, each of which is folded into a square. If one square has triple the area of the other, how long are the pieces?

24 One leg of a right triangle is $\frac{1}{4}$ the length of the hypotenuse, and the area of the triangle is 100 cm^2. How long is the other leg?

25 Show that there do not exist two positive numbers whose sum is 30 and whose product is 240.

26 (cont.) If the sum of two positive numbers is 30, what is the most their product can be?

27 Demand, supply, and price in dollars of a certain sports car are related by

$$Q_s = 2(P - 16000) \qquad Q_d P = 5.25 \times 10^7.$$

Find the equilibrium price.

28 Express x and y in terms of t, assuming $x^2 + y^2 = 1$, $y = t(x+1)$, and $x \neq -1$.

29 The illumination intensity I at distance D from a light bulb of strength S is directly proportional to S and inversely proportional to the square of D. That is, $I = kS/D^2$, where k is a constant that depends on the units of measurement. Suppose two bulbs of intensities S_1 and S_2 lie at distance D apart. Where on the segment joining them are their intensities equal?

30 At distance D from a planet of mass M, the gravitational field strength is GM/D^2, where G is a constant (Newton's law of universal gravitation). Assume the moon's mass is $\frac{1}{81}$ of the earth's and their distance is 382,000 km. How far from the earth to the moon will their gravitational attractions cancel?

31 (cont.) Solve the same problem for the earth and the sun. The sun's mass is 3.33×10^5 that of the earth, and their mutual distance is 1.49×10^8 km.

32 (cont.) The earth's radius is 6.37×10^3 km and the gravitational field strength on the surface of the earth is six times that on the surface of the moon. Estimate the moon's radius. (The gravitation on or above the surface of a planet is the same as if all the planet's mass were concentrated at its center.)

Find the value of the **continued fraction**

33 $$\cfrac{1}{2 + \cfrac{1}{2 + \cfrac{1}{2 + \cfrac{1}{2 + \cdots}}}}$$

34 $$\cfrac{1}{1 + \cfrac{1}{2 + \cfrac{1}{1 + \cfrac{1}{2 + \cfrac{1}{1 + \cfrac{1}{2 + \cdots}}}}}}$$

[*Hint* Let the value of the continued fraction be x, and express the denominator in terms of x.]

9 ORDER

The order relation in the real number system asserts that one number is less than, or greater than, another. Recall the notation

$a < b$	a is less than b
$a \leq b$	a is less than or equal to b
$a > b$	a is greater than b
$a \geq b$	a is greater than or equal to b.

The numbers greater than 0 are the **positive** numbers, and the numbers less than 0 are the **negative** numbers. The relation $a < b$ means that the difference $b - a$ is positive:

$a < b$	if and only if $b - a > 0$
$a \leq b$	if and only if $b - a \geq 0$.

The Number Line

The number line provides an important and extremely useful *geometric* interpretation of the real number system. To construct it, we begin with a line in the euclidean plane (Fig. 1). Step 1: We *direct* the line by placing an arrowhead on it. Then we know right and left. A point P is left of a point Q if going directly from P to Q is in the direction of the arrowhead. Step 2: We choose a point for 0. Step 3: We choose a unit of length.

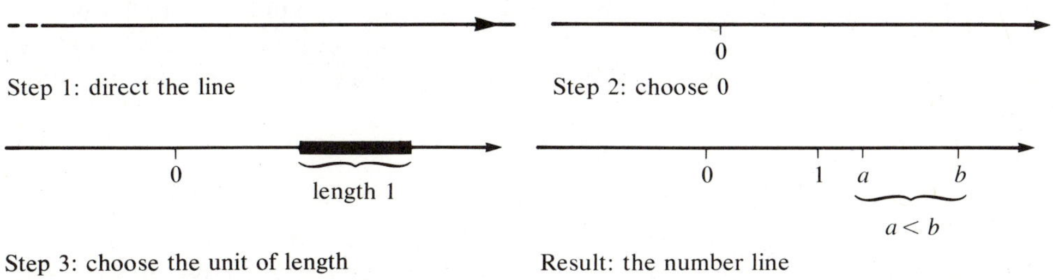

Fig. 1 Construction of the number line

Now a basic mathematical principle states that there is a unique one-to-one correspondence between the set of real numbers and the set of points of the line such that the number 0 corresponds to the point labelled 0, and

(i) $a < b$ if and only if the point to which a corresponds is left of the point to which b corresponds;

(ii) If $a < b$, then the distance between the corresponding points of the line is $b - a$.

Once this one-to-one correspondence between the real number system and a line has been set up, we can *identify* real numbers with points of the line. Thus we refer to the *points* -3, 2.78, $\sqrt{5}$, etc. Now picturing the real number system as a number line helps us to understand order. The notation $a < b$ is mathematical shorthand for saying that a is to the left of b; the notation $a \le b$ for saying that a is left of or equal to b.

Intervals

The set of all numbers between two fixed numbers is an **interval** (or **segment**) on the number line. An interval may include one or both of its end points, or neither, depending on how it is specified. For example, $-2 \le x \le 1$ describes the interval of all numbers between -2 and 1, *including* the end points, while $3 < x < 7$ describes the interval *strictly* between 3 and 7, that is, *excluding* the end points (Fig. 2).

Fig. 2 Examples of intervals
(a) Closed interval $[-2, 1]$: $-2 \le x \le 1$
(b) Open interval $(3, 7)$: $3 < x < 7$

The special names "closed interval" and "open interval" are common for telling whether or not end points are in or out.

Intervals Let $a < b$.

The **closed interval** from a to b, denoted $[a, b]$, is the set of all points x such that $a \le x \le b$.

The **open interval** from a to b, denoted (a, b), is the set of all points x such that $a < x < b$.

Thus in Fig. 2 we have the closed interval $[-2, 1]$ and the open interval $(3, 7)$. Although the notations $[a, b]$ and (a, b) are common in mathematics, there is a possible confusion with ordered pairs, so we won't use this notation much.

Properties of Order

We shall derive five basic properties of order, then summarize them in a box.

Two properties are obvious from the number line. (1) If a is to the left of b and b is to the left of c, then a is to the left of c. In other words, $a < b$ and $b < c$ imply $a < c$. (2) If a is to the left of b and both points are shifted c units to the left or right, the shifted points are in the same order. In other words, if $a < b$ and c is any real number, then $a + c < b + c$.

These two properties can also be verified without geometry:

(1) Suppose $a < b$ and $b < c$. This means $b - a > 0$ and $c - b > 0$. But the sum of positive numbers is positive, hence

$$c - a = (c - b) + (b - a) > 0.$$

Therefore $a < c$.

(2) If c is any real number, then $a + c < b + c$ because

$$(b + c) - (a + c) = b - a > 0.$$

The second property says that you may *add* the same number to both sides of an inequality and get a valid inequality:

$$\begin{array}{c} a < b \\ c = c \\ \hline a + c < b + c. \end{array}$$

Note that subtracting c from both sides is the same as adding $-c$, so that you also may *subtract* the same number from both sides of an inequality. That is, from $a < b$ follows $a - c < b - c$.

Question Can you *multiply* both sides of an inequality by the same number?

Answer Yes, provided the number is positive. If it is negative, you reverse the inequality. For example, suppose we start with $2 < 3$. Then

$$\tfrac{1}{17}\cdot 2 < \tfrac{1}{17}\cdot 3 \qquad 5\cdot 2 < 5\cdot 3 \qquad (8.07)\cdot 2 < (8.07)\cdot 3 \qquad \text{etc.,}$$

but

$$(-\tfrac{1}{17})\cdot 2 > (-\tfrac{1}{17})\cdot 3 \qquad (-5)\cdot 2 > (-5)\cdot 3$$

$$(-8.07)\cdot 2 > (-8.07)\cdot 3 \qquad \text{etc.}$$

In general, we have the properties (3) if $a < b$ and $c > 0$, then $ac < bc$ and (4) if $a < b$ and $c < 0$, then $ac > bc$. The proofs depend on the fact that the product of positive numbers is positive:

(3) $b - a > 0$ and $c > 0$, so $bc - ac = (b - a)c > 0$, hence $ac < bc$.

(4) $b - a > 0$ and $c < 0$, so $-c > 0$ and $ac - bc = (b - a)(-c) > 0$, that is, $bc < ac$.

The same rules hold for division of both sides of an inequality by a non-zero number c. In fact, division by c is the same as multiplication by $1/c$, while c and $1/c$ have the same sign.

Finally we observe that taking reciprocals *reverses* an inequality, provided the numbers involved have the same sign. For example,

$$0 < 2 < 3 \quad \text{but} \quad \frac{1}{2} > \frac{1}{3}; \qquad -4 < -3 < 0 \quad \text{but} \quad \frac{1}{-4} > \frac{1}{-3}.$$

In general, we have property (5) if $0 < a < b$ or $a < b < 0$, then $1/a > 1/b$. The proof is left as an exercise.

Rules for Order

(1) If $a < b$ and $b < c$, then $a < c$.

(2) If $a < b$ and if c is any real number, then $a \pm c < b \pm c$.

(3) If $a < b$ and $c > 0$, then $ac < bc$.

(4) If $a < b$ and $c < 0$, then $ac > bc$.

(5) If $0 < a < b$ or $a < b < 0$, then $\dfrac{1}{b} < \dfrac{1}{a}$.

These rules apply just as well with $\le$ in place of $<$, except in (5) you must have

$$0 < a \le b \quad \text{or} \quad a \le b < 0.$$

EXAMPLE 1 Given $1.4 < \sqrt{2} < 1.5$, show that

(a) $2 + \sqrt{2} < 3.5$ (b) $8.4 < 6\sqrt{2} < 9$ (c) $\dfrac{1}{3 - 2\sqrt{2}} > 5$.

Solution (a) Start with $\sqrt{2} < 1.5$. By Rule (2), you may add 2 to both sides:

$$2 + \sqrt{2} < 2 + 1.5 = 3.5.$$

(b) Start with the two inequalities

$$1.4 < \sqrt{2} \qquad \sqrt{2} < 1.5.$$

By Rule (3), you may multiply both sides of each by 6:

$$8.4 = 6(1.4) < 6\sqrt{2} \qquad 6\sqrt{2} < 6(1.5) = 9.$$

Combine: $8.4 < 6\sqrt{2} < 9$.

(c) One way to show that a number is big is to show that its reciprocal is small. By Rule (3), $2.8 < 2\sqrt{2} < 3$, and it follows by Rule (4) that

$-3 < -2\sqrt{2} < -2.8$. Add 3:

$$0 < 3 - 2\sqrt{2} < 0.2$$

by Rule (2). Finally, by Rule (5),

$$\frac{1}{3 - 2\sqrt{2}} > \frac{1}{0.2} = 5.$$

Several natural extensions of the rules of order are convenient in applications.

(6) If $a < b$, then $-a > -b$.

(7) If $a < A$ and $b < B$, then $a + b < A + B$.

(8) If $0 < a < A$ and $0 < b < B$, then $ab < AB$.

The proofs are left as exercises. Note that Rules (7) and (8) are easily extended to three or more inequalities. For instance, if $a < A$, $b < B$, and $c < C$, then

$$a + b + c < A + B + C.$$

Finally, we recall a basic inequality that has many applications.

(9) If a is any real number, then

$$a^2 \geq 0.$$

In fact, $a^2 > 0$ unless $a = 0$.

Estimates

The rules of order are useful in making estimates with a minimum of computation.

EXAMPLE 2 Verify the estimate, using as little arithmetic as possible

(a) $6 \cdot 8 \cdot 9 \cdot 11 < 5000$

(b) $\pi^2 < 10$

(c) $\frac{5}{9} < \frac{1}{5} + \frac{1}{6} + \frac{1}{7} + \frac{1}{8} + \frac{1}{9} < 1$.

Solution (a) Use Rule (8):

$$6 \cdot 8 \cdot 9 \cdot 11 = 48 \cdot 99 < 50 \cdot 100 = 5000.$$

(b) Use the fact that $\pi < 3.1416 < 3.15 = 315/100 = 63/20$ and Rule (8):

$$\pi^2 = \pi \cdot \pi < \frac{63}{20} \cdot \frac{63}{20} = \frac{3969}{400} < \frac{4000}{400} = 10.$$

(c) Each term after the first is less than $\frac{1}{5}$. Hence, by Rule (7),

$$\frac{1}{5} + \frac{1}{6} + \frac{1}{7} + \frac{1}{8} + \frac{1}{9} < \frac{1}{5} + \frac{1}{5} + \frac{1}{5} + \frac{1}{5} + \frac{1}{5} = 1.$$

Similarly, each term but the last is greater than $\frac{1}{9}$. Hence

$$\frac{1}{5} + \frac{1}{6} + \frac{1}{7} + \frac{1}{8} + \frac{1}{9} > \frac{1}{9} + \frac{1}{9} + \frac{1}{9} + \frac{1}{9} + \frac{1}{9} = \frac{5}{9}.$$

EXAMPLE 3 Find the largest possible error in computing the product of two numbers, $a = 8.4$ and $b = 12.3$, determined by an experiment in which the last digit might be off by 1.

Solution We have

$$8.3 \le a \le 8.5 \quad \text{and} \quad 12.2 \le b \le 12.4.$$

By Rule (8),

$$(8.3)(12.2) \le ab \le (8.5)(12.4),$$

that is, $101.26 \le ab \le 105.40$. Since $(8.4)(12.3) = 103.32$, the larger of

$$105.40 - 103.32 = 2.08 \quad \text{and} \quad 103.32 - 101.26 = 2.06$$

is the largest possible error.

Answer 2.08

Exercises

Abbreviate, using as few symbols as possible

1 x is strictly between 4 and 9.

2 x is between 3 and 5, with 3 and 5 allowed.

3 x is non-negative.

4 x is positive but not more than 10.

5 x is less than y and y is at most z.

6 x is strictly between 1 and 2 and y is strictly between 2 and 3.

Indicate the portion of the number line determined by the inequality

7 $x \le -2$ **8** $x > 4$ **9** $0 < x < 1$

10 $2 \le x < 5$ **11** $x^2 > 9$ **12** $x + 1 > 0$.

Verify the inequality, using as little computation as possible

13 $5667 - 4128 < 6093 - 4128$ **14** $33 \cdot 89 < 36 \cdot 89$

15 $1/\pi > 1/4$ **16** $11/5 < 14/5$

17 $9 \cdot 11 \cdot 13 < 1300$ **18** $97 \cdot 98 \cdot 99 < 1{,}000{,}000$

19 $\frac{1}{3}(8561 + 8774 + 8819) > 8561$ **20** $100\sqrt{2} > 140$

21 $\pi^6 < 1000$ **22** $6.28 < 2\pi < 6.30$

23 $9 + 99 + 999 + 9999 < 11{,}110$ **24** $\frac{1}{2} < \frac{1}{11} + \frac{1}{12} + \frac{1}{13} + \cdots + \frac{1}{20} < \frac{10}{11}$

25 $\dfrac{4}{3} < \dfrac{1}{\sqrt{5}} + \dfrac{1}{\sqrt{6}} + \dfrac{1}{\sqrt{7}} + \dfrac{1}{\sqrt{8}} < 2$ **26** $1 < \frac{1}{4}(1 + \pi) < 1.04$.

27 If $0 < a < b$, show that $a^2 < b^2$.

28 (cont.) Show that the conclusion is not necessarily true if a and b are not both positive.

29 If a and b have opposite signs and $a < b$, find the relation between $1/a$ and $1/b$.

30 If $a < b$, prove that $c - b < c - a$.

31 If $a < b$ and $c < d$, is $ac < bd$?

32 If $a^2 < b^2$, is $a < b$?

Prove

33 Rule (5) **34** Rule (6) **35** Rule (7) **36** Rule (8).

37 Suppose $1.2 < a < 1.3$ and $10.0 < b < 10.5$. Find the set of possible values of ab.

38 Suppose $-1 < a < 1$ and $-3 < b < 5$. Find the set of possible values of ab.

39 I measure five numbers strictly between 12 and 13. Estimate their sum S.

40 If $3 \le a \le 4$ and $5 \le b \le 6$, find the set of possible values of $1/a + 1/b$.

Which is larger (calculator not allowed)

41 $(67)(69)$, 68^2

42 $(1047)(1051)$, 1049^2

43 $(121)(122)(124)(125)$, 123^4

44 $(a - 1)(a + 1)$, a^2?

Shopping. Which is the best and which is the worst buy?

45 toothpaste; *giant* $5\frac{3}{4}$ oz size at 94¢, *large economy* $6\frac{7}{9}$ oz size at \$1.07 or *family* $7\frac{1}{3}$ oz size at \$1.20.

46 corn flakes; *brand* X of 14 oz at 61¢, *brand* Y of 18 oz at 78¢ or *brand* Z of 1 lb at 70¢.

47 detergent; *mammoth* $5\frac{1}{2}$ lb size at \$1.42, *super* $7\frac{3}{4}$ lb size at \$2.06 or *extra savings* $10\frac{1}{4}$ lb size at \$2.70.

48 peanut butter; *small* 4 oz jar at 37¢, *medium* 6.6 oz jar at 58¢ or *large* 12.4 oz jar at \$1.04.

49 Suppose $x > 0$ and $y = \dfrac{1}{1 + \dfrac{1}{1 + \dfrac{1}{x}}}$.

If x increases, does y increase or decrease?

50 Brain teaser. Ten people sit at a round table. Each has an amount of money equal to the average of the amounts held by his two neighbors. All together they have \$100. Find all possible ways the \$100 can be distributed.

10 ABSOLUTE VALUE

Often we need a measure of the size of a number, regardless of its sign. In some sense, -7 seems larger than 2, but we cannot write $-7 > 2$. We introduce absolute values in order to express this feeling precisely. We define the **absolute value** of a real number a, written $|a|$, as follows:

Absolute Value
$$|a| = \begin{cases} a & \text{if } a \ge 0 \\ -a & \text{if } a < 0. \end{cases}$$

Examples

$$|-7| = 7 \qquad |15| = 15 \qquad |-3.1416| = 3.1416 \qquad |7.28| = 7.28$$
$$|0| = 0 \qquad |\tfrac{7}{3}| = \tfrac{7}{3} \qquad |-5| = 5 \qquad |-\tfrac{4}{9}| = \tfrac{4}{9}.$$

Each real number has a non-negative absolute value which measures its "size," or "magnitude." With absolute values, we can say correctly what we were trying to express before, $|-7| > |2|$.

Properties of Absolute Values

Absolute values satisfy some basic rules:

Rules for Absolute Values

(1) $|a| = \begin{cases} a & \text{if } a > 0 \\ 0 & \text{if } a = 0 \\ -a & \text{if } a < 0 \end{cases}$

(2) $|-a| = |a|$ (3) $|ab| = |a| \cdot |b|$ (4) $\left|\dfrac{a}{b}\right| = \dfrac{|a|}{|b|}$ $(b \neq 0)$

(5) $a \leq |a|$ and $-a \leq |a|$

(6) $|a + b| \leq |a| + |b|$ (the triangle inequality).

We have discussed (1), and (2) is obvious by the definition. Rule (3) is pretty obvious also; just test a few cases. However, we'll leave a formal proof to Exs. 60–62. Rule (4) follows from (3) applied to $b(a/b) = a$:

$$\left|b \cdot \frac{a}{b}\right| = |a| \qquad |b| \left|\frac{a}{b}\right| = |a| \qquad \left|\frac{a}{b}\right| = \frac{|a|}{|b|}.$$

Rule (5) is easily proved by considering the three cases $a < 0$, $a = 0$, $a > 0$.

Rule (6) is proved in several steps. First by (5),

$$a \leq |a| \quad \text{and} \quad b \leq |b|, \quad \text{so} \quad a + b \leq |a| + |b|.$$

Likewise

$$-a \leq |a| \quad \text{and} \quad -b \leq |b|, \quad \text{so} \quad -(a + b) \leq |a| + |b|.$$

Hence $\pm(a + b) \leq |a| + |b|$. But for one of the choices of sign, $\pm(a + b) = |a + b|$. Therefore

$$|a + b| \leq |a| + |b|.$$

Examples of Rule (3):

$$|4 \cdot 5| = |20| = 20 = |4| \cdot |5|$$

$$|7 \cdot (-4)| = |-28| = 28 = |7| \cdot |-4|$$

$$|(-6)(-5)| = |30| = 30 = |-6| \cdot |-5|.$$

Examples of Rule (6):

$$|-5 + 3| \leq |-5| + |3|$$

because $|-5 + 3| = |-2| = 2$ and $|-5| + |3| = 5 + 3 = 8$.

$$|(-3) + (-4)| \leq |-3| + |-4|$$

because $|(-3) + (-4)| = |-7| = 7$ and $|-3| + |-4| = 3 + 4 = 7$.

Remark Rules (3) and (6) extend easily to three or more numbers; for instance,

$$|abc| = |a| \cdot |b| \cdot |c| \qquad |a + b + c| \leq |a| + |b| + |c|.$$

For example

$$|3 - 7 + 2| \leq |3| + |-7| + |2|.$$

EXAMPLE 1 Compute (a) $|7 - 30|$ (b) $|1 - \sqrt{2}|$ (c) $|(-2)^5|$.

Solution (a) $|7 - 30| = |-23| = 23$.

(b) $\sqrt{2} > 1.4$ so $1 - \sqrt{2} < 0$. Therefore

$$|1 - \sqrt{2}| = -(1 - \sqrt{2}) = \sqrt{2} - 1.$$

(c) $|(-2)^5| = |-32| = 32$. Alternatively, $|(-2)^5| = |-2|^5 = 2^5 = 32$.

Answer (a) 23 (b) $\sqrt{2} - 1$ (c) 32

Absolute values *and* inequalities frequently occur together. A rather obvious, but very useful rule for such cases is the following.

(7) Let $B \geq 0$. Then $\left.\begin{matrix} |a| < B \\ |a| \leq B \end{matrix}\right\}$ if and only if $\left\{\begin{matrix} -B < a < B \\ -B \leq a \leq B. \end{matrix}\right.$

Geometrically, the (first) statement says that the points with $|a| < B$ are precisely the points of the open interval $-B < x < B$, an interval centered at 0. See Fig. 1.

Fig. 1

EXAMPLE 2 Find all numbers x that satisfy

(a) $|x - 4| = 1$ (b) $\left|\dfrac{1}{x}\right| \leq 2$ (c) $|x - 1| < 5$.

Solution (a) If $|x - 4| = 1$, then either $x - 4 = 1$ or $x - 4 = -1$. Hence $x = 5$ or $x = 3$.

(b) By (4), $|1/x| \leq 2$ if and only if $1/|x| \leq 2$. Clearly $x = 0$ is impossible, so $|x| > 0$, and the inequality is equivalent to $|x| \geq \frac{1}{2}$. Therefore $x \leq -\frac{1}{2}$ or $x \geq \frac{1}{2}$.

(c) By (7), $|x - 1| < 5$ is the same as $-5 < x - 1 < 5$. Adding 1 to these inequalities, we obtain $-4 < x < 6$.

Answer (a) $x = 3, 5$ (b) $x \leq -\frac{1}{2}$ or $x \geq \frac{1}{2}$ (c) $-4 < x < 6$

EXAMPLE 3 Suppose

$$|x + 5| < 10^{-3} \quad \text{and} \quad |y - 2| < 10^{-3}.$$

Show that

$$|(x + y) + 3| < 2 \times 10^{-3}.$$

Solution Write

$$(x + y) + 3 = (x + 5) + (y - 2)$$

and apply the triangle inequality (Rule 6):

$$|(x + y) + 3| \leq |x + 5| + |y - 2| < 10^{-3} + 10^{-3} = 2 \times 10^{-3}.$$

Remark Rule (6) has two useful consequences:

$$(8)\quad |a - b| \le |a| + |b|$$

$$(9)\quad |a| - |b| \le |a - b|.$$

To prove (8), replace b by $-b$ in (6):

$$|a - b| = |a + (-b)| \le |a| + |-b| = |a| + |b|.$$

To prove (9), apply (6) to the sum $a = (a - b) + b$:

$$|a| = |(a - b) + b| \le |a - b| + |b|,$$

hence $|a| - |b| \le |a - b|$.

Geometric Interpretation

Absolute values have an important geometric meaning on the number line. By its very definition, $|a|$ is the distance of the point a from the point 0. Thus the "size" of a is measured by its *distance* from 0, a positive quantity no matter whether a is right or left of 0.

Given distinct points a and b on the number line, the distance between them is $|a - b|$. See Fig. 2. The distance between 5 and 9 is $|5 - 9| = |-4| = 4$. The distance from -2 to 3 is $|-2 - 3| = |-5| = 5$. Since $|a - b| = |b - a|$, it doesn't matter which you subtract from which; the answer comes out the same.

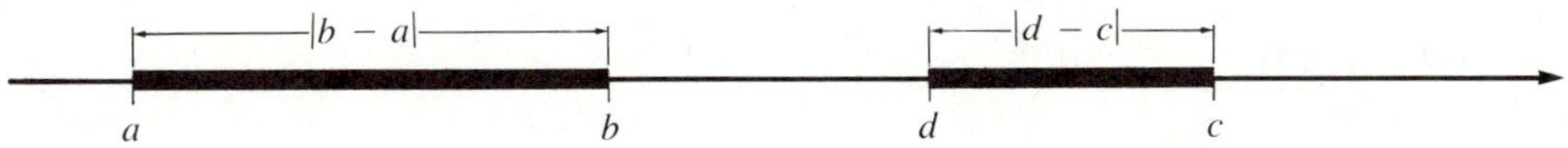

Fig. 2 Distance between points

Using absolute values and inequalities, we can often express geometric facts algebraically, or interpret algebraic relations geometrically. Here are a few examples:

Arithmetic Statement	Geometric Statement
$a > 0$	The point a lies to the right of the point 0.
$a > b$	a lies to the right of b.
$a - b = c > 0$	a lies c units to the right of b.
$a < b < c$	b lies between a and c.
$\|3 - a\| < \frac{1}{2}$	The point a is within $\frac{1}{2}$ unit of the point 3, that is, a lies between $2\frac{1}{2}$ and $3\frac{1}{2}$.
$\|a\| < \|b\|$	The point a closer to 0 than the point b is.

This close relationship between algebra and geometry is extremely important. It can provide two ways of looking at a problem, and hence can increase the chances of solving it. Sometimes we are able to attack geometric problems algebraically, sometimes vice versa.

EXAMPLE 4 Express in algebraic notation using absolute values

(a) The distance from x to -3 is 4.

(b) x is closer to 4 than y is to -6.

(c) x is at most 2 units from 5.

Solution (a) $|x - (-3)| = 4$, that is, $|x + 3| = 4$.

(b) $|x - 4| < |y - (-6)|$, that is, $|x - 4| < |y + 6|$.

(c) The distance from x to 5 is less than or equal to 2, hence $|x - 5| \leq 2$.

Answer (a) $|x + 3| = 4$ (b) $|x - 4| < |y + 6|$ (c) $|x - 5| \leq 2$

EXAMPLE 5 Find all points on the number line that are twice as far from 0 as from 6.

Solution The distance from x to 0 is $|x|$ and the distance from x to 6 is $|x - 6|$. We must find points x for which

$$|x| = 2|x - 6|.$$

Since $|x| = \pm x$ and $|x - 6| = \pm(x - 6)$, there are two possibilities, the same signs or opposite signs:

$x = 2(x - 6)$	$x = -2(x - 6)$
$x = 2x - 12$	$x = -2x + 12$
$x = 12$	$3x = 12 \qquad x = 4$.

Answer 4 and 12.

EXAMPLE 6 Draw on the number line the set of all points x that satisfy

(a) $|x - 4| = 1$ (b) $|x - 2| < 3$ (c) $|x + 2| > 4$.

Solution (a) x lies at distance 1 from 4, hence $x = 3$ or $x = 5$.

(b) The distance from x to 2 is less than 3, hence $-1 < x < 5$.

(c) The distance from x to -2 exceeds 4, hence either $x < -6$ or $x > 2$.

Answer

(a) $x = 3$ or $x = 5$

(b) $-1 < x < 5$

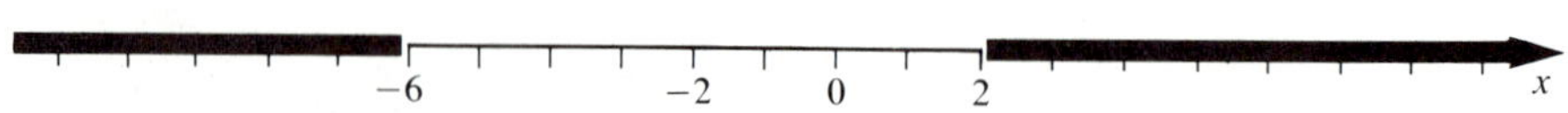

(c) $x < -6$ or $x > 2$

Exercises

Find a simpler expression, not involving absolute values

1 $|-4|$ **2** $-|-6|$ **3** $|4 - 11|$
4 $|8 - 3|$ **5** $|-2| + |-3|$ **6** $|(-2)(-3)|$
7 $|-5| \cdot |-6|$ **8** $|-5|/5$ **9** $|-4/9|$
10 $|-2| - |-3|$ **11** $|-6|^2$ **12** $|(-6)^2|$.

Express without absolute values all numbers x that satisfy the condition

13 $|x| = 0$ **14** $|x| = -1$ **15** $|3x| = 9$

16 $|\frac{1}{2}x| = 3$ **17** $|x - 4| = 1$ **18** $|x + 4| = 1$

19 $|x| \le \frac{1}{2}$ **20** $|x| > \frac{2}{3}$ **21** $|2x| < 8$

22 $|-4x| > 6$ **23** $|x - 1| < 3$ **24** $|2 - x| < 5$

25 $|x - \frac{1}{2}| < 4$ **26** $|x + 8| > 20$ **27** $|2x - 5| \le 1$

28 $|3x + 4| \ge 1$ **29** $|x^2 - 7| = 2$ **30** $|2x^2 - 9| = 1$

31 $\left|\dfrac{1 + 3x}{2}\right| < 5$ **32** $\left|\dfrac{1}{x + 1}\right| < 3$ **33** $\left|\dfrac{2}{x} + 1\right| > 10$

34 $\left|\dfrac{3}{5 - x^2}\right| < \frac{1}{100}$ **35** $|x - 2|\cdot|x + 2| \le 3$ **36** $x + |x| > 12$.

Draw on the number line the set of all points x that satisfy

37 $|x| > 2$ **38** $|x - 5| \le 0.2$ **39** $|x + \frac{3}{2}| < \frac{5}{2}$ **40** $1 < |x| < 3$.

Express in algebraic notation using absolute values

41 x is either 2 or -2

42 x is farther from 0 than from y

43 x is closer to 1 than to 5

44 x lies either to the left of -2 or to the right of 2

45 x is within 3 units of 7

46 x is between 6 and 8.

47 Find all points that are 3 times as far from 5 as from 1.

48 Find all points that are 10 times as far from 5 as from 1.

49 Explain how you can tell at a glance that there is no x for which both $|x - 1| < 2$ and $|x - 12| < 3$.

50 If $a \le b$ and $-a \le b$, show that $|a| \le b$.

51 Explain why $|a| + |b| + |c| > 0$ is algebraic shorthand for "at least one of the numbers a, b, c is different from 0."

52 (cont.) Find an analogous shorthand version of "a, b, c are all different from 0."

53 Suppose $|x - a| < 10^{-6}$. Show that $|7x - 7a| < 10^{-5}$.

54 Suppose $|x - 7| < 10^{-6}$ and $|y - 5| < 10^{-6}$. Show that $|(x + y) - 12| < 10^{-5}$. [*Hint* Use the triangle inequality.]

55 Suppose $|x - 5| < \frac{1}{10}$ and $|y - 7| < \frac{1}{10}$. Prove that $|xy - 35| < 1.3$. [*Hint* $xy - 35 = x(y - 7) + 7(x - 5)$.]

56 (cont.) Suppose $|x - 5| < 10^{-6}$ and $|y - 7| < 10^{-6}$. Prove that $|xy - 35| < 2 \times 10^{-5}$.

57 Suppose $|x - 3| < 10^{-6}$. Prove that $|x^2 - 9| < 10^{-5}$. [*Hint* Factor $x^2 - 9$.]

58* (cont.) Prove $|x^3 - 27| < 5 \times 10^{-5}$.

59 Prove $\big||a| - |b|\big| \le |a - b|$

$$\text{Define } \operatorname{sgn}(a) = \begin{cases} 1 & \text{if } a > 0 \\ 0 & \text{if } a = 0 \\ -1 & \text{if } a < 0. \end{cases}$$

60 Prove $\operatorname{sgn}(ab) = (\operatorname{sgn} a)(\operatorname{sgn} b)$.

61 (cont.) Prove $|a| = (\operatorname{sgn} a)a$.

62 (cont.) Prove Rule (3): $|ab| = |a|\cdot|b|$.

11 INEQUALITIES

Let us look at inequalities involving an unknown. Statements such as $x^2 + 1 > 0$, which are true for all real numbers x, are called **absolute inequalities.** They correspond to identities in the case of equations. Statements such as $x + 1 > 10$, which are not true for all real numbers x, are called **conditional inequalities.** They correspond to conditional equations.

Solving a conditional inequality means finding all real numbers for which the statement is true. For example, the solution of $x + 1 > 10$ is the set of all real numbers greater than 9.

Some techniques for solving conditional inequalities are similar to techniques for solving equations. For example, by (2) on p. 87, you may add or subtract the same quantity from both sides of an inequality. By (3), you may multiply both sides by the same *positive* quantity. If you multiply by a negative quantity, you must *reverse* the inequality.

Example 1 Solve $3x - 7 > x + 1$.

Solution Add 7 to both sides:

$$3x > x + 8.$$

Subtract x from both sides:

$$2x > 8.$$

Now divide by 2, a positive number:

$$x > 4.$$

Answer $x > 4$

Example 2 Solve $\dfrac{x - 3}{2x + 1} < 0$.

Solution The quotient is negative if and only if numerator and denominator have opposite signs. To see where this happens, we make a diagram (Fig. 1).

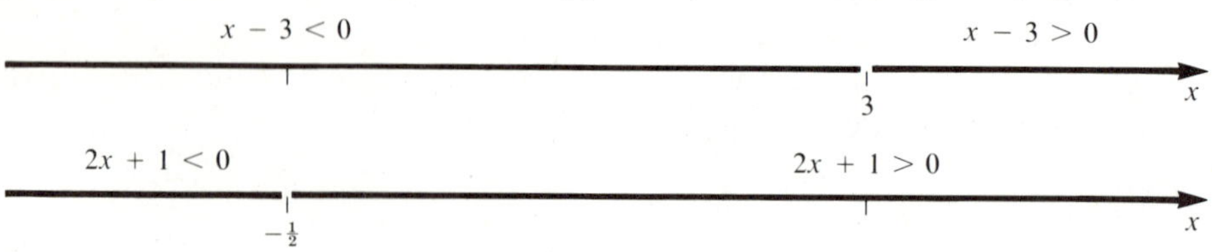

Fig. 1

The signs are opposite only on the segment $-\frac{1}{2} < x < 3$.

Answer $-\frac{1}{2} < x < 3$

Example 3 Solve $\dfrac{13x + 3}{2x + 1} \le 6$.

Solution 1 If this were an equation, we would multiply both sides by $2x + 1$. We can do the same with this inequality *provided* we distinguish two cases, $2x + 1 > 0$ and $2x + 1 < 0$.

Case 1: $2x + 1 > 0$, that is, $x > -\frac{1}{2}$. Multiply both sides by the positive quantity $2x + 1$:

$$(2x + 1)\left(\frac{13x + 3}{2x + 1}\right) \le 6(2x + 1) \qquad 13x + 3 \le 6(2x + 1).$$

Solve this linear inequality:

$$13x + 3 \le 12x + 6 \qquad x \le 3.$$

The solution in Case 1 is $-\frac{1}{2} < x \le 3$.

Case 2: $2x + 1 < 0$, that is, $x < -\frac{1}{2}$. Multiply both sides by the negative quantity $2x + 1$, reversing the inequality:

$$13x + 3 \ge 6(2x + 1) \qquad 13x + 3 \ge 12x + 6 \qquad x \ge 3.$$

Both $x < -\frac{1}{2}$ and $x \ge 3$ together are impossible, so there are no solutions in Case 2.

Solution 2 (*sketch*) We can avoid cases by subtracting 6 from both sides of the conditional inequality, putting the left-hand side over a common denominator, and using the method of the previous example:

$$\frac{13x + 3}{2x + 1} - 6 \le 0 \qquad \frac{13x + 3 - 6(2x + 1)}{2x + 1} \le 0 \qquad \frac{x - 3}{2x + 1} \le 0.$$

This conditional inequality can be solved by the method of Example 2.

Answer $-\frac{1}{2} < x \le 3$

Example 4 Solve $x^2 + 3 \ge 4x$.

Solution Quadratic inequalities are best solved by transposing all terms to one side and factoring:

$$x^2 - 4x + 3 \ge 0 \qquad (x - 1)(x - 3) \ge 0.$$

The product is positive only if both factors have the same sign; the product is 0 only if one of the factors is 0. It is convenient to make a diagram (Fig. 2).

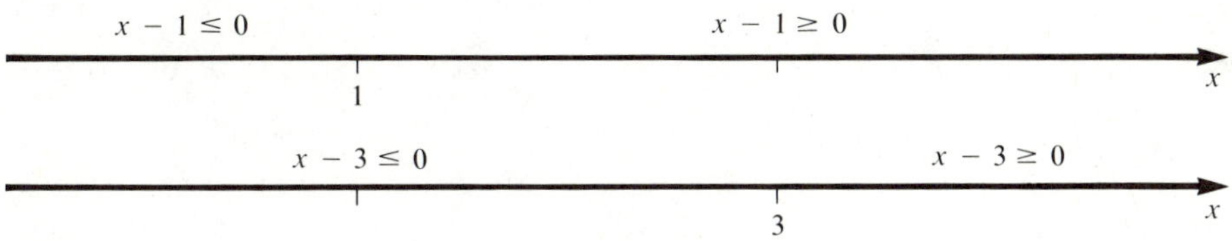

Fig. 2

The signs agree (or the product is 0) for $x \le 1$ or $x \ge 3$.

Answer $x \le 1$ or $x \ge 3$

Example 5 Solve $|5x - 4| < 20$.

Solution This inequality is equivalent to two inequalities without absolute values:

$$-20 < 5x - 4 < 20.$$

Add 4 to each term, then divide by 5:

$$-16 < 5x < 24 \qquad \frac{-16}{5} < x < \frac{24}{5}.$$

Answer $-\frac{16}{5} < x < \frac{24}{5}$

The next example is a typical application of inequalities.

EXAMPLE 6 The assets of an established trust fund are \$200 million and increasing at the rate of \$10 million per year. A new trust fund starts with \$100 million and, under aggressive management, its assets increase at the rate of \$12 million per year. When will its assets exceed 75% of those of the older trust fund?

Solution For convenience, take as the unit of money the megabuck, \$1 million. In t years, the assets of the old trust fund will increase by $10t$. Hence its total assets will be $200 + 10t$. Similarly, those of the new trust fund will be $100 + 12t$. The problem asks for what values of t is

$$100 + 12t > \tfrac{3}{4}(200 + 10t).$$

Multiply both sides by 4 and solve:

$$400 + 48t > 3(200 + 10t) = 600 + 30t$$

$$18t > 200 \qquad t > \tfrac{200}{18} = \tfrac{100}{9} = 11\tfrac{1}{9}.$$

Answer After $11\frac{1}{9}$ years.

Applications of $a^2 \geq 0$

The absolute inequality $a^2 \geq 0$ has many applications. We give two.

EXAMPLE 7 Let a, b, c be constants with $a > 0$. Prove that

$$ax^2 + bx + c \geq \frac{4ac - b^2}{4a}$$

is an absolute inequality (true for all x).

Solution Complete the square:

$$ax^2 + bx + c = a\left(x + \frac{b}{2a}\right)^2 + \frac{4ac - b^2}{4a}.$$

Since $a > 0$ and $(x + b/2a)^2 \geq 0$ we have $a(x + b/2a)^2 \geq 0$, so

$$ax^2 + bx + c = a\left(x + \frac{b}{2a}\right)^2 + \frac{4ac - b^2}{4a} \geq \frac{4ac - b^2}{4a}.$$

EXAMPLE 8 Prove that the inequality

$$\sqrt{ab} \leq \frac{a + b}{2}$$

is true for all $a \geq 0$ and $b \geq 0$.

Solution We start with $(\sqrt{b} - \sqrt{a})^2 \geq 0$ and expand:

$$b - 2\sqrt{ab} + a \geq 0 \qquad a + b \geq 2\sqrt{ab}.$$

Now divide both sides by 2.

Alternative Solution Let $A = \frac{1}{2}(a + b)$ be the arithmetic mean of a and b. Then A is the midpoint of a and b, so we can write $a = A - t$ and $b = A + t$. Then

$$ab = (A - t)(A + t) = A^2 - t^2 \leq A^2 \qquad \sqrt{ab} \leq A.$$

Remark The result is called the **arithmetic-geometric mean inequality.** For other versions see Exs. 45 and 46.

Exercises

Solve

1 $6x - 1 < 3$

2 $2x + 10 < 7$

3 $4x - 5 > 8x + 1$

4 $3(x - 4) > \frac{1}{2}x - 6$

5 $-4 < 2x + 9 < 16$

6 $6 < \frac{1}{2}(x + 3) < 10$

7 $(2x - 1)^2 \le 4x^2 + 2$

8 $(x + 1)(x + 2) \ge (x - 3)(x + 4)$

9 $\dfrac{1}{x + 5} > \frac{1}{8}$

10 $\frac{1}{2} > \dfrac{4}{3 - x}$

11 $x^2 > 6x + 7$

12 $x^2 + 4x \le 12$

13 $\dfrac{2x - 3}{x^2} < 1$

14 $(3x - 5)^2 \ge 0$

15 $\dfrac{x^2}{x^2 + 8} < 1$

16 $|x - 4| < 5$

17 $|2x - 1| < 4$

18 $|3x + 1| > 1$

19 $|x^2 - 1| < 3$

20 $|x| < 2x + 3$

21 $|x| < |x + 5|$

22 $\dfrac{2x + 1}{4x + 1} < 1$

23 $\dfrac{x}{x + 1} > 2$

24 $\dfrac{2x}{3x + 1} \le 1$

25 $\dfrac{x}{2x + 3} > -1$

26 $\dfrac{4x - 1}{3x + 2} > 5$

27 $\dfrac{2x}{x^2 + 1} \le 1$

28 $\dfrac{x}{x - 3} < 0$

29 $\dfrac{x}{8x - 3} > 0$

30 $\dfrac{x^2 - 4}{3x^2 + 1} \ge 0$

31 $(x - 1)(2x - 1) > 0$

32 $(x - 1)(x - 2)(x - 3) > 0$

33 $(x - 2)(x^2 + 1) \ge 0$

34 $(x - 2)^2(x^2 + 1)(x - 3) \ge 0$

35 $\dfrac{x}{(x - 1)(x - 4)} < 0$

36 $\dfrac{(x + 2)(x - 3)}{x^5 - x^4} < 0.$

37 A projectile is fired from ground level. After t sec, its height is $320t - 16t^2$ ft above the ground. During what period is the projectile higher than 1200 ft?

38 Company A will rent a car for \$20 per day. Company B will rent a car for \$12 per day plus an initial fee of \$120. When is it cheaper to rent from Company B?

39 Prove $x(1 - x) \le \frac{1}{4}$ for all x.

40 (cont.) Find the largest possible area of a rectangle of perimeter 2.

41 Find all real numbers c for which there exist two numbers whose sum is c and whose product is 100. [*Hint* Reduce to a quadratic equation; check the sign of its discriminant.]

42 Find all real numbers c for which there exist two numbers whose sum is 10 and the sum of whose squares is c.

43 Prove $\dfrac{x^2 + y^2}{2} \ge \left(\dfrac{x + y}{2}\right)^2$ for all x and y.

44* Prove $(xz + yw)^2 \le (x^2 + y^2)(z^2 + w^2)$ for all x, y, z, w. [*Hint* $(xw - yz)^2 \ge 0$.]

45 Prove $\frac{1}{4}(x + y + z + w) \ge \sqrt[4]{xyzw}$ for all $x \ge 0$, $y \ge 0$, $z \ge 0$, $w \ge 0$. [*Hint* Use the result of Example 8 several times.]

46* (cont.) Prove $\frac{1}{3}(x + y + z) \ge \sqrt[3]{xyz}$ for all $x \ge 0$, $y \ge 0$, $z \ge 0$. [*Hint* Set $w = \frac{1}{3}(x + y + z)$ in Ex. 45.]

Prove the absolute inequalities

47 $x^4 + x^2 - 2x + 3 > 0$

48* $x^4 - 4x^3 + 5x^2 - 2x + 1 > 0.$

49 Prove $x + \dfrac{a^2}{x} \ge 2a$ for $x > 0$.

50 Prove $\dfrac{x}{(x + 1)^2} \le \frac{1}{4}$ for $x > 0$.

REVIEW EXERCISES

Find all solutions

1 $\dfrac{x^2 - 3}{x^2 - 4} = 0$

2 $\dfrac{2}{x} - 3 = \dfrac{2 - 3x}{x}.$

Set up the problem

3 Find 5 consecutive integers whose sum equals the square of the middle one

4 A broker charges a commission on corn contracts of \$15 per contract. A second broker charges a flat transaction fee of \$20 plus \$13 per contract. Find the number of contracts for which the two brokers charge the same total commission.

Solve

5 $\dfrac{3}{2x+1} = \dfrac{1}{x-1}$

6 $\begin{cases} 2x + y = 2 \\ -3x + 2y = -17. \end{cases}$

7 A grain elevator's charge for storing wheat is \$75 plus \$16 per month per 1000 bushels. A second elevator's charge is \$95 plus \$15.85 per month per 1000 bushels. Find the break-even quantity for a farmer who wishes to store his grain for 3 months.

8 How much 97% pure aluminum must be mixed with 99.5% pure aluminum to produce one ton of 98% pure metal?

9 If A can do a job in 8 hours and A and B together can do the same job in 5 hours, how long does it take B to do the job?

10 Find a three digit whole number whose unit digit is the sum of the tens digit and the hundreds digit and such that (1) twice the number plus the number with digits reversed equals 1911 and (2) twice the tens digit is one greater than the hundreds digit.

Solve $x^2 - x - 30 = 0$

11 by factoring

12 by completing the square.

Solve by the quadratic formula

13 $x^2 - 5x + 5 = 0$

14 $3x^2 + 9x + 5 = 0.$

Solve with a calculator

15 $157x^2 - 963x + 101 = 0$

16 $4.38x^2 + 12.6x - 0.481 = 0.$

Solve

17 $(x^2 - 16)(\sqrt{x} - 3) = 0$

18 $\sqrt{x+3} + \sqrt{x-4} = 7$

19 $\dfrac{12 - x}{x - 5} = \dfrac{2x}{x - 4}$

20 $\dfrac{1}{x} - x = \frac{21}{10}.$

21 From a point 100 ft above the ground, a particle is projected straight up with speed 20 ft/sec. It can be shown that after t sec, the height of the particle above the ground is

$$x = -16t^2 + 20t + 100.$$

When does the particle hit the ground?

22 Show that the area of a rectangle of perimeter 4 is at most 1.

23 Indicate the portion of the number line determined by the inequality $x^4 \geq 16$.

24 Suppose $x < 0$ and $y = \dfrac{1}{1 + \dfrac{1}{1 - \dfrac{1}{x}}}$. Does y increase or decrease as x decreases?

25 Draw on the number line all x that satisfy $|2x + 3| < 5$.

26 Suppose $|x - 1| < 10^{-4}$. Show that $|x^2 - 1| < 3 \times 10^{-4}$.

Solve

27 $(x + 2)(x^2 + 4) > 0$

28 $(x - 2)^3(x - 3) > 0.$

29 Solve $x(x - 1)(x - 2)(x - 3)(x - 4) < 0.$

30 An audio amplifier with negative feedback has input x volts and output z volts. Their relation is $z = 100e + 7$, where the *effective input* e is the true input minus 10% of the output. Find the input required for a 12 volt output.

3 FUNCTIONS AND GRAPHS

1 INTRODUCTION

Everyone is familiar with the use of graphs to summarize data. Figure 1 shows three typical graphs. There are many others; one sees graphs depicting length, time, speed, voltage, blood pressure, productivity, profit, etc.

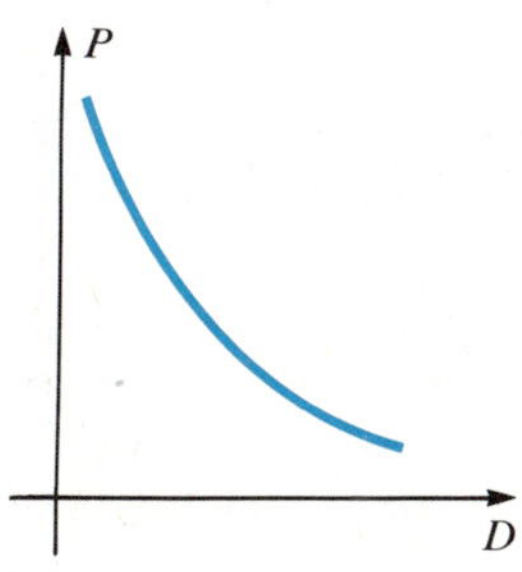

(a) price vs. demand

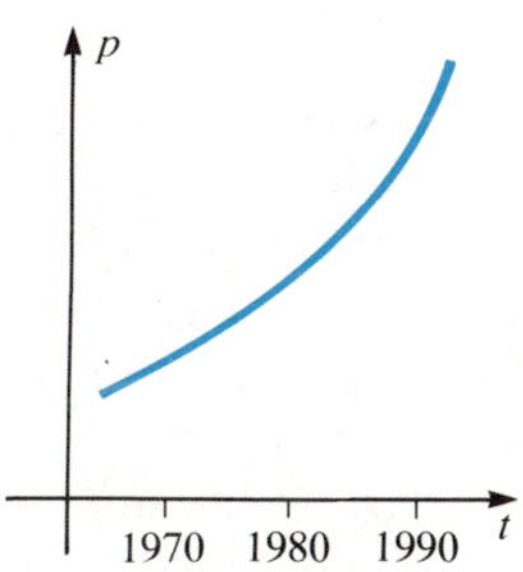

(b) population vs. time

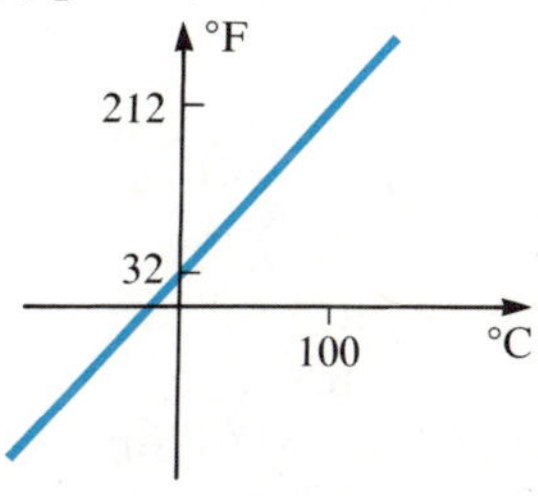

(c) Fahrenheit vs. Celsius

Fig. 1 Graphs

These graphs illustrate visually the way one numerical quantity depends on (or varies with) another. Graph (a) shows how the price of a certain item depends on the demand, (b) shows how the population depends on (varies with) time, and (c) shows how Fahrenheit readings depend on (are related to) Celsius readings.

The graphs in Fig. 1 are pictures of *functions*. The idea of a function is one of the most important in mathematics. Roughly speaking, a function describes the way one quantity depends on (or varies with) another. We say, for instance, that pressure is a function of temperature, or that population is a function of time, etc. Functions lurk everywhere; they are involved in almost every application of mathematics. Therefore they are the objects of a great deal of study.

As Fig. 1 illustrates, a graph is an excellent tool in understanding the nature of a function. For it is a kind of "life history" of a function, to be seen at a glance. That is why there is much emphasis on graphs in this book.

The natural habitat of graphs is the plane. To prepare for the coming discussion of functions and graphs, we'll spend the next section on coordinate systems in the plane.

2 COORDINATES IN THE PLANE

When the points of a line are specified by real numbers, we say that the line is **coordinatized:** each point has a label or **coordinate.** It is possible also to label, or coordinatize, the points of the plane. Let us describe the most common way of doing this.

We start by drawing two perpendicular lines in the plane. We mark their intersection 0 and coordinatize each line as shown in Fig. 1a. We draw one line horizontal and name it the x-axis, the other line vertical and name it the y-axis.

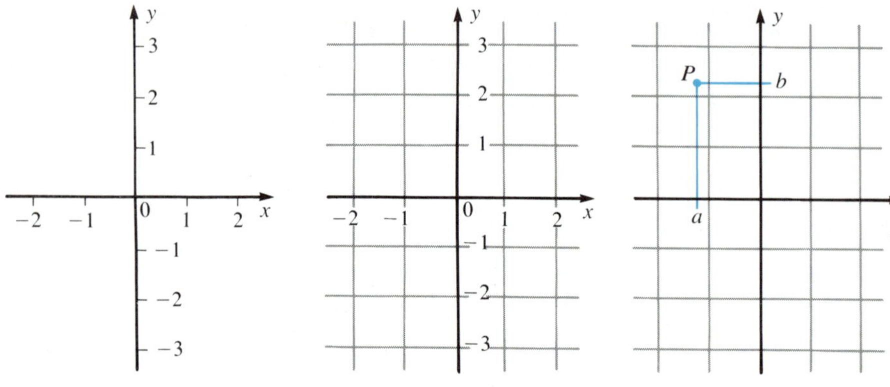

Fig. 1

Consider all lines parallel to the x-axis and all lines parallel to the y-axis (Fig. 1b). These two systems of parallel lines impose a rectangular grid on the whole plane. We use this grid to coordinatize the points of the plane.

Take any point P of the plane. Through P pass one vertical line and one horizontal line (Fig. 1c). They meet the axes in points a and b respectively. We associate with P the ordered pair (a, b); it completely describes the location of P. The number a is the **x-coordinate** of P, and the number b is the **y-coordinate** of P.

Conversely, take any ordered pair (a, b) of real numbers. The vertical line through a on the x-axis and the horizontal line through b on the y-axis meet in a point P whose coordinates are precisely (a, b). Thus there is a one-to-one correspondence

$$P \longleftrightarrow (a, b)$$

between the set of points of the plane and the set of all ordered pairs of real numbers.

A coordinate system in the plane such as we have just described is called a **rectangular** or **cartesian** coordinate system. The point $(0, 0)$ is called the **origin.**

Remark 1 The pair (a, b) is also sometimes called (ungrammatically) the **coordinates** of P. Some writers refer to the x-coordinate of a point as its **abscissa** and the y-coordinate as its **ordinate.**

Remark 2 It is common practice to speak of "the point (a, b)". Logically, this is incorrect since an ordered pair of real numbers is not a point. Still, this slight inaccuracy rarely causes any confusion, but often saves some fuss.

The coordinate axes divide the plane into four **quadrants,** which are numbered as in Fig. 2a. Note that the signs of its coordinates determine the quadrant in which a point lies. Points of the form $(a, 0)$ lie on the x-axis; those of the form

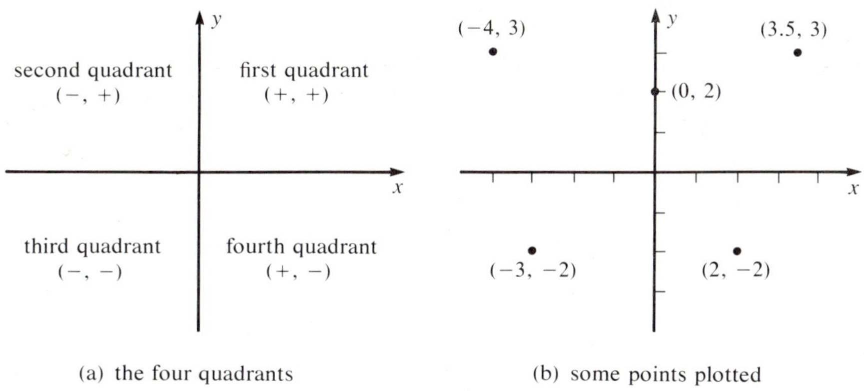

Fig. 2

$(0, b)$ lie on the y-axis. Figure 2b shows a few examples.

It is not compulsory to use the same scales on both axes. For example, in plotting U.S. population for the last 20 years, you would probably choose a reasonable length (say 1 cm) to represent one year on the horizontal axis. But you would use the same length to represent, say, 5,000,000 people on the vertical axis.

Subsets of the Plane

The part of the plane above the x-axis is the set of all points (x, y) for which $y > 0$. It can be described briefly by the simple inequality $y > 0$. Similarly, the part of the plane on and to the right of the y-axis is described by the inequality $x \geq 0$. Often a subset of the plane is conveniently described by an inequality or by a system of inequalities.

EXAMPLE 1 Sketch the set of all points (x, y) for which $y \leq 0$ and $|x| \geq 1$.

Solution First we shade (Fig. 3) the points where $y \leq 0$, that is, the region on and *below* the x-axis. Then we shade the region where $|x| \geq 1$, which consists of two parts, the part on and to the right of the vertical line $x = 1$ (where $x \geq 1$), and the part on and to the left of the vertical line $x = -1$ (where $x \leq -1$). The region where both inequalities are satisfied is shaded twice.

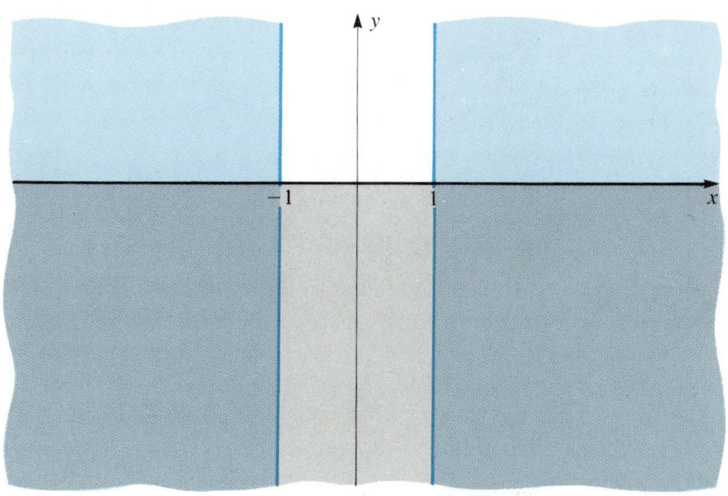

Fig. 3 The set where $y \leq 0$ *and* $|x| \geq 1$ is shaded dark.

The Distance Formula

Given two points (x_1, y_1) and (x_2, y_2) in the coordinate plane, what is the distance between them? We can answer this question easily by the Pythagorean theorem. See Fig. 4.

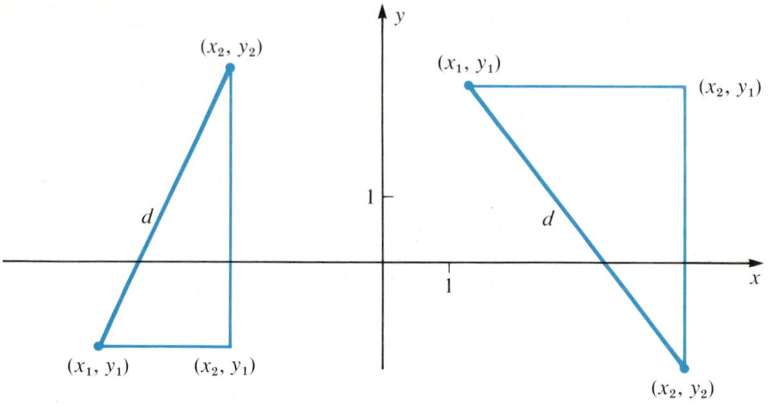

Fig. 4 Distance between two points (equal units on the two axes assumed)

No matter where the two points are located, we introduce the auxiliary point (x_2, y_1) forming a right triangle. Its legs have length $|x_2 - x_1|$ and $|y_2 - y_1|$, so by the Pythagorean theorem,

$$d^2 = |x_2 - x_1|^2 + |y_2 - y_1|^2 = (x_2 - x_1)^2 + (y_2 - y_1)^2.$$

Distance Formula The distance between two points (x_1, y_1) and (x_2, y_2) is

$$\sqrt{(x_2 - x_1)^2 + (y_2 - y_1)^2}.$$

(The same unit of length on *both* axes is assumed.)

If the points (x_1, y_1) and (x_2, y_2) lie on the same horizontal line or the same vertical line, their distance is $|x_1 - x_2|$ or $|y_1 - y_2|$; there is no need to introduce an auxiliary point. Nevertheless, the distance formula still yields the correct answer.

Examples

(a) distance from $(3, -1)$ to $(7, 2)$:

$$d^2 = (7 - 3)^2 + [2 - (-1)]^2 = 4^2 + 3^2 = 25 \qquad d = 5$$

(b) distance from $(2, 4)$ to $(-3, 1)$:

$$d^2 = (-3 - 2)^2 + (1 - 4)^2 = (-5)^2 + (-3)^2 = 34 \qquad d = \sqrt{34}$$

EXAMPLE 2 Show that the three points $P_1 = (1, 3)$, $P_2 = (3, -3)$, and $P_3 = (14, 4)$ are the vertices of an isosceles triangle.

Solution Plot the points (Fig. 5). If the triangle is actually isosceles, then the figure suggests that the equal sides are $\overline{P_1P_3}$ and $\overline{P_2P_3}$. Just check by the distance formula:

$$\overline{P_1P_3} = \sqrt{(14 - 1)^2 + (4 - 3)^2} = \sqrt{169 + 1} = \sqrt{170}$$

$$\overline{P_2P_3} = \sqrt{(14 - 3)^2 + (4 + 3)^2} = \sqrt{121 + 49} = \sqrt{170}.$$

Therefore $\overline{P_1P_3} = \overline{P_2P_3}$, so the triangle is isosceles.

Remark The solution of Example 2 shows how useful an accurate drawing can be.

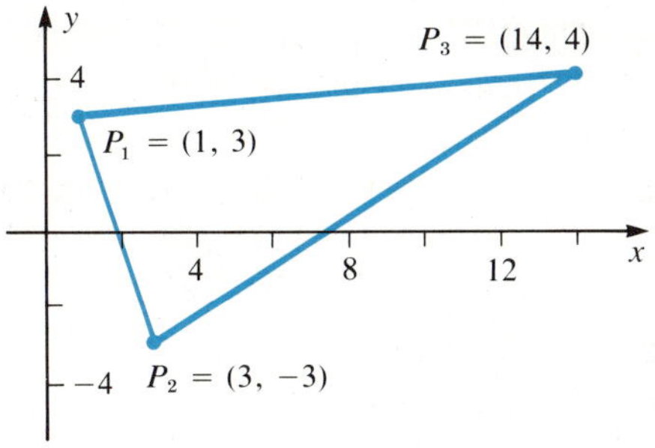

Fig. 5

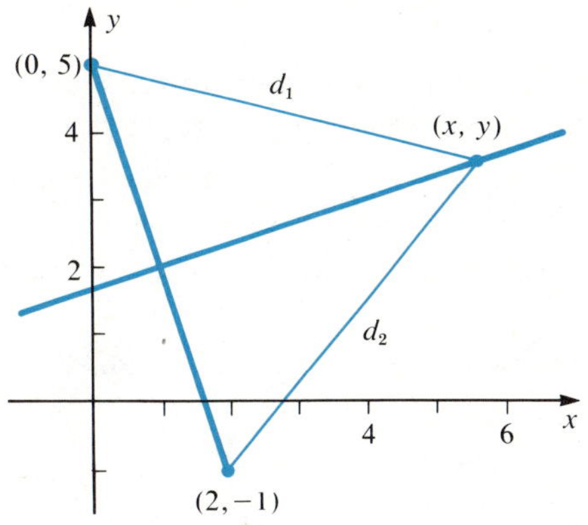

Fig. 6

EXAMPLE 3 Let (x, y) be any point on the perpendicular bisector of the segment joining $(0, 5)$ and $(2, -1)$. Show that $y = \frac{1}{3}x + \frac{5}{3}$.

Solution Points on the perpendicular bisector are equidistant from $(0, 5)$ and $(2, -1)$. Thus $d_1 = d_2$ in Fig. 6. Express d_1 and d_2 using the distance formula, and substitute into this equation. Better yet, to avoid radicals, substitute into the equivalent equation $d_1^2 = d_2^2$:

$$\begin{aligned} d_1^2 &= d_2^2 \\ (x-0)^2 + (y-5)^2 &= (x-2)^2 + (y+1)^2 \\ x^2 + y^2 - 10y + 25 &= x^2 - 4x + 4 + y^2 + 2y + 1 \\ -10y + 25 &= -4x + 4 + 2y + 1 \\ 4x + 20 &= 12y. \end{aligned}$$

Solve for y:

$$y = \tfrac{1}{3}x + \tfrac{5}{3}.$$

Exercises

Plot and label the points on one graph

1 $(-4, 1), (3, 2), (5, -3), (1, 4)$

2 $(0, -2), (3, 0), (-2, 2), (1, -3)$

3 $(0.2, -0.5), (-0.3, 0), (-1.0, -0.1)$

4 $(75, -10), (-15, 60), (95, 40)$.

Choose suitable scales on the axes and label the points

5 $(150, 0.3), (50, 0.6)$

6 $(-0.02, 5), (0.03, 8)$

7 $(0.1, -0.003), (-0.3, 0.007)$

8 $(-0.02, 35), (0.00, -60)$.

Indicate on a suitable diagram all points (x, y) in the plane for which

9 $x = -3$

10 $y = 2$

11 x and y are positive

12 either x or y (or both) is zero

13 $1 \le x \le 3$

14 $-1 \le y \le 2$

15 $-2 \le x \le 2$ and $-2 \le y \le 2$

16 $x > 2$ and $y < 3$

17 both x and y are integers

18 $x^2 > 4$

19 $|x| \ge 1$ and $|y| \le 2$

20 $|x| \ge 2$ and $|y| \ge 2$

21 $xy > 0$ and $|x| \le 3$

22 $|x| + |y| > 0$

23 $|x - 3| < 1$ and $y > 0$

24 $|x - 1| < 2$ and $|y| < 1$

25 $x \ge 0$ and $|y - 2| = 1$

26 $|x + 1| \le \frac{1}{2}$ and $|y - 1| \le 2$.

Write the coordinates (x, y) of the

27 vertices of a square centered at $(0, 0)$, sides of length 2 and parallel to the axes

28 vertices of a square centered at $(1, 3)$, sides of length 2, at 45° angles with the axes

29 vertices of a 3-4-5 right triangle in the first quadrant, right angle at $(0, 0)$, hypotenuse of length 15, longer leg along the x-axis.

30 vertices of an equilateral triangle, sides of length 2, base on the x-axis, vertex on the positive y-axis.

Compute the distance between the points

31 $(4, 1)$ and $(12, 7)$

32 $(-9, 6)$ and $(7, -6)$

33 $(0, 3)$ and $(8, 5)$

34 $(0, 3)$ and $(0, 11)$

35 $(-4, -1)$ and $(3, -3)$

36 $(\frac{1}{2}, \frac{3}{2})$ and $(2, 5)$.

Show that the points are vertices of a right triangle

37 $(2, -3)$ $(-2, 1)$ $(9, 4)$

38 $(-6, 1)$ $(-2, 9)$ $(4, -4)$.

Show that the points are vertices of a square

39 $(0, 0)$ $(3, 4)$ $(-1, 7)$ $(-4, 3)$

40 $(-5, -6)$ $(-12, 18)$ $(12, 25)$ $(19, 1)$.

41 Show that the points are vertices of a regular hexagon with center at the origin
$(2, 0)$ $(1, \sqrt{3})$ $(-1, \sqrt{3})$ $(-2, 0)$ $(-1, -\sqrt{3})$ $(1, -\sqrt{3})$.

42 Verify that the distance between $(3x_1, 3y_1)$ and $(3x_2, 3y_2)$ is three times the distance between (x_1, y_1) and (x_2, y_2).

43 A circle has center $(1, -6)$ and radius 10. Another circle has center $(8, 18)$ and radius 14. Do they intersect?

44 Show that the circle with center $(8, 1)$ and radius 3 is contained in the circle with center $(3, 2)$ and radius 9.

45 Find an equation that a point (x, y) must satisfy if it lies on the perpendicular bisector of the segment connecting $(0, 0)$ and $(4, 6)$.

46 Find all points on the x-axis whose distance from the point $(2, 7)$ is 10.

47 Find all points on the line $y = 2$ that are twice as far from $(3, 0)$ as from the origin.

48 (cont.) Do the same for the line $y = 4$.

49 Find all points at a distance 2 from the origin and having 1 as one of their coordinates.

50 (**Midpoint formula**) Using the distance formula, verify that the midpoint of the segment connecting (x_1, y_1) and (x_2, y_2) is $(\frac{1}{2}(x_1 + x_2), \frac{1}{2}(y_1 + y_2))$.

51 Show that the point $(6, 9)$ is the same distance from the point $(0, 1)$ as from the line $y = -1$.

52 (cont.) Show that in general a point (a, b) has the geometric property described in Ex. 51 provided $a^2 = 4b$.

3 FUNCTIONS

One of the fundamental concepts in mathematics is that of a function. Let us consider some examples.

1. To each city in the U.S. (except Washington, D.C.) corresponds the state in which it is located. This correspondence is called a function from the set of U.S. cities to the set of U.S. states.

2. To each living person is associated his or her height in cm today. This association is a function from the set of living people to the set of positive real numbers.

3. To each non-negative number x is assigned the number $\sqrt{x}$. This assignment is a function from the set of non-negative real numbers to itself.

In general, a **function** f is a rule that assigns to each element of a set **D** an element of a set **Y**. We say that f is a function from **D** to **Y**. (The sets **D** and **Y** can be the same set, as in Example 3 above.) We call **D** the **domain** of the function f.

We emphasize that a function assigns to each x in its domain **D** a *unique* element y in **Y**. For instance, the function in Example 1 assigns to each city the *unique* state that city belongs to. Notice that it assigns both to Los Angeles and to San Francisco the same state, California. That's okay because each city individually gets a single state. What is not okay is assigning to California *both* Los Angeles and San Francisco. That is not a function. However, assigning to each state its unique capital is a legitimate function (from the set of states to the set of cities).

Remark Let us confess that our definition of function is not 100 percent rigorous because we have used the words "rule," "assignment," etc., which are not precise mathematical terms. Mathematical logicians and others who worry about the *foundations* of mathematics (as distinct from mathematics and its applicability) like to define a function as "a set **S** of ordered pairs (x, y) in which each first element x occurs in at most one pair." This indeed bypasses the need for undefined words like "rule" and "assignment" since each ordered pair (x, y) of **S** "assigns" y to x. However, it passes the buck to equally vague words like "collection" and "common property" that go into the "definition" of a *set*.

Although the "set of ordered pairs" approach pleases logicians (for valid reasons), it has nothing to do with real life use of mathematics. In applications, the fruitful concept of functions is as *assignments* or *correspondences*. For example, to each $t > 0$ *corresponds* the distance traveled by a falling body during the first t seconds of fall; to each $r > 0$ *corresponds* the area of a circle of radius r; to each state is *assigned* its population, etc.

Real Functions

In this book we shall deal almost exclusively with functions for which both **D** and **Y** are the real numbers or subsets of the real numbers. So let us agree that unless stated otherwise, a function will mean an assignment of a real number y to each real number x in a set **D** of real numbers. The symbol x, representing the typical number of the domain **D**, is often called the **independent variable.** The symbol y, denoting the number assigned to x, is called the **dependent variable.** (The value of y depends on the value of x). The set of all assigned numbers y is the **range** of the function. Some people like to picture a function f as a machine or black box that converts inputs into outputs (Fig. 1). (Think of a TV receiver, an amplifier, or a synthesizer as a function, or black box, that processes an input into an output.) The set of all acceptable inputs is the domain of f; the set of all possible outputs is the range.

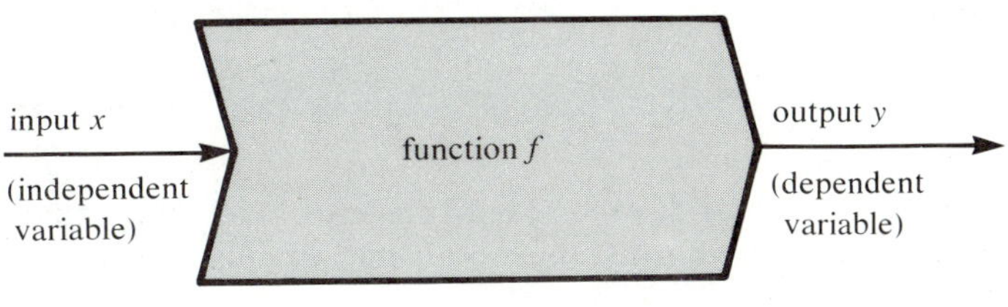

Fig. 1 A function as a "black box"

Functional Notation

The number assigned to x by a function f is often denoted by $f(x)$, read "f of x" or "f at x". If you like inputs and outputs, think of $f(x)$ as the output corresponding to the input x.

We often specify a function f by a formula or an equation that tells how to find $f(x)$ from x. Instead of "the function f that assigns to each real number its square," we write "$f(x) = x^2$" or "$y = x^2$". Although the domain of f is not mentioned, it is understood to be the largest set of x for which the formulas make sense. In this example, the domain of f is the set of all real numbers since x^2 is defined for all real x. The domain of the function $f(x) = \sqrt{x}$ is the set of non-negative real numbers.

Remark It is logically incorrect to say "the function $f(x)$", or "the function x^2", or the function "$y = f(x)$". The symbols "$f(x)$", "y", "x^2" represent numbers, the numbers assigned by the function f to the numbers x. A function is not a *number,* but an *assignment* of a number y or $f(x)$ to each number x in a certain domain. Nevertheless, these slight inaccuracies are very common and harmless; we shall not try to avoid them.

Let us illustrate numerical use of functional notation. Keep in mind that $f(x)$ is the *number* assigned to x by the function f. For example, if $f(x) = 2x + 5$, then no matter what number x is given, the function f doubles it and adds 5.

EXAMPLE 1 Let $f(x) = 2x + 5$. Compute

(a) $f(0)$ (b) $f(3)$ (c) $f(-\sqrt{7})$ (d) $f(1/x)$ (e) $f(x - 8)$.

Solution (a) $f(0) = 2\cdot 0 + 5 = 5$

(b) $f(3) = 2\cdot 3 + 5 = 11$

(c) $f(-\sqrt{7}) = 2(-\sqrt{7}) + 5 = -2\sqrt{7} + 5$

(d) $f(1/x) = 2(1/x) + 5 = 2/x + 5$

(e) $f(x - 8) = 2(x - 8) + 5 = 2x - 11$

Answer (a) 5 (b) 11 (c) $-2\sqrt{7} + 5$ (d) $2/x + 5$ (e) $2x - 11$

EXAMPLE 2 Let $f(x) = x^2 + 1$. Compute

(a) $f(x_1 + x_2)$ (b) $f(5x)$ (c) $f(x^2)$ (d) $\dfrac{f(3 + h) - f(3)}{h}$

Solution (a) $f(x_1 + x_2) = (x_1 + x_2)^2 + 1$

(b) $f(5x) = (5x)^2 + 1 = 25x^2 + 1$

(c) $f(x^2) = (x^2)^2 + 1 = x^4 + 1$

(d) $$\frac{f(3 + h) - f(3)}{h} = \frac{[(3 + h)^2 + 1] - [3^2 + 1]}{h}$$

$$= \frac{6h + h^2}{h} = 6 + h.$$

Answer (a) $(x_1 + x_2)^2 + 1$ (b) $25x^2 + 1$

(c) $x^4 + 1$ (d) $6 + h$

Remark So far we have used only the letter f to denote functions. However, when several functions occur in the same problem, we must denote each one by a different letter.

EXAMPLE 3 Let $f(x) = \dfrac{1}{x^2 + 2}$ and $g(x) = x^3 + 4x$. Show that

(a) $f(-x) = f(x)$ (b) $g(-x) = -g(x)$.

Solution (a) $f(-x) = \dfrac{1}{(-x)^2 + 2} = \dfrac{1}{x^2 + 2} = f(x)$

(b) $g(-x) = (-x)^3 + 4(-x) = -x^3 - 4x = -(x^3 + 4x) = -g(x)$.

Warning In practice, most functions are computed by formulas such as $f(x) = x^2$ or $f(x) = 1 + 7x^4$. Yet there are perfectly good functions not given by formulas. Here are a few examples:

(a) $f(x) =$ the largest integer y for which $y \le x$.

(b) $f(x) = \begin{cases} 1 \text{ if } x > 0 \\ 0 \text{ if } x = 0 \\ -1 \text{ if } x < 0. \end{cases}$

(c) $f(x) = 1$ if x is an integer, $f(x) = -1$ if x is not an integer.

(d) $f(x) =$ number of letters in the English spelling of the rational number x in lowest terms. For example, $f(\frac{1}{2}) = 7$, $f(3) = 5$.

Domains

Most functions arising in practice have simple domains. The most common domains are the whole line, an interval (segment) $a \le x \le b$, a "half-line" such as $x \ge 0$ or $x < 2$ or some simple combination of these. Examples:

FUNCTION	DOMAIN
$f(x) = 2x + 1$	all real x (the whole line)
$f(x) = \sqrt{x + 2}$	$x \ge -2$ (half line)
$f(x) = \sqrt{1 - x^2}$	$-1 \le x \le 1$ (interval)
$f(x) = 1/x$	all x except $x = 0$ (union of two half-lines)

EXAMPLE 4 Find the domain of $f(x) = \sqrt{x} + \sqrt{1 - x^2}$.

Solution The domain is the largest set of real numbers x for which the formula is meaningful. The first term $\sqrt{x}$ is defined for $x \ge 0$. The second term $\sqrt{1 - x^2}$ is defined only for $-1 \le x \le 1$. In order for both terms to be defined, we must have $0 \le x \le 1$. See Fig. 2.

Answer The interval $0 \le x \le 1$.

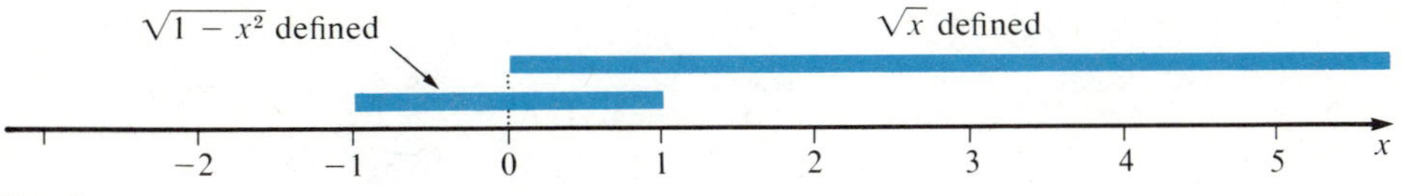

Fig. 2

Exercises

Compute the values of the function

1 $f(x) = 3x - 1$

(a) $f(1)$ (b) $f(-1)$ (c) $f(0.15)$ (d) $f(-12)$

2 $f(x) = |x - 4|$

(a) $f(-2)$ (b) $f(2)$ (c) $f(4 - \sqrt{7})$ (d) $f(\sqrt{7} - 4)$

3 $f(x) = 1/x$

(a) $f(5)$ (b) $f(0.001)$ (c) $f(1/x)$ (d) $f(\frac{1}{2}\sqrt{x})$

4 $f(x) = x^2 + x - 1$

(a) $f(10)$ (b) $f(-0.01)$ (c) $f(-x)$ (d) $f\left(\frac{-1 + \sqrt{5}}{2}\right)$.

5 Which of these functions satisfies $f(-x) = f(x)$?

(a) $f(x) = x^4 + \frac{1}{x^4}$ (b) $f(x) = |x|$ (c) $f(x) = 2x + 1$

(d) $f(x) = (x - 1)^2$.

6 Which of these functions satisfies $f(-x) = -f(x)$?

(a) $f(x) = 1/x$ (b) $f(x) = x^3$ (c) $f(x) = x^3 + 1$

(d) $f(x) = x^5 + 7x^3 - 4x$.

7 Which functions satisfy $f(3x) = 3f(x)$?

(a) $f(x) = x$ (b) $f(x) = 7x$ (c) $f(x) = x + 3$ (d) $f(x) = x^3$.

8 Which functions satisfy $f(x_1 + x_2) = f(x_1) + f(x_2)$?

(a) $f(x) = ax$ (b) $f(x) = x^2$ (c) $f(x) = 1/x$ (d) $f(x) = \sqrt{x}$.

9 Show that $f(x) = x^n$, where n is any integer, satisfies $f(x_1x_2) = f(x_1)f(x_2)$. Give an example of a function that does not satisfy this equation.

10 Find an example of a function that satisfies $f(x^2) = [f(x)]^2$ and an example of a function that does not.

Exercises 11–16 concern the function $f(x) = \begin{cases} 1 \text{ if } x > 0 \\ 0 \text{ if } x = 0 \\ -1 \text{ if } x < 0. \end{cases}$

Compute

11 $f(x) + f(-x)$

12 $\frac{1}{2}[f(x) - f(-x)]$

13 $f(2x)$

14 $f[f(x)]$.

15 Show that $xf(x) = |x|$.

16 Show that $f(x)f(y) = f(xy)$.

Find the domain of $f(x)$

17 $f(x) = 3x - 2$

18 $f(x) = -7x + 6$

19 $f(x) = (4x - 5)^3$

20 $f(x) = |7 - x|$

21 $f(x) = 1/(2x - 3)$

22 $f(x) = x/(x + 2)$

23 $f(x) = x/(3x - 5)$

24 $f(x) = x/(x - 1)(x - 3)$

25 $f(x) = \sqrt{x - 6}$

26 $f(x) = \sqrt{5 - 2x}$

27 $f(x) = \sqrt{4 - 9x^2}$

28 $f(x) = \sqrt{15x^2 + 11}$

29 $f(x) = \sqrt{2x - 3}$

30 $f(x) = \frac{1}{\sqrt{x + 4}}$

31 $f(x) = \sqrt{\frac{1}{4} - x^2}$

32 $f(x) = \sqrt{x^2 - 9} + \sqrt{9 - x^2}$

33 $f(x) = \sqrt{(x - 1)(x - 4)}$

34 $f(x) = \sqrt{x^3 + 1}$

35 $\sqrt{x^2 - x - 12}$

36 $\sqrt{(x - 1)(x - 2)(x - 3)}$.

Find a function $f(x)$ that describes

37 the number of seconds in x hours

38 the Celsius reading corresponding to $x°$ Fahrenheit

39 the diagonal of a square of side x

40 the time required for a plane to fly 1000 miles at x miles/hr if the last 300 miles are against a 40 mph head-wind

41 the value after one year of a \$10,000 investment if x dollars are invested at 6% yearly interest and the rest at 9%

42 the value after 3 years of \$1000 invested at x% annual interest, compounded annually.

4 GRAPHS OF FUNCTIONS

We associate to each function f its graph, a geometric picture of the function. The **graph** of f is the set of all points in the coordinate plane of the form $(x, f(x))$. To each x in the domain of f corresponds one point (x, y) on its graph, where $y = f(x)$. We sometimes think of $f(x)$ as the "height" of the graph at x, above the x-axis if $f(x) > 0$, below if $f(x) < 0$. See Fig. 1.

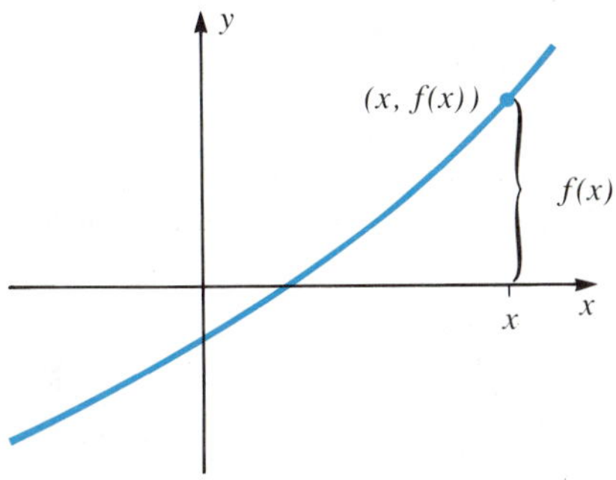

Fig. 1 Graph of f

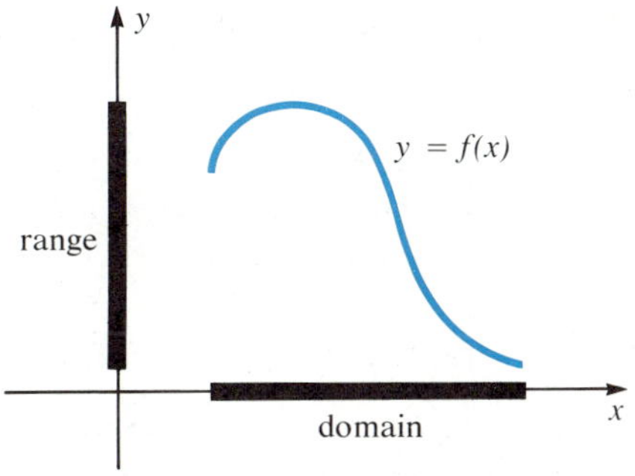

Fig. 2 Domain and range of f

Geometrically, the domain of f is the set of all points on the x-axis at which f is defined. Recall that another set associated with a function f is its **range.** It is the set of all *values* of f:

> **Range** The **range** of f is the set of all real numbers $f(x)$, where x is any real number in the domain of f.

We can see both the domain and range of a function f by inspecting its graph (Fig. 2). The domain of f is the part of the x-axis directly above or below the graph. In other words, the domain is the vertical projection of the graph onto the x-axis. The range of f is the set of all "heights" of the graph. In other words, the range is the horizontal projection of the graph onto the y-axis.

If you think of a function as a machine, then the domain is the set of all inputs the machine will accept, and the range is the set of all outputs it produces.

Terminology It is common to say "the graph of $y = f(x)$" or even "the graph of $f(x)$" instead of "the graph of the function f". For example, if $f(x) = x^2$, we may refer to "the graph of $y = x^2$". Very informal, but common terminology is simply "the graph of x^2".

EXAMPLE 1 Graph the constant function $f(x) = 2$. Show the domain and range of the function.

Solution This extremely simple function assigns to each number the same number, 2. Its graph consists of all points of the form $(x, 2)$. These points form the horizontal line 2 units above the x-axis (Fig. 3, next page).

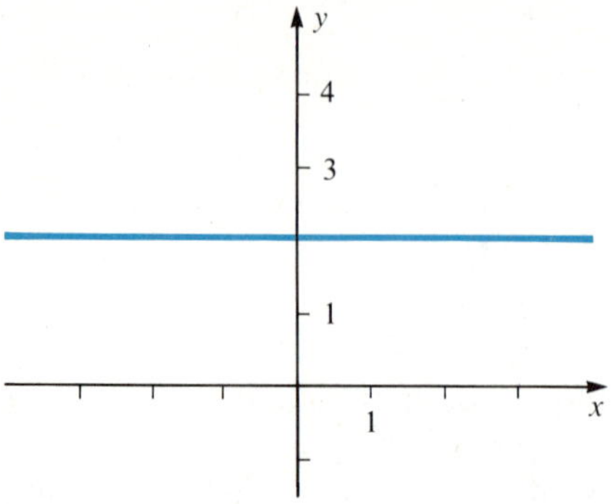

Fig. 3 Graph of $f(x) = 2$
Dom $= \{\text{all } x\}$ Range $= \{2\}$

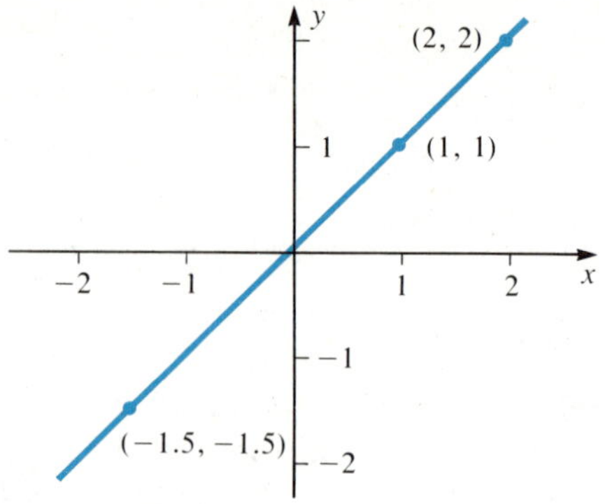

Fig. 4 Graph of $f(x) = x$
Dom $= \{\text{all } x\}$ Range $= \{\text{all } y\}$

EXAMPLE 2 Graph the function $f(x) = x$. Show the domain and range of the function.

Solution The graph consists of all points of the form (x, x). Plotting a few of these, we see that they form a straight line through the origin at an angle of 45° with the positive x-axis (Fig. 4).

EXAMPLE 3 Graph $f(x) = |x|$. Show the domain and range of the function.

Solution If $x \geq 0$, then $|x| = x$. So, on and to the right of the y-axis, the graph is identical to that of $f(x) = x$ in Fig. 4.

If $x < 0$, then $|x| = -x$. So to the left of the y-axis, the graph consists of all points $(x, -x)$, where $x < 0$. Plotting a few of these, we see that they form a straight line through the origin at an angle 135° with the positive x-axis. Combining the two parts, we obtain the complete graph (Fig. 5).

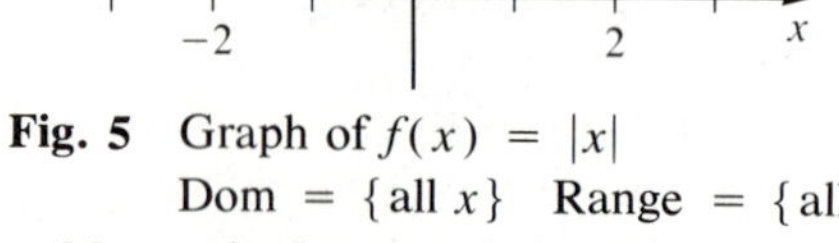

Fig. 5 Graph of $f(x) = |x|$
Dom $= \{\text{all } x\}$ Range $= \{\text{all } y \geq 0\}$

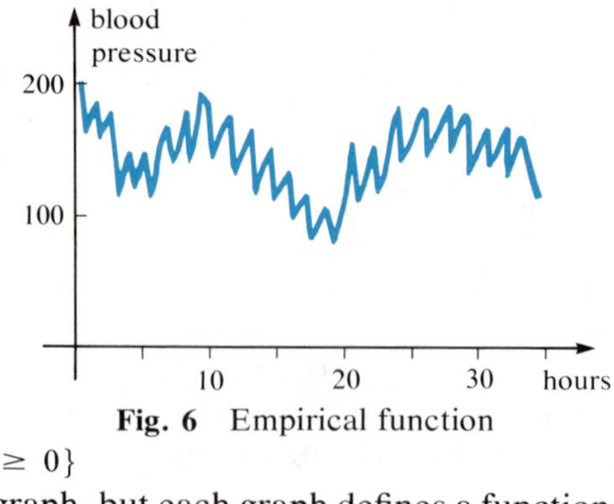

Fig. 6 Empirical function

Not only does each function have a graph, but each graph defines a function. By a graph, we mean here a collection of points (x, y) in the plane such that no two of the points have the same first coordinate (only one point can lie above a point on the x-axis). Such a graph automatically defines a function $f(x)$: to each x that occurs as a first coordinate of a point (x, y), it assigns the second coordinate y. Thus, $f(x)$ is the "height" of the graph above x.

The graphical definition of functions is standard procedure in science. For instance, a scientific instrument recording temperature or blood pressure on a graph is defining a function of time (Fig. 6). There is hardly ever an explicit formula for such a function.

Let us give two examples of functions without explicit formulas. The first is the "nearest integer" function defined by

$$I(x) = \begin{cases} \text{nearest integer to } x \text{ if there is } one \text{ such} \\ x \text{ if there are } two \text{ such} \end{cases}$$

For instance, $I(3.49) = 3$, $I(3.51) = 4$, but $I(3.5) = 3.5$. The graph is shown in Fig. 7a.

The second example is the "saw-tooth" function $s(x)$ whose graph is shown in Fig. 7b.

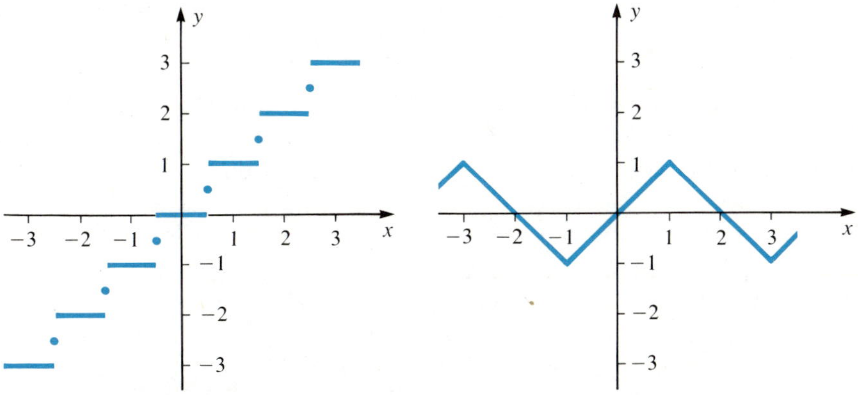

(a) nearest integer function $I(x)$ (b) saw-tooth function $s(x)$

Fig. 7 Functions without explicit formulas

Some Graphing Principles

Suppose we know the graph of $y = f(x)$. Then we can easily find the graph of $y = f(x) + 1$. For each point $(x, f(x) + 1)$ on this graph is one unit higher than the point $(x, f(x))$ on the graph of $y = f(x)$. Therefore, the graph of $y = f(x) + 1$ is one unit higher than that of $y = f(x)$. Similarly, if $c > 0$, then the graph of $y = f(x) + c$ is c units higher and the graph of $y = f(x) - c$ is c units lower (Fig. 8).

> Let $c > 0$. Then the graph of
>
> $$\left.\begin{array}{l} y = f(x) + c \\ y = f(x) - c \end{array}\right\} \text{ is the graph of } y = f(x) \text{ shifted } c \text{ units } \begin{cases} \text{upward} \\ \text{downward.} \end{cases}$$

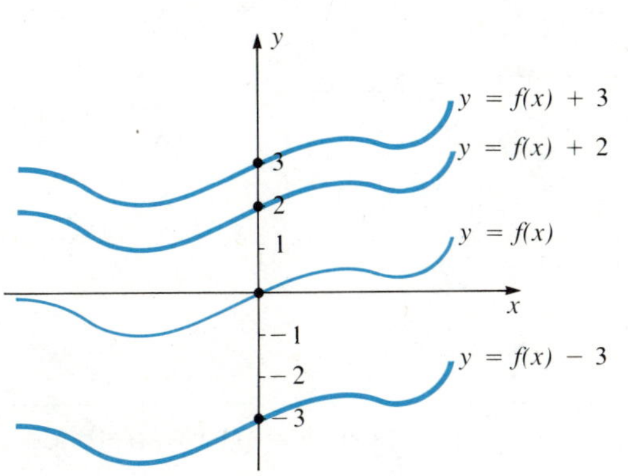

Fig. 8 Graphs of $y = f(x) \pm c$

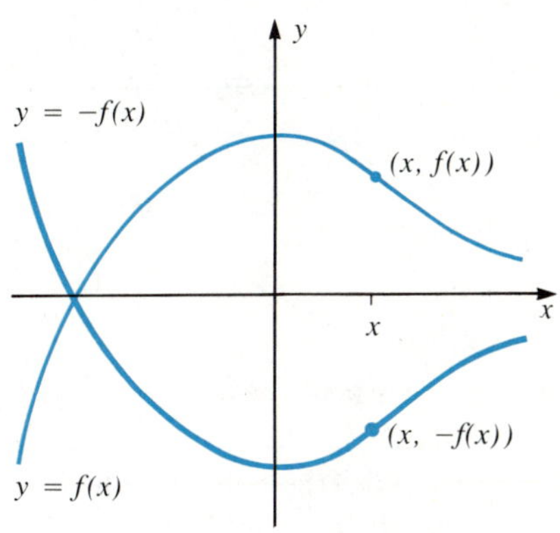

Fig. 9 Graph of $y = -f(x)$

Next we consider the graph of $y = -f(x)$. Each of its points has the form $(x, -y)$, where (x, y) is on the graph of $y = f(x)$. But $(x, -y)$ is the reflection of (x, y) in the x-axis. Therefore the graph of $y = -f(x)$ is the reflection of the graph of $y = f(x)$. See Fig. 9, previous page.

> The graph of $y = -f(x)$ is the reflection of the graph of $y = f(x)$ in the x-axis.

EXAMPLE 4 Graph $y = 2 - |x|$.

Solution Figure 5 shows the graph of $y = |x|$. Turn it upside down (reflect in the x-axis) to obtain (Fig. 10a) the graph of $y = -|x|$. Then shift upward 2 units to obtain (Fig. 10b) the graph of $y = -|x| + 2 = 2 - |x|$.

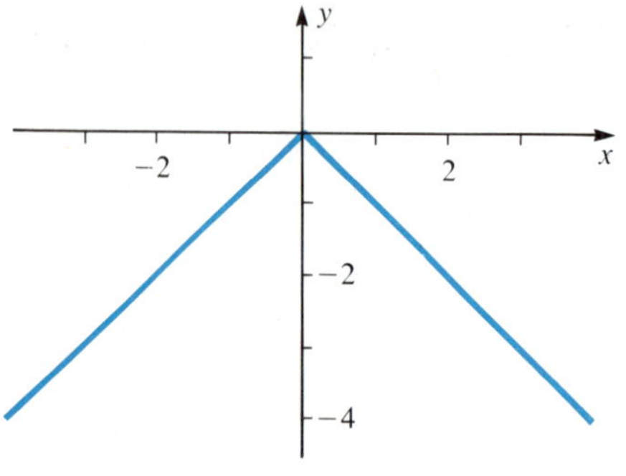

(a) $y = -|x|$

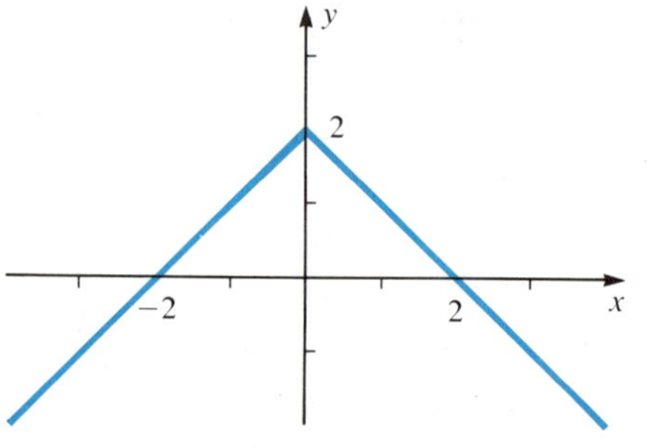

(b) $y = 2 - |x|$

Fig. 10

As our final graphing principle, let us consider the graph of $y = 2f(x)$. Each point of this graph has the form $(x, 2y)$, where (x, y) is on the graph of $y = f(x)$. Therefore, we obtain the graph of $y = 2f(x)$ when we stretch the graph of $y = f(x)$ by a factor of 2 in the y-direction (doubling the "heights"). Similarly, for $c > 0$ the graph of $y = cf(x)$ is the graph of $y = f(x)$ stretched vertically by a factor c. If $c < 1$, then "stretched" must be interpreted as "shrunk." See Fig. 11.

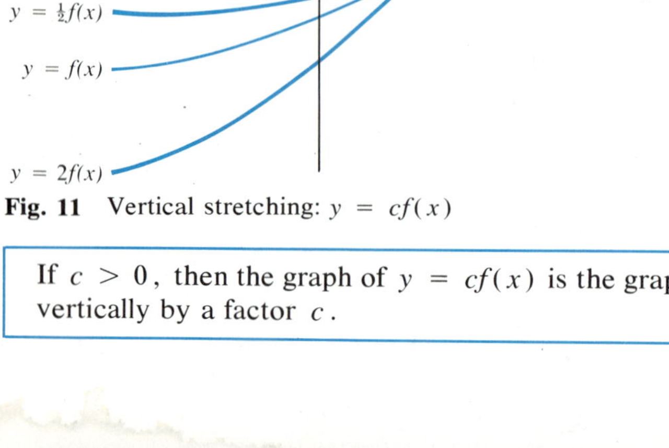

Fig. 11 Vertical stretching: $y = cf(x)$

> If $c > 0$, then the graph of $y = cf(x)$ is the graph of $y = f(x)$ stretched vertically by a factor c.

EXAMPLE 5 Graph $y = 2 + \frac{1}{2}s(x)$, where $s(x)$ is the saw-tooth function in Fig. 7b.

Solution Shrink the graph of $s(x)$ vertically by a factor of $\frac{1}{2}$ to obtain (Fig. 12a) the graph of $y = \frac{1}{2}s(x)$. Then shift upward 2 units to obtain (Fig. 12b) the graph of $y = 2 + \frac{1}{2}s(x)$.

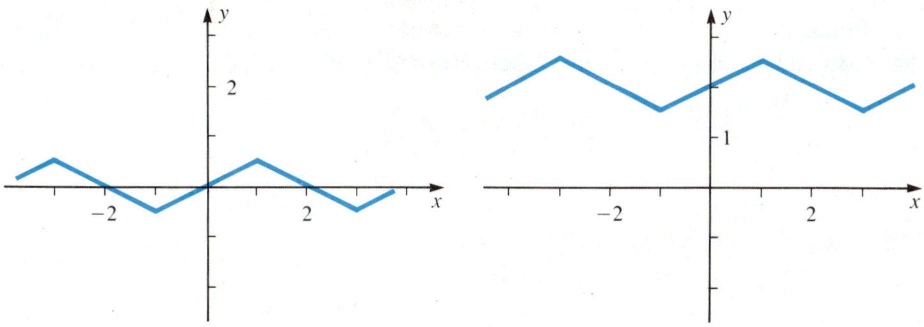

(a) $y = \frac{1}{2}s(x)$ (b) $y = 2 + \frac{1}{2}s(x)$

Fig. 12

Exercises

Graph

1 $f(x) = x + 2$ **2** $f(x) = x - 1$ **3** $f(x) = -x$

4 $f(x) = -x + 1$ **5** $f(x) = -17$ **6** $f(x) = 0.03$

7 $f(x) = 2x$ **8** $f(x) = \frac{1}{2}x - 3$ **9** $f(x) = 2|x|$

10 $f(x) = |x - 1|$ **11** $f(x) = \begin{cases} 0 & x \le 0 \\ 2x & x > 0 \end{cases}$ **12** $f(x) = \begin{cases} x - 1 & x \le 3 \\ 2 & x > 3 \end{cases}$

13 $f(x) = \begin{cases} 1 & x > 0 \\ 0 & x = 0 \\ -1 & x < 0 \end{cases}$ **14** $f(x) = \begin{cases} 1 & \text{if } x \text{ is an integer} \\ -1 & \text{if } x \text{ is not an integer} \end{cases}$

15 $f(x) = \begin{cases} \text{nearest } \textit{even} \text{ integer if there is } \textit{one} \text{ such} \\ x \text{ if there are } \textit{two} \text{ such} \end{cases}$

16 $f(x) = \frac{1}{3}|x| - 1$.

Recall the function $s(x)$ in Fig. 7b. Graph

17 $f(x) = 3s(x)$ **18** $f(x) = -s(x)$

19 $f(x) = |s(x)|$ **20** $f(x) = 1 - 2s(x)$.

21 Graph the function that gives the first class postage on a letter as a function of its weight. (As we go to press, the first ounce or fraction costs 15¢, each additional ounce or fraction, 13¢.)

22 Graph $f(x)$, the distance from the real number x to the nearest integer.

23 A taxi charges 75¢ as soon as you enter, then 10¢ after each $\frac{1}{6}$ mile. Graph $f(x)$, the cost of riding x miles in this cab.

24 A 120-yard high hurdle race has 10 hurdles. The first is 15 yards after the start, and all the rest are spaced at 10-yard intervals. Graph $f(x)$, the number of hurdles remaining after a runner has covered x yards.

Let $[x]$ denote the largest integer n such that $n \le x$. Graph

25 $f(x) = [x]$ **26** $f(x) = [1/x]$ $x > \frac{1}{5}$ **27** $f(x) = [\frac{1}{2}x]$

28 $f(x) = [2x]$ **29** $f(x) = (-1)^{[x]}$

30 $f(x) =$ the remainder when $[x]$ is divided by 3.

5 CONSTRUCTION OF FUNCTIONS

In this section we discuss several standard methods for building new functions out of old ones.

Addition of Functions

Given functions f and g whose domains overlap, their **sum** $f + g$ is the function defined on the common part (intersection) of the domains of f and g by

$$[f + g](x) = f(x) + g(x).$$

Examples

(1) $f(x) = 2x - 3 \qquad g(x) = x^2 - x - 1.$

f and g are defined for all real x; so is $f + g$, and

$$[f + g](x) = (2x - 3) + (x^2 - x - 1) = x^2 + x - 4.$$

(2) $f(x) = \sqrt{x - 2} \qquad g(x) = \sqrt{3 - x}.$

The domain of f is $\{x \geq 2\}$; the domain of g is $\{x \leq 3\}$. The common part of these domains is $\{2 \leq x \leq 3\}$; this set is the domain of $f + g$, and

$$[f + g](x) = \sqrt{x - 2} + \sqrt{3 - x}.$$

A black box interpretation of the sum of functions is shown in Fig. 1.

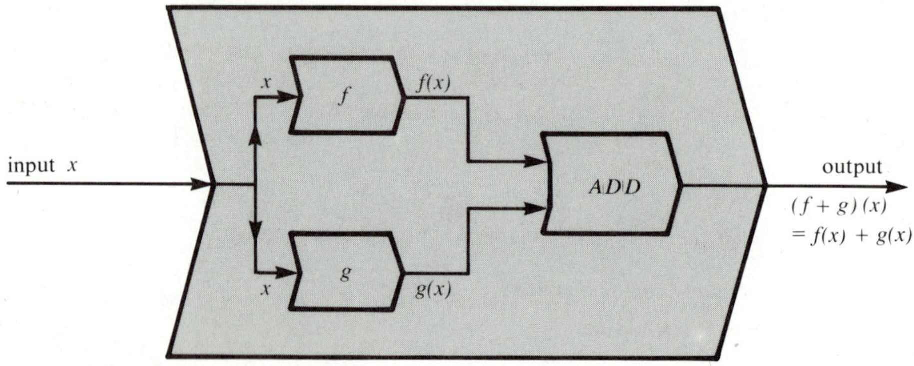

Fig. 1

Multiplication of Functions

Given functions f and g, their **product** fg is defined by

$$[fg](x) = f(x)g(x).$$

For example, if $f(x) = 16x^2$ and $g(x) = (x - 3)^5$, then

$$[fg](x) = 16x^2(x - 3)^5.$$

The domain of fg is the intersection of the domains of f and of g. The reason is precisely the same as for sums: the product is defined wherever both individual factors $f(x)$ and $g(x)$ are defined.

For a black box picture of fg, just replace the adder in Fig. 1 by a multiplier.

Composition of Functions

Given functions f and g, their **composition** $f \circ g$ (also called the **composite function** of f and g) is defined by

$$[f \circ g](x) = f[g(x)].$$

Think of substituting one function into the other, or replacing the variable of f by the function g.

In order for $f[g(x)]$ to be defined, the number $g(x)$ must be in the domain of f. Therefore the domain of $f \circ g$ is the set of all numbers x such that x is in the domain of g *and* $g(x)$ is in the domain of f.

Examples

(1) $f(x) = x^2 + 2x \qquad g(x) = -3x.$

$$[f \circ g](x) = f[g(x)] = [g(x)]^2 + 2[g(x)]$$
$$= (-3x)^2 + 2(-3x)$$
$$= 9x^2 - 6x.$$

Since both g and f are defined for all real x, so is $f \circ g$; the domain of $f \circ g$ is $\{\text{all } x\}$.

(2) $f(x) = 3x - 4 \qquad g(x) = 2x^2 - x + 1:$

$$[f \circ g](x) = f[g(x)] = 3g(x) - 4$$
$$= 3(2x^2 - x + 1) - 4$$
$$= 6x^2 - 3x - 1.$$

Again, the domain of $f \circ g$ is $\{\text{all } x\}$.

The black box interpretation is particularly useful for composite functions (Fig. 2). The black boxes for g and for f are connected in series; the output from g is the input for f. The black boxes for f and g, connected in series, make the new (composite) black box for $f \circ g$.

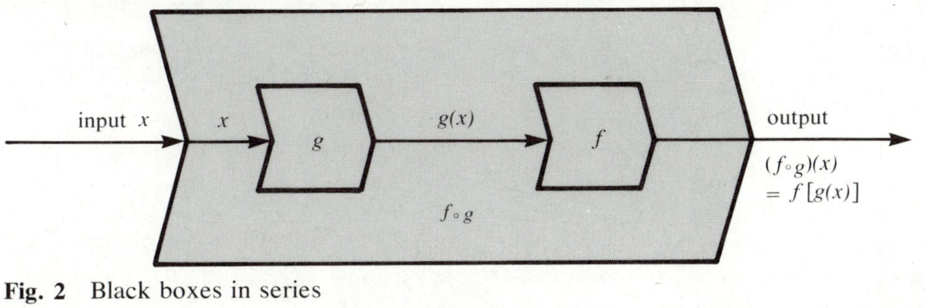

Fig. 2 Black boxes in series

EXAMPLE 1 Let $f(x) = x - 5$ and $g(x) = \sqrt{x}$. Compute and give the domains of

(a) $f \circ g$ (b) $g \circ f$.

Solution (a) g is defined for all $x \geq 0$ and f is defined for *all* x, so $g(x)$ is in the domain of f for each x in the domain of g. Hence $f \circ g$ is defined for all $x \geq 0$ and

$$[f \circ g](x) = g(x) - 5 = \sqrt{x} - 5 \qquad (x \geq 0).$$

(b) f is defined for all x, but g is defined only for all $x \geq 0$. Now $f(x) \geq 0$ only if $x \geq 5$, so the domain of $g \circ f$ is $\{x \geq 5\}$, and

$$[g \circ f](x) = \sqrt{f(x)} = \sqrt{x - 5} \qquad (x \geq 5).$$

Answer (a) $[f \circ g](x) = \sqrt{x} - 5$, domain $\{x \geq 0\}$

(b) $[g \circ f](x) = \sqrt{x - 5}$, domain $\{x \geq 5\}$.

EXAMPLE 2 Let $f(x) = \sqrt{x}$ and $g(x) = \sqrt{25 - x^2} - x$.

(a) Find the domain of $f \circ g$. (b) Compute $f \circ g$.
(c) Find $[f \circ g](3)$ and $[f \circ g](-4)$.

Solution (a) The domain of f is $\{x \geq 0\}$ and the domain of g is $\{-5 \leq x \leq 5\}$. For $g(x)$ to be in the domain of f requires

$$\sqrt{25 - x^2} - x \geq 0 \qquad \sqrt{25 - x^2} \geq x.$$

If $-5 \leq x \leq 0$, this is automatically satisfied. If $0 < x \leq 5$, then $\sqrt{25 - x^2} \geq x$ only if

$$25 - x^2 \geq x^2 \qquad 25 \geq 2x^2 \qquad x^2 \leq \tfrac{25}{2} \qquad x \leq \tfrac{5}{2}\sqrt{2}.$$

The domain of $f \circ g$ is the set of all x in the domain of g such that $g(x)$ is in the domain of f. In this case it is the set $\{-5 \leq x \leq \frac{5}{2}\sqrt{2}\}$.

(b) $[f \circ g](x) = \sqrt{g(x)} = \sqrt{\sqrt{25 - x^2} - x}$.

(c) $[f \circ g](3) = \sqrt{\sqrt{25 - 9} - 3} = \sqrt{4 - 3} = 1$.

$[f \circ g](-4) = \sqrt{\sqrt{25 - 16} + 4} = \sqrt{3 + 4} = \sqrt{7}$.

Answer (a) $\{-5 \leq x \leq \frac{5}{2}\sqrt{2}\}$ (b) $[f \circ g](x) = \sqrt{\sqrt{25 - x^2} - x}$
(c) $1 \quad \sqrt{7}$

Exercises

Find $[f + g](x)$ and $[fg](x)$ and their domains

1 $f(x) = 3x + 1 \quad g(x) = x - 4$
2 $f(x) = 2x - 1 \quad g(x) = 2x + 3$
3 $f(x) = x^2 \quad g(x) = -2x + 1$
4 $f(x) = x^2 + 1 \quad g(x) = -x^2 + x$
5 $f(x) = x \quad g(x) = 1/x$
6 $f(x) = \sqrt{x} \quad g(x) = \sqrt{1 - x}$
7 $f(x) = \sqrt{1 - x} \quad g(x) = \sqrt{x - 2}$
8 $f(x) = \dfrac{1}{x^2 + x} \quad g(x) = \dfrac{1}{x^2 - 4}$.

Compute $[f \circ g](x)$ and $[g \circ f](x)$, and give their domains

9 $f(x) = 3x + 1 \quad g(x) = x - 2$
10 $f(x) = 2x - 1 \quad g(x) = -x^2 + 3x$
11 $f(x) = 2x^2 \quad g(x) = -x - 1$
12 $f(x) = x + 1 \quad g(x) = -x + 1$
13 $f(x) = 2x \quad g(x) = -2x$
14 $f(x) = x + 3 \quad g(x) = 4x + 1$
15 $f(x) = x^2 \quad g(x) = 3$
16 $f(x) = \pi x^2 \quad g(x) = 2x + 5$
17 $f(x) = \sqrt{x} \quad g(x) = x^2 - 9$
18 $f(x) = \sqrt{x} \quad g(x) = \sqrt[3]{x}$
19 $f(x) = 1/x \quad g(x) = x^2 + 2$
20 $f(x) = \sqrt{x} \quad g(x) = |x|$
21 $f(x) = \dfrac{x - 7}{3} \quad g(x) = 3x + 7$
22* $f(x) = \dfrac{2x + 1}{3x + 1} \quad g(x) = \dfrac{x - 1}{-3x + 2}$.

Compute $[f \circ g](x)$

23 $f(x) = x$ and $g(x)$ is any function
24 $g(x) = x$ and $f(x)$ is any function.

Compute $[f \circ f](x)$

25 $f(x) = 1 - x$
26 $f(x) = 1/x \quad x \neq 0$.

Does $f \circ g$ make sense?

27 $f(x) = \sqrt{2x - 5} \quad g(x) = 1 - x^2$
28 $f(x) = g(x) = -\sqrt{x}$.

Suppose the domain of f is the interval $0 \leq x \leq 2$. Find the domain of

29 $f(x - 3)$ **30** $f(-x)$ **31** $f(2x)$ **32** $f(x^2)$.

33 A function f is called **strictly increasing** if whenever x_1 and x_2 are in the domain of f and $x_1 < x_2$, then $f(x_1) < f(x_2)$. Show that the sum of two strictly increasing functions is strictly increasing.

34 (cont.) Show that the composite of two strictly increasing functions is strictly increasing.

35 Let $f(x) = x + 2$. Find a function $g(x)$ such that $[f \circ g](x) = x$.

36 (cont.) Do the same for $f(x) = 2x - 3$.

How do the calculator sequences represent composition of functions? How do the composite functions differ?

37 [x^2] [$\sqrt{\ }$] and [$\sqrt{\ }$] [x^2]

38 [+] 1 [=] [x^2] and [x^2] [+] 1 [=].

In the next four exercises you are asked to find the function represented by a black box that is built out of other black boxes. These are all simplified models of **servomechanisms** because part of the output is returned to the input. We use the building blocks in Fig. 3.

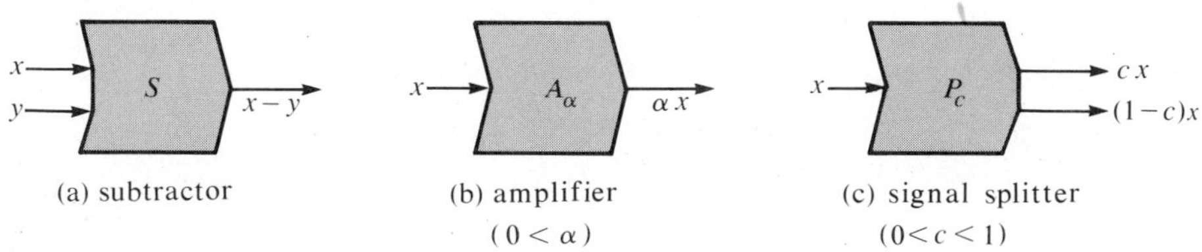

Fig. 3

Express y as a function of x

39

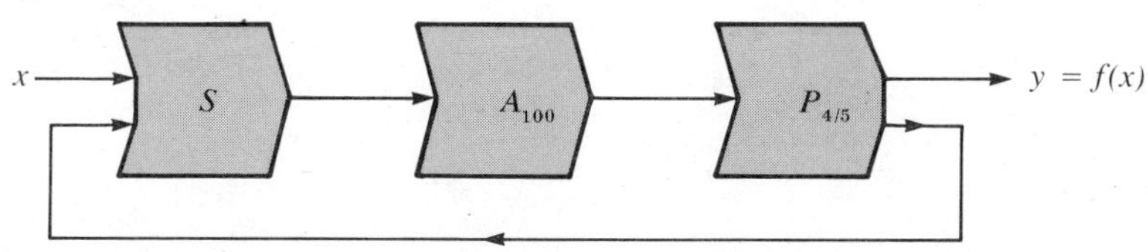

40

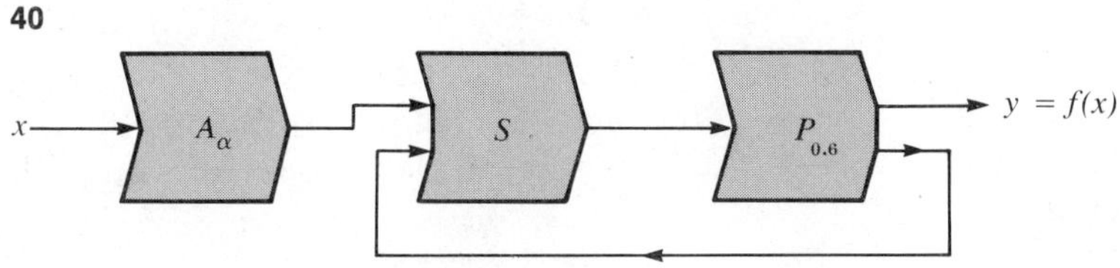

41

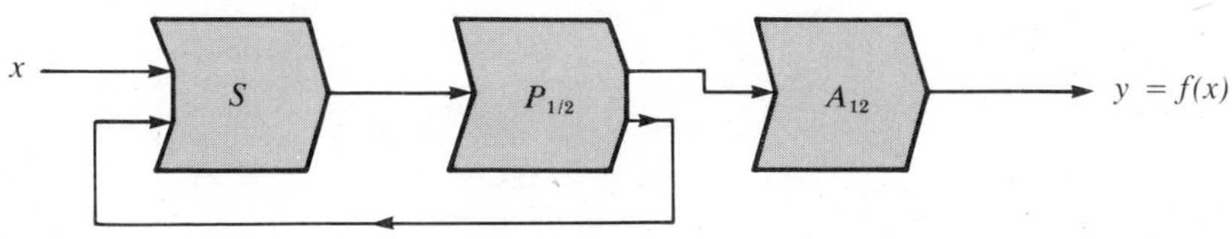

42

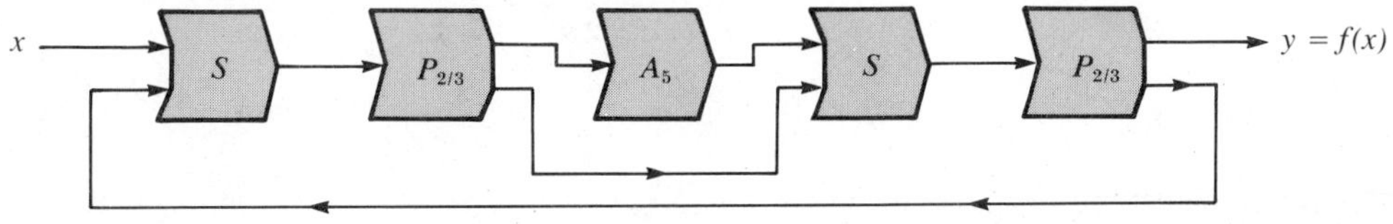

6 LINEAR FUNCTIONS

A function $f(x)$ is called **linear** if

$$f(x) = ax + b$$

for all real values of x, where a and b are constants. The word "linear" is used because the graph of a linear function is a straight line.

> The graph of each linear function $y = ax + b$ is a non-vertical straight line.
>
> Conversely, each non-vertical straight line is the graph of a linear function.

Let us justify these two assertions, taking first the special case $b = 0$. We shall prove: (1) the graph of $y = ax$ is a non-vertical straight line through the origin, and (2) each non-vertical straight line through the origin is the graph of a linear function $y = ax$.

(1) Consider the graph $y = ax$ for $a > 0$. Two points on the graph are $(0, 0)$ and $(1, a)$. Through these points passes a line L that lies in the first and third quadrants (Fig. 1).

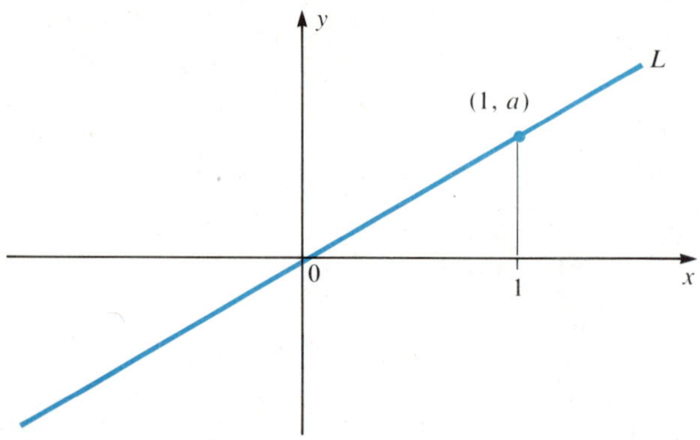

Fig. 1 Line L through $(0, 0)$ and $(1, a)$

We assert that the graph of $y = ax$ is the line L. That means each point of L has the form (x, ax), and conversely, each point in the plane of the form (x, ax) lies on L. See Fig. 2.

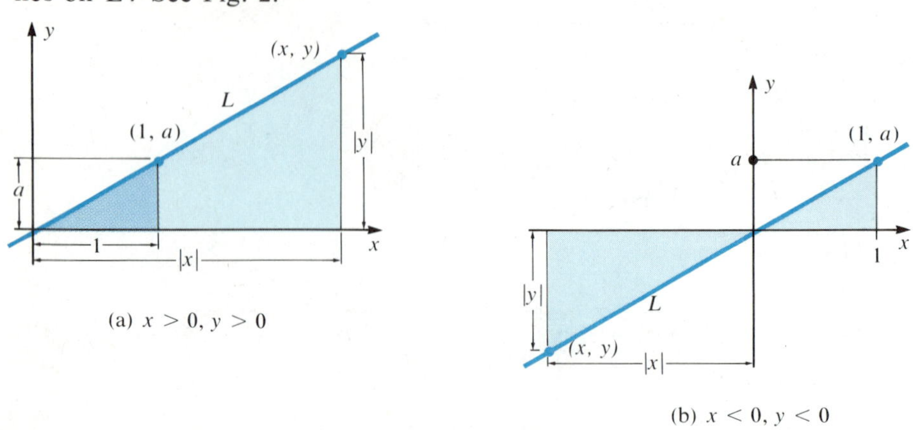

Fig. 2 Similar triangles: $a > 0$

A point (x, y) is on L if and only if the right triangles are similar, that is, the ratios of corresponding legs are equal. For (x, y) in the first quadrant, this condition is

$$\frac{|y|}{|x|} = \frac{y}{x} = \frac{a}{1}, \quad \text{that is,} \quad y = ax.$$

For (x, y) in the third quadrant, x and y are both negative. Hence $|y|/|x| = (-y)/(-x) = y/x$, so the condition is again

$$\frac{y}{x} = \frac{a}{1}, \quad \text{that is,} \quad y = ax.$$

Now consider the graph of $y = ax$ for $a < 0$. Then $a = -|a|$, and we write $y = ax$ as $y = -|a|x$. The graph is the reflection in the x-axis of the graph of $y = |a|x$. Hence it is a line through $(0, 0)$ lying in the second and fourth quadrants (Fig. 3).

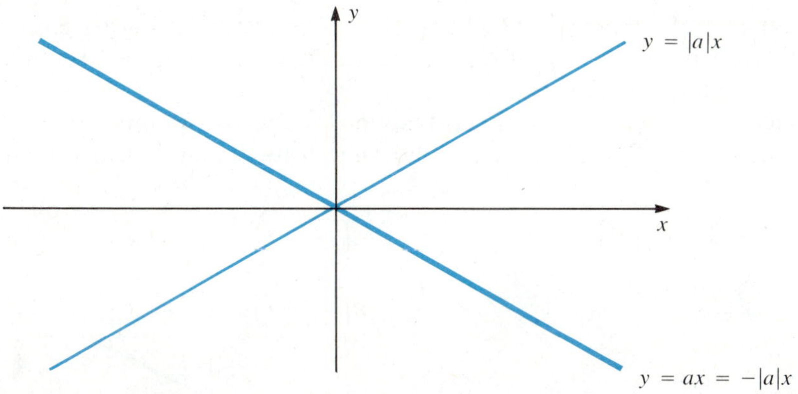

Fig. 3 Graph of $y = ax$ for $a < 0$

If $a = 0$, the graph of $y = ax = 0$ is the x-axis. Thus in all cases the graph of $y = ax$ is a non-vertical straight line through $(0, 0)$.

(2) Conversely, let L be any non-vertical line through $(0, 0)$. Then L passes through a point $(1, a)$, and the same reasoning shows that each point (x, y) on L satisfies $y = ax$.

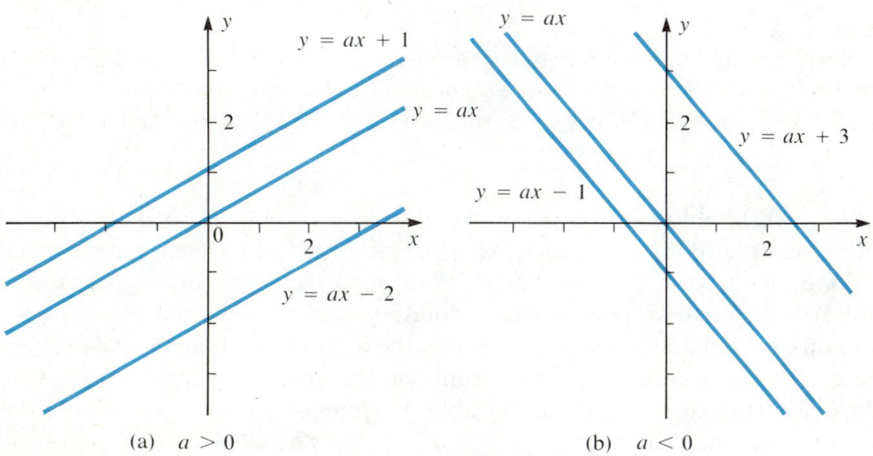

Fig. 4 Examples of $y = ax + b$ for various b

For the general linear function $y = ax + b$, with $b \neq 0$, the graph is just the graph of $y = ax$ moved up or down $|b|$ units, hence a non-vertical straight line. Conversely, each non-vertical straight line is parallel to a non-vertical straight line through $(0, 0)$. Hence it is the graph of $y = ax + b$ for a suitable constant b. See Fig. 4, previous page.

Remark The special case $y = ax$ is often described by saying that "y is **proportional** to x, or **directly proportional** to x." For example, "the number of Twins is proportional to the number of Babies born" means $T = aB$, where a is a constant.

Knowing that the graph of a linear function is a line makes it easy to plot the graph. We just find any two points on the graph, then draw the straight line through them.

EXAMPLE 1 Graph

(a) $y = 2x$ (b) $y = \frac{1}{2}x - 1$.

Solution (a) If $x = 1$, then $y = 2$, hence $(1, 2)$ is one point on the graph. Similarly $(-1, -2)$ is another point on the graph. Plot these two points, then draw the line through them (Fig. 5a).

(b) The values $x = 0$ and $x = 2$ correspond to the two points $(0, -1)$ and $(2, 0)$ on the graph. Plot these points and then draw the line through them (Fig. 5b).

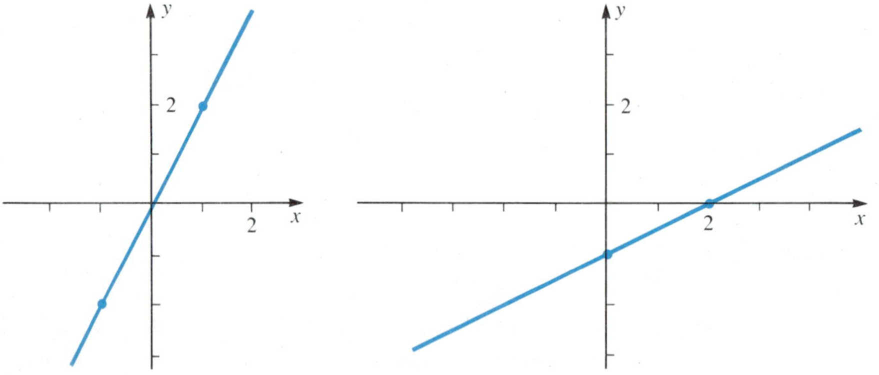

(a) graph of $y = 2x$ (b) graph of $y = \frac{1}{2}x - 1$

Fig. 5

Remark Keep in mind the small difference between the terms linear *function* and linear (first degree) *polynomial*. For $ax + b$ to be a linear *polynomial*, we must have $a \neq 0$. A linear *function* $y = ax + b$ can perfectly well have $a = 0$; its graph is still a line.

Slope

With each non-vertical line in the plane, we associate a measure, called slope, of its steepness relative to the positive x-axis. In Fig. 6, the line D will have a small positive slope, C a large positive slope, B a negative slope of large absolute value, and A a negative slope of small absolute value.

Given a non-vertical line, its slope measures the amount y changes relative to a change in x. In Fig. 7, choose any two points on the line, (x_0, y_0) and (x_1, y_1). As x advances from x_0 to x_1, the variable y changes from y_0 to y_1, so the change is $y_1 - y_0$ while the change in x is $x_1 - x_0$. The **slope** is the ratio of the change in y to the change in x:

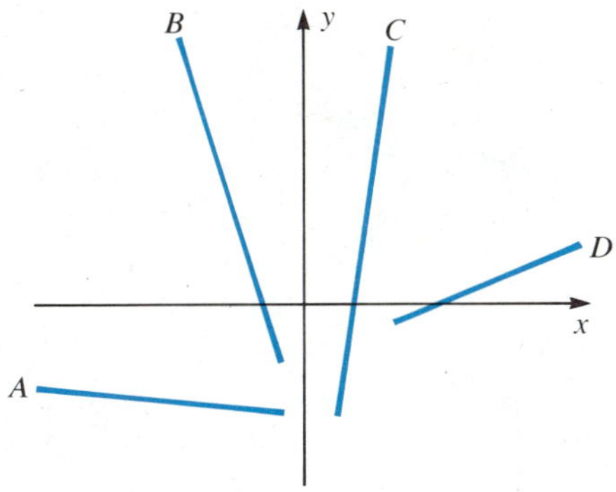

Fig. 6 Various degrees of steepness

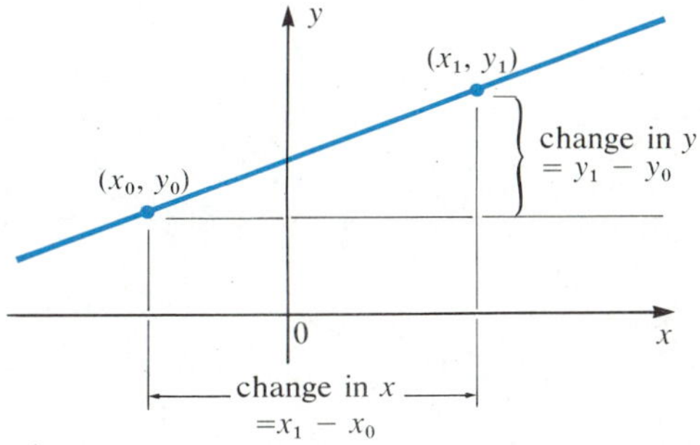

Fig. 7 Changes in x and y

$$\text{slope} = \frac{y_1 - y_0}{x_1 - x_0}.$$

If the line rises steeply, the change in y is greater than the change in x. Hence the slope is a large number. If the line rises slowly, the slope is a small number.

Notice that if the line falls as x increases, $y_1 - y_0$ is negative while $x_1 - x_0$ is positive. Hence the slope is negative.

Note that the slope formula is valid whether (x_1, y_1) is to the right of (x_0, y_0) or to the left of (x_1, y_1). For if we interchange (x_0, y_0) and (x_1, y_1), that only reverses the signs of both numerator and denominator; the ratio remains the same.

If the line in Figure 7 is shifted up or down to a parallel position, then the ratio of $y_1 - y_0$ to $x_1 - x_0$ does not change, hence the slope does not change. If the line is shifted to a non-parallel position the ratio changes, hence the slope changes.

> Two non-vertical lines are parallel if and only if their slopes are equal.

Now let us compute the slope of a line given as the graph of a linear function $y = ax + b$. We choose *any* two points on the line:

$$\text{slope} = \frac{y_1 - y_0}{x_1 - x_0} = \frac{(ax_1 + b) - (ax_0 + b)}{x_1 - x_0} = \frac{a(x_1 - x_0)}{x_1 - x_0} = a.$$

Thus the slope is a.

Next, let us find a meaning for the coefficient b in $y = ax + b$. If $x = 0$, then $y = b$. Hence the line meets the y-axis at $(0, b)$. We call b the ***y*-intercept** of the graph.

> The graph of $y = ax + b$ is a straight line with slope a and y-intercept b.

We can now tell, for instance, that $y = 5x - 2$ represents the line with slope 5 and y-intercept -2. Conversely, given the slope and y-intercept of a line, we can immediately write its equation.

EXAMPLE 2 Graph $x - 3y - 6 = 0$.

Solution Use algebra to work the equation into the form $y = ax + b$:

$$3y = x - 6 \qquad y = \tfrac{1}{3}x - 2$$

The graph is the line of slope $\frac{1}{3}$ with y-intercept -2. See Fig. 8, next page.

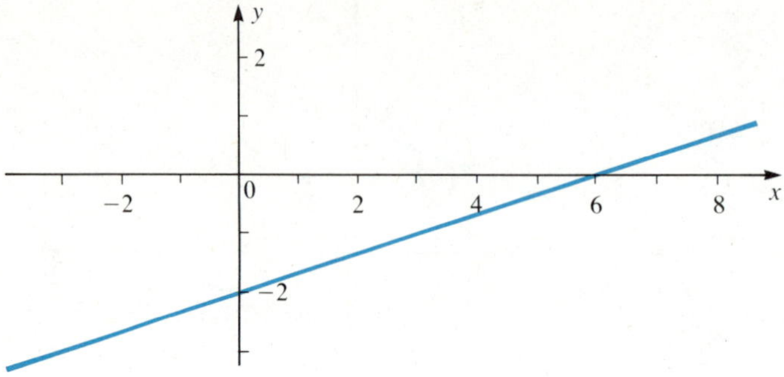

Fig. 8 $x - 3y - 6 = 0$

Formulas for Lines

A non-vertical straight line is determined by its slope (direction) and one point it passes through. The formula $y = ax + b$ illustrates this principle; it determines a line by its slope and the special point $(0, b)$, where it intersects the y-axis.

Now let us find an equation for a line, given its slope a and some point (x_0, y_0) on the line. The equation will have the form $y = ax + b$, where b is the y-intercept. To compute b, we use the fact that (x_0, y_0) satisfies the equation of the line:

$$y_0 = ax_0 + b \qquad b = y_0 - ax_0.$$

Hence

$$y = ax + b = ax + y_0 - ax_0,$$

that is,

$$y - y_0 = a(x - x_0).$$

This equation is called the **point-slope form.**

Now we can easily find the equation of a line given two of its points (x_0, y_0) and (x_1, y_1). We compute the slope

$$a = \frac{y_1 - y_0}{x_1 - x_0},$$

then substitute this value into the point-slope form:

$$y - y_0 = \left(\frac{y_1 - y_0}{x_1 - x_0}\right)(x - x_0).$$

This equation is called the **two-point form.**

A particular case is worth noting when the two points are $(a, 0)$ and $(0, b)$, that is, when we are given the x-intercept a and the y-intercept b. Then the two-point form yields

$$y - 0 = \left(\frac{b - 0}{0 - a}\right)(x - a) = -\frac{b}{a}(x - a) = -\frac{b}{a}x + b.$$

Divide by b and rearrange terms to obtain a convenient form of this equation:

$$\frac{x}{a} + \frac{y}{b} = 1.$$

This equation is called the **two-intercept form.** This form is particularly easy to remember, and is obviously correct. For if $y = 0$, then $x/a = 1$, so $x = a$; hence $(a, 0)$ is on the graph. If $x = 0$, then $y/b = 1$, so $y = b$; hence

$(0, b)$ is on the graph.

We have derived four useful formulas for a straight line:

Slope-intercept form:	$y = ax + b$
Point-slope form:	$y - y_0 = a(x - x_0)$
Two-point form:	$y - y_0 = \dfrac{y_1 - y_0}{x_1 - x_0}(x - x_0)$
Two-intercept form:	$\dfrac{x}{a} + \dfrac{y}{b} = 1.$

These formulas are used to obtain an equation for a given line. Conversely, given an equation in one of these forms, it represents a line that can be easily identified. For example, $y - 3 = 2(x - 1)$ is in point-slope form, hence represents the line through $(1, 3)$ with slope 2.

Remark Each horizontal line has the equation $y = b$, where b is its y-intercept. We have excluded *vertical* lines from this discussion because a vertical line cannot be presented in the form $y = f(x)$. However, each vertical line has the equation $x = a$, where a is its x-intercept.

EXAMPLE 3 Find an equation for the line through $(2, 1)$ and $(3, -7)$.

Solution Substitute into the two-point form:

$$y - 1 = \frac{1 - (-7)}{2 - 3}(x - 2) = -8(x - 2).$$

Answer $y - 1 = -8(x - 2)$

EXAMPLE 4 Find an equation for the line that passes through $(1, 4)$ and is parallel to the line $3x + 2y = 1$.

Solution We are given one point; we need the slope. Since parallel lines have the same slope, the desired slope is the same as that of the line $3x + 2y = 1$. We put this equation into slope-intercept form:

$$2y = -3x + 1 \qquad y = -\tfrac{3}{2}x + \tfrac{1}{2}.$$

The slope is $-\frac{3}{2}$. From the point-slope form, the desired line is

$$y - 4 = -\tfrac{3}{2}(x - 1).$$

Answer $y - 4 = -\frac{3}{2}(x - 1)$

Perpendicular Lines

There is a simple way to tell whether two lines are parallel: their slopes must be equal. There is also a simple way to tell, using slopes, whether two lines are perpendicular. Take two lines through the origin: $y = ax$, where $a > 0$, and $y = bx$, where $b < 0$. See Fig. 9.

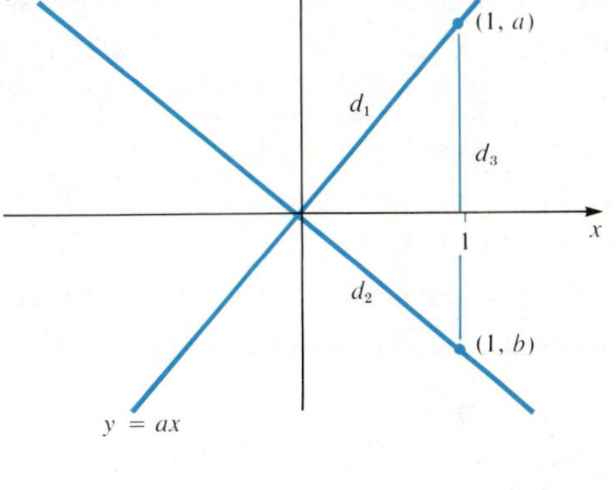

Fig. 9 Perpendicular lines

Clearly $(1, a)$ is a point of $y = ax$ and $(1, b)$ is a point of $y = bx$. The lines are perpendicular if and only if $(1, a)$, $(1, b)$, and $(0, 0)$ are the vertices of a right triangle with its right angle at $(0, 0)$. By the figure, this is equivalent to

$$d_3^2 = d_1^2 + d_2^2.$$

But by the distance formula

$$d_1^2 = 1^2 + a^2 \qquad d_2^2 = 1^2 + b^2 \qquad d_3^2 = (b - a)^2,$$

so the condition for perpendicularity is

$$1 + a^2 + 1 + b^2 = (b - a)^2 = b^2 - 2ab + a^2,$$

that is,

$$2 = -2ab \qquad ab = -1.$$

Thus the product of the slopes must be -1. The same is true for two perpendicular lines that do not intersect at $(0, 0)$. For there is a pair of perpendicular lines with the same slopes that do intersect at $(0, 0)$. Just slide the given lines without turning so that their intersection point moves to $(0, 0)$.

Perpendicular Lines Two non-vertical lines are perpendicular if and only if the product of their slopes is -1.

EXAMPLE 5 Find an equation for the line through $(7, 10)$ that is perpendicular to the line $y = -3x + 5$.

Solution Let a be the slope of the desired line. The line $y = -3x + 5$ has slope -3. In order for the lines to be perpendicular, their slopes must satisfy

$$(-3)(a) = -1, \quad \text{hence} \quad a = \tfrac{1}{3}.$$

Now substitute into the point-slope form:

$$y - 10 = \tfrac{1}{3}(x - 7).$$

Answer $y - 10 = \frac{1}{3}(x - 7)$

Exercises

Graph

1 $y = 2x - 3$ **2** $y = \frac{1}{2}x - 3$ **3** $y = 2x + 9$

4 $y = 3x + 1$ **5** $y = -\frac{1}{4}x + 1$ **6** $y = -2x + 1$

7 $y = 3x + 40$ **8** $y = 9x - 50$ **9** $y = 0.1x + 1.5$

10 $y = -0.3x + 0.2$.

Find the slope of the line through the given points

11 $(0, 0)\ (3, 4)$ **12** $(0, 0)\ (2, 6)$ **13** $(-1, 2)\ (1, 2)$

14 $(6, 5)\ (8, 13)$ **15** $(-4, -3)\ (-1, 7)$ **16** $(2, -\frac{3}{2})\ (\frac{1}{2}, 6)$.

Do these points lie on a straight line?

17 $(3, 1)\ (10, 8)\ (14, 12)$ **18** $(-1, -6)\ (0, -4)\ (2, 0)$

19 $(0, \frac{1}{2})\ (1, \frac{3}{2})\ (5, \frac{11}{2})$ **20** $(9, 7)\ (5, 6)\ (-7, 4)$.

Verify that the points are vertices of a parallelogram

21 $(0, 0)\ (0, 5)\ (3, 9)\ (3, 4)$ **22** $(-4, -5)\ (3, 3)\ (9, 6)\ (2, -2)$.

Are the lines parallel?

23 $x - y = 2$ and $x - y = 3$

24 $3x + 5y + 1 = 0$ and $3x + 5y - 7 = 0$

25 $\dfrac{x}{a} + \dfrac{y}{b} = 1$ and $\dfrac{x}{2a} + \dfrac{y}{2b} = 1$

26 $y - 4 = 3(x - 1)$ and $y - 1 = 3(x - 4)$.

Find the equation and y-intercept of the line with given slope a and passing through the given point

27 $a = 1 \quad (4, 2)$ **28** $a = -1 \quad (2, -1)$ **29** $a = 0 \quad (4, 3)$

30 $a = 2 \quad (1, 3)$ **31** $a = \frac{1}{2} \quad (2, -2)$ **32** $a = \frac{2}{3} \quad (-1, 7)$.

Find the equation of the line through the two given points

33 $(0, 0)\ (1, 2)$ **34** $(1, 0)\ (3, 4)$ **35** $(-1, 0)\ (2, 4)$

36 $(-7, -9)\ (2, 6)$ **37** $(\frac{1}{2}, 1)\ (\frac{3}{2}, 2)$ **38** $(5, 3)\ (-\frac{1}{2}, -1)$

39 $(0, 3)\ (0, -3)$ **40** $(5, -2)\ (8, -2)$.

Find the slope and y-intercept

41 $3x - y - 7 = 0$ **42** $x + 2y + 6 = 0$

43 $3(x - 2) + y + 5 = 2(x + 3)$ **44** $2(x + y + 1) = 3x - 5$.

Find both intercepts and write the equation of the line in two-intercept form

45 $\dfrac{x}{2} + \dfrac{y}{3} = 1$ **46** $x - y = 2$

47 $2x + 3y = 1$ **48** $3x - 5y = 15$.

Plot on the same set of coordinate axes

49 $y = -\frac{1}{2}x + b$ for $b = -1\ \ 0\ \ 1\ \ 2$ **50** $y = ax + 1$ for $a = -3\ \ -1\ \ \frac{1}{3}\ \ 1$

51 $x - y = c$ for $c = -2\ \ -1\ \ 0\ \ 1\ \ 2$ **52** $\dfrac{x}{a} + \dfrac{y}{2} = 1$ for $a = -2\ \ -1\ \ 1\ \ 3$

53 $y - b = 2(x - 1)$ for $b = -2\ \ 0\ \ 1\ \ 2$ **54** $y - 4 = -3(x - a)$ for $a = -2\ \ 0\ \ 1\ \ 2$.

Find an equation for the line

55 through $(3, 4)$ and parallel to $2x - y = 1$

56 parallel to $x + y + 1 = 0$ and intersecting the line $y = 7x - 6$ on the y-axis

57 through $(1, 6)$ and perpendicular to $9x - 5y = 4$

58 through the origin and perpendicular to the line through $(4, 0)$ and $(0, 3)$.

59 Find equations for the sides of the square with vertices $(\pm 1, 0)$ and $(0, \pm 1)$.

60 Show that the parallelogram in Ex. 21 is a rhombus (all sides equal) and verify that its diagonals are perpendicular.

Using slopes, show that the points are vertices of a right triangle

61 $(-3, 5)\ (-1, 3)\ (4, 12)$. **62** $(-6, 15)\ (3, 2)\ (6, 6)$.

63 Show that the lines $ax + by = c$ and $bx - ay = d$ are perpendicular. Assume $ab \neq 0$.

64 Show that the line $y = x$ is the perpendicular bisector of the line segment joining (a, b) and (b, a), where $a \neq b$.

7 QUADRATIC FUNCTIONS

A function $f(x)$ is called **quadratic** if

$$f(x) = ax^2 + bx + c \qquad a \neq 0$$

where a, b, and c are constants. A quadratic function is defined for all values of x.

Quadratic functions occur frequently in applications. For example, if a projectile is shot upwards with muzzle velocity v_0, then its height y at time t is given by the quadratic function

$$y = -\tfrac{1}{2}gt^2 + v_0 t,$$

where g is the constant of gravity.

Graph of $y = ax^2 + c$

We begin by graphing the simplest quadratic, $y = x^2$. We first consider only $x \geq 0$. When x is small, x^2 is very small. For example, $(0.1)^2 = 0.01$, and $(0.001)^2 = 0.000001$. Therefore, as x increases starting from 0, the graph of $y = x^2$ rises very slowly from 0. See Fig. 1.

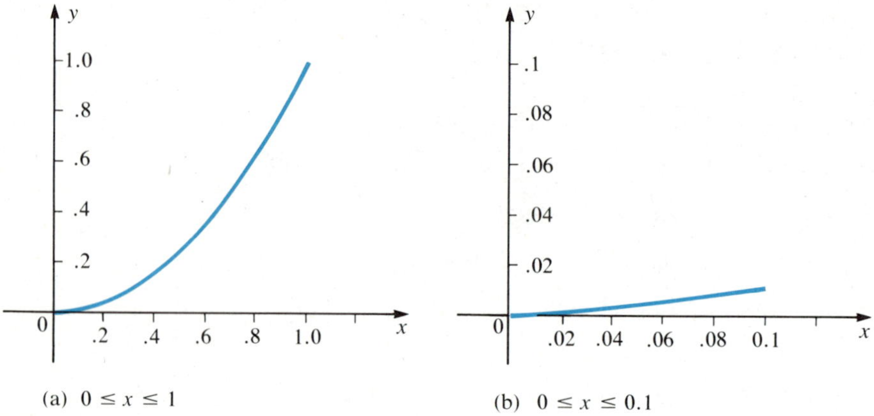

Fig. 1 Graph of $y = x^2$; small x

On the other hand, when x is large, x^2 is very large. For example, $(10)^2 = 100$, and $(1{,}000)^2 = 1{,}000{,}000$. Therefore, as x gets larger and larger, the graph of $y = x^2$ rises very steeply (Fig. 2).

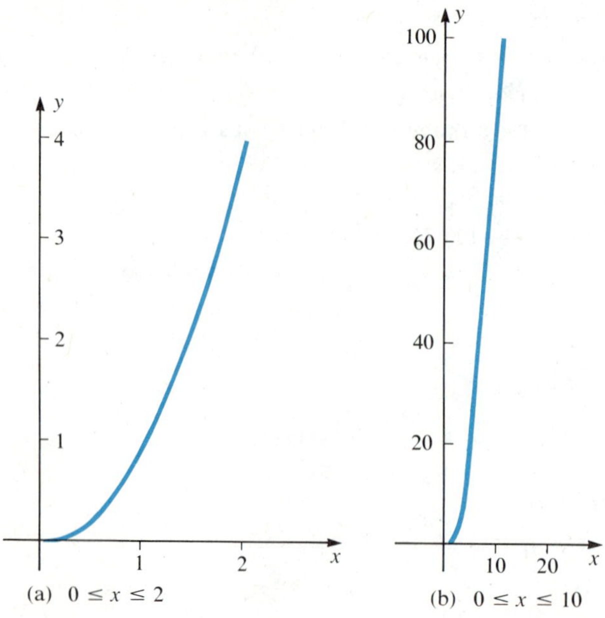

Fig. 2 Graph of $y = x^2$; large x

For $x < 0$, we use the fact that $(-x)^2 = x^2$. That means the value of y at $-x$ is the same as at x. So whenever (x, y) is on the graph, so is $(-x, y)$. Hence the graph of $y = x^2$ for $x < 0$ is the mirror image in the y-axis of the graph for $x > 0$. See Fig. 3.

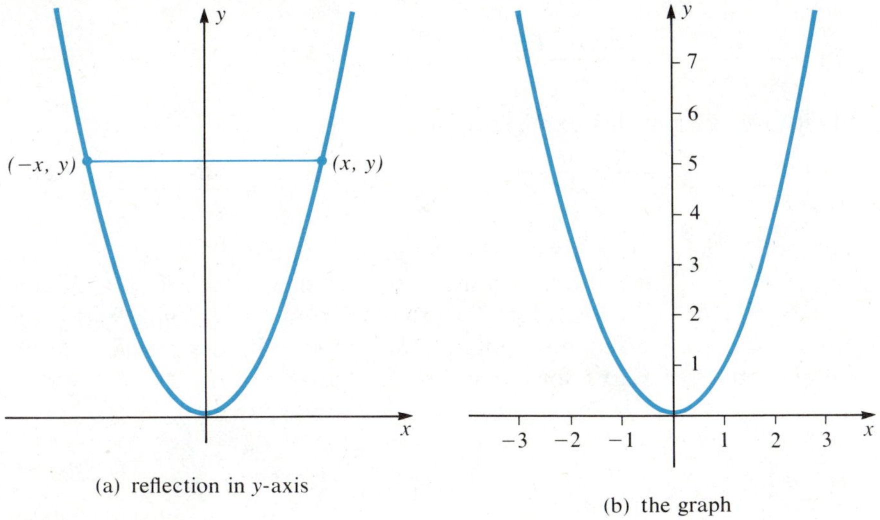

Fig. 3 Graph of $y = x^2$

Once we have the graph of $y = x^2$, the graph of $y = ax^2$ is easy. If $a > 0$, it is the graph of $y = x^2$ stretched vertically by the factor a. If $a < 0$, the graph of $y = ax^2$ is the graph of $y = x^2$ stretched vertically by the factor $|a|$, and then reflected in the x-axis. The curve opens upward if $a > 0$ and has $(0, 0)$ as its lowest point. It opens downward if $a < 0$ and has $(0, 0)$ as its highest point. Finally, we obtain the graph of $y = ax^2 + c$ by shifting the graph of $y = ax^2$ up or down $|c|$ units according to the sign of c. Fig. 4 shows some examples.

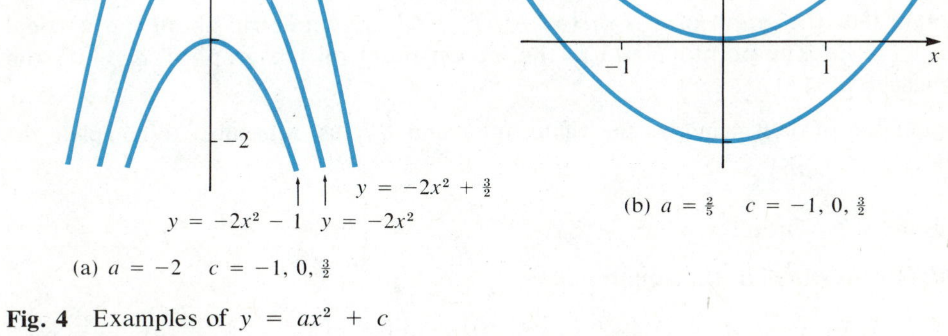

Fig. 4 Examples of $y = ax^2 + c$

Graph of the General Quadratic

To graph the most general quadratic function $y = ax^2 + bx + c$, we complete the square. (We can suppose $a \neq 0$, otherwise the function is linear). We obtain

$$y = ax^2 + bx + c = a\left(x^2 + 2\frac{b}{2a}x\right) + c$$

$$= a\left(x^2 + 2\frac{b}{2a}x + \frac{b^2}{4a^2}\right) + c - \frac{b^2}{4a}$$

$$= a\left(x + \frac{b}{2a}\right)^2 + \frac{4ac - b^2}{4a}.$$

For simplicity we rewrite this expression as

$$y = a(x - h)^2 + d \quad \text{where} \quad h = \frac{-b}{2a} \quad \text{and} \quad d = \frac{4ac - b^2}{4a}.$$

Now for each point (x_0, y_0) that satisfies this equation, the point $(x_0 - h, y_0)$ satisfies $y = ax^2 + d$. This point is h units to the left of (x_0, y_0) if $h > 0$, and $|h|$ units to the right if $h < 0$. Therefore the graph of $y = a(x - h)^2 + d$ is the graph of $y = ax^2 + d$ shifted horizontally $|h|$ units; shifted right if $h > 0$, shifted left if $h < 0$. See Fig. 5 for examples.

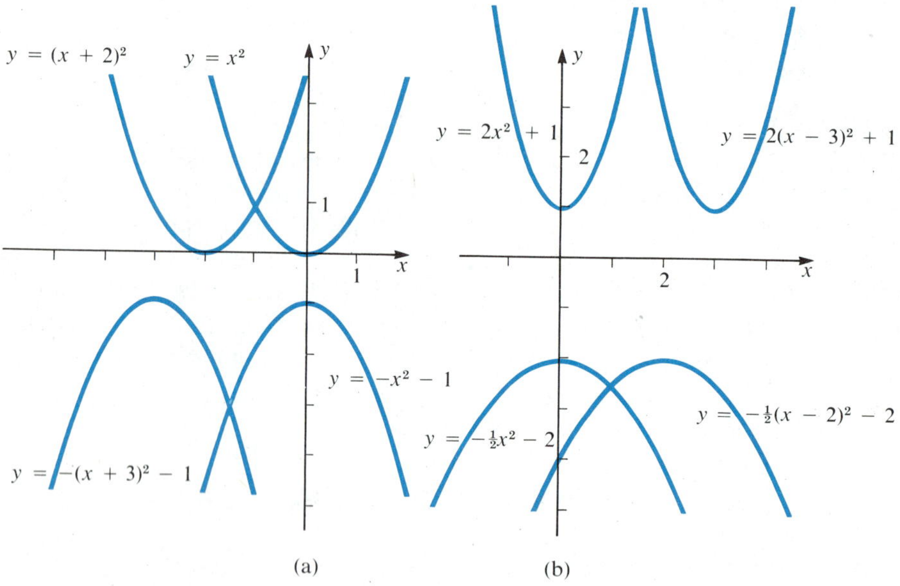

Fig. 5 Graphs of $y = a(x - h)^2 + d$

Note that the graph of $y = a(x - h)^2 + d$ is symmetric about the vertical line $x = h$. The point (h, d) is the lowest point on the graph if $a > 0$, the highest if $a < 0$.

Suggestion Don't memorize the values of h and d. Just remember to complete the square.

EXAMPLE 1 Graph $y = x^2 + 2x + 3$.

Solution Complete the square:

$$y = x^2 + 2x + 3 = (x^2 + 2x + 1) + 2 = (x + 1)^2 + 2.$$

The graph is obtained by shifting the graph of $y = x^2$ one unit to the left, then two units up (Fig. 6).

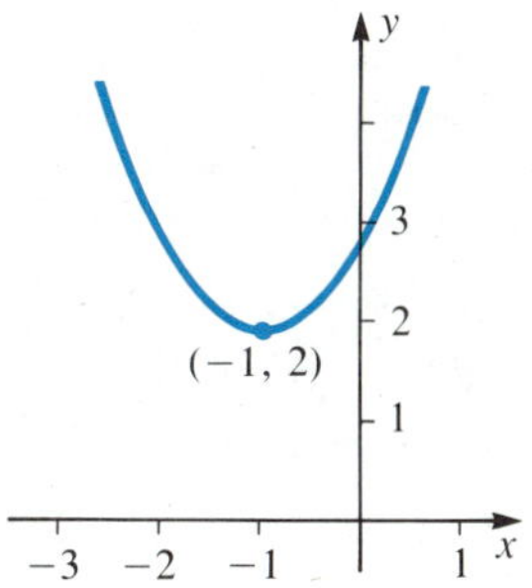

Fig. 6 $y = x^2 + 2x + 3$

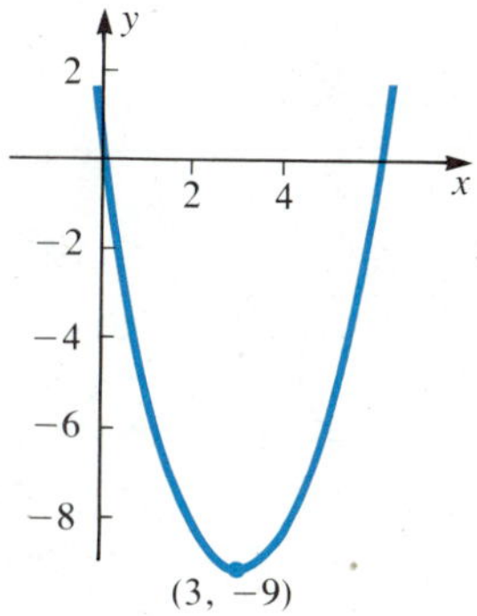

Fig. 7 $y = x^2 - 6x$

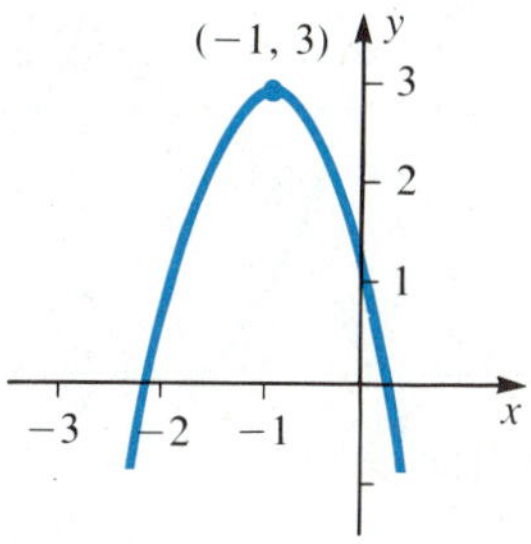

Fig. 8 $y = -2x^2 - 4x + 1$

EXAMPLE 2 Graph $y = x^2 - 6x$ and find the lowest point on the curve.

Solution Complete the square:

$$y = x^2 - 6x = x^2 - 6x + 9 - 9 = (x - 3)^2 - 9.$$

The graph is obtained by shifting the graph of $y = x^2$ to the right 3 units, then down 9 units. By inspection, its lowest point is $(3, -9)$. As a check, note that if $x = 3$, then $y = -9$. But if $x \neq 3$, then $(x - 3)^2 > 0$, so $y > -9$. Hence the lowest point is $(3, -9)$.

Answer $(3, -9)$; Fig. 7

EXAMPLE 3 Graph $y = -2x^2 - 4x + 1$ and find the highest point on the curve.

Solution Complete the square:

$$y = -2x^2 - 4x + 1 = -2(x^2 + 2x) + 1$$
$$= -2(x^2 + 2x + 1) + 2 + 1 = -2(x + 1)^2 + 3.$$

The graph is obtained by shifting the graph of $y = -2x^2$ one unit to the left and three units up. By inspection, its highest point is $(-1, 3)$.

Answer $(-1, 3)$; Fig. 8

Relation to Quadratic Equations

The graph of $y = ax^2 + bx + c$ gives instant information about the solutions of the quadratic equation

$$ax^2 + bx + c = 0.$$

For if x is a solution, then $y = 0$ and conversely. Therefore the solutions of the equation correspond to the points where the graph meets the x-axis. If you have the graph you can actually *see* the solutions.

Because of the shape of the graph, there are only three possibilities (Fig. 9, next page): the graph crosses the x-axis at two points, or it is tangent at one point, or it doesn't touch the axis at all. These observations confirm what we know about real solutions of a quadratic equation; there are either two, one, or none.

We can say even more. Let's assume for the moment that $a > 0$. Then the number of real solutions of $ax^2 + bx + c = 0$ is determined geometrically by

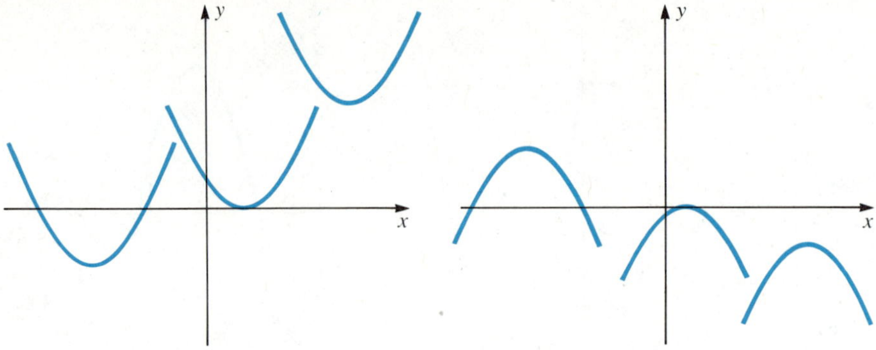

(a) possibilities with $a > 0$ (b) possibilities with $a < 0$

Fig. 9 Does the graph of $y = ax^2 + bx + c$ meet the x-axis?

the position of the lowest point on the graph (Fig. 9a). There are two real solutions if the lowest point is below the x-axis, one if it is on the x-axis, and none if it is above. Algebraically, the number of real solutions is determined by the sign of the discriminant, $D = b^2 - 4ac$: two if $D > 0$, one if $D = 0$, and none if $D < 0$. Now here is a nice agreement between the geometry and the algebra: these two criteria amount to the same thing. By completing the square, we found that the lowest point on the graph is

$$\left(-\frac{b}{2a}, \frac{4ac - b^2}{4a}\right) = \left(-\frac{b}{2a}, -\frac{D}{4a}\right).$$

Since $a > 0$, this point is below the x-axis (two solutions) if $D > 0$, on the x-axis (one solution) if $D = 0$, and above the x-axis (no solutions) if $D < 0$.

In the case $a < 0$, we consider the highest point on the graph, but otherwise the argument is practically the same.

Maximum Problems

Problems involving the maximum (or minimum) value of a quadratic function can be solved by completing the square.

EXAMPLE 4 What is the area of the largest rectangular field that can be enclosed with 2000 ft of fencing?

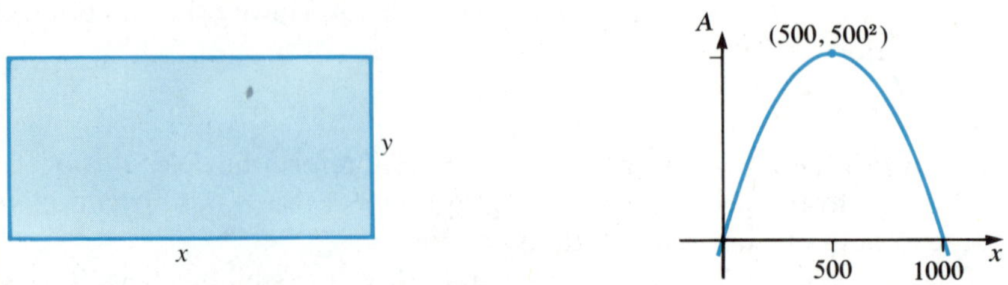

(a) rectangular field (b) graph of $A = -x^2 + 1000x$

Fig. 10

Solution Let x and y denote the length and width of the field (Fig. 10a). Then

$$\text{perimeter} = 2x + 2y = 2000 \qquad y = 1000 - x.$$

Therefore the area is

$$A = xy = x(1000 - x) = -x^2 + 1000x.$$

The problem is equivalent to finding the highest point (x, A) on the graph of $A = -x^2 + 1000x$. See Fig. 10b. Complete the square:

$$A = -(x^2 - 1000x + 500^2) + 500^2 = -(x - 500)^2 + 500^2.$$

Because the first term is negative or zero, the largest possible value of A is 500^2; it occurs only for $x = 500$. Then $y = 1000 - x = 500$, so the rectangle is a square.

Answer 500^2 ft^2; the rectangle is a square.

Example 5 A projectile is fired at time $t = 0$ at a 30 degree angle to the ground. Its height above the ground after t seconds is shown in physics to be $y = \frac{1}{2}v_0 t - 16t^2$ ft, where v_0 is the muzzle velocity (air resistance is neglected). If $v_0 = 1000$ ft/sec, what is the greatest height the projectile reaches?

Solution Complete the square:

$$y = 500t - 16t^2 = -16\left(t^2 - \frac{125}{4}t\right)$$

$$= -16\left[t^2 - \frac{125}{4}t + \left(\frac{125}{8}\right)^2\right] + 16\left(\frac{125}{8}\right)^2$$

$$= -16\left(t - \frac{125}{8}\right)^2 + \frac{125^2}{4}.$$

Therefore the largest possible value of y is $125^2/4$.

Answer $125^2/4 = 3906.25$ ft

Exercises

Graph

1 $y = 2x^2$ **2** $y = -2x^2$ **3** $y = -\frac{1}{2}x^2$
4 $y = \frac{1}{2}x^2$ **5** $y = x^2 + 3$ **6** $y = -x^2 - 3$
7 $y = 2x^2 - 1$ **8** $y = -2x^2 - 1$ **9** $y = 3(x - 5)^2$
10 $y = 1 - 3(x - 5)^2$.

Graph on the indicated domain (use different scales on the axes if necessary)

11 $y = 0.1x^2 \quad 0 \le x \le 100$ **12** $y = -x^2 \quad -0.1 \le x \le 0$.

Graph and find the highest (or lowest) point on each graph

13 $y = x^2 - 4x + 1$ **14** $y = x^2 + 2x - 5$ **15** $y = x^2 + x + 1$
16 $y = x^2 - x + 1$ **17** $y = -x^2 - 2x$ **18** $y = -x^2 + 2x$
19 $y = -x^2 - 4x - 3$ **20** $y = -x^2 + 4x + 1$ **21** $y = 2x^2 - 6x + 1$
22 $y = 2x^2 + 4x$ **23** $y = 3x^2 + 12x - 8$ **24** $y = -3x^2 + 12x - 8$
25 $y = -2x^2 + 8x - 10$ **26** $y = -2x^2 + 12x$ **27** $y = -4x^2 + x$
28 $y = 2(x^2 + x + 1)$ **29** $y = x(2x - 3)$ **30** $y = x^2 - 6x + 2$
31 $y = x^2 + x - 4$ **32** $y = 3x(x + 1)$ **33** $y = -x^2 + x - 2$
34 $y = (x - 2)(x - 6)$ **35** $y = 2(x + 1)(x + 5)$ **36** $y = -(x + 2)(x - 1)$.

37 Show that the graph of $y = ax^2 + bx$ passes through the origin for all choices of a and b.

38 For what value of c does the lowest point of the graph of $y = x^2 + 6x + c$ fall on the x-axis?

39 Under what conditions is the lowest point of the graph of $y = x^2 + 2bx + c$ on the y-axis?

40* What is the relation between the graph of $y = ax^2 + bx + c$ and that of $y = ax^2 - bx + c$?

41 Graph $y = |x^2 - 4|$.

42* Show that the graph of $y = (x - r)(x - s)$ is symmetric about the line $x = \frac{1}{2}(r + s)$.

43 Show that for $0 \leq x \leq 1$, the product $x(1 - x)$ never exceeds $\frac{1}{4}$.

44 A farmer will make a rectangular pen with 100 ft of fencing, using part of a wall of his barn for one side of the pen. What is the largest area he can enclose?

45 A 4-ft line is drawn across a corner of a rectangular room, cutting off a triangular region. Show that its area cannot exceed 4 ft^2. [*Hint:* Use the Pythagorean theorem and work with A^2.]

46 A rectangular solid has a square base, and the sum of its 12 edges is 4 ft. Show that its total surface area (sum of the areas of its 6 faces) is largest if the solid is a cube.

47 Suppose a projectile fired as in Example 5 reaches a maximum height of 2500 ft. What was its muzzle velocity?

48 (cont.) In general, if the muzzle velocity is doubled, is the maximum height of the trajectory also doubled? If not, what does happen?

By completing the square, show that

49 $x^2 + 6x + 10 > 0$ for all x

50 $x^2 + 2xy + 2y^2 \geq 0$ for all x and y

51 $x^2 + 2x + y^2 - 4y + \frac{11}{2} > 0$ for all x and y

52* $x^2 + y^2 + 2z^2 + 2xy + 2yz + 2zx \geq 0$ for all x, y, and z.

53* Let $a > 0$ and $b > 0$. Show that $ax + \dfrac{b}{x} \geq 2\sqrt{ab}$ for all $x > 0$.

54* (cont.) Let $B > 0$. Find the smallest C such that $\dfrac{x}{B + x^2} \leq C$ for all x.

8 GRAPHS OF RELATIONS

Often we graph **relations** of the form $f(x, y) = c$, where $f(x, y)$ is a polynomial in x and y and c is a constant. For example we might graph

$$y^2 - 2x = 0 \quad \text{or} \quad x^2 + 3y^2 - xy + 7x = 4.$$

The **graph** of such a relation is the set of all points (x, y) in the plane that satisfy the relation.

Remark If $f(x)$ is a (polynomial) function, the graph of $y = f(x)$ is the same as the graph of the relation $y - f(x) = 0$.

Linear Relations

The graph of

$$ax + by = c,$$

where a and b are not both 0, is always a straight line. For if $b \neq 0$, we can rewrite the relation in the equivalent form

$$y = -\frac{a}{b}x + \frac{c}{b},$$

whose graph is obviously a straight line. If $b = 0$, then $a \neq 0$ and an equivalent form is

$$x = \frac{c}{a},$$

whose graph is a vertical line.

EXAMPLE 1 Graph

(a) $x + y = 4$ (b) $3x - 2y = 5$.

Solution (a) We write the equivalent relation

$$y = -x + 4$$

whose graph is a line of slope -1 and y-intercept 4. See Fig. 1a.
(b) We write the equivalent relation (solve for y)

$$y = \tfrac{3}{2}x - \tfrac{5}{2}$$

whose graph is a line of slope $\frac{3}{2}$ and y-intercept $-\frac{5}{2}$. See Fig. 1b.

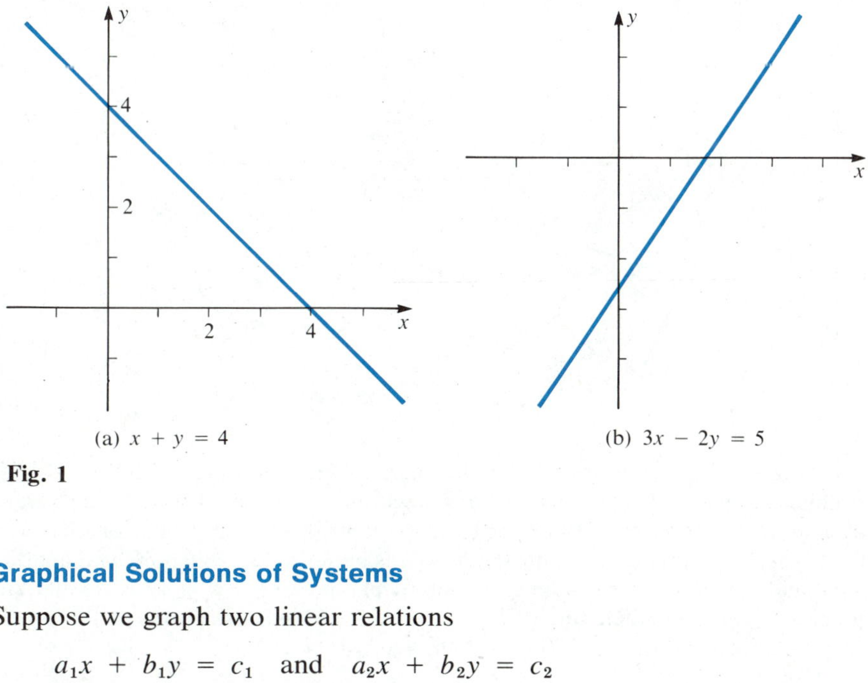

(a) $x + y = 4$ (b) $3x - 2y = 5$

Fig. 1

Graphical Solutions of Systems

Suppose we graph two linear relations

$$a_1x + b_1y = c_1 \quad \text{and} \quad a_2x + b_2y = c_2$$

on the same grid. Then we obtain two straight lines. The intersection of these lines is the solution set of the system

$$\begin{cases} a_1x + b_1y = c_1 \\ a_2x + b_2y = c_2. \end{cases}$$

Generally the two lines intersect in a single point (x, y); then it is the *unique* solution of the system. However, the lines may be parallel (inconsistent system) or the lines may coincide (underdetermined system). Compare the algebraic discussion starting on p. 53.

EXAMPLE 2 Solve graphically

$$\begin{cases} 2x + y = 4 \\ -x + 3y = -9. \end{cases}$$

Solution We can graph each relation very easily by finding *both* intercepts. For $2x + y = 4$, the intercepts are obviously $(2, 0)$ and $(0, 4)$. For $-x + 3y = -9$, the intercepts are $(9, 0)$ and $(0, -3)$. We plot these pairs of intercepts in each case and draw the two lines through them (Fig. 2). The lines appear to intersect at $(3, -2)$, so this ought to be the solution of the system, or close to it. We check:

$$2 \cdot 3 + (-2) = 4 \qquad -(3) + 3(-2) = -9.$$

We have found the solution.

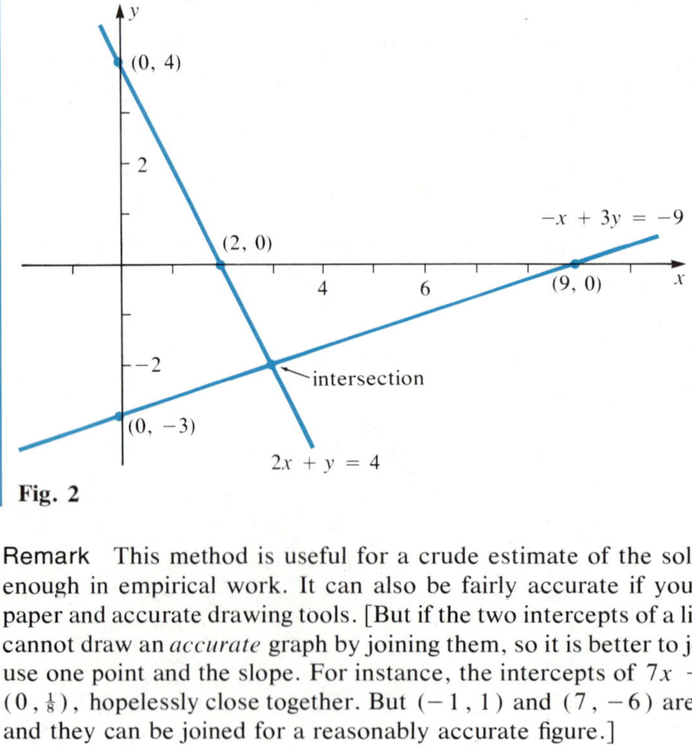

Fig. 2

Remark This method is useful for a crude estimate of the solution of a system—often enough in empirical work. It can also be fairly accurate if you use finely divided graph paper and accurate drawing tools. [But if the two intercepts of a line are close together, you cannot draw an *accurate* graph by joining them, so it is better to join two points far apart or use one point and the slope. For instance, the intercepts of $7x + 8y = 1$ are $(\frac{1}{7}, 0)$ and $(0, \frac{1}{8})$, hopelessly close together. But $(-1, 1)$ and $(7, -6)$ are also points on this graph, and they can be joined for a reasonably accurate figure.]

Circles

Let us find the graph of the single quadratic relation

$$x^2 + y^2 = r^2 \qquad r > 0.$$

It's easy once you see that the relation is a statement about distances. It says that the distance squared from (x, y) to $(0, 0)$ is r^2. Hence a point (x, y) is on the graph if and only if its distance from the origin is r. The graph is a circle about the origin of radius r. See Fig. 3a.

Similarly, the relation

$$(x - h)^2 + (y - k)^2 = r^2$$

describes those points (x, y) whose distance from the fixed point (h, k) is r. The graph is a circle with center (h, k) and radius r. See Fig. 3b.

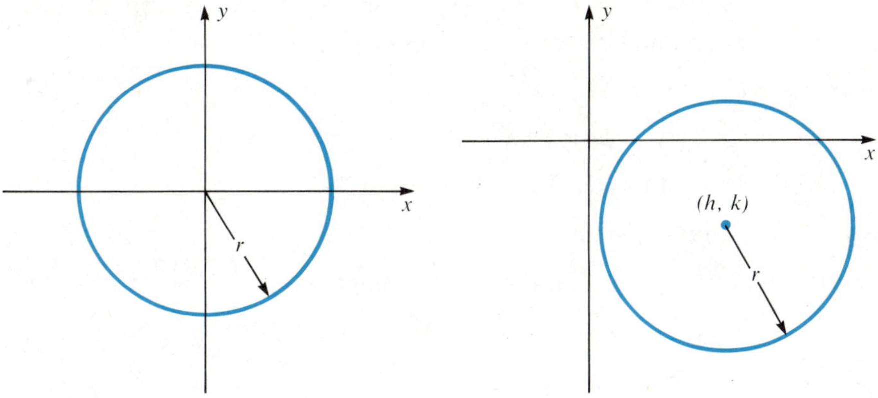

(a) $x^2 + y^2 = r^2$ (b) $(x - h)^2 + (y - k)^2 = r^2$

Fig. 3

> The graph of the relation
>
> $$(x - h)^2 + (y - k)^2 = r^2 \qquad r > 0$$
>
> is a circle with center (h, k) and radius r.

By completing squares, any relation

$$x^2 + y^2 + ax + by + c = 0$$

can be expressed in the form

$$(x - h)^2 + (y - k)^2 = \pm r^2.$$

If the right-hand side is positive, the graph is a circle. If $r^2 = 0$, the graph is the single point (h, k), and if negative, the graph is empty—a sum of squares cannot be negative.

EXAMPLE 3 Graph

(a) $x^2 + y^2 - 4x + 2y = 11$

(b) $3x^2 + 3y^2 - 8x = 1$

(c) $2x^2 + 2y^2 - 12x + 16y = -55$.

Solution (a) Complete the squares in x and in y:

$$(x^2 - 2\cdot 2x) + (y^2 + 2\cdot 1y) = 11$$

$$(x^2 - 2\cdot 2x + 2^2) + (y^2 + 2\cdot 1y + 1^2) = 11 + 4 + 1$$

$$(x - 2)^2 + (y + 1)^2 = 16 = 4^2.$$

The graph is a circle with center $(2, -1)$ and radius 4. See Fig. 4a (next page).

(b) Divide both sides by 3, then complete the square in x:

$$x^2 + y^2 - \tfrac{8}{3}x = \tfrac{1}{3}$$

$$(x^2 - 2\cdot\tfrac{4}{3}x) + y^2 = \tfrac{1}{3}$$

$$(x - \tfrac{4}{3})^2 + y^2 = \tfrac{1}{3} + \tfrac{16}{9} = \tfrac{19}{9} = (\tfrac{1}{3}\sqrt{19})^2.$$

The graph is a circle with center $(\frac{4}{3}, 0)$ and radius $\frac{1}{3}\sqrt{19}$. See Fig. 4b.

(c) Divide by 2 and complete the squares in x and in y:

$$x^2 + y^2 - 6x + 8y = -\tfrac{55}{2}$$

$$(x^2 - 2\cdot 3x) + (y^2 + 2\cdot 4y) = -\tfrac{55}{2}$$

$$(x-3)^2 + (y+4)^2 = -\tfrac{55}{2} + 9 + 16 = -\tfrac{5}{2}.$$

Since $-\frac{5}{2} < 0$, the graph is empty.

Answer (a) Fig. 4a (b) Fig. 4b (c) empty

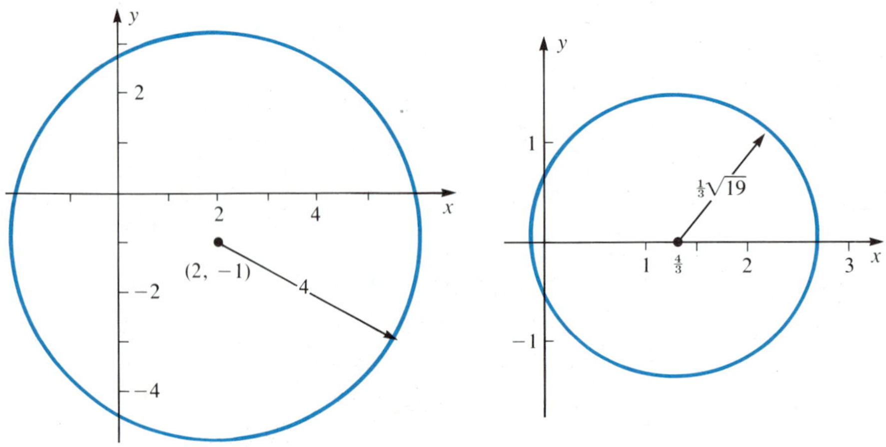

(a) $x^2 + y^2 - 4x + 2y = 11$ (b) $3x^2 + 3y^2 - 8x = 1$

Fig. 4

Inequalities

Relations of the form $f(x, y) = c$ generally lead to curves. The graphs of *inequalities* like

$$f(x, y) \geq c \quad \text{or} \quad f(x, y) < c$$

are generally *regions* bounded by curves.

EXAMPLE 4 Graph the relation $\frac{1}{2}x + \frac{1}{3}y \leq 1$.

Solution First graph the relation $\frac{1}{2}x + \frac{1}{3}y = 1$, a line (Fig. 5a). If you start at any point on this line and move to the *left*, then y remains the same but x *decreases*, hence $\frac{1}{2}x + \frac{1}{3}y < 1$ to the left of the line. Similarly, $\frac{1}{2}x + \frac{1}{3}y > 1$ to the *right* of the line. Therefore the region to the left of and on the line is the graph of $\frac{1}{2}x + \frac{1}{3}y \leq 1$. See Fig. 5b.

We can also graph systems of inequalities. Then we seek the common part (intersection) of the regions determined by the individual inequalities separately.

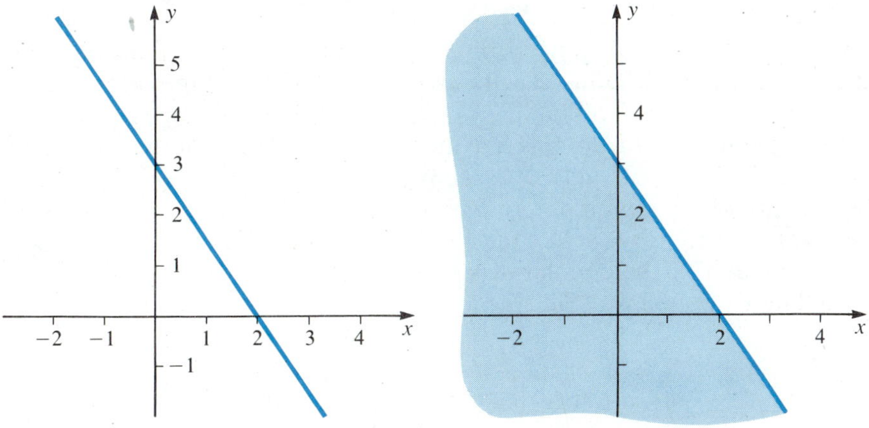

(a) $\frac{1}{2}x + \frac{1}{3}y = 1$ (b) $\frac{1}{2}x + \frac{1}{3}y \leq 1$

Fig. 5

EXAMPLE 5 Find all points (x, y) that satisfy simultaneously

$$y \geq 0 \qquad \tfrac{1}{2}x + \tfrac{1}{3}y \leq 1 \qquad -\tfrac{1}{3}x + \tfrac{1}{5}y \leq 1.$$

(That is, graph this system of inequalities.)

Solution Clearly $y \geq 0$ determines the half-plane above the x-axis, including the x-axis. The inequality $\frac{1}{2}x + \frac{1}{3}y \leq 1$ determines, as seen in Example 4, the region to the left of the line $\frac{1}{2}x + \frac{1}{3}y = 1$ including the line. Similarly, $-\frac{1}{3}x + \frac{1}{5}y \leq 1$ determines the region to the *right* of the line $-\frac{1}{3}x + \frac{1}{5}y = 1$ including the line [which, incidentally, has intercepts $(-3, 0)$ and $(0, 5)$].

The intersection of these three half-planes is shown in Fig. 6; it is a triangular region, including its boundary.

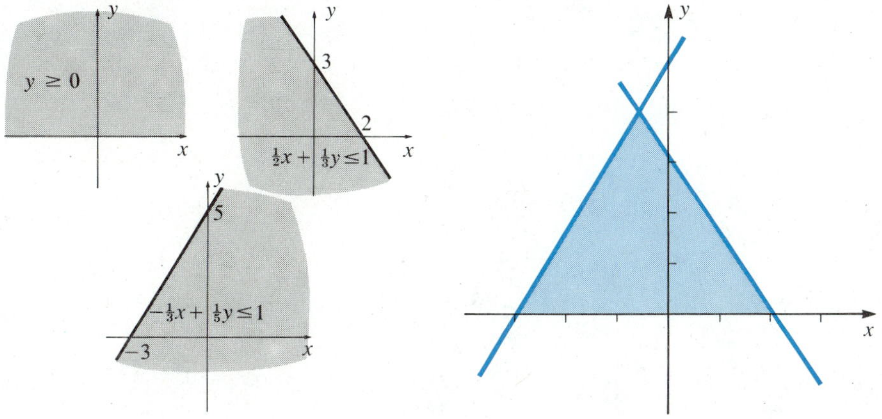

Fig. 6 $\begin{cases} y \geq 0 \\ \frac{1}{2}x + \frac{1}{3}y \leq 1 \\ -\frac{1}{3}x + \frac{1}{5}y \leq 1 \end{cases}$

EXAMPLE 6 Graph the system of inequalities

$$x^2 + y^2 \leq 25 \qquad x - y \geq 0.$$

Solution The inequality $x^2 + y^2 \le 25$ is equivalent to $\sqrt{x^2 + y^2} \le 5$, which determines all points whose distance from the origin is at most 5. This region is clearly the circular disk of radius 5 with center $(0, 0)$. See Fig. 7a.

The graph of $x - y = 0$, that is, $y = x$ is the line of slope 1 through $(0, 0)$. Moving to the *right* from this line increases x and does nothing to y, so $x - y$ increases, hence $x - y > 0$ to the right of $x - y = 0$. Similarly $x - y < 0$ to the left, so the region determined by $x - y \ge 0$ is the half-plane on and to the right of the line $y = x$. See Fig. 7b.

The region common to the circular disk and to the half-plane is a semi-circular region, including its boundary (Fig. 7c).

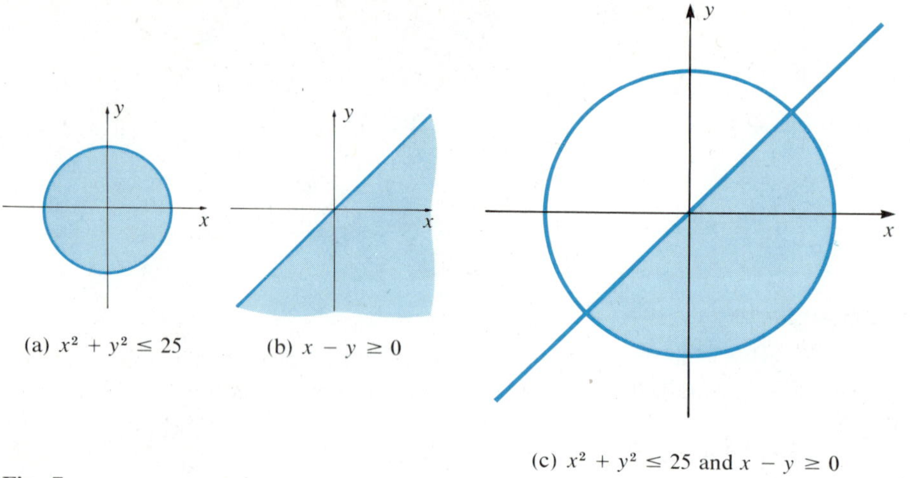

Fig. 7

Exercises

Graph

1 $3x + 2y = 6$ **2** $4x - y = -1$ **3** $x + 3y = -2$ **4** $-3x + 4y = 6$.

Solve the system graphically

5 $\begin{cases} 3x - y = 4 \\ 2x + 3y = -1 \end{cases}$ **6** $\begin{cases} 3x - 4y = 1 \\ x + y = -2 \end{cases}$ **7** $\begin{cases} 2x - y = 4 \\ x + 3y = -5 \end{cases}$

8 $\begin{cases} x + 2y = 4 \\ 2x - 5y = 0 \end{cases}$ **9** $\begin{cases} x + 4y = 1 \\ 3x + 2y = -2 \end{cases}$ **10** $\begin{cases} 2x + 3y = 1 \\ -4x + 3y = -8. \end{cases}$

Graph

11 $x^2 + y^2 - 4x - 4y = 0$ **12** $x^2 + y^2 - 6x = 0$

13 $x^2 + y^2 + 2x + 6y = -6$ **14** $x^2 + y^2 - x + y = \frac{5}{2}$

15 $x^2 + y^2 - x + 2y = 0$ **16** $x^2 + y^2 + 6x - 8y = 25$

17 $2x^2 + 2y^2 - 3x - 5y + \frac{9}{4} = 0$ **18** $3x^2 + 3y^2 - x - y = 0$.

19 Show that the graph of $x^2 + y^2 - 8x - 6y + 16 = 0$ is a circle tangent to the x-axis.

20 Show that the graph of $x^2 + y^2 - 2ax - 2ay + a^2 = 0$ is a circle tangent to both axes.

Graph the system of inequalities

21 $\begin{cases} x \le y \\ y \ge -1 \end{cases}$ **22** $\begin{cases} y \le x + 1 \\ y \ge x - 1 \end{cases}$ **23** $\begin{cases} y \le 2x \\ x \ge 1 \end{cases}$ **24** $\begin{cases} 2y \le x + 2 \\ 2y \ge -x + 2 \end{cases}$

25 $y \geq 0 \quad x + y \leq 1 \quad -x + y \leq 1$

26 $x \leq 0 \quad y \geq -1 \quad 2x + 3y \geq -6$

27 $x + y \geq 1 \quad y \leq x \quad y \geq 5x - 16$

28 $(x - 2)^2 + y^2 \leq 4 \quad y \geq -x \quad y \leq x$

29 $\begin{cases} x^2 + y^2 \geq 1 \\ (x - 2)^2 + y^2 \leq 4 \end{cases}$

30 $\begin{cases} x^2 + (y - 2)^2 \leq 9 \\ (x - 3)^2 + y^2 \leq 9 \end{cases}$

31 $\begin{cases} x^2 + y^2 \leq 9 \\ y \geq -1 \end{cases}$

32 $\begin{cases} x^2 + y^2 \leq 4 \\ x^2 + y^2 \geq 1. \end{cases}$

REVIEW EXERCISES

1 Plot the points $(2, 3)$, $(2, -3)$, $(-2, 3)$, and $(-2, -3)$. Show that they are the vertices of a rectangle.

2 Plot all points (x, y) in the plane for which
(a) $y \geq 0$ (b) $-1 \leq x \leq 1$ and y is an integer.

3 Given $f(x) = 3x + 1$
(a) compute $f(0)$, $f(-2)$, $f[f(x)]$
(b) show that $f(a + b) = f(a) + f(b) - 1$.

4 Find a function $f(x)$ that describes the diameter of a circle of area x.

Graph and show the domain and range

5 $f(x) = x - |x|$

6 $f(x) = \begin{cases} x & \text{if } x \text{ is an integer} \\ -x & \text{otherwise.} \end{cases}$

7 If $f(x) = 2x + 1$ and $g(x) = 3x - 2$, find a formula for $h(x) = 2f(x) - g(x) + f(x)g(x)$.

8 If $f(x) = \sqrt{x}$ and $g(x) = 3 - x$, compute
(a) $[f \circ g](x)$ (b) $[g \circ f](x)$.
In each case state the domain and range of the function.

9 Are the points $(0, 0)$, $(2, 5)$, $(3, 8)$ collinear or not? Explain.

10 Find a linear function $y = ax + b$ whose graph passes through $(0, 3)$ and $(1, 5)$.

11 Find a linear function $y = ax + b$ whose graph passes through $(0, 6)$ and is parallel to the graph of $y = -x$.

12 Find a linear function $y = ax + b$ whose graph is perpendicular to the line through $(3, 1)$ and $(-4, 2)$, and which passes through $(1, -1)$.

13 Graph and find the lowest point of $y = 2x^2 - 12x + 14$.

14 Plot on the same set of coordinate axes
(a) $y = (x - 2)^2$ (b) $y = -3(x - 2)^2$.

15 For what numbers b is the value of $x^2 + 2bx + 1$ positive regardless of the choice of x?

16 A farmer will make a rectangular pen inside his barn with 40 ft of chicken wire, using a corner of the barn for two sides of the pen. Find the maximum area he can enclose.

Graph

17 $x^2 + y^2 - 6x + 6y + 11 = 0$

18 $\begin{cases} x + y \geq 0 \\ y \leq 0 \\ x \leq 2. \end{cases}$

4 POLYNOMIAL AND RATIONAL FUNCTIONS

1 INTRODUCTION

In this chapter we shall study polynomial functions

$$f(x) = a_m x^m + a_{m-1}x^{m-1} + \cdots + a_2 x^2 + a_1 x + a_0$$

and rational functions

$$f(x) = \frac{a_m x^m + \cdots + a_1 x + a_0}{b_n x^n + \cdots + b_1 x + b_0}.$$

Such functions play a central role in mathematics for a number of reasons. (1) They arise naturally in many applications. (2) Their values can be computed using only the simplest operations of arithmetic: addition, subtraction, multiplication, division; hence numerical work with these functions is perfectly suited to high speed computers. (3) They can be used to approximate highly complicated functions, so they are indispensable in applied mathematics.

2 GRAPHS OF MONOMIALS

In this section we shall graph some special, but important, polynomials. We start with the cubic $y = x^3$. First we study its graph for $x \geq 0$. Since $0^3 = 0$, the point $(0, 0)$ is on the graph. As x increases, starting from 0, its cube x^3 also increases, so the graph always goes up. Let's tabulate some values of y for x from 0 to 1:

x	0.0	0.1	0.2	0.3	0.4	0.5	0.6	0.7	0.8	0.9	1.0
x^3	0.0	0.001	0.008	0.027	0.064	0.125	0.216	0.343	0.512	0.729	1.0

This data gives us a good idea of the graph for $0 \leq x \leq 1$. See Fig. 1a. Next, let's tabulate y for some larger values of x:

x	1	2	3	4	5	6	7	8	9	10
x^3	1	8	27	64	125	216	343	512	729	1000

The graph rises rapidly as x grows (Fig. 1b).

Finally we observe that $(-x)^3 = -x^3$, so at a negative value $-x$, the point $(-x, (-x)^3)$ is as far below the x-axis as the point (x, x^3) is above the x-axis. We now have enough information to complete the graph (Fig. 1c).

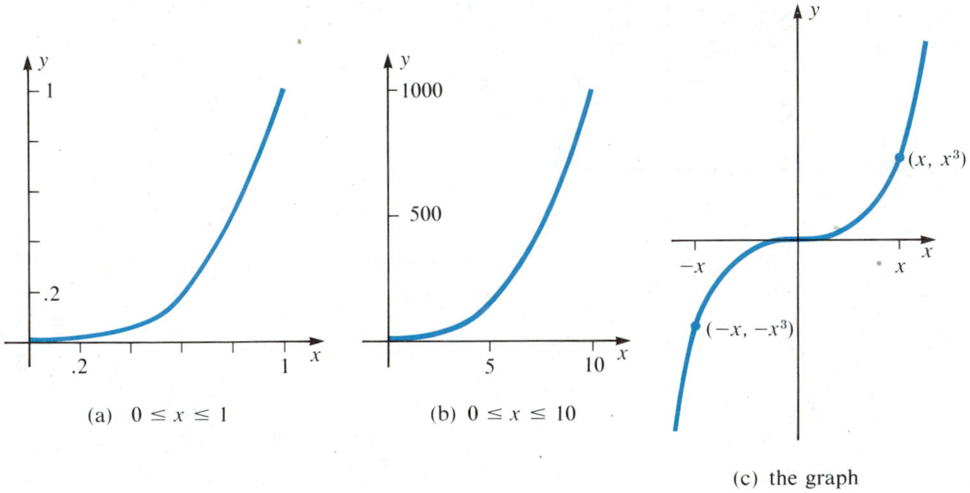

Fig. 1 $y = x^3$

We obtain the graph of $y = x^4$ similarly. If x is small, then y is very, very small: if x is large, then y is very, very large. Thus for $x \geq 0$, the general shape of $y = x^4$ is much like that of $y = x^3$ in Fig. 1b, except flatter near $x = 0$ and steeper for large x. But for x negative, the relation $(-x)^4 = x^4$ implies that $(-x, (-x)^4)$ and (x, x^4) are both the same distance *above* the x-axis. Therefore the graph of $y = x^4$ for $x < 0$ is the mirror image in the y-axis of the graph of $y = x^4$ for $x > 0$. See Fig. 2.

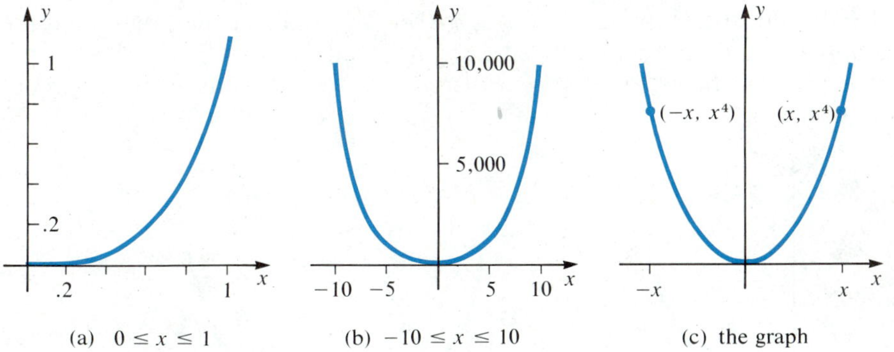

Fig. 2 $y = x^4$

It is interesting to compare the graphs of x^2, x^3, and x^4 for small x and for large x. When x is small, x^2 is very small, x^3 is even smaller, and x^4 is even smaller yet. For instance,

$$(0.1)^2 = 0.01 \qquad (0.1)^3 = 0.001 \qquad (0.1)^4 = 0.0001.$$

But when x is large, x^2 is very large, x^3 is even larger, and x^4 is even larger yet.

For instance,

$$10^2 = 100 \qquad 10^3 = 1000 \qquad 10^4 = 10{,}000.$$

We show the three graphs in Fig. 3.

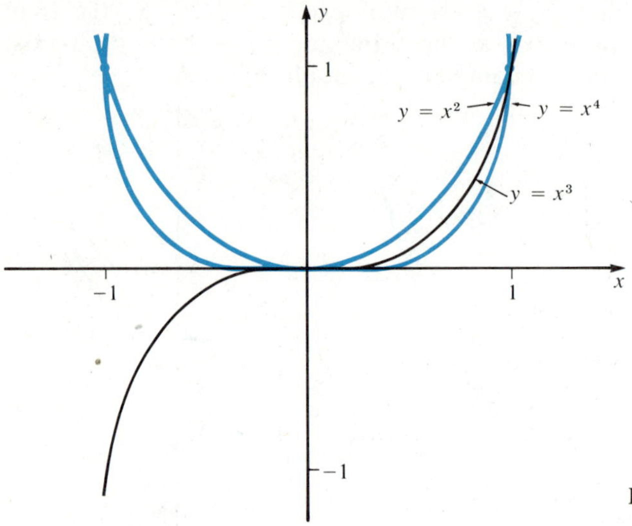

Fig. 3 Graphs of $\begin{cases} y = x^2 \\ y = x^3 \\ y = x^4 \end{cases}$

The graphs of

$$y = x^5 \qquad y = x^7 \qquad y = x^9 \quad \ldots$$

where the exponent is odd are more or less like the graph of $y = x^3$. They are increasingly flat near $x = 0$ and grow increasingly rapidly for x large. Similarly the graphs of

$$y = x^6 \qquad y = x^8 \qquad y = x^{10} \quad \ldots$$

are much like the graph of $y = x^4$ but even flatter still near $x = 0$. See Fig. 4 for examples. Note the important features:

(a) if n is odd, the graph of $y = x^n$ *crosses* the x-axis at the origin (and the x-axis is *tangent* to the graph provided $n \geq 3$), and

(b) if n is even, the graph of $y = x^n$ is *tangent* to the x-axis at the origin but *does not cross* it.

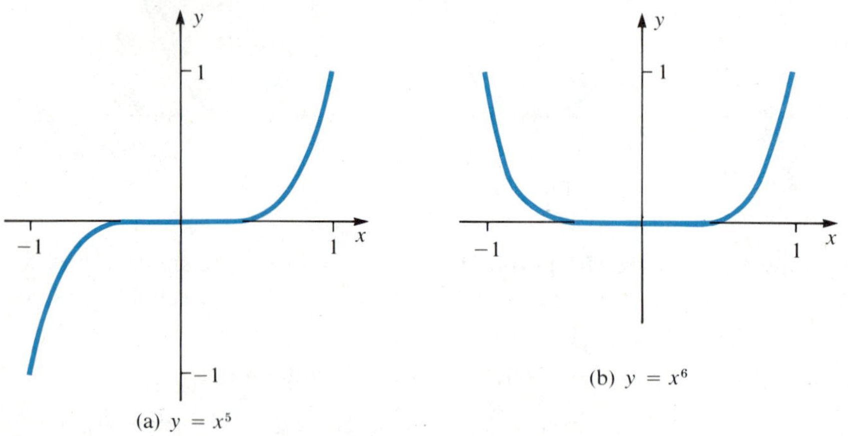

(a) $y = x^5$ (b) $y = x^6$

Fig. 4 Typical graphs of odd and even powers of x

The graph of $y = c(x - r)^n$ is similar to that of $y = x^n$, except that it touches the x-axis at $x = r$ instead of $x = 0$, and it is stretched by the factor $|c|$ in the y-direction (and reflected in the x-axis if $c < 0$). See Fig. 5 for examples.

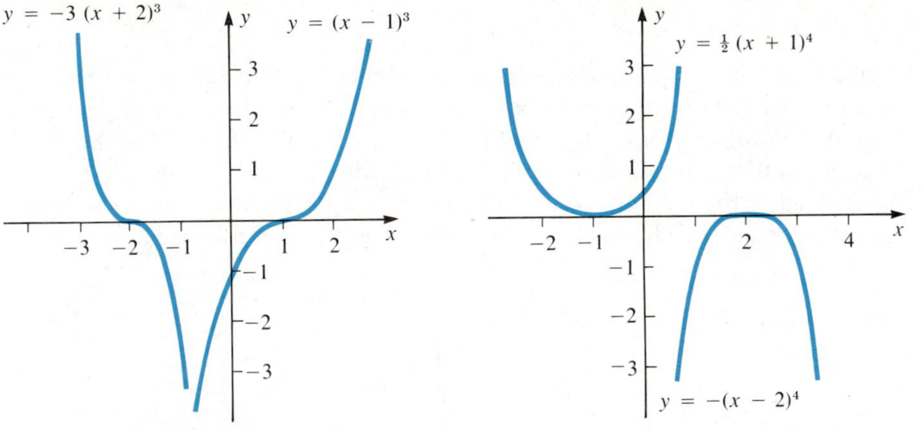

Fig. 5 Examples of $y = c(x - r)^n$

Notation

When x grows larger and larger, we write

$$x \longrightarrow \infty$$

and read "x approaches infinity." This is just a convenient shorthand for saying that x increases without bound, eventually exceeding any preassigned real number. (The notation does *not* imply that there is a "largest" number, infinity.)

The graph of $y = x^n$ rises without bound as x increases without bound. We can abbreviate this behavior by writing

$$x^n \longrightarrow \infty \quad \text{as} \quad x \longrightarrow \infty.$$

If n is even, the graph of $y = x^n$ also rises without bound as x becomes more and more negative without bound. We can write

$$x^n \longrightarrow \infty \quad \text{as} \quad x \longrightarrow -\infty.$$

Finally, if n is odd, the graph of $y = x^n$ falls without bound as x becomes more and more negative. Briefly

$$x^n \longrightarrow -\infty \quad \text{as} \quad x \longrightarrow -\infty.$$

Exercises

Graph roughly to show the general shape

1 $y = \frac{1}{4}x^4$ **2** $y = -x^4$ **3** $y = \frac{1}{8}x^3$
4 $y = -x^3 + 1$ **5** $y = -x^6$ **6** $y = x^3 - 2$
7 $y = x^4 - 1$ **8** $y = x^5 + 1$ **9** $y = -x^5$
10 $y = \frac{1}{16}x^5$ **11** $y = (x + 1)^3$ **12** $y = (x - 1)^4$
13 $y = -(x + 1)^3$ **14** $y = -(x - 1)^4$ **15** $y = (x - 1)^3 + 1$
16 $y = -(x + 1)^4 - 1$ **17** $y = \frac{1}{3}(x + 2)^4$ **18** $y = -\frac{1}{2}(x + 2)^4 - 4$
19 $y = (x - \frac{1}{2})^5 - 1$ **20** $y = (x + \frac{1}{2})^6 - 2$.

Graph accurately; use coordinate paper and plot points computed on your calculator

21 $y = (x - 2)^3 \quad 1 \le x \le 3$ **22** $y = -(x + 1)^3 \quad -2 \le x \le 0$
23 $y = -(x - 1)^4 \quad 0 \le x \le 2$ **24** $y = \frac{1}{2}(x - 2)^4 \quad 1 \le x \le 3$.

3 GRAPHS OF POLYNOMIALS

Graphing the most general polynomial function

$$y = f(x) = a_n x^n + a_{n-1}x^{n-1} + \cdots + a_1 x + a_0 \qquad (a_n \neq 0)$$

is a tough job. In calculus courses tools are developed that help a lot, but here we shall have to work pretty much with our bare hands. About the most we can do is plot a bunch of points and hope for the best. In calculus it is shown that the graph of $f(x)$ changes direction at most $n - 1$ times, which is some help. This means for instance that if Fig. 1a represents the graph of a polynomial, then its degree must be seven or more. The graph in Fig. 1b is typical of polynomial functions.

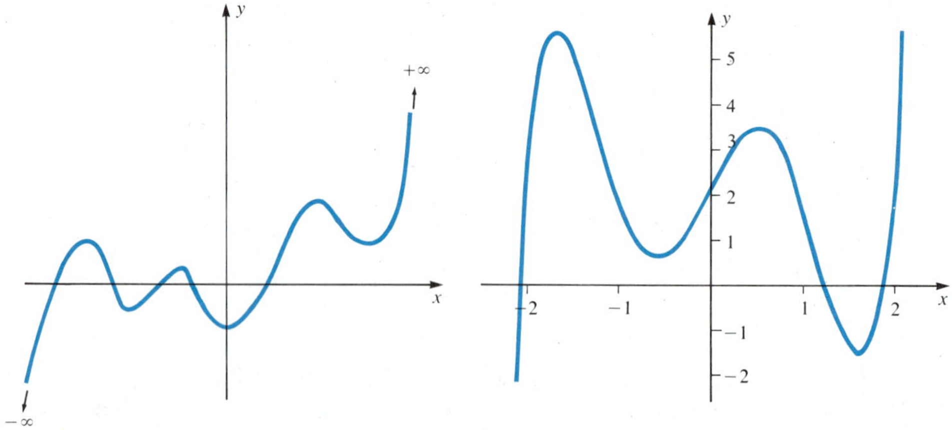

(a) This graph changes direction 6 times, so if it is the graph of a polynomial $f(x)$, then $\deg f(x) \geq 7$.

(b) $y = x^5 - 5x^3 + 4x + 2$

Fig. 1 Graphs of polynomials

We can say something about the behavior of a polynomial graph as $x \longrightarrow \infty$ or $x \longrightarrow -\infty$. Let

$$f(x) = a_n x^n + a_{n-1}x^{n-1} + \cdots + a_1 x + a_0 \qquad (a_n \neq 0).$$

For $|x|$ very large, we expect the term $a_n x^n$ to dominate all other terms. To confirm this, we write

$$f(x) = a_n x^n \left(1 + \frac{a_{n-1}}{a_n x} + \cdots + \frac{a_1}{a_n x^{n-1}} + \frac{a_0}{a_n x^n} \right).$$

For $|x|$ very large, each ratio $a_{n-i}/a_n x^i$ is near 0. Therefore the quantity in parentheses is near 1, and the graph behaves like $y = a_n x^n$. As $x \longrightarrow \infty$ or $-\infty$, it zooms up or down, depending on the sign of a_n and whether n is even or odd. See Fig. 1a again.

Cubics

We know how to graph $y = x^3$ and some modifications of $y = x^3$ like

$$y = a(x - r)^3 \qquad y = a(x - r)^3 + c$$

involving horizontal shift, stretching, reflection in the x-axis, and vertical shift. Unfortunately this does not help us graph the general cubic

$$y = ax^3 + bx^2 + cx + d \qquad (a \neq 0).$$

It turns out, however, that there are only a few possible shapes for the graph of a general cubic, and they are shown in Fig. 2. These shapes are derived from the following properties of $y = ax^3 + bx^2 + cx + d$:

(1) If $a > 0$, then $y \longrightarrow \infty$ as $x \longrightarrow \infty$, and $y \longrightarrow -\infty$ as $x \longrightarrow -\infty$.
(2) If $a < 0$, then $y \longrightarrow -\infty$ as $x \longrightarrow \infty$, and $y \longrightarrow \infty$ as $x \longrightarrow -\infty$.
(3) The graph intersects each horizontal line at least once and at most three times.
(4) The graph either rises steadily, falls steadily, or has one peak and one pit.

We have already discussed (1) and (2); they hold for any polynomial of odd degree, not just cubics. Property (3) will follow from our discussion of the zeros of polynomials in Chapter 9, but Property (4) is beyond our scope, best proved by calculus.

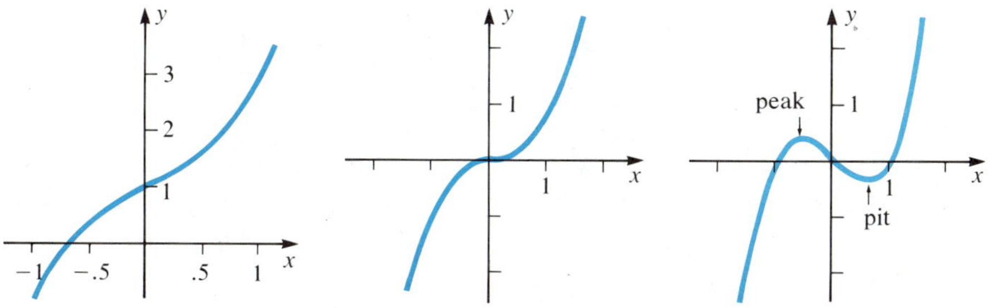

(a) $y = x^3 + x + 1$: rises steadily; no horizontal tangent

(b) $y = x^3$: rises steadily; horizontal tangent at (0, 0)

(c) $y = x^3 - x$

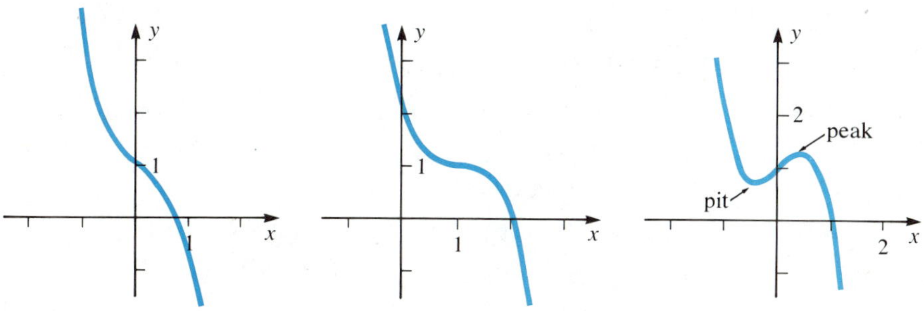

(a′) $y = -x^3 - x + 1$: falls steadily; no horizontal tangent

(b′) $y = -x^3 + 3x^2 - 3x + 2$: falls steadily; horizontal tangent at (1, 1)

(c′) $y = -2x^3 + x + 1$

Fig. 2

When you must graph a cubic, it is best to plot enough points so you can fit a smooth curve and be confident that your curve fits one of the shapes in Fig. 2.

EXAMPLE 1 Graph $y = -x^3 - 3x^2 - x + 2$.

Solution We shall tabulate y for several values of x. A good starting point is $x = 0$. Then we go forwards and backwards until the y's become pretty large (off the map):

x	-4	-3	-2	-1	0	1	2
y	22	5	0	1	2	-3	-20

These points are plotted next, in Fig. 3a. We know that

$$y \longrightarrow \infty \quad \text{as} \quad x \longrightarrow -\infty \qquad \text{and} \qquad y \longrightarrow -\infty \quad \text{as} \quad x \longrightarrow \infty$$

(the coefficient of x^3 is negative), and we know that a cubic has at most one peak and one pit. Therefore the shape of the graph is apparent (see Fig. 2c). For greater accuracy we might also tabulate at the halves:

x	-3.5	-2.5	-1.5	-0.5	0.5	1.5
y	11.6	1.4	0.1	1.9	0.6	-9.6

and plot the corresponding points (Fig. 3b). Finally we join all of these plotted points by a smooth curve (Fig. 3c).

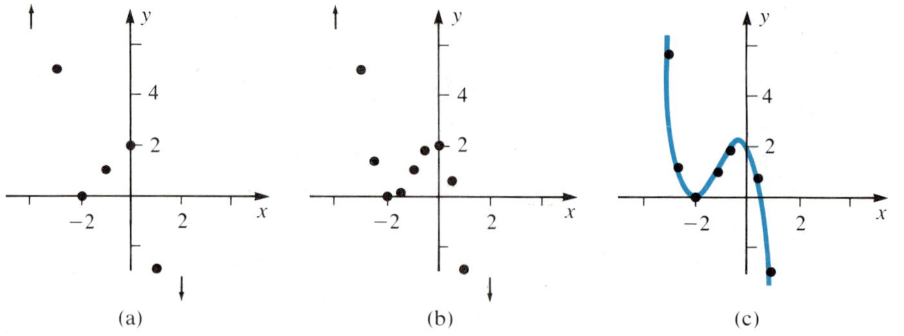

Fig. 3 $y = -x^3 - 3x^2 - x + 2$

Factored Cubics

Cubics of the form

$$y = a(x - r_1)(x - r_2)(x - r_3)$$

are particularly easy to graph because:

(1) The sign of a determines whether $y \longrightarrow \infty$ or $y \longrightarrow -\infty$ as $x \longrightarrow \infty$ or $-\infty$.

(2) $y = 0$ only for $x = r_1$, $x = r_2$, and $x = r_3$.

(3) Say $r_1 \le r_2 \le r_3$. Then y has constant sign on each of the intervals

$$-\infty < x < r_1 \qquad r_1 < x < r_2 \qquad r_2 < x < r_3 \qquad r_3 < x < \infty.$$

This last property holds because a factor $x - r_i$ *changes sign* only when x passes across r_i.

EXAMPLE 2 Graph $y = 2x(x + 1)(x - 2)$.

Solution The leading coefficient 2 is positive, hence

$$y \longrightarrow \infty \quad \text{as} \quad x \longrightarrow \infty \qquad \text{and} \qquad y \longrightarrow -\infty \quad \text{as} \quad x \longrightarrow -\infty.$$

Clearly $y = 0$ only if $x = -1$, 0, and 2. By (3) above, y has constant sign on each of the intervals

$$-\infty < x < -1 \qquad -1 < x < 0 \qquad 0 < x < 2 \qquad 2 < x < \infty.$$

Since $y \longrightarrow -\infty$ as $x \longrightarrow -\infty$, we have $y < 0$ for $-\infty < x < -1$. As x crosses -1 from left to right, only the sign of the factor $x + 1$ changes, so $y > 0$ for $-1 < x < 0$. Thus the signs alternate $- + - +$. We summarize:

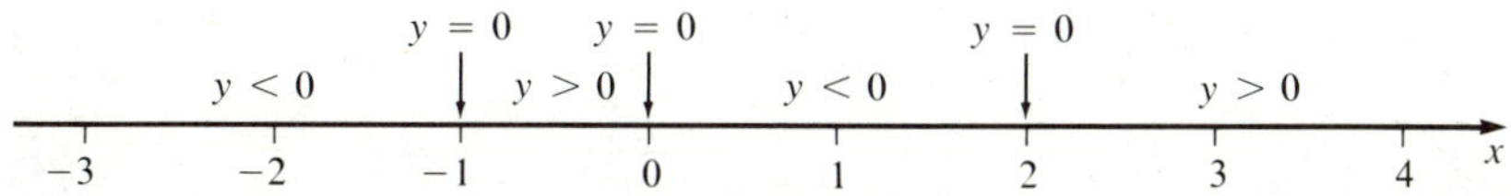

This is enough information for a crude plot; the shape must be that of Fig. 2c. For more accuracy we tabulate some inbetween values and a few beyond:

x	-1.5	-1.0	-0.5	0.0	0.5	1.0	1.5	2.0	2.5
y	-5.2	0	1.2	0	-2.2	-4	-3.8	0	8.8

These points are plotted and joined by a smooth curve in Fig. 4.

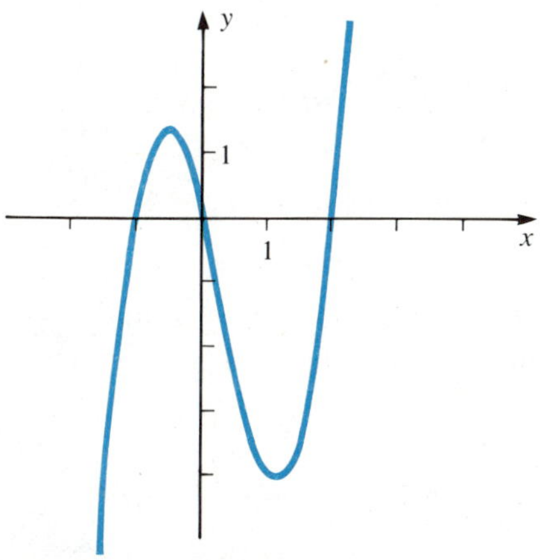

Fig. 4 $y = 2x(x+1)(x+2)$

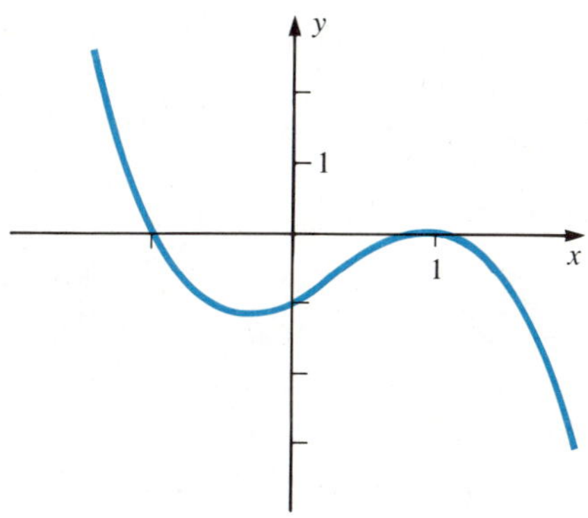

Fig. 5 $y = -(x+1)(x-1)^2$

EXAMPLE 3 Graph $y = -(x+1)(x-1)^2$.

Solution Expanded, $y = -x^3 + \cdots$. The leading coefficient -1 is negative. Therefore

$$y \longrightarrow \infty \quad \text{as} \quad x \longrightarrow -\infty \qquad \text{and} \qquad y \longrightarrow -\infty \quad \text{as} \quad x \longrightarrow \infty.$$

Clearly $y = 0$ only if $x = -1$ or $x = 1$. Therefore y has constant sign on each of the intervals

$$-\infty < x < -1 \qquad -1 < x < 1 \qquad 1 < x < \infty.$$

Clearly $y > 0$ for $-\infty < x < -1$ since $y \longrightarrow \infty$ as $x \longrightarrow -\infty$. As x crosses -1 from left to right, only the sign of the factor $x + 1$ changes, so $y < 0$ for $-1 < x < 1$. As x crosses $+1$ from left to right, the factor $(x-1)^2$ *does not* change sign (because it is a square) so $y < 0$ for $1 < x < \infty$ also. We summarize:

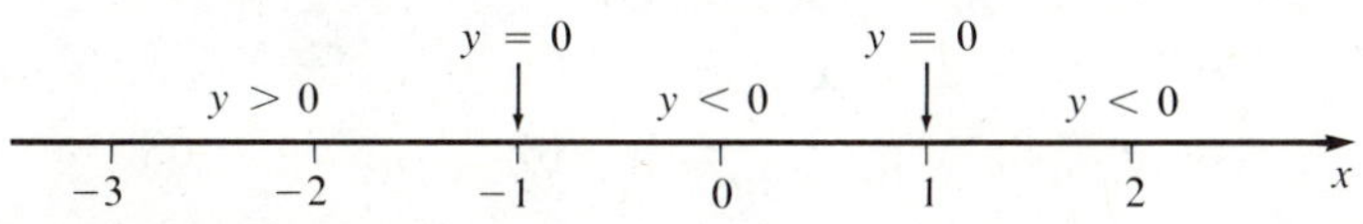

The sign of y changes at $x = -1$, so the graph crosses the x-axis there. The sign of y does not change at $x = 1$, so the graph touches the x-axis, but does not cross it. Thus the graph is *tangent* to the x-axis at $x = 1$. We have enough information for a rough graph (Fig. 5). For greater accuracy, of course, we would plot more points.

Completely Factored Polynomials

The last example is typical of graphing a polynomial that is completely factored into linear factors. In general, it is fairly easy to graph

$$y = a(x - r_1)(x - r_2) \cdots (x - r_n) \qquad (a \neq 0).$$

The basic facts are similar to those for cubics:

(1) The leading term of y is ax^n, so whether $y \longrightarrow \infty$ or $y \longrightarrow -\infty$ as $x \longrightarrow \infty$ or $-\infty$ is determined by the sign of a and the parity of n (whether n is odd or even). For instance, if $a < 0$ and n is even, then $y \longrightarrow -\infty$ as $x \longrightarrow \infty$ and as $x \longrightarrow -\infty$.

(2) $y = 0$ only for $x = r_1$, $x = r_2$, $\cdots$, and $x = r_n$.

(3) Say $r_1 \leq r_2 \leq \cdots \leq r_n$. Then y has constant sign on each of the intervals

$$-\infty < x < r_1 \qquad r_1 < x < r_2 \cdots r_{n-1} < x < r_n \qquad r_n < x < \infty.$$

(Some of these intervals will be empty—whenever $r_i = r_{i+1}$.)

(4) The sign of y changes as x passes through an r_i if the factor $x - r_i$ occurs an odd number of times; the sign does not change if $x - r_i$ occurs an even number of times.

As an illustration of (4), consider

$$y = (x - 1)(x - 2)(x - 3)^5(x - 4)^2.$$

As x increases from slightly less than 3 to slightly more than 3, the factor $(x - 3)^5$ changes sign—because the exponent 5 is odd: $(\text{neg.})^5 = (\text{neg.})$, $(\text{pos.})^5 = (\text{pos.})$. Therefore y itself also changes sign, since the other factor $(x - 1)(x - 2)(x - 4)^2$ has constant sign close to $x = 3$. On the other hand, as x increases through 4, the factor $(x - 4)^2$ *doesn't* change sign; it is always positive (except 0 at $x = 4$). Therefore y also doesn't change sign near $x = 4$.

These considerations allow us to graph any completely factored polynomial.

EXAMPLE 4 Graph

$$y = 4(x - 1)^3(x - 2)^2(x - 3).$$

Solution Clearly $y = 4x^6 + \cdots$. The coefficient 4 of the leading term is positive and its degree 6 is even. Therefore

$$y \longrightarrow \infty \quad \text{as} \quad x \longrightarrow -\infty \qquad \text{and} \qquad y \longrightarrow \infty \quad \text{as} \quad x \longrightarrow \infty.$$

Next, $y = 0$ only if $x = 1$, $x = 2$, or $x = 3$. Therefore y has constant sign on each of the intervals

$$-\infty < x < 1 \qquad 1 < x < 2 \qquad 2 < x < 3 \qquad 3 < x < \infty.$$

Now let us analyze the behavior of y near $x = 1$. We write

$$y = [4(x - 2)^2(x - 3)](x - 1)^3$$

The factor in [ ]'s is not zero at $x = 1$. For x very close to 1 it stays near its value at $x = 1$; this value is

$$4(x - 2)^2(x - 3)\Big|_{x=1} = 4\cdot(-1)^2(-2) = -8.$$

Therefore

$$y \approx -8(x-1)^3 \qquad \text{for} \quad x \approx 1.$$

In other words, the graph we seek is approximated closely near $x = 1$ by the graph of $y = -8(x-1)^3$. Similarly, near $x = 2$,

$$y = [4(x-1)^3(x-3)](x-2)^2 \approx -4(x-2)^2$$

because

$$4(x-1)^3(x-3)\Big|_{x=2} = -4$$

and finally, in the same way,

$$y \approx 32(x-3) \qquad \text{for} \quad x \approx 3.$$

We indicate all of this information in Fig. 6a. Then we fill in the rest by smooth curves to finish a rough graph (Fig. 6b).

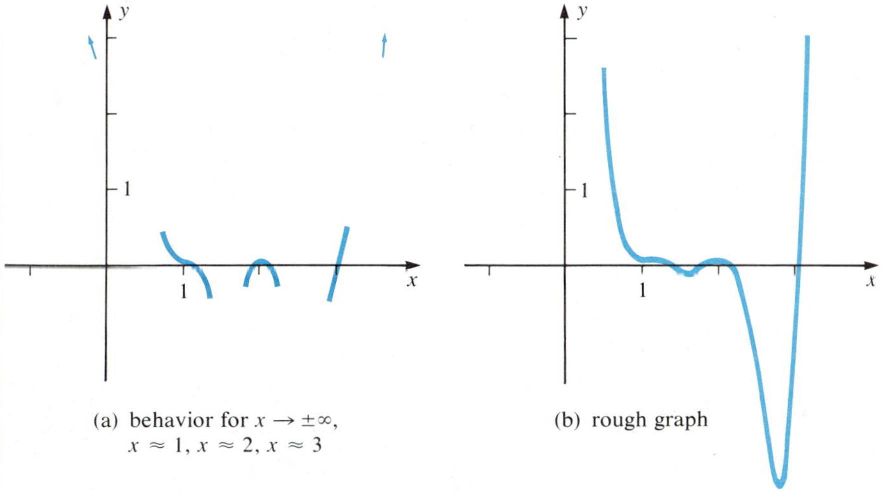

(a) behavior for $x \to \pm\infty$, $x \approx 1$, $x \approx 2$, $x \approx 3$

(b) rough graph

Fig. 6 $y = 4(x-1)^3(x-2)^2(x-3)$

Exercises

Graph roughly; plot for the given values of x

1 $y = x^3 - 4x \quad x = 0 \ \pm 1 \ \pm 2 \ \pm 3$

2 $y = x^3 + 2x^2 - x - 2 \quad x = -3 \ -2 \ -1 \ 0 \ 1 \ 2$

3 $y = \frac{1}{3}x^3 - \frac{1}{2}x^2 - 6x + 1 \quad x = -3 \ -2 \ -1 \ 0 \ 1 \ 2 \ 3 \ 4$

4 $y = x^3 - 3x + 3 \quad x = -3 \ -2 \ -1 \ 0 \ 1 \ 2$

5 $y = -x^3 + 3x \quad x = 0 \ \pm 1 \ \pm 2 \ \pm 3$

6 $y = 2x^3 - 3x^2 - 1 \quad x = 0 \ \pm 1 \ \pm 2$

7 $y = x^4 - 3x^2 \quad x = 0 \ \pm 1 \ \pm 2$

8 $y = 3x^4 + 2x^3 - 3x^2 + 2 \quad x = 0 \ \pm\frac{1}{2} \ \pm 1 \ \pm\frac{3}{2} \ -2.$

Draw a line graph of the signs and sign changes; then sketch the graph roughly

9 $y = x(x-1)(x-2)$

10 $y = (x+2)x(x-2)$

11 $y = (x+2)(x+1)(x-1)$

12 $y = (x+2)(x-1)(x-2)$

13 $y = x^2(x-1)$

14 $y = (x-1)(x-2)^2$

15 $y = -(x+1)(x-1)^2$

16 $y = -x^2(x-1)$

17 $y = \frac{1}{6}(x-1)(x-2)(x-3)(x-4)$

18 $y = \frac{1}{24}x(x-2)(x-3)(x-4)$

19 $y = x^2(x^2-1)$

20 $y = \frac{1}{4}(x-1)^2(x^2-4)$

21 $y = -\frac{1}{4}(x+1)^2(x^2-4)$

22 $y = -\frac{1}{12}x^2(x-2)(x-3)$

23 $y = x(x-1)^3$

24 $y = \frac{1}{8}(x^2-1)^2.$

25 Suppose a polynomial graph $y = f(x)$ passes through $(0, 0)$ $(2, 0)$ $(4, 0)$ $(6, 0)$ and $(1, 5)$ $(3, 1)$ $(5, 5)$. Find the least possible degree of $f(x)$, and sketch such a polynomial graph.

26 Suppose a polynomial graph $y = f(x)$ is tangent to the x-axis at $(0, 0)$ and at $(1, 0)$, and $f \neq 0$. Find the least possible degree of $f(x)$, and sketch such a polynomial graph. [*Hint* **Tangent,** for instance at $(0, 0)$ to the x-axis, means x^2 is a factor of $f(x)$, or maybe even a higher power of x is.]

4 TIPS ON GRAPHING

This section is a guide to graphs, featuring techniques that can reduce the work in graphing.

Symmetry

We have used symmetry in several places in this and the previous chapter. Now we shall examine some of its aspects more systematically.

Consider the graphs of $y = x^2 - 2$ and $y = x^3$ shown in Fig. 1. The one on the left is symmetric in the y-axis. The one on the right is symmetric in the origin; that is, to each point of the graph corresponds an opposite point as seen through a peephole in the origin. In either case we need plot the curve only for $x \geq 0$; we obtain the rest by symmetry. Thus the work is "cut in half."

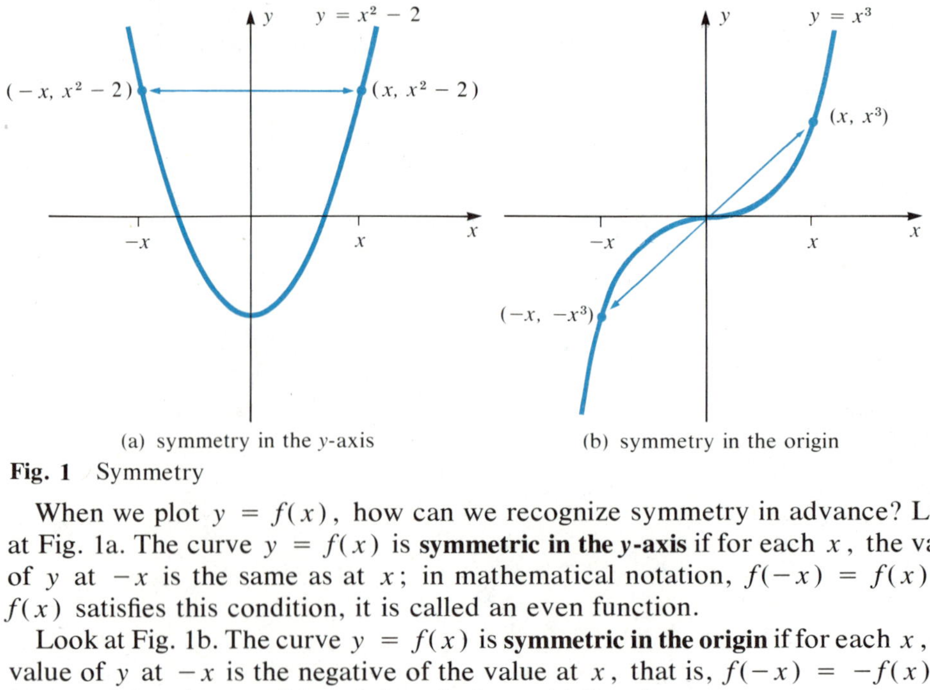

(a) symmetry in the y-axis (b) symmetry in the origin

Fig. 1 Symmetry

When we plot $y = f(x)$, how can we recognize symmetry in advance? Look at Fig. 1a. The curve $y = f(x)$ is **symmetric in the y-axis** if for each x, the value of y at $-x$ is the same as at x; in mathematical notation, $f(-x) = f(x)$. If $f(x)$ satisfies this condition, it is called an even function.

Look at Fig. 1b. The curve $y = f(x)$ is **symmetric in the origin** if for each x, the value of y at $-x$ is the negative of the value at x, that is, $f(-x) = -f(x)$. If $f(x)$ satisfies this condition, it is called an odd function.

> An **even** function $f(x)$ is one for which $f(-x) = f(x)$.
> The graph of an even function is symmetric in the y-axis.
>
> An **odd** function $f(x)$ is one for which $f(-x) = -f(x)$.
> The graph of an odd function is symmetric in the origin.

It can be shown that a polynomial is an *even function* if and only if it involves only *even powers* of the variable. Similarly, a polynomial is an *odd function* if and only if it involves only *odd powers* of the variable.

Examples (1) $f(x) = \dfrac{x^4}{1 + x^2}$

$$f(-x) = \frac{(-x)^4}{1 + (-x)^2} = \frac{x^4}{1 + x^2} = f(x); \text{ even function.}$$

(2) $f(x) = \dfrac{x^3}{1 + x^4}$

$$f(-x) = \frac{(-x)^3}{1 + (-x)^4} = \frac{-x^3}{1 + x^4} = -f(x); \text{ odd function.}$$

Vertical and Horizontal Shifts

We know that adding or subtracting a positive constant c to $f(x)$ shifts the graph of $y = f(x)$ up or down c units. Now let us consider horizontal shifts. How can we shift the graph of $y = f(x)$ three units to the right? More precisely, how can we find a function $g(x)$ whose graph is that of $y = f(x)$ shifted three units to the right?

Consider Fig. 2. For each point (x, y) on the curve $y = g(x)$, there corresponds a point $(x - 3, y)$ on the curve $y = f(x)$. The values of y are the same. But on the first curve $y = g(x)$, on the second, $y = f(x - 3)$. Conclusion: $g(x) = f(x - 3)$. This makes sense. If x represents time, then the value of g "now" is the same as the value of f three seconds ago.

The same reasoning shows the graph of $y = f(x + 3)$ is the graph of $y = f(x)$ shifted three units to the left.

Let $c > 0$. The graph of

$$\left.\begin{aligned} y &= f(x) + c \\ y &= f(x) - c \\ y &= f(x - c) \\ y &= f(x + c) \end{aligned}\right\} \text{ is the graph of } y = f(x) \text{ shifted } c \text{ units } \left\{\begin{array}{l} \text{upward} \\ \text{downward} \\ \text{to the right} \\ \text{to the left.} \end{array}\right.$$

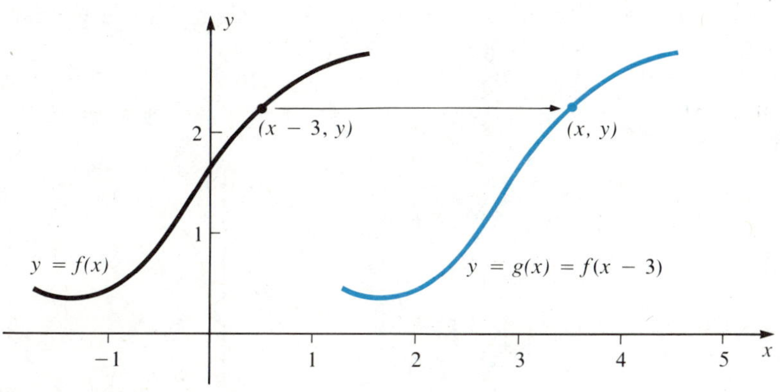

Fig. 2 Horizontal shift

Stretching and Reflecting

If $c > 0$, the graph of $y = cf(x)$ is obtained from that of $y = f(x)$ by stretching by a factor of c in the y-direction. Each point (x, y) is replaced by

(x, cy). Note: "stretching" by a factor less than one is interpreted as shrinking (Fig. 3).

The graph of $y = -f(x)$ is obtained by reflecting the graph of $y = f(x)$ in the x-axis (turning it upside down). That is because each point (x, y) is replaced by the point $(x, -y)$. See Fig. 4.

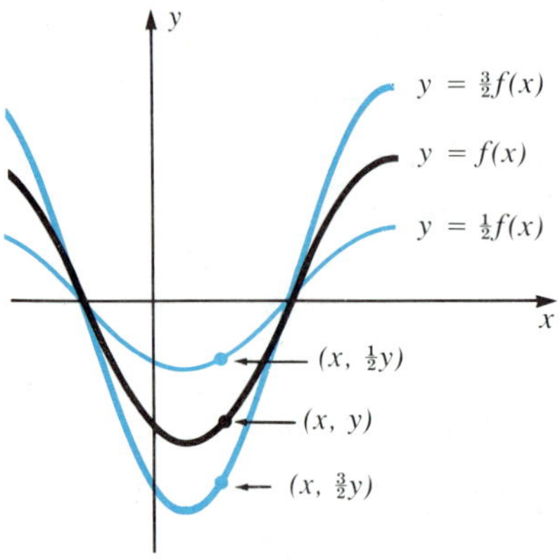

Fig. 3 Stretching in the y-direction

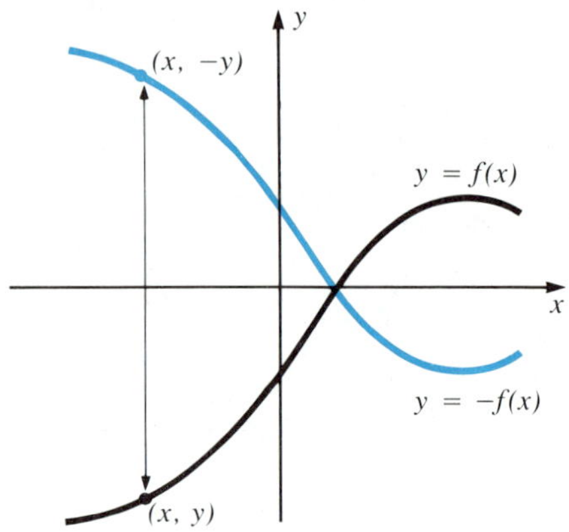

Fig. 4 Reflection in the x-axis

The next example illustrates these techniques.

EXAMPLE 1 Graph

$$y = 2[(x - 1)^3 - (x - 1)].$$

Solution This function resembles

$$f(x) = x^3 - x.$$

Indeed, $2[(x - 1)^3 - (x - 1)] = 2f(x - 1)$, so we must plot $y = 2f(x - 1)$. We'll do this in three steps by plotting

$$y = f(x) \qquad y = 2f(x) \qquad y = 2f(x - 1).$$

Since

$$f(-x) = (-x)^3 - (-x) = -x^3 + x = -(x^3 - x) = -f(x),$$

the function $f(x)$ is odd, so its graph is symmetric in the origin. Hence we can graph $y = f(x)$ for $x \geq 0$, then obtain the rest by symmetry.

Clearly $f(0) = 0$ and $f(1) = 0$. We see also that $f(x) < 0$ for $0 < x < 1$, that $f(x) > 0$ for $x > 1$, and that $f(x) \longrightarrow \infty$ as $x \longrightarrow \infty$. This information gives us Fig. 5a.

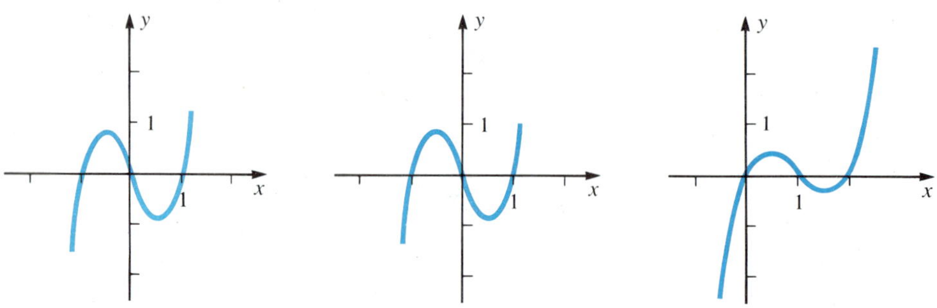

(a) $y = f(x) = x^3 - x$; symmetric in $(0, 0)$

(b) $y = 2f(x) = 2(x^3 - x)$

(c) $y = 2f(x - 1)$
$= 2[(x - 1)^3 - (x - 1)]$

Fig. 5

Next we stretch by the factor 2 to obtain Fig. 5b. Finally we shift one unit to the right for the final graph, Fig. 5c.

Afterthought $f(x) = x(x^2 - 1) = (x + 1)x(x - 1)$, so

$$2f(x - 1) = 2x(x - 1)(x - 2).$$

The methods of the previous section apply!

Free Information

Very often you can get valuable information about a graph for free, just by looking at the equation involved. It's good practice not to start right off plotting points, but to take a minute to think. You should look for symmetry, shifts, behavior as $x \longrightarrow \pm\infty$, and stretching and reflecting. Here are some other things to look for.

(1) *Domain* Is y defined for all real x or is there some restriction on x? For example $y = \sqrt{1 - x^2}$ is defined only for $|x| \le 1$, and $y = 1/(x - 1)(x - 4)$ is not defined for $x = 1$ or $x = 4$.

(2) *Extent* Is there some limitation on y? For example, if $y = 1/(1 + x^2)$, then $0 < y \le 1$ since $1/y = 1 + x^2 \ge 1$. The graph does not extend above the level $y = 1$ or below the level $y = 0$.

(3) *Intercepts* Where does the graph intersect the y-axis $(x = 0)$, and where does it intersect the x-axis $(y = 0)$? For example, the graph of $y = (x^2 - 4)/(x^2 + 5)$ intersects the x-axis at $x = \pm 2$ and the y-axis at $y = -\frac{4}{5}$.

(4) *Sign of y* Can you tell where $y > 0$ or $y < 0$? For example $y = 1/x$ is positive for $x > 0$ and negative for $x < 0$. Also y is never 0.

(5) *Increasing or decreasing?* For example, $y = 1/x$ decreases as x increases through positive values.

You will not always be able to check all these points. At least try to see what you can find easily.

EXAMPLE 2 Graph $y = \dfrac{1}{1 + x^4}$.

Solution The graph is defined for all x. Since $1 + x^4 \ge 1$, we have $0 < y \le 1$. In fact $y = 1$ only at $x = 0$, so the highest point of the curve is $(0, 1)$. As x increases through positive values, y decreases towards 0. (So there is no x-intercept.) The curve is symmetric in the y-axis since the function is even. This free information is enough for a fairly good idea of the curve. Plotting a few points helps fix the shape (Fig. 6).

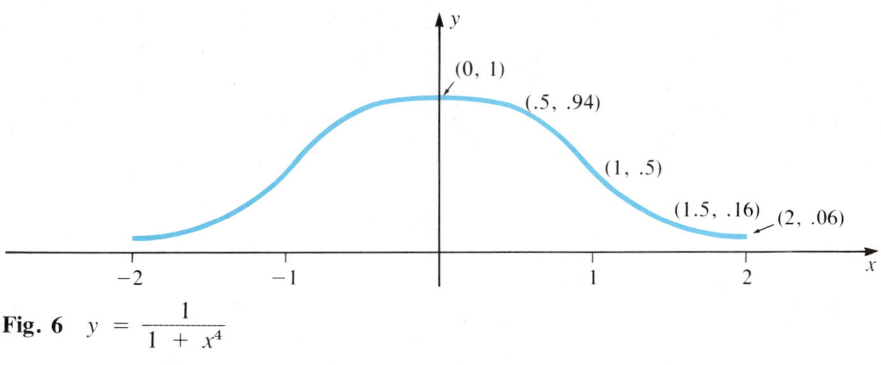

Fig. 6 $y = \dfrac{1}{1 + x^4}$

Exercises

State whether the function is even, odd, neither

1 x^5 **2** x^4 **3** $x|x|$

4 $\dfrac{x^3}{1 + |x|}$ **5** $\dfrac{x^3}{x^2 + 1}$ **6** $\dfrac{x^2}{x^3 + 1}$

7 $x(x^2 + 1)$ **8** $(1 - x)^2 - (1 + x)^2$ **9** $x^3 + x^2 + 1$

10 $(x^3 + 1)(x^3 - 2)$ **11** $\dfrac{x^4 + 1}{x^2 + 1}$ **12** $\sqrt{1 - x^2}$.

Compute $g(x) = \frac{1}{2}[f(x) + f(-x)]$ and show that it is even, where $f(x) =$

13 $3x^3 - 5x^2 + x - 2$ **14** $\dfrac{2x - 1}{x + 3}$ **15** $\dfrac{x^2}{x - 4}$ **16** $\sqrt{1 + x}$.

Compute $h(x) = \frac{1}{2}[f(x) - f(-x)]$ and show that it is odd, where $f(x) =$

17 $x^4 - 5x^3 + 6x^2 - 7x + 8$ **18** $(x^2 + 2)(x^3 - 1)$

19 $\sqrt{1 - x}$ **20** $\dfrac{x^2 - 1}{2x + 1}$.

21 Prove that for any function $f(x)$, the function $g(x) = \frac{1}{2}[f(x) + f(-x)]$ is even.

22 Prove that for any function $f(x)$, the function $h(x) = \frac{1}{2}[f(x) - f(-x)]$ is odd.

23 Write $f(x) = 2x^4 - 7x^3 + x^2 - 3x + 2$ as the sum of an even function and an odd function.

24 Use Ex. 21 and 22 to show that any function $f(x)$ is the sum of an even function and an odd function. Assume a symmetric domain $-c < x < c$.

25 Show that the sum of odd functions is odd.

26 Show that the product of odd functions is even.

27 Show that the product of an odd function and an even function is odd.

28 Suppose $f(x) = a_0 + a_1x + a_2x^2 + \cdots + a_9x^9$ is odd. Find a_6.

29 If f and g are odd, what about $f \circ g$?

30 If f is odd and g even, what about $f \circ g$?

Graph

31 $y = \dfrac{x}{1 + x^4}$ **32** $y = \dfrac{x^2}{1 + x^4}$

33 $y = x^5 - x$ **34** $y = x^4 - x^2$

35 $y = (x + 1)^5 - (x + 1)$ **36** $y = (x - 2)^4 - (x - 2)^2$.

5 RATIONAL FUNCTIONS

One of the ways we construct functions is by dividing two functions $f(x)$ and $g(x)$ to form their quotient $f(x)/g(x)$. The quotient is defined wherever $f(x)$ and $g(x)$ are both defined and $g(x)$ is not 0. A particular case is the quotient of two polynomials $f(x)$ and $g(x)$; such a function is called a **rational function**. Since the domain of any polynomial is $\{\text{all } x\}$, the domain of a rational function $f(x)/g(x)$ is the set of all real x such that $g(x) \neq 0$.

Examples

(1) $\dfrac{x^3 + 1}{(x - 1)(x - 2)}$ domain $\{x \mid x \neq 1, x \neq 2\}$

(2) $\dfrac{x^5 - 3x + 2}{4x^2 + 1}$ domain $\{\text{all } x\}$.

The rational functions include the polynomial functions, because if $p(x)$ is a polynomial, then

$$p(x) = \frac{p(x)}{1}$$

is a quotient of polynomials, hence a rational function.

Identities

Recall what an identity is. When we say two functions $f_1(x)$ and $f_2(x)$ are identically equal and write

$$f_1(x) = f_2(x),$$

we mean that the functions take the same values on the common part of their domains, that is, where both are defined. However, $f_1(x)$ and $f_2(x)$ may have different domains.

Example

$$\frac{2x}{x(x-1)} = \frac{2(x+1)}{(x+1)(x-1)}$$

is an identity. The rational function on the left is defined for all x except $x = 0$ and $x = 1$. The rational function on the right is defined for all x except $x = -1$ and $x = 1$. Therefore the *common* domain of these two rational functions consists of all x except -1, 0, and 1. But if $x \neq -1, 0, 1$, then

$$\frac{2x}{x(x-1)} = \frac{x}{x} \cdot \frac{2}{x-1} = \frac{2}{x-1}$$

and

$$\frac{2(x+1)}{(x+1)(x-1)} = \frac{x+1}{x+1} \cdot \frac{2}{x-1} = \frac{2}{x-1},$$

so the identity is established.

In the preceding example, both sides of the identity are identically equal to the rational function $2/(x-1)$ in which nothing more can be cancelled. In general, each rational function is identically equal to a rational function in lowest terms, that is, where nothing can be cancelled out.

Examples

$$\frac{3x^5}{x^4 - 2x^6} = \frac{3x}{1 - 2x^2} \qquad (x \neq 0, \tfrac{1}{2}\sqrt{2}, -\tfrac{1}{2}\sqrt{2})$$

$$\frac{(x+1)(x+2)(x+3)}{x+2} = (x+1)(x+3) \qquad (x \neq -2).$$

The simplest rational functions that are not polynomials are the functions $y = 1/x^n$. We shall now study these functions and their graphs. For that purpose, we need some notation. In the same spirit that we write $x \longrightarrow \infty$, we shall write

$$x \longrightarrow a+$$

to mean x gets closer and closer to a, but only $x > a$ is allowed. (Read "x approaches a from above", or "from the right".) Similarly, we shall write

$$x \longrightarrow a-$$

to mean x gets closer and closer to a, but only $x < a$ is allowed. (Read "x approaches a from below", or "from the left".)

EXAMPLE 1 Graph $y = \dfrac{1}{x}$.

Solution The function $f(x) = 1/x$ is defined for all x except $x = 0$. It is an odd function since

$$f(-x) = \frac{1}{-x} = -\frac{1}{x} = -f(x).$$

Therefore its graph is symmetric in the origin, so we need plot the graph only for $x > 0$, then construct the part for $x < 0$ by symmetry.

If $x > 0$, then $y = 1/x > 0$. As x increases through positive values, y decreases; in fact $y \longrightarrow 0+$ as $x \longrightarrow \infty$ (for instance, $y = 10^{-6}$ when $x = 10^{6}$). We see also that $y \longrightarrow \infty$ as $x \longrightarrow 0+$ (for instance, $y = 10^{6}$ when $x = 10^{-6}$). This information together with a few plotted points gives a reasonable idea of the graph (Fig. 1).

Remark It happens that the graph of $y = 1/x$ is a *hyperbola*, discussed in Chapter 12, Section 7 and particularly in Example 4 of Chapter 12, Section 8.

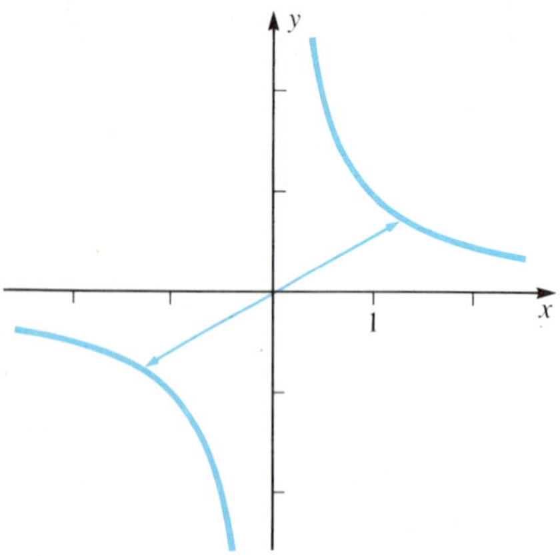

Fig. 1 $y = \dfrac{1}{x}$

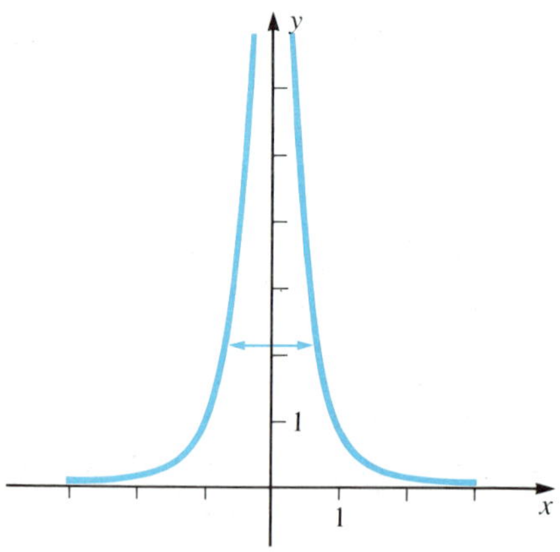

Fig. 2 $y = \dfrac{1}{x^2}$

EXAMPLE 2 Graph $y = \dfrac{1}{x^2}$.

Solution The function $f(x) = 1/x^2$ is defined for all $x \neq 0$. It is an even function since

$$f(-x) = \frac{1}{(-x)^2} = \frac{1}{x^2} = f(x).$$

Therefore its graph is symmetric in the y-axis. We need plot it only for $x > 0$, then obtain the rest by reflection in the y-axis.

As x increases through positive values, $f(x)$ decreases. In fact $f(x) \longrightarrow 0+$ as $x \longrightarrow \infty$ [for instance, $f(1000) = 10^{-6}$]. Also $f(x) \longrightarrow \infty$

as $x \longrightarrow 0+$ [for instance, $f(0.001) = 10^6$].

We use this information and a few plotted points to construct a rough graph (Fig. 2).

Figures 1 and 2 are typical of the graphs of $y = 1/x^n$. If n is odd, then $1/x^n$ is an odd function and we have symmetry with respect to the origin. If n is even, then $1/x^n$ is an even function and we have symmetry with respect to the y-axis. See Fig. 3. Note that the larger the exponent n, the faster the curve $y = 1/x^n$ approaches 0 as $x \longrightarrow \infty$, and the faster it zooms up to ∞ as $x \longrightarrow 0+$.

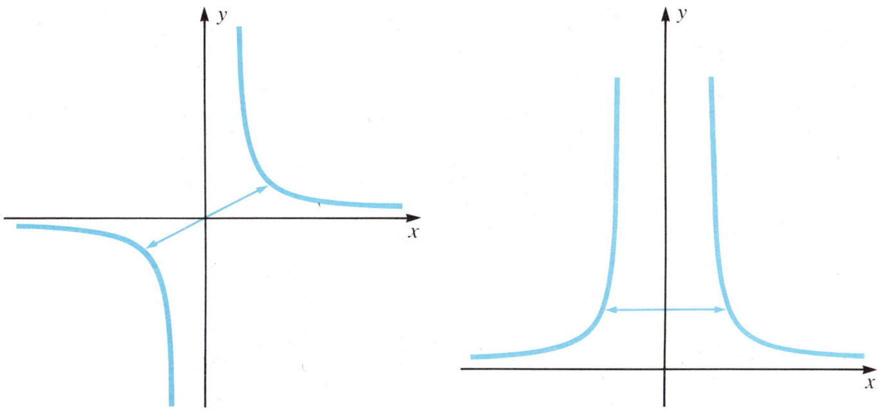

(a) n odd (b) n even

Fig. 3 $y = \dfrac{1}{x^n}$

By shifting and reflecting we can easily use the graphs we know to graph any function

$$y = \frac{a}{(x - r)^n} + b.$$

EXAMPLE 3 Graph

(a) $y = \dfrac{1}{x - 1}$ (b) $y = \dfrac{-1}{(x + 2)^2} + 1$.

Solution (a) This is the graph of $y = f(x - 1)$ where $f(x) = 1/x$. Just shift the graph of $y = 1/x$ one unit to the right (Fig. 4a).

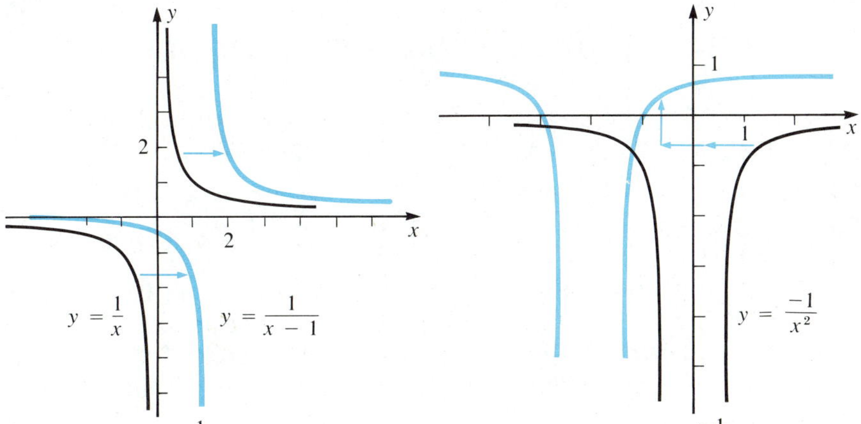

(a) $y = \dfrac{1}{x - 1}$ (b) $y = \dfrac{-1}{(x + 2)^2} + 1$

Fig. 4

(b) This is the graph of $y = f(x+2) + 1$, where $f(x) = -1/x^2$. Just turn the graph of $y = 1/x^2$ upside down, then shift it two units to the left and one unit up (Fig. 4b).

Exercises

Find the domain of

1 $\dfrac{x+3}{x-4}$ **2** $\dfrac{x^2+2}{3x^2+2}$ **3** $\dfrac{x^2+2x+1}{x+1}$

4 $\dfrac{x^2+2x+6}{5x+2}$ **5** $\dfrac{x^2+3x+2}{x^2-4}$ **6** $\dfrac{x^2-1}{x^2+3x+2}$

7 $\dfrac{x^2}{x^6-8x^3}$ **8** $\dfrac{x^2-1}{x^3-1}$ **9** $\dfrac{x+1}{x^3-x^2+x-1}$

10 $\dfrac{5x+8}{x^4-3x^2+2}$ **11** $\dfrac{1}{x^4-2x^2+1}$ **12** $\dfrac{x+1}{x^3+125}$

13 $\dfrac{x^2-1}{3x^2-5x-3}$ **14** $\dfrac{(x+1)(x-2)}{x^6+2x^3+4}$.

Is the relation an identity? If so, where is it valid?

15 $\dfrac{x^3+x}{x^2+1} = x$ **16** $\dfrac{2x+1}{2x^2+3x+1} = \dfrac{x}{x^2+x}$

17 $\dfrac{x^2-1}{x^2-x-2} = \dfrac{x^2+x-2}{x^2-4}$ **18** $\dfrac{x+1}{x^3+1} = \dfrac{1}{x^2+x+1}$

19 $\dfrac{x^2-2x-3}{x^2-9} = \dfrac{x^2-x}{x^2+3x}$ **20** $\dfrac{x^3-x^2-4x+1}{x^2-1} = \dfrac{x^2-4}{x-1}$.

21 Graph $y = 1/x^3$ accurately for $-2 \le x \le 2$, $x \ne 0$.

22 Graph $y = 1/x^4$ accurately for $-2 \le x \le 2$, $x \ne 0$.

Graph

23 $y = \dfrac{1}{x^2} - 1$ **24** $y = \dfrac{1}{x} + 3$ **25** $y = -\dfrac{1}{x} + 1$ **26** $y = -\dfrac{1}{x^2} - 2$

27 $y = \dfrac{\frac{1}{2}}{x-3}$ **28** $y = \dfrac{1}{x-2} + 3$ **29** $y = \dfrac{1}{2x+5}$ **30** $y = \dfrac{\frac{3}{2}}{(2x-1)^2}$

31 $y = \dfrac{2}{x+1} - 1$ **32** $y = \dfrac{-2}{(x-2)^2} - 1$.

6 GRAPHS OF RATIONAL FUNCTIONS

In this section we shall learn some of the important features of the graphs of rational functions. First, let's look at some more examples.

EXAMPLE 1 Graph (a) $y = \dfrac{1}{x^2+1}$ (b) $y = \dfrac{2x}{x^2+1}$

Solution (a) The function $1/(x^2+1)$ is even, so the graph is symmetric about the y-axis. Since $x^2 + 1 \ge 1$, we see that $0 < y \le 1$ for all values of x. If x starts at 0 and increases, then y starts at 1 and decreases; if $x \longrightarrow \infty$, then $y \longrightarrow 0+$. This information plus a few plotted points yields a graph (Fig. 1).

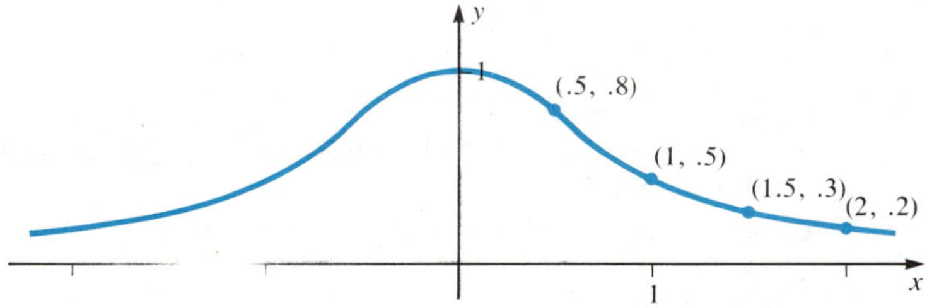

Fig. 1 $y = \dfrac{1}{x^2 + 1}$

(b) The function is odd so the graph is symmetric with respect to the origin.
Certainly $(0, 0)$ is on the graph, and $y > 0$ when $x > 0$. Near $x = 0$, the denominator $x^2 + 1$ is close to 1, hence the graph is closely approximated by the graph of $y = 2x$.

To study the curve for large values of x, divide numerator and denominator by x^2, the highest power of x in the denominator:

$$y = \frac{2x}{x^2 + 1} = \frac{1}{x} \cdot \frac{2}{1 + 1/x^2}.$$

For large values of x, the second factor is near 2. For instance, if $x = 1000$, then

$$\frac{2}{1 + 1/x^2} = \frac{2}{1.000001} \approx 1.999998$$

Therefore $y \approx 2/x$ for large x, and $y \longrightarrow 0+$ as $x \longrightarrow \infty$. We summarize all of this information in Fig. 2a. Apparently the graph starts upwards from the origin at about slope 2, but soon starts to decline, ultimately dying out towards zero. Plotting a few points confirms this (Fig. 2b).

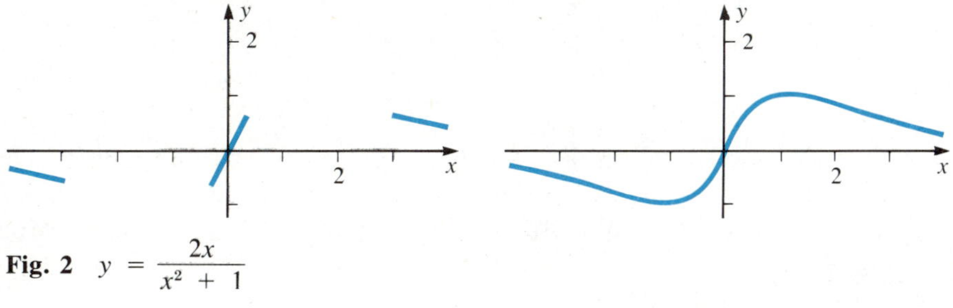

Fig. 2 $y = \dfrac{2x}{x^2 + 1}$

EXAMPLE 2 Graph (a) $y = \dfrac{x}{x + 1}$ (b) $y = \dfrac{x^2}{x^2 + 1}$.

Solution (a) The function is neither odd nor even, so there is no apparent symmetry. Clearly $y = 0$ at $x = 0$ and $y \approx x$ when $x \approx 0$ because the denominator $x + 1 \approx 1$. The denominator equals 0 at $x = -1$ so we expect the graph to blow up as $x \longrightarrow -1$. It does because

$$y = \frac{x}{x + 1} \approx \frac{-1}{x + 1} \quad \text{for} \quad x \approx -1.$$

Hence $y \longrightarrow -\infty$ as $x \longrightarrow -1+$ and $y \longrightarrow \infty$ as $x \longrightarrow -1-$.

To find what happens as $x \longrightarrow \infty$ or $x \longrightarrow -\infty$, we divide both numerator and denominator by x:

$$y = \frac{x}{x+1} = \frac{1}{1 + 1/x}.$$

This form of y shows us that $y \longrightarrow 1-$ as $x \longrightarrow \infty$ and $y \longrightarrow 1+$ as $x \longrightarrow -\infty$. It also shows that y increases as x increases [$1/x$ decreases, $1 + 1/x$ decreases, $y = 1/(1 + 1/x)$ increases].

We summarize this information in Fig. 3a and complete a rough graph in Fig. 3b.

(a) Alternative Solution

$$y = \frac{x}{x+1} = \frac{x+1-1}{x+1} = \frac{-1}{x+1} + 1.$$

Think about it, and compare Fig. 1, page 158.

Remark Like $y = 1/x$, the graph of $y = x/(x+1)$ is also a hyperbola. See the remark on p. 158 for details.

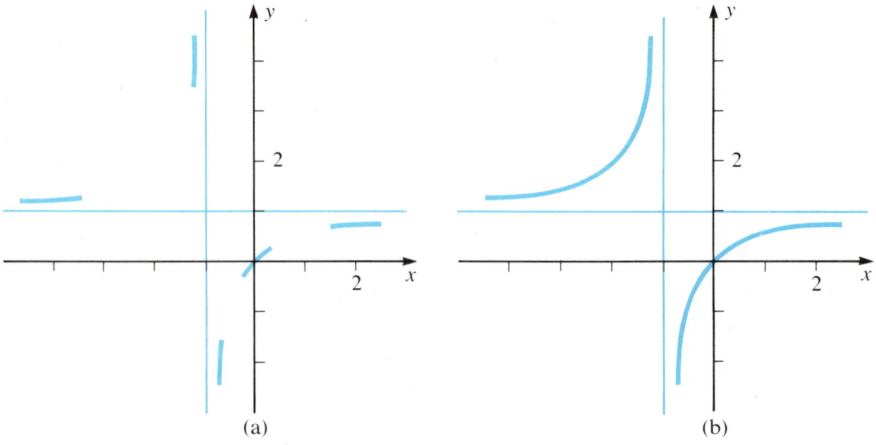

Fig. 3 $y = \dfrac{x}{1+x}$

(b) The function is even so we can plot the graph for $x \geq 0$ and reflect in the y-axis. Note that $y = 0$ at $x = 0$, $y \approx x^2$ for $x \approx 0$, and $y \longrightarrow 1-$ as $x \longrightarrow \infty$. Also y increases as x increases from 0. These facts lead to the rough graph in Fig. 4.

(b) Alternative Solution

$$y = \frac{x^2}{x^2+1} = \frac{x^2+1-1}{x^2+1} = \frac{-1}{x^2+1} + 1, \text{ etc.}$$

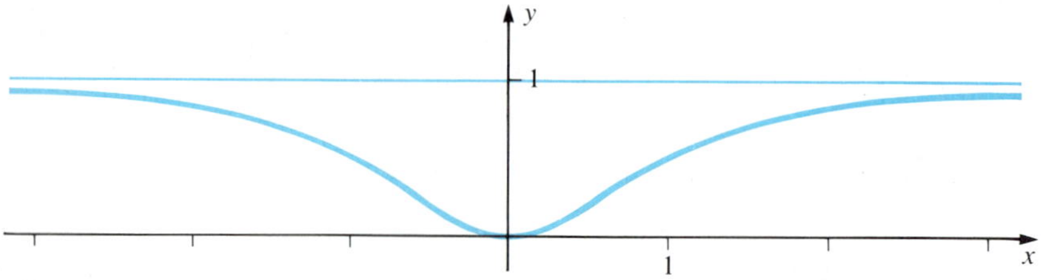

Fig. 4 $y = \dfrac{x^2}{x^2+1}$

Horizontal Asymptotes

Observe the special roles of the horizontal lines $y = 0$ in Figs. 1 and 2 and $y = 1$ in Figs. 3 and 4. The corresponding graphs approach closer and closer to these lines as $x \longrightarrow \infty$ or $x \longrightarrow -\infty$.

In general, a horizontal line $y = c$ is called a **horizontal asymptote** of a graph $y = f(x)$ if $y \longrightarrow c$ as $x \longrightarrow -\infty$ or as $x \longrightarrow \infty$ (or both).

The examples in this and the preceding section all illustrate the following statements:

> **Horizontal Asymptotes** Suppose
>
> $$r(x) = \frac{a_m x^m + \cdots + a_1 x + a_0}{b_n x^n + \cdots + b_1 x + b_0} \qquad (a_m \neq 0, b_n \neq 0)$$
>
> (1) If $m = n$, then the line $y = a_m/b_n$ is a horizontal asymptote of $y = r(x)$. In fact, $r(x) \longrightarrow a_m/b_n$ as $x \longrightarrow \infty$ and as $x \longrightarrow -\infty$.
>
> (2) If $m < n$ (bottom-heavy case), then the x-axis $(y = 0)$ is a horizontal asymptote of $y = r(x)$. In fact, $r(x) \longrightarrow 0$ as $x \longrightarrow \infty$ and as $x \longrightarrow -\infty$.

Instead of proving these statements, we shall give examples that show the idea of the proofs. In each case we divide both numerator and denominator by the highest power of x in the denominator.

Examples

(1) $r(x) = \dfrac{2x^3 + 1}{5x^3 + x - 4} \qquad (m = n = 3, a_m = 2, b_n = 5).$

Divide by x^3: $\quad r(x) = \dfrac{2 + \dfrac{1}{x^3}}{5 + \dfrac{1}{x^2} - \dfrac{4}{x^3}}.$

For large values of $|x|$, the numerator is approximately 2, the denominator approximately 5, so $r(x) \longrightarrow \frac{2}{5} = a_m/b_n$ as $x \longrightarrow \infty$ or $x \longrightarrow -\infty$. Therefore the line $y = \frac{2}{5}$ is a horizontal asymptote.

(2) $r(x) = \dfrac{x^2 + 5x - 3}{x^3 + x^2 + 1} \qquad (m = 2, n = 3).$

Divide numerator and denominator by x^3:

$$r(s) = \frac{\dfrac{1}{x} + \dfrac{5}{x^2} - \dfrac{3}{x^3}}{1 + \dfrac{1}{x} + \dfrac{1}{x^3}}.$$

For large values of $|x|$, the numerator is near 0, the denominator near 1. Therefore $r(x) \longrightarrow 0$ as $x \longrightarrow \infty$ or $x \longrightarrow -\infty$, so the x-axis is a horizontal asymptote.

Vertical Asymptotes

Now observe the special role of the vertical line $x = -1$ in Fig. 3. The graph approaches this line "at infinity" as $x \longrightarrow -1$ from either side.

In general a vertical line $x = a$ is called a **vertical asymptote** of a graph $y = f(x)$ if $y \longrightarrow \infty$ or $-\infty$ as $x \longrightarrow a+$ or $x \longrightarrow a-$.

> **Vertical Asymptote** Let $r(x) = f(x)/g(x)$ be a quotient of two polynomials and let a be a real number such that $g(a) = 0$ but $f(a) \neq 0$. Then the line $x = a$ is a vertical asymptote of the graph $y = r(x)$.

EXAMPLE 3 Locate the asymptotes and graph

$$y = \frac{x^2 + 4}{x^2 - 4}.$$

Solution First we have

$$y = \frac{x^2 + 4}{x^2 - 4} = \frac{1 + 4/x^2}{1 - 4/x^2} \longrightarrow 1 \quad \text{as} \quad x \longrightarrow \pm\infty$$

so the horizontal line $y = 1$ is an asymptote. Next, factor the denominator:

$$y = \frac{x^2 + 4}{(x + 2)(x - 2)}$$

Clearly $x = -2$ and $x = 2$ are zeros of the denominator, but not of the numerator, so the lines $x = -2$ and $x = 2$ are vertical asymptotes.

The numerator $x^2 + 4$ is always positive. If $x < -2$ or $x > 2$ the denominator $x^2 - 4$ is positive; if $-2 < x < 2$, the denominator is negative. Therefore

$$\left.\begin{matrix} y > 0 \\ y < 0 \\ y > 0 \end{matrix}\right\} \text{ for } \left\{\begin{matrix} -\infty < x < -2 \\ -2 < x < 2 \\ 2 < x < \infty. \end{matrix}\right.$$

This is almost enough information for a graph. If we add that $y = -1$ when $x = 0$ and note that y is an *even* function of x, then we are in business (Fig. 5).

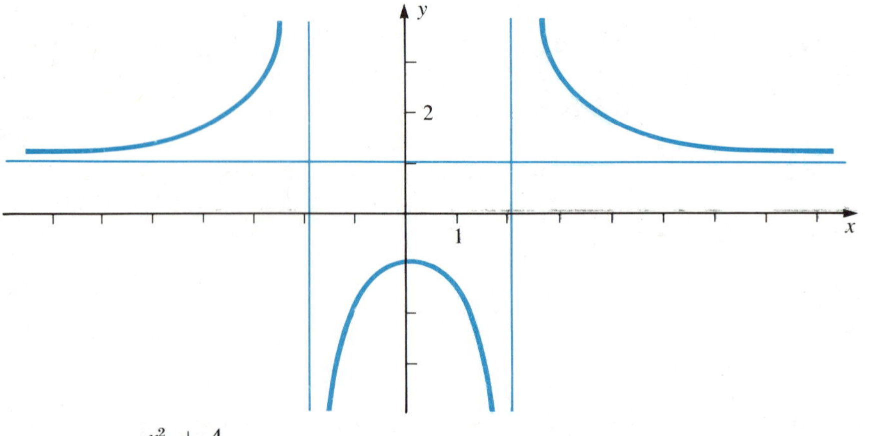

Fig. 5 $y = \dfrac{x^2 + 4}{x^2 - 4}$

Remark To fine-tune the graph, write

$$y = \frac{x^2 + 4}{x^2 - 4} = \frac{x^2 - 4 + 8}{x^2 - 4} = 1 + \frac{8}{x^2 - 4}.$$

It follows easily that y decreases as x increases from 0 to 2 and y decreases as x increases from 2 to ∞.

Behavior as $x \longrightarrow \infty$ or $x \longrightarrow -\infty$

Again consider a rational function

$$r(x) = \frac{a_m x^m + \cdots + a_1 x + a_0}{b_n x^n + \cdots + b_1 x + b_0} \qquad (a_m \neq 0, b_n \neq 0).$$

If the numerator and denominator have equal degree $(m = n)$, then $r(x) \longrightarrow a_m/b_n$ as $x \longrightarrow \infty$ or $x \longrightarrow -\infty$, as already noted. If the rational function is bottom-heavy $(m < n)$, then $r(x) \longrightarrow 0$ as $x \longrightarrow \infty$ or $x \longrightarrow -\infty$, also already noted.

There remains the case of a top-heavy rational function $(m > n)$. In that case $r(x) \longrightarrow \infty$ or $-\infty$ as $x \longrightarrow \infty$ or $x \longrightarrow -\infty$. We shall illustrate this with an example, but omit the proof in general.

Example $r(x) = \dfrac{x^4 + 3x}{x^2 + 12} \qquad (m = 4, n = 2).$

Divide both numerator and denominator by x^2:

$$r(x) = \frac{x^2 + \dfrac{3}{x}}{1 + \dfrac{12}{x^2}}.$$

For large values of $|x|$, the denominator is approximately 1, the numerator approximately x^2. Hence $r(x)$ behaves like $y = x^2$ as $|x| \longrightarrow \infty$. Therefore $r(x) \longrightarrow \infty$.

Exercises

Graph; identify all horizontal and vertical asymptotes

1 $y = \dfrac{x^3}{x^4 + 1}$ **2** $y = \dfrac{x^2}{x^4 + 1}$ **3** $y = \dfrac{5(x - 1)}{(x - 1)^2 + 4}$

4 $y = \dfrac{x^2 + 1}{x^2 + 2}$ **5** $y = \dfrac{x}{2x - 3}$ **6** $y = \dfrac{-x}{2x + 4}$

7 $y = \dfrac{x - 1}{x^2}$ **8** $y = \dfrac{2x + 3}{x^2}$ **9** $y = \dfrac{1}{x + 1} - \dfrac{1}{x - 1}$

10 $y = \dfrac{1}{x} - \dfrac{1}{x - 1}$ **11** $y = x^2 - \dfrac{1}{x^2}$ **12** $y = \dfrac{1}{x - 1} - \dfrac{2}{x} + \dfrac{1}{x + 1}.$

Describe the behavior of the rational function as $x \to \infty$ and $-\infty$

13 $\dfrac{x^5 - x^3}{3x^5 + x - 1}$ **14** $\dfrac{(x^2 - 3)(2x - 1)}{3x^3 - x}$

15 $\dfrac{2x^6 + x^5 - x - 3}{5x^5 - 4x^4 + x}$ **16** $\dfrac{(x^3 - x + 1)(x^3 - x^2 + 1)}{(x^2 - x - 1)(x^2 + x - 1)}.$

7 COMPLETELY FACTORED RATIONAL FUNCTIONS

It is fairly easy to sketch the graph of a rational function in lowest terms that is completely factored into linear factors:

$$y = a\frac{(x - r_1)(x - r_2) \cdots (x - r_m)}{(x - s_1)(x - s_2) \cdots (x - s_n)}$$

Clearly $y = 0$ if $x = r_1$ or $x = r_2$, etc., so the graph intersects the x-axis at each $x = r_i$, and only at these points. The denominator is 0 at $x = s_1$, $x = s_2$, etc., so each line $x = s_j$ is a vertical asymptote of the graph.

EXAMPLE 1 Graph $y = \dfrac{x}{(x+1)(x-2)}$

Solution The numerator is 0 only at $x = 0$; the denominator is 0 at $x = -1$ and $x = 2$. Therefore the graph intersects the x-axis only at $x = 0$, and the lines $x = -1$ and $x = 2$ are the vertical asymptotes. The function is bottom-heavy, so the x-axis is a horizontal asymptote.

To study the graph near $x = 0$ we write

$$y = \frac{x}{(x+1)(x-2)} = \left[\frac{1}{(x+1)(x-2)}\right] x \approx -\tfrac{1}{2}x \quad \text{for} \quad x \approx 0$$

because $1/(x+1)(x-2) = -\frac{1}{2}$ if $x = 0$. Similarly

$$y \approx \frac{\frac{1}{3}}{x+1} \quad \text{for} \quad x \approx -1 \qquad \text{and} \qquad y \approx \frac{\frac{2}{3}}{x-2} \quad \text{for} \quad x \approx 2.$$

We show all of these facts about the graph in Fig. 1a, and fill in the rough graph in Fig. 1b.

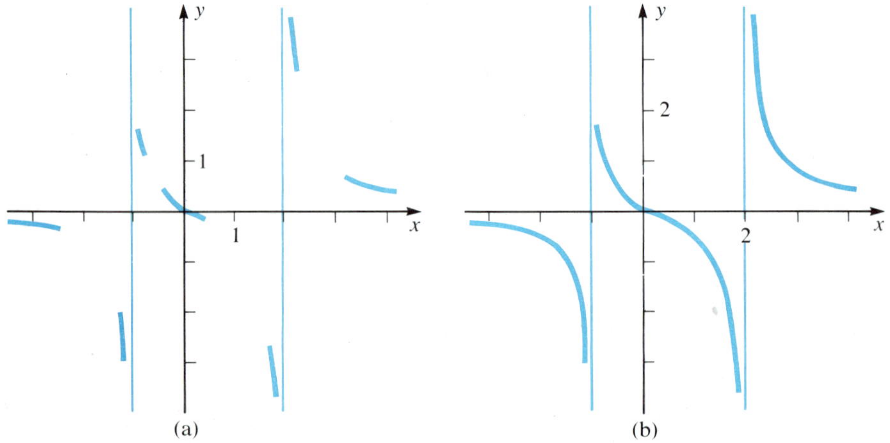

Fig. 1 $y = \dfrac{x}{(x+1)(x-2)}$

EXAMPLE 2 Graph $y = 2\dfrac{(x-1)(x+1)}{(x-2)(x+2)}$.

Solution The numerator is 0 at $x = 1$ and $x = -1$; the denominator is 0 at $x = 2$ and $x = -2$. Therefore the graph intersects the x-axis only at $x = 1$ and $x = -1$, and the lines $x = 2$ and $x = -2$ are the vertical asymptotes.

Since the numerator and the denominator have the same degree, there is a horizontal asymptote (see p. 163). Since

$$y = \frac{2(x^2-1)}{x^2-4} = 2\frac{1-1/x^2}{1-4/x^2} \longrightarrow 2 \quad \text{as} \quad x \longrightarrow \pm\infty,$$

the horizontal asymptote is the line $y = 2$. Note, incidentally, the free information: y is an even function of x, so the graph is symmetric in the y-axis.

We need the behavior of y for x near $1, -1, 2$, and -2. Near $x = 1$,

$$y = 2\left[\frac{x+1}{(x-2)(x+2)}\right](x-1) \approx -\tfrac{4}{3}(x-1)$$

because $2(x+1)/(x-2)(x+2) = -\frac{4}{3}$ when $x = 1$. Similarly

$$y \approx \tfrac{4}{3}(x+1) \quad \text{for} \quad x \approx -1,$$

and

$$y \approx \frac{\frac{3}{2}}{x - 2} \quad \text{for} \quad x \approx 2 \qquad y \approx \frac{-\frac{3}{2}}{x + 2} \quad \text{for} \quad x \approx -2.$$

We sketch all of the facts in Fig. 2a, as well as the y-intercept $(0, \frac{1}{2})$. Then we fill in the rough graph in Fig. 2b.

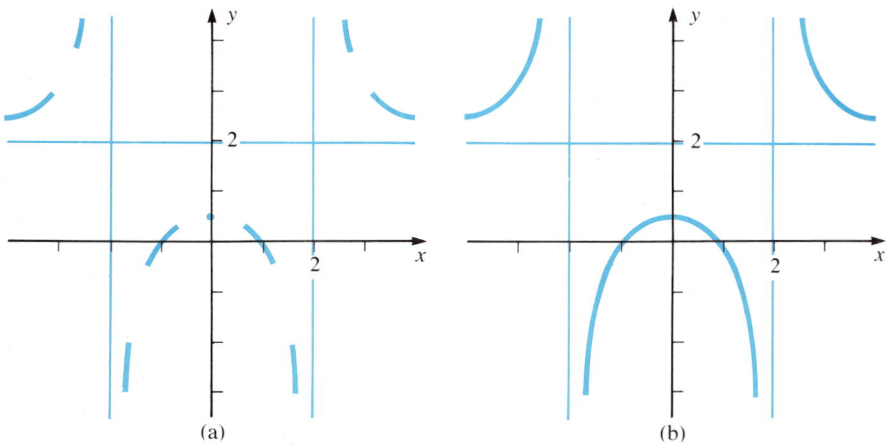

Fig. 2 $y = 2\dfrac{(x - 1)(x + 1)}{(x - 2)(x + 2)}$

EXAMPLE 3 Graph $y = \dfrac{(x + 1)(x - 2)}{(x + 2)(x - 1)}$.

Solution The graph intersects the x-axis at two points, $x = -1$ and $x = 2$, and has two vertical asymptotes, $x = -2$ and $x = 1$. In Example 2 there were also two points of intersection with the x-axis and two vertical asymptotes, so you might expect the graphs to be similar. But there is an important difference: in Example 2 both x-intercepts are between the vertical asymptotes, whereas here only one is between them.

Anyhow, as before we find the horizontal asymptote $y = 1$, the y-intercept $(0, 1)$ and the local behavior:

$$y \approx \tfrac{3}{2}(x + 1) \quad \text{for} \quad x \approx -1 \qquad y \approx \tfrac{3}{4}(x - 2) \quad \text{for} \quad x \approx 2$$

$$y \approx \frac{-\frac{4}{3}}{x + 2} \quad \text{for} \quad x \approx -2 \qquad y \approx \frac{-\frac{2}{3}}{x - 1} \quad \text{for} \quad x \approx 1.$$

We sketch these facts in Fig. 3a, then fill in the graph in Fig. 3b.

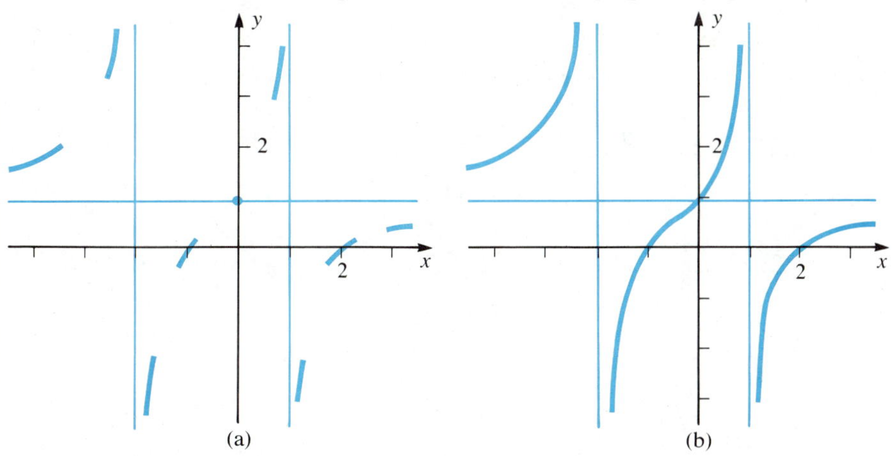

Fig. 3 $y = \dfrac{(x + 1)(x - 2)}{(x + 2)(x - 1)}$

We close this section with two examples showing what happens when there is a repeated factor.

EXAMPLE 4 Graph $y = \dfrac{2x}{(x-1)^2}$.

Solution The graph intersects the x-axis at $x = 0$, and near there $y \approx 2x$. The denominator is 0 at $x = 1$, and near there $y \approx 2/(x-1)^2$. Therefore the line $x = 1$ is a vertical asymptote. The quotient is bottom-heavy, so $y = 0$ is a horizontal asymptote. We sketch the graph in Fig. 4.

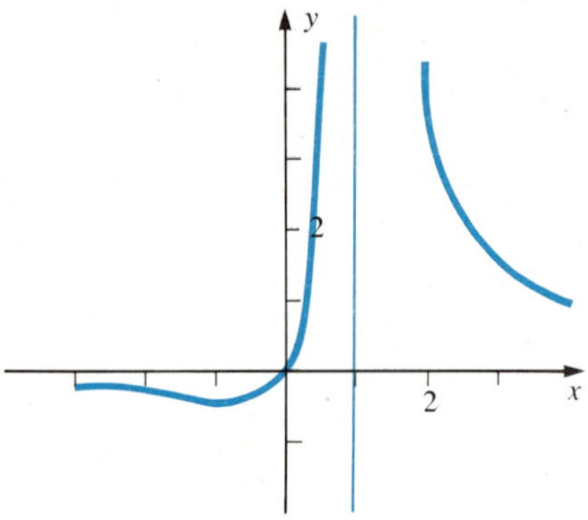

Fig. 4 $y = \dfrac{2x}{(x-1)^2}$

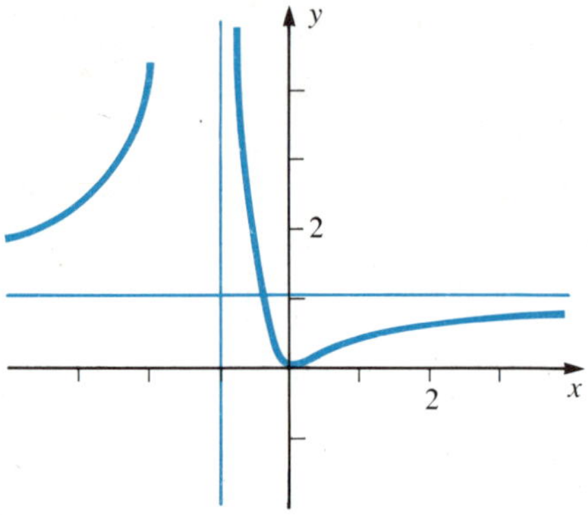

Fig. 5 $y = \dfrac{x^2}{(x+1)^2}$

EXAMPLE 5 Graph $y = \dfrac{x^2}{(x+1)^2}$.

Solution The graph intersects the x-axis at $x = 0$ and near there $y \approx x^2$. The line $x = -1$ is a vertical asymptote, and $y \approx 1/(x+1)^2$ near $x = -1$. Since

$$y = \frac{x^2}{(x+1)^2} = \frac{x^2}{x^2 + 2x + 1}$$

we conclude as usual that $y = 1$ is a horizontal asymptote. We sketch the graph in Fig. 5.

Exercises

Graph; identify asymptotes and other distinctive features

1 $y = \dfrac{x}{(x+1)(x-1)}$ **2** $y = \dfrac{x}{(x+2)(x-1)}$ **3** $y = \dfrac{x}{(x-1)(x-3)}$

4 $y = \dfrac{x+1}{x(x-2)}$ **5** $y = \dfrac{(x-1)(x-2)}{(x-3)(x-4)}$ **6** $y = \dfrac{x(x+2)}{(x-1)(x-2)}$

7 $y = \dfrac{x(x+2)}{(x+1)(x-1)}$ **8** $y = \dfrac{x(x-2)}{(x+1)(x-1)}$ **9** $y = \dfrac{(x+1)(x-1)}{(x+2)(x-2)}$

10 $y = \dfrac{x(x-3)}{(x-1)(x-2)}$ **11** $y = \dfrac{x+1}{(x-1)^2}$ **12** $y = \dfrac{x-1}{(x-2)^2}$

13 $y = \dfrac{(x-1)^2}{x^2}$ **14** $y = \left(\dfrac{x+1}{x-1}\right)^2$ **15** $y = \dfrac{x^2}{(x+1)(x-1)}$

16 $y = \dfrac{-x^2}{(x-1)(x-2)}$ **17** $y = -\dfrac{(x+1)(x+2)}{x^2}$ **18** $y = \dfrac{(x-1)(x-3)}{(x-2)^2}$

19 $y = \dfrac{(x+1)(x-1)}{x^3}$ **20** $y = \dfrac{x^3}{(x+1)(x-1)}$ **21** $y = \dfrac{(x+2)^2}{x^3}$

22 $y = \dfrac{x^3}{(x+1)^3}$ **23** $y = \dfrac{x^3}{(x-1)^2}$ **24** $y = \dfrac{-x^2}{(x+1)^2}$

25 $y = \dfrac{(x+1)(x-1)}{(x+2)x(x-2)}$ **26** $y = \dfrac{(x+2)x(x-2)}{(x+1)(x-1)}$.

REVIEW EXERCISES

Graph roughly

1 $y = -\frac{1}{2}(x+2)^3 - 1$ **2** $y = \frac{1}{3}(x-2)^4 - 3$.

Draw a line graph showing signs and sign changes; then graph roughly

3 $y = (x-1)(x-2)(x-4)$ **4** $y = \frac{1}{2}x^2(x-2)$

5 $y = x^3(x+1)(x+2)$ **6** $y = x^3(x-1)^2(x+1)$.

Graph

7 $y = -\dfrac{2}{x} + 3$ **8** $y = \dfrac{1}{3x^2} + 4$.

Graph, showing all horizontal and vertical asymptotes

9 $y = \dfrac{x+1}{x^2+2}$ **10** $y = \dfrac{(x+1)^4}{x^2(x^2+1)}$ **11** $y = \dfrac{x^3+1}{x^3-1}$

12 $y = \dfrac{x^3-1}{x^3+1}$ **13** $y = \dfrac{x}{(x-2)(x+1)}$ **14** $y = \dfrac{x}{(x+2)(x+3)}$

15 $y = \dfrac{-x^2}{(x+1)(x+3)}$ **16** $y = \dfrac{x^4}{(x+1)(x+2)}$.

5 EXPONENTIAL AND LOGARITHM FUNCTIONS

1 EXPONENTIAL FUNCTIONS

In this chapter, we first study exponential functions, then turn them around to obtain logarithm functions. Both exponential and logarithm functions are new kinds of animals. They cannot be expressed solely in terms of addition, subtraction, multiplication, division, and taking roots (radicals). This sets them apart from simpler functions, such as polynomials and rational functions.

In Chapter 1, we discussed exponential expressions a^p, where p is an integer or a rational number, and $a > 0$. Now we introduce exponential *functions* $f(x) = a^x$, where x takes *all real values*.

A strict definition of a^x for all real values is technical. It requires defining such numbers as $a^{\sqrt{2}}$. Rather than attempt a definition, let us see what properties an exponential function ought to have. For example, suppose $f(x) = 2^x$ were defined for *all* real x. What would it be like?

If $x = n$, an integer, then 2^x should agree with our former definition of 2^n. We can tabulate some values of the function:

x	0	1	2	3	4	5	6	7	8	9	10
2^x	1	2	4	8	16	32	64	128	256	512	1024

The values increase rapidly; joining the plotted points by a smooth curve, we get a sketch of the graph of $y = 2^x$ for $x \geq 0$. See Fig. 1a.

If $x = -n$, where n is a positive integer, then 2^x should mean $2^{-n} = 1/2^n$. We tabulate some values, using 2-place accuracy:

x	−10	−9	−8	−7	−6	−5	−4	−3	−2	−1	0
$f(x) = 2^x$	0.00	0.00	0.00	0.01	0.02	0.03	0.06	0.12	0.25	0.50	1.00

The values decrease very rapidly. They suggest the graph shown in Fig. 1b. (Note that the x-axis is a horizontal asymptote.)

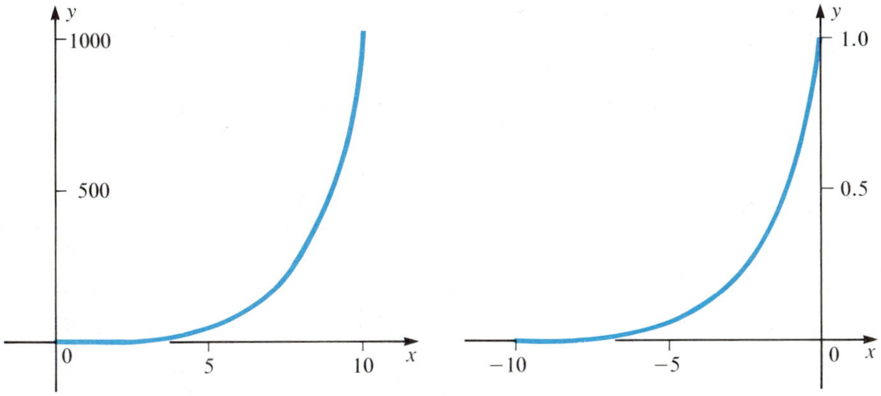

(a) $0 \leq x \leq 10$ (b) $-10 \leq x \leq 0$

Fig. 1 $y = 2^x$

Now let's plot $y = 2^x$ for "reasonable" values of x, say $-3 \leq x \leq 3$ (Fig. 2). We see that the graph always rises as x increases. It rises very fast as x increases through positive values, and it approaches the x-axis very fast as x decreases through negative values.

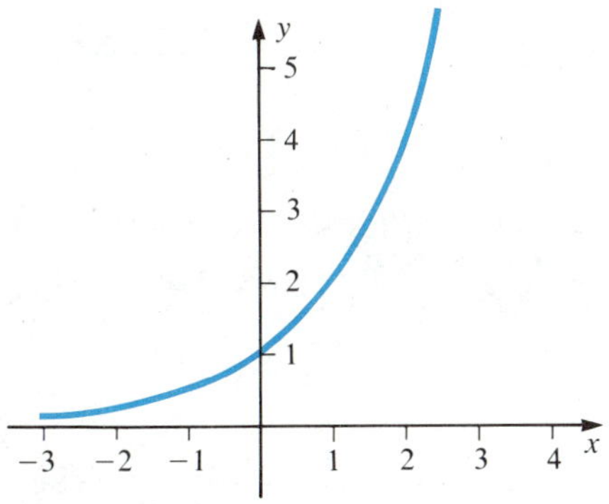

Fig. 2 $y = 2^x$

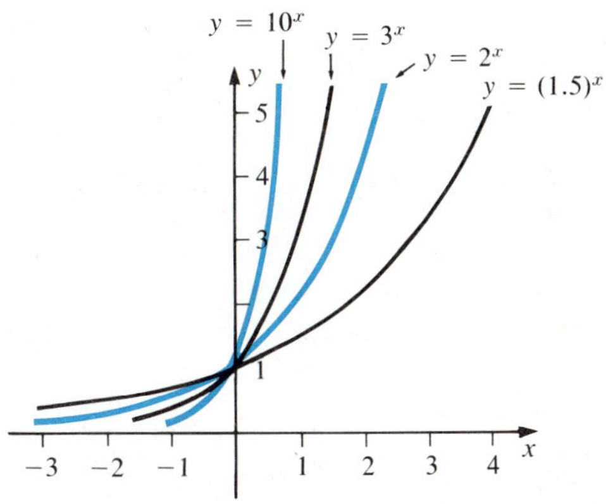

Fig. 3 $y = a^x$ for various $a > 1$

The graph of $y = a^x$, for any $a > 1$, is similar. Figure 3 shows some examples. The larger a is, the faster the curve zooms up as x increases, and the faster it decays as x decreases.

Properties of Exponential Functions

We leave the actual construction of exponential functions to more advanced courses. In this course, we simply accept their existence and list their properties, based on experimental evidence.

For each number $a > 0$, there exists an exponential function a^x with the following properties:

(1) a^x is defined for all real x, and $a^x > 0$.

(2) a^x takes all positive real values.

(3) $a^n = \underbrace{a \cdot a \cdot a \cdots a}_{n \text{ factors}}$ for each positive integer n.

(4) If $a > 1$, then a^x is a strictly increasing function ($a^x < a^y$ whenever $x < y$). Also $a^x \longrightarrow \infty$ as $x \longrightarrow \infty$ and $a^x \longrightarrow 0+$ as $x \longrightarrow -\infty$.

(5) The rules of exponents hold:

$$a^{x+y} = a^x a^y \qquad a^{x-y} = \frac{a^x}{a^y} \qquad a^{-x} = \frac{1}{a^x} \qquad a^0 = 1$$

$$(a^x)^y = a^{xy} \qquad a^x b^x = (ab)^x \qquad 1^x = 1.$$

The number a is called the **base** of the exponential function a^x.

Calculating Exponential Functions

Use your calculator to approximate values of an exponential function. For instance, to find $(3.47)^{2.538}$ on most calculators you key in

3 . 4 7 [y^x] 2 . 5 3 8 [=] 23.5 15608

Further examples:

$(297.6)^{-6.84}$ 2 9 7 . 6 [y^x] 6 . 8 4 [+/−] [=] 1.2032236 ×10⁻¹⁷

$(0.8712)^{1.97}$. 8 7 1 2 [y^x] 1 . 9 7 [=] 0.7621355

Graph of $y = a^x$ for $a < 1$

So far we have sketched the graphs of exponential functions $y = a^x$ only for $a > 1$. Now let us graph $y = a^x$ for $0 < a < 1$. To do so we write $a = 1/b$, where $b > 1$. By the rules of exponents

$$a^x = b^{-x} = \frac{1}{b^x}$$

Since b^x is an increasing function, a^x is a decreasing function. We can say even more: the graph of $y = a^x$ is the mirror image in the y-axis of the graph of $y = b^x$. Why? Because the height of $y = a^x$ at $-x$ is the height of $y = b^x$ at x. For example, the graph of $y = (\frac{1}{2})^x$ is the mirror image in the y-axis of the graph of $y = 2^x$. See Fig. 4.

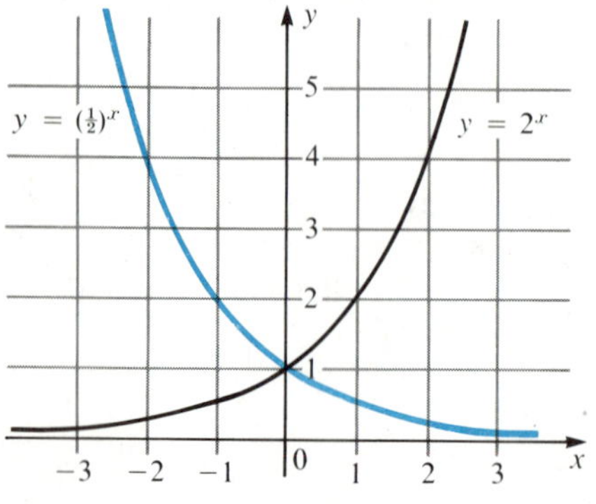

Fig. 4 $y = (\frac{1}{2})^x$

Use of Exponential Functions

Exponential functions occur frequently in applications concerning natural growth. We give a few examples now; others later when we have more technique.

EXAMPLE 1 A culture has 5000 bacteria now. In t hours it will have

$$N = 5000 \times 2^{t/k}$$

bacteria, where k is the time it takes the culture to double. If $k = 7$ hours, how many bacteria will there be in 16.5 hours?

Solution $N = 5000 \times 2^{16.5/7}$

Calculate:

1 6 . 5 [÷] 7 [=] [STO] 2 [y^x] [RCL] [×] 5 0 0 0 [=] 25617.733

Answer 25,618

EXAMPLE 2 A container of hot water is at temperature 90 °C at time $t = 0$. It is surrounded by air at constant temperature 20 °C. If stirred steadily, its temperature after t minutes will be

$$u(t) = 20 + 70 \times 1.071^{-t} \text{ °C}.$$

Find its temperature after (a) 15 min. (b) 30 min. (c) 60 min.

Solution A calculator routine for computing $u(t)$ is

1 . 0 7 1 [y^x] t [+/−] [×] 7 0 [+] 2 0 [=].

For $t = 15$, the calculation yields $u(15) \approx 45.02$. Similarly $u(30) \approx 28.94$ and $u(60) \approx 21.14$.

Answer (a) 45.02 °C (b) 28.94 °C (c) 21.14 °C.

EXAMPLE 3 If \$1000 is invested at $9\frac{1}{2}\%$ annual interest, compounded annually, for t years, it will be worth

$$(1000)(1 + 0.095)^t$$

dollars. Show that it will triple in approximately 12.1 years.

Solution In 12.1 years the value of the investment will be

$$1000 \times 1.095^{12.1}$$

By calculator this is approximately \$2998.55, pretty close to \$3000.

Exercises

Graph

1 $y = 3^x \quad -3 \le x \le 0$

2 $y = 3^x \quad 0 \le x \le 3$

3 $y = (1.5)^x \quad -3 \le x \le 3$

4 $y = 10^x \quad 0 \le x \le 6$

5 $y = 10^{-x} \quad 0 \le x \le 6$

6 $y = (0.4)^x \quad -3 \le x \le 3$

7 $y = 2^{x-1}$

8 $y = \frac{1}{2}(2^x + 2^{-x})$

9 $y = \frac{1}{2}(2^x - 2^{-x})$

10 $y = 2^{-x^2}$

11 $y = 2^{|x|}$

12 $y = 2^{-|x|}$

13 $y = 1 - 2^{-x}$

14 $y = 2^x - 3$

15 $y = 3^x - 2^x$

16 $y = 2^x - x$.

If you graph $y = f(x)$ and $y = g(x)$ on the same coordinate plane, each intersection of the graphs corresponds to a solution of $f(x) = g(x)$, and vice versa. Use this technique to determine the number of solutions of

17 $x = 5^{-x}$

18 $x = 2^x$

19 $10^x = x + 1$

20 $5^x = 2 - x^2$.

21 Find a positive function $f(x)$ for which $f(2x) = f(x)^2$.

22 Find a positive function $f(x)$ for which $f(x + 1) = 3f(x)$.

23 Suppose a certain culture of bacteria doubles in 4.5 hours. By how much does it multiply in 10 hours? [See Example 1, p. 173.]

24 A culture of bacteria contains 8500 bacteria now, and we know that it doubles in 10 hours. How many bacteria were there 3 hours ago? [See Example 1, p. 173.]

25 The temperature of a certain container of hot oil being cooled is

$$u(t) = 25 + 130 \times 1.208^{-t}\ ^\circ\mathrm{C}$$

after t minutes. Find (a) its initial temperature, (b) its ultimate temperature (t very large), (c) its temperature after 12 minutes.

26 (cont.) Show that the temperature drops half way from 155 °C to 25 °C after approximately 3.67 minutes.

27 Show that money invested at 16% annual interest, compounded annually, will double in about 4 years 8 months. [See Example 3, p. 173.]

28 Show that money invested at $11\frac{1}{2}$% annual interest, compounded annually, will triple in about 10 years 1 month. [See Example 3, p. 173.]

29 Which is worth more now, \$1000 invested at 4% eleven years ago, or \$1000 invested at 8% six years ago? [See Example 3, p. 173.]

30 How much invested at 5% twelve years ago would be worth today exactly as much as \$2000 invested at 10% six years ago? [See Example 3, p. 173.]

2 INVERSE FUNCTIONS

Our next goal is to define a new function, $x = \log_a y$, called the logarithm function to the base a. We'll do this by "inverting" the exponential function $y = a^x$. First, let's examine the idea of inverse functions.

Inverse Functions

We start with an easy example, $f(x) = 2x$. Suppose x is given. The function f assigns to x the number $y = 2x$. Now suppose that not x, but y is given. Then to y corresponds $x = \frac{1}{2}y$. This correspondence is a function, $g(y) = \frac{1}{2}y$. It is obtained by solving $y = 2x$ for x in terms of y.

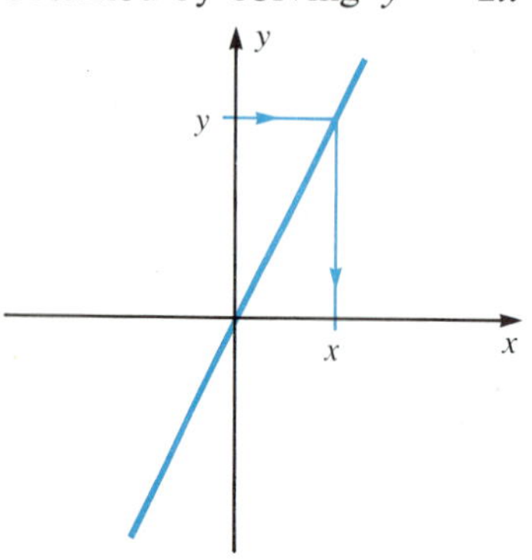

Fig. 1 The graph of $y = 2x$ assigns a y to each x. But it also assigns an x to each y!

Look at the graph of $y = 2x$ in Fig. 1. To each x on the x-axis corresponds a unique y on the y-axis. But seen from the y-axis, the graph assigns to each y a unique x on the x-axis. Thus, related to the function $y = f(x)$ there is a function $x = g(y)$. The line $y = 2x$ is the graph of either of these functions; which one depends on your point of view.

The function f takes you from x to y. Then g takes you from this y back to x. Thus

$$g[f(x)] = \tfrac{1}{2}f(x) = \tfrac{1}{2}(2x) = x.$$

Similarly, g takes you from y to x, then f takes you from x back to y:

$$f[g(y)] = 2g(y) = 2(\tfrac{1}{2}y) = y.$$

The actions of f and g neutralize each other. Apply one after the other and you are back where you started. In the language of composite functions,

$$(g \circ f)(x) = x \qquad (f \circ g)(y) = y.$$

Inverse Function Suppose $f(x)$ and $g(y)$ are functions such that

$$g[f(x)] = x \text{ for all } x \text{ in the domain of } f$$

$$f[g(y)] = y \text{ for all } y \text{ in the domain of } g.$$

Then g is the **inverse function** of f and f is the **inverse function** of g.

For a black box interpretation, see Fig. 2.

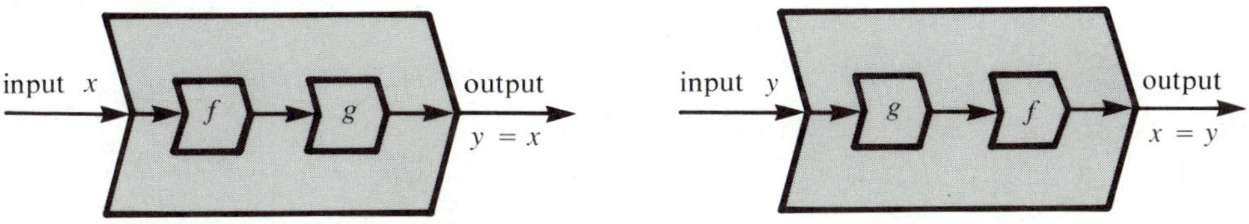

Fig. 2 Inverse functions: $\begin{cases} (g \circ f)(x) = x \\ (f \circ g)(y) = y \end{cases}$

Examples

(1) $f(x) = 2x - 5 \qquad g(y) = \tfrac{1}{2}(y + 5)$.

Then $g[f(x)] = \tfrac{1}{2}[f(x) + 5] = \tfrac{1}{2}[(2x - 5) + 5] = x$

and $f[g(y)] = 2g(y) - 5 = 2 \cdot \tfrac{1}{2}(y + 5) - 5 = y$.

(2) $f(x) = x^3 \qquad g(y) = y^{1/3}$.

Then $g[f(x)] = [f(x)]^{1/3} = (x^3)^{1/3} = x$

and $f[g(y)] = [g(y)]^3 = (y^{1/3})^3 = y$.

In our examples so far, we have found the inverse function $x = g(y)$ by solving $y = f(x)$ explicitly for x in terms of y. But difficulties can arise. We may not be able to solve for x, as in the case of $y = x^7 + 3x$. Or, as in the case of $y = x^2$, there may be more than one value of x for a given y, which is not allowed in the definition of a function. Thus, it may not be clear whether a given function has an inverse. We need a closer look at the question.

Domains and Ranges

We shall have to be very precise about the domains and ranges of our function.

The **domain** of a function is the set of points on which it is defined. When specifying a function, we sometimes explicitly state its domain. Other times, we just give a formula for the function without mentioning a domain. Then it is understood that the domain is the set of all real numbers for which the formula makes sense. For instance, we can define $f_1(x) = \sqrt{x}$ on the domain $\{1 \le x < 2\}$. Then f_1 is a function whose domain is $\{1 \le x < 2\}$. We can also define f_2 by $f_2(x) = \sqrt{x}$ without giving a domain. Then it is understood that the domain of f_2 is the set of all non-negative real numbers, since that is where $\sqrt{x}$ makes sense. The functions f_1 and f_2 are *not* equal because they have different domains.

We haven't used the range of a function much, so let's recall that the **range** of f is the set of all values of f, that is, the set of all real numbers $f(x)$, where x is in the domain of f.

It is useful to have the notation

dom(f)	domain of f
range(f)	range of f.

Intervals

We'll be dealing with functions whose domains and ranges are intervals. Recall that an **interval** is the set of all real numbers between two given real numbers, possibly including one or both of the end points. Examples are

$$\{x \mid -2 < x < \sqrt{19}\} \qquad \{x \mid -2 \le x \le \sqrt{19}\}$$
$$\{x \mid -2 \le x < \sqrt{19}\} \qquad \{x \mid -2 < x \le \sqrt{19}\}.$$

We'll use the shorter set notation

$$\{-2 < x < \sqrt{19}\} \qquad \{-2 \le x \le \sqrt{19}\} \qquad \text{etc.}$$

We allow also infinite intervals where the "end points" may be $-\infty$ or $+\infty$. Examples are

$$\{-\infty < x < \infty\} \qquad \{-\infty < x < 2\} \qquad \{10 \le x < \infty\}$$

which we abbreviate as

$$\{\text{all } x\} \qquad \{x < 2\} \qquad \{10 \le x\}.$$

Betweenness Property One property distinguishes intervals from all other sets on the line: if an interval contains two points a and b, then it contains all points between a and b.

This is proved in more advanced courses.

Existence of Inverse Functions

Examine the graphs in Fig. 3. Each of the functions shown has as domain an interval **I** and as range an interval **J** . In (a) and (b), it is clear that to each y in **J** corresponds a unique x in **I**. Hence an inverse function exists. But in (c), there is a y in **J** to which corresponds several points x in **I**. Hence no inverse function can exist.

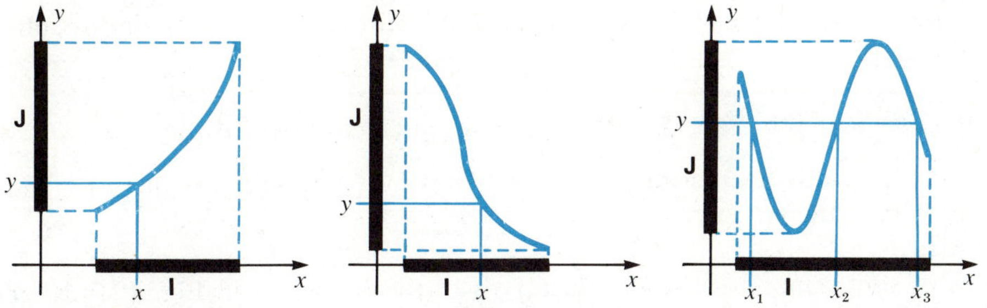

(a) one x for each y (b) one x for each y (c) several x for some y

Fig. 3 The domain **I** is the set on which the function is *defined*.
The range **J** is the set of *values* of the function.

With these examples in mind, we now consider functions $f(x)$ with three properties:

(1) The domain of $f(x)$ is an interval **I**.

(2) $f(x)$ is **strictly increasing,** that is, whenever x_1 and x_2 are points of **I** and $x_1 < x_2$, then $f(x_1) < f(x_2)$.

(3) the range of $f(x)$ is an interval **J**.

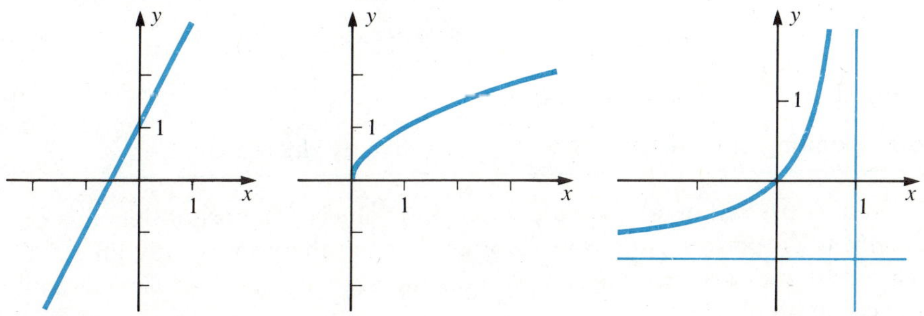

(a) $f(x) = 2x + 1$
$\text{dom}(f) = \{\text{all } x\}$
$\text{range}(f) = \{\text{all } y\}$

(b) $f(x) = \sqrt{x}$
$\text{dom}(f) = \{x \geq 0\}$
$\text{range}(f) = \{y \geq 0\}$

(c) $f(x) = \dfrac{x}{1 - x}$
$\text{dom}(f) = \{x < 1\}$
$\text{range}(f) = \{y > -1\}$

Fig. 4 Strictly increasing functions

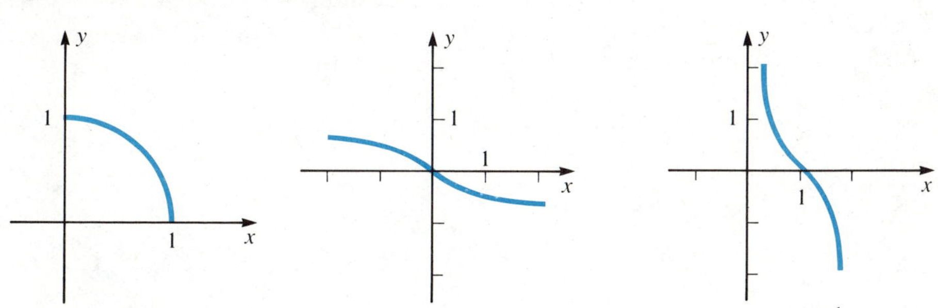

(a) $f(x) = \sqrt{1 - x^2}$
$\text{dom}(f) = \{0 \leq x \leq 1\}$
$\text{range}(f) = \{0 \leq y \leq 1\}$

(b) $f(x) = \begin{cases} \dfrac{-x}{x + 1} & x \geq 0 \\ \dfrac{x}{x - 1} & x < 0 \end{cases}$
$\text{dom}(f) = \{\text{all } x\}$
$\text{range}(f) = \{-1 < y < 1\}$

(c) $f(x) = \dfrac{x - 1}{x(x - 2)}$
$\text{dom}(f) = \{0 < x < 2\}$
$\text{range}(f) = \{\text{all } y\}$

Fig. 5 Strictly decreasing functions

Fig. 4 on the previous page shows examples of such functions. Note their domains and ranges.

We also consider functions for which condition (2) is replaced by

(2′) $f(x)$ is **strictly decreasing,** that is, whenever x_1 and x_2 are points of **I** and $x_1 < x_2$, then $f(x_1) > f(x_2)$.

Examples of such functions are the negatives of the functions in Fig. 4. Further examples are shown in Fig. 5, previous page.

We now come to the main point of this section. It is a theorem that guarantees the existence of an inverse function for any function that satisfies (1), (2) or (2′), and (3).

Existence of Inverse Functions
Let $f(x)$ be a strictly increasing function whose domain and range are intervals **I** and **J** respectively. Then there exists a unique inverse function $g(y)$ with domain **J** and range **I** such that

$$f[g(y)] = y \text{ for all } y \text{ in } \mathbf{J},$$

and

$$g[f(x)] = x \text{ for all } x \text{ in } \mathbf{I}.$$

The function $g(y)$ is strictly increasing.

The corresponding statements hold if $f(x)$ is strictly decreasing.

This theorem is proved in advanced courses in analysis. The proof depends on the Betweenness Property of intervals. Notice that the theorem is only an existence statement; it does not give any formula for the inverse function $g(y)$ or any method for computing it.

Example 1 Let $f(x) = ax + b \quad (a > 0)$. Show that $f(x)$ has an inverse function $g(y)$. Compute $g(y)$ and give its range and domain.

Solution Obviously $f(x)$ is strictly increasing and $\operatorname{dom}(f) = \operatorname{range}(f) = \{\text{all reals}\}$. Therefore it has an inverse function $g(y)$ with $\operatorname{dom}(g) = \operatorname{range}(g) = \{\text{all reals}\}$. To compute $g(y)$, solve $y = ax + b$ for x:

$$x = \frac{1}{a}y - \frac{b}{a} \qquad g(y) = \frac{1}{a}y - \frac{b}{a}.$$

Check

$$f[g(y)] = ag(y) + b = a\left(\frac{1}{a}y - \frac{b}{a}\right) + b = y$$

$$g[f(x)] = \frac{1}{a}f(x) - \frac{b}{a} = \frac{1}{a}(ax + b) - \frac{b}{a} = x.$$

Answer $g(y) = \frac{1}{a}y - \frac{b}{a} \qquad \operatorname{dom}(g) = \operatorname{range}(g) = \{\text{all reals}\}$

Example 2 Let $f(x) = x^2 \quad \operatorname{dom}(f) = \{0 \le x\}$. Show that $f(x)$ has an inverse function $g(y)$. Compute $g(y)$ and give its range and domain.

Solution If $0 \leq x_1 < x_2$, then $x_1^2 < x_2^2$. Hence $f(x)$ is strictly increasing. Its range is the interval $\{0 \leq y\}$. Therefore $f(x)$ has an inverse $g(y)$ with domain $\{0 \leq y\}$ and range $\{0 \leq x\}$. Solve $y = x^2$ for x:

$$x = \sqrt{y}, \quad \text{that is,} \quad g(y) = \sqrt{y} \qquad y \geq 0.$$

Check

$$f[g(y)] = [g(y)]^2 = (\sqrt{y})^2 = y$$

$$g[f(x)] = \sqrt{f(x)} = \sqrt{x^2} = x.$$

Answer $g(y) = \sqrt{y} \qquad \text{dom}(g) = \text{range}(g) = \{\text{reals} \geq 0\}$

Remark The natural domain of $f(x) = x^2$ is $\{\text{all } x\}$. But on this domain, $f(x)$ has no inverse function. That's because $y = x^2$ implies $x = \pm\sqrt{y}$, two values of x for each positive y. To remove the difficulty, we restricted the domain of $f(x) = x^2$ to $\{x \geq 0\}$. Then $x = +\sqrt{y}$ only, so an inverse function exists. Geometrically speaking, we cut down the domain of $f(x)$ to the largest sub-domain on which the function is strictly increasing.

Example 3 Let $f(x) = \dfrac{1}{x}$ $\quad \text{dom}(f) = \{x > 0\}$. Show that $f(x)$ has an inverse function $g(y)$. Compute $g(y)$ and give its range and domain.

Solution If $0 < x_1 < x_2$, then $1/x_1 > 1/x_2$ so $f(x)$ is strictly decreasing. Its range is $\{0 < y\}$. Hence $f(x)$ has an inverse function $g(y)$ with $\text{dom}(g) = \{0 < y\}$ and $\text{range}(g) = \{0 < x\}$. Solve $y = 1/x$ for x:

$$x = g(y) = 1/y.$$

Check

$$f[g(y)] = \frac{1}{g(y)} = \frac{1}{1/y} = y \qquad g[f(x)] = \frac{1}{f(x)} = \frac{1}{1/x} = x.$$

Answer $g(y) = \dfrac{1}{y} \qquad \text{dom}(g) = \text{range}(g) = \{\text{positive reals}\}$

Example 4 Let $f(x) = \dfrac{x}{1 - x}$ $\quad \text{dom}(f) = \{x < 1]$. Show that $f(x)$ has an inverse function $g(y)$. Compute $g(y)$ and give its range and domain.

Solution Rewrite $f(x)$:

$$f(x) = \frac{(x - 1) + 1}{1 - x} = -1 + \frac{1}{1 - x}.$$

As x increases on the interval $-\infty < x < 1$, then $1 - x$ decreases from ∞ to 0. Hence $1/(1 - x)$ increases from 0 to ∞. As a result, $f(x)$ is a strictly increasing function and its values run from -1 to ∞. See Fig. 4c, page 177.

According to the theorem, $f(x)$ has an inverse function $g(y)$ with $\text{dom}(g) = \{-1 < y\}$ and $\text{range}(g) = \{x < 1\}$. To compute $g(y)$, solve $y = f(x)$ for x:

$$y = \frac{x}{1 - x} \qquad y(1 - x) = x \qquad y - yx = x.$$

Hence

$$y = yx + x = (y + 1)x \qquad x = g(y) = \frac{y}{y + 1}.$$

Check

$$f[g(y)] = \frac{g(y)}{1 - g(y)} = \frac{\dfrac{y}{y+1}}{1 - \dfrac{y}{y+1}} = \frac{y}{(y+1) - y} = y$$

$$g[f(x)] = \frac{f(x)}{f(x) + 1} = \frac{\dfrac{x}{1-x}}{\dfrac{x}{1-x} + 1} = \frac{x}{x + (1 - x)} = x.$$

Answer $g(y) = \dfrac{y}{y+1}$ $\quad \text{dom}(g) = \{-1 < y\}$ $\quad \text{range}(g) = \{x < 1\}$

In the examples above, $f(x)$ was simple enough that we could solve $y = f(x)$ for $x = g(y)$. But that is not always the case. For example, take $y = x^7 + 3x$. This function is strictly increasing, so our theorem guarantees the existence of an inverse function $g(y)$. But there is no formula for x in terms of y. Even approximating $g(1)$ isn't easy. It amounts to approximating the real solution of

$$x^7 + 3x = 1.$$

(Actually $x \approx 0.3331814032$.) We simply have to swallow the fact that $g(y)$ is a legitimate function but there is no algebraic formula for it.

Some Questions of Notation

The function f that assigns to each real number its square can be written

$$f(x) = x^2 \quad \text{or} \quad f(t) = t^2 \quad \text{or} \quad f(y) = y^2.$$

It doesn't matter what symbol represents the independent variable.

We have been writing $x = g(y)$ for the inverse of $f(x)$. But this makes us a little uncomfortable because we are used to x as the independent variable and y as the dependent variable. However, it doesn't matter what symbols we use for the variables; there is nothing to stop us from writing $y = g(x)$. For instance, we can write $y = \sqrt{x}$ as the inverse of $y = x^2$. Here $x \geq 0$ for both functions. But in general $y = f(x)$ and $y = g(x)$ may have different domains. Therefore, when we write $y = f(x)$, then x is in the domain of f; when we write $y = g(x)$, then x is in the domain of g.

Graphs

Suppose we have the graph of a strictly increasing (or decreasing) function $y = f(x)$. Then we can easily find the graph of its inverse function $y = g(x)$. We just flip the graph of $y = f(x)$ over the line $y = x$.

Here is why it works. Each point (x, y) on the graph of $y = f(x)$ satisfies $x = g(y)$. In other words, the graph is also the graph of $x = g(y)$. But we want the graph of $y = g(x)$, so we interchange x and y. Now interchanging the coordinates of a point simply reflects that point in the line (mirror) $y = x$. See Fig. 6a. Therefore the graph of $y = g(x)$ is the reflection of the graph of $x = g(y)$, that is, the reflection of the graph of $y = f(x)$. See Fig. 6b.

Graphs of Inverse Functions Let f and g be a pair of inverse functions. Then the graphs of $y = f(x)$ and $y = g(x)$ are mirror images of each other in the line $y = x$.

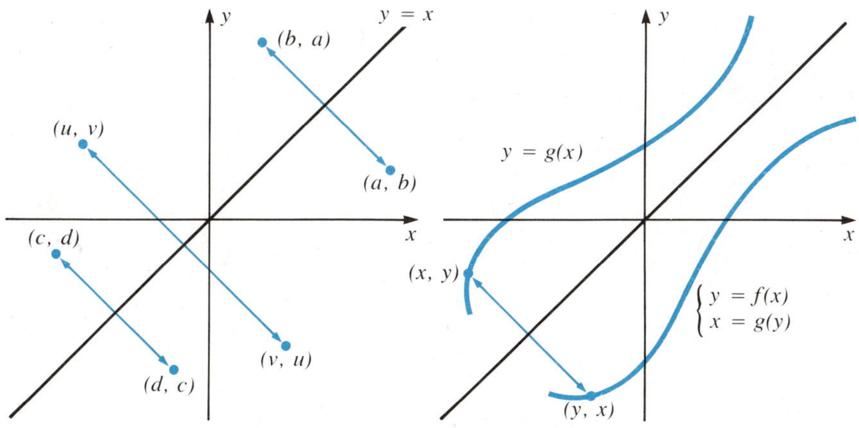

(a) reflection in $y = x$ (b) reflection in $y = x$ of $y = f(x)$ yields the inverse function $y = g(x)$

Fig. 6 Graph of inverse function

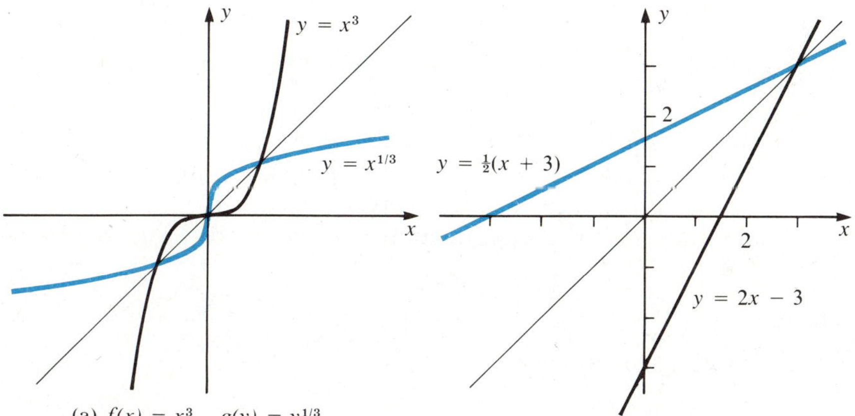

(a) $f(x) = x^3 \quad g(y) = y^{1/3}$ (b) $f(x) = 2x - 3 \quad g(y) = \frac{1}{2}(y + 3)$

Fig. 7 Pairs of inverse functions

Some examples are shown in Fig. 7.

Exercises

Explain why the function has an inverse, and find the domain of the inverse function $g(y)$

1 $f(x) = x^3 \quad \{\text{all } x\}$

2 $f(x) = -1/x \quad \{x < 0\}$

3 $f(x) = -1/x \quad \{x > 0\}$

4 $f(x) = 1/x^2 \quad \{x < 0\}$

5 $f(x) = x^2 \quad \{3 < x < 4\}$

6 $f(x) = 2x^2 - 5 \quad \{3 < x < 4\}$

7 $f(x) = x^5 + x^3 \quad \{\text{all } x\}$

8 $f(x) = x^7 + 3x \quad \{\text{all } x\}$

9 $f(x) = 1/x^3 \quad \{x > 0\}$

10 $f(x) = 1/x \quad \{x < 0\}$

11 $f(x) = x^2 \quad \{x \le 0\}$

12 $f(x) = x^4 + x^2 \quad \{x \le 0\}$.

Find the inverse function and its domain

13 $f(x) = \frac{1}{8}x^3 \quad \{\text{all } x\}$

14 $f(x) = 1/x \quad \{x < 0\}$

15 $f(x) = -1/x \quad \{x > 0\}$

16 $f(x) = 4x^2 \quad \{x \le 0\}$

17 $f(x) = 5/x^2 \quad \{x > 0\}$

18 $f(x) = 1 - 3x^2 \quad \{x \ge 0\}$

19 $f(x) = \sqrt{x + 1} \quad \{x \ge -1\}$

20 $f(x) = \sqrt[3]{x} \quad \{\text{all } x\}$

21 $f(x) = \sqrt{x^2 + 4} \quad \{x \ge 0\}$

22 $f(x) = \sqrt{x^3 + 27} \quad \{x \ge -3\}$

23 $f(x) = \dfrac{1}{x^4 + 10} \quad \{x \le 0\}$

24* $f(x) = x + \dfrac{1}{x} \quad \{x \ge 1\}$

25 $f(x) = \dfrac{-1}{x^2}\quad \{x < 0\}$

26* $f(x) = \dfrac{x}{1 + |x|}\quad \{\text{all } x\}.$

Compute $g(4)$ where g is the inverse function of f

27 $f(x) = x^7 + 3x + 4\quad \{\text{all } x\}$

28 $f(x) = x^3 - 23\quad \{\text{all } x\}$

29 $f(x) = \dfrac{x}{1 - x}\quad \{x < 1\}$

30 $f(x) = x^4 + x^2\quad \{x \geq 0\}.$

On the same graph plot $y = f(x)$ and $y = g(x)$, where g is the inverse function of f

31 $f(x) = \frac{1}{2}x + 1$

32 $f(x) = -3x + 2$

33 $f(x) = x^2\quad \{x \geq 0\}$

34 $f(x) = 1/x\quad \{x > 0\}$

35 $f(x) = 1/x^2\quad \{x > 0\}$

36 $f(x) = \sqrt{1 + x}\quad \{x \geq -1\}$

37 $f(x) = \begin{cases} x + 1 & 0 \leq x \leq 1 \\ x^2 + 1 & 1 \leq x < \infty \end{cases}$

38 $f(x) = \begin{cases} x^2 & 0 \leq x \leq 1 \\ \sqrt{x} & 1 \leq x < \infty. \end{cases}$

39 Suppose $g(y)$ is the inverse of $f(x)$. Find the domain and inverse of $f(-x)$.

40* Suppose $f_1(x)$ and $f_2(x)$ have inverses $g_1(x)$ and $g_2(x)$ respectively. Suppose the range of f_1 lies in the domain of f_2. Show that $f_2 \circ f_1$ has the inverse $g_1 \circ g_2$.

3 LOGARITHM FUNCTIONS

We are ready for the inverse of the exponential function $f(x) = a^x$. If $a > 1$, then $y = a^x$ is strictly increasing. Its domain is $\{\text{all } x\}$, and its range is $\{y > 0\}$. By the discussion of Section 2, there exists an inverse function $x = g(y)$ with domain $\{y > 0\}$ and range $\{\text{all } x\}$. The function g is called the **logarithm function** to the **base** a. We write it as

$$x = \log_a y.$$

By the nature of inverse functions,

$$g[f(x)] = x \qquad f[g(y)] = y.$$

In this case, these relations mean

$$\log_a (a^x) = x \qquad a^{\log_a y} = y.$$

As usual, we prefer to interchange x and y. So let us consider the function

$$y = \log_a x.$$

Its graph is the reflection of the graph of $y = a^x$ in the line $y = x$. See Fig. 1.

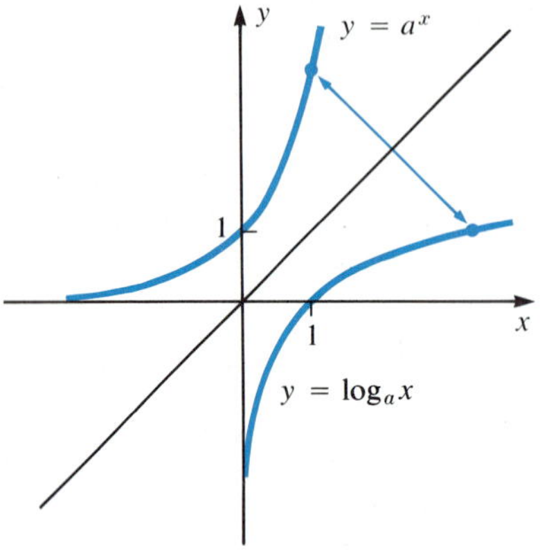

Fig. 1 Inverse functions: $y = a^x$ and $y = \log_a x$

We see that $\log_a x$ is defined for all $x > 0$. It is a strictly increasing function taking each real value exactly once. In fact,

$$\log_a x \longrightarrow \infty \quad \text{as} \quad x \longrightarrow \infty.$$

The y-axis is a vertical asymptote of its graph,

$$\log_a x \longrightarrow -\infty \quad \text{as} \quad x \longrightarrow 0+.$$

Logarithm to base *a*
Let $a > 1$. There is a unique function $y = \log_a x$ defined for all $x > 0$ and satisfying

$$a^{\log_a x} = x \quad (x > 0) \qquad \log_a a^x = x \quad (\text{all } x).$$

The function $\log_a x$ is strictly increasing.

$$\log_a x \longrightarrow \infty \quad \text{as} \quad x \longrightarrow \infty$$

and

$$\log_a x \longrightarrow -\infty \quad \text{as} \quad x \longrightarrow 0+.$$

Since $a^0 = 1$, it follows that $\log_a 1 = 0$. The function $\log_a x$ is strictly increasing. Therefore if $0 < x < 1$, then $\log_a x < \log_a 1 = 0$. Similarly if $x > 1$, then $\log_a x > \log_a 1 = 0$.

Let $a > 1$. Then $\begin{cases} \log_a x < 0 & \text{for } 0 < x < 1 \\ \log_a 1 = 0 & \\ \log_a x > 0 & \text{for } 1 < x. \end{cases}$

Since $\log_a x_1 < \log_a x_2$ if $x_1 < x_2$, the function $\log_a x$ takes different values for different values of x. This fact is important in applications, especially in the following form:

Let $a > 1$. If $\log_a x_1 = \log_a x_2$, then $x_1 = x_2$.

Exponential and Logarithmic Relations

Because the functions a^x and $\log_a x$ are inverse to each other, each relation of the form $y = a^x$ is equivalent to the relation $x = \log_a y$.

Examples

Exponential Relation	Equivalent Logarithmic Relation
$3^0 = 1$	$\log_3 1 = 0$
$3^1 = 3$	$\log_3 3 = 1$
$3^2 = 9$	$\log_3 9 = 2$
$3^3 = 27$	$\log_3 27 = 3$
$3^{-1} = \frac{1}{3}$	$\log_3 \frac{1}{3} = -1$
$3^{-2} = \frac{1}{9}$	$\log_3 \frac{1}{9} = -2$
$3^{1/2} = \sqrt{3}$	$\log_3 \sqrt{3} = \frac{1}{2}$

To find $x = \log_a y$, it suffices to find the number x that satisfies $a^x = y$.

EXAMPLE 1 Find

(a) $\log_2 32$ (b) $\log_5 \frac{1}{25}$ (c) $\log_4 8$ (d) $7^{\log_7 12}$

Solution (a) Set $x = \log_2 32$. Then x must satisfy $2^x = 32$. But $2^5 = 32$, therefore $x = 5$.

(b) Set $x = \log_5 \frac{1}{25}$. Then x must satisfy $5^x = \frac{1}{25}$. But $5^{-2} = \frac{1}{25}$, therefore $x = -2$.

(c) Set $x = \log_4 8$. Then x must satisfy $4^x = 8$. But $4^{3/2} = 2^3 = 8$, therefore $x = \frac{3}{2}$.

(d) Set $x = \log_7 12$. Then x must satisfy $7^x = 12$. Therefore $7^{\log_7 12} = 12$.

Answer (a) 5 (b) -2 (c) $\frac{3}{2}$ (d) 12

EXAMPLE 2 Solve for x

(a) $\log_2(4x - 1) = -3$ (b) $\log_5(x^2 - 6) = \log_5 x$.

Solution (a) Convert the equation to an equivalent exponential form:

$$4x - 1 = 2^{-3} = \frac{1}{8}.$$

Solving for x yields $x = \frac{9}{32}$.

(b) The equation $\log_5(x^2 - 6) = \log_5 x$ holds only if

$$x^2 - 6 = x.$$

Solve this quadratic:

$$x^2 - x - 6 = 0 \qquad (x - 3)(x + 2) = 0 \qquad x = -2, 3.$$

The solution $x = -2$ is unacceptable because $\log_5(-2)$ is undefined. The other solution is OK because if $x = 3$, then $x^2 - 6 = 3$ so $\log_5(x^2 - 6) = \log_5 3$.

Answer (a) $\frac{9}{32}$ (b) 3

Properties of Logarithm Functions

Logarithms satisfy certain rules (algebraic properties) of great importance for both theory and computation:

Rules for Logarithm Functions

(1) $\log_a(x_1x_2) = \log_a x_1 + \log_a x_2$

(2) $\log_a\left(\dfrac{x_1}{x_2}\right) = \log_a x_1 - \log_a x_2$

(3) $\log_a x^r = r \cdot \log_a x$.

These properties are practically restatements of corresponding properties of a^x. Take (1), for instance. Suppose $y_1 = \log_a x_1$ and $y_2 = \log_a x_2$, that is, $x_1 = a^{y_1}$ and $x_2 = a^{y_2}$. Then

$$x_1x_2 = a^{y_1}a^{y_2} = a^{y_1+y_2},$$

so

$$\log_a(x_1x_2) = y_1 + y_2 = \log_a x_1 + \log_a x_2.$$

There are similar verifications of the other two properties; they are left as exercises.

EXAMPLE 3 Given

$$\log_7 2 \approx 0.3562 \qquad \log_7 3 \approx 0.5646 \qquad \log_7 5 \approx 0.8271,$$

estimate (a) $\log_7 15$ (b) $\log_7 \frac{32}{9}$ (c) $\log_7 \sqrt[3]{0.4}$ (d) $\log_7 35$.

Solution (a) By rule (1),

$$\log_7 15 = \log_7(3 \times 5) = \log_7 3 + \log_7 5$$
$$\approx 0.5646 + 0.8271 = 1.3917.$$

(b) By rules (2) and (3),

$$\log_7 \tfrac{32}{9} = \log_7 32 - \log_7 9 = \log_7(2^5) - \log_7(3^2)$$
$$= 5\log_7 2 - 2\log_7 3 \approx 5(0.3562) - 2(0.5646) = 0.6518.$$

(c) By rules (3) and (2),

$$\log_7 \sqrt[3]{0.4} = \log_7(0.4)^{1/3} = \tfrac{1}{3}\log_7 0.4 = \tfrac{1}{3}\log_7 \tfrac{2}{5}$$
$$= \tfrac{1}{3}(\log_7 2 - \log_7 5) \approx \tfrac{1}{3}(0.3562 - 0.8271) \approx -0.1570.$$

(d) By rule (1),

$$\log_7 35 = \log_7(7 \times 5) = \log_7 7 + \log_7 5$$
$$= 1 + \log_7 5 \approx 1 + 0.8271 = 1.8271$$

Answer (a) 1.3917 (b) 0.6518 (c) -0.1570 (d) 1.8271

EXAMPLE 4 Write as $\log_a x$

(a) $-3(\log_a 4 + \log_a 5)$ (b) $2\log_a 7 + \log_a 3$

(c) $\frac{1}{2}\log_a 2 + \frac{1}{2}\log_a 3 - 2\log_a 5$.

Solution

(a) $-3(\log_a 4 + \log_a 5) = -3\log_a(4 \times 5)$

$$= -3\log_a 20 = \log_a 20^{-3} = \log_a \tfrac{1}{8000}.$$

(b) $2\log_a 7 + \log_a 3 = \log_a 7^2 + \log_a 3 = \log_a(7^2 \times 3) = \log_a 147$.

(c) $\frac{1}{2}\log_a 2 + \frac{1}{2}\log_a 3 - 2\log_a 5 = \log_a 2^{1/2} + \log_a 3^{1/2} - \log_a 5^2$

$$= \log_a \sqrt{2} + \log_a \sqrt{3} - \log_a 25 = \log_a(\sqrt{2} \times \sqrt{3}) - \log_a 25$$
$$= \log_a \sqrt{6} - \log_a 25 = \log_a\left(\frac{\sqrt{6}}{25}\right).$$

Answer (a) $\log_a \frac{1}{8000}$ (b) $\log_a 147$ (c) $\log_a \frac{1}{25}\sqrt{6}$.

EXAMPLE 5 Solve $\log_2 x - \log_2(x - 2) = 4$.

Solution By the rules of logarithms,

$$\log_2 x - \log_2(x - 2) = \log_2 \frac{x}{x - 2} = 4.$$

Hence $\dfrac{x}{x - 2} = 2^4 = 16$.

Solve for x:

$$x = 16(x-2) \qquad 15x = 32 \qquad x = \tfrac{32}{15}.$$

This answer checks. Both $\log_2 x$ and $\log_2(x-2)$ are defined for $x = \frac{32}{15}$.

Answer $\frac{32}{15}$

Exercises

Graph

1 $y = \log_2 x$ **2** $y = \log_3 x$ **3** $y = \log_{10} x$
4 $y = \log_2 |x|$ **5** $y = \log_2 x^2$ **6** $y = \log_2(x^2+1)$.

The graph of $y = \log_{10} x$ rises very slowly.

7 Where does it reach the level $y = 1$? 2? 10?

8 How much does it rise as x increases from 1,000,000 to 10,000,000?

Express as a statement about logarithms

9 $2^{10} = 1024$ **10** $5^4 = 625$
11 $8^{-2/3} = \frac{1}{4}$ **12** $\sqrt{961} = 31$

Find (without a calculator)

13 $\log_2 128$ **14** $\log_2 1/(2\sqrt{2})$ **15** $\log_2 \sqrt[5]{16}$
16 $\log_3 81$ **17** $\log_3 9\sqrt{3}$ **18** $\log_4 4$
19 $\log_4 2$ **20** $\log_4 1024$ **21** $\log_5 5^{12}$
22 $\log_5 0.008$ **23** $\log_{10} 0.0001$ **24** $\log_8 64^{15}$
25 $\log_{24} 1$ **26** $10^{\log_{10} 17}$ **27** $\log_2 0.03125$
28 $\log_{16} \frac{1}{2}$ **29** $2^{3 \log_2 5}$ **30** $\log_{a^2} a$.

Solve for x

31 $\log_{12} x = 0$ **32** $\log_5 x = 2$
33 $\log_3(x+5) = 4$ **34** $\log_3(x+7) = -1$
35 $\log_4(2x+3) = \frac{1}{2}$ **36** $\log_{27} x^2 = \frac{1}{3}$
37 $\log_9(x^2-10) = 1$ **38** $\log_5(x^2-7x+7) = 0$
39 $\log_3 3^{2x+1} = 15$ **40** $\log_3 27^x = 4.5$
41 $\log_x 8 = 3$ **42** $\log_x 8 = -\frac{3}{2}$

Use the approximations $\log_{11} 2 \approx 0.2891$ $\log_{11} 3 \approx 0.4582$ $\log_{11} 5 \approx 0.6712$ to estimate

43 $\log_{11} 6$ **44** $\log_{11} 48$ **45** $\log_{11} \frac{9}{16}$
46 $\log_{11} \sqrt{12}$ **47** $\log_{11} 45$ **48** $\log_{11} 225$
49 $\log_{11}(\frac{1}{96}\sqrt{5})$ **50** $\log_{11} \sqrt[3]{\frac{36}{5}}$ **51** $\log_{11} \frac{3}{25}$
52 $\log_{11}(\frac{6}{125})^{1/5}$.

Write as $\log_a x$

53 $\log_a 6 + 3\log_a 2$ **54** $\log_a \sqrt{2} - \frac{5}{2}\log_a 4$

55 $\log_a 2\cdot 3 - \log_a 3\cdot 4 + \log_a 4\cdot 5 - \log_a 5\cdot 6$

56 $2\log_a 2 + 4\log_a 4 - 8\log_a 8$.

Solve for x

57 $\log_6 x + \log_6(x-1) = 1$ **58** $\log_6(2x+1) - \log_6(2x-1) = 1$

59 $\log_2 \sqrt{x} + \log_2 \dfrac{1}{x} = 3$ **60** $\frac{1}{2}\log_3 x - \frac{1}{3}\log_3 x^2 = 1$

61 $\log_2 x^2 - \log_2(3x+8) = 1$ **62** $\log_{10}(\log_{10} x) = 1$.

63 Prove Rule (2) on page 184. **64** Prove Rule (3) on page 184.

65 Given an example showing that $\log_a(x_1+x_2) \neq \log_a x_1 + \log_a x_2$.

66 Find a non-zero function $f(x)$ satisfying $f(x_1x_2) = f(x_1) + f(x_2)$.

67 Find all x such that $-2 < \log_{10} x < -1$.

68 Find the domain of $\log_2 \log_2 x$.

4 COMPUTING LOGARITHM FUNCTIONS

For practical computations, the most useful base of logarithms is 10. This is because we write numbers in decimal notation. A key feature of decimal notation is the way we multiply a number by powers of 10. We merely shift the decimal point. For logarithms to the base 10, there is a corresponding feature:

$$\log_{10}(10^n x) = \log_{10} 10^n + \log_{10} x = n + \log_{10} x.$$

Thus multiplying x by a power of 10 (shifting the decimal point) just adds an integer to $\log_{10} x$. For instance,

$$\log_{10}(1057.398) = 3 + \log_{10}(1.057398),$$

$$\log_{10}(0.000092) = -5 + \log_{10} 9.2.$$

The logarithm of x to the base 10 is called the **common logarithm** of x and is written $\log x$. When you see the expression $\log x$ without an explicit base, then base 10 is understood.

Until recently, computations with logarithms were done using tables of common logs. Now hand-held calculators have made log table calculations obsolete. Most scientific calculators have keys for $\log x$ and 10^x. To find $\log x$ on your calculator, just key in x and follow by [log].

Examples

		Keys	Display
(1)	$\log(1057.398)$	1 0 5 7 . 3 9 8 [log]	3.0242388
(2)	$\log(0.000092)$	. 0 0 0 0 9 2 [log]	-4.0362121

Exponential Equations

Common logarithms are useful in equations involving exponents. The calculator is a magnificent tool for finding numerical solutions.

Example 1 Solve for x $\quad 7^x = 54$.

Solution Since 54 is a little larger than 7^2, the answer should be a little more than 2. Now the exponential statement $7^x = 54$ is equivalent to the logarithmic statement

$$x = \log_7 54.$$

Theoretically at least, this solves the problem. But this theoretical solution does not give the numerical value of x.

To estimate the number x, apply the function log to both sides of the original equation $7^x = 54$:

$$\log 7^x = \log 54 \qquad x \log 7 = \log 54 \qquad x = \frac{\log 54}{\log 7}.$$

Now find log 54 and log 7, and divide. This can all be done on your calculator by a short sequence of steps:

5 4 [log] [÷] 7 [log] [=] $\qquad$ 2.049932

The result is slightly more than 2, as predicted.

Answer 2.049932

EXAMPLE 2 Estimate the solution of

(a) $11^x = 5000$ (b) $5^{-x} = 0.038$.

Solution (a) Apply the function log to both sides:

$$x \log 11 = \log 5000 \qquad x = \frac{\log 5000}{\log 11}.$$

Now estimate x by the key sequence

5 0 0 0 [log] [÷] 1 1 [log] [=] 3.551945

(b) Apply log to both sides:

$$-x \log 5 = \log 0.038 \qquad x = -\frac{\log 0.038}{\log 5}.$$

Now use the sequence

. 0 3 8 [log] [÷] 5 [log] [=] [+/−] 2.031870

Answer (a) 3.551945 (b) 2.031870

EXAMPLE 3 Estimate the solution of $5^x = 9 \times 4^x$.

Solution 1 Apply log to both sides and use the rules for logarithms:

$$\log 5^x = \log(9 \times 4^x) = \log 9 + \log 4^x$$

$$x \log 5 = \log 9 + x \log 4$$

$$x(\log 5 - \log 4) = \log 9.$$

Therefore

$$x = \frac{\log 9}{\log 5 - \log 4} = \left(\frac{1}{\log 5 - \log 4}\right)(\log 9).$$

Now calculate x by the sequence

5 [log] [−] 4 [log] [=] [1/x] [×] 9 [log] [=] 9.846686

Solution 2 Divide both sides by 4^x and use laws of exponents:

$$\frac{5^x}{4^x} = 9 \qquad (\tfrac{5}{4})^x = 9 \qquad (1.25)^x = 9.$$

By the usual technique

$$x \log 1.25 = \log 9 \qquad x = \frac{\log 9}{\log 1.25} \approx 9.846686$$

The answer is the same as in Solution 1 because

$$\log 1.25 = \log \tfrac{5}{4} = \log 5 - \log 4.$$

Answer 9.846686

Inverse Logarithms (Antilogs)

Suppose you are given $\log x = 1.816$. How do you find x? Remember that $\log x$ and 10^x are inverse functions. Therefore

$$x = 10^{\log x} = 10^{1.816}$$

If your calculator has a [10^x] key, use the sequence

1 . 8 1 6 [10^x] 65.463617

If it doesn't have a [10^x] key, then the sequence [INV] [log] will do the same thing. Another route is

1 0 [y^x] 1 . 8 1 6 [=] 65.463617

Logs to Other Bases

For practical computations with logarithms, the most useful base is 10 because it goes so well with decimals. However, for theoretical work the most natural base is a number called e. To 10 places,

$$e = 2.71828\,18285.$$

A strict definition of e requires calculus. There is a nice geometric interpretation of e, however. The graph of each exponential function passes through the point $(0, 1)$. Among all of these, only the graph of $y = e^x$ has slope 1 at that point (Fig. 1).

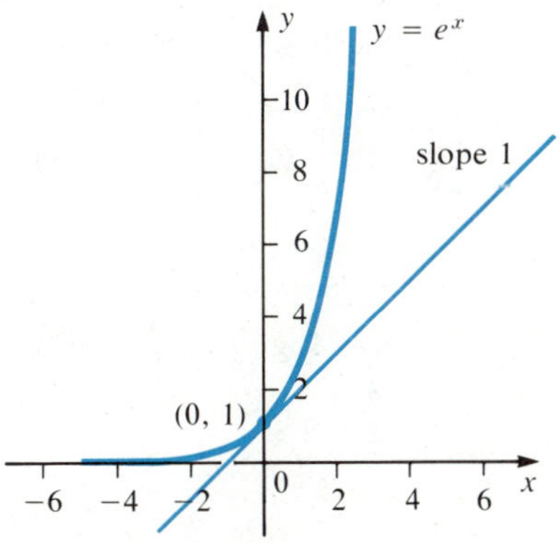

Fig. 1 $y = e^x$

The function $\log_e x$ is called the **natural logarithm** function. It is written $\ln x$. You can easily obtain $\ln x$ and e^x on any scientific calculator. Many have keys for each of these functions. Others may not have a key for e^x, but you can calculate e^x as the inverse of $\ln x$. For example, you can find $e^{3.86}$ by the sequence

3 . 8 6 [INV] [ln] 47.46535

Natural Logarithms

Suppose we need a numerical value for $\log_a x$ and a is not 10 or e. There is no calculator key for this, so what can we do? Actually we have already run into this problem. Back in Example 1, we found that the *theoretical* solution to $7^x = 54$ was $\log_7 54$. For a *numerical* solution, we applied log to both sides of the equation and found $x = (\log 54)/(\log 7)$. The answers must be the same. Therefore

$$\log_7 54 = \frac{\log 54}{\log 7}.$$

There is a general principle here. To find $y = \log_a x$, write the equivalent exponential relation,

$$a^y = x.$$

Apply log to both sides:

$$\log a^y = \log x$$

$$y \cdot \log a = \log x\,.$$

Now solve for y.

$$\boxed{\log_a x = \frac{\log x}{\log a} = \left(\frac{1}{\log a}\right) \log x\,.}$$

This formula shows that each function $\log_a x$ is just a constant multiple of $\log x$. For example

$$\log_2 x \approx (3.321928) \log x$$

because $1/\log 2 \approx 3.321928$.

Suggestion Don't memorize the formula in the box. It's better to remember the way it was derived: by applying log to both sides of $a^y = x$.

EXAMPLE 4 Let $1 < a < b$. Show that $(\log_a b)(\log_b a) = 1$.

Solution 1 Apply the boxed formula twice, once to $\log_a$ with $x = b$ and once to $\log_b$ with $x = a$:

$$\log_a b = \frac{\log b}{\log a} \qquad \log_b a = \frac{\log a}{\log b}.$$

Clearly the product $(\log_a b)(\log_b a)$ equals 1.

Solution 2 Set $y = \log_a b$ so $a^y = b$. Apply the function $\log_b$ to both sides:

$$\log_b a^y = \log_b b = 1 \qquad y \cdot \log_b a = 1\,.$$

But $y = \log_a b$, so we are done.

EXAMPLE 5 Estimate an x for which $\log_3 x = \log x + 1$.

Solution Replace $\log_3 x$ by an equivalent expression in terms of common logs:

$$\frac{\log x}{\log 3} = \log x + 1\,.$$

Solve for $\log x$:

$$\left(\frac{1}{\log 3} - 1\right) \log x = 1 \qquad \log x = \frac{1}{\dfrac{1}{\log 3} - 1}.$$

Calculate the number on the right, then press [10ˣ] (or [INV] [log]) to obtain x:

3 [log] [1/x] [−] 1 [=] [1/x] [10ˣ] 8.175029

Answer 8.175029

Exercises

Calculate to 6 places

1 $\log 2.385$ **2** $\log 0.01105$ **3** $\log(2.385 \times 10^{26})$

4 $\log(4.7 \times 10^{-13})$ **5** $\ln 3.14159$ **6** $\ln 178{,}000$.

Estimate the solution to 6 significant figures

7 $2^x = 10$ **8** $3^x = 77$ **9** $3^{-x} = 0.002$
10 $10^{-x} = 1.7$ **11** $1066^x = 1492$ **12** $1492^x = 1066$
13 $(3.5)^x = 4$ **14** $(9.1)^x = 5.8$ **15** $(1.001)^x = 10$
16 $(0.999)^x = 0.1$ **17** $(0.99)^x = 1.01$ **18** $(1.001)^x = 0.999$
19 $10^x = e$ **20** $e^{-x} = 3.82$ **21** $\log\log x = 1.013$
22 $\log_x 10 = 3$ **23** $\log_x 10 = -1.54$ **24** $8^x = 9^x$.

Estimate to 6 places

25 $\log_7 5$ **26** $\log_{11} 9$ **27** $\log_8 2$
28 $\log_{13} 17$ **29** $\log_{1.48} 2.2$ **30** $\log_{6.331} 1.049$.

Estimate the solution to 6 significant figures

31 $\log_4 x = 2 + \log_5 x$ **32** $\ln x - \log x = 1$
33 $(\log x)^2 = \ln x$ **34** $\log_2 x + \log_3 x + \log_4 x = 3$.

Estimate the base a to 5 places

35 $\log_a x = \frac{3}{4}\log x \quad \text{all } x > 0$ **36** $\log_a x = \frac{1}{2}\log x \quad \text{all } x > 0$.

5 APPLICATIONS

Natural Growth

We shall discuss a growth law that occurs frequently in nature. To illustrate the idea, let us imagine an experiment. A group of laboratory mice are split randomly into a small group of 100 and a large group of 500. After two months, the first group has increased to 150. What is a reasonable guess for the size of the second group?

Let $S(t)$ denote the population of the smaller group after t months, and $L(t)$ the same for the larger group. We are given that

$$S(0) = 100 \qquad L(0) = 500 \qquad \text{and that} \quad S(2) = 150.$$

We want an estimate for $L(2)$.

One "solution" is to assume that both populations grow at the *same constant rate*. Then

$$S(t) = 100 + kt \qquad \text{and} \qquad L(t) = 500 + kt,$$

where k is the rate of increase. But this is nonsense: in t months, each population would increase by the same amount, kt. Obviously the larger group should have a larger increase.

Let's think. If a population of 100 increases to 150 in two months, that is a 50% increase. Hence a population of 500 also should increase by 50% in the same time, so $L(2) = 750$. This is the key. The populations do not grow at the same rate, but they do increase by the same percentage of themselves in a fixed time. So, regardless of its actual size, the population should increase by 50% in any period of two months. If so, then

$$S(2) = S(0) \times 1.5$$

$$S(4) = S(2) \times 1.5 = [S(0) \times 1.5] \times 1.5 = S(0) \times (1.5)^2$$

$$S(6) = S(4) \times 1.5 = [S(0) \times (1.5)^2] \times 1.5 = S(0) \times (1.5)^3$$

and in general,

$$S(2n) = S(0) \times (1.5)^n = S(0) \times (1.5)^{2n/2}.$$

Thus we obtain an exponential function on the domain $\{2, 4, 6, \cdots, 2n, \cdots\}$. It seems reasonable to interpolate and guess that

$$S(t) = S(0) \times (1.5)^{t/2} = 100 \times (1.5)^{t/2}$$

for any time t. For instance

$$S(3) = 100 \times (1.5)^{3/2} \approx 184$$

$$S(5.5) = 100 \times (1.5)^{5.5/2} \approx 305.$$

The function $L(t)$ has the same growth law (50% increase in population for 2 months) except for a different initial value, so

$$L(t) = L(0) \times (1.5)^{t/2} = 500 \times (1.5)^{t/2}.$$

For instance

$$L(2) = 500 \times 1.5 = 750$$

$$L(3) = 500 \times (1.5)^{3/2} \approx 919.$$

We have been led to consider a growth law of the form

$$F(t) = c \cdot a^t \qquad c \quad \text{a constant}.$$

For such a law, the percentage increase over any time interval of duration d depends only on d, not on the time or the size of $F(t)$. For

$$F(t + d) = c \cdot a^{t+d} = c \cdot a^t \cdot a^d = F(t) \cdot a^d.$$

Thus the population grows by a fixed factor a^d, that is, by a fixed percentage of itself.

For very short time intervals, we can almost imagine "instantaneous rate of growth." This idea is made precise in calculus. It is proved that at each instant, the growth rate of $F(t) = c \cdot a^t$ is proportional to $F(t)$ itself. Conversely, if $F(t)$ is a function growing this way, then $F(t) = c \cdot a^t$.

We now state what we shall call the Natural Growth Law.

Natural Growth Law Let $F(t)$ be a positive function with domain $\{\text{all } t\}$. Suppose the rate of increase of $F(t)$ at any instant of time t is proportional to $F(t)$. Then

$$F(t) = F_0 \cdot a^t$$

where a is a positive constant and $F_0 = F(0)$.

We say that $F(t)$ obeys the **natural growth law.**

EXAMPLE 1 A function $F(t)$ obeys the natural growth law. Suppose its value triples in 5 minutes. (a) Find the function. (b) By what factor does the function increase in 12 minutes? (c) How long does it take the function to multiply six-fold?

Solution (a) We are given

$$F(t) = F_0 \cdot a^t \quad \text{and} \quad F(5) = 3F_0.$$

Therefore

$$F_0 \cdot a^5 = 3F_0 \qquad a^5 = 3 \qquad a = 3^{1/5}$$

$$F(t) = F_0 \cdot a^t = F_0 \cdot (3^{1/5})^t = F_0 \cdot 3^{t/5}.$$

(b) $F(12) = F_0 \cdot 3^{12/5}$

so the function increases to $3^{12/5}$ times its initial value in the first 12 minutes.

(c) We must solve $F(t) = 6F_0$ for t:

$$F_0 \cdot 3^{t/5} = 6F_0 \qquad 3^{t/5} = 6 \qquad \frac{t}{5} \cdot \log 3 = \log 6$$

$$t = 5\frac{\log 6}{\log 3} \approx 8.155 \text{ minutes}.$$

Answer (a) $F(t) = F_0 \cdot 3^{t/5}$ (b) $3^{12/5} \approx 13.97$

(c) $5\dfrac{\log 6}{\log 3} \approx 8.155$ minutes

Population Growth

Under certain conditions the rate of growth of a population is proportional to the current population. Therefore the population obeys the natural growth law.

Example 2 A yeast culture has initial weight 10 g. The weight doubles in 2 hours. Assuming the growth rate is proportional to current weight, find

(a) the growth law (b) the weight after $3\frac{1}{2}$ hours

(c) when the weight reaches 24 g.

Solution (a) Let $W(t)$ denote the weight at t hours. Then

$$W(t) = W_0 \cdot a^t = 10 \times a^t \quad \text{and} \quad W(2) = 20.$$

Therefore

$$10 \times a^2 = 20 \qquad a^2 = 2 \qquad a = 2^{1/2}$$

$$W(t) = 10 \times (2^{1/2})^t = 10 \times 2^{t/2}.$$

(b) $W(3.5) = 10 \times 2^{3.5/2} = 10 \times 2^{1.75} \approx 33.64$ g

(c) Solve $W(t) = 24$ for t:

$$10 \times 2^{t/2} = 24 \qquad 2^{t/2} = 2.4 \qquad \frac{t}{2} \log 2 = \log 2.4$$

$$t = 2\,\frac{\log 2.4}{\log 2} \approx 2.526 \text{ hr}$$

Answer (a) $W(t) = 10 \times 2^{t/2}$ (b) 33.64 g (c) 2.526 hr

Example 3 On 1 Jan. 1980 the population of city A was 1,772,000 and the population of city B was 2,091,000. The population of A is increasing at the rate of 3.13% per year and the population of B is increasing at the rate of 1.66% per year. When will A catch up to B?

Solution The population $A(t)$ of the first city adds $3.13\% = 0.0313$ times its value to itself each year, that is, it *multiplies* by the factor $1 + 0.0313 = 1.0313$ each year. By the natural growth law,

$$A(t) = 1{,}772{,}000 \times (1.0313)^t.$$

Similarly for the second city

$$B(t) = 2{,}091{,}000 \times (1.0166)^t.$$

We must solve $A(t) = B(t)$ for t:

$$1{,}772{,}000 \times (1.0313)^t = 2{,}091{,}000 \times (1.0166)^t$$

$$\frac{(1.0313)^t}{(1.0166)^t} = \frac{2091}{1772} \qquad \left(\frac{1.0313}{1.0166}\right)^t = \frac{2091}{1772} \qquad (1.01446)^t \approx 1.18002$$

$$t\log(1.01446) \approx \log(1.18002) \qquad t \approx \frac{\log(1.18002)}{\log(1.01446)} \approx 11.530 \text{ years}.$$

Now 0.530 years = 0.530×12 months = 6.36 months, and 0.36 months = 0.36×31 days $\approx$ 11 days, so the populations will be equal approximately 11 yr 6 mo 11 days after 1 Jan. 1980.

Answer mid-July 1991.

Natural Decay

If a positive function $F(t)$ *decreases* (decays) at a rate proportional to the value of the function, then

$$F(t) = F_0 \cdot b^t$$

where $0 < b < 1$. This is a particular case of natural growth. Since we are generally more comfortable with bases greater than one, we can replace b by $1/a = a^{-1}$, where $a > 1$. Thus

$$F(t) = F_0 \cdot a^{-t} \qquad a > 1$$

is the law of natural decay.

For instance, a poor investment might decrease in value at the rate of 6% per year. Assuming the natural decay law, its value after t years would be

$$V(t) = V_0 \cdot (1 - 0.06)^t = V_0 \cdot (0.94)^t.$$

Since $1/0.94 \approx 1.063830$, we can also write

$$V(t) = V_0 \cdot (1.063830)^{-t}.$$

The law of natural decay applies to the disintegration of radioactive elements. With each such element is associated an important constant, its **half-life.** This is the time in which a sample decays to half of its mass.

Example 4 A 1 kg sample of Thorium X decays to 1 g in 36.2755 days. Find its half-life to two places.

Solution The decay law is of the form

$$X(t) = 1000 \cdot a^{-t} \text{ g},$$

where t is in days. To apply it, we need the value of a. We are given

$$X(36.2755) = 1,$$

hence

$$1000 \cdot a^{-36.2755} = 1 \qquad a^{36.2755} = 1000$$

$$a = (1000)^{1/36.2755} \approx 1.20976.$$

Therefore the decay law is

$$X(t) = 1000 \times (1.20976)^{-t}.$$

Now the half-life is the time in which the sample decays to 500 g. So we solve $X(t) = 500$ for t:

$$1000 \times (1.20976)^{-t} = 500 \qquad (1.20976)^{t} = 2$$

$$t = \frac{\log 2}{\log(1.20976)} \approx 3.64 \text{ days}$$

Answer 3.64 days

The long half-life of ^{14}C (carbon-14) has been applied to determine the age of very old objects. The next example is a typical application of carbon-14 dating.

EXAMPLE 5 All living wood contains the same concentration of ^{14}C. When wood dies, the ^{14}C decays at a rate proportional to its mass; its half-life is 5568 years.

In 1950, archeologists excavated wood from a city built in the time of King Hammurabi, the law giver. Analysis showed that this wood contained 61.28% of the normal concentration of ^{14}C. Find an approximate date for King Hammurabi.

Solution Let $m(t)$ be the mass of ^{14}C at time t years from King H. Then

$$m(t) = m_0 \cdot a^{-t}.$$

Since $m(5568) = \frac{1}{2}m_0$, we have

$$m_0 \cdot a^{-5568} = \tfrac{1}{2}m_0 \qquad a^{-5568} = \tfrac{1}{2} \qquad a^{5568} = 2.$$

Therefore $a = 2^{1/5568}$ and

$$m(t) = m_0 \cdot 2^{-t/5568}.$$

We must solve $m(t) = (0.6128)m_0$ for t:

$$m_0 \cdot 2^{-t/5568} = (0.6128)m_0 \qquad 2^{t/5568} = \frac{1}{0.6128}$$

$$\frac{t}{5568}\log 2 = \log\left(\frac{1}{0.6128}\right) = -\log(0.6128)$$

$$t = -5568\,\frac{\log(0.6128)}{\log 2} \approx 3934 \text{ yr}.$$

The decay started about 3934 years before 1950, that is, in 1985 BC. (There is no year 0; the year before 1 AD is 1 BC.)

Answer 1985 BC.

Cooling

Suppose that an object has been heated to temperature T_0. At time $t = 0$ it is plunged into a cooling bath that is kept at the constant temperature B. Let $T(t)$ denote the temperature of the object at time t. It is known that the difference

$$T(t) - B$$

obeys the decay law; it decreases at a rate proportional to itself. Hence

$$T(t) - B = (T_0 - B)a^{-t} \qquad a > 1.$$

EXAMPLE 6 During heat treatment, a piece of steel at 400 °C is plunged into an oil bath kept at 25 °C. The steel cools to 200 °C in 68 sec. (a) Find the cooling law for the steel. (b) Find the time needed to cool to 50 °C.

Solution (a) By the decay law

$$T(t) - 25 = (400 - 25)a^{-t} = 375 \cdot a^{-t}.$$

Since $T(68) = 200$ we have

$$375 \cdot a^{-68} = T(68) - 25 = 200 - 25 = 175$$

$$a^{68} = \frac{375}{175} \approx 2.1429 \qquad a \approx (2.1429)^{1/68} \approx 1.0113.$$

Therefore

$$T(t) = 25 + 375 \times (1.0113)^{-t}.$$

(b) We must solve $T(t) = 50$ for t:

$$25 + 375 \times (1.0113)^{-t} = 50 \qquad (1.0113)^{-t} = \frac{50 - 25}{375} \approx 0.066667$$

$$t \approx -\frac{\log(0.066667)}{\log(1.0113)} \approx 241 \text{ sec}.$$

Answer (a) $T(t) = 25 + 375 \times (1.0113)^{-t}$ (b) 241 sec

Magnitudes of Stars

The magnitude of a star is a measure of its brightness (intensity). Bright stars have magnitudes around 0; the faintest star that can be seen with the naked eye has magnitude about 6. Thus magnitude is a decreasing function of brightness. Precisely, the **magnitude** of a star is

$$m = -2.5 \log \frac{B}{B_0},$$

where B is the brightness of the star and B_0 is a reference point chosen so that a star of brightness B_0 has magnitude 0. The constant 2.5 is chosen so that a star of brightness $\frac{1}{100}B_0$ has magnitude 5. (Check this.)

EXAMPLE 7 The brightest star in the sky is Sirius, with magnitude -1.5. How many times brighter is Sirius than a star (Eri A) of magnitude 4.4?

Solution We want the ratio B_S/B, where B_S is the brightness of Sirius and B is the brightness of the other star. We are given

$$-2.5 \log \frac{B_S}{B_0} = -1.5 \quad \text{and} \quad -2.5 \log \frac{B}{B_0} = 4.4.$$

To find the relation between B_S and B, we must eliminate B_0 from these equations. To do so, rewrite the equations as

$$-2.5(\log B_S - \log B_0) = -1.5 \quad \text{and} \quad -2.5(\log B - \log B_0) = 4.4$$

and subtract:

$$-2.5(\log B_S - \log B) = -1.5 - 4.4 = -5.9$$

$$-2.5 \log \frac{B_S}{B} = -5.9 \qquad \log \frac{B_S}{B} = \frac{5.9}{2.5} = 2.36.$$

Therefore $\frac{B_S}{B} = 10^{2.36} \approx 229$.

Answer 229 times brighter

Semi-log and Log-log Paper

These are graph papers that use logarithmic scales on one or both axes. On semi-log paper (Fig. 1a) the horizontal axis is scaled as usual. But the vertical axis is scaled logarithmically. This means that the vertical distance from the horizontal axis to the level marked y is not y, but $\log y$.

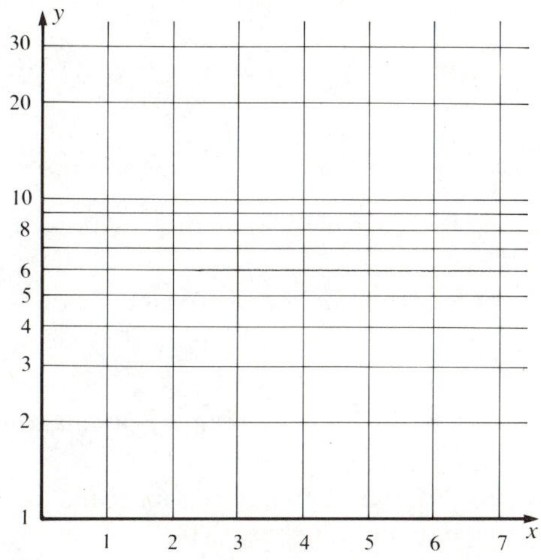

(a) semi-log paper

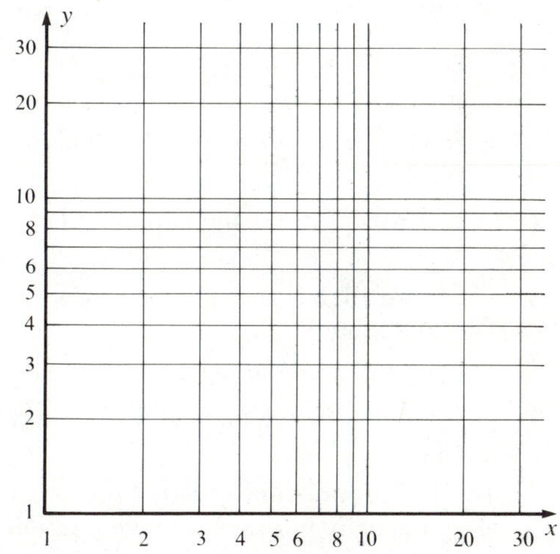

(b) log-log paper

Fig. 1

Imagine a rectangular grid superimposed on the semi-log paper, with both sets of axes coinciding and the same scale on the horizontal axes. Then a given point will have two sets of coordinates: (x, y) on the semi-log grid and (u, v) on the rectangular grid. These coordinates are related by

$$u = x \qquad v = \log y.$$

In log-log paper (Fig. 1b) both axes have logarithmic scales. The coordinates (x, y) are related to rectangular coordinates (u, v) by

$$u = \log x \qquad v = \log y.$$

EXAMPLE 8 Find the equation of the general straight line plotted on semi-log paper.

Solution The rectangular equation of the line is

$$v = au + b.$$

But $u = x$ and $v = \log y$. So the equivalent equation in semi-log coordinates is

$$\log y = ax + b.$$

It can be written as

$$y = 10^{ax+b},$$

which is equivalent to

$$y = 10^b \cdot 10^{ax} = kc^x,$$

where $k = 10^b$ and $c = 10^a$ are positive constants.

Answer $y = kc^x$ where $k > 0$ and $c > 0$

Remark Example 8 illustrates the convenience of semi-log paper for dealing with exponential functions. For example, suppose a scientist suspects an exponential relation between two quantities in an experiment. He can plot data points on semi-log paper. If they show an approximate straight line pattern, that is strong evidence for his suspicion.

Exercises

In exercises concerning growth (decay), always assume the natural growth (decay) law unless otherwise stated.

1 A bacteria colony doubles in 3 hours. How long does it take to increase 5-fold? 10-fold?

2 (cont.) After 7 hours the population of the colony is 10^7. What was its original population?

3 There are 10^5 bacteria in a culture at the start of an experiment and 10^6 after 5 hours. Find the growth law.

4 A bacteria colony grows by 4% in 10 minutes. How long will it take to double?

5 See Example 3. When will the population of city A be 20% greater than the population of city B?

6 The 1970 population of San Antonio, Texas was 707,500. At what rate must the city grow annually in order to have one million inhabitants by the year 2000?

7 Find an interpretation for k in the growth law $N(t) = N_0 3^{t/k}$.

8 Cities C and D presently have equal populations. The population of C has been growing at the rate of 3.1% per year and the population of D has been growing at the rate of 1.7% per year. When was C's population 15% lower than D's population?

9 The estimated 1970 world population was 3.63×10^9. Assuming a steady growth rate of doubling in 35 years, estimate the world population in the years 2000, 2100, 2200.

10 (cont.) If this growth rate continues too long we will run out of space to hold all these people. When will there be one person for each square foot of land? The land area of the earth is about 1.6×10^{15} ft^2.

Exercise 10 shows that the basic growth law is unrealistic over a long period. A more sophisticated model takes into consideration the struggle for food and space as the population grows. It proposes the formula

$$p(t) = \frac{ak}{bk + e^{-at}},$$

where a and b are constants and

$$k = \frac{p_0}{a - bp_0} \qquad p_0 = \text{population at time } t = 0.$$

11 Taking $a = 0.0290$ and $b = 2.94 \times 10^{-12}$, find k. Then estimate world population in the years 2000, 2100, 2200 based upon $p_0 = 3.63 \times 10^9$ in 1970. What is the ultimate world population?

12 (cont.) The U.S. population in 1790 was 3.93×10^6. Taking $a = 3.054 \times 10^{-2}$ and $b = 1.189 \times 10^{-10}$, find k. Then find what population the formula predicts for 1950 and for 1970. What does it predict as the ultimate U.S. population?

13 Thorium-X has a half-life of 3.64 days. How long will it take for a sample to decay to $\frac{1}{10}$ of its original mass?

14 A certain radioactive substance loses $\frac{1}{5}$ of its original mass in 3 days. What is its decay law?

15 A substance decays according to the law $M(t) = M_0e^{-t}$, where t is in seconds. What is its half-life?

16 Under ideal conditions the rate of *decrease* of atmospheric pressure with increasing altitude above sea level is proportional to the pressure. If the barometer reads 30 in. at sea level and 25 in. at 4000 ft., find the pressure at 20,000 ft.

17 A 5-lb. sample of radioactive material contains 2 lb. of radium-F, half-life 138.3 days, and 3 lb. of thorium-X, half-life 3.64 days. When will the sample contain equal amounts of radium-F and thorium-X?

18 Wood from the Lascaux caves in France was found in 1950 to contain 14.52% of the normal concentration of carbon-14. Estimate the date of the famous Lascaux cave paintings. [See Example 5, p. 195.]

19 A metal object in air at 20 °C cools from 85 °C to 40 °C in 13 minutes. How much longer will it take to cool to 21 °C?

20 In a superconductivity experiment, a piece of metal at 25 °C must be cooled to −268 °C. The metal is immersed in liquified gas held at −271 °C, and cools to −100 °C in 6.2 min. What is the total time required to reach −268 °C?

21 A teapot is cooling in a room at 68 °F. Its temperature decreases from 200 °F to 175 °F in the first minute. How much longer will it take to reach the desirable drinking temperature of 130 °F?

22 Molten lava at 900 °C is discharged from a volcano into air at 25 °C. If it cools to 780 °C in three hours, find its temperature after 35 hours.

23 How many times brighter is the sun, magnitude −26.8, than Sirius, magnitude −1.5?

24 Show that the difference in magnitude between two stars depends only on the *ratio* of their brightness.

25 Loudness of sounds is measured in decibels. The loudness in decibels of a sound of intensity I is

$$10 \log \frac{I}{I_0},$$

where I_0 is a standard very faint sound. The loudness of normal speech is 60 decibels. How many times more intense is the noise of a jet landing at 118 decibels?

26 If two sounds differ by one decibel, by what percentage does the intensity of the louder sound exceed the intensity of the softer one?

27 (a) Show that $y = 3x^2$ is a straight line in log-log coordinates.
(b) What is the equation of the general straight line in log-log coordinates? See Fig. 1b.

28* The sequence $1, 1, 2, 3, 5, 8, 13, 21, 34, \cdots$ is the sequence of **Fibonacci numbers.** Each is the sum of the preceding two. If we denote the sequence by $F_1, F_2, F_3, \cdots$ then $F_1 = F_2 = 1$ and $F_{n+2} = F_{n+1} + F_n$. It is known that

$$F_n = \frac{1}{\sqrt{5}}(\alpha^n - \beta^n) \quad \text{where } \alpha = \tfrac{1}{2}(1+\sqrt{5}),\ \beta = \tfrac{1}{2}(1-\sqrt{5}).$$

Find the n for which F_n is closest to 10^{12}.

29* In calculus it is shown that for large values of n,

$$1 + \tfrac{1}{2} + \tfrac{1}{3} + \tfrac{1}{4} + \cdots + \frac{1}{n} \approx \ln n + \gamma,$$

where $\gamma \approx 0.5772$. Estimate how many terms are needed until the sum $1 + \frac{1}{2} + \frac{1}{3} + \cdots$ exceeds 10.

30* (cont.) Estimate the sum

$$\frac{1}{1001} + \frac{1}{1002} + \frac{1}{1003} + \cdots + \frac{1}{2000}.$$

6 RATES OF GROWTH [Optional]

In some applications it is important to understand the growth rate as $x \longrightarrow \infty$ of a^x, and the decay rate of a^{-x}, where $a > 1$. Our standard of comparison is powers of 10. This is psychological; we feel in our hearts that a billion (10^9) is pretty large, but we are not so sure at first sight if, say, 2^{25} is or isn't so large.

We'll first make some rough estimates of the growth rate of 2^x as $x \longrightarrow \infty$, starting with

$$2^{10} = 1024 > 10^3.$$

From this inequality follow, for instance,

$$2^{20} = (2^{10})^2 > (10^3)^2 = 10^6$$

$$2^{30} = (2^{10})^3 > (10^3)^3 = 10^9$$

$$2^{100} = (2^{10})^{10} > (10^3)^{10} = 10^{30} \qquad \text{etc.}$$

In general,

$$2^{10n} > 10^{3n}.$$

This gives a pretty good idea of the extremely rapid growth of 2^x. Actually, we can also express the inequality in the form $2^x > 10^{0.3x}$, another way of looking at it.

We are convinced that 2^x grows rapidly. But so do x^5 and x^{10}. How does 2^x stack up against these functions for rapid growth? Just for fun let's compare 2^x and x^{10} for some large values of x, using our estimates for 2^x. First we'll try $x = 100$:

$$2^{100} > 10^{30} \quad \text{and} \quad 100^{10} = (10^2)^{10} = 10^{20},$$

$$\text{hence} \quad \frac{2^{100}}{100^{10}} > \frac{10^{30}}{10^{20}} = 10^{10}.$$

Thus if $x = 100$, then 2^x is more than 10^{10} times as large as x^{10}. Next, we'll try $x = 1000$:

$$2^{1000} = 2^{10\times100} > 10^{3\times100} = 10^{300} \quad \text{and} \quad 1000^{10} = (10^3)^{10} = 10^{30},$$

$$\text{hence} \quad \frac{2^{1000}}{1000^{10}} > \frac{10^{300}}{10^{30}} = 10^{270}.$$

Thus if $x = 1000$, then 2^x is more than 10^{270} times as large as x^{10}. Wow! Even though $2^x < x^{10}$ for small values of x (up to about 58.77), 2^x is overwhelmingly larger than x^{10} as $x \longrightarrow \infty$.

Similar reasoning applies to any power x^n, not just x^{10}, and to any exponential function a^x with $a > 1$, not just 2^x.

Growth of a^x Let $a > 1$ and let n be any whole number. Then

$$\frac{a^x}{x^n} \longrightarrow \infty \quad \text{as} \quad x \longrightarrow \infty .$$

In words, the exponential function a^x grows more rapidly than any power function x^n.

Example Again just for fun, let's try to find an x such that $(1.01)^x > x^{100}$. Since $(1.01)^x$ doesn't grow very rapidly at first:

$$(1.01)^5 \approx 1.051 \qquad (1.01)^{10} \approx 1.105 \qquad (1.01)^{25} \approx 1.282$$

this seems difficult. It isn't actually once you get a start. Play around on your calculator; you soon will find an x such that $(1.01)^x > 10$. For instance, $(1.01)^{250} \approx 12.0 > 10$. This is the start we need. Now we have for instance

$$(1.01)^{250{,}000} > 10^{1000}$$

But

$$(250{,}000)^{100} < (1{,}000{,}000)^{100} = (10^6)^{100} = 10^{600},$$

so
$$\frac{(1.01)^{250{,}000}}{(250{,}000)^{100}} > \frac{10^{1000}}{10^{600}} = 10^{400}.$$

Therefore for $x = 250{,}000$, the number $(1.01)^x$ is more than 10^{400} times as large as x^{100}. Convinced?

Note Let us try briefly to see *why* $2^x/x^{10} \longrightarrow \infty$ as $x \longrightarrow \infty$. To do so, we analyze what happens when x increases to $x + 1$. The numerator of $2^x/x^{10}$ increases from 2^x to $2^{x+1} = 2 \cdot 2^x$, that is, it doubles. The denominator increases from x^{10} to $(x + 1)^{10}$. But

$$\frac{(x+1)^{10}}{x^{10}} = \left(1 + \frac{1}{x}\right)^{10}$$

is little larger than 1 when x is large. For instance $(\frac{1001}{1000})^{10} \approx 1.01$. So while the numerator of $2^x/x^{10}$ is doubling, the denominator is increasing by a tiny percentage; therefore the fraction itself is practically doubling.

Now let's look at the decay of a^{-x} as $x \longrightarrow \infty$. If $a > 1$, then $a^{-x} \longrightarrow 0+$ as $x \longrightarrow \infty$ because

$$a^{-x} = \frac{1}{a^x} \quad \text{and} \quad a^x \longrightarrow \infty .$$

How rapidly does a^{-x} decay to 0? More rapidly than any function x^{-n}, where $n > 0$.

Decay of a^{-x} Let $a > 1$ and $n > 0$. Then

$$\frac{a^{-x}}{x^{-n}} = x^n a^{-x} \longrightarrow 0+ \quad \text{as} \quad x \longrightarrow \infty .$$

As an example let us compare 2^{-x} and x^{-10} for $x = 1000$:

$$\frac{2^{-1000}}{(1000)^{-10}} = \frac{(1000)^{10}}{2^{1000}} < \frac{1}{10^{270}} = 10^{-270}.$$

The ratio is indeed small.

Instead of using a^{-x} where $a > 1$, we can use b^x where $0 < b < 1$; we just take $b = a^{-1}$.

EXAMPLE 1 Estimate an x such that

(a) $(1.001)^x > 10$ (b) $(0.999)^x < 0.1$.

Solution (a) By calculator, $(1.001)^{10} > 1.01$. But $(1.01)^{250} > 10$ was observed earlier, hence

$$(1.001)^{2500} = [(1.001)^{10}]^{250} > (1.01)^{250} > 10.$$

(b) $\dfrac{1}{0.999} = 1.001001001 \cdots > 1.001$,

therefore $\left(\dfrac{1}{0.999}\right)^{2500} > (1.001)^{2500} > 10$. It follows that

$$(0.999)^{2500} < 0.1.$$

Answer (a) 2500 [actually $(1.001)^{2304} > 10 > (1.001)^{2303}$]

(b) 2500 [actually $(0.999)^{2302} < 0.1 < (0.999)^{2301}$]

EXAMPLE 2 Estimate an x such that

(a) $(1.001)^x > x^{100}$ (b) $(0.999)^x < x^{-1000}$.

Solution (a) From Example 1, $(1.001)^{2500} > 10$. It follows that

$$(1.001)^{2,500,000} > 10^{1000}.$$

But $(2,500,000)^{100} < (10^7)^{100} = 10^{700} < 10^{1000}$,

so $x = 2.5 \times 10^6$ is large enough.

(b) From Example 1, $(0.999)^{2500} < 0.1 = 10^{-1}$. It follows that

$$(0.999)^{25,000,000} < 10^{-10,000}.$$

But $(25,000,000)^{-1000} > (10^8)^{-1000} = 10^{-8000} > 10^{-10,000}$,

so $x = 2.5 \times 10^7$ is large enough.

Answer (a) 2.5×10^6 (b) 2.5×10^7

Remark We have emphasized the rapid growth as $x \longrightarrow \infty$ of a^x, where $a > 1$. Does anything grow even more rapidly? Sure, b^x, where $b > a$, because the ratio

$$\frac{b^x}{a^x} = \left(\frac{b}{a}\right)^x$$

is itself an exponential function with base $b/a > 1$, hence grows rapidly.

Exercises

1 Compare the values of 2^x and x^{10} for $x = 10, 20, 30, \cdots, 100$ by computing their ratio.

2 Compare the values of 3^x and x^3 for $x = 1, 2, 3, \cdots, 10$ by computing their ratio.

3 Compare the values of 2^{-x} and x^{-2} for $x = 1, 2, 3, \cdots, 10$ by computing their ratio.

4 Compare the values of 3^{-x} and x^{-4} for $x = 1, 2, 3, \cdots, 10$ by computing their ratio.

5 Find a value of n for which $2^n > 10^{50}$.

6 Find a value of x for which $2^x > x^{100}$.

7 Find a value of x for which $(1.01)^x > x^{10}$.

8 Find a value of x for which $(1.0001)^x > x^5$.

9 Compare 2^x and x^{100} for $x = 10^3$ and $x = 10^6$ by examining their ratio.

10 Compare $(1.1)^x$ and x^{10} for $x = 10^3$ and $x = 10^6$ by examining their ratio. Use $2 < (1.1)^{10}$.

11 Give convincing numerical evidence that

$$\frac{\log_{10} x}{x} \longrightarrow 0 \quad \text{as} \quad x \longrightarrow \infty.$$

12 $\log_2 x$ grows slowly as $x \longrightarrow \infty$. So does $\sqrt[100]{x}$. Which grows more slowly in the long run?

13 $\log_{10} x \longrightarrow -\infty$ as $x \longrightarrow 0+$. How about $x \log x$ as $x \longrightarrow 0+$? [As usual, $\log = \log_{10}$.]

14 (cont.) The same question for $\sqrt[10]{x} \log x$.

15 Which is larger for $x > 1$, $\log_2 x$ or $\log_3 x$? [*Hint* Compare the graphs of the corresponding exponential functions.]

16 (cont.) The same question for $0 < x < 1$.

17 Find the smallest number x on your calculator so that x [log] [log] [log] [log] does not result in **Error.**

18 Find the largest number x on your calculator so that x [e^x] [e^x] [e^x] [e^x] does not result in **Error.**

7 PRACTICAL CALCULATION

How lucky you are! You belong to the first generation of students not burdened by the use of logarithm tables. We would like to show you what you missed, and also point out some aspects of calculation worth knowing.

Logarithms were introduced as a labor saving device in computation. The rules

$$\log(x_1x_2) = \log x_1 + \log x_2 \qquad \log\frac{x_1}{x_2} = \log x_1 - \log x_2 \qquad \log x^{1/n} = \frac{1}{n}\log x$$

convert tedious multiplication, division, and root extraction into less tedious addition, subtraction, and division. The trouble is that before calculators, this method required tables of logarithms. To divide x_1 by x_2, you looked up $\log x_1$ and $\log x_2$ in a table and subtracted them, obtaining $\log(x_1/x_2)$. Then you looked in the table again to find what number had this logarithm.

Common log tables tabulate the function $\log x$ from $x = 1$ to $x = 10$ at some fixed increment, depending on the accuracy of the tables. A portion of a 4-place table looks like this:

N	0	1	2	3	4	5	6	7	8	9	Proportional Parts								
											1	2	3	4	5	6	7	8	9
55	7404	7412	7419	7427	7435	7443	7451	7459	7466	7474	1	2	2	3	4	5	5	6	7
56	7482	7490	7497	7505	7513	7520	7528	7536	7543	7551	1	2	2	3	4	5	5	6	7
57	7559	7566	7574	7582	7589	7597	7604	7612	7619	7627	1	2	2	3	4	5	5	6	7
58	7634	7642	7649	7657	7664	7672	7679	7686	7694	7701	1	1	2	3	4	4	5	6	7
59	7709	7716	7723	7731	7738	7745	7752	7760	7767	7774	1	1	2	3	4	4	5	6	7

We interpret the main part of the table at the left as giving values of $\log x$ from

$$\log 5.50 \approx 0.7404 \quad \text{to} \quad \log 5.99 \approx 0.7774$$

in increments of 0.01. Then we can read off, for instance, that $\log 5.74 \approx 0.7589$.

The "Proportional Parts" section is for finding logs of numbers between listed values. For example, the fourth row of the table gives

$$\log 5.86 \approx 0.7679 \qquad \log 5.87 \approx 0.7686 .$$

Suppose we want $\log 5.868$. Now 5.868 is 8 tenths of the way from 5.86 to 5.87. We look under 8 in the fourth row of Proportional Parts and find 6. This 6 must be added to the last digit of 0.7679 in order to go (approximately) 8 tenths of the way from 0.7679 to 0.7686. Thus

$$\log 5.868 \approx 0.7679 + 0.0006 = 0.7685 .$$

With a little practice, you can do the addition mentally while looking at the table and only write down the result,

$$\log 5.868 \approx 0.7685 .$$

This technique of going part way between listed values is called **linear interpolation.** We'll say more about it shortly.

In a typical calculation with logarithms, we usually end up with the log of the answer. Then we use the table backwards (inverse function 10^x) to find the answer. For instance, given

$$\log x = 0.7716 \quad \text{we read} \quad x \approx 5.91 .$$

But suppose we are given

$$\log x = 0.7600 .$$

From the third row of the table,

$$\log 5.75 \approx 0.7597 \qquad \log 5.76 \approx 0.7604 .$$

From 7597 to 7600 is 3. Over the 3 in the *third* row of Proportional Parts is 4. So we go 4 tenths of the way from 5.75 to 5.76. Therefore

$$x \approx 5.75 + 0.004 = 5.754$$

If a number is not in the range $1 \leq x < 10$, you can still find its log from the table. Write the number in scientific notation and use the laws of logarithms. For example,

$$\begin{aligned} \log 5754 &= \log(5.754 \times 10^3) = \log 5.754 + \log 10^3 \\ &\approx 0.7600 + 3 = 3.7600 , \\ \log 0.05754 &= \log(5.754 \times 10^{-2}) = \log 5.754 + \log 10^{-2} \\ &\approx 0.7600 - 2 = -1.2400 . \end{aligned}$$

EXAMPLE 1 Use 4-place tables to estimate $x = \dfrac{(551.2)(0.5937)}{5.706}$.

Solution By the laws of logarithms,

$$\log x = \log 551.2 + \log 0.5937 - \log 5.706 .$$

Look up the logs in the table (p. 203), arrange them in a column, and sum the column with the indicated signs. Then look up the antilog (inverse function, 10^x):

$$\begin{array}{rl|c}
\log 551.2 \approx & 0.7414 + 2 & + \\
\log 0.5937 \approx & 0.7736 - 1 & + \\
\log 5.706 \approx & 0.7564 & - \\
\hline
\log x \approx & 0.7586 + 1 & \\
x \approx & 5.735 \times 10^1 = 57.35\,. &
\end{array}$$

Answer 57.35 (By calculator: 57.351462)

Note The data was given to 4 significant figures, so only 4 figures in the answer are meaningful. See p. 206.

Well, that's the way calculations used to be done, and still are done in less affluent parts of the world. These calculations were tedious, and there were many possibilities for error. Hand-held calculators have changed all that. They have eliminated the need for logarithms as computational aids, although not as important functions in theoretical work. Before discussing use of calculators, we mention three aspects of numerical work: interpolation, accuracy, and round-off.

Linear Interpolation

It is important to know how to read tables. Many useful functions are not built into calculators but do exist in tables. Many results of highly accurate measurements are tabulated in handbooks.

Generally, tables do not have Proportional Parts supplements. To "read between the lines", we need linear interpolation. For instance, suppose we want $\log 3.1517$. From a 5-place table we find

$$\log 3.151 \approx 0.49845 \qquad \log 3.152 \approx 0.49859\,.$$

We pretend that the function $\log x$ is linear for $3.151 \le x \le 3.152$. If so, since 3.1517 is $\frac{7}{10}$ of the way from 3.151 to 3.152, then $\log 3.1517$ is $\frac{7}{10}$ of the way from 0.49845 to 0.49859. See Fig. 1. But

$$\tfrac{7}{10}(0.49859 - 0.49845) = \tfrac{7}{10}(0.00014) \approx 0.00010\,.$$

Therefore

$$\log 3.1517 \approx 0.49845 + 0.00010 = 0.49855\,.$$

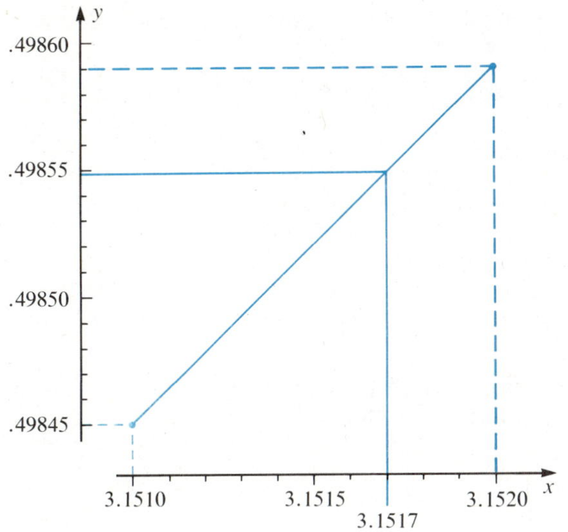

Fig. 1 Example of linear interpolation

Accuracy

What does it mean that a measurement is accurate to 3 decimal places? It means that the error made is at most $\pm 5 \times 10^{-4}$.

For example, consider a chemist's analytical balance that weighs anything up to 150 grams with one-milligram accuracy. Since 1 mg = 0.001 g, the read-out is accurate to 3 decimal places. If the read-out is 0.493 g, then the sample actually weighs between 0.4925 and 0.4935 g. If the read-out is 104.228 g, then the sample actually weighs between 104.2275 and 104.2285 g.

We say that 0.493 has 3 significant figures and 104.228 has 6 significant figures. In general, if a number is written with a decimal point, its number of **significant figures** is the number of digits from the left-most non-zero digit to the right-most digit.

Examples

NUMBER	SIG. FIGS.
12.8	3
12.80	4
1500.0	5
1.5×10^3	2
10^9	1

NUMBER	SIG. FIGS.
0.04	1
1.336	4
4.38×10^{-6}	3
3.1416	5
3.14159	6

The number of significant digits indicates how accurate the data is. For example, compare the readings 0.493 and 104.228. Both imply a maximum error of $\pm 5 \times 10^{-4}$. For the first (3 significant figures), this amounts to about 5 parts in 5000. But for the second (6 significant figures), it amounts to about 5 parts in 1,000,000. Thus more significant figures imply greater relative accuracy.

Note that 12.80 implies greater accuracy than 12.8. For 12.80 indicates a maximum error of $\pm 5 \times 10^{-3}$, whereas 12.8 indicates a maximum error of 5×10^{-2}.

Don't confuse the number of decimal places of accuracy with the number of significant figures. A weighing of 104.228 is accurate to 3 decimal places but has 6 significant figures.

Round-off

Suppose we have a 5-place read-out, but we only require 2-place accuracy. Then we must **round off** each 5-place reading to 2 places (and this means decimal places).

Examples

NUMBER	ROUNDED-OFF NUMBER
0.48265	0.48
0.48701	0.49
0.49013	0.49
0.49501	0.50
0.49500	?

The last entry is a problem. Do we round off to 0.49 or 0.50? The convention varies, but we shall adopt the rule "make the last digit even". So we round off to 0.50 because 0 is even and 9 is odd.

Rules for Rounding Off

(1) If the discarded portion is less than $5000 \cdots$, then drop it.

(2) If the discarded portion is greater than $5000 \cdots$, then drop it and add 1 to the last digit kept.

(3) If the discarded portion is exactly $5000 \cdots$, then drop it; if the last digit kept is even, do nothing; if the last digit kept is odd, add 1 to it.

Each time we round off, we introduce an error. But we feel these rules are fair and hope that in a series of calculations with round-off, the errors will more or less average out, not pile up.

Example Round off 9.86507 to 0, 1, 2, 3, and 4 decimal places. Solution: 10, 9.9, 9.87, 9.865, 9.8651.

Example Round off 9.865 to 2 decimal places. Solution: 9.86.

Remark: Note the different results:

9.86507 —(2 places)→ 9.87

9.86507 —(3 places)→ 9.865 —(2 places)→ 9.86

The one-step round-off is more accurate than the two-step procedure.

There is a lesson to be learned here: if you want 4 places it is better to use a 4-place table rather than round off from a 5-place table. The reason is that the entries in the tables are already rounded off by the table-makers; you may lose accuracy in rounding off again. For example, in 4-place and 5-place common log tables we find

$$\log 1.19 \approx 0.0755 \quad \text{and} \quad \log 1.19 \approx 0.07555.$$

If we round off the second entry to 4 places we get 0.0756. But a 6-place table shows $\log 1.19 \approx 0.075547$ so that 0.0755 is more accurate.

Exercises

Use the table on p. 203 to estimate

1 log 5.52 **2** log 5.98 **3** log 0.598

4 log 552 **5** log 58700 **6** log 0.000581

7 log 5503 **8** log 0.05977 **9** log 564500

10 log 59.89.

N	0	1	2	3	4	5	6	7	8	9
15	1761	1790	1818	1847	1875	1903	1931	1959	1987	2014
16	2041	2068	2095	2122	2148	2175	2201	2227	2253	2279
17	2304	2330	2355	2380	2405	2430	2455	2480	2504	2529
18	2553	2577	2601	2625	2648	2672	2695	2718	2742	2765
19	2788	2810	2833	2856	2878	2900	2923	2945	2967	2989

Use the portion of a log table and linear interpolation to estimate

11 log 1.503 **12** log 1.987 **13** log 18.45

14 log 172.3 **15** log 0.001677 **16** log 175100.

Use the same table and linear interpolation to find 10^x (the antilog of x) where x =

17 0.2566 **18** 2.2260 **19** 5.2537

20 1.2891 **21** 0.2504 − 1 **22*** −0.7824.

Round off to 2 decimal places

23 0.4444 0.3128 0.1075 0.2555 **24** 6.411 10.91 2.0041 3.0095.

Round off to 3 decimal places.

25 0.0005 0.00049 16.2445 3.7855 **26** 1.8125 3.14159265 0.9997 0.99946.

27 (a) Compute 1.255 + 0.395 + 2.116 + 1.336, then round off to 2 places.
(b) Round off each term first, then add. Compare the answers. Which answer is more accurate?

28 (cont.) Do the same for 0.255 + 0.365 + 0.166 + 0.823.

Round off to 3 significant figures

29 1046.0 55.521 10.05 **30** 9.095 9.094 9.0949.

An inexperienced technician uses a voltmeter with 2 significant figure readings. Rewrite accurately his data

31 0.4 **32** 12.0 **33** 2.3 **34** 9.

8 USE OF CALCULATORS

Many models of hand-held calculators are available, each with its own special features. We cannot deal with all models and features here. Rather, we shall discuss some features common to all scientific calculators and some basic techniques of using them.

For the purposes of this course, and even more advanced courses, a fairly inexpensive scientific calculator will serve you well. It should display 5 to 8 significant figures; more are really not necessary. It should have the following types of keys:

(a) [x^2] [$\sqrt{\ }$] [$1/x$] [y^x]

(b) [log] [ln] [10^x] [e^x]

(c) One memory with

[STO] (also called [Min] [x→M] etc.)

[RCL] (also called [MR] [RM] etc.)

[M+] (also called [SUM] etc.)

(d) [( ] [ )] (to at least one level, more if possible)

(e) [sin] [cos] [tan]

(f) [INV] (also called [ARC] [F] [2nd] etc.)

(g) [π]

As pointed out in Section 4, you don't really need all four keys of type (b). The key sequence [INV] [log] and [INV] [ln] are equivalent to [10^x] and [e^x] respectively.

Lots of other keys are available on various models. For example, there are statistical keys, [e], [$x^{1/y}$], [n!], [° ′ ″], [→rθ], and others, but we don't advise paying extra for them. Parentheses [( ] and [ )] are now standard on most scientific calculators, but many perfectly good older models lack them.

Since the main source of errors is hitting the wrong key, an uncluttered keyboard is preferable. Also you might like a calculator with a liquid crystal display.

You can't read it in the dark, but the batteries last 500–2000 hours.

A final suggestion: Whatever calculator you buy, read the instructions carefully and do enough numerical experiments to familiarize yourself with its capabilities and peculiarities.

Left-to-Right Logic

This is the system usually found in the least expensive calculators. In our examples, we usually assume left-to-right logic unless noted otherwise.

On any calculators, the keys

$\boxed{+}$, $\boxed{-}$, $\boxed{\times}$, $\boxed{\div}$, $\boxed{y^x}$

are called **binary** keys because they operate on *two* numbers. In left-to-right logic, each binary key completes all previous pending operations.

Examples

$3 \boxed{+} 4 \boxed{\times} 2 \boxed{=}$ produces $(3 + 4) \times 2 = 14$

$1\ 0 \boxed{-} 3 \boxed{y^x} 2 \boxed{=}$ produces $(10 - 3)^2 = 49$.

Each binary key acts as if it were preceded by $\boxed{=}$. For example the calculation

$3 \boxed{+} 4 \boxed{\times} 2 \boxed{=}$ is equivalent to $3 \boxed{+} 4 \boxed{=} \boxed{\times} 2 \boxed{=}$.

The keys

$\boxed{x^2}$ $\boxed{\sqrt{\ }}$ $\boxed{10^x}$ $\boxed{\log}$ $\boxed{\sin}$ etc.

are called **unary** keys because they operate on a single number. For example

$x \boxed{\sqrt{\ }}$ produces $\sqrt{x}$, $x \boxed{\log}$ produces $\log x$, etc.

The action of a unary key is completed *before* pending operations.

Examples

$7 \boxed{+} 9 \boxed{\sqrt{\ }} \boxed{=}$ produces $7 + \sqrt{9} = 10$

$5 \boxed{x^2} \boxed{-} 3 \boxed{x^2} \boxed{=} \boxed{\sqrt{\ }}$ produces $\sqrt{5^2 - 3^2} = 4$.

If you forget $\boxed{=}$ in either example, the answer will be 3. Why?

Example 1 Assuming left-to-right logic, find the calculation performed by the key sequence

(a) $6 \boxed{+} 5 \boxed{-} 2 \boxed{\times} 3 \boxed{\div} 7 \boxed{=}$

(b) $5 \boxed{x^2} \boxed{x^2} \boxed{-} 1\ 6 \boxed{\sqrt{\ }} \boxed{=} \boxed{\log}$.

Solution (a) Move from left to right performing each operation on the previous total. The result is

$$\frac{(6 + 5 - 2) \times 3}{7}.$$

(b) Similarly, proceeding from left to right we obtain

$$(5^2)^2 - \sqrt{16}$$

at the next to last step. The last step yields the log of this number.

Answer (a) $\dfrac{(6 + 5 - 2) \times 3}{7}$ (b) $\log[(5^2)^2 - \sqrt{16}]$.

Algebraic Logic

In this system, some operations have higher priority than others. The usual order of priority is

highest ⟶ 1 unary operations

2 $\boxed{y^x}$

3 $\boxed{\times}$ $\boxed{\div}$

4 $\boxed{+}$ $\boxed{-}$

lowest ⟶ 5 $\boxed{=}$.

An operation of higher priority is always completed before any pending previous operations of lower priority.

Examples

3 $\boxed{+}$ 4 $\boxed{\times}$ 2 $\boxed{=}$ produces $3 + (4 \times 2) = 11$

1 0 $\boxed{-}$ 3 $\boxed{y^x}$ 2 $\boxed{=}$ produces $10 - 3^2 = 1$

3 $\boxed{\times}$ 5 $\boxed{-}$ 2 4 $\boxed{\div}$ 8 $\boxed{=}$ produces $(3 \times 5) - (24 \div 8) = 12$.

Calculators with algebraic logic often have parentheses. Many have parentheses within parentheses. All work on one basic rule: closing an open parenthesis automatically completes all operations pending since the last $\boxed{(}$.

Examples

$\boxed{(}$ 4 $\boxed{+}$ 3 $\boxed{\times}$ 2 $\boxed{)}$ produces $4 + (3 \times 2) = 10$

$\boxed{(}$ 4 $\boxed{+}$ 3 $\boxed{)}$ $\boxed{\times}$ $\boxed{(}$ 5 $\boxed{-}$ 3 $\boxed{)}$ produces $5 - 3 = 2$

$\boxed{(}$ 4 $\boxed{+}$ 3 $\boxed{)}$ $\boxed{\times}$ $\boxed{(}$ 5 $\boxed{-}$ 3 $\boxed{)}$ $\boxed{=}$ produces

$$(4 + 3) \times (5 - 3) = 14$$

Note that there is no $\boxed{=}$ in the first example; closing the parentheses completes all operations inside. In the second example, closing the last parentheses simply completes the operation inside that set of parentheses, $5 - 3 = 2$. If you want the product $(4 + 3) \times (5 - 3)$, you must complete the $\boxed{\times}$ operation by hitting $\boxed{=}$ at the end (third example).

EXAMPLE 2 Assuming algebraic logic, find the calculation set up by the key sequence

(a) 4 $\boxed{-}$ 5 $\boxed{\div}$ 6 $\boxed{+}$ 7 $\boxed{\times}$ 8 $\boxed{=}$

(b) $\boxed{(}$ 5 $\boxed{\times}$ $\boxed{(}$ 3 $\boxed{-}$ 1 $\boxed{)}$ $\boxed{y^x}$ 4 $\boxed{+}$ 1 $\boxed{)}$.

Solution (a) Since $\boxed{\div}$ and $\boxed{\times}$ have priority over $\boxed{-}$ and $\boxed{+}$, the sequence yields

$$4 - (5 \div 6) + (7 \times 8).$$

(b) Closing the outer parentheses completes the sequence inside. The operation $\boxed{y^x}$ has highest priority, then $\boxed{\times}$, and last $\boxed{+}$. The result is

$$[5 \times (3 - 1)^4] + 1.$$

Answer (a) $4 - (5 \div 6) + (7 \times 8)$ (b) $[5 \times (3 - 1)^4] + 1$

Warning Be very careful using [=]. On some models, [=] will not only complete all pending operations within ( ), but also all pending operations before the (. For example, the logical set-up for calculating

$$(21 - 5)\sqrt{59 + 22} \qquad (= 144)$$

would seem to be

[(] 2 1 [−] 5 [)] [×] [(] 5 9 [+] 2 2 [=] [√‾] [)] [=] .

However, on a TI-58 we obtain 36. The joker is the first [=]. We thought it would complete only the addition $59 + 22$. It did, but it also completed the pending multiplication, giving $16 \cdot 81 = 1296$. The next step gave $\sqrt{1296} = 36$. Then nothing remained pending, so [)] and [=] had no effect.

Reverse Polish Logic

At first these calculators seem strange; they have no [=] key! Yet the reverse Polish logic is highly efficient. In fact, we believe it is easier to use an RPL calculator than other types. When you are ready for an expensive calculator, consider a Hewlett-Packard; they are the Cadillacs among calculators.

This system uses an "operational stack," which consists of the displayed number plus other numbers stacked "above" the display. Pressing [ENTER ↑] moves the whole stack up one space. Pressing a binary key combines the bottom two numbers (the displayed number and the one right "above" it) and moves the rest of the stack down one space.

Examples

3 [ENTER ↑] 4 [+] produces $3 + 4 = 7$.

5 [ENTER ↑] 2 [ENTER ↑] 7 [+] [×] produces $(7 + 2) \times 5 = 45$.

Store and Recall

The key [STO] stores the displayed number in the memory (without changing the display). The key [RCL] recalls the stored number to the display (without changing the contents of the memory).

These features make for efficient computing. You can store a number in the memory and bring it back whenever needed. For example, [+] [RCL] [=] adds the stored number to the display; [y^x] [RCL] [=] raises the displayed number to the power of the stored number.

Examples (left-to-right logic)

5 [STO] [x^2] [+] [RCL] [=] produces $5^2 + 5 = 30$

8 3 [+] 7 9 [=] [STO] 4 3 6 [−] 8 7 [×] [RCL] [=]
produces $(83 + 79) \times (436 - 87) = 56538$.

If the first [=] is omitted in the second example, then only 79 (the displayed number) goes into the memory.

Keying in numbers is a major source of errors. When the same number occurs several times in a calculation, try to cut down the number of times you must key it in, hopefully to one time.

Example 3 Let $a = 9.69699$. Compute $\frac{1}{a}(\log a)^a + \sqrt{a}$.

Solution There is great danger of an error when keying in such a confusing number. The most obvious solution requires entering a four times. But that is poor technique and very risky. With efficient use of the memory, it's only necessary once.

Enter a at the beginning and store it. Then whenever you need it, just hit [RCL]. With this strategy, a reasonable key sequence is

9.69699 [STO] [log] [y^x] [RCL] [÷] [RCL] [+] [RCL] [√] [=]

Answer 3.204511

Example 4 Let $a = 4.0767$. Compute $\frac{\sqrt{a}}{a^2 + 5}$.

Solution Enter a and store it. Compute the denominator and take its reciprocal. Then multiply by $\sqrt{a}$:

4.0767 [STO] [x^2] [+] 5 [=] [1/x] [×] [RCL] [√] [=]

One step shorter:

4.0767 [STO] [x^2] [+] 5 [÷] [RCL] [√] [=] [1/x]

Answer 0.09339187

Memory Plus

The key [M+] adds the number in the display to the memory. Usually it *completes pending operations* before making this addition.

Example (starting with memory clear)

2 [×] 3 [M+] 4 [×] 5 [M+] 6 [×] 7 [M+] [RCL]

produces $(2 \times 3) + (4 \times 5) + (6 \times 7) = 68$.

Constant Factors

Suppose you need to multiply each of a whole slew of numbers by the same constant. Some calculators have a "constant factor" feature for doing this without reentering the factor each time. The sequence c [×] x [=] produces cx, as usual. But when you follow by y [=], you get cy. Then z [=] produces cz, etc. (Some models use a [K] key to activate the constant factor feature.)

Example 5 A weekend special at your record store announces a $27\frac{1}{2}\%$ discount on all marked prices. Allowing for a 4% sales tax, what is your actual cost on records and tapes at

\$1.98, 2.37, 3.19, 3.89, 4.89, 5.57, 7.29?

Round up each price to the nearest cent.

Solution First key in

1 [−] .275 [×] 1.04 [×].

This is your actual discount factor, taking tax into account. It is set up in the calculator as a constant factor. Now continue

1 . 9 8 [=] 2 . 3 7 [=] 3 . 1 9 [=] 3 . 8 9 [=] 4 . 8 9 [=]

5 . 5 7 [=] 7 . 2 9 [=] .

After each [=] round up to get the corresponding cost.

Answer \$1.50, 1.79, 2.41, 2.94, 3.69, 4.20, 5.50

Overflow

The largest number the calculator holds is $9.999 \cdots \times 10^{99}$. A calculation that results in a larger number at any stage puts the calculator into an error state called **overflow.** Similarly, numbers smaller than 10^{-99} yield **underflow** and the calculator probably displays 0. Numbers beyond the capacity of the calculator can sometimes be handled indirectly by using logarithms.

EXAMPLE 6 Estimate

(a) 2^{3000} (b) 3^{-5000}

Solution (a) The number is so large that

2 [y^x] 3 0 0 0 [=]

results in **Error.** However, we can calculate

$$\log 2^{3000} = 3000 \cdot \log 2$$

by

2 [log] [×] 3 0 0 0 [=]

resulting in 903.089988. Therefore

$$2^{3000} \approx 10^{903.089988} = 10^{0.089988} \times 10^{903}.$$

The factor $10^{0.089988}$ is within the capacity of the calculator; its value is approximately 1.2302, hence

$$2^{3000} \approx 1.23 \times 10^{903}.$$

(The last digits are doubtful so we round.) The following routine does the whole calculation:

2 [log] [×] 3 0 0 0 [−] 9 0 3 [=] [10^x] .

You must pause at [−] to read 903, the integer part of the logarithm, then subtract this integer part. Also you must multiply the result by 10^{903}.

(b) Similarly, first we do

3 [log] [×] 5 0 0 0 [+/−] [+] ,

read the integer part −2385, and finish with

2 3 8 6 [=] [10^x] ,

to obtain 2.4759×10^{-2386}. (Why do we add 2386 rather than 2385; what would have happened had we added 2385 instead?)

Answer (a) 1.23×10^{903} (b) 2.48×10^{-2386}

Accuracy and Hidden Digits

Most calculators work internally with one to three more significant digits than are displayed. Usually the number displayed is a round-off of the internal number. Not always, however; in some models the final digits are truncated (chopped off) rather than rounded. Thus the internal number

· · · 68 may appear as · · · 6.

On some models you can see the hidden final digits by using a little trick. Let's illustrate the trick for $\sqrt{2}$ on three calculators:

2 [√]	1.4 142 135	Casio fx-2000
	1.4 142 135	Sharp EL-5805
	1.4 142 13562	Texas Instr. TI-59.

We multiply by 1000, then subtract 1414. This deletes the first 4 significant digits, making space in the display for the hidden digits. The results:

0.2 13562	fx-2000
0.2 1357	EL-5805
0.2 13562373	TI-59.

Thus we get the following estimates for $\sqrt{2}$:

1.41421 3562	fx-2000
1.41421 357	EL-5805
1.41421 35623 73	TI-59.

Exercises

Find the result of the calculation in left-to-right logic

1 a [+] b [√] [+] c [√] [=]
2 a [+] b [√] [+] c [=] [√]
3 a [×] b [1/x] + c [1/x] [=]
4 a [×] b [=] [1/x] [+] c [=] [1/x]
5 a [×] b [÷] c [+] a [÷] b [×] c [=]
6 a [−] b [−] c [×] 3 [+] a [×] b [=]
7 a [+] b [10^x] [+] c [y^x] 2 [−] a [=]
8 a [+] 1 [y^x] b [−] 1 [÷] a [=] [1/x] [×] c [=]
9 a [+] 1 [=] [1/x] [+] 1 [=] [1/x] [+] 1 [=] [1/x]
10 a [1/x] [y^x] a [+] a [y^x] a [1/x] [+] a [1/x] [=] [1/x]

Find the result of the calculation in algebraic logic

11 a [×] b [÷] c [+] a [÷] b [×] c [=]
12 a [−] b [−] c [×] 3 [+] a [×] b [=]
13 a [+] b [y^x] c [×] d [=]
14 a [×] b [y^x] c [×] d [+] a [y^x] b [1/x] [=]
15 [(] [(] a [−] b [)] [×] c [−] [(] b [−] c [)] [×] a [)]
16 [(] a [×] [(] a [+] b [×] c [)] [−] c [)] [×] a [=]
17 a [+] b [y^x] [(] a [−] b [)] [=] [×] [(] a [−] b [)] [=]
18 [(] a [−] b [)] [÷] a [+] b [÷] [(] a [+] b [)] [=]
19 a [+] [1/x] [×] b [=]
20 (cont.) a [+] [1/x] [×] b [+] [1/x] [×] c [=]

Set up a routine on a left-to-right calculator with memory, keying a only once

21 $a \log(3 + a)$ **22** $\sqrt{1 + a^2} + a^4$ **23** $a - a^{1/a}$

24 $a^{(1+a)}$ **25** $\dfrac{1 + a}{1 - a}$ **26** $\dfrac{3 + a}{2 + a^2}$

27 $(\sqrt[4]{1 + 5a^4} - a)^a - a$ **28** $(1 + 2a)^{\sqrt{1+a}}$

Set up a routine on an algebraic calculator with memory and parentheses, keying a only once

29 $\dfrac{2 + a}{3 - a}$ **30** $\dfrac{a^2 + 4}{a^3 + 7}$

31 $(a^{1/a} + 1)[(1/a)^a + 1]$ **32** $a^{a^a} + a^a + a$

33 $a^2 + \log(a^3 - 2)$ **34** $\dfrac{10}{3a^2 - 1} + (a^5 - 2) \times 10^a$.

Calculate

35 $\dfrac{2a + 1}{(a - 3)^5}$ for $a = 4.1891$

36 $[1 + \log(4a + 3)]^a$ for $a = 2.5887$

37 $\sqrt{\dfrac{a}{a^2 + 6}}$ for $a = 0.6077$

38 $\sqrt{a + \sqrt{a + \sqrt{a}}}$ for $a = 3.9399$.

39 In 1971 a computation proved that $2^{19937} - 1$ is a prime number (the largest prime known at that time). How large is it?

40 In 1978 it was announced that $2^{21701} - 1$ is a prime number. How large is it? The same question for $2^{23209} - 1$, the largest known prime as of April, 1979, and for $2^{44497} - 1$, the largest one as of June, 1979.

Estimate to 3 significant figures.

41 $(0.99)^{1,000,000}$ **42** $\left(\dfrac{1}{\sqrt{2}}\right)^{10,000}$.

Try to find one additional hidden digit

43 $\sqrt{3}$ **44** $(1.56)^{2.81}$.

45 Suppose a is in the display and b is in the memory. Find a routine that interchanges a and b without reentering either (and without use of a memory exchange key if you happen to have one).

46 (cont.) With the same data, end up with a in the display and b^2 in the memory.

9 ADDITIONAL CALCULATOR APPLICATIONS [Optional]

Polynomial Evaluation

Suppose we need the value of a polynomial $f(x)$ for a specific number x. Computing $f(x)$ can involve a lot of arithmetic, so it is worthwhile learning how to do it on a calculator.

For example, suppose we want the value of

$$f(x) = 5x^2 + 2x - 4$$

for $x = 2.7727$. We immediately store x in the memory to avoid entering it again. But then what? If we compute $5x^2$, we will have to store it someplace while we compute $2x - 4$. But the memory is already in use holding x. We could copy down $5x^2$, but that is poor technique.

The solution is to write $f(x)$ in a more convenient form. We factor the first two terms:

$$f(x) = (5x + 2)x - 4.$$

This form suggests a routine that requires no storage of partial results. To evaluate $f(2.7727)$, we use the key sequence (left to-right logic)

2 . 7 7 2 7 [STO] [×] 5 [+] 2 [×] [RCL] [−] 4 [=]

with result 39.98473. (For algebraic logic, insert [=] between 2 and [×].)

We can apply this technique also to higher degree polynomials.

EXAMPLE 1 Estimate on a calculator

$$f(x) = 3x^4 - 7x^3 + 5x^2 + 2x - 4$$

for $x =$ 1.80634 1.80734 1.80834 1.80934.

Solution Factor x out of the first 4 terms:

$$f(x) = (3x^3 - 7x^2 + 5x + 2)x - 4.$$

Repeat the operation on the cubic in parentheses:

$$f(x) = ((3x^2 - 7x + 5)x + 2)x - 4.$$

Once more:

$$f(x) = (((3x - 7)x + 5)x + 2)x - 4.$$

This form suggests the routine

x [STO] [×] 3 [−] 7 [×] [RCL] [+] 5 [×] [RCL] [+] 2
[×] [RCL] [−] 4 [=]

Now apply this sequence to the given values of x.

Answer 6.60895 6.63125 6.65360 6.67600

Remark The four values of x in Example 1 are equally spaced at steps of 0.001. In such a case, there is a nice shortcut. After the polynomial is evaluated for the first x, use the key sequence . 0 0 1 [M+] [RCL]. This stores the next x and gets it ready in the display for the next evaluation of $f(x)$.

The routine of Example 1 can be adapted to any polynomial.

Polynomial Evaluation A routine for evaluating the polynomial

$$f(x) = a_n x^n + a_{n-1}x^{n-1} + \cdots + a_2x^2 + a_1x + a_0$$

is the key sequence (left-to-right logic)

x [STO] [×] a_n

[+] a_{n-1} [×] [RCL] [+] a_{n-2} [×] [RCL] · · · [+] a_1 [×] [RCL]

[+] a_0 [=].

If $a_k < 0$, replace [+] a_k by [−] $|a_k|$. If $a_k = 0$, simply omit [+] a_k.

For an algebraic logic calculator, modify the routine by inserting [=] before each [×] after the first one.

EXAMPLE 2 Verify that $x = 29$ is a solution of the equation

$$2x^6 - 59x^5 + 26x^4 + 88x^3 - 24x^2 - 144x - 29 = 0.$$

Solution Evaluate the polynomial on the left-hand side for $x = 29$. Use the key sequence

2 9 [STO] [×] 2 [−] 5 9 [×] [RCL] [+] 2 6 [×] [RCL] [+] 8 8 [×] [RCL]

[−] 2 4 [×] [RCL] [−] 1 4 4 [×] [RCL] [−] 2 9 [=].

The result is 0; hence $x = 29$ is a solution.

Remark 1 When some coefficients are 0, shortcuts may be possible. For example, if

$$f(x) = x^9 - 8x^5 + 7x^3 + 1,$$

You can write

$$f(x) = ((x^4 - 8)x^2 + 7)x^3 + 1.$$

This form suggests the key sequence

x [STO] [x^2] [x^2] [−] 8 [×] [RCL] [x^2] [+] 7

[×] [RCL] [x^2] [×] [RCL] [−] 1 [=].

Remark 2 On a calculator with [(] and [)], rational functions can be evaluated easily. Suppose $r(x) = f(x)/g(x)$, where $f(x)$ and $g(x)$ are polynomials. Let [$f(x)$] and [$g(x)$] denote the *polynomial* routines described above. Then $r(x)$ is evaluated by

x [STO] [$f(x)$] [÷] [(] [$g(x)$] [)] [=].

Solutions of Equations

Complicated equations such as

$$x^4 = x + 1 \qquad e^x = 2 - x^3 \qquad x^5 + 2x^3 + x^2 - 1 = 0$$

usually cannot be solved exactly. However, solutions can be approximated by numerical methods. These require lots of computation, so the calculator is a valuable aid. We are going to illustrate some basic methods that might be considered ''educated'' trial-and-error.

EXAMPLE 3 Estimate to 3 places the positive solution of $x^4 = x + 1$.

Solution First graph $y = x^4$ and $y = x + 1$ on the same grid (Fig. 1).

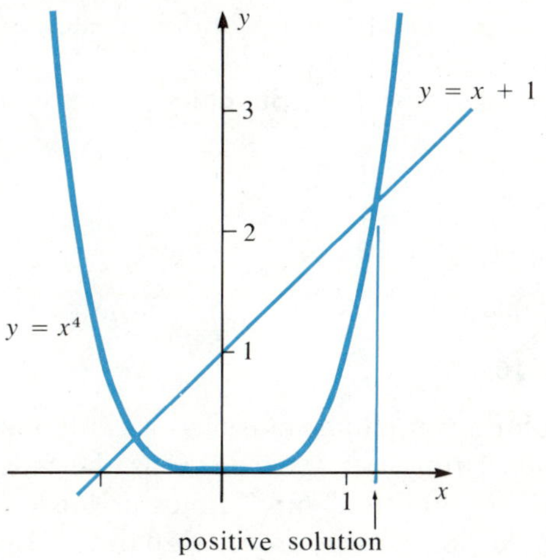

Fig. 1 Graphical estimate of the positive root of $x^4 = x + 1$

The figure shows that the graphs cross for x around 1.1 or 1.2. This gives a first, crude approximation to the solution. For more accurate information, let us tabulate x, x^4, and $x + 1$ by tenths, starting at 1.0. The calculator sequence x [x^2] [x^2] gives x^4 easily.

x	1.0	1.1	1.2	1.3
x^4	1.0000	1.4641	2.0736	2.8561
$x + 1$	2.0	2.1	2.2	2.3

The table shows that $x^4 < x + 1$ for $x = 1.2$ and $x^4 > x + 1$ for $x = 1.3$. Hence the positive solution is trapped between 1.2 and 1.3. To pin it down tighter, we tabulate by hundredths, starting at 1.2, working towards 1.3 until $x^4 > x + 1$.

	x	1.20	1.21	1.22	1.23
(to 4 places)	x^4	2.0736	2.1436	2.2153	2.2889
	$x + 1$	2.20	2.21	2.22	2.23

We see that $x^4 < x + 1$ for $x = 1.22$ and $x^4 > x + 1$ for $x = 1.23$, so the solution is between 1.22 and 1.23. We continue:

x	1.220	1.221
x^4	2.2153	2.2226
$x + 1$	2.220	2.221

The solution is between 1.220 and 1.221. Which is closer? To decide, we just check the midpoint 1.2205:

$$(1.2205)^4 \approx 2.2190 < 1.2205 + 1 = 2.2205.$$

Therefore, the solution is between 1.2205 and 1.2210, so it is closer to 1.221 than to 1.220.

Answer 1.221

Use of Linear Interpolation

Suppose Example 3 called for 6 places instead of 3. The way we were going, that would require much more work. At the next step, we would calculate x^4 and $x + 1$ for $x = 1.2201, 1.2202, 1.2203, \cdots$. We might have to go all the way to 1.2209 before $x^4 > x + 1$. Then we might need 9 calculations at steps of 0.00001, etc.

There is a way to cut down the work considerably. For convenience, we write the equation $x^4 = x + 1$ in the form

$$f(x) = 0, \quad \text{where} \quad f(x) = x^4 - x - 1.$$

From the last table in the solution of Example 3, we find

$$f(1.220) \approx 2.2153 - 2.2200 = -0.0047,$$

$$f(1.221) \approx 2.2226 - 2.2210 = 0.0016.$$

Since the second value is closer to 0, the solution should be closer to 1.221 than to 1.220. In fact, the value $f(1.220)$ is about 3 times as far from 0 as the value $f(1.221)$ is from 0. So we suspect that the solution is about 3 times as far from 1.220 as from 1.221. That would place it about $\frac{3}{4}$ of the way from 1.220 to 1.221, at 1.22075. Therefore, instead of testing 1.2201, 1.2202, $\cdots$, we jump right to

1.2207. If we feel especially confident, we could jump right to 1.22075 and start testing at intervals of 10^{-5}. That would pay off since, to 6 places, the solution turns out to be 1.220744.

This method worked so well that we should examine what is really going on. Basically we assumed that as x increased from 1.220 to 1.221, the values of $f(x)$ varied proportionally from $f(1.220)$ to $f(1.221)$. In other words, we assumed that $f(x)$ was *linear* and estimated the solution by linear interpolation (Fig. 2a).

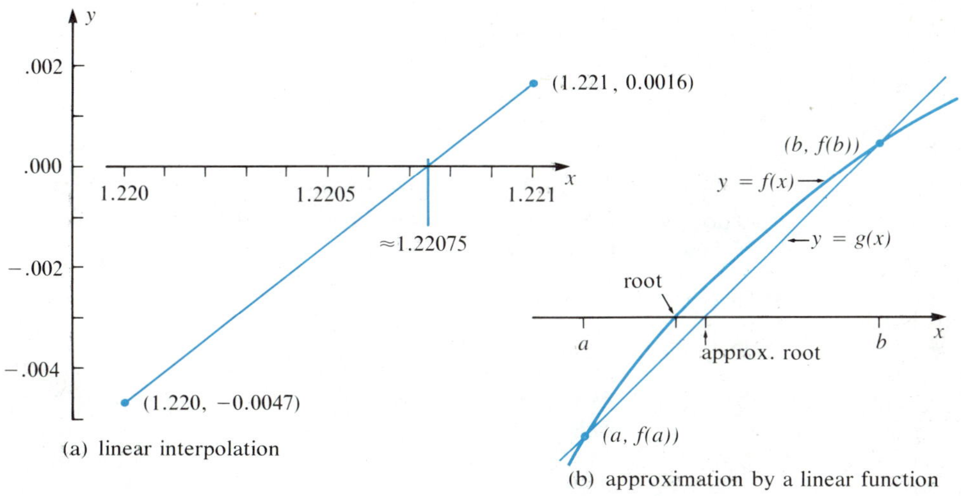

Fig. 2 Root location

The technique of linear interpolation is useful for approximating solutions of equations $f(x) = 0$ in general. We assume that the graph of $y = f(x)$ is a smooth curve. By graphical or numerical methods, we find a small interval $[a, b]$ such that $f(a)$ and $f(b)$ are of opposite sign. Then there is a solution of $f(x) = 0$ inside $[a, b]$. We replace $f(x)$ by a linear function $g(x)$ on $[a, b]$, then approximate the solution of $f(x) = 0$ by the solution of $g(x) = 0$. See Fig. 2b. This linear interpolation can be done graphically. With a piece of millimeter graph paper and a straight edge you can get accurate estimates.

Example 4 Estimate to 5 places the solution of

$$e^x = 2 - x^3.$$

Solution We start by graphing $y = e^x$ and $y = 2 - x^3$ on the same grid (Fig. 3).

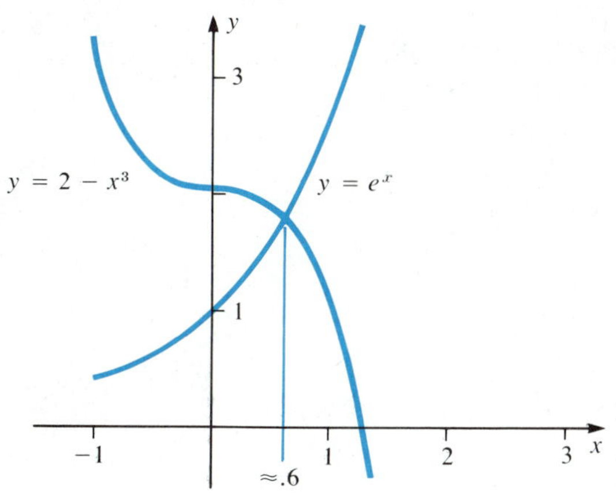

Fig. 3 To solve: $e^x = 2 - x^3$

Even a fairly crude figure indicates that there is only one solution, around $x = 0.6$.

The equation is equivalent to $f(x) = 0$, where

$$f(x) = e^x + x^3 - 2.$$

(Note that $f(x)$ can be evaluated by

x [STO] [y^x] 3 [−] 2 [+] [RCL] [e^x] [=] .)

We tabulate $f(x)$ to three places, starting at $x = 0.5$.

x	0.5	0.6
$f(x)$	−0.226	0.038

There is a solution between 0.5 and 0.6, close to 0.6. We estimate it graphically by linear interpolation (Fig. 4a). A good guess is $x \approx 0.586$, so let us start again at $x = 0.585$ and work forward by increments of 0.001:

x	0.585	0.586	0.587
$f(x)$	−0.00481	−0.00198	0.00085

The solution lies between 0.586 and 0.587. We interpolate again (Fig. 4b).

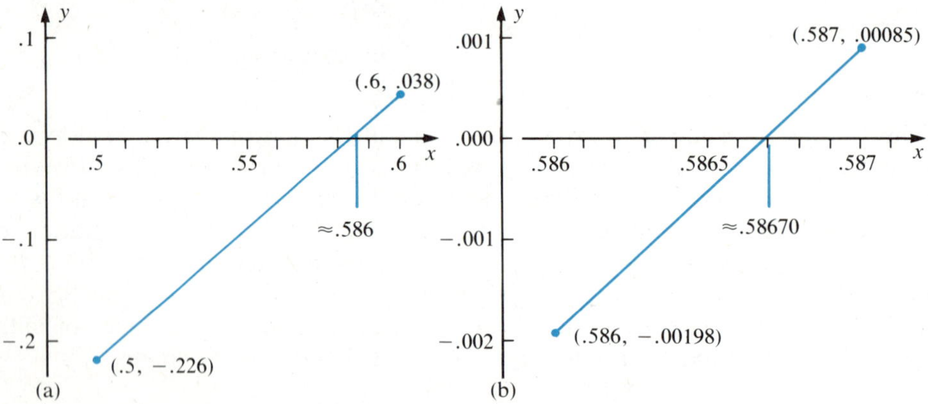

Fig. 4

The graph indicates $x \approx 0.58670$. We calculate values:

x	0.58669	0.58670	0.58671
$f(x)$	-3.12×10^{-5}	-2.89×10^{-6}	2.54×10^{-5}

The solution is between 0.58670 and 0.58671, closer to 0.58670.

Answer 0.58670

Remark The linear interpolations can be done numerically on a calculator rather than graphically. Assume that $a < b$ and that $f(a)$ and $f(b)$ are of opposite sign. The linear function interpolating

$$(a, f(a)) \quad \text{and} \quad (b, f(b))$$

is

$$g(x) = \frac{f(b) - f(a)}{b - a}(x - a) + f(a).$$

The solution of $g(x) = 0$ is

$$x = a + \frac{-f(a)}{f(b) - f(a)}(b - a).$$

One can check that

$$x = a + \frac{|f(a)|}{|f(b)| + |f(a)|}(b - a),$$

which is easy to calculate. Actually only the multiplier of $b - a$ must be calculated, since $b - a = 10^{-n}$ at each step. For instance, in the last graphical interpolation above, $a = 0.586$, $b - a = 0.001$, and

$$\frac{|f(a)|}{|f(a)| + |f(b)|} = \frac{0.00198}{0.00198 + 0.00085} \approx 0.70.$$

Therefore $(0.70)(0.001) = 0.00070$ must be added to $a = 0.586$ for the next approximation.

The shortcut

1 9 8 [STO] [+] 8 5 [÷] [RCL] [=] [1/x] 0.70

suggests itself.

Programmable Calculators

A programmable calculator can remember a sequence of instructions and execute the sequence repeatedly. This capability is useful whenever a function $f(x)$ must be evaluated for many values of x, as in Example 4. You load into the program memory a sequence of steps that computes the function (for example, p. 220, line 6). Then if you enter any number x and press the run/stop [R/S] key, the calculator will compute $f(x)$.

But a programmable calculator can do more. It can make decisions (do one thing if the display is positive, another if it is negative) and it can loop (repeat a sequence of steps a prescribed number of times). These are tremendous capabilities. For example, they enable the entire process of searching for a solution of $f(x) = 0$ as in Example 4, interpolation and all, to be loaded into the program. (Any particular function $f(x)$ must be programmed and loaded as a subroutine.) Once this is set up, you tell the calculator where to start looking, and what accuracy you want. Then you just press [R/S], sit back, and wait for the answer.

Exercises

Evaluate the polynomial at the indicated values

1 $x^3 + 5x^2 + 7x + 2$ 1.4 -32

2 $x^3 + 15x^2 + 3x + 10$ 0.35 13

3 $5x^3 + 2x^2 + 6x + 1$ 17.8 -7.36

4 $2x^3 + 19x^2 + x + 4$ -5.56 29.2

5 $6x^4 - 2x^3 - x^2 + 3x + 2$ 1.11015 -4.37298

6 $3x^4 + 3x^3 + 7x^2 - 4x - 8$ -0.01473 21.159

7 $3x^6 + 2x^5 + 2x^3 + x^2 + 5$ 3.14159 -2.71828

8 $4x^7 + 2x^3 - 7x + 1$ 52.803 -0.15961

9 $-5x^6 + 9x^3 + 7x^2 - 3$ 3.0102 -2.23607

10 $7x^6 + 6x^5 + 3x^2 + 2x + 1$ 0.36602 -2.56155.

Which of the given numbers is a solution of the equation?

11 $x^5 - 7x^4 + 3x^3 - 17x + 90 = 0 \quad -2 \quad 2 \quad 3$

12 $2x^6 + 3x^5 + 4x^4 + 5x^3 + 6x^2 + 7x - 951 = 0 \quad -2 \quad -3 \quad -4$

13 $x^5 + x^4 - 9x^3 - 8x^2 + 3x - 18 = 0 \quad -2 \quad 2 \quad 3$

14 $x^6 - 9x^5 + 26x^4 - 23x^3 - 9x^2 + 26x - 24 = 0 \quad -1 \quad 2 \quad 3 \quad 4.$

Solve approximately to 6 places

15 $x^3 + x^2 = 1$

16 $x = \dfrac{1}{2 + x^4}$

17 $10^x = x + 1 \quad x < 0$

18 $x \cdot 10^x = 100$

19 $\log x = -x$

20 $2 - \log x = x$

21 $x^5 + 2x^3 + x^2 - 1 = 0$ [*Hint* Test $x = 0, 1, 2, 3$, etc.]

22 $2x^3 - 21x^2 + 72x - 55 = 0$

23 $e^x = 10x \quad x > 2$

24 $2^{-x} = x$

25 $2^x + x^3 = 0$

26 $2^x + 3^x = 15.$

10 BUSINESS APPLICATIONS [Optional]

As further applications of exponential functions, we shall discuss several basic problems involving the investment of money. This section might be considered as a very short course in "math of finance."

Compound Interest

Suppose you deposit an amount of money P (principal) in a savings account. At regular time intervals, it draws interest at interest rate i per period (for example, $\frac{1}{2}\%$ per month). What is the value A of the account after n interest periods?

This question leads immediately to an exponential function. After one period the interest on P is Pi. This amount is added to the account, whose value becomes

$$P + Pi = P(1 + i).$$

Thus the principal is multiplied by $1 + i$. You start the second period with this new principle, $P(1 + i)$. In turn it is multiplied by $1 + i$ at the end of the second period, so the value of the account then is

$$[P(1 + i)](1 + i) = P(1 + i)^2.$$

The pattern is clear: each interest payment multiplies the current amount by $(1 + i)$. Hence powers of $(1 + i)$ build up.

> **Compound Interest** The value of a sum P invested for n periods at interest rate i per period is
>
> $$A = P(1 + i)^n.$$

EXAMPLE 1 I deposit \$6,000 in a savings account at 8% annual interest. Find the value of my account after 7 years if interest is compounded (a) annually, (b) quarterly.

Solution (a) Apply the formula with $P = 6000$, $i = 0.08$, and $n = 7$:

$$A = 6000(1.08)^7.$$

Use the calculator sequence

1 . 0 8 [y^x] 7 [×] 6 0 0 0 [=] 10282.95

(b) With 4 payments per year, $i = \frac{1}{4}(0.08) = 0.02$ and $n = 7 \times 4 = 28$. Hence

$$A = 6000(1.02)^{28}.$$

Use the sequence

1 . 0 2 [y^x] 2 8 [×] 6 0 0 0 [=] 10446.15

Answer (a) \$10,282.95 (b) \$10,446.15

EXAMPLE 2 At what annual interest rate r will money double in 8 years if compounded monthly?

Solution The monthly interest rate is $i = r/12$ and the number of periods is $n = 8 \times 12 = 96$. After 8 years, we must have $A = 2P$, so the set-up is

$$2P = P(1 + r/12)^{96} \qquad (1 + r/12)^{96} = 2$$

$$1 + r/12 = 2^{1/96} \qquad r = 12(2^{1/96} - 1).$$

Calculate r by the key sequence (left-to-right logic)

2 [y^x] 9 6 [1/x] [−] 1 [×] 1 2 [=] 0.08696

Answer 8.696%

Suppose a loan at interest rate i per period is repaid in a fractional time. It is established business practice that the formula $A = P(1 + i)^n$ still applies, even though n is not an integer.

EXAMPLE 3 A company buys a treasury bill for \$500,000 at $8\frac{5}{8}\%$ annual interest, compounded monthly. It redeems the bill after 18 months and 10 days. Find the amount it receives.

Solution Use the formula $A = P(1 + i)^n$ with

$$P = 500{,}000 \qquad i = \tfrac{1}{12}(0.08625) = 0.0071875 \quad \text{and}$$

$$n = 18\tfrac{1}{3} \approx 18.333333$$

Thus

$$A \approx (500{,}000)(1.0071875)^{18.333333} \approx 570{,}154.65$$

Answer \$570,154.65

Remark There are inexpensive calculators specifically for business applications which do the calculations in these and the following examples directly and easily.

Annuity

"Annuity" originally meant a yearly payment of a fixed amount, but now it refers to any periodic payment of a fixed amount. Suppose you deposit at the *end* of each time period the same amount P into a savings account paying interest i per period and you do this for n time periods. What will be the value of your account immediately after the n-th payment? (For instance, you deposit \$10 at the end of each month in a savings account paying $\frac{1}{2}\%$ per month, and you close the account immediately after your 24-th payment. How much do you withdraw?)

Let A_k denote the value of the account after k payments, that is, at the end of the k-th time period. Clearly $A_1 = P$. The amount A_k will consist of three parts: (1) the k-th payment P, (2) the amount A_{k-1} of the account after the $(k - 1)$-th payment, and (3) the interest $A_{k-1}i$ earned by A_{k-1} in the k-th period. Therefore

$$A_k = P + A_{k-1} + A_{k-1}i = P + A_{k-1}(1 + i).$$

Let's work out the first few:

$$\begin{aligned} A_1 &= P \\ A_2 &= P + A_1(1 + i) = P + P(1 + i) = P[1 + (1 + i)] \\ A_3 &= P + A_2(1 + i) = P + P[1 + (1 + i)](1 + i) \\ &= P[1 + (1 + i) + (1 + i)^2]. \end{aligned}$$

Continuing we obtain

$$A_n = P[1 + (1 + i) + (1 + i)^2 + \cdots + (1 + i)^{n-1}].$$

To sum the bracketed expression, we recall the polynomial identity

$$x^n - y^n = (x - y)(x^{n-1} + x^{n-2}y + x^{n-3}y^2 + \cdots + xy^{n-2} + y^{n-1})$$

given in the box on p. 30. We set $x = 1 + i$ and $y = 1$, so $x - y = i$, and the identity becomes

$$(1 + i)^n - 1 = i[(1 + i)^{n-1} + \cdots + (1 + i)^2 + (1 + i) + 1].$$

Dividing by i we obtain

$$(1 + i)^{n-1} + \cdots + (1 + i)^2 + (1 + i) + 1 = \frac{(1 + i)^n - 1}{i}.$$

Therefore

$$A_n = P\frac{(1 + i)^n - 1}{i}.$$

This is the basic law of annuities. Obviously the important quantity is

$$\frac{(1 + i)^n - 1}{i},$$

the amount of an annuity of \$1 per period. It is commonly denoted $s_{\overline{n}|i}$ in tables or on business calculators.

> **Annuity** The value A of an account after the n-th periodic payment of P at interest rate i per period is
>
> $$A = Ps_{\overline{n}|i} \quad \text{where} \quad s_{\overline{n}|i} = \frac{(1 + i)^n - 1}{i}.$$

EXAMPLE 4 A company started an annuity for an employee 35 years ago, depositing \$12.50 at the end of each month in an account paying $4\frac{1}{2}\%$ annually, compounded monthly. How much is the annuity now worth?

Solution In this example

$$i = \tfrac{1}{12}(0.045) = 0.00375 \qquad n = 35 \times 12 = 420.$$

Compute $s_{\overline{n}|i}$:

$$s_{\overline{n}|i} = \frac{(1 + i)^n - 1}{i} = \frac{(1.00375)^{420} - 1}{0.00375} \approx 1017.7420$$

Consequently

$$A = Ps_{\overline{n}|i} \approx (12.50)(1017.7420) \approx 12721.78$$

(A direct routine for a left-to-right calculator is

.0 4 5 [÷] 1 2 [=] [STO] [+] 1 [y^x] 4 2 0 [−] 1 [÷] [RCL] [=].)

Answer $12,721.78

Let's try a more complicated version of the same example.

EXAMPLE 5 I deposited $10 at the end of each month in a savings plan paying 4% annual interest, compounded monthly. At the end of 8 years the bank increased its interest rate to $5\frac{1}{4}\%$ and I increased my deposits to $15. What is the value of my account now, 7 years later?

Solution At the end of the first 8 years the value was $V_1 = 10 \times s_{\overline{n}|i}$, where

$$n = 8 \times 12 = 96 \quad \text{and} \quad i = \frac{0.04}{12} \approx 0.0033333333,$$

that is,

$$V_1 = 10\frac{(1+i)^{96} - 1}{i} \approx 1129.1853.$$

Over the next 7 years this amount earned interest, increasing to

$$V_2 = V_1(1+i)^n,$$

where now

$$n = 7 \times 12 = 84 \quad \text{and} \quad i = \frac{0.0525}{12} = 0.004375,$$

so

$$V_2 \approx (1129.1853)(1.004375)^{84} \approx 1629.372.$$

Meanwhile, the $15 monthly deposits accumulated to

$$V_3 = 15\frac{(1+i)^n - 1}{i} = 15\frac{(1.004375)^{84} - 1}{0.004375} \approx 1518.729.$$

The total value of the account is

$$V = V_2 + V_3 \approx 1629.372 + 1518.729 \approx 3148.10$$

Answer $3148.10

Sinking Fund

The value of an annuity at the end of its duration is called a **sinking fund.** It is often used in business to meet a fixed future obligation. Sometimes we must find the periodic payment P needed to produce a sinking fund A. Then we solve the formula $A = Ps_{\overline{n}|i}$ for P:

$$P = \frac{A}{s_{\overline{n}|i}} = A\frac{i}{(1+i)^n - 1}.$$

EXAMPLE 6 A corporation floats a $28,000,000 bond issue, due in 12 years. How much money must be deposited monthly in a 7% account (compounded monthly) to retire the bonds at maturity?

Solution Here

$$n = 12 \times 12 = 144 \quad \text{and} \quad i = \frac{0.07}{12} \approx 0.0058333333.$$

Therefore

$$s_{\overline{n}|i} = \frac{(1 + i)^{144} - 1}{i} \approx 224.69498$$

$$P = \frac{A}{s_{\overline{n}|i}} = \frac{2.8 \times 10^7}{s_{\overline{n}|i}} \approx 124{,}613.37$$

Answer \$124,613.37

Let us consider an example in which the interest rate is unknown.

Example 7 I deposit \$125 monthly in a savings account for 9 years. During that time the bank's interest rate fluctuates. However, when I close the account, its value is \$17,962.07. Estimate the effective annual interest rate, compounded monthly, that is, the constant interest rate that would produce the same result.

Solution Let r be the unknown annual rate. Then

$$i = \frac{r}{12} \qquad n = 9 \times 12 = 108 \qquad P = 125 \quad \text{and} \quad A = 17962.07$$

Therefore

$$s_{\overline{n}|i} = \frac{A}{P} \approx 143.6966$$

$$\frac{(1 + i)^{108} - 1}{i} = 143.6966\,.$$

We cannot solve this equation algebraically, so we try to squeeze down on i by guessing. We evaluate the left-hand side for various values of i using the routine

i [STO] [+] 1 [y^x] 1 0 8 [−] 1 [÷] [RCL] [=] .

We find

i	.01	.005	.006
$s_{\overline{108}|i}$	192.89	142.74	151.34

The solution is between .005 and .006, probably closer to .005. For a closer estimate, we interpolate:

$$\frac{143.70 - 142.74}{151.34 - 142.74} \approx 0.11$$

This suggests trying $i = 0.0051$. The corresponding $s_{\overline{n}|i}$ is slightly too small, so we go to $i = 0.0052$:

i	.0051	.0052
$s_{\overline{108}|i}$	143.5713	144.4089

We interpolate again:

$$\frac{143.6966 - 143.5713}{144.4089 - 143.5713} \approx 0.15$$

Our next guess is $i = 0.005115$ which gives $s_{\overline{108}|i} \approx 143.6966$, a bulls-eye! Therefore $r = 12i = 12 \times 0.005115 = 0.06138$.

Answer 6.138%

Payment of Debts

What amount A can you borrow if you propose to repay the loan by paying amount P at the end of each period, for n periods? Assume the interest rate is i per period.

We wish to express A in terms of P, n, and i. There are several ways to solve this problem. Here is one based on our previous work and nice logical reasoning. Suppose the lender deposits each payment, as fast as he receives it, in a sinking fund (savings account at interest rate i). Then after the n-th payment this money is worth

$$Ps_{\overline{n}|i} = P\frac{(1+i)^n - 1}{i}$$

He reasons that this must be precisely the value of his original investment A after n periods, that is, $A(1+i)^n$. Hence

$$P\frac{(1+i)^n - 1}{i} = A(1+i)^n.$$

We solve this equation for A:

$$A = P\frac{(1+i)^n - 1}{i(1+i)^n} = P\frac{1-(1+i)^{-n}}{i}$$

This is the basic law for the payment of a debt. The important quantity is

$$a_{\overline{n}|i} = \frac{1-(1+i)^{-n}}{i},$$

known as the **present value of an annuity** (of future payments) of amount 1.

Payment of Debts A payment of P per period for n periods at interest rate i per period will repay a debt whose present value is

$$A = Pa_{\overline{n}|i} \qquad a_{\overline{n}|i} = \frac{1-(1+i)^{-n}}{i}.$$

EXAMPLE 8 I qualify for a 30-year home mortgage at $10\frac{3}{4}\%$. I can afford monthly payments of \$280. How much can I borrow?

Solution

$$n = 30 \times 12 = 360 \qquad i = \frac{0.1075}{12} \approx 0.0089583333$$

$$a_{\overline{n}|i} = \frac{1-(1+i)^{-n}}{i} \approx 107.12587$$

$$A = Pa_{\overline{n}|i} \approx (280)(107.12587) \approx 29{,}995.24$$

A left-to-right calculator routine for the whole problem is

0 . 1 0 7 5 [÷] 1 2 [=] [STO] [+] 1 [y^x] 3 6 0 [+/−]

[−] 1 [÷] [RCL] [+/−] [×] 2 8 0 [=]

Answer \$29,995.24

The next example contains what is called a **balloon payment,** the premature retirement of a debt.

Example 9 Suppose that I repay the mortgage in Example 8 for 13 years, when I win a large cash prize and decide to pay off the remaining debt. How much is that?

Solution Think of the periodic payments as a sinking fund, whose value after 13 years is $Ps_{\overline{n}|i}$, where $P = 280$, $n = 13 \times 12 = 156$, and $i = \frac{1}{12}(0.1075)$. At that time the lender's original investment of \$29,995.24 has grown by a factor of $(1 + i)^n$. What is still owed (the balloon payment) is the difference

$$(29{,}995.24)(1 + i)^n - 280\frac{(1 + i)^n - 1}{i} \approx 26188.33$$

Answer \$26,188.33

Exercises

1 To what amount does \$100 grow in 2 years at 10% annual interest (a) compounded monthly, (b) compounded daily?

2 How long does it take money to double at interest compounded monthly (a) at 6% annual interest, (b) at 12% annual interest?

3 Bank B offers 7% compounded quarterly while bank C offers $6\frac{3}{4}$% compounded daily. Which is better for the investor?

4 At 12% compounded monthly, money will triple in how long?

5 What annual interest rate, compounded monthly, is equivalent to 25% compounded annually?

6 What annual interest rate, compounded daily, is equivalent to 100% compounded annually?

7 At 9% compounded monthly, what should be the initial investment in a trust fund to be worth \$25,000 in 21 years?

8 Ben Franklin said that a penny saved is a penny earned. Suppose Ben had saved a penny on his 21-st birthday in 1727 and invested it at 10% interest compounded quarterly. By 1982, how much would that penny have earned?

9 A short-term loan of \$125,000 at $11\frac{1}{4}$% annual interest compounded monthly is repaid by a single payment after 14 months and 6 days. How much is due?

10 A short-term loan of \$200,000 is repaid after 7 months and 20 days. The amount repaid is \$214,492.55. Find the annual interest rate if compounded monthly.

11 To what amount will a monthly deposit of \$15 accumulate in 10 years at $9\frac{3}{4}$% annual interest compounded monthly?

12 My ordinary life insurance policy puts 10% of my \$160 annual premium into a $5\frac{1}{2}$% annuity. How much is it worth after 37 years?

13 How much must be deposited monthly at $\frac{1}{2}$% per month to save \$10,000 in 17 years?

14 To repay a loan of \$1,200,000 in $6\frac{1}{2}$ years, how much must be deposited monthly into a sinking fund paying $8\frac{3}{8}$% compounded monthly?

15 How long will it take monthly deposits of \$50 to accumulate to \$12,500 if the annual interest rate is $7\frac{1}{2}$% compounded monthly?

16 How long will it take monthly deposits of \$50 to accumulate to \$12,500 if the annual interest rate is 11% compounded monthly?

17 I deposit \$50 per month in a savings plan for 20 years. The annual interest rate changes from 5% to 6% after 8 years; the compounding is always monthly. Find the final value.

18* I deposit \$15 per month in a savings plan for 18 years. The interest rate is $4\frac{1}{2}$% annually the first 4 years, $5\frac{3}{4}$% the next 7 years, and $5\frac{1}{4}$% the remaining 7 years, always compounded monthly. Find the final value.

19 Find the annual interest rate, compounded monthly, if \$100 deposits each month accumulate to \$26,046.78 in 13 years.

20 Find the annual interest rate, compounded monthly, if \$10 deposits each month accumulate to \$2405.08 in 12 years.

21 How large a mortgage can be paid off in 40 years at $9\frac{3}{4}\%$ annual interest compounded monthly with payments of \$79.35 per month?

22 I take a mortgage of \$28,000 for 18 years at $8\frac{1}{2}\%$, compounded monthly. Find my monthly payment.

23 I take a mortgage of \$90,000 for 28 years at $10\frac{1}{4}\%$, compounded monthly. Find my monthly payment.

24 (cont.) Suppose I decide to pay off the mortgage after 17 years. How much is due?

25 I borrow \$3500 from a loan shark at 2% per month. I repay \$90 a month for 6 years. How much do I then owe?

26 I borrow \$2500 from a loan shark at 3% per month. I repay \$40 per month for 5 years. How much do I then owe?

27 A loan company offers cheap loans up to \$2500 for only 9% simple interest per year. I enter and request \$2500 for one year. The lender deducts the interest, $(2500)(0.09) = \$225$ and hands me the remaining \$2275, reminding me to repay $\frac{1}{12}(2500) = \$208.34$ per month for 12 months. (They always round up to the nearest cent in these deals.) Estimate the honest annual interest rate, compounded monthly.

28 (cont.) The next customer wants \$2000 for 2 years, so the lender deducts the $(2000)(0.18)$, etc. Now work out the true interest rate.

29 In order to sell his house, seller S gives buyer B a second mortgage of \$9000 at 11% compounded monthly. What are B's monthly payments if he pays off the \$9000 in 7 years?

30 (cont.) Suppose that S really needs cash, so he immediately sells the second mortgage to an investor I for \$8600. What is I's annual rate of return (compounded monthly) on this investment?

REVIEW EXERCISES

Graph

1 $y = (0.8)^x$

2 $y = (1.2)^x$.

Find the inverse function; give its domain and range

3 $y = \begin{cases} 3x \\ 2x \end{cases}$ for $\begin{cases} x \le 0 \\ x \ge 0 \end{cases}$

4 $y = x^4 + 2x^2$ for $x \ge 0$

5 $y = 1 + 2^x$

6* $y = \frac{1}{2}(2^x - 2^{-x})$.

Graph

7 $y = \log_{10}(100x)$

8 $y = \log_2(x - 1)$

9 $y = \log_{10}\left(1 + \dfrac{1}{x}\right)$

10 $y = \log_{10}\left(\dfrac{1}{x} - 1\right)$.

Calculate

11 $\log 8 + \log(0.125)$

12 $(\log 3.000)/(\ln 3.000)$.

Solve to 5 significant figures

13 $(1.1432)^x = 17.611$

14 $\log(5^x + 2) = 2$.

How accurate are the approximations

15 $\log \ln 2 \approx -\dfrac{1}{2\pi}$

16 $\dfrac{4}{\sqrt{t}} \approx \pi$ where $t = \frac{1}{2}(1 + \sqrt{5})$?

17 How long will it take a bacteria colony to triple if it doubles in 7 hr 35 min?

18 A radioactive chemical loses $\frac{1}{10}$ of its original mass in $5\frac{1}{2}$ hours. Give its decay law.

Which grows faster as $x \longrightarrow \infty$

19 $2^{x+\log_2 x}$ or 3^x

20 $\log_3(x + 3^x)$ or x?

Use a calculator to estimate to 6 significant figures

21 $\dfrac{3}{\sqrt{6} - \dfrac{2}{\sqrt{5} - 1}}$

22 $\dfrac{5 \ln a}{7 + e^a} + \sqrt[4]{a}$ where $a = 0.414392$

Evaluate

23 $f(-1.02394)$ where $f(x) = 3x^4 - 5x^3 + 6x^2 - x + 2$

24 $f(10.1386)$ where $f(x) = 1.7x^5 - 3.9x^4 + 2.1x^2 + 5.2x - 3.0$.

Solve to 6 places

25 $3^x - 2^x = 2$

26 $x(x - 1)(x - 2) = \frac{1}{4}$.

27 Find the monthly deposit, at 6% per year compounded monthly, that will accumulate \$5000 in 5 years.

28 Find the monthly payment on a 25-year mortgage for \$92,000 at 11.375% annual interest, compounded monthly.

6 TRIGONOMETRIC FUNCTIONS

1 ANGLES, SINES, AND COSINES

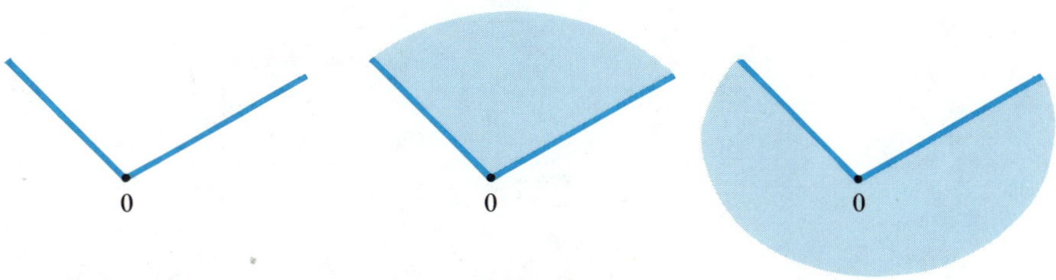

(a) two rays (b) one angle they determine (c) the other angle

Fig. 1 Two rays from 0 cut the plane into two angles.

Let us review some geometry. Two rays (half lines) from a point 0 cut the plane into two **angles** (Fig. 1). The point 0 is called the **vertex** of each angle. There is one exception, which we agree to include anyhow: when the rays coincide (Fig. 2a). Then they determine a **whole angle** (Fig. 2b) and a **zero angle** (Fig. 2c).

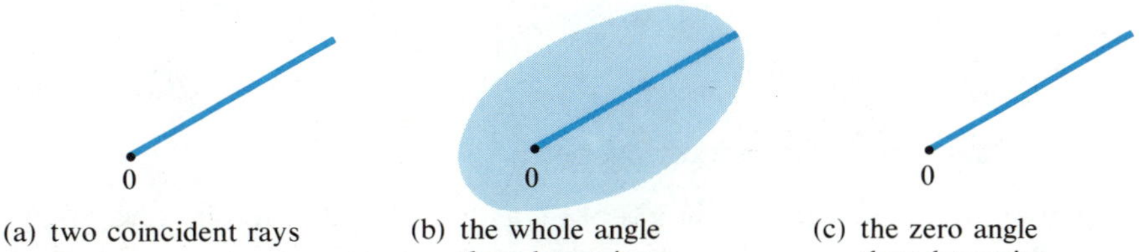

(a) two coincident rays (b) the whole angle they determine (c) the zero angle they determine

Fig. 2 The "exception", included anyhow

Each angle has an **initial side** and a **terminal side.** If you stand on the vertex 0 of the angle facing into the angle, its initial side is on your right and its terminal side is on your left (Fig. 3a). Alternatively, imagine a ray from 0 that sweeps the angle *counterclockwise,* from its initial to its terminal side (Fig. 3b).

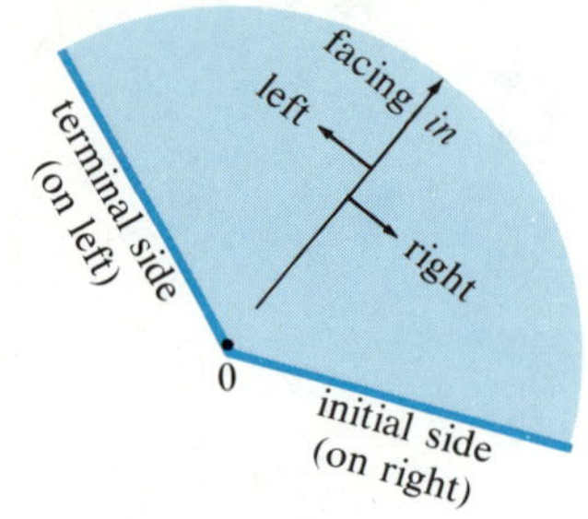

(a) The observer looks into the angle from the vertex.

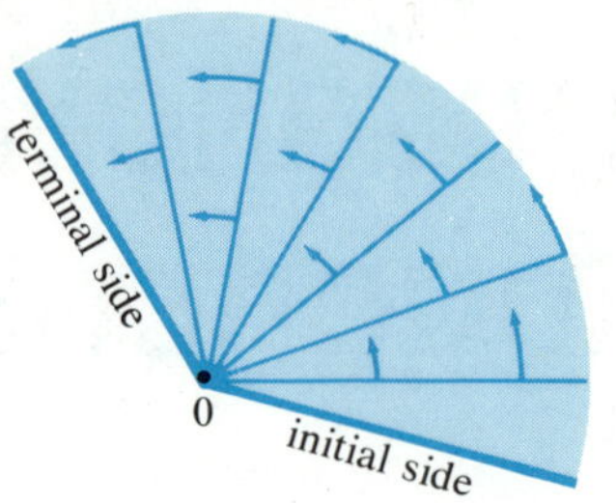

(b) counterclockwise sweep of the angle

Fig. 3

The common unit of angle *measure* is the **degree** (°). The whole angle is divided into 360 degrees (Fig. 4a). Examples are shown in Figs. 4b and 4c.

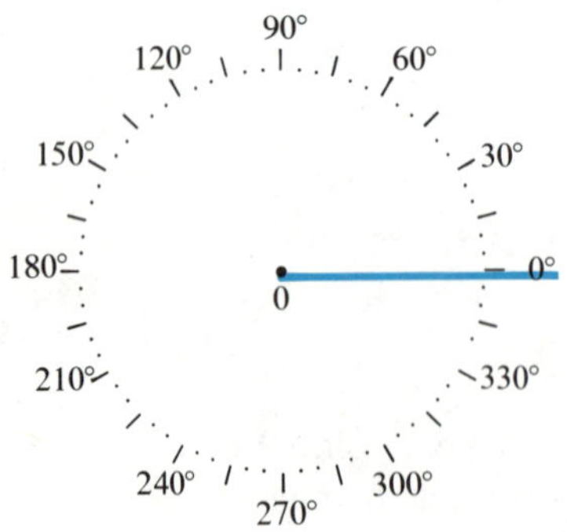

(a) whole angle divided into 360°

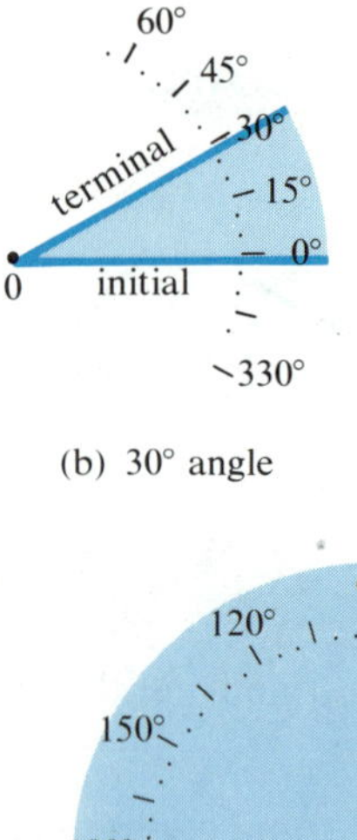

(b) 30° angle

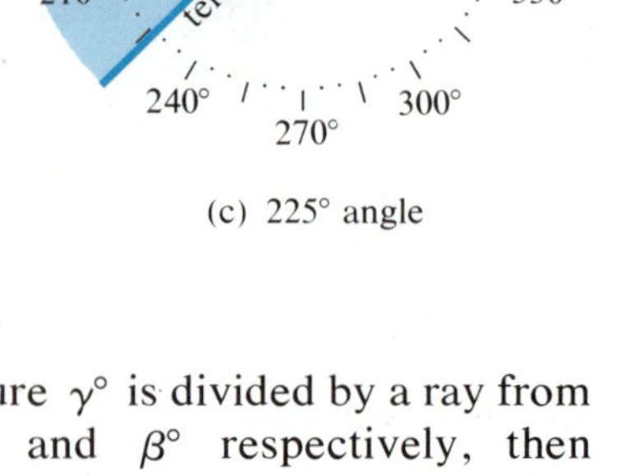

Fig. 4

(c) 225° angle

Angle measure has several important properties.

(1) Angle measure is *additive:* if an angle of measure $\gamma°$ is divided by a ray from its vertex into two subangles of measure $\alpha°$ and $\beta°$ respectively, then $\gamma° = \alpha° + \beta°$.

(2) Congruent angles have equal measure. That is, if a transformation of the euclidean plane that preserves all distances (**congruence** transformation) takes an angle of measure $\alpha°$ to an angle of measure $\beta°$, then $\alpha° = \beta°$.

(3) Similar angles have equal measure. That is, if a transformation of the euclidean plane that stretches all distances by the same positive factor k (a **similarity** transformation) takes an angle of measure $\alpha°$ to an angle of measure $\beta°$, then $\alpha° = \beta°$.

Recall the theorems: (1) Congruent triangles have equal corresponding angles. (2) Similar triangles have equal corresponding angles.

Terminology It is customary to refer to two different angles of equal *measure* as "equal angles." This is convenient and harmless. It goes along with the common usage of saying "equal segments" when we mean segments of equal length.

We shall often say things like "α is an angle and $\alpha = 30°$", when we mean "α is an angle and the *measure* of α equals $30°$". It is just too tedious to drag along the word measure when the context is perfectly clear.

A **straight angle** is half of a whole angle, that is, an angle of 180°. See Fig. 5a. A **right angle** is half a straight angle, or a quarter of a whole angle, that is, an angle of 90°. See Fig. 5b. Three quarters of a whole angle has no special name; it is an angle of 270°. An example is the angle from the positive x-axis, counterclockwise to the negative y-axis (Fig. 5c).

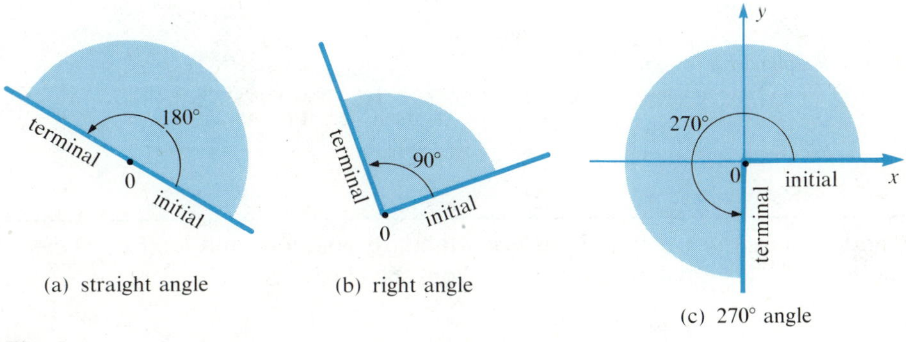

Fig. 5

Sine and Cosine

An angle is in **standard position** with respect to coordinate axes when its vertex is at the origin and its initial side is along the positive x-axis.

Let θ be an angle in standard position. Let (x, y) be *any* point on its terminal side different from its vertex (Fig. 6a). Let r be the distance from the origin to (x, y). By the distance formula (really the Pythagorean theorem),

$$r = \sqrt{x^2 + y^2}.$$

Of course $r > 0$ because $(x, y) \neq (0, 0)$. Now consider the two ratios

$$\frac{x}{r} \quad \text{and} \quad \frac{y}{r}.$$

These ratios depend *only* on θ, not on the choice of (x, y) on the terminal side.

Why? Consider the *point*

$$\left(\frac{x}{r}, \frac{y}{r}\right)$$

whose coordinates are the ratios x/r and y/r in question. On the one hand this point lies on the terminal side because its coordinates are just the coordinates of (x, y) stretched by the positive factor $1/r$. On the other hand, it lies on the **unit circle,** the circle of radius 1 and center $(0, 0)$. This is so because

$$\left(\frac{x}{r}\right)^2 + \left(\frac{y}{r}\right)^2 = \frac{x^2 + y^2}{r^2} = \frac{r^2}{r^2} = 1.$$

Therefore, the point $(x/r, y/r)$ is determined geometrically (Fig. 6b) as the unique intersection of the terminal side with the unit circle.

Thus x/r and y/r depend only on θ, that is, they are *functions* of θ. We call them the **cosine** of θ and the **sine** of θ respectively.

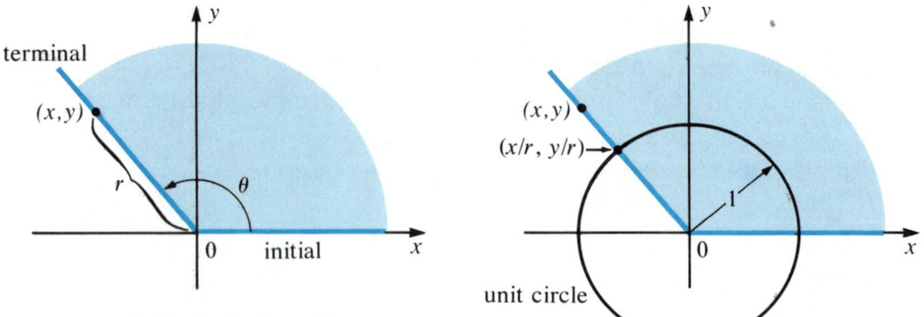

(a) θ is in standard position, (x,y) on its terminal side, $r = \sqrt{x^2 + y^2} > 0$.

(b) $(x/r, y/r)$ is the intersection of the terminal side of θ with the unit circle.

Fig. 6

> **Sine and Cosine** Let θ be an angle in standard position, and let (x, y) be any point on its terminal side, different from the origin. Set $r = \sqrt{x^2 + y^2}$. Then
>
> $$\cos\theta = \frac{x}{r} \qquad \sin\theta = \frac{y}{r}.$$

Thus we have defined two functions of θ with domain $0 \le \theta \le 360°$.

Examples

(1) $0°$ terminal side = positive x-axis,

$(x, y) = (x, 0)$ with $x > 0$ $\quad r = x$

$$\frac{x}{r} = \frac{x}{x} = 1 \qquad \frac{y}{r} = \frac{0}{r} = 0 \qquad \cos 0° = 1 \qquad \sin 0° = 0$$

(2) $90°$ terminal side = positive y-axis

$(x, y) = (0, y)$ with $y > 0$ $\quad r = y$

$$\frac{x}{r} = \frac{0}{r} = 0 \qquad \frac{y}{r} = \frac{y}{y} = 1 \qquad \cos 90° = 0 \qquad \sin 90° = 1$$

(3) $180°$ terminal side $=$ negative x-axis

$$(x, y) = (x, 0) \text{ with } x < 0 \qquad r = |x| = -x$$

$$\frac{x}{r} = \frac{x}{-x} = -1 \qquad \frac{y}{r} = \frac{0}{r} = 0 \qquad \cos 180° = -1 \qquad \sin 180° = 0$$

(4) $270°$ terminal side $=$ negative y-axis

$$(x, y) = (0, y) \text{ with } y < 0 \qquad r = |y| = -y$$

$$\frac{x}{r} = \frac{0}{r} = 0 \qquad \frac{y}{r} = \frac{y}{-y} = -1 \qquad \cos 270° = 0 \qquad \sin 270° = -1$$

(5) $360°$ same as $0°$ $\qquad \cos 360° = 1 \qquad \sin 360° = 0$

The functions sine and cosine have many interesting properties which we shall study soon. For the moment we note only the following.

$$-1 \le \sin\theta \le 1 \qquad -1 \le \cos\theta \le 1$$

$$\sin^2\theta + \cos^2\theta = 1$$

[Here $\cos^2\theta$ is an abbreviation for $(\cos\theta)^2$, etc.] They are consequences of the fact that $(\cos\theta, \sin\theta) = (x/r, y/r)$ is a point on the *unit* circle, which means $\cos^2\theta + \sin^2\theta = 1$. It is clear geometrically that $|\cos\theta| \le 1$ and $|\sin\theta| \le 1$, or analytically from $\cos^2\theta \le \cos^2\theta + \sin^2\theta = 1$, etc.

Further Examples

EXAMPLE 1 Find the cosine and sine of (a) $45°$ (b) $135°$.

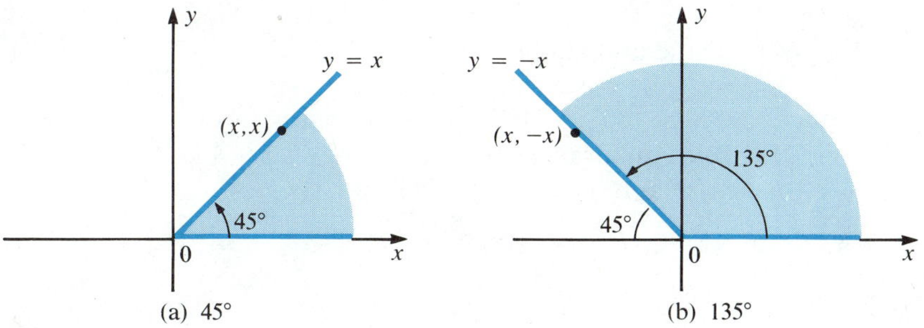

Fig. 7

Solution (a) The terminal side of a $45°$ angle in standard position is the portion of the line $y = x$ in the first quadrant (Fig. 7a). Its typical point is $(x, y) = (x, x)$ where $x > 0$. Therefore

$$r = \sqrt{x^2 + y^2} = \sqrt{x^2 + x^2} = x\sqrt{2}$$

$$\cos 45° = \frac{x}{r} = \frac{x}{x\sqrt{2}} = \frac{1}{\sqrt{2}} = \tfrac{1}{2}\sqrt{2}$$

$$\sin 45° = \frac{y}{r} = \frac{x}{x\sqrt{2}} = \frac{1}{\sqrt{2}} = \tfrac{1}{2}\sqrt{2}.$$

(b) Since $135° = 180° - 45°$, the terminal side of a $135°$ angle in standard position is the second quadrant portion of the line $y = -x$. See Fig. 7b. Its typical point is $(x, y) = (x, -x)$ where $x < 0$. Therefore

$$r = \sqrt{x^2 + y^2} = \sqrt{x^2 + x^2} = |x|\sqrt{2} = -x\sqrt{2}$$

$$\cos 135° = \frac{x}{r} = \frac{x}{-x\sqrt{2}} = \frac{-1}{\sqrt{2}} = -\tfrac{1}{2}\sqrt{2}$$

$$\sin 135° = \frac{y}{r} = \frac{-x}{-x\sqrt{2}} = \frac{1}{\sqrt{2}} = \tfrac{1}{2}\sqrt{2}.$$

Answer (a) $\cos 45° = \sin 45° = \tfrac{1}{2}\sqrt{2}$

(b) $\cos 135° = -\tfrac{1}{2}\sqrt{2}$ $\sin 135° = \tfrac{1}{2}\sqrt{2}$

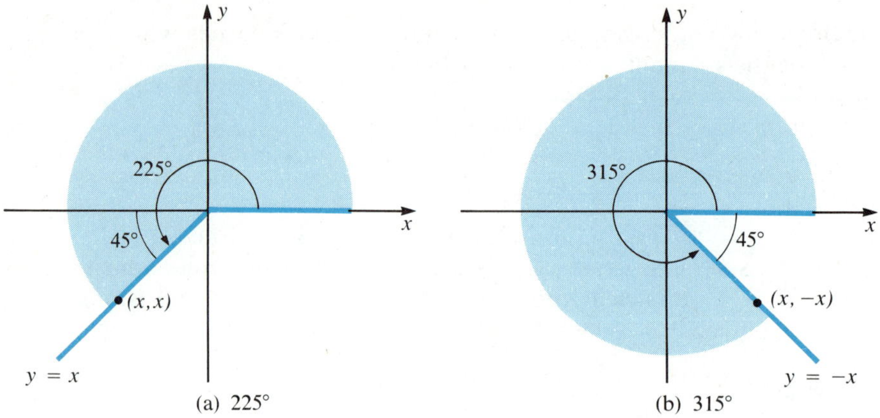

Fig. 8

EXAMPLE 2 Find the cosine and sine of (a) 225° (b) 315°.

Solution (a) Since $225° = 180° + 45°$, the terminal side of a 225° angle in standard position is the third quadrant portion of $y = x$. See Fig. 8a. Its typical point is $(x, y) = (x, x)$ where $x < 0$. Therefore

$$r = \sqrt{x^2 + y^2} = \sqrt{x^2 + x^2} = |x|\sqrt{2} = -x\sqrt{2}$$

$$\cos 225° = \frac{x}{r} = \frac{x}{-x\sqrt{2}} = \frac{-1}{\sqrt{2}} = -\tfrac{1}{2}\sqrt{2}$$

$$\sin 225° = \frac{y}{r} = \frac{x}{-x\sqrt{2}} = \frac{-1}{\sqrt{2}} = = -\tfrac{1}{2}\sqrt{2}.$$

(b) Since $315° = 360° - 45°$, the terminal side of a 315° angle in standard position is the fourth quadrant portion of $y = -x$. See Fig. 8b. Its typical point is $(x, y) = (x, -x)$ where $x > 0$. Therefore

$$r = \sqrt{x^2 + y^2} = \sqrt{x^2 + x^2} = x\sqrt{2}$$

$$\cos 315° = \frac{x}{r} = \frac{x}{x\sqrt{2}} = \frac{1}{\sqrt{2}} = \tfrac{1}{2}\sqrt{2}$$

$$\sin 315° = \frac{y}{r} = \frac{-x}{x\sqrt{2}} = \frac{-1}{\sqrt{2}} = -\tfrac{1}{2}\sqrt{2}.$$

Answer (a) $\cos 225° = \sin 225° = -\frac{1}{2}\sqrt{2}$

(b) $\cos 315° = \frac{1}{2}\sqrt{2} \qquad \sin 315° = -\frac{1}{2}\sqrt{2}$

Graphs

Let us tabulate all the values we have found, rounding to three places ($\frac{1}{2}\sqrt{2} \approx 0.707$):

$\theta°$	0	45	90	135	180	225	270	315	360
$\cos\theta$	1	.707	0	−.707	−1	−.707	0	.707	1
$\sin\theta$	0	.707	1	.707	0	−.707	−1	−.707	0

This gives us enough data for rough graphs of $y = \cos\theta$ and $y = \sin\theta$. See Fig. 9. We shall make a detailed study of these graphs in a later section.

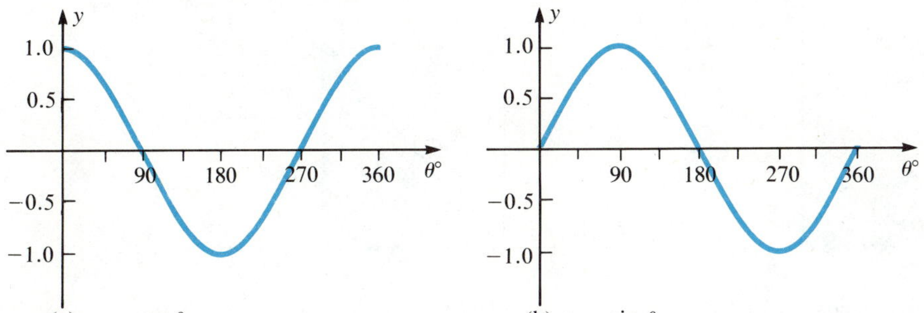

(a) $y = \cos\theta$ (b) $y = \sin\theta$

Fig. 9

Calculating Trigonometric Functions

All scientific calculators can estimate $\sin\theta°$ and $\cos\theta°$. They often calculate also in other units of angle measure (*radians*, sometimes *grads*) and you must make sure before calculating that the mode switch (or key) is set to [deg]. Then you simply enter $\theta°$ and press [sin] or [cos].

Examples

(1) $\sin 60°$ 6 0 [sin] 0.8660254

(2) $\cos 135°$ 1 3 5 [cos] −0.7071068

Exercises

Draw in standard position

1 10° **2** 80° **3** 120° **4** 210°
5 240° **6** 300° **7** 330° **8** 190°.

Find $\sin\theta$ and $\cos\theta$ exactly for an angle θ in standard position whose terminal side passes through

9 $(4, 3)$ **10** $(-3, 4)$ **11** $(-5, -12)$ **12** $(12, -5)$
13 $(7, 24)$ **14** $(-24, -7)$ **15** $(40, -9)$ **16** $(-9, 40)$
17 $(-11, -60)$ **18** $(-60, 11)$ **19** $(1, \sqrt{2})$ **20** $(-\sqrt{2}, -1)$
21 $(\sqrt{2}, \sqrt{7})$ **22** $(-\sqrt{7}, -\sqrt{2})$ **23** $(\sqrt{3}, -\sqrt{5})$ **24** $(\sqrt{5}, -\sqrt{3})$.

Estimate by calculator to 4 places

25	sin 17.61°	**26**	sin 48.92°	**27**	cos 5.21°	**28**	cos 78.33°
29	sin 100.62°	**30**	sin 164.55°	**31**	cos 93.75°	**32**	cos 171.82°
33	sin 206.17°	**34**	sin 251.39°	**35**	cos 199.02°	**36**	cos 254.10°
37	sin 279.03°	**38**	sin 294.16°	**39**	cos 280.40°	**40**	cos 355.01°

2 RIGHT TRIANGLES AND TANGENTS

The angles of a right triangle are its right angle, 90°, and its two complementary acute angles, α and β. Remember that **acute** means $0 < \alpha < 90°$ and $0 < \beta < 90°$, and **complementary** means $\alpha + \beta = 90°$. The angles α and β are complementary because the sum of the three angles of (any) triangle is 180°, so $\alpha + \beta + 90° = 180°$, hence $\alpha + \beta = 90°$.

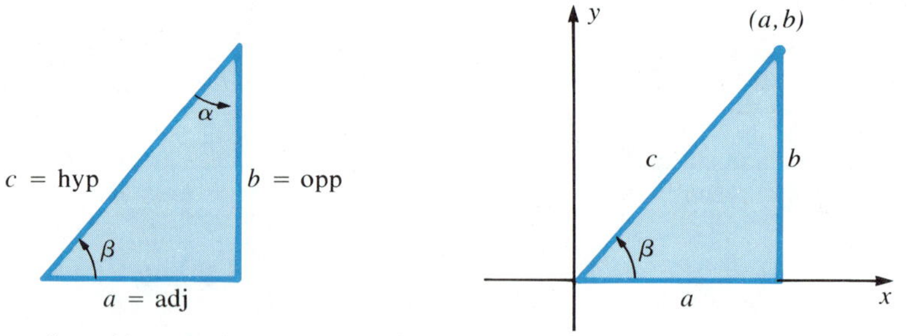

(a) b = side opposite β
a = side adjacent β
c = hypotenuse

(b) fit axes

Fig. 1 Right triangle

Take a right triangle, labelled as in Fig. 1a. To find $\sin\beta$ and $\cos\beta$, we fit axes as in Fig. 1b. (It may be necessary to turn the triangle upside-down first.) Then $(x, y) = (a, b)$ is a point on the terminal side of β, now in standard position, and $r = c$. Hence

$$\sin\beta = \frac{b}{c} \quad \text{and} \quad \cos\beta = \frac{a}{c}.$$

Since b is the length of the leg of the triangle opposite β and a is the length of the leg adjacent to α, we may write:

> For an acute angle β of a right triangle,
>
> $$\sin\beta = \frac{\text{opposite}}{\text{hypotenuse}} \qquad \cos\beta = \frac{\text{adjacent}}{\text{hypotenuse}}.$$

Look at α in Fig. 1a. Its opposite side is a, and its adjacent side is b, so

$$\sin\alpha = \frac{a}{c} = \cos\beta \quad \text{and} \quad \cos\alpha = \frac{b}{c} = \sin\beta.$$

Since $\beta = 90° - \alpha$, this proves an important result:

> If α is any acute angle, then
>
> $$\sin\alpha = \cos(90° - \alpha) \quad \text{and} \quad \cos\alpha = \sin(90° - \alpha).$$

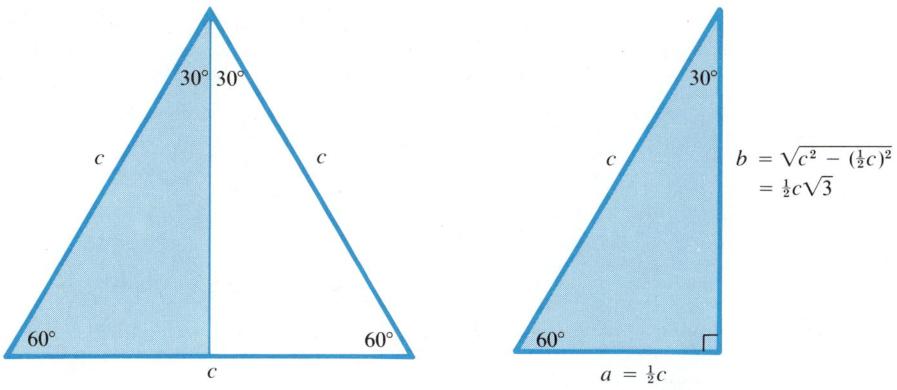

(a) equilateral triangle (b) 30-60-90 triangle

Fig. 2

EXAMPLE 1 Find the cosine and sine of

(a) $60°$ (b) $30°$.

Solution (a) Drop a perpendicular from one vertex of an equilateral triangle of side c to its opposite side (Fig. 2a). It splits the triangle into two 30-60-90 triangles. The side adjacent to the $60°$ angle (Fig. 2b) has length $a = \frac{1}{2}c$. By the Pythagorean theorem, the opposite side is

$$b = \sqrt{c^2 - a^2} = \sqrt{c^2 - (\tfrac{1}{2}c)^2} = \sqrt{\tfrac{3}{4}c^2} = \tfrac{1}{2}c\sqrt{3}.$$

Therefore

$$\sin 60° = \frac{\text{opp}}{\text{hyp}} = \frac{b}{c} = \frac{\frac{1}{2}c\sqrt{3}}{c} = \tfrac{1}{2}\sqrt{3}$$

$$\cos 60° = \frac{\text{adj}}{\text{hyp}} = \frac{a}{c} = \frac{\frac{1}{2}c}{c} = \tfrac{1}{2}$$

(b) Since $30°$ and $60°$ are complementary acute angles,

$$\sin 30° = \cos 60° = \tfrac{1}{2} \quad \text{and} \quad \cos 30° = \sin 60° = \tfrac{1}{2}\sqrt{3}.$$

Answer (a) $\sin 60° = \frac{1}{2}\sqrt{3}$ $\cos 60° = \frac{1}{2}$

(b) $\sin 30° = \frac{1}{2}$ $\cos 30° = \frac{1}{2}\sqrt{3}$

The 30-60-90 triangle can be used to find sines and cosines of many non-acute angles. Remember that the sides are

short leg: $\frac{1}{2}c$ long leg: $\frac{1}{2}c\sqrt{3}$ hypotenuse: c.

It helps to take $c = 2$: the sides of a 30-60-90 triangle are in the proportion 1 to $\sqrt{3}$ to 2.

EXAMPLE 2 Find $\sin 150°$ and $\cos 150°$.

Solution Since $150° = 180° - 30°$, the terminal side of a $150°$ angle in standard position lies in the second quadrant, at $30°$ to the negative x-axis (Fig. 3, next page). Take the point P on this terminal side at distance 2 from $(0, 0)$, and drop a perpendicular from P to the x-axis. A 30-60-90 triangle is formed; its hypotenuse is 2, so its short leg is 1 and its long leg $\sqrt{3}$. Therefore

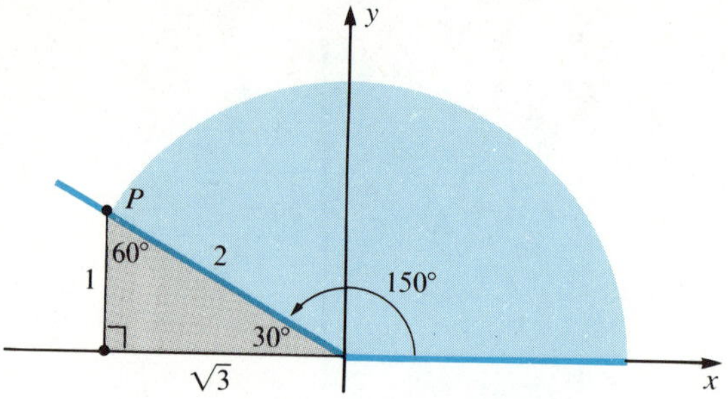

Fig. 3

$P = (-\sqrt{3}, 1)$, being in the *second* quadrant. Consequently

$$\cos 150° = \frac{x}{r} = \frac{-\sqrt{3}}{2} = -\tfrac{1}{2}\sqrt{3} \qquad \sin 150° = \frac{y}{r} = \frac{1}{2} = \tfrac{1}{2}.$$

Answer $\cos 150° = -\tfrac{1}{2}\sqrt{3} \qquad \sin 150° = \tfrac{1}{2}$

EXAMPLE 3 Express in terms of $v = \sin 35°$

(a) $\cos 35°$ (b) $\sin 125°$ (c) $\cos 235°$.

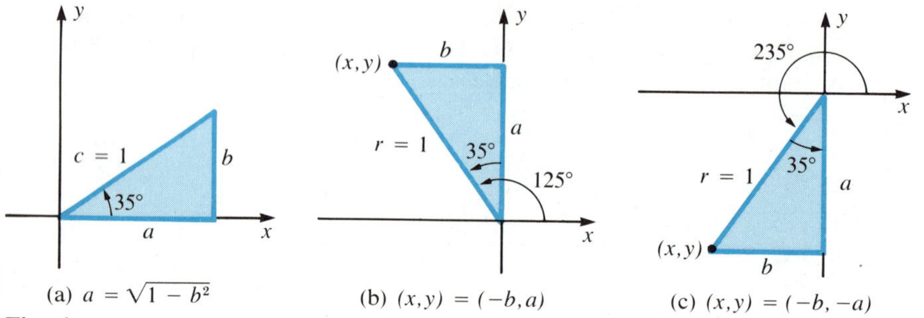

Fig. 4

Solution (a) Draw 35° in standard position and a corresponding 35-55-90 right triangle (Fig. 4a). Take hyp $= 1$. Then

$$v = \sin 35° = \frac{\text{opp}}{\text{hyp}} = \frac{b}{1} = b.$$

By the Pythagorean theorem,

$$a = \text{adj} = \sqrt{1 - b^2} = \sqrt{1 - v^2}$$

so

$$\cos 35° = \frac{\text{adj}}{\text{hyp}} = \frac{\sqrt{1 - v^2}}{1} = \sqrt{1 - v^2}.$$

[Of course, we could have used $\cos^2 35° = 1 - \sin^2 35°$, but we need the triangle for (b) and (c).]

(b) In standard position, the terminal side of 125° lies in the second quadrant, 35° forward of the y-axis. Place the triangle from (a) accordingly (Fig. 4b). Clearly $x = -b = -v$ and $y = a = \sqrt{1 - v^2}$, so

$$\sin 125° = \frac{y}{r} = \frac{\sqrt{1 - v^2}}{1} = \sqrt{1 - v^2}.$$

(c) The terminal side of 235° lies in the third quadrant, 35° short of the negative y-axis (270°). Place the triangle from (a) accordingly (Fig. 4c). Clearly $x = -b = -v$ and $y = -a = -\sqrt{1 - v^2}$, so

$$\cos 235° = \frac{x}{r} = \frac{-v}{1} = -v.$$

Answer (a) $\sqrt{1 - v^2}$ (b) $\sqrt{1 - v^2}$ (c) $-v$

Tangent

The tangent function is another useful trigonometric function.

> **Tangent** Let θ be an angle in standard position, but not 90° or 270°. Let (x, y) be any point on its terminal side, different from $(0, 0)$. Then
>
> $$\tan\theta = \frac{y}{x}.$$
>
> Alternatively,
>
> $$\tan\theta = \frac{\sin\theta}{\cos\theta}.$$
>
> In words, $\tan\theta$ is the *slope* of the terminal side.

Note that

$$\frac{\sin\theta}{\cos\theta} = \frac{y/r}{x/r} = \frac{y}{x},$$

so the ratio y/x depends only on θ, not on the choice of the point (x, y) on the terminal side. The exceptions, 90° and 270°, are angles whose terminal sides coincide with one side or the other of the y-axis, hence the denominator $x = 0$, and y/x is undefined. (Alternatively, $\cos\theta = 0$.) Finally, recall that the slope of the terminal side, a non-vertical line through $(0, 0)$, is $(y - 0)/(x - 0) = y/x = \tan\theta$.

Examples

(1) $\tan 45° = \dfrac{\sin 45°}{\cos 45°} = \dfrac{\frac{1}{2}\sqrt{2}}{\frac{1}{2}\sqrt{2}} = 1$

(2) $\tan 150° = \dfrac{\sin 150°}{\cos 150°} = \dfrac{\frac{1}{2}}{-\frac{1}{2}\sqrt{3}} = \dfrac{-1}{\sqrt{3}} = -\frac{1}{3}\sqrt{3}$

If β is an acute angle of a right triangle (Fig. 1, page 238), then

$$\tan\beta = \frac{\text{opposite}}{\text{adjacent}}.$$

EXAMPLE 4 Express in terms of $T = \tan 35°$

(a) $\cos 35°$ (b) $\sin 145°$ (c) $\tan 305°$.

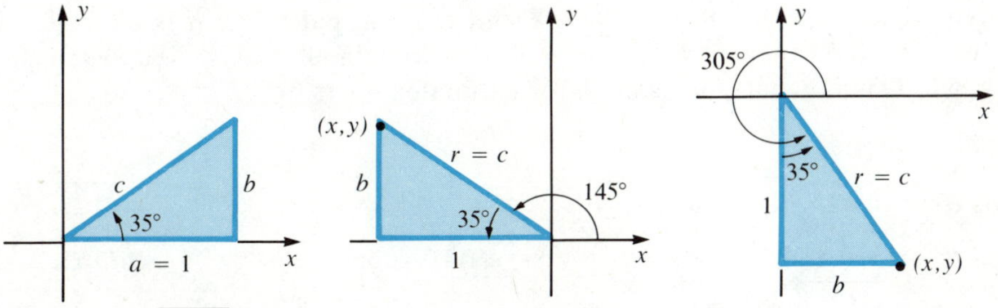

(a) $c = \sqrt{1 + b^2}$ (b) $(x,y) = (-1, b)$ (c) $(x,y) = (b, -1)$

Fig. 5

Solution (a) Draw 35° in standard position and a corresponding 35-55-90 right triangle (Fig. 5a, previous page). Take adj = 1. Then

$$T = \tan 35° = \frac{\text{opp}}{\text{adj}} = \frac{b}{1} = b.$$

By the Pythagorean theorem,

$$c = \text{hyp} = \sqrt{1 + b^2} = \sqrt{1 + T^2}$$

so

$$\cos 35° = \frac{\text{adj}}{\text{hyp}} = \frac{1}{\sqrt{1 + T^2}}.$$

(b) The terminal side of 145° lies in the second quadrant, 35° short of the negative x-axis. Place the triangle from (a) accordingly (Fig. 5b). Clearly $x = -1$ and $y = b = T$, so

$$\sin 145° = \frac{y}{r} = \frac{T}{\sqrt{1 + T^2}}.$$

(c) The terminal side of 305° lies in the fourth quadrant, 35° forward of the negative y-axis. Place the triangle from (a) accordingly (Fig. 5c). Clearly $x = b = T$ and $y = -1$,, so

$$\tan 305° = \frac{y}{x} = \frac{-1}{T}.$$

Answer (a) $\dfrac{1}{\sqrt{1 + T^2}}$ (b) $\dfrac{T}{\sqrt{1 + T^2}}$ (c) $\dfrac{-1}{T}$

To estimate tan $\theta°$ on a calculator, use the [tan] key. But first set the mode to [deg].

Examples

(1)	tan 27.42°	2 7 . 4 2 [tan]	0.5188
(2)	tan 310°	3 1 0 [tan]	−1.1917536

Calculating Angles

Each angle θ of a triangle satisfies $0 < \theta < 180°$. When you solve a triangle (as we do in Chapter 8) for one of its angles θ, you usually find $\sin\theta$ or $\cos\theta$, from which you must calculate θ. Your calculator does this by the sequences

[INV] [sin] and [INV] [cos],

but some care must be taken. First of all, be sure the calculator is in the [deg] mode. Now let's look at [INV] [cos]. From Fig. 9a, page 237, it is clear that if $-1 < y < 1$, then there is a unique θ in the domain $0 < \theta < 180°$ such that $\cos\theta = y$. This is what your calculator estimates.

Examples

(1)	$\cos\theta° = 0.42$	. 4 2 [INV] [cos]	65.17
(2)	$\cos\theta° = -0.834$	. 8 3 4 [+/−] [INV] [cos]	146.51

Thus $\cos 65.17° \approx 0.42$ and $\cos 146.51° \approx -0.834$, as you can check directly on your calculator.

Finding θ from $\sin\theta = y$ is more complicated. As the graph in Fig. 9b, page 237, indicates, if $0 < y < 1$, there are *two* values of θ satisfying $0 < \theta < 180°$ and $\sin\theta = y$. Your calculator finds the *smaller* one only, the one satisfying $0 < \theta < 90°$. The larger one is $180° - \theta$, as is clear from Fig. 6. (Of course, there is only *one* value of θ satisfying $0 < \theta < 180°$ and $\sin\theta = 1$; namely $\theta = 90°$.)

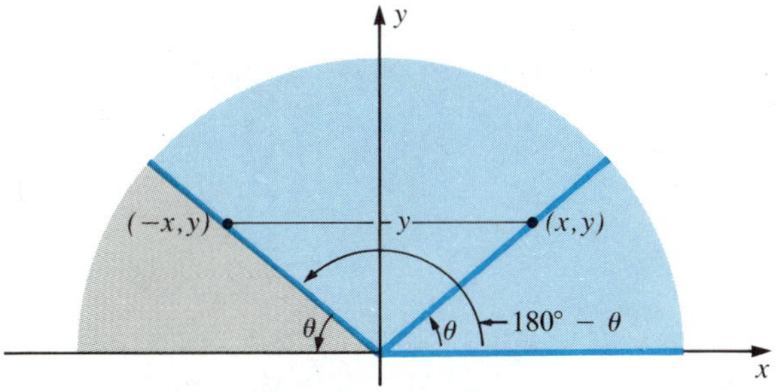

Fig. 6 $\sin(180° - \theta) = \dfrac{y}{r} = \sin\theta$

EXAMPLE 5 Estimate to two places all θ satisfying $0 < \theta < 180°$ and $\sin\theta = 0.6413$.

Solution Set the [deg] mode, then key in

. 6 4 1 3 [INV] [sin] → 39.89

Hence one answer is 39.89°. The other is

$$180 - 39.89 = 140.11°$$

Check 3 9 . 8 9 [sin] → .6413

1 4 0 . 1 1 [sin] → .6413

Answer 39.89° 140.11°

Remark Evidently $\cos\theta$ is a strictly decreasing function for $0 < \theta < 180°$. The sequence [INV] [cos] calculates its *inverse* function. This function is commonly denoted arc cos y. Similarly arc sin y denotes the inverse function of the strictly increasing function $\sin\theta$ on the interval $0 < \theta < 90°$. We shall study these inverse trigonometric functions in the next chapter.

Exercises

Find (exact values)

1 sin 120°	**2** cos 120°	**3** sin 210°	**4** cos 210°
5 sin 240°	**6** cos 240°	**7** sin 300°	**8** cos 300°
9 sin 330°	**10** cos 330°	**11** tan 0°	**12** tan 180°
13 tan 30°	**14** tan 60°	**15** tan 120°	**16** tan 135°
17 tan 210°	**18** tan 225°	**19** tan 240°	**20** tan 300°
21 tan 315°	**22** tan 330°		

Express in terms of $K = \cos 20°$

23 sin 20°	**24** sin 70°	**25** cos 160°	**26** sin 200°
27 cos 290°	**28** cos 340°	**29** tan 160°	**30** tan 290°.

Express in terms of $T = \tan 75°$

31 tan 15°	**32** cos 15°	**33** sin 165°	**34** tan 165°
35 tan 195°	**36** sin 195°	**37** sin 255°	**38** cos 255°.

Estimate to 4 places by calculator

39 tan 17.61° **40** tan 72.41° **41** tan 100.61° **42** tan 275.73°.

Estimate to two places the angle θ satisfying $0 < \theta < 180°$ and $\cos\theta =$

43 0.3333 **44** 0.6667 **45** 0.8984 **46** 0.75
47 −0.2 **48** −0.31 **49** −0.2179 **50** −0.3964.

Estimate to two places all angles satisfying $0 < \theta < 180°$ and $\sin\theta =$

51 0.3333 **52** 0.6667 **53** 0.9243 **54** 0.6.

3 CIRCULAR FUNCTIONS

Radian Measure

The degree is the unit of angle measure used in practical work, like surveying, navigation, and machine design. Another unit of angle measure called the radian is better for scientific work. Recall what a degree is: it is the unit of angle measure taken so that the whole angle measures 360°. The **radian** is the unit of angle measure taken so that the whole angle measures 2π radians. Of course π denotes the real number such that a circle of radius r has circumference $2\pi r$. Let's look at a circle of radius r in Fig. 1a. The whole angle is 2π radians and the circumference is $2\pi r$. It follows (Fig. 1b) that a central angle of θ radians subtends (determines) an arc of length $s = r\theta$. In particular (Fig. 1c) a central angle of 1 radian in a circle of radius r subtends an arc of length precisely r.

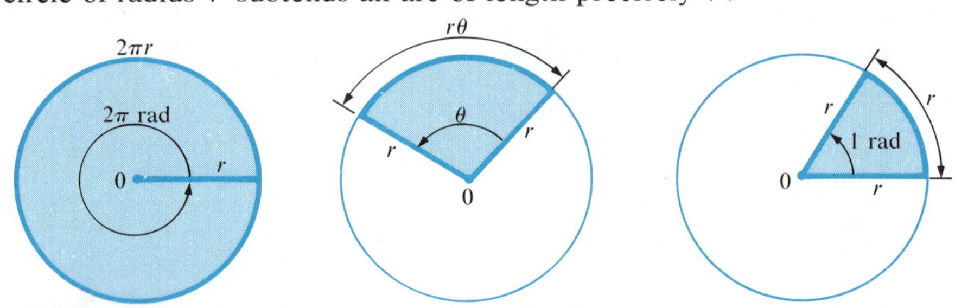

(a) central angle = 2π rad subtended arc = $2\pi r$

(b) central angle = θ rad subtended arc = $r\theta$

(c) central angle = 1 rad subtended arc = r

Fig. 1 Radian measure

Since 2π radians equals 360°, we have π radians equal 180° (easier to remember). Thus 1 radian equals $180/\pi$ degrees and 1° equals $\pi/180$ radians.

> **Radian Measure** A central angle of 1 radian in a circle of radius r subtends an arc of length r.
>
> $$\pi \text{ radians} = 180° \qquad 2\pi \text{ radians} = 360°$$
>
> $$1 \text{ rad} = \frac{180°}{\pi} \approx 57.29578° \qquad 1° = \frac{\pi}{180} \text{ rad} \approx 1.745329 \times 10^{-2} \text{ rad}$$

Examples

(1) $18° = 18 \cdot \dfrac{\pi}{180} = \frac{1}{10}\pi$ radians

(2) $47.69° = 47.69 \cdot \dfrac{\pi}{180} \approx 0.8323$ radians

(3) $\frac{2}{9}\pi \text{ radians} = \frac{2}{9}\pi \cdot \dfrac{180°}{\pi} = 40°$

(4) $1.3645 \text{ radians} = 1.3645 \cdot \dfrac{180°}{\pi} \approx 78.180°$

It is common practice to omit the unit "radian". Thus we write $\theta = \frac{1}{10}\pi$ as an abbreviation for $\theta = \frac{1}{10}\pi$ radians. (This convention is justified because the radian is, in certain respects, a natural unit of angle measure, whereas the degree is man-made.)

Radian Measures of Common Angles

$15° = \frac{1}{12}\pi$	$30° = \frac{1}{6}\pi$	$45° = \frac{1}{4}\pi$	$60° = \frac{1}{3}\pi$	$75° = \frac{5}{12}\pi$
$90° = \frac{1}{2}\pi$	$105° = \frac{7}{12}\pi$	$120° = \frac{2}{3}\pi$	$135° = \frac{3}{4}\pi$	$150° = \frac{5}{6}\pi$
$165° = \frac{11}{12}\pi$	$180° = \pi$	$210° = \frac{7}{6}\pi$	$225° = \frac{5}{4}\pi$	$240° = \frac{4}{3}\pi$
$270° = \frac{3}{2}\pi$	$300° = \frac{5}{3}\pi$	$315° = \frac{7}{4}\pi$	$330° = \frac{11}{6}\pi$	$360° = 2\pi$

The $\boxed{\pi}$ key on your calculator is a shortcut for keying in π. Since $\theta°$ equals $(\pi/180)\theta$ radians, the key sequence

$$\boxed{\pi}\ \boxed{\div}\ 1\ 8\ 0\ \boxed{\times}\ \theta\ \boxed{=}$$

converts $\theta°$ to radians. If your calculator has a constant factor capability (p. 212), you can keep $\pi/180$ as a constant factor and easily do a batch of conversions, like Exs. 21–24 below.

Similarly

$$1\ 8\ 0\ \boxed{\div}\ \boxed{\pi}\ \boxed{\times}\ \beta\ \boxed{=}$$

converts β radians to degrees.

Trig Functions

From now on (unless otherwise noted) when we write $\sin\theta$ or $\cos\theta$ or $\tan\theta$, we mean that θ is in radians.

Examples

$$\sin\tfrac{1}{2}\pi = 1 \qquad \sin\tfrac{1}{6}\pi = \tfrac{1}{2} \qquad \cos\tfrac{3}{4}\pi = -\tfrac{1}{2}\sqrt{2}$$

$$\cos\pi = -1 \qquad \tan\tfrac{1}{6}\pi = \tfrac{1}{3}\sqrt{3} \qquad \tan\tfrac{1}{4}\pi = 1.$$

These all come from previous calculations. For instance, $\tan\frac{1}{4}\pi = \tan 45° = 1$. As quickly as possible, you should learn to think in radians and toss away the crutch of passing through degrees. You should think of a right angle as $\frac{1}{2}\pi$, of half of a right angle as $\frac{1}{4}\pi$, of one-third of a right angle as $\frac{1}{6}\pi$, etc.

The Sine Function

Let us recall the graph of $y = \sin\theta$, only this time with θ in radians, $0 \le \theta \le 2\pi$. The function y increases from 0 to 1 as θ increases from 0 to $\frac{1}{2}\pi$. Then y decreases from 1 to -1 as θ increases from $\frac{1}{2}\pi$ to $\frac{3}{2}\pi$. Finally y increases from -1 to 0 as θ increases from $\frac{3}{2}\pi$ to 2π. See Fig. 2, next page.

Now we are going to do a construction we have never done before. We are going to extend this function $y = \sin\theta$ to be a function on the whole real line, not just on the interval $0 \le \theta \le 2\pi$. We do it simply by shifting the graph left and right in steps of 2π. See Fig. 3, next page. The pieces match because $\sin 2\pi = \sin 0$, so the next piece always takes off from where the previous one stops.

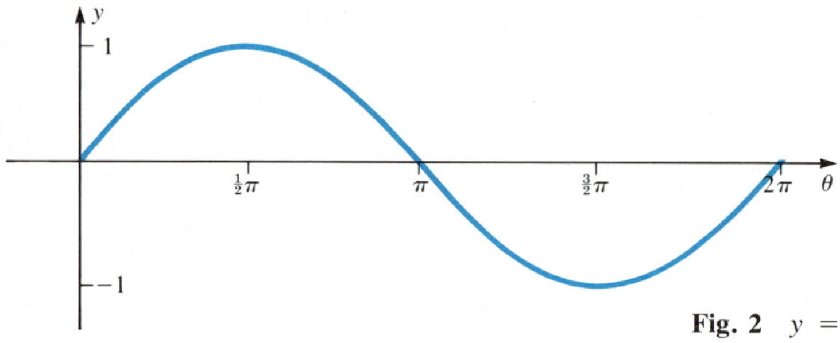

Fig. 2 $y = \sin\theta \qquad 0 \le \theta \le 2\pi$

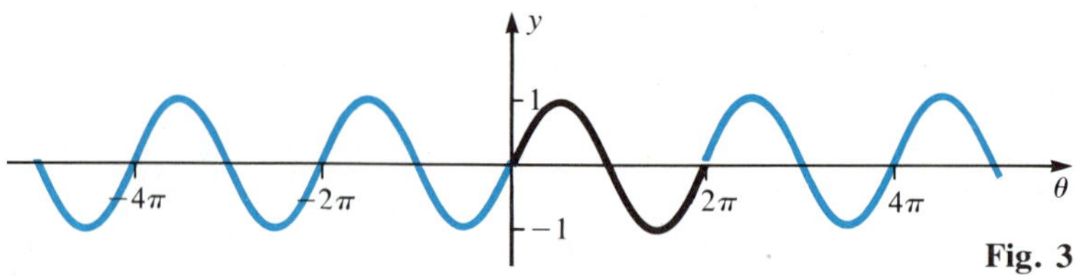

Fig. 3 $y = \sin\theta \qquad$ all θ

The result is a function $y = \sin\theta$ defined for all θ, and satisfying the relation

$$\sin(\theta + 2\pi n) = \sin\theta \qquad n \text{ any integer}.$$

This expresses analytically what we did geometrically with the graph. The piece from $2\pi n$ to $2\pi(n + 1)$ is obtained by shifting the original piece (from 0 to 2π) to the right $2\pi n$ units (left if n is negative).

Examples

(1) $\sin\frac{27}{4}\pi = \sin(\frac{3}{4}\pi + 6\pi) = \sin\frac{3}{4}\pi = \frac{1}{2}\sqrt{2}$

(2) $\sin\frac{34}{3}\pi = \sin(\frac{4}{3}\pi + 10\pi) = \sin\frac{4}{3}\pi = -\frac{1}{2}\sqrt{3}$

(3) $\sin(-\frac{19}{6}\pi) = \sin(\frac{5}{6}\pi - 4\pi) = \sin\frac{5}{6}\pi = \frac{1}{2}$.

Your calculator will handle, with reasonable accuracy, $\sin\theta$ for any value of θ, negative, or greater than 2π, provided of course that it fits on the calculator.

Examples (First set [rad] mode.)

(1)	$\sin\frac{34}{3}\pi$	3 4 [×] [π] [÷] 3 [=] [sin]	−0.866025
(2)	$\sin(-27.68)$	2 7 . 6 8 [+/−] [sin]	−0.559957

Note Many calculators have a [grad] mode. Some countries on the metric system prefer the **grad** as the practical unit of angle measure. There are 100 grads in a right angle, so $\frac{1}{6}\pi = 30° = 33\frac{1}{3}$ grad.

Angles Again

We have defined $\sin\theta$ for each real number θ. Does that mean we have defined angles of all possible sizes, even negative? Yes and no. We can't change geometry; a geometric angle is what it was in Fig. 1, p. 231: one of the regions determined by two rays. But let's think dynamically instead of statically. Let's think of a ray, pivoted at the vertex, that starts on the initial side and sweeps counterclockwise to the terminal side. Figure 4a shows an angle of $\frac{1}{6}\pi$ swept out this way. Now let's suppose that the ray, like a lighthouse beam, sweeps a full revolution and then stops at the same terminal side (Fig. 4b). Then it has swept through an angle of 2π plus $\frac{1}{6}\pi$ more, altogether $\frac{13}{6}\pi$. If instead it rotates two full revolutions, then stops at the terminal side (Fig. 4c), the result is a swept angle of $4\pi + \frac{1}{6}\pi = \frac{25}{6}\pi$. If it sweeps backwards (*clockwise*), that counts as a negative angle (Fig. 4d).

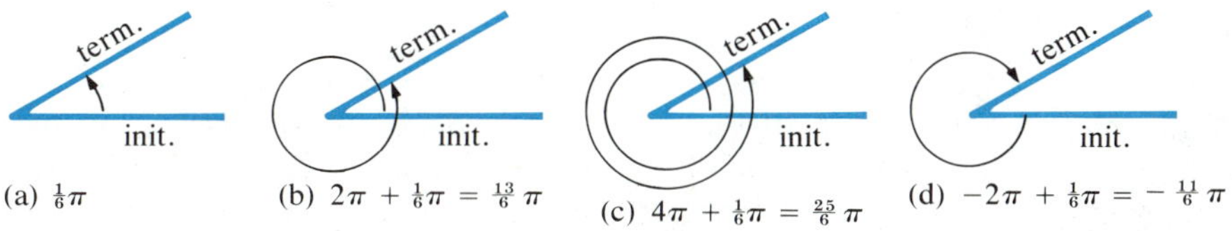

Fig. 4 Swept angles—same initial and terminal sides

Our new (dynamic) angle consists of an initial side, a terminal side, and a history of how we got there: counterclockwise or clockwise, and how many full revolutions before stopping. The initial side and one single real number describes all this. For instance, $\theta = -\frac{38}{3}\pi$ tells you the rotation is *clockwise* through 6 full revolutions ($-6 \times 2\pi = -12\pi = -\frac{36}{3}\pi$) and $-\frac{2}{3}\pi$ more. Its terminal side is in the 3-rd quadrant at angle $\frac{1}{3}\pi$ with the negative x-axis. (Check this.)

Remark The **revolution** is another unit of angle measurement, commonly used with rotating machinery. It is a whole angle, so

$$1 \text{ rev} = 360° = 2\pi \text{ radians}.$$

For instance, a certain electric hand drill idles at 1150 rpm (revolutions per minute). That means in each minute it turns (sweeps) through an angle of $1150 \text{ rev} = 2300\pi$.

Properties of the Sine Function

Let θ_1 and θ_2 be two real numbers. Suppose the corresponding angles they determine have the same terminal side. Then they differ by an integer multiple of a whole angle, that is, $\theta_2 - \theta_1 = 2\pi n$. Conversely if θ_1 and θ_2 differ by an integer multiple of 2π, then the corresponding angles they sweep out have the same terminal side.

Since $\sin\theta$ was cooked up to satisfy $\sin(\theta + 2\pi n) = \sin\theta$, we conclude that if θ_1 and θ_2 determine the same terminal side, then $\sin\theta_1 = \sin\theta_2$.

Now look again at the graph of $y = \sin\theta$ in Fig. 3. It looks like an odd function; can we prove it is? We want to prove $\sin(-\theta) = -\sin\theta$. To reach the terminal side of $-\theta$, we do the same motion *clockwise* that we did *counterclockwise* to reach the terminal side of θ. See Fig. 5a. It is clear from this that the terminal

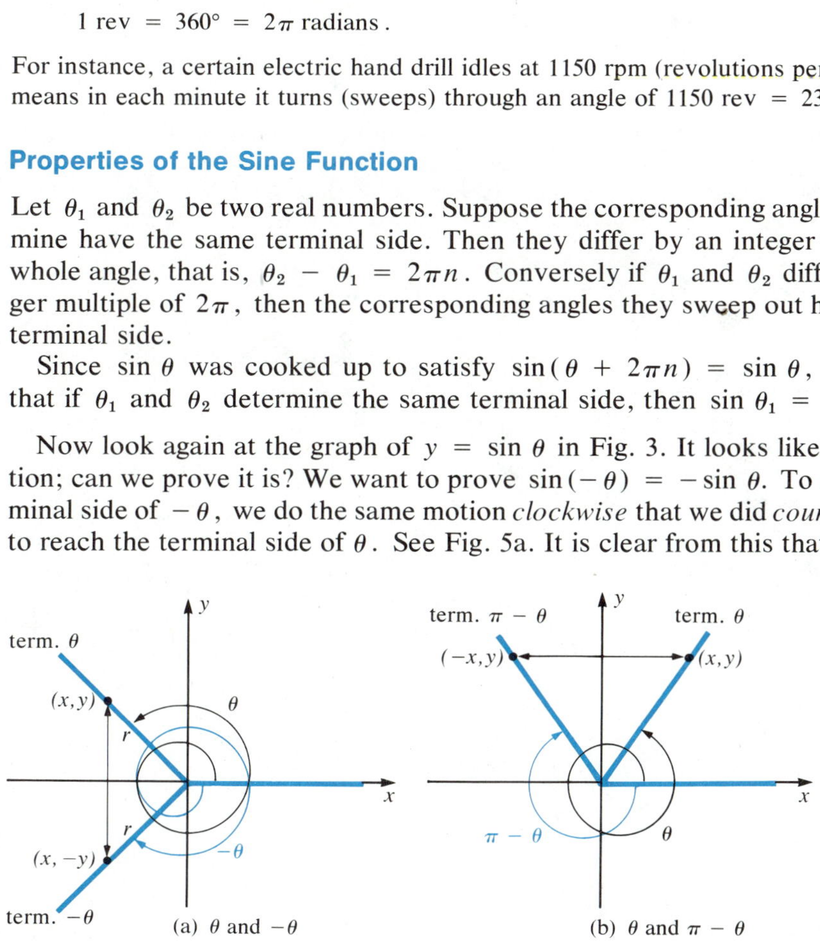

Fig. 5

side of $-\theta$ is the reflection in the x-axis of the terminal side of θ. Choose (x, y) on the terminal side of θ, at distance r from $(0, 0)$. Then $(x, -y)$ is on the terminal side of $-\theta$, hence

$$\sin(-\theta) = \frac{-y}{r} = -\frac{y}{r} = -\sin\theta.$$

Look again at the graph of $y = \sin\theta$. It seems to be symmetric in the vertical line $x = \frac{1}{2}\pi$, that is, apparently $\sin(\pi - \theta) = \sin\theta$. We have already verified this relation for the case $0 < \theta < \pi$. But it is true for all θ. Indeed, you rotate as much *clockwise* starting from the negative x-axis to reach the terminal side of $\pi - \theta$ as you rotate *counterclockwise* from the positive x-axis to reach the terminal side of θ. See Fig. 5b. But the negative x-axis is π radians forward of the positive x-axis. Geometrically, this means that the terminal sides of $\pi - \theta$ and θ are symmetric in the y-axis. Now an argument like that of the last paragraph establishes $\sin(\pi - \theta) = \sin\theta$.

$$\sin(-\theta) = -\sin\theta \qquad \sin(\pi - \theta) = \sin\theta$$

From these relations, or by further geometric arguments, other properties of the sine function follow, for example

$$\sin(\theta + \pi) = -\sin\theta \qquad \sin(2\pi - \theta) = -\sin\theta.$$

The Cosine Function

Now let us recall the graph of $y = \cos\theta$, only with θ in radians, $0 \le \theta \le 2\pi$. The function decreases from 1 to -1 as θ increases from 0 to π. Then y increases from -1 back to 1 as θ increases from π to 2π. See Fig. 6.

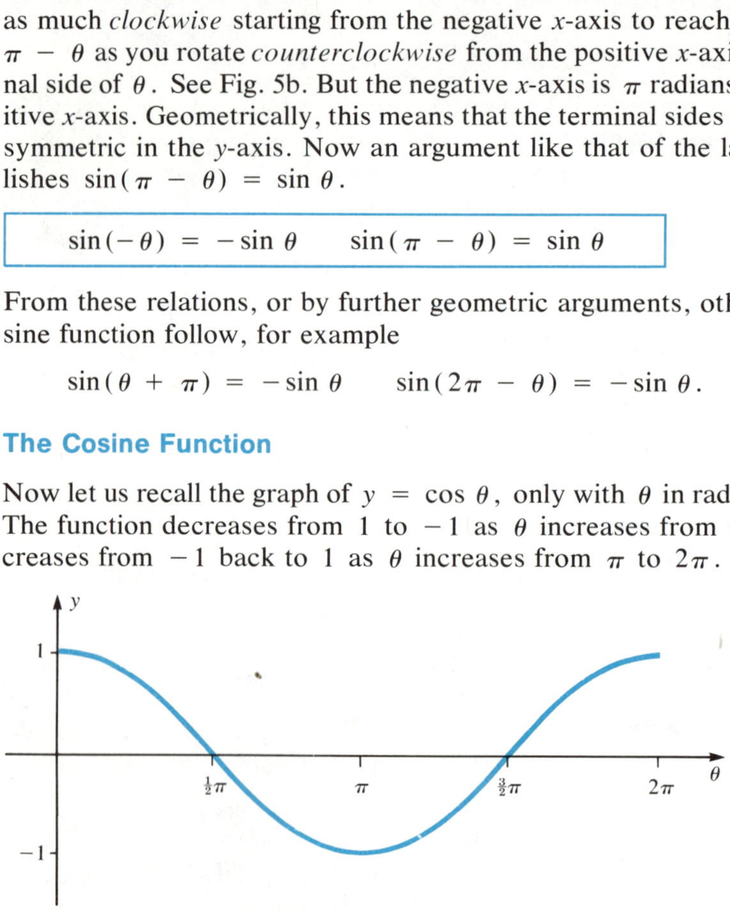

Fig. 6 $y = \cos\theta \qquad 0 \le \theta \le 2\pi$

We extend this function to a function on the whole real line by shifting the graph left and right in steps of 2π. See Fig. 7. The result is a function $y = \cos\theta$ defined for all θ, and satisfying the relation

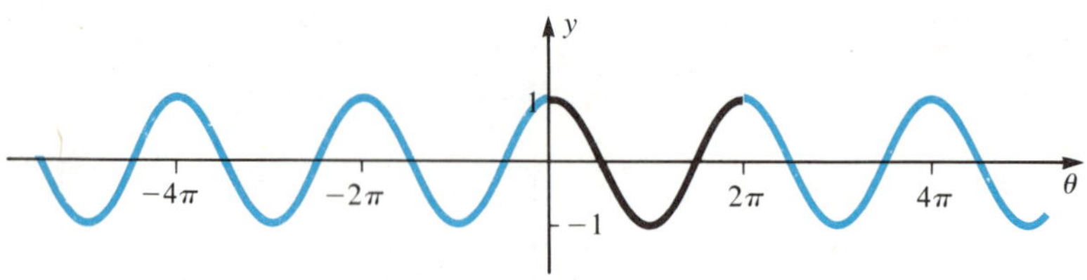

Fig. 7 $y = \cos\theta \qquad$ all θ

$$\cos(\theta + 2\pi n) = \cos\theta.$$

The graphs of $y = \sin\theta$ and $y = \cos\theta$ look remarkably alike! Apparently if you shift the graph of $y = \sin\theta$ to the left by $\frac{1}{2}\pi$, you get the graph of $y = \cos\theta$. Analytically,

$$\sin(\theta + \tfrac{1}{2}\pi) = \cos\theta.$$

We shall establish this by a geometric argument. Study Fig. 8 and its legend carefully. (You may wish to draw it with the terminal side of θ in other quadrants to be convinced.) If $P = (x, y)$ is on the terminal side of θ at distance r from $(0, 0)$, then $Q = (-y, x)$ is on the terminal side of $\theta + \frac{1}{2}\pi$, at the same distance. Therefore

$$\sin(\theta + \tfrac{1}{2}\pi) = \frac{x}{r} = \cos\theta.$$

From the same set-up we also have

$$\cos(\theta + \tfrac{1}{2}\pi) = \frac{-y}{r} = -\frac{y}{r} = -\sin\theta.$$

This also can be derived analytically from the boxed relation above and properties of $\sin\theta$:

$$\begin{aligned}\cos(\theta + \tfrac{1}{2}\pi) &= \sin[(\theta + \tfrac{1}{2}\pi) + \tfrac{1}{2}\pi] \\ &= \sin(\theta + \pi) = -\sin\theta.\end{aligned}$$

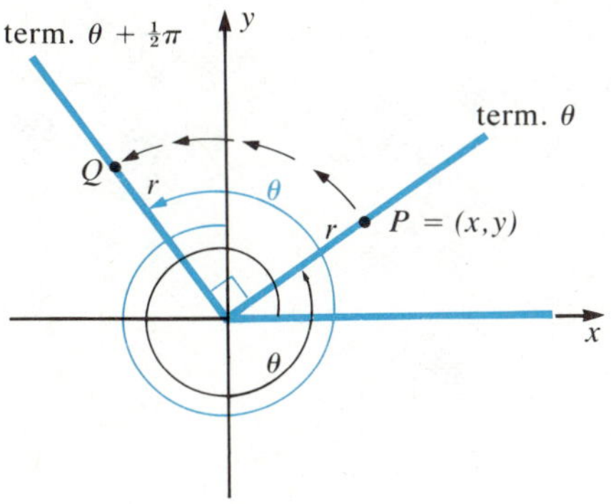

Fig. 8 The terminal side of $\theta + \frac{1}{2}\pi$ bears the same relation to the positive y-axis that the terminal side of θ bears to the positive x-axis. It also bears the same relation to the *negative* x-axis that term. θ bears to the positive y-axis. Therefore $Q = (-y, x)$.

Because of the relation $\cos\theta = \sin(\theta + \frac{1}{2}\pi)$ expressing $\cos\theta$ in terms of sine, every property of the sine function implies a property of the cosine function. For instance, $\sin(\pi - \theta) = \sin\theta$ can be used as follows:

$$\begin{aligned}\cos(-\theta) &= \sin(-\theta + \tfrac{1}{2}\pi) = \sin[\pi - (\tfrac{1}{2}\pi + \theta)] \\ &= \sin(\tfrac{1}{2}\pi + \theta) = \cos\theta.\end{aligned}$$

Therefore $\cos\theta$ is an even function. Similarly, we can derive other relations involving the cosine. Let us summarize the main such properties of both sine and cosine.

Properties of Sine and Cosine

(1) $-1 \le \sin\theta \le 1 \qquad -1 \le \cos\theta \le 1$

(2) $\sin^2\theta + \cos^2\theta = 1$

(3) $\sin(\theta + 2\pi n) = \sin\theta \qquad \cos(\theta + 2\pi n) = \cos\theta$

(4) sin is odd, cosine is even:

$\sin(-\theta) = -\sin\theta \qquad \cos(-\theta) = \cos\theta$

(5) $\sin(\pi - \theta) = \sin\theta \qquad \cos(\pi - \theta) = -\cos\theta$

(6) $\sin(\theta + \frac{1}{2}\pi) = \cos\theta \qquad \cos(\theta + \frac{1}{2}\pi) = -\sin\theta$

(7) $\sin(\theta + \pi) = -\sin\theta \qquad \cos(\theta + \pi) = -\cos\theta$

(8) $\sin(\frac{1}{2}\pi - \theta) = \cos\theta \qquad \cos(\frac{1}{2}\pi - \theta) = \sin\theta$

Remark You should memorize (1)–(4). The first two are easy; the next two are easy if you remember the graphs, which you should. Relations (5)–(8) are a little harder to remember, but you only need (8); the others all follow easily, for example

$$\sin(\theta + \tfrac{1}{2}\pi) = \sin[\tfrac{1}{2}\pi - (-\theta)] = \cos(-\theta) = \cos\theta, \text{ etc.}$$

Actually there is only one relation (8), since both say that if $\alpha + \beta = \frac{1}{2}\pi$, then $\sin\alpha = \cos\beta$. In the case of acute angles, this is an easy-to-remember statement about complementary angles (in a right triangle). Actually, (5)–(8) are all statements about symmetries in the graphs of sine and cosine; they can be recalled as needed by looking at the graphs, or by a quick sketch of angles and their terminal sides. The big box isn't as big a deal as it first appears.

EXAMPLE 1 Express in terms of $y = \sin\frac{1}{8}\pi$

(a) $\sin\frac{9}{8}\pi$ (b) $\cos(\frac{51}{8}\pi)$ (c) $\cos(-\frac{23}{8}\pi)$

Solution (a) Use (7):

$$\sin\tfrac{9}{8}\pi = \sin(\tfrac{1}{8}\pi + \pi) = -\sin\tfrac{1}{8}\pi = -y.$$

(b) Use (3) and (8):

$$\cos(\tfrac{51}{8}\pi) = \cos(\tfrac{3}{8}\pi + 6\pi) = \cos\tfrac{3}{8}\pi$$
$$= \cos(\tfrac{1}{2}\pi - \tfrac{1}{8}\pi) = \sin\tfrac{1}{8}\pi = y.$$

(c) Use (3), (4), (5), and (2):

$$\cos(-\tfrac{23}{8}\pi) = \cos(-\tfrac{7}{8}\pi - 2\pi) = \cos(-\tfrac{7}{8}\pi) = \cos\tfrac{7}{8}\pi$$
$$= \cos(\pi - \tfrac{1}{8}\pi) = -\cos\tfrac{1}{8}\pi$$
$$= -\sqrt{1 - \sin^2\tfrac{1}{8}\pi} = -\sqrt{1 - y^2}.$$

Answer (a) $-y$ (b) y (c) $-\sqrt{1 - y^2}$

Exercises

Convert to exact degrees

1 $\frac{1}{10}\pi$ **2** $\frac{1}{15}\pi$ **3** $\frac{3}{5}\pi$ **4** $\frac{3}{8}\pi$ **5** $\frac{2}{9}\pi$
6 $\frac{1}{90}\pi$ **7** $\frac{13}{12}\pi$ **8** $\frac{17}{12}\pi$ **9** $\frac{19}{12}\pi$ **10** $\frac{23}{12}\pi$.

Convert to radians; express in the form $b\pi$, where b is rational

11 10° **12** 12° **13** 36° **14** 67.5° **15** 54°

16 202.5° **17** 357° **18** 275° **19** 228° **20** 108°.

Convert to 4-place radians on a calculator

21 12.85° **22** 10.34° **23** 172.91° **24** 0.0751°.

Convert to 2-place degrees on a calculator

25 0.5400 **26** 5.1774 **27** 0.03161 **28** 1.0003.

Find the arc length of an arc of radius 5 subtended by a central angle $\theta =$

29 2.604 **30** 0.1313 **31** 11.65° **32** 104.61°.

Find (exactly) $\sin\theta$ and $\cos\theta$ for $\theta =$

33 3π **34** $-\pi$ **35** 51π **36** -7π
37 $\frac{19}{4}\pi$ **38** $\frac{73}{4}\pi$ **39** $-\frac{19}{4}\pi$ **40** $-\frac{73}{4}\pi$
41 $\frac{28}{3}\pi$ **42** $\frac{67}{6}\pi$ **43** $-\frac{28}{3}\pi$ **44** $-\frac{67}{6}\pi$.

Find (exactly) all θ in the interval $-\pi \le \theta \le \pi$ that satisfy

45 $\sin\theta = -\frac{1}{2}$ **46** $\cos\theta = -\frac{1}{2}$ **47** $\cos\theta = \frac{1}{2}\sqrt{2}$ **48** $\sin\theta = \frac{1}{2}\sqrt{2}$
49 $\sin\theta = -1$ **50** $\cos\theta = -1$ **51** $\cos\theta = \frac{1}{2}\sqrt{3}$ **52** $\sin\theta = \frac{1}{2}\sqrt{3}$.

Find all real numbers θ that satisfy

53 $\sin\theta = 0$ **54** $\cos\theta = 0$ **55** $\sin\theta = 1$ **56** $\cos\theta = -1$.

Find all θ in the interval $0 \le \theta \le 2\pi$ that satisfy

57 $\cos\theta \ge 0$ **58** $\sin\theta \ge \frac{1}{2}$ **59** $\sin\theta > \cos\theta$
60 $\cos\theta \ge \sin\theta$ **61*** $\sin\theta + \cos\theta \ge 1$ **62*** $\sin\theta - \cos\theta \ge 1$.

Express in terms of $\sin\theta$ or $\cos\theta$

63 $\sin(\theta - 3\pi)$ **64** $\cos(\theta + 5\pi)$ **65** $\cos(\theta + \frac{5}{2}\pi)$
66 $\sin(\theta - \frac{7}{2}\pi)$ **67** $\sin(\frac{9}{2}\pi - \theta)$ **68** $\cos(\theta - \frac{17}{2}\pi)$.

Express in terms of $y = \sin\frac{1}{12}\pi$

69 $\sin\frac{13}{12}\pi$ **70** $\sin\frac{5}{12}\pi$ **71** $\cos(-\frac{11}{12}\pi)$ **72** $\cos(-\frac{37}{12}\pi)$.

Express in terms of $x = \cos\frac{1}{10}\pi$

73 $\cos\frac{11}{10}\pi$ **74** $\cos\frac{19}{10}\pi$ **75** $\sin\frac{7}{5}\pi$ **76** $\sin(-\frac{12}{5}\pi)$.

77 Prove $\cos(\theta + \frac{1}{2}\pi) + \cos(\theta - \frac{1}{2}\pi) = 0$.

78* Prove $\sin\frac{1}{15}\pi + \sin\frac{3}{15}\pi + \sin\frac{5}{15}\pi + \cdots + \sin\frac{29}{15}\pi = 0$.

79 (A true story) The instruction manual of a certain calculator gives these directions for calculating $\sin 63°$ and $\cos\frac{1}{4}\pi$:

6 3 [sin] 0.8910065

[π] [÷] 4 [=] [cos] 0.9999061

Explain why the writer of the manual should be fired.

80* Find all real numbers θ such that θ radians and θ degrees have the same terminal side. Equivalently, such that both $\cos\theta = \cos\theta°$ and $\sin\theta = \sin\theta°$.

81 I am not sure what mode my calculator is in, so I key 3 0 [sin]. The result is a negative number. Conclusion?

82 (cont.) The result is a positive number, not 0.5. Conclusion?

83 The graph of $y = \sin\theta$ appears symmetric with respect to the line $\theta = \frac{9}{2}\pi$. Prove it so.

84 The graph of $y = \sin\theta$ appears symmetric with respect to the point $(3\pi, 0)$. Prove it so.

4 THE TANGENT; GRAPHING METHODS

Let's return to the tangent function, which we have rather neglected. Recall that $\tan\theta°$ was defined for $0 \le \theta° \le 360°$, not $90°$ or $270°$. Translating to radians, $\tan\theta$ is defined (so far) for $0 \le \theta \le 2\pi$, $\theta \ne \frac{1}{2}\pi$, $\theta \ne \frac{3}{2}\pi$. We have three ways to look at $\tan\theta$ when θ is in standard position. First, $\tan\theta = y/x$ where (x, y) is any point on the terminal side of θ different from $(0, 0)$. Second, $\tan\theta$ is the slope of its terminal side. Third, $\tan\theta = (\sin\theta)/(\cos\theta)$.

Any of the three can be used to extend $\tan\theta$ to the whole real axis, except for those θ whose terminal sides are vertical; they are the odd multiples of $\frac{1}{2}\pi$:

$$\pm\tfrac{1}{2}\pi \quad \pm\tfrac{3}{2}\pi \quad \pm\tfrac{5}{2}\pi \quad \cdots$$

Tangent The function $\tan\theta$ is defined for all real θ except $\pm\frac{1}{2}\pi$, $\pm\frac{3}{2}\pi$, $\pm\frac{5}{2}\pi$, $\cdots$. If θ is in standard position, then $\tan\theta$ is the slope of the terminal side of θ, that is, $\tan\theta = y/x$ where (x, y) is on the terminal side and $x \ne 0$. Alternatively,

$$\tan\theta = \frac{\sin\theta}{\cos\theta}.$$

Examples

(1) $$\tan\tfrac{17}{3}\pi = \frac{\sin\frac{17}{3}\pi}{\cos\frac{17}{3}\pi} = \frac{\sin(-\frac{1}{3}\pi + 6\pi)}{\cos(-\frac{1}{3}\pi + 6\pi)} = \frac{\sin(-\frac{1}{3}\pi)}{\cos(-\frac{1}{3}\pi)} = \frac{-\frac{1}{2}\sqrt{3}}{\frac{1}{2}} = -\sqrt{3}.$$

(2) $$\tan(-\tfrac{3}{4}\pi) = \frac{\sin(-\frac{3}{4}\pi)}{\cos(-\frac{3}{4}\pi)} = \frac{-\sin\frac{1}{4}\pi}{-\cos\frac{1}{4}\pi} = 1.$$

The tangent function has two immediate properties, both obvious from the slope definition, but we'll prove them analytically anyhow. First, $\tan\theta$ is an odd function:

$$\tan(-\theta) = \frac{\sin(-\theta)}{\cos(-\theta)} = \frac{-\sin\theta}{\cos\theta} = -\tan\theta.$$

Second, $\tan\theta$ is unchanged when θ is replaced by $\theta \pm \pi$:

$$\tan(\theta \pm \pi) = \frac{\sin(\theta \pm \pi)}{\cos(\theta \pm \pi)} = \frac{-\sin\theta}{-\cos\theta} = \tan\theta.$$

More generally, if n is a whole number, then

$$\tan(\theta \pm n\pi) = \tan\theta.$$

For instance,

$$\tan(\theta + 2\pi) = \tan(\theta + \pi + \pi) = \tan(\theta + \pi) = \tan\theta$$

$$\tan(\theta - 3\pi) = \tan(\theta - \pi - \pi - \pi) = \tan(\theta - \pi - \pi) = \tan(\theta - \pi) = \tan\theta.$$

Remark $\tan(\theta + \pi) = \tan\theta$ is a surprise; our first guess would be only $\tan(\theta + 2\pi) = \tan\theta$, the way $\sin\theta$ and $\cos\theta$ behave.

Examples

(1) $\tan \frac{17}{3}\pi = \tan(-\frac{1}{3}\pi + 6\pi) = \tan(-\frac{1}{3}\pi) = -\tan\frac{1}{3}\pi = -\sqrt{3}.$

(2) $\tan(-\frac{3}{4}\pi) = \tan(-\frac{3}{4}\pi + \pi) = \tan\frac{1}{4}\pi = 1.$

$$\tan(-\theta) = -\tan\theta \qquad \tan(\theta \pm n\pi) = \tan\theta$$

Graph of tan θ

Let us start on the interval $-\frac{1}{2}\pi < \theta < \frac{1}{2}\pi$. There $y = \tan\theta$ is a strictly increasing odd function. By interpreting $\tan\theta$ as the slope of the terminal side of θ, we see that as θ increases from $-\frac{1}{2}\pi+$ to 0 to $\frac{1}{2}\pi-$, the function $y = \tan\theta$ increases from $-\infty$ to 0 to $+\infty$. This gives us a fair picture (Fig. 1a). Next, because $\tan(\theta + n\pi) = \tan\theta$, we can shift this portion left and right in steps of π to obtain the complete graph (Fig. 1b).

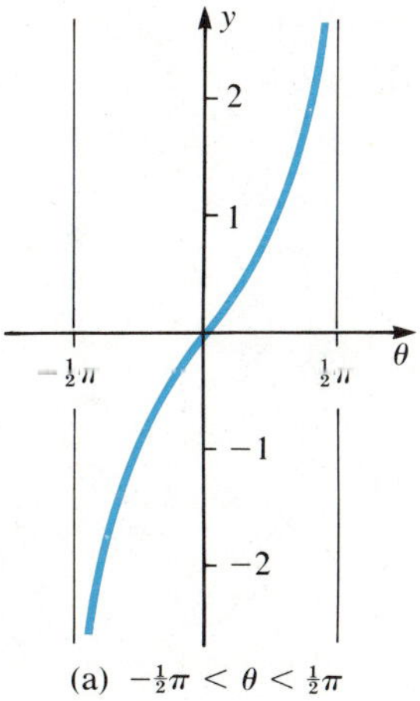

(a) $-\frac{1}{2}\pi < \theta < \frac{1}{2}\pi$

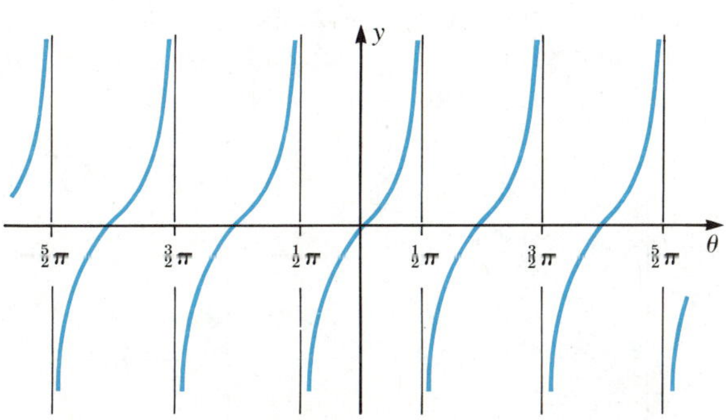

(b) complete graph

Fig. 1 $y = \tan\theta$

EXAMPLE 1 Find all solutions of $\tan\theta = 1$.

Solution We know one solution: $\theta = \frac{1}{4}\pi$. From the graph, every other solution differs from this one by an integer multiple of π.

Answer $\frac{1}{4}\pi + n\pi$ $\quad n$ any integer

Complementary Angles

The formulas (8) on p. 250 for sin and cos of $\frac{1}{2}\pi - \theta$ imply something about tan:

$$\tan(\tfrac{1}{2}\pi - \theta) = \frac{1}{\tan\theta}.$$

This is true because

$$\tan(\tfrac{1}{2}\pi - \theta) = \frac{\sin(\frac{1}{2}\pi - \theta)}{\cos(\frac{1}{2}\pi - \theta)} = \frac{\cos\theta}{\sin\theta} = \frac{1}{\tan\theta}.$$

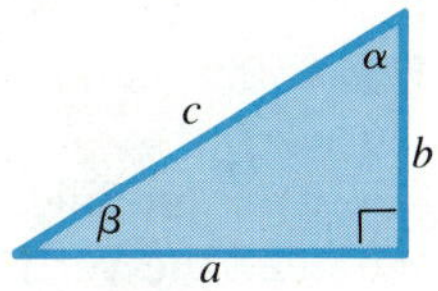

Fig. 2

For acute angles, the formula has a direct right triangle interpretation (Fig. 2). In the triangle, α and β are complementary, and

$$\tan\beta = \frac{\text{opp}}{\text{adj}} = \frac{b}{a} = \frac{1}{a/b} = \frac{1}{\tan\alpha}.$$

Graph Construction

We now describe geometric constructions for the graph of $y = \sin\theta$, then for $y = \tan\theta$. Of course, the $\cos\theta$ graph is free once we have the sine. These constructions are quite accurate for practical purposes; needed are graph paper, compass, ruler, and protractor (not really needed).

First we draw the θ- and y-axes and mark scales on each. Given θ, we wish to construct the corresponding point $(\theta, \sin\theta)$ on the graph of $y = \sin\theta$.

To the left, out of the way, we draw an auxiliary y-axis, an auxiliary x-axis colinear with the θ-axis, and a unit circle (Fig. 3). Given θ, we draw a corresponding angle in standard position. Its terminal side intersects the unit circle in $(\cos\theta, \sin\theta)$. We project this point horizontally; where it meets the vertical through θ (on the right-hand graph) is the desired point $(\theta, \sin\theta)$.

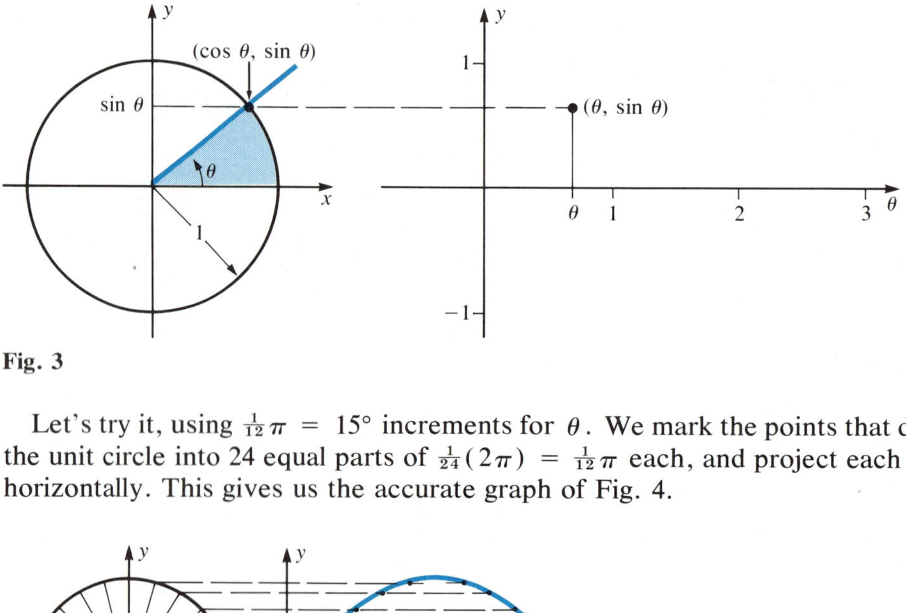

Fig. 3

Let's try it, using $\frac{1}{12}\pi = 15°$ increments for θ. We mark the points that divide the unit circle into 24 equal parts of $\frac{1}{24}(2\pi) = \frac{1}{12}\pi$ each, and project each point horizontally. This gives us the accurate graph of Fig. 4.

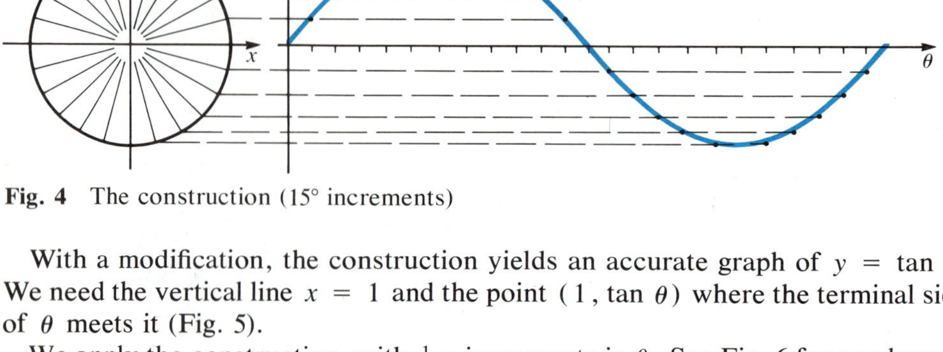

Fig. 4 The construction (15° increments)

With a modification, the construction yields an accurate graph of $y = \tan\theta$. We need the vertical line $x = 1$ and the point $(1, \tan\theta)$ where the terminal side of θ meets it (Fig. 5).

We apply the construction, with $\frac{1}{12}\pi$ increments in θ. See Fig. 6 for one branch of $y = \tan\theta$.

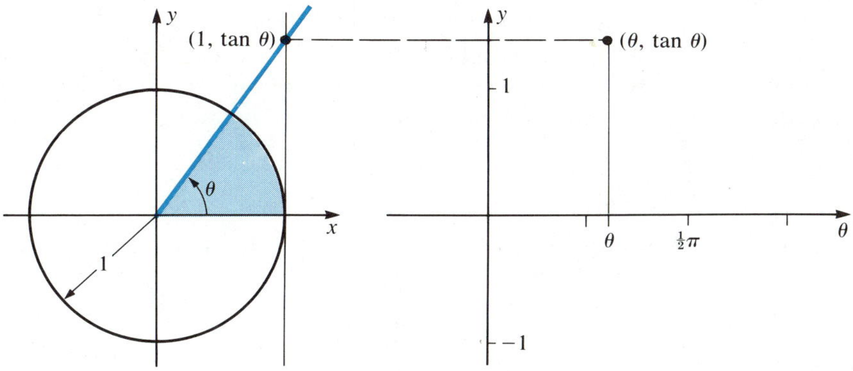

Fig. 5

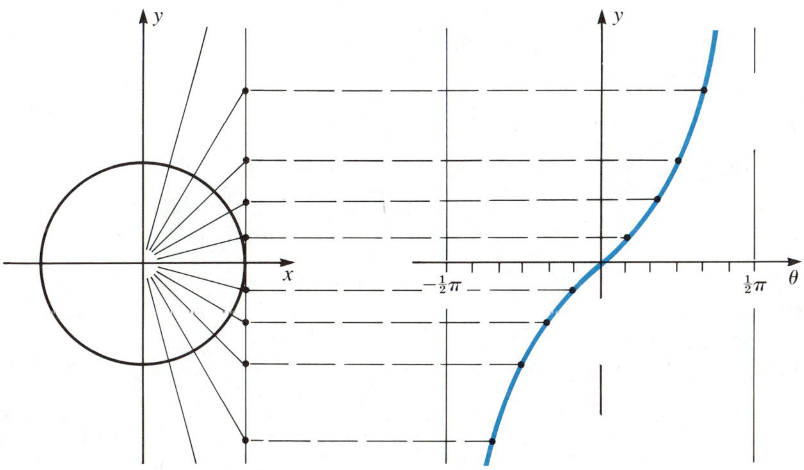

Fig. 6 The construction (15° increments)

Approximations Near $\theta = 0$

We now examine $y = \sin\theta$, $y = \cos\theta$, and $y = \tan\theta$ with a microscope near $\theta = 0$. We begin with a drawing (Fig. 7a) of a small θ. It is geometrically clear that

$$\sin\theta \longrightarrow 0 \quad\text{and}\quad \cos\theta \longrightarrow 1 \quad\text{as}\quad \theta \longrightarrow 0.$$

Thus for our first approximation we can write

$$\sin\theta \approx 0 \quad\text{and}\quad \cos\theta \approx 1 \quad\text{for}\quad \theta \approx 0.$$

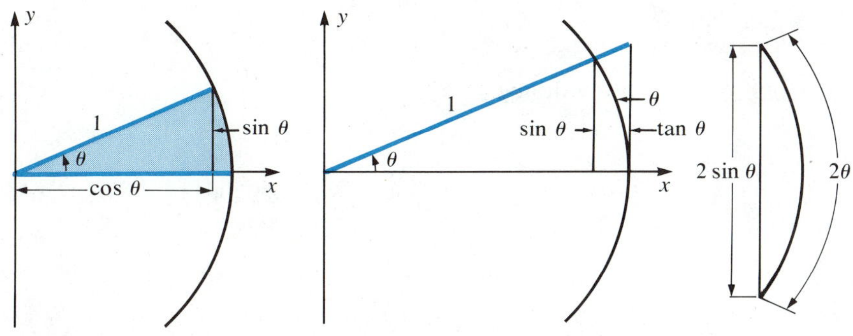

(a) $\left.\begin{matrix}\sin\theta \to 0\\ \cos\theta \to 1\end{matrix}\right\}$ as $\theta \to 0$ (b) more detail (c) $2\sin\theta < 2\theta$

Fig. 7 Small θ

We fine-tune these estimates, starting with a more detailed drawing (Fig. 7b). The arc has length θ; the outside vertical has length $\tan\theta$. Look at the arc (Fig. 7c, left) along the unit circle from the terminal sides of $-\theta$ and θ. Its length 2θ certainly exceeds the corresponding chord length $2\sin\theta$, hence

$$2\sin\theta < 2\theta \qquad \sin\theta < \theta.$$

Now look at the outer right triangle; its legs are 1 and $\tan\theta$, so its area is $\frac{1}{2}\tan\theta$. It includes the sector (of radius 1 and central angle θ) of area $\frac{1}{2}r^2\theta = \frac{1}{2}\theta$. Hence

$$\tfrac{1}{2}\theta < \tfrac{1}{2}\tan\theta \qquad \theta < \tan\theta = \frac{\sin\theta}{\cos\theta} \qquad \theta\cos\theta < \sin\theta$$

All together,

$$\theta\cos\theta < \sin\theta < \theta.$$

Since $\cos\theta \longrightarrow 1$ as $\theta \longrightarrow 0$, and $\sin\theta$ is trapped between $\theta\cos\theta$ and θ, we conclude that

$$\sin\theta \approx \theta \quad \text{as} \quad \theta \approx 0.$$

We have pulled ourselves up by our bootstraps from $\sin\theta \approx 0$ to $\sin\theta \approx \theta$. (Our figures have $\theta > 0$, but the conclusion holds for $\theta < 0$ also; this is clear from the geometry, or because both $\sin\theta$ and θ are odd functions.)

Let us test $\sin\theta \approx \theta$ for some small θ:

θ	.00	.05	.10	.15	.20	.25	.30
$\sin\theta$	.0000	.0500	.0998	.1494	.1987	.2474	.2955

These values support our estimate. Now let us test smaller θ, this time to 8 places.

θ	.01	.02	.03	.04	.05
$\sin\theta$	.00999983	.01999867	.02999550	.03998933	.04997917

The numerical evidence for $\sin\theta \approx \theta$ is very strong!

Now we can improve our approximation $\cos\theta \approx 1$ as $\theta \approx 0$. First

$$\cos\theta = \sqrt{1-\sin^2\theta} \approx \sqrt{1-\theta^2} \qquad \text{for} \quad \theta \approx 0.$$

For a simpler estimate, we use the approximation*

$$\sqrt{1-\theta^2} \approx 1 - \tfrac{1}{2}\theta^2 \qquad \text{for} \quad \theta \approx 0.$$

From this,

$$\cos\theta \approx \sqrt{1-\theta^2} \approx 1 - \tfrac{1}{2}\theta^2 \qquad \text{for} \quad \theta \approx 0.$$

Here is some strong numerical evidence:

θ	.01	.02	.03	.04	.05
$\cos\theta$	.99995000	.99980001	.99955003	.99920011	.99875026
$1-\frac{1}{2}\theta^2$	.99995	.99980	.99955	.99920	.99875

---

* It follows from

$$(1 - \tfrac{1}{2}\theta^2)^2 = 1 - \theta^2 + \tfrac{1}{4}\theta^4 \approx 1 - \theta^2$$

because $\frac{1}{4}\theta^4$ is negligible compared to θ^2. Hence

$$\sqrt{1-\theta^2} \approx \sqrt{(1-\tfrac{1}{2}\theta^2)^2} = 1 - \tfrac{1}{2}\theta^2.$$

Now let's estimate $\tan\theta$ for $\theta \approx 0$. It's easy to start:

$$\tan\theta = \frac{\sin\theta}{\cos\theta} \approx \frac{0}{1} = 0 \qquad \text{for} \quad \theta \approx 0.$$

Better is

$$\tan\theta = \frac{\sin\theta}{\cos\theta} \approx \frac{\theta}{1} = \theta \qquad \text{for} \quad \theta \approx 0.$$

(It doesn't improve the estimate perceptibly to use $\cos\theta \approx 1 - \frac{1}{2}\theta^2$ in the denominator.) Let's check some calculator values:

θ	.01	.05	.10
$\tan\theta$	.01000033	.05004171	.10033467

The agreement is excellent. Now we summarize our findings:

Approximations Near $\theta = 0$

$$\sin\theta \approx \theta \qquad \cos\theta \approx 1 - \tfrac{1}{2}\theta^2 \qquad \tan\theta \approx \theta$$

Remark It is easy to deduce from the discussion that $(\sin\theta)/\theta \longrightarrow 1$ as $\theta \longrightarrow 0$ but $\theta \neq 0$. This result is important in calculus.

Approximation Near $\theta = \frac{1}{2}\pi$

We know that $\tan\theta \longrightarrow \infty$ as $\theta \longrightarrow \frac{1}{2}\pi-$. How fast? No problem. As $\theta \longrightarrow \frac{1}{2}\pi-$, then $\frac{1}{2}\pi - \theta \longrightarrow 0+$. But (complementary angles)

$$\tan\theta = \frac{1}{\tan(\frac{1}{2}\pi - \theta)} \approx \frac{1}{\frac{1}{2}\pi - \theta} \qquad \text{as} \quad \theta \longrightarrow \tfrac{1}{2}\pi-.$$

It may be clearer to write this estimate in the form (substituting $\frac{1}{2}\pi - \theta$ for θ):

$$\tan(\tfrac{1}{2}\pi - \theta) \approx \frac{1}{\theta} \qquad \text{as} \quad \theta \longrightarrow 0+$$

We test some calculator values:

θ	.01	.005	.001
$\tan(\frac{1}{2}\pi - \theta)$	99.996667	199.99833	999.99967
$1/\theta$	100	200	1000

Not bad!

Exercises

Find $\tan\theta$ (exact) for $\theta =$

1 $-\pi$ **2** 7π **3** $-\frac{1}{4}\pi$ **4** $-\frac{1}{6}\pi$
5 $-\frac{1}{3}\pi$ **6** $-\frac{4}{3}\pi$ **7** $\frac{13}{3}\pi$ **8** $\frac{11}{4}\pi$
9 $\frac{100}{3}\pi$ **10** $-\frac{101}{3}\pi$.

Find all real θ satisfying

11 $\tan\theta = 0$ **12** $\tan\theta = -1$ **13** $\tan\theta = -\dfrac{1}{\sqrt{3}}$
14 $\tan\theta = -\sqrt{3}$ **15** $\tan\theta > 0$ **16** $\tan\theta > 1$.

Prove

17 $\tan(\pi - \theta) = -\tan\theta$ **18** $\tan(\theta + \frac{1}{2}\pi) = \dfrac{-1}{\tan\theta}$
19 $(1 + \tan^2\theta)\cos^2\theta = 1$ **20** $\dfrac{1 - \tan^2\theta}{1 + \tan^2\theta} = \cos^2\theta - \sin^2\theta$.

Find all possible values for $\tan\theta$ if

21 $\sin\theta = \frac{3}{5}$

22 $\cos\theta = -\frac{5}{13}$

23 $\cos\theta = \frac{1}{2}$

24 $\sin\theta = -\frac{1}{2}$.

Estimate by calculator to 6 significant figures

25 $\tan 4.000$

26 $\tan(-3.521)$

27 $\tan 10.03$

28 $\tan(-15.19)$

29 $\tan\frac{11}{7}$

30 $\tan\frac{128}{81}$.

Tabulate for $\theta = .002, .004, .006, .008, .010$

31 θ and $\sin\theta$

32 θ, $\cos\theta$, and $1 - \frac{1}{2}\theta^2$.

33 Up to approximately what positive value of θ is the estimate $\sin\theta \approx \theta$ accurate to 3 places?

34 (cont.) Answer the same question for the estimate $\cos\theta \approx 1 - \frac{1}{2}\theta^2$.

Justify the estimate

35 $\sin 1° \approx \dfrac{\pi}{180}$

36 $\cos\theta \approx 1 - \frac{1}{2}\sin^2\theta$, θ small.

37 Show, for $\theta°$ small and positive, that

$$\tan(90° - \theta°) \approx \frac{180}{\pi\theta}.$$

38 (cont.) Find approximately a first quadrant angle $\theta°$ such that $\tan\theta° = 100$.

39 Show that $\tan(\theta - \frac{1}{2}\pi) \approx -1/\theta$ for $\theta \longrightarrow 0+$.

40 Show that $\tan(\theta + \frac{1}{2}\pi) \approx -1/\theta$ for $\theta \longrightarrow 0+$.

41 Prove the estimate

$$\tfrac{1}{2}(\sin\theta + \tan\theta) \approx \theta \quad \text{for} \quad \theta \approx 0.$$

Verify that it is more accurate than both $\sin\theta \approx \theta$ and $\tan\theta \approx \theta$ for $\theta = 0.1, 0.05, 0.01, 0.005$. Use 8 places if possible.

42 (cont.) Now compare also the estimate

$$\tfrac{2}{3}\sin\theta + \tfrac{1}{3}\tan\theta \approx \theta$$

for the same values. Which is better?

5 COTANGENT, SECANT, AND COSECANT

We now define three more trigonometric functions. They, in a sense, complete the story on what are called direct trigonometric functions.

Cotangent, Secant, Cosecant Let θ be an angle in standard position and (x, y) a point on its terminal side, different from $(0, 0)$. Set $r = \sqrt{x^2 + y^2}$. Then

$$\cot\theta = \frac{x}{y} \quad (y \neq 0) \qquad \sec\theta = \frac{r}{x} \quad (x \neq 0) \qquad \csc\theta = \frac{r}{y} \quad (y \neq 0).$$

Equivalently,

$$\cot\theta = \frac{\cos\theta}{\sin\theta} = \frac{1}{\tan\theta} \qquad \theta \neq n\pi, \quad n \text{ an integer}$$

$$\sec\theta = \frac{1}{\cos\theta} \qquad \theta \neq (n + \tfrac{1}{2})\pi, \quad n \text{ an integer}$$

$$\csc\theta = \frac{1}{\sin\theta} \qquad \theta \neq n\pi, \quad n \text{ an integer}$$

Examples

(1) $\cot \frac{1}{6}\pi = \dfrac{1}{\tan \frac{1}{6}\pi} = \dfrac{1}{\frac{1}{3}\sqrt{3}} = \sqrt{3}$

(2) $\sec \frac{3}{4}\pi = \dfrac{1}{\cos \frac{3}{4}\pi} = \dfrac{1}{-\frac{1}{2}\sqrt{2}} = -\sqrt{2}$

(3) $\csc(-\frac{5}{6}\pi) = \dfrac{1}{\sin(-\frac{5}{6}\pi)} = \dfrac{1}{-\frac{1}{2}} = -2$

Properties

Since cot, sec, csc are the reciprocals of tan, cos, sin respectively, properties of the latter functions translate readily into properties of the former. We first note a batch of relations concerning shifting and parity (even or odd).

$$\begin{cases} \cot(\theta + n\pi) = \cot \theta \\ \cot(-\theta) = -\cot \theta \end{cases} \qquad \begin{cases} \sec(\theta \pm 2\pi n) = \sec \theta \\ \sec(-\theta) = \sec \theta \end{cases}$$

$$\begin{cases} \csc(\theta \pm 2\pi n) = \csc \theta \\ \csc(-\theta) = -\csc \theta \end{cases}$$

Next, we know that $\tan(\frac{1}{2}\pi - \theta) = 1/\tan \theta$ (complementary angles), hence $\cot \theta = \tan(\frac{1}{2}\pi - \theta)$. Also $\sin(\frac{1}{2}\pi - \theta) = \cos \theta$, so taking reciprocals, $\csc(\frac{1}{2}\pi - \theta) = \sec \theta$. Replacing θ by $\frac{1}{2}\pi - \theta$ in these relations leads to two similar ones; all together:

$$\begin{cases} \tan(\frac{1}{2}\pi - \theta) = \cot \theta \\ \cot(\frac{1}{2}\pi - \theta) = \tan \theta \end{cases} \qquad \begin{cases} \sec(\frac{1}{2}\pi - \theta) = \csc \theta \\ \csc(\frac{1}{2}\pi - \theta) = \sec \theta \end{cases}$$

Examples $\frac{1}{3}\pi + \frac{1}{6}\pi = \frac{1}{2}\pi$, so

(1) $\tan \frac{1}{3}\pi = \cot \frac{1}{6}\pi = \sqrt{3}$ (2) $\sec \frac{1}{3}\pi = \csc \frac{1}{6}\pi = 2$

(3) $\tan \frac{1}{6}\pi = \cot \frac{1}{3}\pi = \frac{1}{3}\sqrt{3}$ (4) $\sec \frac{1}{6}\pi = \csc \frac{1}{3}\pi = \dfrac{2}{\sqrt{3}} = \frac{2}{3}\sqrt{3}$.

Graphs

We construct the graph of $y = \cot \theta$ by using the relation

$$\cot \theta = \tan(\tfrac{1}{2}\pi - \theta) = \tan[-(\theta - \tfrac{1}{2}\pi)].$$

We first draw $y = \tan \theta$, then reflect it in the y-axis; the result is $y = \tan(-\theta)$. Then we shift $\frac{1}{2}\pi$ to the right. The result is $y = \tan(\frac{1}{2}\pi - \theta) = \cot \theta$. See Fig. 1, next page.

The complete graph is drawn in Fig. 2, next page. Note the vertical asymptote at $\theta = n\pi$ for each integer n.

To graph $y = \sec \theta$ and $y = \csc \theta$ we use the relations

$$\sec \theta = \frac{1}{\cos \theta} \qquad \text{and} \qquad \csc \theta = \frac{1}{\sin \theta}$$

and plot points (Fig. 3, next page, and Fig. 4, page 261). Note the vertical asymptotes where the functions are undefined. Note also that $\csc(\theta + \frac{1}{2}\pi) = \sec\theta$ because $\sin(\theta + \frac{1}{2}\pi) = \cos\theta$, so the curve $y = \sec\theta$ is obtained by shifting the curve $y = \csc\theta$ to the left by $\frac{1}{2}\pi$.

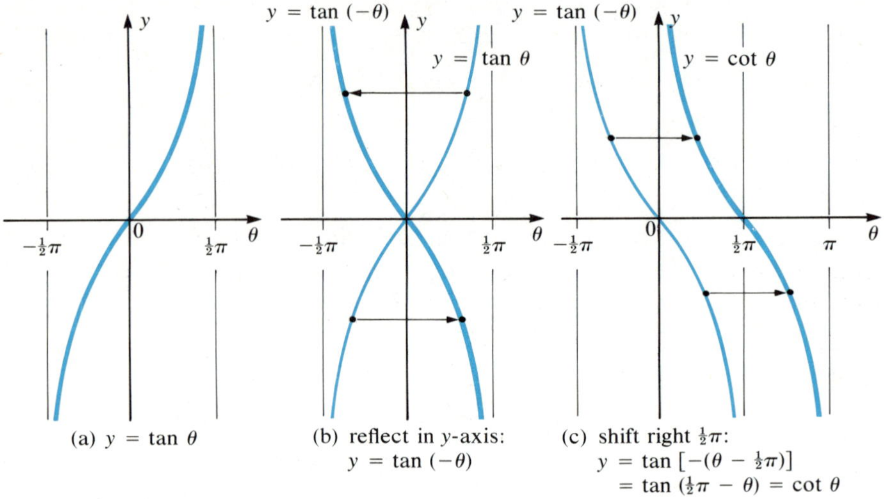

Fig. 1 Construction of the graph of $y = \cot\theta$

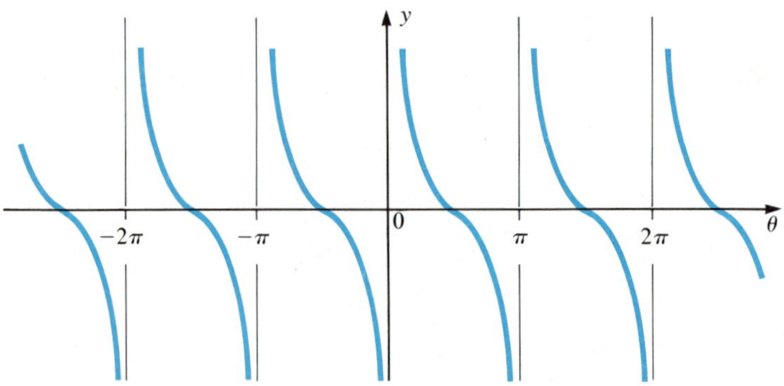

Fig. 2 $y = \cot\theta$

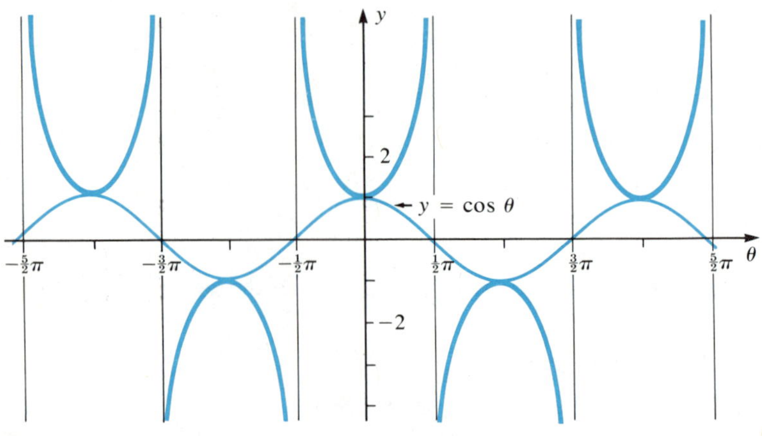

Fig. 3 $y = \sec\theta$

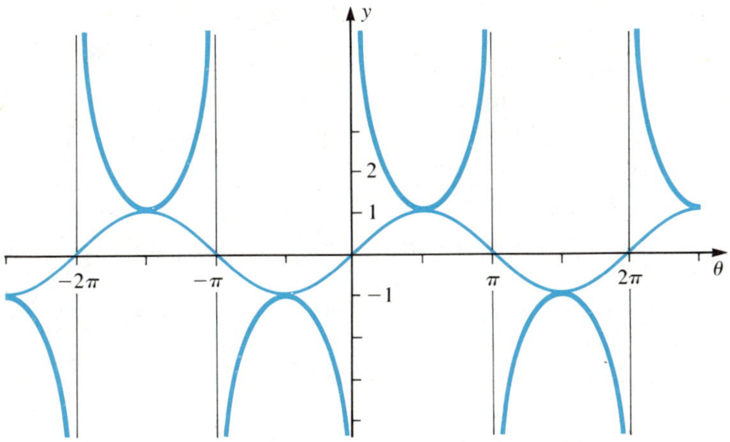

Fig. 4 $y = \csc\theta$

EXAMPLE 1 Find all values of θ for which $\csc\theta = 2$.

Solution If $\csc\theta = 2$, then

$$\frac{1}{\sin\theta} = 2 \qquad \sin\theta = \tfrac{1}{2}.$$

For $0 \le \theta \le 2\pi$, there are two such values: $\theta = \frac{1}{6}\pi$ and $\theta = \frac{5}{6}\pi$. By the graph in Fig. 4, the only other possibilities are these plus multiples of 2π.

Answer $\frac{1}{6}\pi + 2\pi n \qquad \frac{5}{6}\pi + 2\pi n \qquad$ (n any integer)

Values of cot θ, sec θ, csc θ

Calculators do not have keys for these functions. But $\cot\theta$, $\sec\theta$, $\csc\theta$ are the reciprocals of $\tan\theta$, $\cos\theta$, $\sin\theta$, so their values can be estimated using [tan], [cos], [sin], and the reciprocal key [1/x].

Examples (Set [rad] mode).

(1)	cot 0.483	. 4 8 3 [tan] [1/x]	1.9068324
(2)	csc(−0.107)	. 1 0 7 [+/−] [sin] [1/x]	−9.3636516
(3)	csc 3π	3 [×] [π] [=] [sin] [1/x]	ERROR

(There is a grain of salt in the last example. On a TI-59, we indeed get an error condition. But on a Sharp EL-5806, we get $\sin 3\pi \approx -4 \times 10^{-10}$, so $\csc 3\pi \approx -2.5 \times 10^{9}$. Small machine errors are likely with calculators; don't forget it.)

Exercises

Find $\cot\theta$, $\sec\theta$, and $\csc\theta$ (exact) for $\theta =$

1 $\frac{1}{4}\pi$ **2** $\frac{2}{3}\pi$ **3** $\frac{5}{3}\pi$ **4** $\frac{7}{6}\pi$
5 $-\frac{1}{3}\pi$ **6** $-\frac{2}{3}\pi$ **7** $-\frac{3}{4}\pi$ **8** $-\frac{13}{6}\pi$
9 $\frac{29}{3}\pi$ **10** $\frac{53}{6}\pi$.

Find all θ for which

11 $\sec\theta = 2$ **12** $\csc\theta = -2$ **13** $\cot\theta = \sqrt{3}$

14 $\cot\theta = \frac{1}{3}\sqrt{3}$ **15** $\sec\theta = -1$ **16** $\csc\theta = 1$.

Estimate on a calculator to 6 significant figures

17 $\cot 0.3113$ **18** $\cot(-1.0144)$ **19** $\sec(2.500)$

20 $\sec(-0.0439)$ **21** $\csc(-3.4124)$ **22** $\csc(0.1158)$.

23 Justify the estimate $\cot\theta \approx 1/\theta$ for θ small.

24 (cont.) Tabulate $1/\theta - \cot\theta$ for $\theta = 0.10, 0.08, 0.06, 0.04, 0.02; 0.01; 0.005$.

25 (cont.) Examine the table for 0.1 and 0.01. Do you see a pattern? If so, test it on the other entries in the table.

26 Justify the estimate $\sec\theta \approx 1 + \frac{1}{2}\theta^2$ for θ small. Tabulate $\sec\theta$ and $1 + \frac{1}{2}\theta^2$ for $\theta = 0.1, 0.05, 0.01, 0.005, 0.001$.

27 Justify the estimate $\csc\theta \approx 1/\theta$ for θ small. Test for $\theta = 0.1, 0.05, 0.01, 0.005, 0.001$.

28 In view of Exs. 23 and 27, you would guess that $\csc\theta - \cot\theta$ is not too large if θ is small. Show numerically that $\csc\theta - \cot\theta \approx \frac{1}{2}\theta$ appears to be true.

Find all θ satisfying

29 $\tan\theta \geq \cot\theta$ **30*** $\tan\theta + \cot\theta \geq 2$.

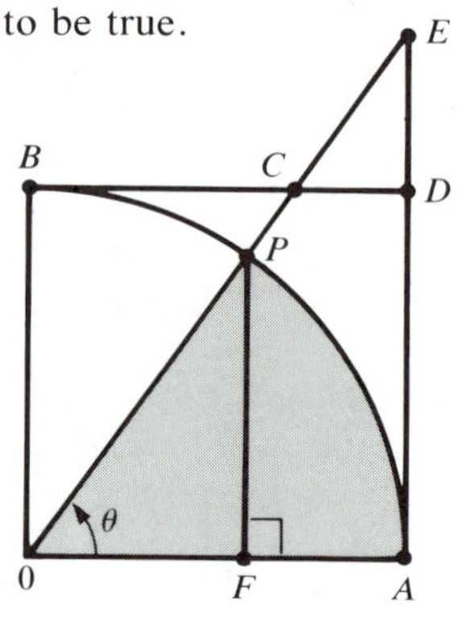

The quarter circle has radius 1. Express each length in terms of θ

31 $\overline{AE}$ **32** $\overline{BC}$

33 $\overline{OE}$ **34** $\overline{OC}$

35 $\overline{OF}$ **36** $\overline{FP}$

37 $\overparen{AP}$ **38** $\overparen{PB}$.

6 PERIODIC FUNCTIONS AND FURTHER GRAPHS

A function $f(\theta)$ with domain {all θ} is called **periodic of period p** provided p is a positive constant and

$$f(\theta + p) = f(\theta)$$

for all θ.

Examples

(1) $f(\theta) = \sin\theta$ is periodic of period 2π

(2) $f(\theta) = \cos\theta$ is periodic of period 10π (also of period 2π)

(3) $f(\theta) = 5\cos^2 3\theta - \sin^3 7\theta$ is periodic of period 2π.

If $f(\theta)$ is periodic of period p, then

$$f(\theta + 2p) = f(\theta + p + p) = f(\theta + p) = f(\theta)$$

$$f(\theta + 3p) = f(\theta + 2p + p) = f(\theta + 2p) = f(\theta)$$

and in general

$$f(\theta + np) = f(\theta) \qquad n \text{ any whole number.}$$

On the other side,

$$f(\theta - p) = f(\theta - p + p) = f(\theta),$$

and as with positive multiples,

$$f(\theta - np) = f(\theta).$$

If $f(\theta)$ is periodic of period p, then

$$f(\theta + np) = f(\theta) \qquad n \text{ any integer}.$$

Constructing Periodic Functions

Let $f(\theta)$ be periodic of period p. The graph of $f(\theta)$ on an interval of length p such as $a \le \theta < a + p$ or $a < \theta \le a + p$ is called a **cycle** of the function. (Note that exactly one end-point is included.) For instance, the graph of $y = \sin\theta$ on $0 \le \theta < 2\pi$ is a cycle of $\sin\theta$; so is its graph on $-\pi < \theta \le \pi$. What is important about a cycle is that it completely determines the periodic function. This is so because given any θ, there is an integer n such that $\theta + np$ is in the interval; then $f(\theta) = f(\theta + np)$.

This reasoning can be turned around; what we have in mind is the process by which we defined $\sin\theta$ once we had it for $0 \le \theta < 2\pi$. Given any interval $a \le \theta < a + p$ (or $a < \theta \le a + p$) of length p and a function $f(\theta)$ on this interval, there is a unique extension of this function to a periodic function of period p on the whole θ-axis. Its graph is simply the graph on $a \le \theta < a + p$ shifted left and right by steps of p.

Examples (1) $f(\theta) = \theta$ for $0 \le \theta < 1$, $p = 1$.

The corresponding periodic function of period 1 is graphed in Fig. 1.

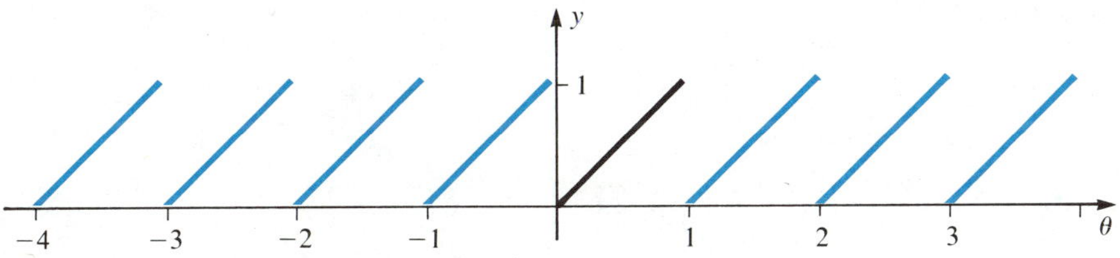

Fig. 1 $y = \theta$ for $0 \le \theta < 1$ and $y(\theta)$ periodic of period 1

(2) $f(\theta) = \theta^2$ for $-1 \le \theta < 1$, $p = 2$.

The corresponding periodic function of period 2 is graphed in Fig. 2.

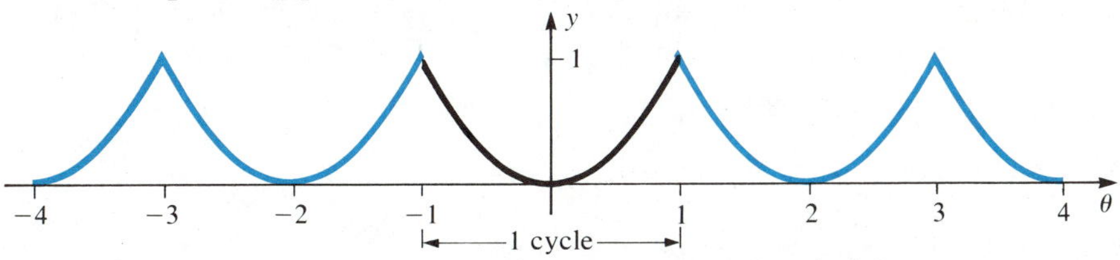

Fig. 2 $y = \theta^2$ for $-1 \le \theta < 1$ and $y(\theta)$ periodic of period 2

(3) $f(\theta) = \begin{cases} \sin\theta & \text{for } 0 \le \theta \le \pi \\ 0 & \text{for } \pi < \theta < 2\pi \end{cases} \qquad p = 2\pi.$

The corresponding periodic function is graphed in Fig. 3. It is what a rectified alternating electric current (half wave rectifier) looks like, current vs. time.

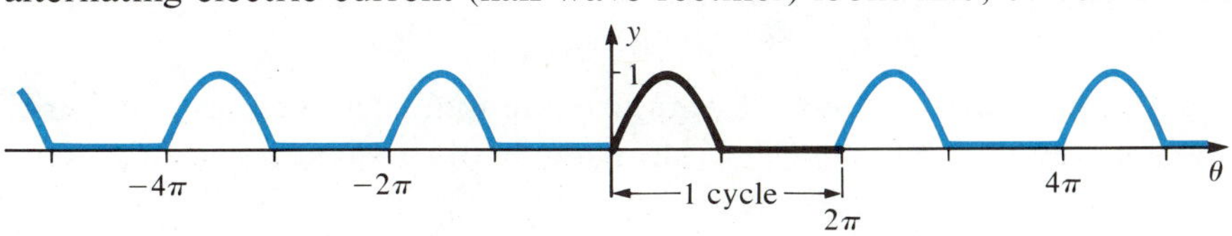

Fig. 3 $y = \begin{cases} \sin\theta & \text{for } 0 \le \theta < \pi \\ 0 & \text{for } \pi \le \theta < 2\pi \end{cases}$ and $y(\theta)$ periodic of period 2π

Graphs of $y = \sin b\theta$

Let's start with $y = \sin 2\theta$. This is just like $y = \sin \theta$, except that we have changed the scale (stretched) on the θ-axis. As θ goes from 0 to π, then 2θ goes from 0 to 2π, so $y = \sin 2\theta$ has a cycle on $0 \le \theta < \pi$. It is periodic of period π because

$$\sin 2(\theta + \pi) = \sin(2\theta + 2\pi) = \sin 2\theta.$$

The graph is sketched in Fig. 4.

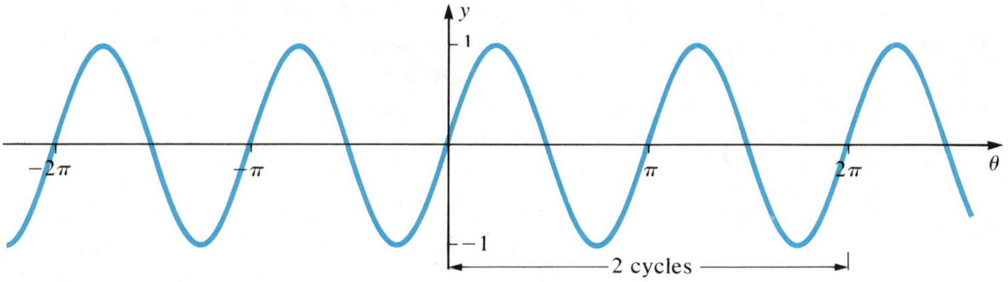

Fig. 4 $y = \sin 2\theta$

Similarly, the graph of $y = \sin 3\theta$ makes one cycle on the interval $0 \le \theta < \frac{2}{3}\pi$ and three cycles on the interval $0 \le \theta < 2\pi$. The graph of $y = \sin\frac{1}{2}\theta$ makes one cycle on the interval $0 \le \theta < 4\pi$, a half-cycle on the interval $0 \le \theta < 2\pi$. See Fig. 5.

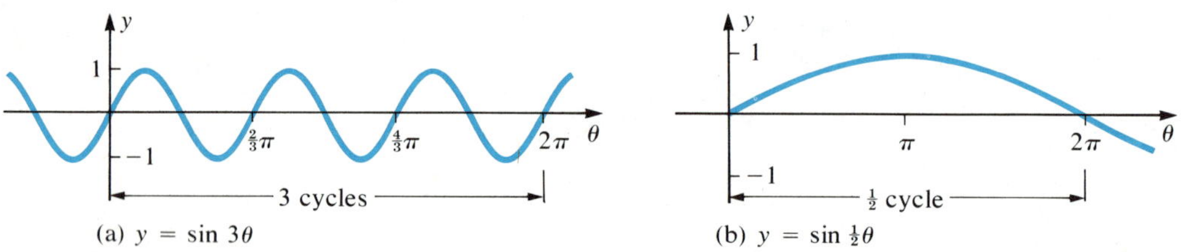

Fig. 5

In general, for each $b > 0$, the graph of $y = \sin b\theta$ is the graph of $y = \sin \theta$ compressed (or stretched) horizontally. The period of $y = \sin b\theta$ is $2\pi/b$:

$$\sin b(\theta + 2\pi/b) = \sin(b\theta + 2\pi) = \sin b\theta.$$

Similar statements apply to $y = \cos b\theta$.

Graph of $y = a\sin(b\theta - c) + d$

Once we note that

$$\sin(b\theta - c) = \sin b\left(\theta - \frac{c}{b}\right),$$

it is clear that the graph of

$$y = a\sin(b\theta - c) + d$$

comes from the graph of $y = \sin b\theta$ by our usual routine of stretching, reflecting, and shifting horizontally and vertically. The same holds if sin is replaced by cos.

EXAMPLE 1 Graph

(a) $y = -2\cos 3\theta$ (b) $y = 5\sin(\frac{1}{3}\theta - \pi) + 10$.

Solution (a) Start with $y = \cos\theta$. First compress the θ-axis by a factor of 3, then stretch the y-axis by a factor of 2, and finally reflect in the θ-axis for the sign change (Fig. 6a).

(b) Rewrite the equation in the form

$$y = 5\sin\tfrac{1}{3}(\theta - 3\pi) + 10.$$

The graph is obtained from that of $y = 5\sin\frac{1}{3}\theta$ by shifting 3π units to the right and 10 units up. Now $\sin\frac{1}{3}\theta$ oscillates $\frac{1}{3}$ as fast as $\sin\theta$; a cycle has length $3\cdot 2\pi = 6\pi$. Also $y = 5\sin\frac{1}{3}\theta$ oscillates between the levels $y = 5$ and $y = -5$. The shifted graph oscillates between the levels $y = 5 + 10 = 15$ and $y = -5 + 10 = 5$. See Fig. 6b.

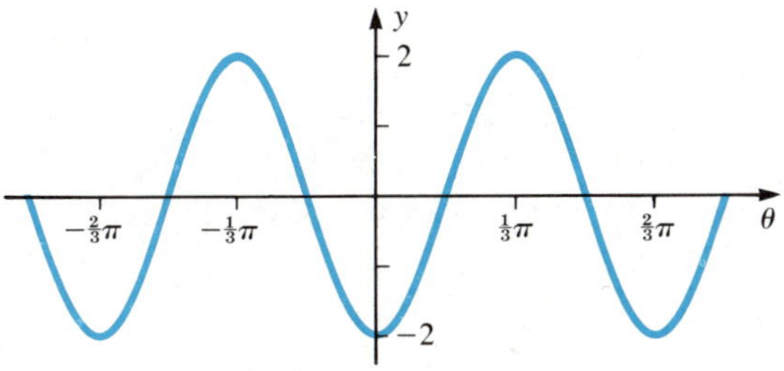

(a) $y = -2\cos 3\theta$

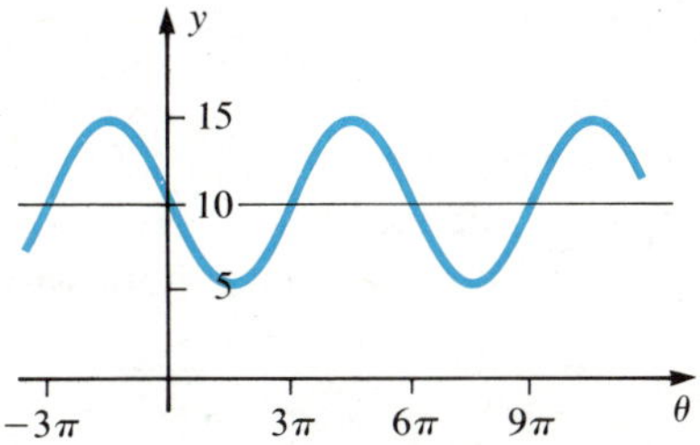

(b) $y = 5\sin(\frac{1}{3}\theta - \pi) + 10$

Fig. 6

EXAMPLE 2 A point P moves uniformly (with constant angular speed) counterclockwise around the circle of radius 4 and center $(0, 0)$. It starts at $(0, 4)$ at time $t = 0$, and its angular speed is 10 revolutions per second. Express its x-coordinate in terms of t (in seconds) and graph the resulting function in the t,x-plane.

Solution From Fig. 7a, we see that $x = 4\cos\theta$. But we are asked for x in terms of t. So we should try to express θ in terms of t.

Given: the motion is uniform. That means θ is a linear function of t. Because $\theta = \frac{1}{2}\pi$ when $t = 0$, we may write $\theta = at + \frac{1}{2}\pi$. The constant a is the constant rate of θ (angular speed) in rad/sec. Thus

$$a = 10 \text{ rev/sec} = 10(2\pi)\text{ rad/sec} = 20\pi \text{ rad/sec}.$$

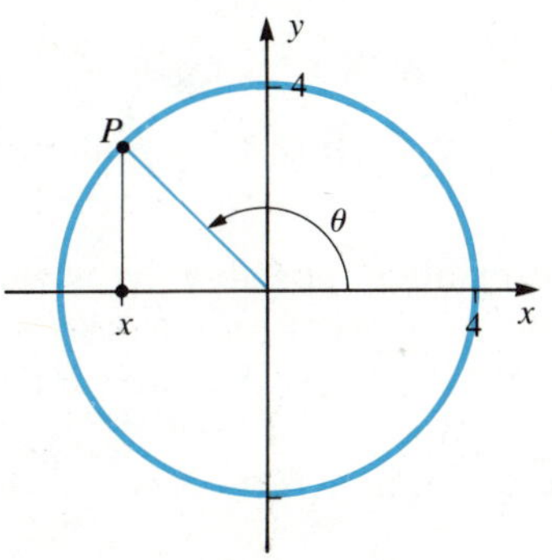

(a) the set-up

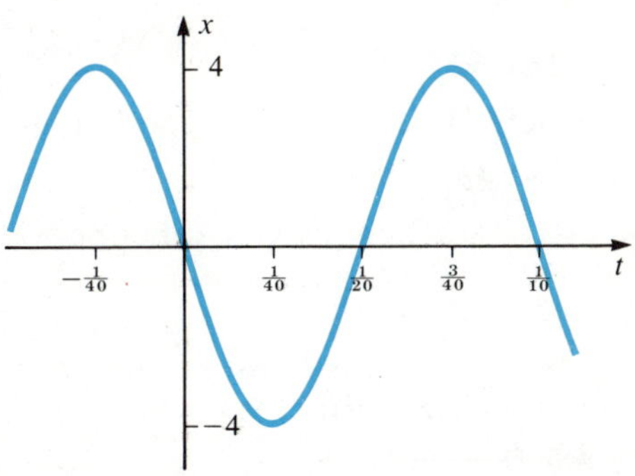

(b) the graph: $x = 4\cos[20\pi(t + \frac{1}{40})]$

Fig. 7

Therefore $\theta = 20\pi t + \frac{1}{2}\pi$,

$$x = 4\cos\theta = 4\cos(20\pi t + \tfrac{1}{2}\pi) = 4\cos[20\pi(t + \tfrac{1}{40})].$$

This is the desired function; its graph (Fig. 7b) is a cosine curve, shifted left $\frac{1}{40}$ and with the scales adjusted.

Remark Suppose a point moves uniformly around a circle, like P in Example 2. The oscillatory motion of its projection on any line is called **simple harmonic motion.** Such motion is very common in nature. For example, imagine a weight hanging on the end of a spring attached to the ceiling. If pulled down a little from its rest position and released, the spring will oscillate up and down. The motion is simple harmonic motion along a vertical line.

Suppose you must graph a periodic function $y = f(\theta)$ of period p. Then graph it carefully on any segment of the θ-axis of length p. After that, or before that, it simply repeats itself.

EXAMPLE 3 Graph $y = \cos 2\theta + 2\cos\theta$.

Solution The function is periodic of period 2π. We tabulate enough values for $-\pi \le \theta < \pi$ to make an accurate graph on that interval. Actually, we can cut our calculations in half by noting that $y(\theta)$ is an even function. Therefore, we need only tabulate values for $0 \le \theta \le \pi$:

θ	0	$\frac{1}{10}\pi$	$\frac{2}{10}\pi$	$\frac{3}{10}\pi$	$\frac{4}{10}\pi$	$\frac{5}{10}\pi$	$\frac{6}{10}\pi$	$\frac{7}{10}\pi$	$\frac{8}{10}\pi$	$\frac{9}{10}\pi$	π
y	3.0	2.7	1.9	0.9	−0.2	−1.0	−1.4	−1.5	−1.3	−1.1	−1.0

We plot these points, reflect in the y-axis, then join by a smooth curve. By periodicity we can draw as much of the graph as we want (Fig. 8).

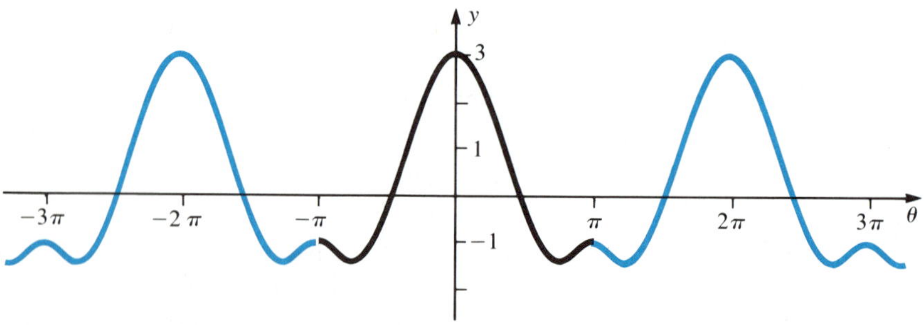

Fig. 8 $y = \cos 2\theta + 2\cos\theta$

Fundamental Period

Let $f(\theta)$ be a periodic function of period p. Then $f(\theta)$ is periodic of period $2p$ since

$$f(\theta + 2p) = f(\theta).$$

It is also periodic of period $3p$, or period $4p$, etc. Thus each of the numbers

$$p, 2p, 3p, 4p, \cdots$$

is a period of $f(\theta)$. There may be other periods, for instance, possibly $\frac{1}{3}p$ is also a period of $f(\theta)$. For instance $p = 6\pi$ is a period of $f(\theta) = \cos\theta$. So are $12\pi, 18\pi, \cdots$. But so also is $\frac{1}{3}p = 2\pi$. This number 2π is the *least* (positive) period of $f(\theta) = \cos\theta$, as is obvious from its graph.

If it exists, the *least* (positive) period p of a periodic function $f(\theta)$ is called the **fundamental period** of $f(\theta)$.

A function can be periodic without having a fundamental period: any constant function is so, and there are others (Ex. 62). While a certain function may have a fundamental period, it may not be easy to find it and prove it so. The advantage of having the fundamental period rather than some multiple is that you have to study the function in detail on a smaller interval.

For instance, consider $f(\theta) = \sin^3\theta\cos^5\theta$. Obviously 2π is a period of $f(\theta)$. It is less obvious, but true, that π is a period of $f(\theta)$. That π is *the fundamental period* of $f(\theta)$ is far from obvious; we probably need a graph to be convinced. [Actually, we can do it this way. First we note that $f(\theta) = 0$ only for $\theta = \frac{1}{2}\pi n$. Therefore the only candidate for a period smaller than π is $\frac{1}{2}\pi$. But $f(\frac{1}{4}\pi) > 0$ and $f(\frac{1}{4}\pi + \frac{1}{2}\pi) < 0$, so $\frac{1}{2}\pi$ is not a period; π is the fundamental period.]

Generally, if you are asked to find a period of a function, you should find one as small as you possibly can. But finding the fundamental period, if there is one, might be very difficult. Try it for

$$f(\theta) = 1.4\sin 3\theta + 1.2\sin 5\theta - 2.3\cos 6\theta$$

for instance. Here 2π is surely a period; is there a smaller one? That's hard!

Singularities

We would like to say that $\tan\theta$ is periodic of period π because

$$\tan(\theta \pm n\pi) = \tan\theta.$$

We are not supposed to do so because the domain of $\tan\theta$ falls short of the whole θ-axis. The points $\theta = (n + \frac{1}{2})\pi$ are singularities of $\tan\theta$, points where it cannot be defined.

Usually one says so anyhow, taking a looser attitude towards periodic functions. Some points can be omitted, but whenever θ_0 is omitted, then all $\theta_0 + np$, where p is the period, must also be omitted.

So don't be surprised to hear that $f(\theta) = \sec\theta$ is periodic of period 2π, etc.

Other Trigonometric Graphs

Functions such as $f(\theta) = \theta\sin\theta$ and $f(\theta) = \theta + \cos\theta$ involve trigonometric functions but are not periodic. Let us graph one such function.

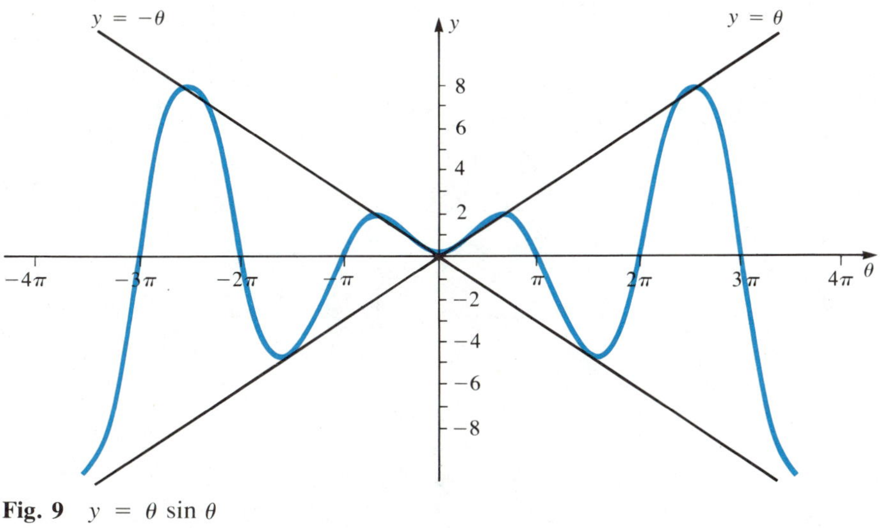

Fig. 9 $y = \theta\sin\theta$

EXAMPLE 4 Graph $y = \theta \sin \theta$.

Solution Since $\theta \sin \theta$ is even, we can concentrate on $\theta \geq 0$. Ordinarily $\sin \theta$ oscillates between the horizontal lines $y = 1$ and $y = -1$. But here the values of $\sin \theta$ are magnified by the factor θ. Therefore the graph oscillates between the lines $y = \theta$ and $y = -\theta$. We draw these lines as guides.

Now we plot some crucial points. Since $y = 0$ where $\sin \theta = 0$, the graph meets the θ-axis at $\theta = 0, \pi, 2\pi, \cdots$. Halfway between these values, at $\theta = \frac{1}{2}\pi, \frac{3}{2}\pi, \frac{5}{2}\pi, \cdots$ we have $\sin \theta = \pm 1$, so $y = \pm\theta$. For these values of θ, the curve alternately reaches its upper and lower guidelines. One final piece of information: $\sin \theta \approx \theta$ near $\theta = 0$, so $y \approx \theta^2$ for $\theta \approx 0$.

This is enough information to sketch the graph for $\theta \geq 0$. We reflect in the y-axis to extend the graph to $\theta < 0$. See Fig. 9, previous page.

Exercises

Graph

1 $y = \begin{cases} 1 & 0 \leq \theta < 1 \\ -1 & 1 \leq \theta < 2 \end{cases}$ period 2

2 $y = \sqrt{1 - \theta^2}$ $-1 \leq \theta < 1$ period 2

3 $y = \begin{cases} \theta & 0 \leq \theta < 2 \\ 4 - \theta & 2 \leq \theta < 4 \end{cases}$ period 4

4 $y = |\theta|$ $-1.5 \leq \theta < 1.5$ period 3.

Show that $f(\theta)$ is periodic by finding a period

5 $\sin \frac{1}{2}\theta - \cos \frac{1}{5}\theta$

6 $\sin \frac{1}{3}\theta + \cos \frac{1}{4}\theta$

7 $\cos 2\pi\theta$

8 $\sin \pi\theta$

9 $|\cos \theta|$

10 $\sin^2 \theta$

11 $\sin 2\theta \cos^3 2\theta$

12 $\sin(\theta - \frac{1}{6}\pi) + \cos(\theta - \frac{1}{5}\pi)$.

Graph $y =$

13 $\sin 4\theta$

14 $\sin 3\theta$

15 $\cos 3\theta$

16 $\cos 4\theta$

17 $\sin 2\pi\theta$

18 $\cos \pi\theta$

19 $3 \sin 2\theta$

20 $-3 \cos 2\theta$

21 $1 + \cos \theta$

22 $2 - \sin \theta$

23 $11 + 3 \cos \theta$

24 $-6 + 2 \sin \theta$

25 $|\sin \theta|$

26 $|\cos \theta|$

27 $\sin^2 \theta$

28 $\cos^2 \theta$

29 $\sin(\theta + \frac{1}{6}\pi)$

30 $\cos(\theta + \frac{1}{4}\pi)$

31 $\frac{1}{2} \sin(\theta + \frac{1}{3}\pi)$

32 $2 \cos(\theta - \frac{1}{6}\pi)$

33 $\sin(3\theta - \frac{1}{3}\pi)$

34 $2 \cos 4(\theta - \frac{1}{6}\pi)$

35 $-4 \sin 3\theta$

36 $-2 \cos(\theta + \frac{1}{4}\pi)$.

Solve Example 2 for the following data

	Center	Radius	Initial position	Angular speed
37	(0, 0)	3	(−3, 0)	1 rev/sec
38	(0, 0)	3	(0, −3)	30 rev/min
39	(1, 0)	1	(2, 0)	10 rev/sec
40	(−3, 0)	2	(−3, 2)	5 rev/sec

Graph $y =$

41 $\sin 2\theta - \sin \theta$

42 $\sin \theta + \cos 2\theta$

43 $\sin \theta + \cos \theta$

44 $\sin \theta - 2 \cos \theta$

45 $3 \cot \theta$

46 $\csc 2\theta + 1$

47 $\tan\theta + \cot\theta$

48 $\tan\theta - \cot\theta$

49 $\sec\theta - \csc\theta$

50 $\sec\theta + \csc\theta$

51 $\theta\cos\theta$

52 $\theta^2 \sin\theta$

53 $\theta + \cos\theta$

54 $\theta + \sin\theta$

55 $|\theta|\sin\theta$

56 $2^{-\theta/2\pi}\sin\theta$.

57 Show graphically that $\cos\theta = \theta$ has only one solution, and locate it approximately.

58 Show graphically that $\pi\cos 2\theta = \theta$ has exactly five solutions, and locate them approximately.

59 Locate graphically all solutions of $\sin 2\theta = \cos\theta$ in the interval $0 \le \theta \le 2\pi$.

60 Locate graphically all solutions of $\theta = 3\pi\sin\theta$.

61 Suppose that, to graph $y = \sin 20\pi\theta$, you plot points for $\theta = 0, 0.1, 0.2, 0.3, \cdots, 2.0$. What happens and why?

62* Suppose $f(\theta) = 0$ if θ is irrational and $f(\theta) = 1$ if θ is rational. Show that $f(\theta)$ is periodic of period p for each positive rational number p, but $f(\theta)$ has no fundamental period.

REVIEW EXERCISES

Find exact values of $\sin\theta$ and $\cos\theta$ for θ in standard position with its terminal side through

1 $(\sqrt{5}, \sqrt{7})$

2 $(\sqrt{7}, -\sqrt{5})$.

Estimate by calculator to 5 places

3 $\sin 32.756°$

4 $\cos 4.1012$.

Express in terms of $S = \sin 27°$

5 $\cos 117°$

6 $\tan 243°$.

7 Convert $\frac{3}{7}\pi$ to degrees.

8 Convert $113°$ to $b\pi$ radians, b rational.

Find all θ with $0 \le \theta < 2\pi$ such that

9 $\sin\theta = \cos\theta$

10 $\sin\theta = -\cos\theta$.

Find

11 $\tan(-\frac{23}{4}\pi)$ exactly

12 $\tan\frac{2}{7}\pi$ to 6 places.

Calculate to 6 places

13 $\sec^2 1.385 - \tan^2 1.385$

14 $\csc^2 4.998 - \cot^2 4.998$.

Find all θ for which

15 $\sec\theta = \frac{2}{3}\sqrt{3}$

16 $\cot\theta = -1$.

Graph

17 $y = 2\sin\theta - \cos 2\theta$

18 $y = \sin\theta + \tan\theta$.

19 Find the smallest positive angle θ on your calculator such that $\cos\theta \ne 1$.

20 Find the largest angle $\theta° < 90°$ on your calculator such that $\sin\theta° \ne 1$.

7 IDENTITIES AND INVERSE FUNCTIONS

1 BASIC IDENTITIES

Because the six trigonometric functions are closely related to each other, there are a multitude of identities interconnecting them; we begin with three very basic ones:

$$\sin^2\theta + \cos^2\theta = 1$$
$$1 + \tan^2\theta = \sec^2\theta$$
$$1 + \cot^2\theta = \csc^2\theta .$$

The first identity, already discussed on p. 235, is an immediate consequence of the definition of $\sin\theta$ and $\cos\theta$:

$$\cos^2\theta + \sin^2\theta = \left(\frac{x}{r}\right)^2 + \left(\frac{y}{r}\right)^2 = \frac{x^2 + y^2}{r^2} = \frac{r^2}{r^2} = 1 .$$

To derive the second identity, divide both sides of the first by $\cos^2\theta$:

$$\frac{\sin^2\theta}{\cos^2\theta} + \frac{\cos^2\theta}{\cos^2\theta} = \frac{1}{\cos^2\theta} \qquad \tan^2\theta + 1 = \sec^2\theta .$$

To derive the third, divide both sides of the first by $\sin^2\theta$:

$$\frac{\sin^2\theta}{\sin^2\theta} + \frac{\cos^2\theta}{\sin^2\theta} = \frac{1}{\sin^2\theta} \qquad 1 + \cot^2\theta = \csc^2\theta .$$

Remark The identity $1 + \tan^2\theta = \sec^2\theta$ is valid wherever both sides are defined, that is, for all θ except $\pm\frac{1}{2}\pi, \pm\frac{3}{2}\pi, \pm\frac{5}{2}\pi, \cdots$. Remember that in general an identity is an equation that is valid for all values of the variable(s) for which all terms in the equation are defined.

In addition to the three relations in the box, there are the relations that define tan, cot, sec, and csc:

$$\tan\theta = \frac{\sin\theta}{\cos\theta} \qquad \cot\theta = \frac{1}{\tan\theta} = \frac{\cos\theta}{\sin\theta}$$

$$\sec\theta = \frac{1}{\cos\theta} \qquad \csc\theta = \frac{1}{\sin\theta}.$$

Now we have the tools for deriving and proving many other identities.

EXAMPLE 1 Show that

(a) $\cos\theta\tan\theta = \sin\theta$ (b) $\cot\theta\sec\theta = \csc\theta$.

Solution (a) $\cos\theta\tan\theta = \cos\theta\dfrac{\sin\theta}{\cos\theta} = \sin\theta$.

(b) $\cot\theta\sec\theta = \dfrac{\cos\theta}{\sin\theta}\dfrac{1}{\cos\theta} = \dfrac{1}{\sin\theta} = \csc\theta$.

EXAMPLE 2 Show that

(a) $\tan^2\theta + \tan^4\theta = \tan^2\theta\sec^2\theta$ (b) $\dfrac{\cos^2\theta}{1-\cos^2\theta} = \cot^2\theta$.

Solution (a) Factor the left-hand side and use $1 + \tan^2\theta = \sec^2\theta$:

$$\tan^2\theta + \tan^4\theta = \tan^2\theta(1+\tan^2\theta) = \tan^2\theta\sec^2\theta.$$

(b) From $\sin^2\theta + \cos^2\theta = 1$ follows $1 - \cos^2\theta = \sin^2\theta$. Hence

$$\frac{\cos^2\theta}{1-\cos^2\theta} = \frac{\cos^2\theta}{\sin^2\theta} = \left(\frac{\cos\theta}{\sin\theta}\right)^2 = \cot^2\theta.$$

Functions in Terms of Other Functions

Suppose we are given an acute θ with $\sin\theta = \frac{3}{4}$. Can we find $\cos\theta$? Yes; from $\cos^2\theta = 1 - \sin^2\theta$,

$$\cos\theta = \sqrt{1-\sin^2\theta} = \sqrt{(1-(\tfrac{3}{4})^2} = \sqrt{\tfrac{7}{16}} = \tfrac{1}{4}\sqrt{7}.$$

Watch out, though. If θ is in the *second* quadrant, then $\cos\theta$ is *negative*, so

$$\cos\theta = -\sqrt{1-\sin^2\theta} = -\tfrac{1}{4}\sqrt{7}.$$

In general,

$$\sin\theta = \pm\sqrt{1-\cos^2\theta} \quad\text{and}\quad \cos\theta = \pm\sqrt{1-\sin^2\theta},$$

where the choice of + or − depends on the quadrant of θ. Similarly,

$$\tan\theta = \pm\sqrt{\sec^2\theta - 1} \qquad \sec\theta = \pm\sqrt{1+\tan^2\theta}$$

$$\cot\theta = \pm\sqrt{\csc^2\theta - 1} \qquad \csc\theta = \pm\sqrt{1+\cot^2\theta}.$$

Again, the choice of sign depends on the quadrant of θ.

EXAMPLE 3 Suppose $\tan\theta = -2$ and θ is an angle in the second quadrant. Find

(a) $\sec\theta$ (b) $\cos\theta$ (c) $\sin\theta$.

Solution (a) In the second quadrant, $\sec\theta < 0$. Therefore,

$$\sec\theta = -\sqrt{1+\tan^2\theta} = -\sqrt{1+(-2)^2} = -\sqrt{5}.$$

(b) $\cos\theta = \dfrac{1}{\sec\theta} = -\dfrac{1}{\sqrt{5}}$.

(c) In the second quadrant, $\sin\theta > 0$. Therefore

$$\sin\theta = \sqrt{1-\cos^2\theta} = \sqrt{1-\tfrac{1}{5}} = \sqrt{\tfrac{4}{5}} = \frac{2}{\sqrt{5}}.$$

Another way:

$$\sin\theta = \cos\theta\,\frac{\sin\theta}{\cos\theta} = \cos\theta\tan\theta = \left(-\frac{1}{\sqrt{5}}\right)(-2) = \frac{2}{\sqrt{5}}.$$

Answer (a) $-\sqrt{5}$ (b) $-\dfrac{1}{\sqrt{5}}$ (c) $\dfrac{2}{\sqrt{5}}$.

EXAMPLE 4 For θ in the second quadrant, express in terms of $\tan\theta$:

(a) $\sec\theta$ (b) $\cos\theta$ (c) $\sin\theta$.

Solution This is the same problem as Example 3, except that $\tan\theta$ is not given a specific value. We solve it in the same way.

(a) $\sec\theta = -\sqrt{1+\tan^2\theta}$

(b) $\cos\theta = \dfrac{1}{\sec\theta} = \dfrac{-1}{\sqrt{1+\tan^2\theta}}$

(c) $\sin\theta = \cos\theta\tan\theta = \dfrac{-\tan\theta}{\sqrt{1+\tan^2\theta}}$

Answer (a) $-\sqrt{1+\tan^2\theta}$ (b) $\dfrac{-1}{\sqrt{1+\tan^2\theta}}$ (c) $\dfrac{-\tan\theta}{\sqrt{1+\tan^2\theta}}$

EXAMPLE 5 Suppose θ is in the fourth quadrant. Express the other five trig functions in terms of $\csc\theta$.

Solution Note that $\csc\theta < 0$ in the fourth quadrant (actually $\csc\theta < -1$). We have first

$$\sin\theta = \frac{1}{\csc\theta}.$$

Since $\cos\theta > 0$ in the fourth quadrant,

$$\cos\theta = \sqrt{1-\sin^2\theta} = \sqrt{1-\frac{1}{\csc^2\theta}} = \sqrt{\frac{\csc^2\theta-1}{\csc^2\theta}} = \frac{\sqrt{\csc^2\theta-1}}{-\csc\theta}.$$

The minus sign is chosen so $\cos\theta > 0$. (Don't forget that $\csc\theta < 0$ in this example.)

Having $\sin\theta$ and $\cos\theta$, it is now easy to express the remaining three functions in terms of $\csc\theta$:

$$\tan\theta = \frac{\sin\theta}{\cos\theta} = \left(\frac{1}{\csc\theta}\right)\Big/\left(\frac{\sqrt{\csc^2\theta-1}}{-\csc\theta}\right) = \frac{-1}{\sqrt{\csc^2\theta-1}}$$

$$\cot\theta = \frac{1}{\tan\theta} = -\sqrt{\csc^2\theta-1} \qquad \sec\theta = \frac{1}{\cos\theta} = \frac{-\csc\theta}{\sqrt{\csc^2\theta-1}}$$

Answer

$\sin\theta$	$\cos\theta$	$\tan\theta$	$\cot\theta$	$\sec\theta$
$\dfrac{1}{\csc\theta}$	$-\dfrac{\sqrt{\csc^2\theta-1}}{\csc\theta}$	$\dfrac{-1}{\sqrt{\csc^2\theta-1}}$	$-\sqrt{\csc^2\theta-1}$	$\dfrac{-\csc\theta}{\sqrt{\csc^2\theta-1}}$

A Geometric Method

Let $f(\theta)$ be any one of the six functions $\sin\theta, \cdots, \csc\theta$. Then $f(\theta)$ equals a quotient of two of the three numbers x, y, r, where (x, y) lies on the terminal side of θ. By setting the denominator equal to ± 1, according to the quadrant of θ, we easily express x, y, r all in terms of $f(\theta)$. Then any other of the six trig functions, another quotient, is readily expressed in terms of $f(\theta)$.

EXAMPLE 6 Express $\sec\theta$ in terms of $\csc\theta$ for θ in the first quadrant.

Solution For the first quadrant, all we need is a right triangle (Fig. 1). Since $\csc\theta = \text{hyp}/\text{opp}$, we choose $\text{opp} = 1$. Then $\text{hyp} = \csc\theta$ and

$$\text{adj} = \sqrt{\text{hyp}^2 - \text{opp}^2} = \sqrt{\csc^2\theta - 1}.$$

Therefore $$\sec\theta = \frac{\text{hyp}}{\text{adj}} = \frac{\csc\theta}{\sqrt{\csc^2\theta - 1}}.$$

Answer $\sec\theta = (\csc\theta)/\sqrt{\csc^2\theta - 1}$.

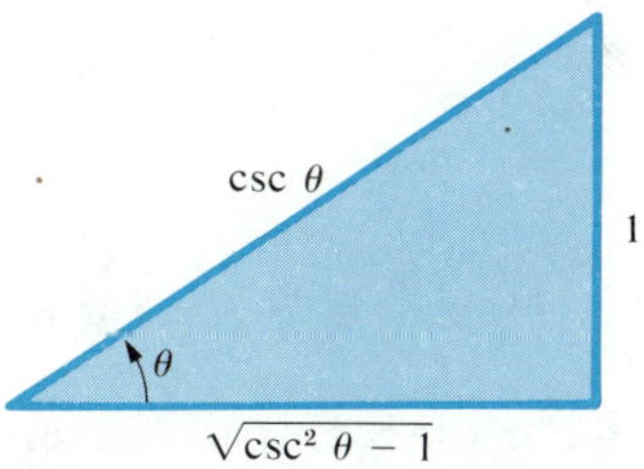

Fig. 1

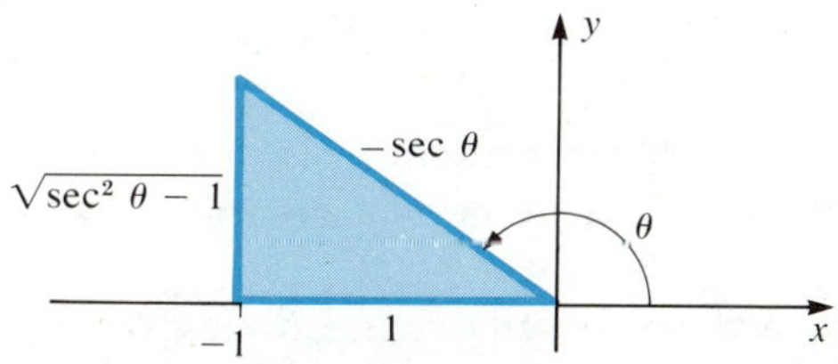

Fig. 2

EXAMPLE 7 Express $\csc\theta$ in terms of $\sec\theta$ for θ in the second quadrant.

Solution In the second quadrant, $x < 0$ and $y > 0$. Since $\sec\theta = r/x$, we choose $x = -1$. Then $r = -\sec\theta$ and

$$y = \sqrt{r^2 - x^2} = \sqrt{\sec^2\theta - 1}.$$

See Fig. 2. Therefore $$\csc\theta = \frac{r}{y} = \frac{-\sec\theta}{\sqrt{\sec^2\theta - 1}}.$$

Answer $\csc\theta = -(\sec\theta)/\sqrt{\sec^2\theta - 1}$

Exercises

Show that

1 $\sin\theta\sec\theta = \tan\theta$

2 $\sec\theta\cot\theta = \csc\theta$

3 $\tan\theta\csc\theta = \sec\theta$

4 $\cos\theta\csc\theta = \cot\theta$

5 $\dfrac{\sin^2\theta}{1 - \sin^2\theta} = \tan^2\theta$

6 $\dfrac{1 - \cos^2\theta}{\tan^2\theta} = \cos^2\theta$

7 $\dfrac{\sec^2\theta}{\sec^2\theta - 1} = \dfrac{1}{\sin^2\theta}$

8 $\dfrac{\cot^2\theta + 1}{\cot^2\theta} = \dfrac{1}{\cos^2\theta}$

9 $\dfrac{\sin\theta}{1 - \sin^2\theta} = \tan\theta\sec\theta$

10 $\dfrac{1 - \cos^2\theta}{\cos\theta} = \sin\theta\tan\theta$

11 $\sec^4\theta - \sec^2\theta = \sec^2\theta\tan^2\theta$

12 $\cos^3\theta - \cos\theta = -\sin^2\theta\cos\theta$

13 $1 - (\cos\theta - \sin\theta)^2 = 2\sin\theta\cos\theta$

14 $1 - \sin\theta\cos\theta\tan\theta = \cos^2\theta$

15 $\dfrac{\cos^2\theta}{1 + \sin\theta} = 1 - \sin\theta$

16 $\dfrac{\sec\theta - 1}{\tan^2\theta} = \dfrac{1}{\sec\theta + 1}$.

Find the exact values of the remaining trigonometric functions

17 $\cos\theta = -\frac{4}{5}$ quadrant 3
18 $\sin\theta = -\frac{3}{5}$ quadrant 4
19 $\tan\theta = 3$ quadrant 1
20 $\cot\theta = \frac{1}{2}$ quadrant 1
21 $\sec\theta = -2$ quadrant 2
22 $\csc\theta = 3$ quadrant 2
23 $\cot\theta = -\frac{3}{2}$ quadrant 4
24 $\tan\theta = 4$ quadrant 3.

Express the first function in terms of the second; adjust the sign so the formula is correct in the given quadrant

25 $\sin\theta$ in terms of $\cot\theta$ quadrant 2
26 $\sin\theta$ in terms of $\cot\theta$ quadrant 4
27 $\csc\theta$ in terms of $\cot\theta$ quadrant 1
28 $\csc\theta$ in terms of $\cot\theta$ quadrant 3
29 $\sec\theta$ in terms of $\cot\theta$ quadrant 3
30 $\cos\theta$ in terms of $\cot\theta$ quadrant 2
31 $\tan\theta$ in terms of $\sin\theta$ quadrant 4
32 $\sec\theta$ in terms of $\sin\theta$ quadrant 1
33 $\csc\theta$ in terms of $\cos\theta$ quadrant 2
34 $\tan\theta$ in terms of $\cos\theta$ quadrant 3.

2 PROVING IDENTITIES

Because there are so many relations among the trigonometric functions, two trigonometric expressions may look quite different, yet be identical. For example, $\tan\theta$ is identical to

$$\frac{\sin\theta}{\cos\theta} \qquad \frac{1}{\cot\theta} \qquad \sin\theta\sec\theta \qquad \cot\theta(\sec^2\theta - 1)$$

and to many other expressions. If, for instance, the expression $(\sec^2\theta - 1)\cot\theta$ were involved in a numerical computation, it would save work to replace it by $\tan\theta$.

Suppose we are given two algebraic expressions A and B in the basic trigonometric functions $\sin\theta$, $\cos\theta$, etc., and we want to prove the identity $A = B$ (for all values of θ for which the expressions are defined). If the identity is simple enough, we can sometimes convert A into B by a sequence of steps.

EXAMPLE 1 Prove $\cos\theta + \sin\theta\tan\theta = \sec\theta$.

Solution

$$\begin{aligned}\cos\theta + \sin\theta\tan\theta &= \cos\theta + \sin\theta\,\frac{\sin\theta}{\cos\theta}\\ &= \cos\theta + \frac{\sin^2\theta}{\cos\theta} = \frac{\cos^2\theta + \sin^2\theta}{\cos\theta}\\ &= \frac{1}{\cos\theta} = \sec\theta.\end{aligned}$$

In a more complicated identity, it may be hard to transform one side into the other. A systematic procedure for proving such an identity is to express everything in A and B in terms of sines and cosines only, then to simplify. If the results are the same, then $A = B$ is proved.

EXAMPLE 2 Prove $\dfrac{\csc\theta + 1}{\cot\theta} = \dfrac{1}{\sec\theta - \tan\theta}$.

Solution Express the left-hand side in terms of sines and cosines:

$$A = \frac{\csc\theta + 1}{\cot\theta} = \frac{\dfrac{1}{\sin\theta} + 1}{\dfrac{\cos\theta}{\sin\theta}} = \frac{1 + \sin\theta}{\cos\theta}$$

The right-hand side is

$$B = \frac{1}{\sec\theta - \tan\theta} = \frac{1}{\dfrac{1}{\cos\theta} - \dfrac{\sin\theta}{\cos\theta}} = \frac{\cos\theta}{1 - \sin\theta}.$$

To prove $A = B$, we must show that

$$\frac{1+\sin\theta}{\cos\theta} = \frac{\cos\theta}{1-\sin\theta}, \quad \text{that is,} \quad (1+\sin\theta)(1-\sin\theta) = \cos^2\theta.$$

But $(1+\sin\theta)(1-\sin\theta) = 1 - \sin^2\theta = \cos^2\theta$, so indeed $A = B$.

One way to prove an identity $A = B$ is to prove a string of equalities

$$A = B_1 = B_2 = \cdots = B_n = B.$$

Actually, this is a string of equivalent identities

$$A = B_1 \qquad A = B_2 \cdots A = B_n \qquad A = B$$

that transform A into B one step at a time. This is how we did Example 1.

This method is fine if you can do it. However, for more complicated identities, as in Example 2, it can be difficult to apply. In general, you can just as well prove any equivalent relation such as $A + C = B + C$ or $(A + C)D = (B + C)D$, provided $D \neq 0$. In fact, if you can find a chain of equivalent relations

$$A = B, \quad A_1 = B_1, \quad A_2 = B_2, \quad \cdots, \quad A_n = B_n,$$

and the last one is an identity, then so is the first, and the proof is complete.

Warning Be sure that each step is *reversible*. Whatever reasoning leads from $A_j = B_j$ to $A_{j+1} = B_{j+1}$ must work in reverse gear; that is, the truth of $A_{j+1} = B_{j+1}$ must imply the truth of $A_j = B_j$, not just the other way. In particular, squaring both sides is invalid; for instance the chain

$$-\sin\theta = \sqrt{1-\cos^2\theta}, \qquad \sin^2\theta = 1 - \cos^2\theta, \qquad \sin\theta = \sqrt{1-\cos^2\theta}$$

certainly requires some qualification.

A convenient shorthand is to abbreviate $\cos\theta$ by c and $\sin\theta$ by s. This not only saves a lot of writing, but also decreases the chance of error.

Example 3 Prove $\dfrac{1-\cos\theta}{1+\cos\theta} = (\csc\theta - \cot\theta)^2$.

Solution Express everything in terms of $c = \cos\theta$ and $s = \sin\theta$; then form a chain of equivalent relations:

$$\frac{1-c}{1+c} = \left(\frac{1}{s} - \frac{c}{s}\right)^2 \qquad \frac{1-c}{1+c} = \frac{(1-c)^2}{s^2}$$

$$s^2(1-c) = (1+c)(1-c)^2 = (1+c)(1-c)(1-c)$$
$$= (1-c^2)(1-c) = s^2(1-c).$$

Converting to sines and cosines is not the only approach, nor necessarily the quickest in every case.

Example 4 Prove $\dfrac{\tan\alpha - \tan\beta}{\cot\alpha - \cot\beta} = -\tan\alpha\tan\beta$.

Solution Here it is easiest to express the cotangents in terms of tangents. The chain of relations is

$$\begin{aligned}\tan\alpha - \tan\beta &= -\tan\alpha\tan\beta(\cot\alpha - \cot\beta)\\ &= -\tan\alpha\tan\beta\left(\frac{1}{\tan\alpha} - \frac{1}{\tan\beta}\right)\\ &= -\tan\beta + \tan\alpha.\end{aligned}$$

Each student develops his or her own style for proving identities. Some like to start with the left-hand side and transform it into the right-hand side. Others start with the more complicated looking side and modify it. We recommend a middle course; fiddle around a little with both sides and try to find a middle meeting ground. Sometimes if you want to prove $A/B = C/D$, it pays to clear of fractions first, then attempt to prove the *equivalent* identity $AD = BC$.

Exercises

Simplify

1 $\sin^4\theta - \cos^4\theta$

2 $(\sin\theta + \cos\theta)^2 - 2\sin\theta\cos\theta$

3 $\dfrac{1 - \cos^2\theta}{\sin\theta}$

4 $\dfrac{1 + \tan^2\theta}{\tan^2\theta}$.

5 $(\sec^2\theta - \tan^2\theta)^3$

6 $\tan^4\theta\cot^5\theta$

7 $\cos^2\theta + \cos^2(\frac{1}{2}\pi - \theta)$

8 $\cot\theta\sec\theta$.

Prove the identity

9 $\tan\theta + \cot\theta = \sec\theta\csc\theta$

10 $2\cos^2\theta - 1 = 1 - 2\sin^2\theta$

11 $(1 - \sin\theta)(1 + \csc\theta) = \cos\theta\cot\theta$

12 $(\cos\theta + 1)(\sec\theta - 1) = \sin\theta\tan\theta$

13 $\sec\theta - \cos\theta = \sin\theta\tan\theta$

14 $\csc\theta = \sin\theta + \cos\theta\cot\theta$

15 $\sec^2\theta - \csc^2\theta = \tan^2\theta - \cot^2\theta$

16 $\sin^2\theta\cot^2\theta + \cos^2\theta\tan^2\theta = 1$

17 $\sec^2\theta + \csc^2\theta = \sec^2\theta\csc^2\theta$

18 $\tan^2\theta - \sin^2\theta = \tan^2\theta\sin^2\theta$

19 $\sec^4\theta - \tan^4\theta = 1 + 2\tan^2\theta$

20 $\sin^4\theta + \cos^4\theta + 2\sin^2\theta\cos^2\theta = 1$

21 $\cot^4\theta + \cot^2\theta = \csc^4\theta - \csc^2\theta$

22 $\cos^6\theta + 3\cos^2\theta\sin^2\theta + \sin^6\theta = 1$

23 $(\cos\theta - \sin\theta)^2 + (\cos\theta + \sin\theta)^2 = 2$

24 $(x\cos\theta + y\sin\theta)^2 + (-x\sin\theta + y\cos\theta)^2 = x^2 + y^2$

25 $\dfrac{1 - \tan\theta}{1 + \tan\theta} = \dfrac{\cot\theta - 1}{\cot\theta + 1}$

26 $\dfrac{1}{1 - \sin\theta} - \dfrac{1}{1 + \sin\theta} = 2\tan\theta\sec\theta$

27 $\dfrac{\tan\theta}{\sec\theta - \cos\theta} = \dfrac{\sec\theta}{\tan\theta}$

28 $\dfrac{\sin\theta}{\csc\theta + \cot\theta} = 1 - \cos\theta$

29 $\sec\theta + \csc\theta = (\tan\theta + \cot\theta)(\cos\theta + \sin\theta)$

30 $\dfrac{1 + \cos\theta}{\sin\theta} + \dfrac{\sin\theta}{1 + \cos\theta} = 2\csc\theta$

31 $\dfrac{\tan\theta - \sin\theta\cos\theta}{\sec\theta} = \sin^3\theta$

32 $\dfrac{1 + 2\sin\theta\cos\theta}{\cos^2\theta} = (1 + \tan\theta)^2$

33 $\cos^4\theta - \cos^2\theta = \sin^4\theta - \sin^2\theta$

34 $\dfrac{\cos^3\theta - \sin^3\theta}{\cos\theta - \sin\theta} = 1 + \sin\theta\cos\theta$

35 $(1 - \tan\theta)(1 - \cot\theta) = 2 - \dfrac{1}{\sin\theta\cos\theta}$

36 $\tan\theta - \cot\theta = \dfrac{1 - 2\cos^2\theta}{\sin\theta\cos\theta}$

37 $\dfrac{\cos\theta}{1 - \tan\theta} + \dfrac{\sin\theta}{1 - \cot\theta} = \sin\theta + \cos\theta$

38 $\dfrac{\tan^2\theta + 1}{\tan^2\theta - 1} = \dfrac{\sec^2\theta + \csc^2\theta}{\sec^2\theta - \csc^2\theta}$

39 $\dfrac{\cos\alpha - \sin\beta}{\cos\beta - \sin\alpha} = \dfrac{\cos\beta + \sin\alpha}{\cos\alpha + \sin\beta}$

40 $\dfrac{\tan\alpha + \tan\beta}{\cot\alpha + \cot\beta} = \dfrac{\tan\alpha\tan\beta - 1}{1 - \cot\alpha\cot\beta}$

3 THE ADDITION LAWS

The **addition laws** for the sine and cosine are formulas that express $\sin(\alpha + \beta)$ and $\cos(\alpha + \beta)$ in terms of $\sin\alpha$, $\sin\beta$, $\cos\alpha$, and $\cos\beta$:

$$\sin(\alpha + \beta) = \sin\alpha\cos\beta + \cos\alpha\sin\beta$$
$$\cos(\alpha + \beta) = \cos\alpha\cos\beta - \sin\alpha\sin\beta.$$

These addition formulas are very basic in mathematics and should be memorized. Before we discuss a proof, let us note the equivalent forms, obtained by substituting $-\beta$ for β and using $\cos(-\beta) = \cos\beta$ and $\sin(-\beta) = -\sin\beta$:

$$\sin(\alpha - \beta) = \sin\alpha\cos\beta - \cos\alpha\sin\beta$$
$$\cos(\alpha - \beta) = \cos\alpha\cos\beta + \sin\alpha\sin\beta.$$

(The reasoning is reversible: the same substitution $\beta \longrightarrow -\beta$ takes the formulas of the second box into those of the first.)

Another point to note is that the addition law for the cosine implies that for the sine (and vice versa). For suppose $\cos(\alpha - \beta) = \cos\alpha\cos\beta + \sin\alpha\sin\beta$ is true for all α and β. Replace α by $\frac{1}{2}\pi - \alpha$ and use the relations

$$\cos(\tfrac{1}{2}\pi - \theta) = \sin\theta \qquad \sin(\tfrac{1}{2}\pi - \theta) = \cos\theta.$$

We have

$$\cos(\tfrac{1}{2}\pi - \alpha - \beta) = \cos(\tfrac{1}{2}\pi - \alpha)\cos\beta + \sin(\tfrac{1}{2}\pi - \alpha)\sin\beta$$

hence

$$\sin(\alpha + \beta) = \sin\alpha\cos\beta + \cos\alpha\sin\beta.$$

Thus we shall have proved both addition laws when we have proved the cosine one.

Proof of the Cosine Addition Law

Take α and β in standard position (Fig. 1a, next page). Their terminal sides meet the unit circle in

$$A = (\cos\alpha, \sin\alpha) \quad \text{and} \quad B = (\cos\beta, \sin\beta).$$

By the distance formula, the chord length $\overline{AB}$ satisfies

$$\begin{aligned}\overline{AB}^2 &= (\cos\alpha - \cos\beta)^2 + (\sin\alpha - \sin\beta)^2\\ &= (\cos^2\alpha + \sin^2\alpha) + (\cos^2\beta + \sin^2\beta)\\ &\quad - 2(\cos\alpha\cos\beta + \sin\alpha\sin\beta),\end{aligned}$$

hence

$$\overline{AB}^2 = 2 - 2(\cos\alpha\cos\beta + \sin\alpha\sin\beta).$$

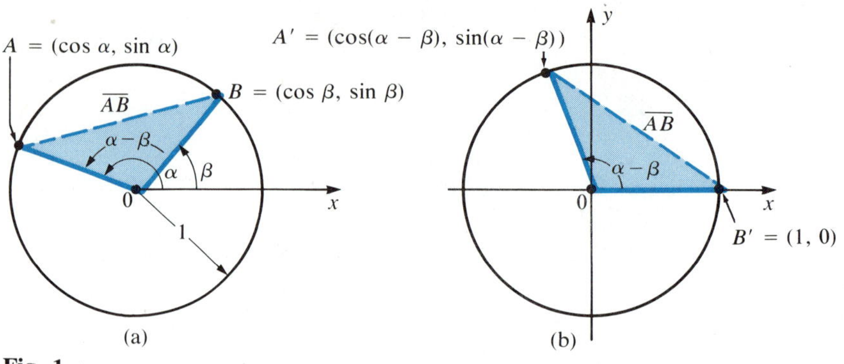

Fig. 1

Now let us *rotate* the triangle $0AB$ about 0 through angle $-\beta$. The resulting triangle $0A'B'$, congruent to $0AB$, has

$$A' = (\cos(\alpha - \beta), \sin(\alpha - \beta)) \quad \text{and} \quad B' = (1, 0).$$

In other words, $0A'$ is the terminal side of $\alpha' = \alpha - \beta$ and $0B'$ is the terminal side of $\beta' = 0$. By the formula for chord length, with α and β replaced by α' and β',

$$\overline{A'B'}^2 = 2 - 2\cos(\alpha - \beta).$$

But $\overline{A'B'} = \overline{AB}$ because rotation preserves length. Hence

$$2 - 2\cos(\alpha - \beta) = 2 - 2(\cos\alpha\cos\beta - \sin\alpha\sin\beta),$$

so

$$\cos(\alpha - \beta) = \cos\alpha\cos\beta - \sin\alpha\sin\beta.$$

Addition Law for the Tangent

Addition formulas for the other functions can be derived from those for sine and cosine. The most important is the formula for tangent:

$$\tan(\alpha + \beta) = \frac{\tan\alpha + \tan\beta}{1 - \tan\alpha\tan\beta} \qquad (\tan\alpha\tan\beta \neq 1).$$

Proof

$$\tan(\alpha + \beta) = \frac{\sin(\alpha + \beta)}{\cos(\alpha + \beta)} = \frac{\sin\alpha\cos\beta + \cos\alpha\sin\beta}{\cos\alpha\cos\beta - \sin\alpha\sin\beta}.$$

Divide numerator and denominator by $\cos\alpha\cos\beta$:

$$\tan(\alpha + \beta) = \frac{\dfrac{\sin\alpha}{\cos\alpha} + \dfrac{\sin\beta}{\cos\beta}}{1 - \dfrac{\sin\alpha\sin\beta}{\cos\alpha\cos\beta}} = \frac{\tan\alpha + \tan\beta}{1 - \tan\alpha\tan\beta}.$$

EXAMPLE 1 Find exact values for

(a) $\cos 75°$ (b) $\tan 75°$.

Solution (a) Since $75 = 45 + 30$, we can use the addition laws to express $\cos 75°$ in terms of $\sin 45°$, $\cos 45°$, $\sin 30°$, and $\cos 30°$:

$$\cos 75^\circ = \cos(45^\circ + 30^\circ) = \cos 45^\circ \cos 30^\circ - \sin 45^\circ \sin 30^\circ$$
$$= (\tfrac{1}{2}\sqrt{2})(\tfrac{1}{2}\sqrt{3}) - (\tfrac{1}{2}\sqrt{2})(\tfrac{1}{2}) = \tfrac{1}{4}\sqrt{6} - \tfrac{1}{4}\sqrt{2}.$$

(b) By the addition law for tangent,

$$\tan 75^\circ = \frac{\tan 45^\circ + \tan 30^\circ}{1 - \tan 45^\circ \tan 30^\circ} = \frac{1 + \frac{1}{3}\sqrt{3}}{1 - 1\cdot\frac{1}{3}\sqrt{3}} = \frac{3 + \sqrt{3}}{3 - \sqrt{3}}.$$

For a neater answer, we rationalize the denominator:

$$\tan 75^\circ = \frac{3 + \sqrt{3}}{3 - \sqrt{3}} = \frac{3 + \sqrt{3}}{3 - \sqrt{3}}\,\frac{3 + \sqrt{3}}{3 + \sqrt{3}} = \frac{9 + 6\sqrt{3} + 3}{9 - 3}$$
$$= \frac{12 + 6\sqrt{3}}{6} = 2 + \sqrt{3}.$$

Answer (a) $\frac{1}{4}(\sqrt{6} - \sqrt{2})$ (b) $2 + \sqrt{3}$

Example 2 Suppose α and β are angles in the first quadrant, $\tan\alpha = \frac{1}{2}$ and $\tan\beta = \frac{1}{3}$. Find $\alpha + \beta$.

Solution By the addition law for tangent,

$$\tan(\alpha + \beta) = \frac{\tan\alpha + \tan\beta}{1 - \tan\alpha\tan\beta} = \frac{\frac{1}{2} + \frac{1}{3}}{1 - \frac{1}{2}\cdot\frac{1}{3}} = \frac{\frac{5}{6}}{\frac{5}{6}} = 1.$$

Thus $\tan(\alpha + \beta) - 1$. Therefore $\alpha + \beta$ is one of the angles $\frac{1}{4}\pi + 2\pi n$ or $\frac{5}{4}\pi + 2\pi n$. Which one? Well, α and β are first quadrant angles, so $0 < \alpha + \beta < \pi$. The only possibility in this range is $\alpha + \beta = \frac{1}{4}\pi$.

Answer $\frac{1}{4}\pi$

Double-Angle Formulas

There are important special cases of the addition formulas that express $\sin 2\theta$, $\cos 2\theta$, and $\tan 2\theta$ in terms of $\sin\theta$, $\cos\theta$, and $\tan\theta$. In the formulas for $\sin(\alpha + \beta)$, $\cos(\alpha + \beta)$, and $\tan(\alpha + \beta)$, set $\alpha = \beta = \theta$. The results are the **double-angle formulas:**

$$\sin 2\theta = 2\sin\theta\cos\theta \qquad \cos 2\theta = \cos^2\theta - \sin^2\theta$$
$$\tan 2\theta = \frac{2\tan\theta}{1 - \tan^2\theta}.$$

The second formula has two useful alternative forms. First we replace $\cos^2\theta$ by $1 - \sin^2\theta$:

$$\cos 2\theta = \cos^2\theta - \sin^2\theta = (1 - \sin^2\theta) - \sin^2\theta = 1 - 2\sin^2\theta.$$

Next we replace $\sin^2\theta$ by $1 - \cos^2\theta$:

$$\cos 2\theta = \cos^2\theta - \sin^2\theta = \cos^2\theta - (1 - \cos^2\theta) = 2\cos^2\theta - 1.$$

$$\cos 2\theta = 1 - 2\sin^2\theta = 2\cos^2\theta - 1.$$

Example 3 Suppose θ is an angle in the second quadrant and $\sin\theta = \frac{3}{5}$. Find

(a) $\cos 2\theta$ (b) $\sin 2\theta$ (c) the quadrant of 2θ.

Solution (a) Use the double angle formula for cosine, preferably the form that involves only $\sin\theta$:

$$\cos 2\theta = 1 - 2\sin^2\theta = 1 - 2(\tfrac{3}{5})^2 = 1 - \tfrac{18}{25} = \tfrac{7}{25}.$$

(b) The double angle formula for sine requires both $\sin\theta$ and $\cos\theta$. Since θ is in the second quadrant, $\cos\theta < 0$. Therefore

$$\cos\theta = -\sqrt{1 - \sin^2\theta} = -\sqrt{1 - (\tfrac{3}{5})^2} = -\sqrt{1 - \tfrac{9}{25}}$$
$$= -\sqrt{\tfrac{16}{25}} = -\tfrac{4}{5}.$$

Now apply the double angle formula:

$$\sin 2\theta = 2\sin\theta\cos\theta = 2(\tfrac{3}{5})(-\tfrac{4}{5}) = -\tfrac{24}{25}.$$

(c) Since $\cos 2\theta > 0$ and $\sin 2\theta < 0$, the terminal side of 2θ lies in the fourth quadrant.

Answer (a) $\tfrac{7}{25}$ (b) $-\tfrac{24}{25}$ (c) 4-th quadrant

EXAMPLE 4 Express $\sin 3\theta$ in terms of $\sin\theta$ and $\cos\theta$.

Solution Write $3\theta = \theta + 2\theta$, and use the addition laws to express $\sin 3\theta$ in terms of $\sin\theta$, $\cos\theta$, $\sin 2\theta$, and $\cos 2\theta$,

$$\sin 3\theta = \sin(\theta + 2\theta) = \sin\theta\cos 2\theta + \cos\theta\sin 2\theta.$$

Now apply the double-angle formulas for $\sin 2\theta$ and $\cos 2\theta$:

$$\sin 3\theta = \sin\theta[\cos^2\theta - \sin^2\theta] + \cos\theta[2\sin\theta\cos\theta]$$
$$= 3\sin\theta\cos^2\theta - \sin^3\theta.$$

If you prefer the answer in terms of $\sin\theta$ alone, you can replace $\cos^2\theta$ by $1 - \sin^2\theta$.

Answer $\sin 3\theta = 3\sin\theta\cos^2\theta - \sin^3\theta = 3\sin\theta - 4\sin^3\theta.$

Remarkable! The addition laws for sine and cosine are perhaps the most striking formulas in elementary mathematics. The only formulas you have seen before vaguely resembling them are the addition laws for exponential and logarithm functions:

$$a^{x+y} = a^x a^y \qquad \log_a x + \log_a y = \log_a(xy).$$

But the sine and cosine addition laws express $\sin(\alpha + \beta)$ and $\cos(\alpha + \beta)$ in terms of *both* sines and cosines of α and β.

Note that all the formulas of this section are consequences of the addition laws.

Exercises

Use the addition laws to compute

1 $\cos(\theta + 2\pi)$ and $\sin(\theta + 2\pi)$

2 $\cos(\theta + \pi)$ and $\sin(\theta + \pi)$

3 $\cos(\theta + \frac{1}{2}\pi)$ and $\sin(\theta + \frac{1}{2}\pi)$

4 $\cos(\frac{1}{2}\pi - \theta)$ and $\sin(\frac{1}{2}\pi - \theta)$.

Express

5 $\cos(\theta + \frac{1}{4}\pi)$ and $\sin(\theta + \frac{1}{4}\pi)$ in terms of $\cos\theta$ and $\sin\theta$

6 $\cos(\theta + \frac{1}{6}\pi)$ and $\sin(\theta + \frac{1}{6}\pi)$ in terms of $\cos\theta$ and $\sin\theta$

7 $\cot(\alpha + \beta)$ in terms of $\cot\alpha$ and $\cot\beta$

8 $\tan(\alpha - \beta)$ in terms of $\tan\alpha$ and $\tan\beta$.

Find exactly

9 $\sin 75°$ **10** $\tan 105°$ **11** $\cos 15°$ **12** $\sin 15°$

13 $\sin 9° \cos 21° + \cos 9° \sin 21°$

14 $\sin 81° \cos 36° - \cos 81° \sin 36°$

15 $\cos \frac{3}{10}\pi \cos \frac{1}{20}\pi + \sin \frac{3}{10}\pi \sin \frac{1}{20}\pi$

16 $\cos 0.3\pi \cos 0.7\pi - \sin 0.3\pi \sin 0.7\pi$.

17 Suppose $0 < \alpha < \frac{1}{2}\pi$, $\pi < \beta < \frac{3}{2}\pi$, $\tan \alpha = \frac{1}{3}$ and $\tan \beta = \frac{1}{2}$. Find $\alpha + \beta$.

18 Suppose α, β, and γ are in the first quadrant, $\tan \alpha = \frac{1}{3}$, $\tan \beta = \frac{1}{7}$, and $\tan \gamma = \frac{1}{2}$, Show that $\alpha + \beta = \gamma$.

19 Suppose $0 < \alpha < \frac{1}{2}\pi$, $\cos \alpha = \frac{1}{4}$, $\frac{1}{2}\pi < \beta < \pi$, and $\sin \beta = \frac{1}{5}$. Find $\sin(\beta - \alpha)$.

20 Suppose $\frac{1}{2}\pi < \alpha < \pi$, $\sin \alpha = \frac{2}{3}$, $\pi < \beta < \frac{3}{2}\pi$, and $\tan \beta = 2$. Find $\cos(\alpha + \beta)$.

21 Suppose θ is in the second quadrant and $\cos \theta = -\frac{1}{3}$. Find $\sin 2\theta$ and $\cos 2\theta$.

22 Suppose θ is in the fourth quadrant and $\sin \theta = -\frac{2}{3}$. Find $\sin 2\theta$ and $\cos 2\theta$.

23 Suppose $\tan \theta = 4$. Find $\cos 2\theta$ and $\tan 2\theta$.

24 Suppose $\tan \theta = 2$. Find $\sin 2\theta$ and $\tan 2\theta$.

Express

25 $\cos 3\theta$ in terms of $\cos \theta$

26 $\tan 3\theta$ in terms of $\tan \theta$

27 $\csc 3\theta$ in terms of $\csc \theta$
[Use Example 3.]

28 $\sec 3\theta$ in terms of $\sec \theta$
[Use Example 25.]

Use double-angle formulas to prove

29 $\sin 4\theta = 4 \sin \theta \cos \theta - 8 \sin^3 \theta \cos \theta$

30 $\cos 4\theta = 8 \cos^4 \theta - 8 \cos^2 \theta + 1$

31 $\cot 2\theta + \csc 2\theta = \cot \theta$

32 $\csc 2\theta - \cot 2\theta = \tan \theta$

33 $\tan \theta + \cot \theta = 2 \csc 2\theta$

34 $\sin \theta \cos \theta \cos 2\theta \cos 4\theta = \frac{1}{8} \sin 8\theta$.

35 If $0 < \theta < \frac{1}{4}\pi$, which is larger, $\tan 2\theta$ or $2 \tan \theta$?

36* If $\frac{1}{4}\pi < \theta < \frac{1}{3}\pi$, show that $\sin \theta < \sin 2\theta < \sqrt{2} \sin \theta$.

Prove

37 $\cot 2\theta = \dfrac{\cot^2 \theta - 1}{2 \cot \theta}$

38 $\sec 2\theta = \dfrac{\sec^2 \theta}{2 - \sec^2 \theta}$

39 $\cot(\alpha + \beta) = \dfrac{\cot \alpha \cot \beta - 1}{\cot \alpha + \cot \beta}$

40 $\tan 4\theta = \dfrac{4 \tan \theta - 4 \tan^3 \theta}{1 - 6 \tan^2 \theta + \tan^4 \theta}$

41 $\dfrac{1 + \cos 2\theta}{1 - \cos 2\theta} = \cot^2 \theta$

42 $\dfrac{\sin 2\theta + \sin \theta}{\cos 2\theta + \cos \theta + 1} = \tan \theta$

43 $\dfrac{\sec \theta - \csc \theta}{\sec \theta + \csc \theta} = \tan 2\theta - \sec 2\theta$

44 $\dfrac{\cot \theta - \tan \theta}{\cot \theta + \tan \theta} = \cos 2\theta$.

4 FURTHER IDENTITIES

Half-Angle Formulas

From the double-angle formulas, we can obtain half-angle formulas that express $\sin \frac{1}{2}\theta$, $\cos \frac{1}{2}\theta$, and $\tan \frac{1}{2}\theta$ in terms of $\sin \theta$ and $\cos \theta$. We start with the double-angle formulas

$$\cos 2\theta = 2 \cos^2 \theta - 1 = 1 - 2 \sin^2 \theta$$

and solve for $\sin \theta$ and $\cos \theta$:

$$\sin \theta = \pm\sqrt{\frac{1 - \cos 2\theta}{2}} \qquad \cos \theta = \pm\sqrt{\frac{1 + \cos 2\theta}{2}},$$

where the sign depends on the quadrant. Replacing θ by $\frac{1}{2}\theta$, we obtain the half-angle formulas:

$$\sin \tfrac{1}{2}\theta = \pm\sqrt{\frac{1-\cos\theta}{2}} \qquad \cos \tfrac{1}{2}\theta = \pm\sqrt{\frac{1+\cos\theta}{2}},$$

where the sign of each is chosen according to the quadrant of $\frac{1}{2}\theta$.

We could obtain a formula for $\tan \frac{1}{2}\theta$ by dividing the expressions for $\sin \frac{1}{2}\theta$ and $\cos \frac{1}{2}\theta$, but we avoid square roots by using a little trick:

$$\tan \tfrac{1}{2}\theta = \frac{\sin \frac{1}{2}\theta}{\cos \frac{1}{2}\theta} = \frac{\sin \frac{1}{2}\theta}{\cos \frac{1}{2}\theta} \cdot \frac{2\cos \frac{1}{2}\theta}{2\cos \frac{1}{2}\theta}.$$

By the double-angle formulas, the numerator is $\sin 2(\frac{1}{2}\theta) = \sin\theta$ and the denominator is

$$2\cos^2(\tfrac{1}{2}\theta) = 1 + [2\cos^2(\tfrac{1}{2}\theta) - 1] = 1 + \cos 2(\tfrac{1}{2}\theta) = 1 + \cos\theta.$$

Therefore

$$\tan \tfrac{1}{2}\theta = \frac{\sin\theta}{1+\cos\theta}.$$

This half-angle formula for the tangent has a geometric interpretation. In Fig. 1a, the inscribed angle is $\frac{1}{2}\theta$, half the corresponding central angle. From the right triangle (Fig. 1b), $\tan \frac{1}{2}\theta = \sin\theta/(1+\cos\theta)$.

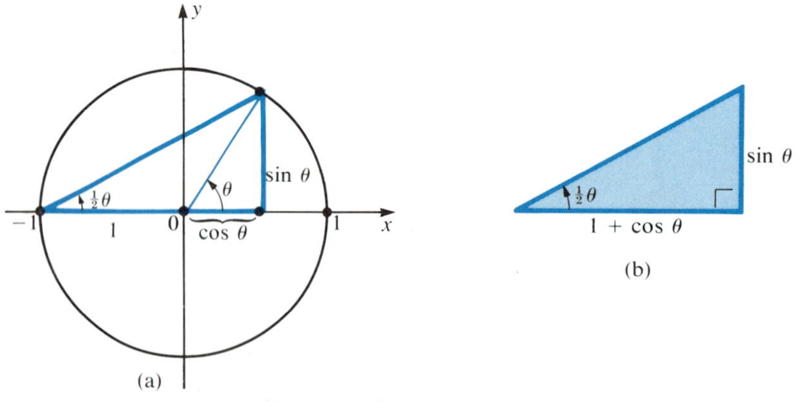

Fig. 1 $\tan \frac{1}{2}\theta = \dfrac{\text{opposite}}{\text{adjacent}} = \dfrac{\sin\theta}{1+\cos\theta}$

Identities expressing $\sin\theta$ and $\cos\theta$ in terms of $\tan \frac{1}{2}\theta$ are useful in calculus:

$$\sin\theta = \frac{2\tan \frac{1}{2}\theta}{1+\tan^2 \frac{1}{2}\theta} \qquad \cos\theta = \frac{1-\tan^2 \frac{1}{2}\theta}{1+\tan^2 \frac{1}{2}\theta}.$$

The proofs are easy:

$$\sin 2\theta = 2\sin\theta\cos\theta = \frac{2\sin\theta\cos\theta}{\sin^2\theta + \cos^2\theta} = \frac{2\dfrac{\sin\theta}{\cos\theta}}{\dfrac{\sin^2\theta}{\cos^2\theta} + 1}$$

$$= \frac{2\tan\theta}{1+\tan^2\theta}.$$

$$\cos 2\theta = \cos^2\theta - \sin^2\theta = \frac{\cos^2\theta - \sin^2\theta}{\cos^2\theta + \sin^2\theta} = \frac{1 - \dfrac{\sin^2\theta}{\cos^2\theta}}{1 + \dfrac{\sin^2\theta}{\cos^2\theta}}$$

$$= \frac{1 - \tan^2\theta}{1 + \tan^2\theta}.$$

Now replace θ by $\frac{1}{2}\theta$.

EXAMPLE 1 Find exact values for (a) $\cos 105°$ (b) $\tan \frac{1}{12}\pi$.

Solution (a) Since $105° = \frac{1}{2}(210°)$ and $210° = 180° + 30°$, use the half-angle formula for cosine, with the negative sign because $105°$ is in the second quadrant:

$$\cos 105° = -\sqrt{\frac{1 + \cos 210°}{2}} = -\sqrt{\frac{1 - \frac{1}{2}\sqrt{3}}{2}} = -\tfrac{1}{2}\sqrt{2 - \sqrt{3}}.$$

(b) Since $\frac{1}{12}\pi = \frac{1}{2}(\frac{1}{6}\pi)$, use the formula for $\tan \frac{1}{2}\theta$:

$$\tan \tfrac{1}{12}\pi = \frac{\sin \frac{1}{6}\pi}{1 + \cos \frac{1}{6}\pi} = \frac{\frac{1}{2}}{1 + \frac{1}{2}\sqrt{3}} = \frac{1}{2 + \sqrt{3}}.$$

Rationalizing the denominator gives us $\tan \frac{1}{12}\pi = 2 - \sqrt{3}$.

Answer (a) $-\frac{1}{2}\sqrt{2 - \sqrt{3}}$ (b) $2 - \sqrt{3}$

Sums of Sines and Cosines

We give two useful applications of the addition formulas. The first is to express any function

$$f(\theta) = a\cos\theta + b\sin\theta \qquad [(a, b) \neq (0, 0)]$$

as a constant times a single cosine or sine.

To do this, we choose an angle θ_0 in standard position whose terminal side passes through (a, b). Then

$$\cos\theta_0 = \frac{a}{\sqrt{a^2 + b^2}} \qquad \sin\theta_0 = \frac{b}{\sqrt{a^2 + b^2}}.$$

(The angle θ_0 is only determined up to a multiple of 2π. We can insist that $0 \leq \theta_0 < 2\pi$; then θ_0 is unique.)

Now we have

$$a\cos\theta + b\sin\theta$$

$$= \sqrt{a^2 + b^2}\left(\frac{a}{\sqrt{a^2 + b^2}}\cos\theta + \frac{b}{\sqrt{a^2 + b^2}}\sin\theta\right)$$

$$= \sqrt{a^2 + b^2}(\cos\theta_0\cos\theta + \sin\theta_0\sin\theta) = \sqrt{a^2 + b^2}\cos(\theta - \theta_0).$$

If $a^2 + b^2 > 0$, then

$$a\cos\theta + b\sin\theta = \sqrt{a^2 + b^2}\cos(\theta - \theta_0),$$

where θ_0 is the angle satisfying $0 \leq \theta_0 < 2\pi$ and

$$\cos\theta_0 = \frac{a}{\sqrt{a^2 + b^2}} \qquad \sin\theta_0 = \frac{b}{\sqrt{a^2 + b^2}}.$$

EXAMPLE 2 Express $-\cos\theta + \sqrt{3}\sin\theta$ as a constant times a cosine.

Solution Here $a = -1$, $b = \sqrt{3}$, and $a^2 + b^2 = 4$, so we write

$$-\cos\theta + \sqrt{3}\sin\theta = 2(-\tfrac{1}{2}\cos\theta + \tfrac{1}{2}\sqrt{3}\sin\theta)$$

We choose θ_0 to satisfy $0 \le \theta_0 < 2\pi$ and

$$\cos\theta_0 = -\tfrac{1}{2} \qquad \sin\theta_0 = \tfrac{1}{2}\sqrt{3}$$

Clearly $\theta_0 = \frac{2}{3}\pi$.

Answer $2\cos(\theta - \frac{2}{3}\pi)$.

Remark Suppose instead we choose $\theta_0 = \frac{1}{6}\pi$. Then $\sin\theta_0 = \frac{1}{2}$ and $\cos\theta_0 = \frac{1}{2}\sqrt{3}$, so the given function is

$$2(-\sin\theta_0\cos\theta + \cos\theta_0\sin\theta) = 2\sin(\theta - \theta_0).$$

In general, each function $a\cos\theta + b\sin\theta$ can be expressed as a constant multiple of a cosine or a sine.

EXAMPLE 3 Prove that $|-\cos\theta + \sqrt{3}\sin\theta| \le 2$ for all values of θ.

Solution By the result of Example 2,

$$|-\cos\theta + \sqrt{3}\sin\theta| = |2\cos(\theta - \tfrac{2}{3}\pi)| = 2|\cos(\theta - \tfrac{2}{3}\pi)| \le 2\cdot 1 = 2.$$

Products of Sines and Cosines

The second application of the addition formulas is to express products of sines and cosines as sums. First we add the formulas

$$\sin(\alpha + \beta) = \sin\alpha\cos\beta + \cos\alpha\sin\beta$$
$$\sin(\alpha - \beta) = \sin\alpha\cos\beta - \cos\alpha\sin\beta,$$

to obtain

$$\sin(\alpha + \beta) + \sin(\alpha - \beta) = 2\sin\alpha\cos\beta.$$

Next we add and subtract the formulas

$$\cos(\alpha + \beta) = \cos\alpha\cos\beta - \sin\alpha\sin\beta$$
$$\cos(\alpha - \beta) = \cos\alpha\cos\beta + \sin\alpha\sin\beta,$$

to obtain

$$\cos(\alpha + \beta) + \cos(\alpha - \beta) = 2\cos\alpha\cos\beta$$
$$\cos(\alpha - \beta) - \cos(\alpha + \beta) = 2\sin\alpha\sin\beta.$$

We interpret the results as formulas for products of sines and cosines:

$$\sin\alpha\sin\beta = \tfrac{1}{2}[\cos(\alpha - \beta) - \cos(\alpha + \beta)]$$
$$\cos\alpha\cos\beta = \tfrac{1}{2}[\cos(\alpha + \beta) + \cos(\alpha - \beta)]$$
$$\sin\alpha\cos\beta = \tfrac{1}{2}[\sin(\alpha + \beta) + \sin(\alpha - \beta)].$$

Examples

(1) $\sin 75^\circ \sin 15^\circ = \frac{1}{2}(\cos 60^\circ - \cos 90^\circ) = \frac{1}{2}(\frac{1}{2} - 0) = \frac{1}{4}$

(2) $\sin 105^\circ \cos 75^\circ = \frac{1}{2}(\sin 180^\circ + \sin 30^\circ) = \frac{1}{2}(0 + \frac{1}{2}) = \frac{1}{4}$

(3) $\cos 105° \cos 75° = \frac{1}{2}(\cos 180° + \cos 30°) = \frac{1}{2}(-1 + \frac{1}{2}\sqrt{3})$

Sometimes we go the other way and express the sum or difference of two sines or two cosines as a product. We write

$$\alpha = \tfrac{1}{2}(\alpha + \beta) + \tfrac{1}{2}(\alpha - \beta) = \gamma + \delta$$
$$\beta = \tfrac{1}{2}(\alpha + \beta) - \tfrac{1}{2}(\alpha - \beta) = \gamma - \delta.$$

where $\gamma = \frac{1}{2}(\alpha + \beta)$ and $\delta = \frac{1}{2}(\alpha - \beta)$. Then

$$\begin{aligned}\sin\alpha + \sin\beta &= \sin(\gamma + \delta) + \sin(\gamma - \delta)\\ &= 2\sin\gamma\cos\delta\\ &= 2\sin\tfrac{1}{2}(\alpha + \beta)\cos\tfrac{1}{2}(\alpha - \beta).\end{aligned}$$

In this way we derive four formulas:

$$\begin{aligned}\sin\alpha + \sin\beta &= 2\sin\tfrac{1}{2}(\alpha + \beta)\cos\tfrac{1}{2}(\alpha - \beta)\\ \sin\alpha - \sin\beta &= 2\cos\tfrac{1}{2}(\alpha + \beta)\sin\tfrac{1}{2}(\alpha - \beta)\\ \cos\alpha + \cos\beta &= 2\cos\tfrac{1}{2}(\alpha + \beta)\cos\tfrac{1}{2}(\alpha - \beta)\\ \cos\alpha - \cos\beta &= -2\sin\tfrac{1}{2}(\alpha + \beta)\sin\tfrac{1}{2}(\alpha - \beta).\end{aligned}$$

The last two batches of formulas have important applications in situations where one vibration is imposed on another, such as in the modulation of radio signals, and in the phenomenon of beats in acoustics. See Section 8.

Exercises

Show that

1 $\sin 22.5° = \frac{1}{2}\sqrt{2 - \sqrt{2}}$ **2** $\tan 22.5° = \sqrt{2} - 1$ **3** $\tan 67.5° = \sqrt{2} + 1$.

4 Compute $\cos 15°$ two ways: by half-angle formulas and by addition laws. Conclude that $\sqrt{2 + \sqrt{3}} = \frac{1}{2}(\sqrt{6} + \sqrt{2})$.

Prove

5 $\tan\frac{1}{2}\theta = \dfrac{1 - \cos\theta}{\sin\theta}$ **6** $\cot\frac{1}{2}\theta = \dfrac{1 + \cos\theta}{\sin\theta}$.

Express each combination as a positive constant times a cosine

7 $4\cos\theta + 3\sin\theta$ **8** $3\cos\theta - 4\sin\theta$

9 $\cos\theta - 2\sin\theta$ **10** $\cos\theta + 2\sin\theta$

11 $2\cos\theta + 3\sin\theta$ **12** $-2\cos\theta - 3\sin\theta$

13 $3\cos\theta - \sin\theta$ **14** $5\sin\theta$.

15 Prove that $|\sin\theta + \cos\theta| \le \sqrt{2}$.

16 (cont.) Find all θ for which $|\sin\theta + \cos\theta| = \sqrt{2}$.

17 Find the maximum value of $\cos\theta - 3\sin\theta$.

18 Find the minimum value of $4\sin\theta + 3\cos\theta$.

Express each function as half a sum (difference) of sines or cosines

19 $\sin\theta\sin 2\theta$ **20** $\cos 2\theta\cos\theta$

21 $\sin 3\theta\cos 4\theta$ **22** $\sin 101\theta\sin 100\theta$.

Express each function as a product of constants, sines, and cosines

23 $\sin\theta + \sin 2\theta$ **24** $\cos 2\theta - \cos\theta$

25 $\sin 5\theta - \sin 4\theta$ **26** $\sin 5\theta + \sin 4\theta$

27 $\cos 6\theta - \cos 5\theta$ **28** $\cos\theta + \cos 3\theta$.

Prove

29 $\sin 75° + \sin 15° = \frac{1}{2}\sqrt{6}$

30 $\sin 75° - \sin 15° = \frac{1}{2}\sqrt{2}$

31 $\cos 105° + \cos 15° = \frac{1}{2}\sqrt{2}$

32 $\cos 105° - \cos 15° = -\frac{1}{2}\sqrt{6}$.

33 From Exs. 29 and 30, find $\sin 75°$ and $\sin 15°$.

34 From Exs. 31 and 32, find $\cos 105°$ and $\cos 15°$.

Prove

35 $\tan\alpha + \tan\beta = \sin(\alpha+\beta)/\cos\alpha\cos\beta$

36 $\cos^4\theta = \frac{1}{8}(3 + 4\cos 2\theta + \cos 4\theta)$

37 $\sin^4\theta = \frac{1}{8}(3 - 4\cos 2\theta + \cos 4\theta)$

[*Hint* Use half-angle formulas twice.]

38 $\sin^2\alpha - \sin^2\beta = \sin(\alpha+\beta)\sin(\alpha-\beta)$

39 $\cos^2\alpha - \sin^2\beta = \cos(\alpha+\beta)\cos(\alpha-\beta)$

40 $1 + 2\cos\theta + 2\cos 2\theta + 2\cos 3\theta = \dfrac{\sin\frac{7}{2}\theta}{\sin\frac{1}{2}\theta}$.

[*Hint* Start by multiplying the left side by $\sin\frac{1}{2}\theta$.]

41 $\sin\theta\sin 2\theta\cos 2\theta = \frac{1}{4}(\cos 3\theta - \cos 5\theta)$

42 $\sin\theta\sin 2\theta\sin 3\theta = \frac{1}{4}(\sin 2\theta + \sin 4\theta - \sin 6\theta)$.

43 Set $t = \tan\frac{1}{2}\theta$. Express $\sin 2\theta$ in terms of t. [*Hint* Use half-angle formulas.]

44 Set $t = \tan\frac{1}{2}\theta$. Express $\cos 2\theta$ in terms of t. [*Hint* Use half-angle formulas.]

Prove

45 $\cot\alpha - \cot\beta = \dfrac{\sin(\beta-\alpha)}{\sin\alpha\sin\beta}$

46 $\dfrac{\tan\frac{1}{2}(\alpha+\beta)}{\tan\frac{1}{2}(\alpha-\beta)} = \dfrac{\sin\alpha+\sin\beta}{\sin\alpha-\sin\beta}$

47 $\dfrac{\sin\alpha-\sin\beta}{\cos\alpha+\cos\beta} = \tan\frac{1}{2}(\alpha-\beta)$

48 $\dfrac{\sin\alpha+\sin\beta}{\cos\alpha-\cos\beta} = \cot\frac{1}{2}(\beta-\alpha)$

49 $\dfrac{\sin\alpha+\sin\beta}{\cos\alpha+\cos\beta} = \tan\frac{1}{2}(\alpha+\beta)$

50 $\dfrac{\sin\alpha-\sin\beta}{\cos\alpha-\cos\beta} = -\cot\frac{1}{2}(\alpha+\beta)$.

51 Verify numerically that $\sin 40° + \sin 20° = \cos 10°$.

52 (cont.) Now prove the relation is true, not just true to 8 or 10 decimal places.

53 From our work on p. 256, we know that $\sin\theta < \theta < \tan\theta$ for $0 < \theta < \frac{1}{2}\pi$. Apply this with $\theta = \frac{1}{6}\pi$ to prove $3 < \pi < 3.47$.

54* (cont.) Use $\theta = \frac{1}{12}\pi$ and half angle formulas to prove

$$12\sqrt{\tfrac{1}{2}(1 - \tfrac{1}{2}\sqrt{3})} < \pi < 12/(2+\sqrt{3}),$$

hence $3.105 < \pi < 3.216$.

55* (cont.) Continue estimating π with $\theta = \frac{1}{24}\pi$.

56* (cont.) Use the same method to estimate π, choosing $\theta = \frac{1}{16}\pi$. (You should conclude that on a 4-banger calculator with $\boxed{\sqrt{\ }}$, you can calculate π with high accuracy. Also, $\sqrt{x}$ can be calculated by hand to any degree of accuracy. Hence so can π!)

5 INVERSE TRIGONOMETRIC FUNCTIONS

Let us recall the basic facts about inverse functions, discussed in detail in Chapter 5, Section 2. Suppose $f(x)$ has domain on interval **I** and range on interval **J**, and $f(x)$ is either strictly increasing or strictly decreasing on **I**. Then $f(x)$ has a unique inverse function $g(y)$ with domain **J** and range **I**, and

$$f[g(y)] = y \qquad g[f(x)] = x.$$

Also $g(y)$ is strictly increasing or decreasing, the way $f(x)$ is.

We now apply this theory to the functions sine, cosine, and tangent.

Arc Sine

The function $y = \sin\theta$, defined for all real θ, has no inverse because it is neither strictly increasing nor strictly decreasing. In fact, if $-1 \le c \le 1$, then there are infinitely many values of θ for which $\sin\theta = c$. But the very point of an inverse is to provide a *unique* θ for each given c.

The situation is not hopeless, however. We can limit ourselves to a smaller domain where $\sin\theta$ is either strictly increasing or strictly decreasing. A natural choice is the interval $-\frac{1}{2}\pi \le \theta \le \frac{1}{2}\pi$. There $\sin\theta$ increases strictly and takes all values from -1 to 1. See Fig. 1.

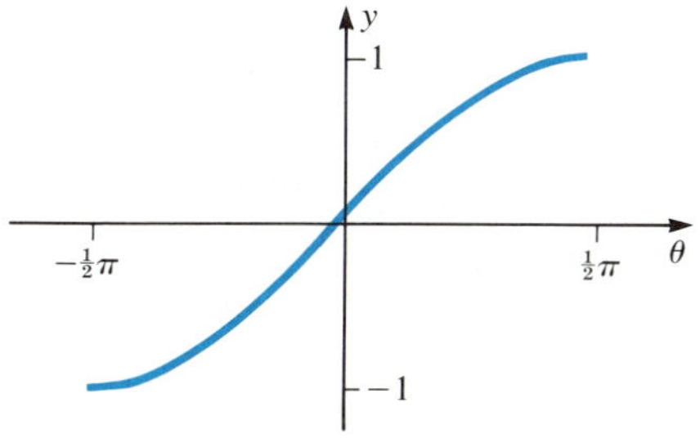

Fig. 1 $y = \sin\theta \quad -\frac{1}{2}\pi \le \theta \le \frac{1}{2}\pi$; y is strictly increasing.

Now we have all the ingredients for an inverse: the function $y = \sin\theta$, defined only on the interval $\mathbf{I} = \{-\frac{1}{2}\pi \le \theta \le \frac{1}{2}\pi\}$, is strictly increasing on $\mathbf{I}$, and has range the interval $\mathbf{J} = \{-1 \le y \le 1\}$. Hence there exists a unique inverse function called the **arc sine** and written

$$\theta = \text{arc sin } y.$$

Stated briefly, $\text{arc sin } y$ is that unique angle for which $\sin\theta = y$ and $-\frac{1}{2}\pi \le \theta \le \frac{1}{2}\pi$. Figure 2 illustrates the idea geometrically.

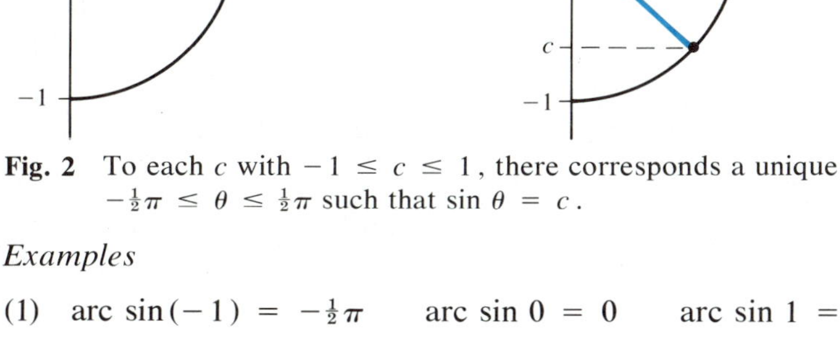

Fig. 2 To each c with $-1 \le c \le 1$, there corresponds a unique θ with $-\frac{1}{2}\pi \le \theta \le \frac{1}{2}\pi$ such that $\sin\theta = c$.

Examples

(1) $\text{arc sin}(-1) = -\frac{1}{2}\pi \qquad \text{arc sin } 0 = 0 \qquad \text{arc sin } 1 = \frac{1}{2}\pi.$

(2) $\text{arc sin } \frac{1}{2} = \frac{1}{6}\pi$ because $\sin\frac{1}{6}\pi = \frac{1}{2}$ and $-\frac{1}{2}\pi \le \frac{1}{6}\pi \le \frac{1}{2}\pi$.

(3) $\text{arc sin } \frac{1}{2}\sqrt{2} = \frac{1}{4}\pi$ because $\sin\frac{1}{4}\pi = \frac{1}{2}\sqrt{2}$ and $-\frac{1}{2}\pi \le \frac{1}{4}\pi \le \frac{1}{2}\pi$.

(4) $\text{arc sin}(-\frac{1}{2}\sqrt{3}) = -\frac{1}{3}\pi$

because $\sin(-\frac{1}{3}\pi) = -\frac{1}{2}\sqrt{3}$ and $-\frac{1}{2}\pi \le -\frac{1}{3}\pi \le \frac{1}{2}\pi$.

Arc Sine There is a unique function $\theta = \text{arc sin } y$ with domain $\{-1 \le y \le 1\}$ and satisfying

$$\sin(\text{arc sin } y) = y \qquad (-1 \le y \le 1)$$

$$\text{arc sin}(\sin\theta) = \theta \qquad (-\tfrac{1}{2}\pi \le \theta \le \tfrac{1}{2}\pi).$$

The function arc sin y is strictly increasing.*

We obtain the graph of $\theta = \text{arc sin } y$ from the graph of $y = \sin\theta$ shown in Fig. 1. Since

$$y = \sin\theta \quad \text{and} \quad \theta = \text{arc sin } y$$

are equivalent statements (provided $-\frac{1}{2}\pi \le \theta \le \frac{1}{2}\pi$), Figure 1 is the graph of $\theta = \text{arc sin } y$. However, the independent variable is on the vertical axis and the dependent variable is on the horizontal axis. For a more conventional graph, we interchange the axes and reflect the graph in the line $y = \theta$. The result (Fig. 3) is the graph of $\theta = \text{arc sin } y$.

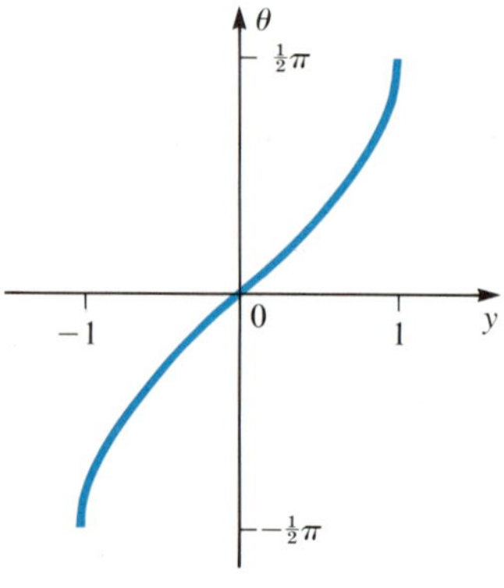

Fig. 3 $\theta = \text{arc sin } y$

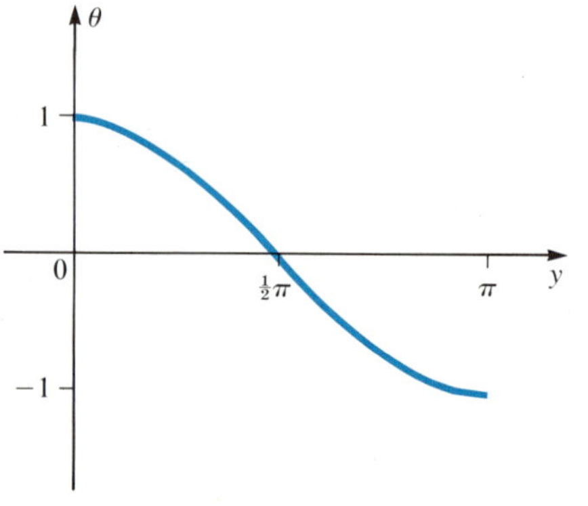

Fig. 4 $y = \cos\theta \qquad 0 \le \theta \le \pi$; y is strictly decreasing.

Arc Cosine

Like $\sin\theta$, the function $\cos\theta$ defined for all real θ has no inverse. But we can limit $\cos\theta$ to a smaller domain. A natural choice is the interval $0 \le \theta \le \pi$. There $\cos\theta$ decreases strictly and takes all values from 1 to -1. See Fig. 4.

We are now set up for an inverse: the function $y = \cos\theta$, defined only on the interval $\mathbf{I} = \{0 \le \theta \le \pi\}$, is strictly decreasing on $\mathbf{I}$, and has range the interval $\mathbf{J} = \{-1 \le y \le 1\}$. Hence there exists a unique inverse function called the **arc cosine** and written

$$\theta = \text{arc cos } y.$$

Note that arc cos y is that angle θ for which $\cos\theta = y$ and $0 \le \theta \le \pi$.

Examples

(1) $\text{arc cos } 1 = 0$ because $\cos 0 = 1$ and $0 \le 0 \le \pi$.

(2) $\text{arc cos } 0 = \frac{1}{2}\pi$ because $\cos\frac{1}{2}\pi = 0$ and $0 \le \frac{1}{2}\pi \le \pi$.

(3) $\text{arc cos}(-1) = \pi$ because $\cos\pi = -1$ and $0 \le \pi \le \pi$.

(4) $\text{arc cos } \frac{1}{2} = \frac{1}{3}\pi$ because $\cos\frac{1}{3}\pi = \frac{1}{2}$ and $0 \le \frac{1}{3}\pi \le \pi$.

* The notations arcsin y and $\sin^{-1} y$ are also used for arc sin y.

(5) $\arccos(-\frac{1}{2}\sqrt{2}) = \frac{3}{4}\pi$ because $\cos\frac{3}{4}\pi = -\frac{1}{2}\sqrt{2}$ and $0 \le \frac{3}{4}\pi \le \pi$.

Arc Cosine There is a unique function $\theta = \arccos y$ with domain $-1 \le y \le 1$ and satisfying

$$\cos(\arccos y) = y \qquad (-1 \le y \le 1)$$

$$\arccos(\cos\theta) = \theta \qquad (0 \le \theta \le \pi).$$

The function $\arccos y$ is strictly decreasing.

We obtain the graph of $\theta = \arccos y$ from Fig. 4 by interchanging the axes and reflecting the graph in the line $y = \theta$. See Fig. 5.

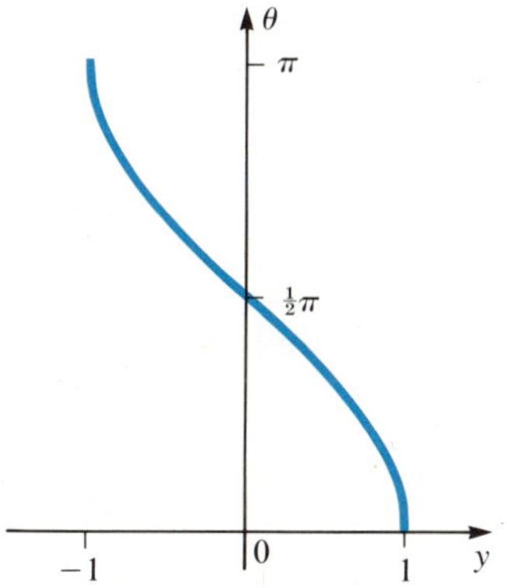

Fig. 5 $\theta = \arccos y$

Notation Since we are used to calling the horizontal axis the x-axis, let us replace y by x and write $\theta = \arcsin x$ and $\theta = \arccos x$.

Relation Between Arc Sin and Arc Cos

From the examples above we have

$$\arcsin(-1) + \arccos(-1) = -\tfrac{1}{2}\pi + \pi = \tfrac{1}{2}\pi$$

$$\arcsin 0 + \arccos 0 = 0 + \tfrac{1}{2}\pi = \tfrac{1}{2}\pi$$

$$\arcsin\tfrac{1}{2} + \arccos\tfrac{1}{2} = \tfrac{1}{6}\pi + \tfrac{1}{3}\pi = \tfrac{1}{2}\pi.$$

These relations suggest the following identity:

$$\arcsin x + \arccos x = \tfrac{1}{2}\pi \qquad (-1 \le x \le 1).$$

To prove the identity, let $\theta = \arcsin x$. Then $x = \sin\theta$ and $-\frac{1}{2}\pi \le \theta \le \frac{1}{2}\pi$. It follows that $-\theta$ is in the same interval, $-\frac{1}{2}\pi \le -\theta \le \frac{1}{2}\pi$. Add $\frac{1}{2}\pi$ to the three terms in this (double) inequality:

$$0 \le \tfrac{1}{2}\pi - \theta \le \pi.$$

But

$$\cos(\tfrac{1}{2}\pi - \theta) = \sin\theta = x,$$

hence $\frac{1}{2}\pi - \theta = \arccos x$. Therefore

$$\begin{aligned}\arcsin x + \arccos x &= \theta + (\tfrac{1}{2}\pi - \theta)\\ &= \tfrac{1}{2}\pi.\end{aligned}$$

The identity has a simple geometric interpretation in the case $0 \le x \le 1$. Figure 6 shows a right triangle in which $\sin\alpha = \cos\beta = x$. That means $\alpha = \arcsin x$ and $\beta = \arccos x$. But $\alpha + \beta = \frac{1}{2}\pi$.

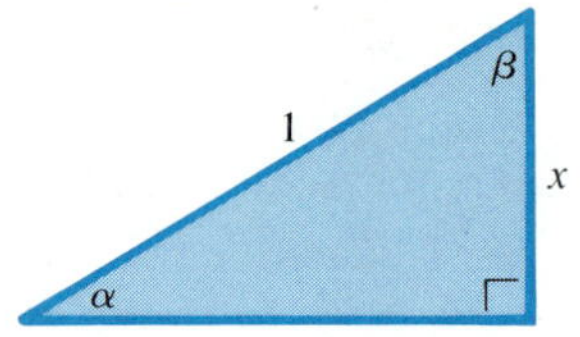

Fig. 6

EXAMPLE 1 Show that $\arcsin x = \arccos\sqrt{1 - x^2}$ for $0 \le x \le 1$.

Solution Set $\theta = \arcsin x$. Then $0 \le \theta \le \frac{1}{2}\pi$ since $0 \le x \le 1$, and $x = \sin\theta$. Since θ is in the first quadrant, $\cos\theta \ge 0$. Therefore

$$\cos\theta = \sqrt{1 - \sin^2\theta} = \sqrt{1 - x^2} \qquad \theta = \arccos\sqrt{1 - x^2}$$

But $\theta = \arcsin x$, so $\arcsin x = \arccos\sqrt{1 - x^2}$.

Arc Tangent

The graph of $\tan\theta$ (Fig. 1, p. 253) shows that $\tan\theta$ is strictly increasing for $-\frac{1}{2}\pi < \theta < \frac{1}{2}\pi$ and takes on every possible real value. Therefore an inverse function exists.

Arc Tangent There is a unique function $\theta = \arctan x$ with domain $\{\text{all } x\}$ and satisfying

$$\tan(\arctan x) = x \qquad (-\infty < x < \infty)$$

$$\arctan(\tan\theta) = \theta \qquad (-\tfrac{1}{2}\pi < \theta < \tfrac{1}{2}\pi).$$

The function $\arctan x$ is strictly increasing.

Note that $\arctan x$ is the angle θ for which $\tan\theta = x$ and $-\frac{1}{2}\pi < \theta < \frac{1}{2}\pi$.

Examples

(1) $\arctan 1 = \frac{1}{4}\pi$ since $\tan\frac{1}{4}\pi = 1$ and $-\frac{1}{2}\pi < \frac{1}{4}\pi < \frac{1}{2}\pi$

(2) $\arctan(-1) = -\frac{1}{4}\pi$ since $\tan(-\frac{1}{4}\pi) = -1$ and $-\frac{1}{2}\pi < -\frac{1}{4}\pi < \frac{1}{2}\pi$

(3) $\arctan(-\sqrt{3}) = -\frac{1}{3}\pi$ since $\tan(-\frac{1}{3}\pi) = -\sqrt{3}$ and $-\frac{1}{2}\pi < -\frac{1}{3}\pi < \frac{1}{2}\pi$

We obtain the graph of $\theta = \arctan x$ by reflecting the graph of $x = \tan\theta$. See Fig. 7. The lines $\theta = \frac{1}{2}\pi$ and $\theta = -\frac{1}{2}\pi$ are horizontal asymptotes of the graph. That checks: for example, as $x \longrightarrow \infty$, the first quadrant angle whose tangent is x approaches $\frac{1}{2}\pi$.

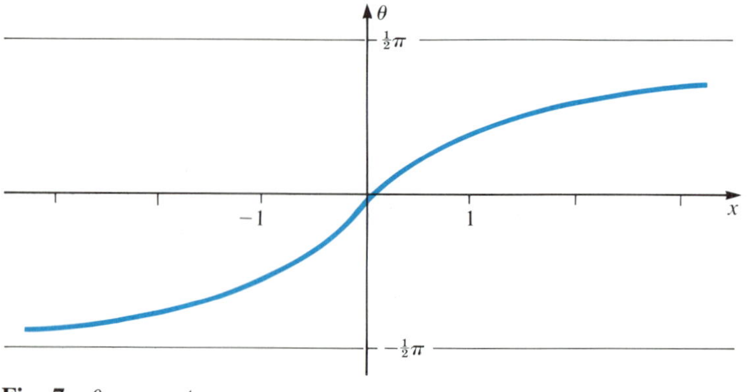

Fig. 7 $\theta = \arctan x$

Example 2 Prove $\arcsin x = \arctan \dfrac{x}{\sqrt{1-x^2}} \quad (-1 < x < 1)$.

Solution This identity is a disguised version of a statement that looks easier: if θ is in the first or fourth quadrants and if $\sin\theta = x$, then $\tan\theta = x/\sqrt{1-x^2}$.

Let $\theta = \arcsin x$, so $\sin\theta = x$. Since θ is in the first or fourth quadrant, $\cos\theta \geq 0$, hence

$$\tan\theta = \frac{\sin\theta}{\cos\theta} = \frac{\sin\theta}{\sqrt{1-\sin^2\theta}} = \frac{x}{\sqrt{1-x^2}}.$$

Therefore $\arctan \dfrac{x}{\sqrt{1-x^2}} = \theta = \arcsin x$.

Example 3 Find exactly

(a) $\arcsin(\cos 15°)$ (b) $\tan[\arccos(-\frac{1}{2})]$ (c) $\sin[\arctan(-\frac{3}{4})]$.

Solution (a) $\cos 15° = \sin 75° = \sin\frac{5}{12}\pi$ and $-\frac{1}{2}\pi \leq \frac{5}{12}\pi \leq \frac{1}{2}\pi$, hence

$$\arcsin(\cos 15°) = \tfrac{5}{12}\pi.$$

(b) $\arccos(-\frac{1}{2}) = \frac{2}{3}\pi$ because $\cos\frac{2}{3}\pi = -\frac{1}{2}$ and $0 \leq \frac{2}{3}\pi \leq \pi$. Therefore

$$\tan[\arccos(-\tfrac{1}{2})] = \tan\tfrac{2}{3}\pi = -\sqrt{3}.$$

(c) Let $\theta = \arctan(-\frac{3}{4})$. Then $-\frac{1}{2}\pi < \theta < \frac{1}{2}\pi$ and $\tan\theta = -\frac{3}{4} < 0$, so $-\frac{1}{2}\pi < \theta < 0$. Put θ in standard position; its terminal side must be in the fourth quadrant, and $(x, y) = (4, -3)$ is a point on this terminal side. Clearly $r^2 = 4^2 + (-3)^2 = 25$, so $r = 5$ and

$$\sin[\arctan(-\tfrac{3}{4})] = \sin\theta = \frac{y}{r} = -\tfrac{3}{5}.$$

Answer (a) $\frac{5}{12}\pi$ (b) $-\sqrt{3}$ (c) $-\frac{3}{5}$

Calculating Inverse Trig Functions

Your calculator has a key

[INV] or [ARC] or [F]

It is used in combination with [sin], [cos], and [tan] to estimate the corresponding inverse functions. (These are sometimes written $\sin^{-1}$, $\cos^{-1}$, etc.)

Examples (First set [rad] mode.)

		Keys	Display
(1)	$\arcsin 0.628$	. 6 2 8 [INV] [sin]	0.678981
(2)	$\arccos(-0.922)$	. 9 2 2 [+/−] [INV] [cos]	2.74401
(3)	$\arcsin(1.485)$	1 . 4 8 5 [INV] [sin]	ERROR

(First set [deg] mode.)

		Keys	Display
(4)	$\arctan 0.505$	. 5 0 5 [INV] [tan]	26.7938
(5)	$\arctan(-3.190)$	3 . 1 9 [+/−] [INV] [tan]	-72.5949
(6)	$\arcsin\frac{3}{17}$	3 [÷] 1 7 [=] [INV] [sin]	10.1642

Exercises

Give exact values

1 $\arcsin(-\frac{1}{2})$ **2** $\arcsin(\frac{1}{2}\sqrt{3})$ **3** $\arcsin(-\frac{1}{2}\sqrt{2})$
4 $\arccos(\frac{1}{2}\sqrt{2})$ **5** $\arccos(-\frac{1}{2})$ **6** $\arccos(\frac{1}{2}\sqrt{3})$
7 $\arccos(-\frac{1}{2}\sqrt{3})$ **8** $\arctan 0$ **9** $\arctan(\frac{1}{3}\sqrt{3})$
10 $\arctan(-\frac{1}{3}\sqrt{3})$ **11** $\arctan(\sqrt{3})$ **12** $\arccos(\cos\frac{1}{7}\pi)$
13 $\sin(\arcsin 0.3)$ **14** $\tan(\arctan 35)$ **15** $\arctan(\tan 1)$
16 $\cos[\arccos(-\frac{1}{3})]$ **17** $\arccos[\cos(-\frac{1}{3})]$ **18** $\arctan[\tan(-1)]$.

Prove

19 $\arcsin(-x) = -\arcsin x$

20 $\arccos(-x) = \pi - \arccos x$

21 $\arctan(-x) = -\arctan x$

22 $\cos(\arcsin x) = \sqrt{1 - x^2}$

23 $\cot(\arctan x) = \dfrac{1}{x}$ for $x \neq 0$

24 $\arctan x = \arcsin \dfrac{x}{\sqrt{1 + x^2}}$

25 $\arcsin(\cos\theta) = \frac{1}{2}\pi - \theta$ for $0 \leq \theta \leq \pi$

26 $\arctan(\cot\theta) = \frac{1}{2}\pi - \theta$ for $0 < \theta < \pi$

27 $2\arccos x = \arccos(2x^2 - 1)$ for $0 \leq x \leq 1$

28 $2\arctan x = \arctan \dfrac{2x}{1 - x^2}$ for $-1 < x < 1$

29 $\arctan\frac{1}{2} + \arctan\frac{1}{3} = \frac{1}{4}\pi$

30 $\arctan\frac{1}{3} + \arctan\frac{1}{7} = \arctan\frac{1}{2}$.

31 By Example 1, the functions $\arcsin x$ and $\arccos\sqrt{1 - x^2}$ are equal for $0 \leq x \leq 1$. Find their relation for $-1 \leq x \leq 0$.

32 According to Ex. 27, the functions $2\arccos x$ and $\arccos(2x^2 - 1)$ are equal for $0 \leq x \leq 1$. Find their relation for $-1 \leq x \leq 0$.

Find exactly

33 $\arccos(\sin 35°)$

34 $\cos[\arctan(-\frac{1}{3})]$

35 $\cos[\arcsin(-0.7)]$

36 $\tan[\arcsin(-\frac{2}{3})]$.

Use trigonometric identities to find exactly

37 $\sin(2\arcsin\frac{1}{3})$

38 $\cos(2\arcsin\frac{1}{3})$

39 $\tan(2\arctan 5)$

40 $\sin(2\arctan 5)$

41 $\cos(\arcsin\frac{3}{5} + \arccos\frac{5}{13})$

42 $\tan(\arcsin\frac{12}{13} - \arcsin\frac{5}{13})$.

43 Show that $x = \cot\theta$ on the domain $0 < \theta < \pi$ has an inverse function $\theta = \operatorname{arc\,cot} x$, with domain $\{\text{all } x\}$, and strictly decreasing.

44 (cont.) Graph $\theta = \operatorname{arc\,cot} x$.

45 (cont.) Prove $\arctan x + \operatorname{arc\,cot} x = \frac{1}{2}\pi$.

46 (cont.) Give a sequence of calculator steps for estimating $\operatorname{arc\,cot} x$.

47 Graph $x = \sec\theta$ for $0 \leq \theta \leq \pi$, $\theta \neq \frac{1}{2}\pi$. Show that $\sec\theta$ restricted to this domain has an inverse function, $\operatorname{arc\,sec} x$.

48 (cont.) Show that $\operatorname{arc\,sec} x$ is defined for $|x| \geq 1$, and sketch its graph.

49 Graph $x = \csc\theta$ for $-\frac{1}{2}\pi \leq \theta \leq \frac{1}{2}\pi$, $\theta \neq 0$. Show that $\csc\theta$ restricted to this domain has an inverse function, $\operatorname{arc\,csc} x$.

50 (cont.) Show that $\operatorname{arc\,csc} x$ is defined for $|x| \geq 1$, and sketch its graph.

Estimate on a calculator; answer in radians to 4 places

51 $\arcsin\frac{1}{4}$ **52** $\arccos\frac{1}{6}$ **53** $\arctan\frac{10}{3}$
54 $\arctan(-10)$ **55** $\arcsin(-\frac{9}{11})$ **56** $\arccos(-\frac{13}{29})$.

Estimate on a calculator; answer in degrees to 2 places

57 $\arcsin 0.1203$ **58** $\arccos 0.8177$ **59** $\arctan(-294.7)$
60 $\arccos(0.1212)$

61 arc tan 1.352 + arc tan $\dfrac{1}{1.352}$ **62** arc sin(−0.944) + arc cos(−0.944).

63 Tabulate x, arc sin x, and $x + \frac{1}{6}x^3$ for $x = 0.1, 0.05, 0.01$. Conclusion?

64 Tabulate x, arc tan x, and $x - \frac{1}{3}x^3$ for the same values as in Ex. 63. Conclusion?

65 There are 3 squares in Fig. 8. Show that $\alpha + \beta = \gamma$. [*Hint* See Ex. 29.]

66 P bisects one leg of the 45-45-90 triangle in Fig. 9. Find tan β. [*Hint* See Ex. 29.]

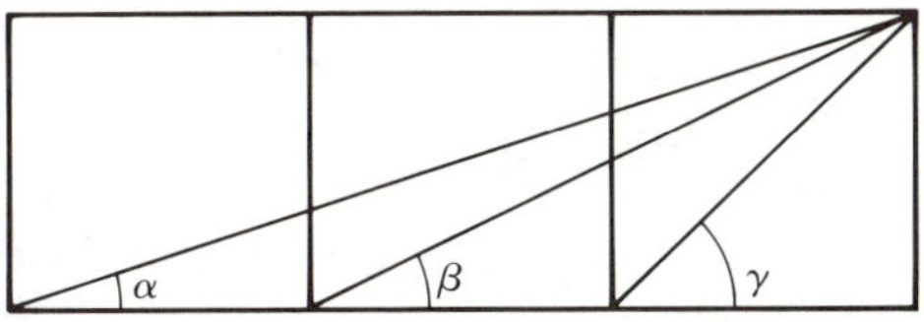

Fig. 8

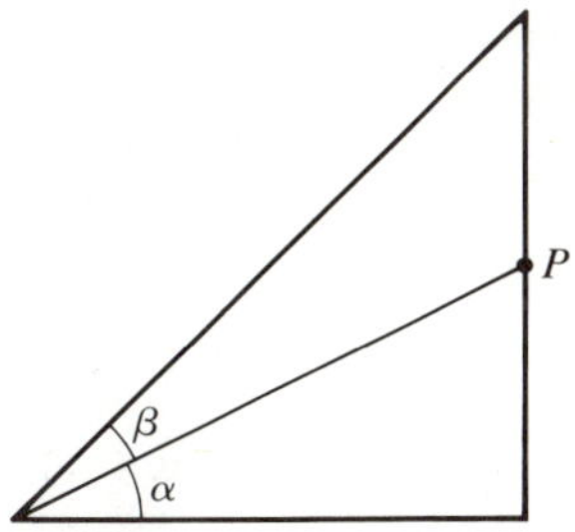

Fig. 9

6 TRIGONOMETRIC EQUATIONS

An equation involving trigonometric functions is called a **trigonometric equation.** In this section we study equations that involve *only* trigonometric functions of θ such as $\sin\theta$, $\cos\theta$, $\cdots$ and perhaps also $\sin 2\theta$, $\cos 3\theta$, etc. Examples are

$$\sin 2\theta = \cos\theta \qquad \csc\theta = 3\sin\theta \qquad \sec^2\theta = 4\tan\theta - 2$$

$$\cot 3\theta = 5 + \cos 2\theta$$

Whenever θ is a root of a trigonometric equation of the type considered here, so is $\theta + 2\pi n$ for each integer n. Therefore it suffices to find all the roots in a conveniently chosen interval of length 2π, such as $0 \le \theta < 2\pi$ or $-\pi < \theta \le \pi$.

We shall always choose the interval

$$-\pi < \theta \le \pi$$

for the simple and practical reason that the values of arc sin x, arc cos x, and arc tan x all fall into this interval.

Thus when you are asked for *all* solutions of a trigonometric equation, you must first find *all* solutions $\theta_1, \theta_2, \cdots, \theta_r$ in the interval $-\pi < \theta \le \pi$, and then write

$$\theta_1 + 2\pi n \qquad \theta_2 + 2\pi n \quad \cdots \quad \theta_r + 2\pi n \qquad (n \text{ an integer})$$

for *all solutions* of the equation. Remember that θ must appear *only* in the form $\sin m\theta$, $\cos m\theta$, $\cdots$, $\csc m\theta$, where m is an integer, but it can appear so more than once.

We begin our work with the simplest type of trigonometric equation.

EXAMPLE 1 Find all solutions in $-\pi < \theta \le \pi$ of

(a) $\sin\theta = \frac{1}{2}\sqrt{3}$ (b) $\cos\theta = 0.2598$ (c) $\tan\theta = 3.9810$.

Solution (a) $\sin\theta$ is positive in the first and second quadrants. There is a solution in each (Fig. 1a, next page):

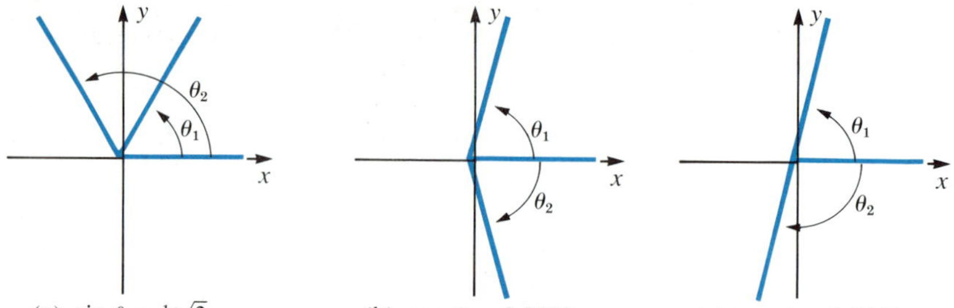

Fig. 1

$$\theta_1 = \text{arc}\sin(\tfrac{1}{2}\sqrt{3}) = \tfrac{1}{3}\pi \qquad \text{in the 1-st quadrant}$$

$$\theta_2 = \pi - \text{arc}\sin(\tfrac{1}{2}\sqrt{3}) = \tfrac{2}{3}\pi \qquad \text{in the 2-nd quadrant.}$$

(b) $\cos\theta$ is positive in the first and fourth quadrants. There is a solution in each (Fig. 1b):

$$\theta_1 = \text{arc}\cos 0.2598 \approx 1.30798 \qquad \text{in the 1-st quadrant}$$

$$\theta_2 = -\text{arc}\cos 0.2598 \approx -1.30798 \qquad \text{in the 4-th quadrant.}$$

(c) $\tan\theta$ is positive in the first and third quadrants. There is a solution in each (Fig. 1c):

$$\theta_1 = \text{arc}\tan 3.9810 \approx 1.32470 \qquad \text{in the 1-st quadrant}$$

$$\theta_2 = \text{arc}\tan 3.9810 - \pi \approx -1.81690 \qquad \text{in the 3-rd quadrant}$$

(We subtract (rather than add) π to make θ fall into the interval $-\pi < \theta \leq \pi$.)

Answer (a) $\frac{1}{3}\pi \quad \frac{2}{3}\pi$

(b) $\theta_1 = \text{arc}\cos 0.2598 \approx 1.30798 \qquad \theta_2 = -\theta_1 \approx -1.30798$

(c) $\theta_1 = \text{arc}\tan 3.9810 \approx 1.32470 \qquad \theta_2 = \theta_1 - \pi \approx -1.81690$

In the example we were given positive values for sin, cos, and tan. We handle negative values of cos and tan similarly; sin must be treated a bit differently. Suppose we seek all θ in the interval $-\pi < \theta \leq \pi$ such that

$$\sin\theta = a \quad \text{where} \quad -1 < a < 0.$$

The 4-th quadrant solution is $\theta_1 = \text{arc}\sin a$, and

$$-\tfrac{1}{2}\pi < \theta_1 < 0$$

The 3-rd quadrant solution θ_2 lies in the interval $-\pi < \theta < -\frac{1}{2}\pi$. It satisfies $\theta_2 + \theta_1 = -\pi$. See Fig. 2. Therefore

$$\theta_2 = -\theta_1 - \pi = -\text{arc}\sin a - \pi.$$

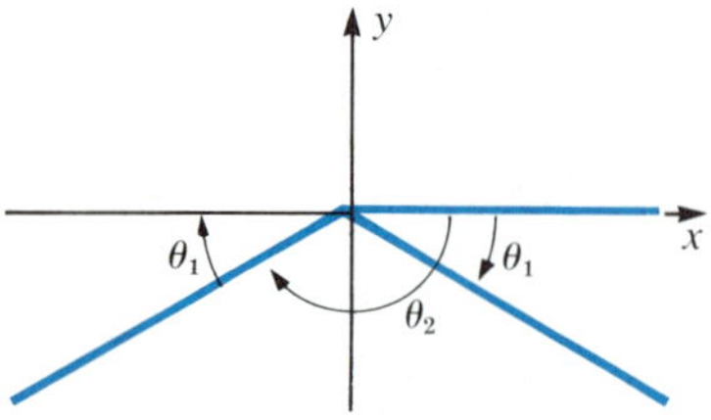

Fig. 2 $-1 < \sin\theta < 0$: $\theta_1 + \theta_2 = -\pi$

EXAMPLE 2 Find all solutions in the interval $-\pi < \theta \le \pi$ of

(a) $\sin\theta = -0.3122$ (b) $\cos\theta = -\frac{1}{2}$ (c) $\tan\theta = -0.2419$

Solution (a) $\theta_1 = \arcsin(-0.3122) \approx -0.31751$ (4-th quadrant)

$\theta_2 = -\arcsin(-0.3122) - \pi \approx -2.82408$ (3-rd quadrant)

(b) $\theta_1 = \arccos(-\frac{1}{2}) = \frac{2}{3}\pi$ (2-nd quadrant)

$\theta_2 = -\arccos(-\frac{1}{2}) = -\frac{2}{3}\pi$ (3-rd quadrant)

(c) $\theta_1 = \arctan(-0.2419) \approx -0.23734$ (4-th quadrant)

$\theta_2 = \arctan(-0.2419) + \pi \approx 2.90425$ (2-nd quadrant)

We add (rather than subtract) π to fall into the interval $\frac{1}{2}\pi < \theta < \pi$.

Answer

(a) $\theta_1 = \arcsin(-0.3122) \approx -0.31751$ $\theta_2 = -\theta_1 - \pi \approx -2.82408$

(b) $\theta_1 = \frac{2}{3}\pi$ $\theta_2 = -\frac{2}{3}\pi$

(c) $\theta_1 = \arctan(-0.2419) \approx -0.23734$ $\theta_2 = \theta_1 + \pi \approx 2.90425$

EXAMPLE 3 Solve $\sin 2\theta = \frac{1}{2}$.

Solution Clearly, if $-\pi < 2\theta \le \pi$ and $\sin 2\theta = \frac{1}{2}$, then $2\theta = \frac{1}{6}\pi$ or $2\theta = \pi - \frac{1}{6}\pi = \frac{5}{6}\pi$. Thus *any* solution θ of $\sin 2\theta = \frac{1}{2}$ must satisfy

$$2\theta = \tfrac{1}{6}\pi + 2\pi n \quad \text{or} \quad 2\theta = \tfrac{5}{6}\pi + 2\pi n$$

that is,

$$\theta = \tfrac{1}{12}\pi + n\pi \quad \text{or} \quad \theta = \tfrac{5}{12}\pi + n\pi,$$

where n is an arbitrary integer. This is a correct answer as it stands; however, it is usually preferable to display all solutions θ in the interval $-\pi < \theta \le \pi$, and then add $2\pi n$ to each. They are (choose $n = 0$ and $n = -1$)

$$\tfrac{1}{12}\pi \qquad \tfrac{1}{12}\pi - \pi = -\tfrac{11}{12}\pi \qquad \tfrac{5}{12}\pi \qquad \tfrac{5}{12}\pi - \pi = -\tfrac{7}{12}\pi.$$

Answer $\frac{1}{12}\pi + 2\pi n$ $-\frac{11}{12}\pi + 2\pi n$ $\frac{5}{12}\pi + 2\pi n$ $-\frac{7}{12}\pi + 2\pi n$

Now we can handle equations of the form $\sin m\theta = c$, etc., in which the value of a trigonometric function is known. This generally represents the final step in the solution of a (more general) trigonometric equation. Before that we try to use algebra technique and trigonometric identities to solve for a trigonometric function of θ, from which we find θ.

EXAMPLE 4 Find all solutions of $\sin 2\theta = \cos\theta$.

Solution Use the double angle formula for $\sin 2\theta$:

$$2\sin\theta\cos\theta = \cos\theta \qquad (2\sin\theta - 1)\cos\theta = 0.$$

The product is zero only if one of the factors is zero, that is, only if

$$2\sin\theta - 1 = 0 \quad \text{or} \quad \cos\theta = 0.$$

In the first case, $\sin\theta = \frac{1}{2}$. The solutions of this equation in the interval

$-\pi < \theta \leq \pi$ are $\frac{1}{6}\pi$ and $\frac{5}{6}\pi$.

In the second case, $\cos\theta = 0$, and the solutions of this equation in the interval $-\pi < \theta \leq \pi$ are $\frac{1}{2}\pi$ and $-\frac{1}{2}\pi$.

Answer $\frac{1}{6}\pi + 2\pi n \quad \frac{5}{6}\pi + 2\pi n \quad \frac{1}{2}\pi + 2\pi n \quad -\frac{1}{2}\pi + 2\pi n$

EXAMPLE 5 Solve $\sec^2\theta = 4\tan\theta - 2$ in the interval $-\pi < \theta \leq \pi$.

Solution Replace $\sec^2\theta$ by $1 + \tan^2\theta$. The result is a quadratic equation for $\tan\theta$:

$$1 + \tan^2\theta = 4\tan\theta - 2 \qquad \tan^2\theta - 4\tan\theta + 3 = 0.$$

The quadratic factors:

$$(\tan\theta - 1)(\tan\theta - 3) = 0 \qquad \text{hence} \quad \tan\theta = 1 \quad \text{or} \quad \tan\theta = 3.$$

If $\tan\theta = 1$, then $\theta = \frac{1}{4}\pi$ or $\frac{1}{4}\pi - \pi = -\frac{3}{4}\pi$. If $\tan\theta = 3$, then

$$\theta = \arctan 3 \approx 1.24905 \quad \text{or} \quad \arctan 3 - \pi \approx -1.89255.$$

Answer $\frac{1}{4}\pi \quad -\frac{3}{4}\pi \quad \arctan 3 \approx 1.24905 \quad \arctan 3 - \pi \approx -1.89255$

EXAMPLE 6 Solve $\tan\theta° = 2\cos\theta°$ in the interval $-180° < \theta° \leq 180°$.

Solution Since $\tan\theta° = \sin\theta°/\cos\theta°$,

$$\frac{\sin\theta°}{\cos\theta°} = 2\cos\theta° \qquad \sin\theta° = 2\cos^2\theta°.$$

Replace $\cos^2\theta°$ by $1 - \sin^2\theta°$; the result is a quadratic equation for $\sin\theta°$:

$$2\sin^2\theta° + \sin\theta° - 2 = 0.$$

Now solve for $\sin\theta°$ by the quadratic formula:

$$\sin\theta° = \frac{-1 \pm \sqrt{17}}{4} = \begin{cases} \frac{1}{4}(-1 + \sqrt{17}) \approx 0.78078 \\ \frac{1}{4}(-1 - \sqrt{17}) \approx -1.28. \end{cases}$$

Since $\sin\theta° < -1$ is impossible, only $\sin\theta° = \frac{1}{4}(\sqrt{17} - 1)$ yields solutions. In this case $\sin\theta° > 0$, so there is one solution in the first quadrant and one in the second:

$$\theta_1° = \arcsin\tfrac{1}{4}(\sqrt{17} - 1) \approx 51.3317° \qquad \theta_2° = 180° - \theta_1° \approx 128.6683°$$

Answer 51.3317° 128.6683°

Exercises

Solve on the interval $-\pi < \theta \leq \pi$, exactly if possible, otherwise to 4-place accuracy

1 $\sin\theta = \frac{1}{2}\sqrt{2}$
2 $\sin\theta = -\frac{1}{2}\sqrt{2}$
3 $\cos\theta = -0.8121$
4 $\cos\theta = 0.8121$
5 $\tan\theta = \sqrt{3}$
6 $\tan\theta = -\sqrt{3}$
7 $\sin 2\theta = \frac{1}{2}\sqrt{3}$
8 $\sin 2\theta = -\frac{1}{2}\sqrt{3}$
9 $\cos 3\theta = \frac{1}{2}\sqrt{2}$
10 $\cos 3\theta = -\frac{1}{2}\sqrt{2}$
11 $\tan 4\theta = -\sqrt{3}$
12 $\tan 4\theta = \sqrt{3}$
13 $\cos 3\theta = -0.6917$
14 $\tan 2\theta = 5.000$
15 $\sin 2\theta = 0.8162$
16 $\tan 3\theta = -0.8164$.

Solve exactly on the interval $0 \leq \theta° < 360°$

17 $\sin\theta° = -\frac{1}{2}$
18 $\cos\theta° = -\frac{1}{2}$
19 $\tan\theta° = \frac{1}{3}\sqrt{3}$
20 $\tan\theta° = -\frac{1}{3}\sqrt{3}$.

Solve on the interval $-\pi < \theta \le \pi$, exactly if possible, otherwise to 4-place accuracy

21 $\sin^2\theta = \cos^2\theta$

22 $16\sin^4\theta = \cos^4\theta$

23 $3\sin^2\theta + 7\cos\theta = 5$

24 $6\cos^4\theta - 5\cos^2\theta + 1 = 0$

25 $\cos 2\theta = 2\cos\theta$

26 $\sin 2\theta = \sin\theta$

27 $\cot 2\theta = 1 + \tan\theta$

28 $\sec^2\theta - 5\tan\theta - 1 = 0$

29 $\sec\theta + \cos\theta = 4$

30 $12\sin\theta + 5 = 2\csc\theta$

31 $\sin^2\theta + \sin\theta = 3$

32 $2\sin\theta - 2\cos\theta = 3$

33 $2\cos^2\theta + \sin\theta = 2$

34 $2\cos^3\theta = \cos\theta$

35 $\tan\theta = \sin\theta$

36 $\tan\theta = \cot\theta$

37 $\sin^6\theta = \sin^2\theta$

38 $\sec^3\theta - 2\sec\theta = 0$

39 $\tan\theta = \cot^2\theta$

40 $\tan^2\theta + \tan\theta = 0$

41 $2\sin^2\theta = -3\cos\theta$

42 $2\sin\theta\cos\theta = \cos 2\theta$

43 $\cos\theta = \cos 2\theta$

44 $\tan\theta = \tan 2\theta$

45 $\tan\theta + \cot\theta = \sec\theta\csc\theta$

46 $\tan^2\theta\sec^2\theta = 9 + \tan^2\theta$

47 $\cos\theta - \cos 3\theta = 0$

48 $\cos\theta - \sqrt{3}\sin\theta = 1$

49 $\tan\theta + \sec\theta = 1$

50 $\tan\theta = \sin 2\theta$

51 $\sin 3\theta + \sin 5\theta = 0$

52 $\cos 3\theta + \cos 5\theta = 0$.

7 EQUATIONS INVOLVING θ [Optional]

Now we consider equations involving both trigonometric functions of θ and also θ in other forms, for example,

$$\sin\theta = \theta^2 \qquad \theta\tan\theta = 1 \qquad \theta + \sin\theta = 2.5.$$

Such equations may have no roots, a finite number of roots, or infinitely many. Periodicity doesn't apply to them in general. If θ_0 is a root, there is no reason whatsoever for $\theta_0 + 2\pi n$ to be a root also.

There are no algebraic methods for solving such equations. You can only hope to approximate a root by numerical techniques. A good way to start is by drawing a graph. That should show the number of roots and their approximate locations.

EXAMPLE 1 Locate roughly the roots of

(a) $\sin\theta = \theta^2$ (b) $\theta\tan\theta = 1$.

Solution (a) Graph $y = \sin\theta$ and $y = \theta^2$ on the same grid (Fig. 1). The curves cross at two points, indicating two roots. One is $\theta = 0$ and the other is near $\theta = 1$.

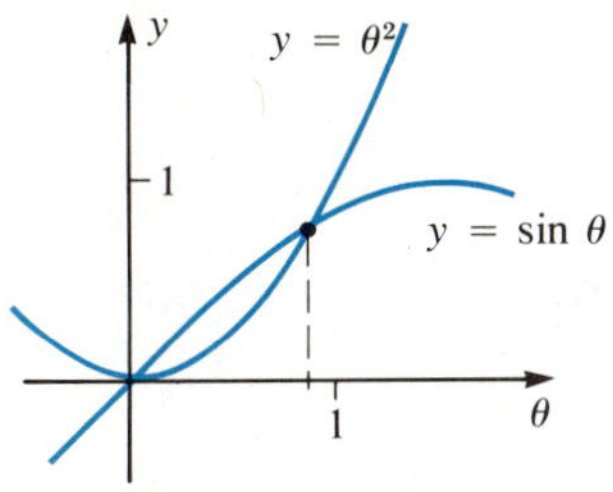

Fig. 1 Graphical solution of $\sin\theta = \theta^2$

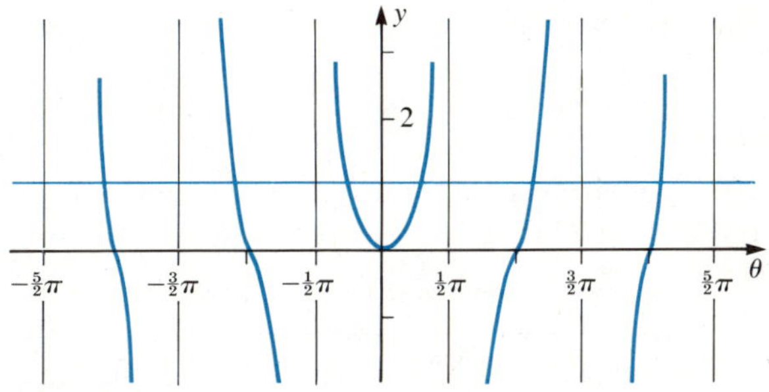

Fig. 2 Graphical solution of $\theta\tan\theta = 1$

(b) Graph the even function $y = \theta\tan\theta$. One way is to start with a graph of $y = \tan\theta$ (Fig. 1, p. 253), superimpose the line $y = \theta$, and multiply corresponding values (Fig. 2). Even the roughest sketch shows infinitely many branches and the graph crossing the horizontal line $y = 1$ infinitely often. The intersections correspond to roots $\pm\theta_0, \pm\theta_1, \pm\theta_2, \cdots$ of $\theta \tan\theta = 1$, where

$$\tfrac{1}{4}\pi < \theta_0 < \tfrac{1}{2}\pi \qquad \pi < \theta_1 < \tfrac{3}{2}\pi \cdots n\pi < \theta_n < (n + \tfrac{1}{2})\pi.$$

As we move to the right, the branches become steeper and steeper at $\theta = n\pi$, so we expect θ_n to be quite close to $n\pi$ if n is large.

The graphical method is a good start, but for more accurate approximations to roots, computational techniques are necessary. We'll use the method developed in Chapter 5, Section 9, pp. 217–221.

EXAMPLE 2 Estimate to 6 places the positive root of $\sin\theta = \theta^2$.

Solution Set

$$f(\theta) = \theta^2 - \sin\theta.$$

We'll tabulate $f(\theta)$; for this we need a calculator routine. For example, in the [rad] mode:

θ [STO] [x^2] [−] [RCL] [sin] [=].

(One step shorter:

θ [x^2] [−] [$\sqrt{\ }$] [sin] [=].)

We start at $\theta = 1.0$ and work backwards by tenths:

θ	.8	.9	1.0
$f(\theta)$	−.077	.027	.159

The root lies between 0.8 and 0.9, probably closer to 0.9. To estimate at what fraction of the way from 0.8 to 0.9 it lies, we interpolate:

$$\text{fractional part} = \frac{77}{77 + 27} \approx 0.7$$

Hence we now suspect that $\theta \approx 0.8 + 0.07 = 0.87$. We tabulate again, starting at 0.87 and moving by steps of 0.01:

θ	.86	.87	.88
$f(\theta)$		−.00743	.00366

The root lies in the interval $0.87 < \theta < 0.88$. We interpolate again.

$$\text{fractional part} = \frac{743}{743 + 366} \approx 0.67$$

Now we are in a tiny interval so we can safely move ahead *two* more places; we try $\theta \approx 0.8767$ and tabulate:

θ	.8766	.8767	.8768
$f(\theta)$		-2.92×10^{-5}	8.22×10^{-5}

One more interpolation:

$$\text{fractional part} = \frac{292}{292 + 822} \approx 0.26 \qquad \theta \approx 0.876726$$

Answer 0.876726

EXAMPLE 3 Estimate to 4 places the root of $\theta \tan \theta = 1$ near $\theta = -2\pi$.

Solution Set $f(\theta) = \theta \tan \theta - 1$. Since $-2\pi \approx -6.28$ we start tabulating $f(\theta)$ at -6.3:

θ	-6.5	-6.4	-6.3
$f(\theta)$	.432	$-.249$	$-.894$

The root is in the interval $-6.5 < \theta < -6.4$. Interpolate:

$$\text{fractional part} = \frac{432}{432 + 249} \approx 0.6.$$

We must go $\frac{6}{10}$ of the way from -6.5 to -6.4, that is, to -6.44. (Since we are working with negative numbers, it would have been wiser to have computed the fractional part from -6.4 to -6.5:

$$\text{fractional part} = \frac{249}{249 + 432} \approx 0.4 \qquad \theta \approx -6.44.)$$

Tabulate values again:

θ	-6.45	-6.44	-6.43
$f(\theta)$		.0182	$-.0491$

$$\text{fractional part} = \frac{491}{491 + 182} \approx 0.73 \qquad \theta \approx -6.4373$$

Answer -6.4373

Iteration

The next example will illustrate an entirely different method of solution.

EXAMPLE 4 Estimate to 4 places the root of $\cos \theta = \theta$.

Solution First graph $y = \cos \theta$ and $y = \theta$ on the same grid (Fig. 3). Clearly $\cos \theta = \theta$ only for one value of θ, about 0.8.

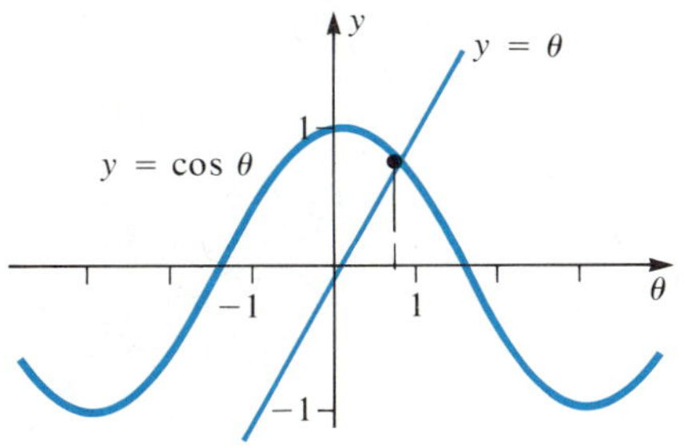

Fig. 3 $\cos \theta = \theta$

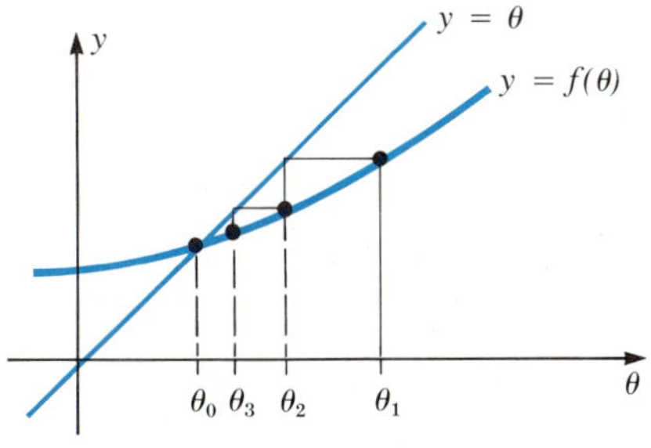

Fig. 4 Assume $f(\theta)$ increases *slowly* near θ_0. Set $\theta_{n+1} = f(\theta_n)$. Then $\theta_n \longrightarrow \theta_0$.

On your calculator (set [rad] mode), watch closely as you press

. 8 [cos] [cos] [cos]

and continue until 4 places stabilize. After pressing [cos] 18 times the display is 0.739136, and after one more press, 0.739051. Since the values alternate up down up down, we conclude that all further values will round to 0.7391.

Answer 0.7391

This iteration method is remarkable. Unfortunately, it does not always work. The equation in Example 4 has the form $\theta = f(\theta)$. Near a root, computing $f(\theta)$ practically gives you back θ again. The theory of such equations says that if you start sufficiently close to a root θ_0 with a guess θ_1 and if $f(\theta)$ is changing sufficiently slowly near θ_0, then the sequence of iterations

$$\theta_2 = f(\theta_1) \qquad \theta_3 = f(\theta_2) \qquad \theta_4 = f(\theta_3)$$

will approach closer and closer to θ_0. Perhaps Fig. 4, previous page, helps show why.

EXAMPLE 5 Find the root of $\tan\theta = \theta$ near $\frac{3}{2}\pi$ to 5 places.

Solution A graph (Fig. 5a) of $y = \tan\theta$ and $y = \theta$ helps us see what is going on. The root is just a little less than $\frac{3}{2}\pi \approx 4.71$, so let us set [rad] mode, then start at 4.6 and iterate the tangent:

4.6 [tan] [tan] [tan] [tan] . . .

The results are

8.86 $\quad -0.633 \quad -0.734 \quad -0.903 \cdots$

Most unsatisfactory; we have drifted far away from the root near $\frac{3}{2}\pi$.

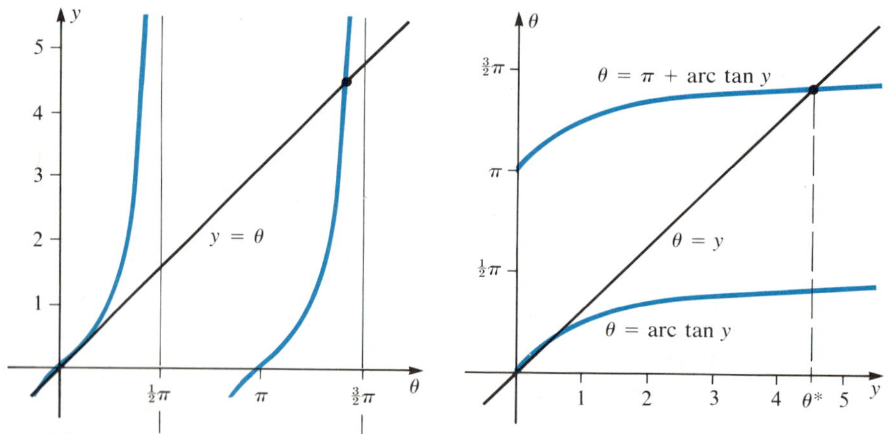

(a) $y = \tan\theta$ and $y = \theta$ (b) $\theta = \pi + \text{arc tan}\, y$ and $\theta = y$

Fig. 5

The trouble is that the function $\tan\theta$ increases rapidly near the root (Fig. 5a). We can remedy this rapid growth by using the *inverse* function to the function $y = \tan\theta$ on the domain $\frac{1}{2}\pi < \theta < \frac{3}{2}\pi$. Where the function increases rapidly, its inverse function increases slowly. By Fig. 5b, the inverse function is

$$g(y) = \text{arc tan}\, y + \pi \qquad -\infty < y < \infty.$$

Clearly, if $\tan\theta_0 = \theta_0$, then

$$\theta_0 = g(\tan\theta_0) = g(\theta_0),$$

so the problem of solving $\tan\theta = \theta$ near $\frac{3}{2}\pi$ is equivalent to the problem of solving $g(y) = y$ near $\frac{3}{2}\pi$. We iterate, again in [rad] mode and starting at 4.6:

4.6 [g] [g] [g] . . . ,

where

[g] denotes* [INV] [tan] [+] [π] [=].

The results are

4.4983 4.4936 4.49342 4.493410 4.493409 4.493409 · · ·

After only 6 iterations, 6 decimal places are fixed,

Answer 4.49341

Exercises

Estimate to 4 places the root of

1 $\theta + \sin\theta = 2.5$

2 $\cos\theta = \sqrt{\theta}$

3 $\theta = 1 + \sin\theta$

4 $\sin\theta = \theta^3 \qquad \theta > 0$

5 $\theta\tan\theta = 0.5 \qquad 0 < \theta < \frac{1}{2}\pi$

6 $\cos\theta = e^{-\theta} \qquad 0 < \theta < \frac{1}{2}\pi$

7 $\cos 2\theta = \frac{1}{6}\theta^{3/2} \qquad \frac{1}{2}\pi < \theta < \pi$

8 $\sin\pi\theta^2 = 1 - \theta \qquad 0 < \theta < 1$

9 $e^{\theta} = \sin 3\theta \qquad -\frac{1}{2}\pi < \theta < 0$

10 $\tan\theta = \frac{1}{5}\theta^3 \qquad \frac{1}{2}\pi < \theta < \frac{3}{2}\pi$.

Solve by iteration to 4 places

11 $\sin\theta + 0.5 = \theta$

12 $\tan\theta = \theta \qquad \theta \approx 7$

13 $\cos\theta = 2\theta$

14* $4\cos\theta = \theta$

15 $x = 3\ln x \qquad x \approx 5$

16* $x = 3\ln x \qquad x \approx 2$.

8 VIBRATIONS [Optional]

In this section we shall describe some real life applications of trigonometric functions and identities.

Sines and cosines describe many of the important types of vibrations that occur in the physical world, such as vibrations of strings, fluctuations in pressures, currents, voltages, etc. Mathematically, such vibrations are given by functions of the type

$$P = A\sin(2\pi\nu t - \phi).$$

The constants A, ν, and ϕ have names:

A **amplitude** $\qquad \nu$ **frequency** $\qquad \phi$ **phase angle**

A fourth constant of the vibration is $T = \dfrac{1}{\nu}$ **period**

The function P is periodic with period T. Its frequency ν is measured in cycles per second (called **hertz,** and abbreviated Hz). Note that P can also be expressed as

$$P = A\cos(2\pi\nu t - \phi - \tfrac{1}{2}\pi),$$

so a vibration can be described by either a sine or a cosine depending on how the phase angle ϕ is adjusted. We won't worry much about ϕ, although in some applications it is important. Usually, we can start the clock so that $\phi = 0$.

To see a vibration, clamp a couple of inches of the tip of a bread knife onto a table with the heel of one hand, most of the blade and the handle projecting from the edge of the table (Fig. 1, next page). With your other hand, pull the handle down a little and release it sharply, with a "ping." Then the knife will vibrate, and

* On a programmable calculator with "user definable" keys, this [g] might indeed be a single key stroke.

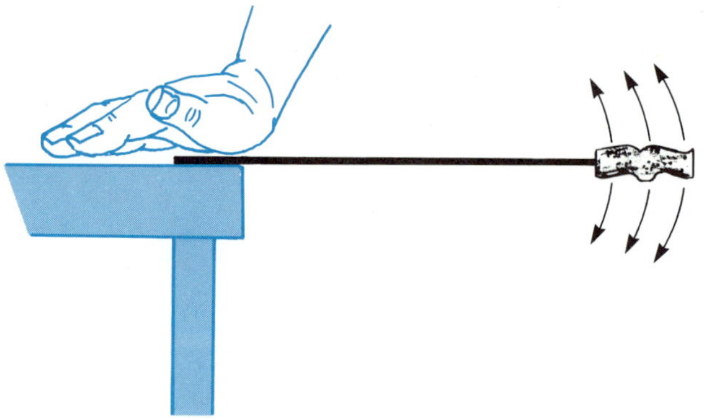

Fig. 1 Vibrating bread knife

the position of the end of the handle will vary from its rest position according to a formula of the type

$$y = A\sin(2\pi\nu t - \phi).$$

For a typical knife, ν will be between 8 and 12 Hz, depending on how much of the knife projects. Hence an up-down-up cycle will take between $\frac{1}{12}$ and $\frac{1}{8}$ sec.

Beats

When a tuning fork is set in motion, it causes a rapid vibration in air pressure given by the formula

$$P = A\sin 2\pi\nu t.$$

The ear detects the disturbance, which it hears as a musical note. For a middle C fork, $\nu \approx 261.63$ Hz, a much faster vibration than that of our bread knife.

Actually, a note produced by a musical instrument is not "pure" but a combination (superposition) of several pure notes. The corresponding formula is of the type

$$P = A_1\sin 2\pi\nu_1 t + A_2\sin 2\pi\nu_2 t + A_3\sin 2\pi\nu_3 t + \cdots$$

where $\nu_1 < \nu_2 < \nu_3 < \cdots$, and the numbers $A_2, A_3, \cdots$ are small compared to A_1. Thus, to the basic pure tone $A_1 \sin 2\pi\nu_1 t$ are added small amounts of higher tones (overtones or harmonics). These overtones give the note its particular quality; they help the ear distinguish between a note on a violin and the same note played on a clarinet.

Suppose two pure musical notes of equal loudness are played together. What do we hear? If their frequencies μ and ν are sufficiently different, we hear two distinct notes (a chord). But if μ and ν are close together, say $|\mu - \nu| \leq 10$, our ears cannot distinguish between the notes, and we hear one note. But we hear that note in an unexpected way; its volume is not steady but pulses up and down several times a second. This is called the phenomonon of **beats.**

To hear beats, try this experiment on a guitar. Tune the low E-string up to F# and tune the A-string down to F#. Then slightly flatten either of these. Pluck the two open strings together and the beats will be unmistakable. The closer in tune with each other the strings are, the slower the beats.

How can we explain beats? Believe it or not, by a trigonometric identity. When the two notes are sounded together, they cause a vibration in air pressure described by

$$P = A\sin 2\pi\mu t + A\sin 2\pi\nu t.$$

(We assume equal amplitudes for simplicity.) Now, there is an identity that expresses a sum of sines as a product (p. 285):

$$P = A(\sin 2\pi\mu t + \sin 2\pi\nu t) = 2A\cos 2\pi\sigma t \sin 2\pi\rho t,$$

where

$$\rho = \tfrac{1}{2}(\mu + \nu) \qquad \sigma = \tfrac{1}{2}(\mu - \nu).$$

Let us write

$$P = V(t)\sin 2\pi\rho t, \quad \text{where} \quad V(t) = 2A\cos 2\pi\sigma t.$$

The ear hears a note of frequency ρ and loudness $V(t)$. Since μ and ν are close, σ is small so $V(t)$ is a slowly fluctuating function. In fact, $V(t)$ varies from $2A$ (loud) to 0 (silence) to $-2A$ (loud) to 0 (silence) and back to $2A$ (loud) as t varies over a full period $0 \le t \le 1/\sigma$. See Fig. 2. The ear does not distinguish between positive and negative loudness; hence in one period of $V(t)$ there are two complete cycles of the beat:

loud soft loud soft loud

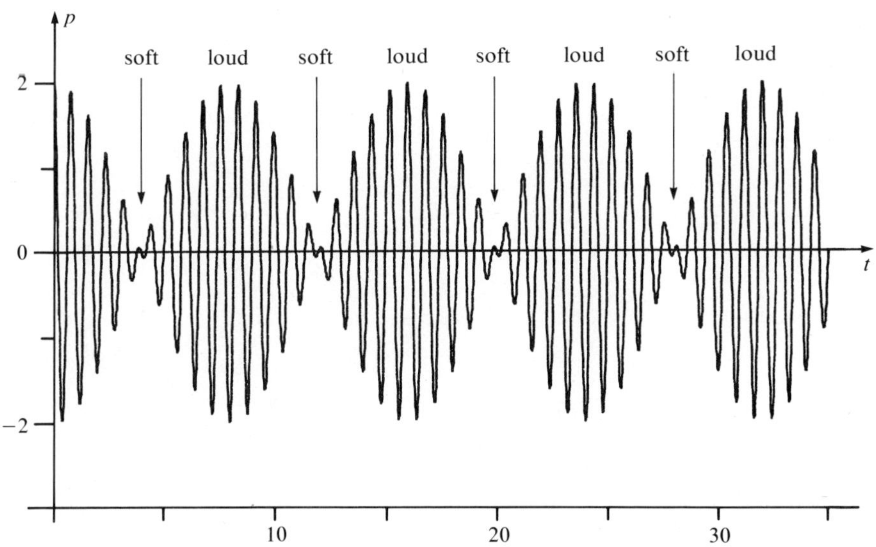

Fig. 2 $P = A(\cos 2\pi\mu t + \cos 2\pi\nu t)$ where $\mu \approx \nu$.
(In this computer-graphics example, $A = 1$, $\mu = \frac{21}{16}$, and $\nu = \frac{19}{16}$.)

Thus the **beat frequency** is $|2\sigma| = |\mu - \nu|$.

To illustrate these ideas, suppose you tune the A-string of a guitar to 92.5 Hz and the E-string to 90.5 Hz. Then

$$\rho = \tfrac{1}{2}(92.5 + 90.5) = 91.5 \qquad \sigma = \tfrac{1}{2}(92.5 - 90.5) = 1$$

and

$$P = A[\sin 2\pi(92.5)t + \sin 2\pi(90.5)t]$$
$$= (2A\cos 2\pi t)\sin 2\pi(91.5)t.$$

The ear will hear the note with pitch given by 91.5 Hz and loudness varying from $2A$ to 0 to $2A$, with two beats per second.

Remark Here is an experiment that makes the beat phenomenon visual. You will need two C-clamps (or a vise), two hacksaw blades, and a small rubber band (Fig. 3). Clamp one blade so the part projecting can be pinged with an audible note. (About 3–4 inches will project.) Clamp the second blade parallel to the first, about 2 inches away, and adjust it to exactly the same note. Then snap the rubber band over the two blades, about halfway. Finally, ping one blade and watch the "beat" move back and forth from one blade to the other.

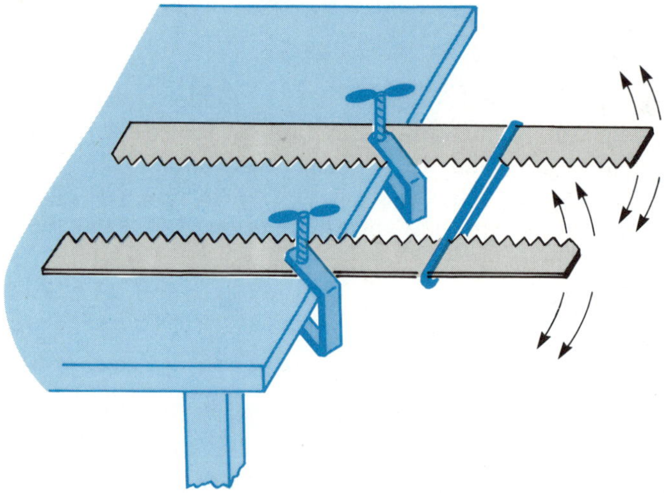

Fig. 3 "Beating" hacksaw blades

Amplitude Modulation

Let us describe an application of trigonometric identities to AM radio. The current generated by an AM transmitter is a pure vibration

$$i = A \sin 2\pi\mu t.$$

The **carrier frequency** (for commercial broadcast stations) lies in the interval

$$550 \text{ khz} \le \mu \le 1600 \text{ khz},$$

where khz means kilohertz, one thousand Hz. When the station is broadcasting its radio signal, but no sound, the amplitude A is constant, $A = A_0$. The constant A_0 is proportional to the power of the station. Sound information is added to the signal by varying (modulating) A, hence the name **amplitude modulation** or AM. To broadcast a sound signal $s(t)$, the amplitude A_0 is modulated by a fraction of its value proportional to $s(t)$. Thus the amplitude of the modulated signal is

$$A = A(t) = A_0[1 + ms(t)] \qquad (0 \le m \le 1)$$

For example, suppose middle C is being broadcast by a station at 980 on your radio dial. Then the modulated current is

$$i = i(t) = A_0(1 + m \sin 2\pi\nu t) \sin 2\pi\mu t,$$

where

$$\mu = 980{,}000 \text{ Hz} \quad \text{and} \quad \nu \approx 261.63 \text{ Hz}.$$

(Note that ν is very small compared to μ.) The modulated signal is actually the superposition (sum) of three pure vibrations—this is where trigonometric identities come in. We write

$$i = A_0[\sin 2\pi\mu t + m \sin 2\pi\nu t \sin 2\pi\mu t].$$

By the formulas for products on p. 284,

$$\sin 2\pi\nu t \sin 2\pi\mu t = \tfrac{1}{2}\cos 2\pi(\mu - \nu)t - \tfrac{1}{2}\cos 2\pi(\mu + \nu)t,$$

hence

$$i = A_0[\sin 2\pi\mu t + \tfrac{1}{2}m \cos 2\pi(\mu - \nu)t - \tfrac{1}{2}m \cos 2\pi(\mu + \nu)t].$$

Thus the radio signal broadcasting a pure musical note of frequency ν is the sum of three pure signals:

$A_0 \sin 2\pi\mu t$	at the carrier frequency μ
$\tfrac{1}{2}A_0 m \cos 2\pi(\mu - \nu)t$	the **left sideband** at frequency $\mu - \nu$
$\tfrac{1}{2}A_0 m \cos 2\pi(\mu + \nu)t$	the **right sideband** at frequency $\mu + \nu$.

The U.S. Federal Communications Commission requires each AM station to broadcast within a bandwidth of 10 khz centered at its assigned frequency μ. This means

$$\mu - 5000 < \mu - \nu \quad \text{and} \quad \mu + \nu < \mu + 5000,$$

hence, $0 \le \nu < 5000$ Hz. For all speech and most music this range of audio frequencies is adequate. (The highest piano note is about 4186 Hz.) However, the music transmitted is not HiFi; overtones are filtered out because of the bandwidth restrictions.

The graph of a (greatly distorted) modulated radio signal is shown in Fig. 4. You would hardly guess the decomposition of this graph into the carrier signal and the two sidebands.

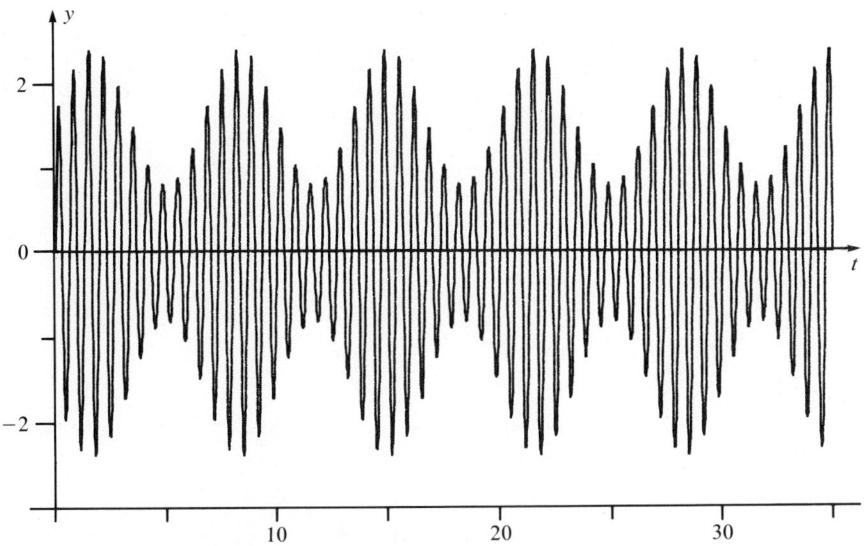

Fig. 4 Modulated AM signal $y = A(t) \sin 2\pi\mu t$ where $A(t) = A_0(1 + m \sin 2\pi\nu t)$. (In this computer-graphics example, $A_0 = 1.6$, $m = 0.5$, $\mu = 1.5$, and $\nu = 0.15$.)

REVIEW EXERCISES

For θ in the fourth quadrant and $\sec\theta = 3$, find

1 $\sin\theta$

2 $\cot\theta$.

Prove the identity

3 $\tan^2\theta - \sin^2\theta = \sin^2\theta\tan^2\theta$

4 $\sin\theta\cos\theta = \dfrac{1-\cos^2\theta}{\tan\theta}$

5 $\dfrac{1}{\sec\theta + \tan\theta} = \dfrac{1-\sin\theta}{\cos\theta}$

6 $\sec^4\theta + \tan^4\theta = 1 + 2\sec^2\theta\tan^2\theta$

7 $\cos\frac{1}{2}\theta\cos\frac{1}{4}\theta = \dfrac{\sin\theta}{4\sin\frac{1}{4}\theta}$

8 $\cos\frac{1}{2}\theta\cos\frac{1}{4}\theta\cos\frac{1}{8}\theta = \dfrac{\sin\theta}{8\sin\frac{1}{8}\theta}$.

9 Use $105^\circ = 60^\circ + 45^\circ$ to find $\sin 105^\circ$.

10 Use $165^\circ = 135^\circ + 30^\circ$ to find $\tan 165^\circ$.

11 Express $-5\cos\theta + 12\sin\theta$ in the form $a\cos(\theta - \theta_0)$.

12 Express $\sin 5\theta\cos 3\theta$ as a sum of sines.

13 Express $\cos\theta - \cos 3\theta$ as a product of a constant and two sines.

14 Prove $\dfrac{\cos(\alpha - \beta)}{\cos\alpha\sin\beta} = \tan\alpha + \cot\beta$.

Find exactly

15 $\arcsin(\cos\frac{5}{7}\pi)$

16 $\sin[\arccos(-0.3)]$.

Graph (for all x)

17 $y = \arcsin(\sin x)$

18 $y = \arccos(\cos x)$.

Solve on the interval $-\pi < \theta \le \pi$

19 $\sin 3\theta = -\frac{1}{2}\sqrt{3}$

20 $\tan 3\theta = -1$.

Find all solutions

21 $\sin 2\theta = -\sin\theta$

22 $\tan\theta = -\sin\theta$.

Solve to 4 places

23 $\sin^3\theta = \theta^2 - 1$

24 $\tan\theta = 2\theta \quad 0 < \theta < \frac{1}{2}\pi$.

8 SOLVING TRIANGLES

1 RIGHT TRIANGLES

This chapter concerns applications of the trigonometric functions to problems of practical geometry. In the fields that use these applications: astronomy, navigation, surveying, civil engineering, machine shop, architecture, construction trades, mechanical engineering, etc., angles are always measured in degrees, not in radians. So in this chapter we shall talk degrees almost exclusively.

Remark Recall that the complete circle is divided into 360 degrees. This division is very old; perhaps it originated in the ancient world from the mistaken belief that the Earth's orbit was a circle traced in a year of 360 days, so one degree represented one day's arc.

Trigonometry

The word **trigonometry** comes from Greek and means the measurement of triangles. The general problem of trigonometry is this: A triangle has three sides and three angles. Given some of these six quantities, measure (compute) the others. We discuss this problem for right triangles here, and for oblique triangles in Sections 3 and 4.

Recall that the trigonometric functions of an acute angle can be expressed as ratios of the sides of a right triangle (Fig. 1). Using these ratios, we can solve problems about right triangles: given two sides, or given one side and one acute angle, find the other sides and angles. (''Side'' is short for side length, and ''angle'' for angle measure.)

$$\sin\theta = \frac{\text{opp}}{\text{hyp}} \qquad \csc\theta = \frac{\text{hyp}}{\text{opp}}$$

$$\cos\theta = \frac{\text{adj}}{\text{hyp}} \qquad \sec\theta = \frac{\text{hyp}}{\text{adj}}$$

$$\tan\theta = \frac{\text{opp}}{\text{adj}} \qquad \cot\theta = \frac{\text{adj}}{\text{opp}}$$

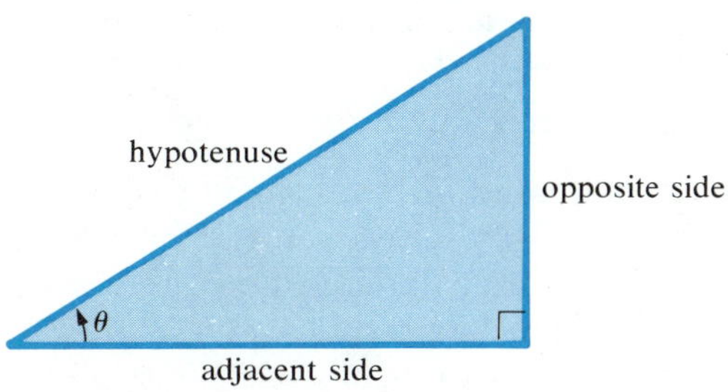

Fig. 1

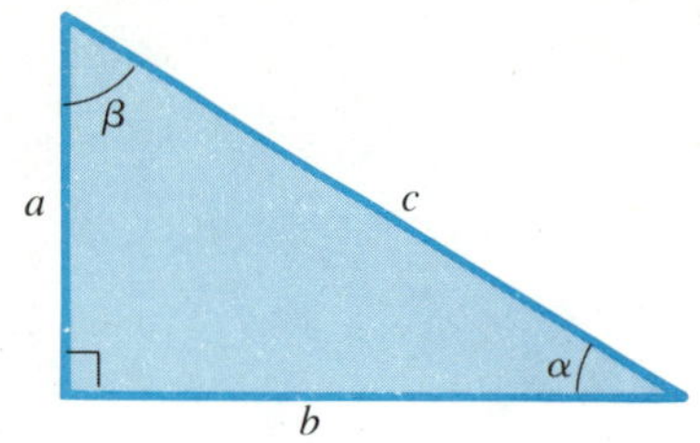

Fig. 2

Example 1 The sides and angles of a right triangle are labeled as in Figure 2.

(a) Given $a = 5.12$ and $c = 8.97$, find b, α, and β.

(b) Given $a = 1.483$ and $\beta = 53.92°$, find b, c, and α.

Solution (a) First find b by the Pythagorean Theorem:

$$b = \sqrt{c^2 - a^2} = \sqrt{(8.97)^2 - (5.12)^2} \approx 7.37.$$

Next find α:

$$\sin\alpha = \frac{a}{c} \qquad \alpha = \arcsin\frac{a}{c} = \arcsin\frac{5.12}{8.97} \approx 34.8°$$

Finally, note that $\alpha + \beta = 90°$. Therefore

$$\beta = 90° - \alpha \approx 90° - 34.8° \approx 55.2°.$$

(b) First find b:

$$\tan\beta = \frac{b}{a} \qquad b = a\tan\beta = 1.483\tan 53.92° \approx 2.035.$$

Next find c:

$$\cos\beta = \frac{a}{c} \qquad c = \frac{a}{\cos\beta} = \frac{1.483}{\cos 53.92°} \approx 2.518$$

Finally,

$$\alpha = 90° - \beta = 90° - 53.92° = 36.08°.$$

Answer (a) $b \approx 7.37$ $\alpha \approx 34.8°$ $\beta \approx 55.2°$

(b) $b \approx 2.035$ $c \approx 2.518$ $\alpha = 36.08°$

Remark Accuracy. The answer to a problem cannot be more accurate than the data. For example, given that the two legs of a right triangle are 8.43 and 9.16, our calculator shows the hypotenuse to be $\sqrt{8.43^2 + 9.16^2} \approx 12.44871479$. But the extra decimal places are unjustified. The only reasonable answer consistent with the accuracy of the data is 12.45. We'll say more about measurement and accuracy in Section 6.

Degrees, Minutes, and Seconds

For modern computation, it is convenient to express fractions of degrees in decimal parts, such as 31.742°. However, there is a long tradition of using sixtieths of a degree and sixtieths of a sixtieth of a degree as fractional units. This tradition goes back to the ancient Sumerians (*ca*. 3000 B.C.), who introduced the sexagesimal (base 60) system of computation. This system was taken over by the Babylonians and then the Greek astronomers. For better or for worse, today most devices that measure fractions of degrees are calibrated in these sixtieth units.

> **Degrees, Minutes, and Seconds** An angle of one **degree** (1°) is $\frac{1}{360}$ of a complete circle angle.
>
> An angle of one **minute** (1′) is $\frac{1}{60}$ of one degree.
>
> An angle of one **second** (1″) is $\frac{1}{60}$ of one minute.

We write $D°M'S''$ for an angle of D degrees plus M minutes plus S seconds. Thus

$$10°15'30'' = (10 + \tfrac{1}{60}\cdot 15 + (\tfrac{1}{60})^2\cdot 30)° \approx 10.258333°.$$

In general,

$$D°M'S'' = (D + \tfrac{1}{60}M + (\tfrac{1}{60})^2 S)°.$$

In order to convert $D°M'S''$ to decimal degrees, we express the right-hand side as

$$(S/60 + M)/60 + D.$$

This form suggests a (left-to-right) calculator routine:

S [÷] 6 0 [+] M [÷] 6 0 [+] D [=].

Example (set [deg] mode) 35°22′41″

4 1 [÷] 6 0 [+] 2 2 [÷] 6 0 [+] 3 5 [=] 35.37806

To calculate a trig function of an angle given as $D°M'S''$, convert to decimal degrees first, then press the appropriate trig key.

Example (set [deg] mode) sin 12°17′23″

2 3 [÷] 6 0 [+] 1 7 [÷] 6 0 [+] 1 2 [=] [sin] 0.212855

Now let's look at the reverse problem: given an angle in decimal degrees, express it in degrees, minutes, and seconds.

EXAMPLE 2 Convert 35.37806° to the form $D°M'S''$.

Solution The problem is to find D, M, and S so that

$$35.37806 = D + \tfrac{1}{60}M + (\tfrac{1}{60})^2 S.$$

Obviously $D = 35$. To find M, subtract 35 and then multiply by 60:

$$0.37806 = \tfrac{1}{60}M + (\tfrac{1}{60})^2 S$$

$$22.6836 = M + \tfrac{1}{60}S.$$

Hence $M = 22$. To find S, we subtract 22 and then multiply by 60:

$$0.6836 = \tfrac{1}{60}S \qquad S = 41.016.$$

Answer 35°22′41″

EXAMPLE 3 Express in degrees, minutes, and seconds the acute angle θ satisfying

$$\cos\theta = 0.343434.$$

Solution First calculate (in [deg] mode) the desired angle in decimal degrees

. 3 4 3 4 3 4 [INV] [cos] 69.913769

To convert to $D°M'S''$, continue the sequence of key strokes:

[−] 6 9 [×] 6 0 [=] 54.82614

Finally,

[−] 5 4 [×] 6 0 [=] 49.5684

Answer 69°54′49.6″

Remark Many calculators have keys for converting DMS degrees to decimal degrees and vice versa. Their use varies from instrument to instrument, so you must read the instruction manual for details.

On one calculator, you enter 123°43′27.5″ as

1 2 3 . 4 3 2 7 5

and press [→DEG] for the decimal form 123.7243°. To reverse this, you press [→D.MS] and the display reads 123.43275, which is interpreted as 123°43′27.5″.

Negative angles The notation $-D°M'S''$ is understood to mean $-(D°M'S'')$, or

$$-D - \tfrac{1}{60}M - (\tfrac{1}{60})^2 S \text{ deg}.$$

For instance

$$-12°17'24'' = -12.2900°.$$

Exercises

In the following exercises you will be given two parts of the right triangle of Fig. 3. Find the remaining sides and angles.

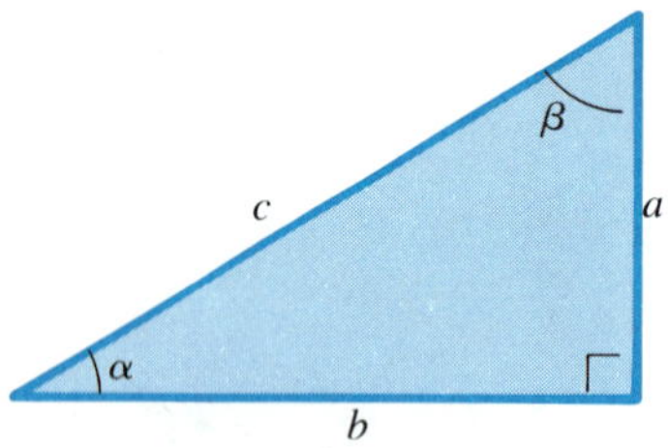

Fig. 3

1 $a = 12 \quad c = 13$

2 $a = 6 \quad c = 13$

3 $a = 107.3 \quad c = 175.5$

4 $a = 24.03 \quad b = 58.81$

5 $b = 15 \quad c = 22$

6 $b = 8.0 \quad c = 12.5$

7 $b = 45.80 \quad c = 66.75$

8 $b = 237.6 \quad c = 289.7$

9 $a = 4 \quad b = 5$

10 $a = 16 \quad b = 9$

11 $a = 27.12 \quad b = 9.315$

12 $a = 18.06 \quad c = 143.1$

13 $c = 30.0 \quad \alpha = 14.5°$

14 $c = 30.0 \quad \alpha = 71.4°$

15 $a = 15.0 \quad \alpha = 25.6°$

16 $a = 9.50 \quad \alpha = 40.8°$

17 $c = 235.5 \quad \beta = 34.2°$

18 $c = 1293 \quad \beta = 68.8°$

19 $b = 56.03 \quad \beta = 43.9°$

20 $b = 526.2 \quad \beta = 33.7°$

21 $c = 1.02570 \quad \alpha = 12.43°$

22 $a = 2.00143 \quad \beta = 51.25°$

23 $a = 3.11701 \quad \alpha = 5.08°$

24 $b = 138.212 \quad \beta = 72.21°$

Express in decimal degrees

25 1″ **26** 1′ **27** 59″ **28** 59′

29 12′12″ **30** 6′3″ **31** $\tfrac{1}{2}$° **32** $\tfrac{1}{10}$°

33 17°3′39″ **34** 59°17′39″ **35** 73°0′23″ **36** 89°59′52″

37 71°57′51″ **38** 1°13′47″ **39** 157′ **40** 10693″.

Express in degrees, minutes, and seconds

41 19.1082° **42** 57.1131° **43** 72.9499° **44** 61.7431°
45 0.0200° **46** 137.1933° **47** 0.000478° **48** 3.00015°

Express in the form $D°M'S''$ the acute θ satisfying

49 $\sin\theta = 0.35162$ **50** $\sin\theta = 0.82141$
51 $\cos\theta = 0.35162$ **52** $\cos\theta = 0.82141$
53 $\tan\theta = 3.10482$ **54** $\tan\theta = 5.00000$
55 $\sin\theta = 4.84814 \times 10^{-6}$ **56** $\tan\theta = 2.90888 \times 10^{-4}$.

Compute

57 $\sin 15°12'19''$ **58** $\sin 9'13''$ **59** $\cos 73°51'39''$
60 $\cos 5°0'17''$ **61** $\tan 53°9'48''$ **62** $\tan 1''$
63 $\cos 89°59'59''$ **64** $\tan 89°59'59''$ **65** $\sin(223°34'12'')$
66 $\sin(349°12'0'')$ **67** $\cos(163°57'55'')$ **68** $\tan(300°12'23'')$.

Compute

69 $\sin 2' - 2\sin 1'$ **70** $\sin 30°1'' - \sin 30°$
71 $\sin 88°1'' - \sin 88°$ **72** $\sin 89°50'1'' - \sin 89°50'$
73 $\tan 1°1'' - \tan 1°$ **74** $\tan 89°1'' - \tan 89°$.

A hand of an analog clock sweeps an angle of $D°M'S''$. Find the elapsed time if the hand is

75 the hour hand **76** the minute hand **77** the second hand.

78 It is between 2 PM and 3 PM, and the minute hand is exactly 193° clockwise from the hour hand. What time is it?

2 APPLICATIONS

We begin with the type of geometric problem that is typical of applications of right triangle trigonometry.

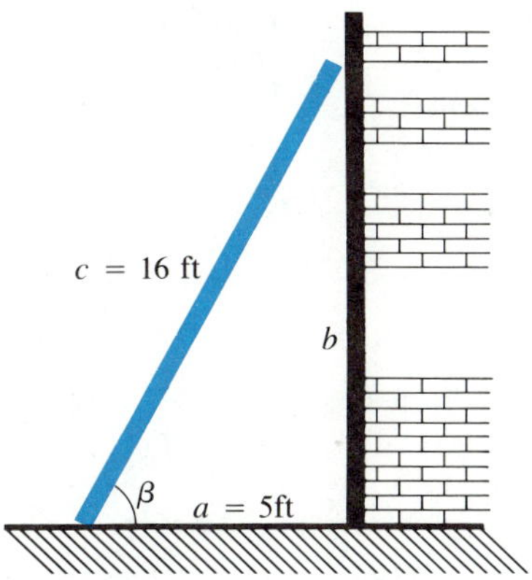

Fig. 1 Ladder

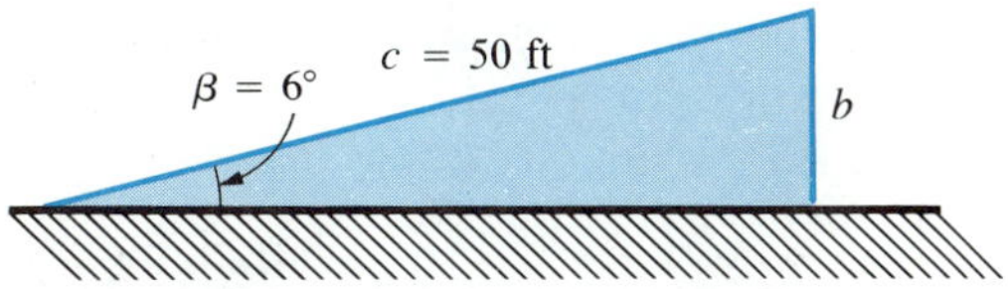

Fig. 2 Ramp (angle exaggerated)

EXAMPLE 1 A 16-ft ladder leans against a wall, with its base 5 ft from the wall. Find the angle the ladder makes with the ground and find how far up the wall the ladder extends.

Solution Draw a figure (Fig. 1). We want β and b; we are given $a = 5$ ft and $c = 16$ ft. Let us assume these measurements are accurate to one-half inch, and write $a = 5.0$, $c = 16.0$. To find β, we use

$$\cos\beta = \frac{a}{c}, \quad \text{so} \quad \beta = \text{arc cos}\,\frac{a}{c} = \text{arc cos}\,\frac{5.0}{16.0} \approx 71.8^\circ.$$

By the Pythagorean theorem,

$$b = \sqrt{c^2 - a^2} = \sqrt{16^2 - 5^2} \approx 15.2 \text{ ft}$$

Answer 71.8° 15.2 ft

Query Are we justified in writing 71.8°, or should we round to 72°? Try varying 5.0 and 16.0 by about $\frac{1}{2}$ inch, say by ± 0.05, and see what happens to β.

Example 2 A 50-ft ramp is inclined at an angle of 6° with the ground. How high does it rise above the ground?

Solution We seek b in Fig. 2, previous page; we are given $c = 50.0$ ft and $\beta = 6.0°$. Since

$$\sin\beta = \frac{b}{c}, \quad \text{we have} \quad b = c\sin\beta = 50\sin 6^\circ \approx 5.2 \text{ ft}.$$

Answer 5.2 ft

Multi-step Problems

Example 3 Estimate θ in Fig. 3a.

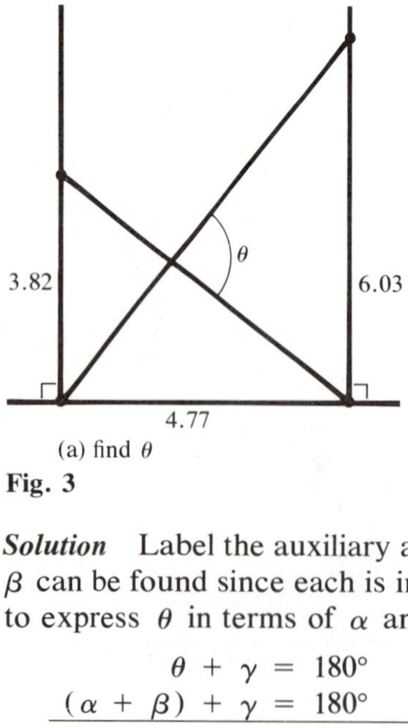

(a) find θ

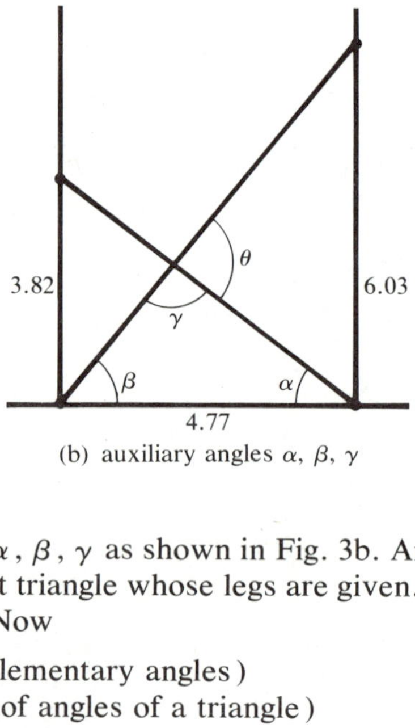

(b) auxiliary angles α, β, γ

Fig. 3

Solution Label the auxiliary angles α, β, γ as shown in Fig. 3b. Angles α and β can be found since each is in a right triangle whose legs are given. The idea is to express θ in terms of α and β. Now

$$\begin{aligned} \theta + \gamma &= 180^\circ && \text{(supplementary angles)} \\ (\alpha + \beta) + \gamma &= 180^\circ && \text{(sum of angles of a triangle)} \\ \hline \text{so} \qquad \theta &= \alpha + \beta && \text{(subtracting)}. \end{aligned}$$

From the figure

$$\tan\alpha = \frac{3.82}{4.77} \qquad \tan\beta = \frac{6.03}{4.77}$$

$$\alpha = \text{arc tan}\,\frac{3.82}{4.77} \approx 38.7^\circ \qquad \beta = \text{arc tan}\,\frac{6.03}{4.77} \approx 51.7^\circ$$

$$\theta = \alpha + \beta \approx 90.4^\circ$$

Answer 90.4°

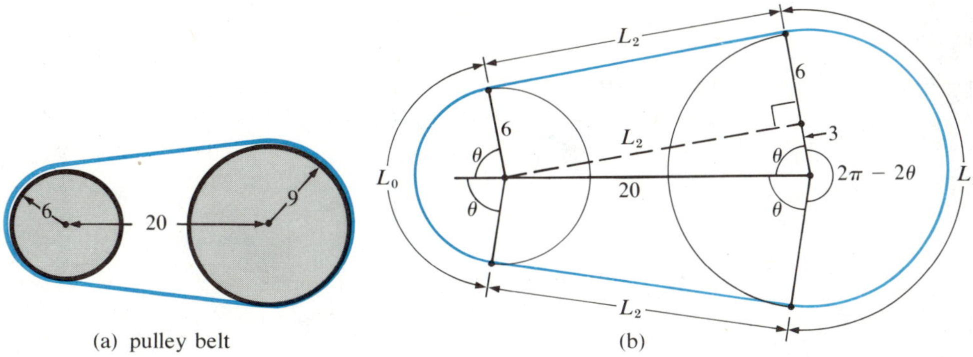

Fig. 4

EXAMPLE 4 Two pulleys of radii 6.00 in. and 9.00 in. are connected by a belt (Fig. 4a). The distance between their centers is 20.00 in. Find the length of the belt.

Solution The belt consists of a small circular arc of length L_0, a large arc of length L_1, and two straight segments, each of length L_2. Thus the total length of the belt is

$$L = L_0 + L_1 + 2L_2.$$

Each straight piece is a common tangent to the two circles. Draw radii to the two upper points of tangency (Fig. 4b). These radii are parallel, since each is perpendicular to the common tangent. Therefore they determine equal central angles θ as shown.

Now draw an auxiliary line, through the center of the smaller circle and parallel to the upper common tangent. It forms a rectangle and a right triangle. From the right triangle,

$$L_2 = \sqrt{20^2 - 3^2} = \sqrt{391} \approx 19.77$$

and

$$\cos\theta = \tfrac{3}{20} = 0.15 \qquad \theta = \operatorname{arc}\cos 0.15 \approx 1.42 \text{ rad}.$$

The smaller circular arc is subtended by the central angle 2θ. Therefore its length (radius times central angle in *radians*) is

$$L_0 = 6 \cdot 2\theta = 12\theta \approx 17.04$$

The larger arc is subtended by the central angle $2\pi - 2\theta$, so its length is

$$L_1 = 9 \cdot (2\pi - 2\theta) \approx 30.99.$$

Finally

$$L = L_0 + L_1 + 2L_2 \approx 17.04 + 30.99 + 2 \cdot 19.77 = 87.57$$

Answer 87.57 in.

Example 5 A surveyor measures (Fig. 5, next page) from A to B, using two intermediate stations S and T in the same vertical plane as A and B. Estimate the horizontal and vertical distances, h and v, from A to B.

Solution The horizontal distance h from A to B is the sum of three horizontal distances: from A to S, from S to T, and from T to B. You must imagine that

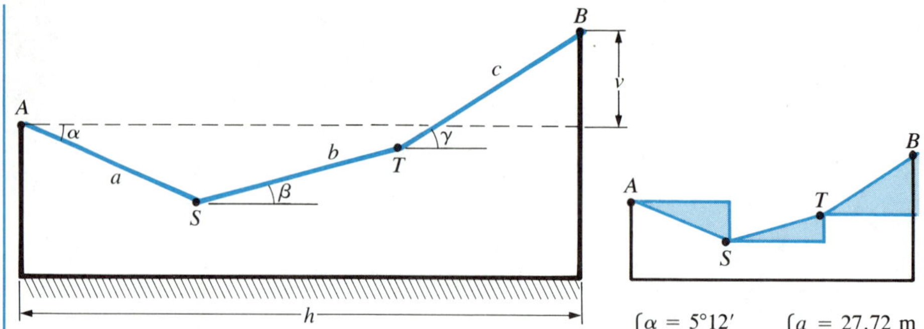

$$\begin{cases} \alpha = 5°12' \\ \beta = 3°43' \\ \gamma = 7°35' \end{cases} \qquad \begin{cases} a = 27.72 \text{ m} \\ b = 34.39 \text{ m} \\ c = 23.41 \text{ m} \end{cases}$$

Fig. 5 (angles exaggerated)

three right triangles with horizontal and vertical legs have been added to the figure. Then

$$\begin{aligned} h &= a \cos \alpha + b \cos \beta + c \cos \gamma \\ &= 27.72 \cos 5°12' + 34.39 \cos 3°43' + 23.41 \cos 7°35' \\ &\approx 27.606 + 34.318 + 23.205 \approx 85.13 \text{ m}. \end{aligned}$$

The vertical distance v from A to B consists of a negative contribution (downward from A to S) followed by two positive contributions (upward from S to T to B). Thus

$$\begin{aligned} v &= -a \sin \alpha + b \sin \beta + c \sin \gamma \\ &= -27.72 \sin 5°12' + 34.39 \sin 3°43' + 23.41 \sin 7°35' \\ &\approx -2.512 + 2.229 + 3.089 \approx 2.81 \text{ m} \end{aligned}$$

Answer $h \approx 85.13$ m $\quad v \approx 2.81$ m

Remark The computation goes briskly if you use [→DEG] and [M+]. For h, with slight variations for different calculators, it should be

5 . 1 2 [→DEG] [cos] [×] 2 7 . 7 2 [M+]
3 . 4 3 [→DEG] [cos] [×] 3 4 . 3 9 [M+]
7 . 3 5 [→DEG] [cos] [×] 2 3 . 4 1 [M+] [RCL] 85.13

Be sure to clear the memory before you start calculating v.

Isosceles Triangles

An isosceles triangle (Fig. 6a) has equal base angles α, apex angle β, base b, and equal slant sides a. By drawing the altitude (Fig. 6b) of length h, we cut the isosceles triangle into two congruent right triangles. Several relations are clear from the figure:

$$\frac{\frac{1}{2}b}{a} = \cos \alpha = \sin \tfrac{1}{2}\beta \qquad h^2 + (\tfrac{1}{2}b)^2 = a^2$$

$$2\alpha + \beta = 180° \qquad h = a \sin \alpha.$$

With these formulas we can express various sides or angles in terms of others.

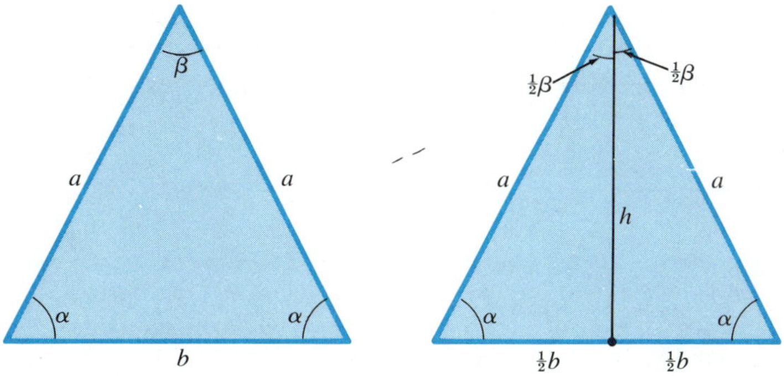

(a) isosceles triangle (b) altitude h

Fig. 6

EXAMPLE 6 The tent in Fig. 7a has a rectangular floor, and its peak is centered. Find the base angles α and β of the front and side triangular flaps respectively, and their common slant height s.

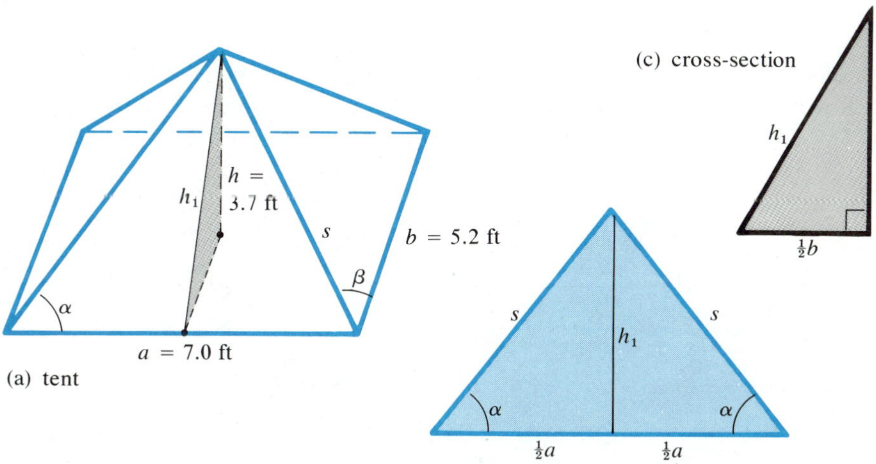

(a) tent (b) front flap

Fig. 7

Solution The front flap is an isosceles triangle (Fig. 7b) with base $a = 7.0$ ft. We could find α and s from this triangle if we knew its altitude h_1. But h_1 is the hypotenuse (Fig. 7c) of a right triangle with legs h and $\frac{1}{2}b$, so

$$h_1^2 = h^2 + (\tfrac{1}{2}b)^2 = (3.7)^2 + (2.6)^2 = 20.45 \qquad h_1 \approx 4.5222$$

Now, returning to Fig. 7b, we have

$$\tan\alpha = \frac{h_1}{\frac{1}{2}a} \approx \frac{4.5222}{3.5} \approx 1.2921 \qquad \alpha \approx \arctan 1.2921 \approx 52.26°$$

and

$$s^2 = h_1^2 + (\tfrac{1}{2}a)^2 = 20.45 + (3.5)^2 = 32.7 \qquad s \approx 5.7184$$

We treat the right side flap similarly:

$$h_2^2 = h^2 + (\tfrac{1}{2}a)^2 = (3.7)^2 + (3.5)^2 = 25.94 \qquad h_2 \approx 5.0931$$

and

$$\tan\beta = \frac{h_1}{\frac{1}{2}b} \approx \frac{5.0931}{2.6} \approx 1.9589 \qquad \beta \approx \arctan 1.9589 \approx 62.96°$$

There is an easier way to find β. Once we have found s, we can use

$$\cos\beta = \frac{\frac{1}{2}b}{s} \approx \frac{2.6}{5.7184} \approx 0.45467 \qquad \beta \approx \text{arc}\cos 0.45467 \approx 62.96°$$

This is a good check!

Answer $\alpha \approx 52°$ $\quad \beta \approx 63°$ $\quad s = 5.7$ ft

Remark We rounded off the answer to two significant figures, precisely what the data (and common sense) of the problem justifies. For the purpose of illustration, we carried 4 or 5 significant figures through the work. Actually, with a calculator, you never should do *any* intermediate rounding; only round your final answer.

For instance, if you want $\sqrt{5.7^2 + 3.8^2}$ to two significant figures, then

$$\sqrt{5.7^2 + 3.8^2} = \sqrt{32.49 + 14.44} = \sqrt{46.93} = 6.850547 \cdots \approx 6.9$$

is correct. However, rounding to two significant figures at each step gives

$$\sqrt{5.7^2 + 3.8^2} \approx \sqrt{32 + 14} = \sqrt{46} = 6.7823 \approx 6.8$$

Area

The area of a triangle with base b and height h equals $\frac{1}{2}bh$. From this fact follows a formula for the area of a triangle in terms of two sides and their included angle:

$$A = \tfrac{1}{2}ab\sin\gamma.$$

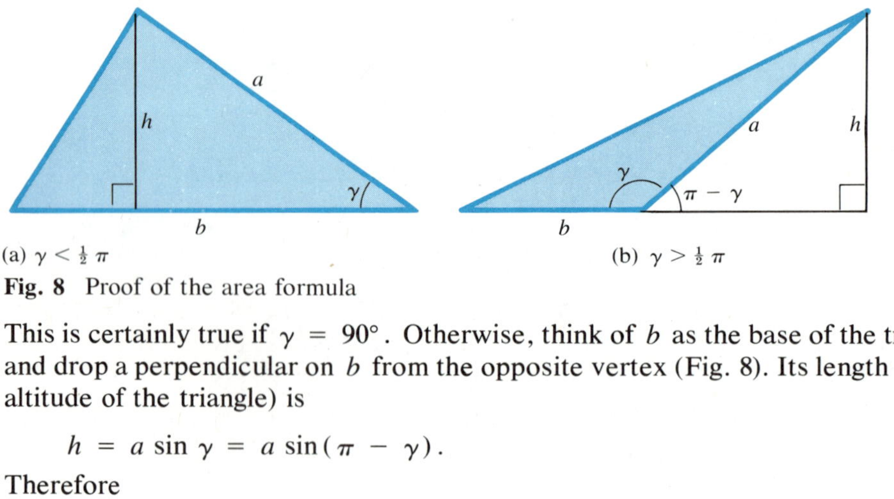

(a) $\gamma < \frac{1}{2}\pi$ (b) $\gamma > \frac{1}{2}\pi$

Fig. 8 Proof of the area formula

This is certainly true if $\gamma = 90°$. Otherwise, think of b as the base of the triangle and drop a perpendicular on b from the opposite vertex (Fig. 8). Its length h (the altitude of the triangle) is

$$h = a\sin\gamma = a\sin(\pi - \gamma).$$

Therefore

$$A = \tfrac{1}{2}bh = \tfrac{1}{2}ba\sin\gamma.$$

EXAMPLE 7 Express the area of an isosceles triangle in terms of its base b and slant side a.

Solution In Fig. 6b, we have

$$a^2 = h^2 + (\tfrac{1}{2}b)^2 \qquad h = \sqrt{a^2 - \tfrac{1}{4}b^2}.$$

Therefore

$$A = \tfrac{1}{2}bh = \tfrac{1}{2}b\sqrt{a^2 - \tfrac{1}{4}b^2} = \tfrac{1}{4}b\sqrt{4a^2 - b^2}.$$

Answer $A = \frac{1}{4}b\sqrt{4a^2 - b^2}$.

Exercises

1 When the elevation of the Sun is 60.5°, a flagpole casts a 44-ft shadow. Find the height of the flagpole (Fig. 9).

2 (cont.) At what elevation will the shadow be twice as long?

3 A 15-ft ladder leans against a wall making a 70° angle with the ground. At what height does its upper end touch the wall?

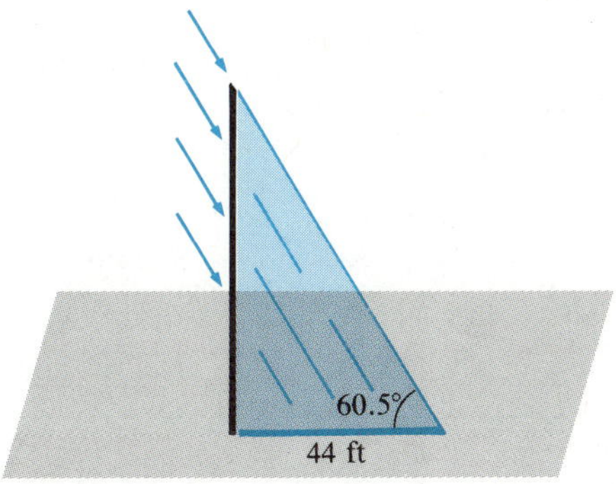

Fig. 9

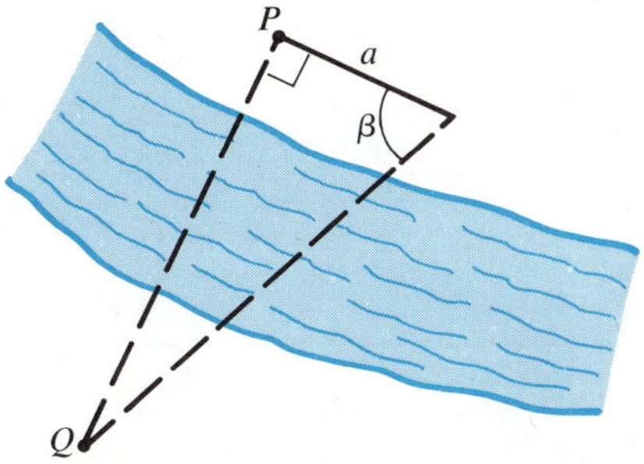

Fig. 10

4 You have a tape rule and a protractor and you wish to estimate the distance from P to an inaccessible point Q. See Fig. 10. You lay out a right triangle with leg $a = 50$ ft and measure $\beta = 67°$. What is the distance?

5 A road makes an angle of 4.5° with the horizontal. I start at sea level and drive exactly 5 km. What is my elevation above sea level?

6 I observe that the angle of elevation of a certain peak is 20°. I drive 2.0 km closer along a level road and find that now the angle is 32°. What is the height of the peak?

7 During the first few seconds of its flight, a rocket rises straight up. Five seconds after launch, an observer 1.5 km away notes that its angle of elevation is 7.5°. Four seconds later the angle is 58.1°. How far did the rocket rise during these four seconds?

8 A plane flying at an altitude of 300 m passes directly overhead. Two seconds later its angle of elevation is 41°. Compute its speed.

9 In Fig. 11 find ac/b^2.

10 In Fig. 12 find $x + y$.

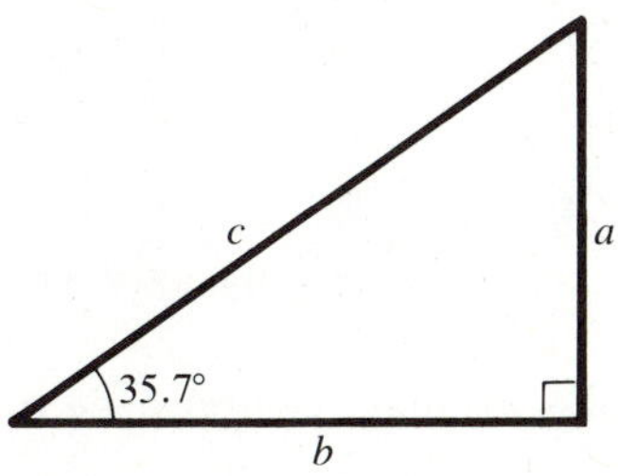

Fig. 11

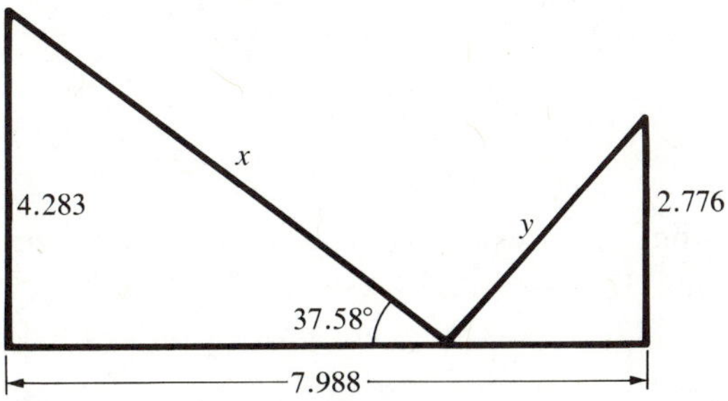

Fig. 12

11 Show that the length of the belt in Fig. 13, next page, is

$$L = 2\sqrt{D^2 - (R_1 - R_0)^2} + 2\pi R_1 - 2(R_1 - R_0)\,\text{arc cos}\left(\frac{R_1 - R_0}{D}\right).$$

12 (cont.) Verify this formula for $R_0 = 0$.

13 (cont.) Verify this formula for $R_0 = R_1$.

14* (cont.) Suppose $R_0 = 3.00$, $R_1 = 6.00$, and $L = 60.00$. Estimate D numerically to two places.

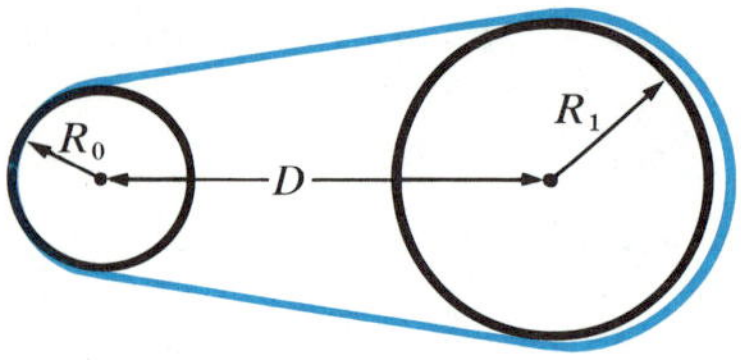

Fig. 13

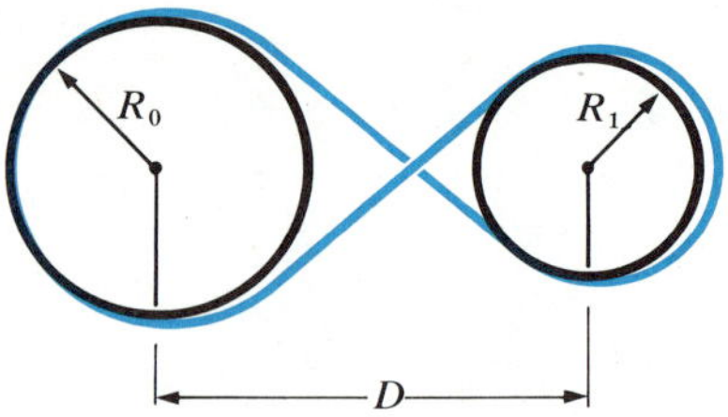

Fig. 14

15 Find the length of the belt in Fig. 14 if $R_0 = R_1$.

16* (cont.) Find the length if $R_0 < R_1$.

Find the horizontal and vertical distances from A to B.

	17 (Fig. 15)	**18** (Fig. 15)	**19** (Fig. 15)	**20** (Fig. 15)
a	17.83 m	29.51 m	61.3 m	104.7 m
b	32.91 m	12.22 m	52.6 m	281.2 m
α	9.014°	4.112°	7.52°	3.104°
β	4.771°	2.790°	3.46°	1.892°

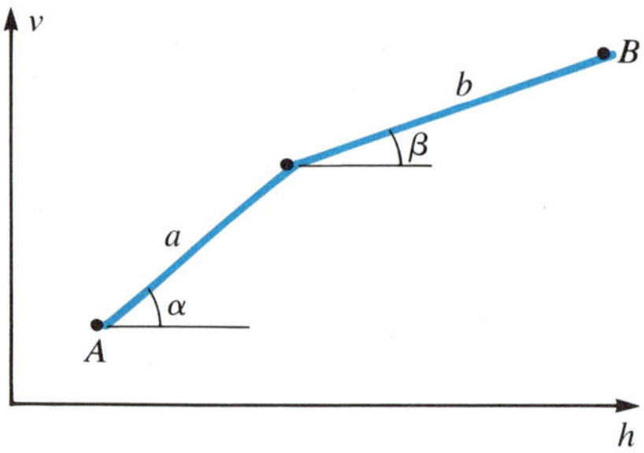

Fig. 15

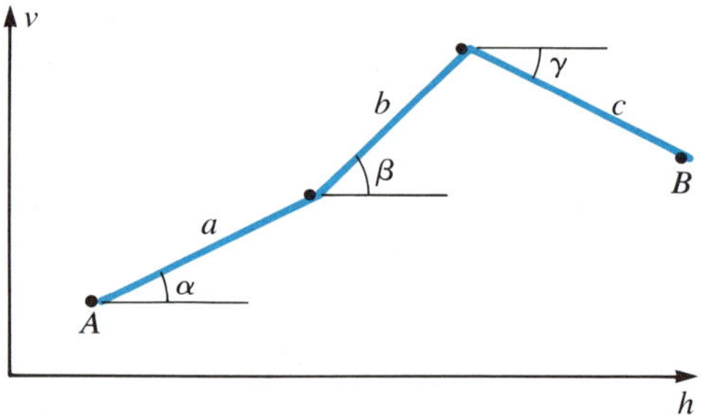

Fig. 16

	21 (Fig. 16)	**22** (Fig. 16)	**23** (Fig. 16)	**24** (Fig. 16)
a	10.2 m	58.39 m	21.69 m	1056 m
b	8.6 m	41.11 m	31.82 m	1932 m
c	3.4 m	19.38 m	31.24 m	2158 m
α	5.1°	2.49°	3.52°	6.603°
β	6.6°	7.17°	6.14°	2.417°
γ	4.2°	3.94°	1.43°	3.528°

25 A room measures $12 \times 15 \times 8$ ft. Compute the angles that the diagonal makes with each of the edges.

26 In Fig. 17, express the chord length L in terms of the radius r and the central angle θ.

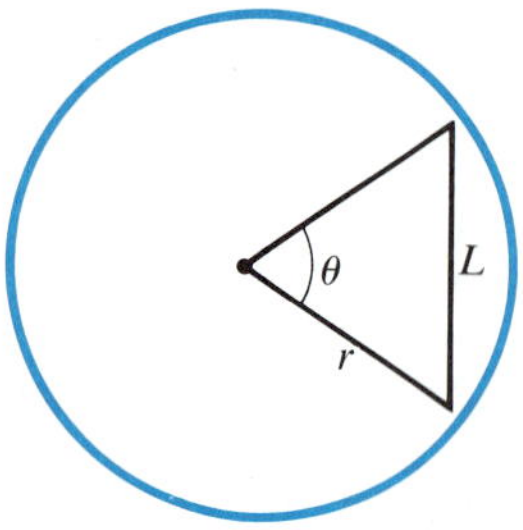

Fig. 17

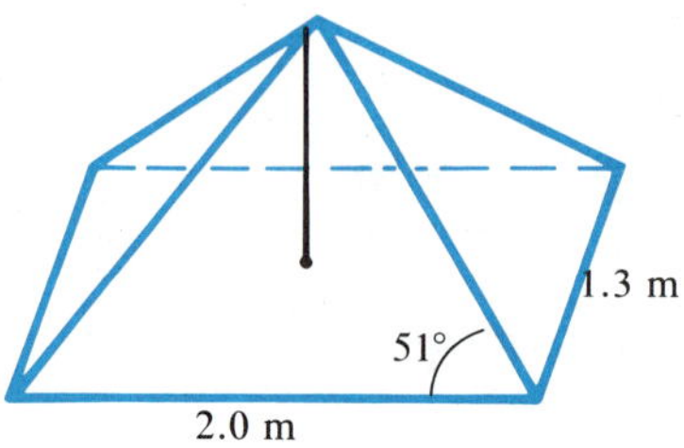

Fig. 18

27 (cont.) Find the perimeter of an equilateral triangle inscribed in a circle of radius r.

28 (cont.) Find the perimeter of a regular hexagon inscribed in a circle of radius r.

29 (cont.) Find the perimeter of a regular octagon inscribed in a circle of radius r.

30 Find the height of the tent in Fig. 18. The floor is rectangular and the peak centered.

31 Express the arc length s in terms of a and y (Fig. 19).

32 Estimate θ in Fig. 20 to three places.

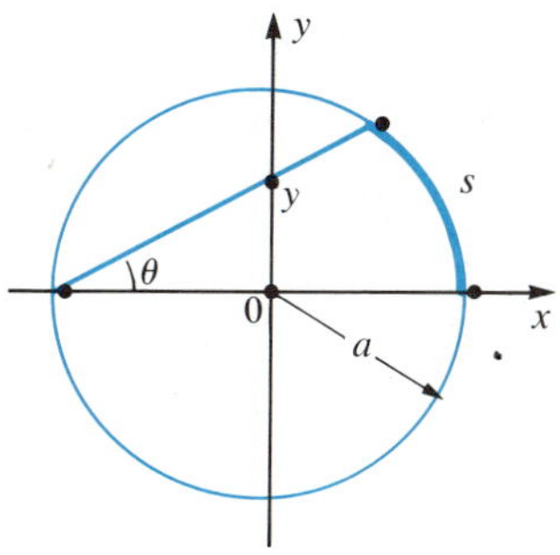

Fig. 19

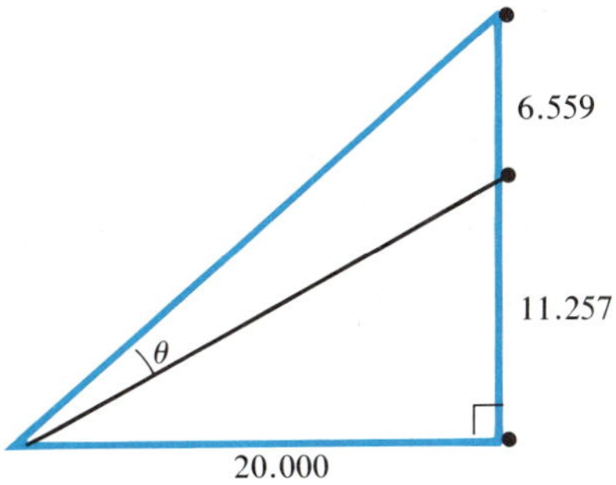

Fig. 20

33 Express the area of an isosceles triangle in terms of its base b and its apex angle β.

34 Express the area of an isosceles triangle in terms of its slant side a and its apex angle β.

35 (cont.) Find the area of a regular octagon inscribed in a circle of radius r.

36 (cont.) Find the area of a regular dodecagon (12 sides) inscribed in a circle of radius r.

37 Find the apex angle of the isosceles triangle whose equal sides are 10 ft and area is 25 ft^2.

38 An isosceles triangle has base 1 and apex angle 20°. For what apex angle would its area be twice as much?

39 The pyramid (Fig. 21) with square base a has height h. Express its edge length b and apex angle in terms of a and h.

40* The 20-ft pipe will never turn the corner in the 6-ft hall (Fig. 22). Approximate the angle at which it gets stuck.

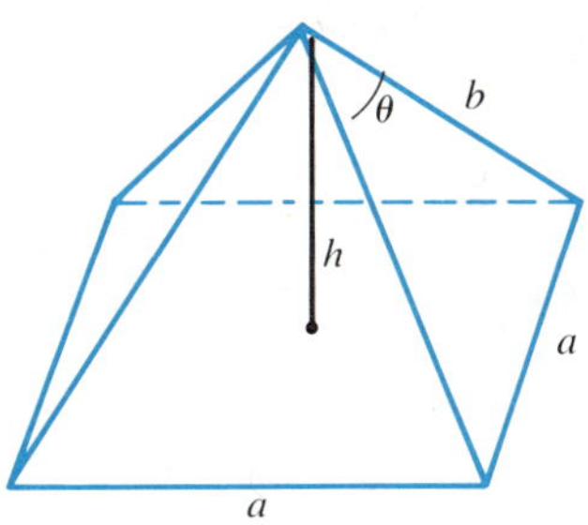

Fig. 21

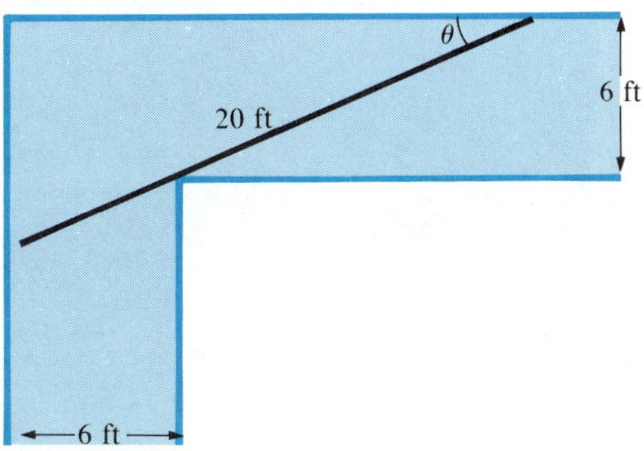

Fig. 22

The **sine bar** (Fig. 23, next page) is a precision tool used in laboratories and machine shops to measure angles. It consists of two cylinders of equal radii, placed parallel, with their axes 25.0000 cm apart. They are connected to a flat surface, parallel to the plane that passes through their axes. In practice, the cylinders rest on highly accurate gauge blocks on a horizontal table (Fig. 24, next page). The sine bar is then tilted to the horizontal through an angle θ.

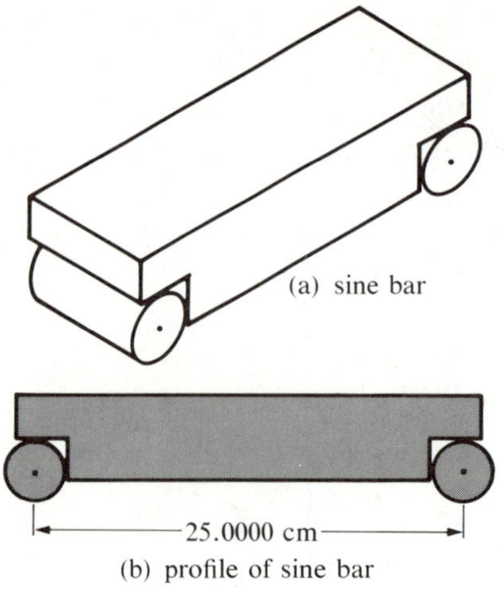

Fig. 23

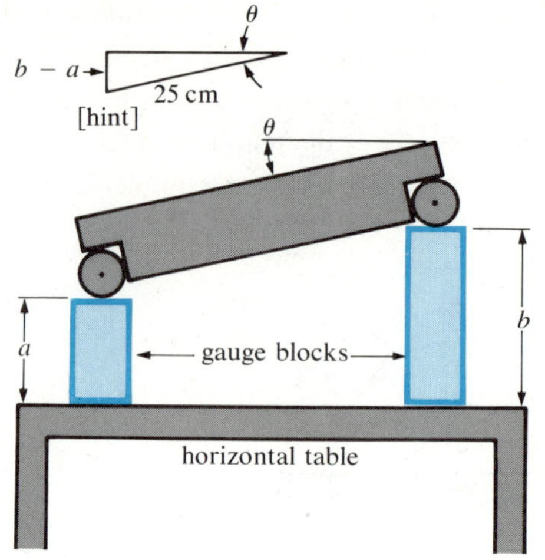

Fig. 24

Find θ in °′″

	41	42	43	44
a	1.0010 cm	2.0000 cm	2.0000 cm	1.0000 cm
b	3.3705 cm	6.0050 cm	12.5120 cm	7.3596 cm

Find b in cm

	45	46	47	48
a	1.0042 cm	2.0000 cm	1.5000 cm	1.3872 cm
θ	3°12′52″	5°0′17″	1°52′43″	25°5′21″.

The remaining exercises are related to practical surveying. The figures are stripped of instruments, terrain, and surveyors, giving only the essentials of the math.

49 (Fig. 25) Find the angle θ in °′″ that gives 100 power magnification.

50 (Fig. 26) Find h if $\alpha = 1°17'29''$

51 (Fig. 26) Find α if $h = 37.804$ m

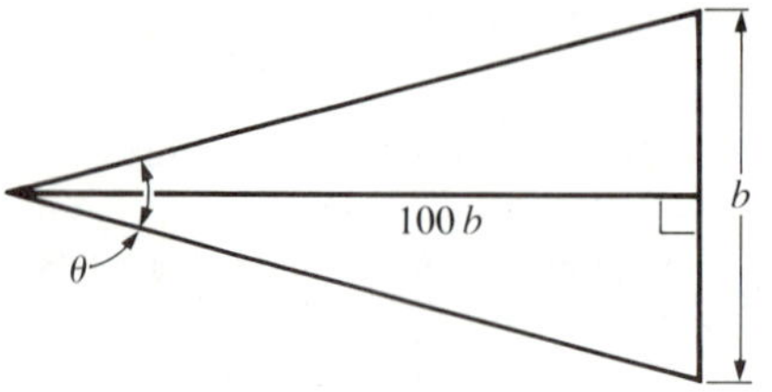

Fig. 25 (Isosceles triangle)

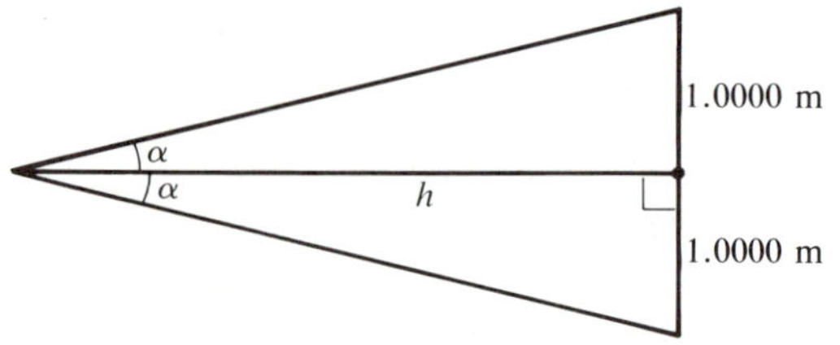

Fig. 26

52 (Fig. 27) Find h and v if $\theta = 13°17'5''$ and $\alpha = 48'19''$.

53 (Fig. 27) Find θ and h if $v = 19.762$ m and $\alpha = 38'17''$.

54 (Fig. 27) Find θ and v if $h = 82.78$ m and $\alpha = 32'54''$.

55 (Fig. 28) Find v and h if $d = 1.4820$ m and $\theta = 16°2'27''$.

56* (Fig. 28) Find v and θ if $h = 100.00$ m and $d = 2.0000$ m.

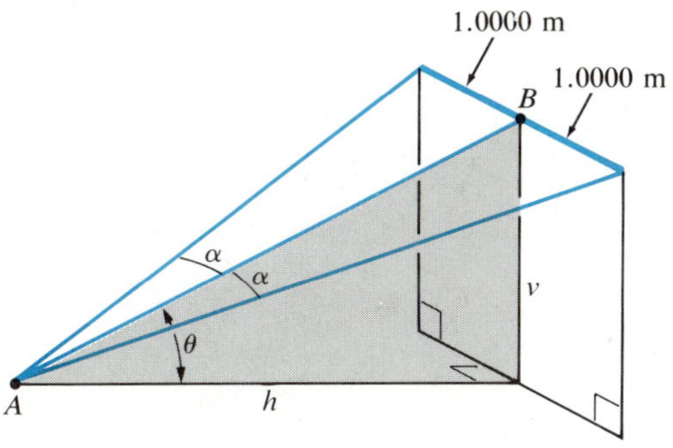

Fig. 27 (Horizontal subtense bar; vertical plane shaded)

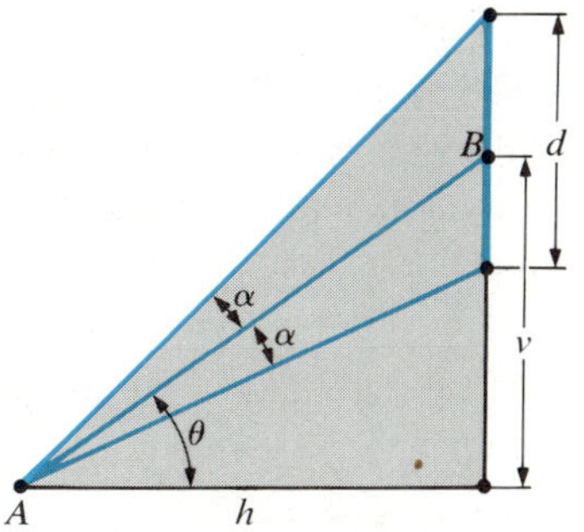

Fig. 28 (Vertical subtense bar) $\alpha = 17'\ 11''$

3 LAW OF COSINES

We now take up oblique triangles, that is, non-right triangles. In this section we develop and apply a formula called the law of cosines, and in the next section another formula called the law of sines. Together these are adequate for all numerical problems concerning oblique triangles.

Generally we shall label the six "parts" of a triangle as in Fig. 1. Each side has a Latin label and its opposite angle has the corresponding Greek label.

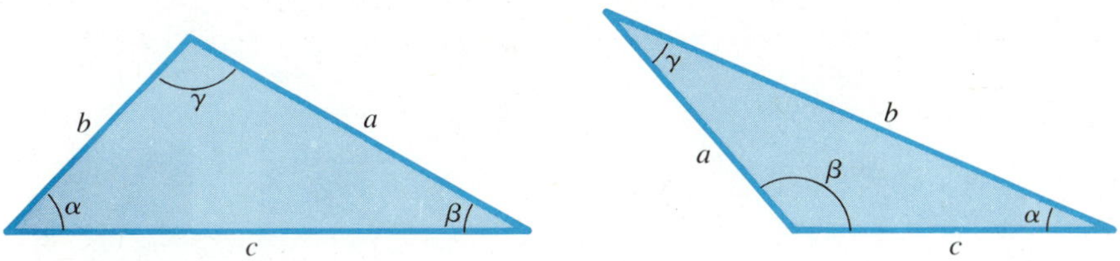

Fig. 1 The 6 parts of a triangle: 3 sides and 3 angles

Now we can state the law of cosines.

> **Law of Cosines**
>
> $$c^2 = a^2 + b^2 - 2ab\cos\gamma.$$

(The law of cosines is a generalization of the Pythagorean Theorem. If γ is a right angle, then $\cos\gamma = 0$, and the formula in the preceding box says that $c^2 = a^2 + b^2$.)

In words, the square of any side of a triangle equals the sum of the squares of the other two sides minus twice their product by the cosine of the included angle. Thus, for the triangle in Fig. 1, we also have

$$a^2 = b^2 + c^2 - 2bc\cos\alpha \quad \text{and} \quad b^2 = c^2 + a^2 - 2ca\cos\beta.$$

To prove the law of cosines, choose axes so that the angle γ is in standard position with side a along the positive x-axis (Fig. 2, next page). The upper vertex of the triangle is at $(b \cos \gamma, b \sin \gamma)$. By the distance formula,

$$\begin{aligned} c^2 &= (b \cos \gamma - a)^2 + (b \sin \gamma)^2 \\ &= (b^2 \cos^2 \gamma - 2ab \cos \gamma + a^2) + b^2 \sin^2 \gamma \\ &= b^2(\cos^2 \gamma + \sin^2 \gamma) + a^2 - 2ab \cos \gamma \\ &= a^2 + b^2 - 2ab \cos \gamma, \end{aligned}$$

so the formula is proved.

The law of cosines is an effective tool when we are given either all three sides of a triangle or two sides and their included angle. Let us consider these two cases.

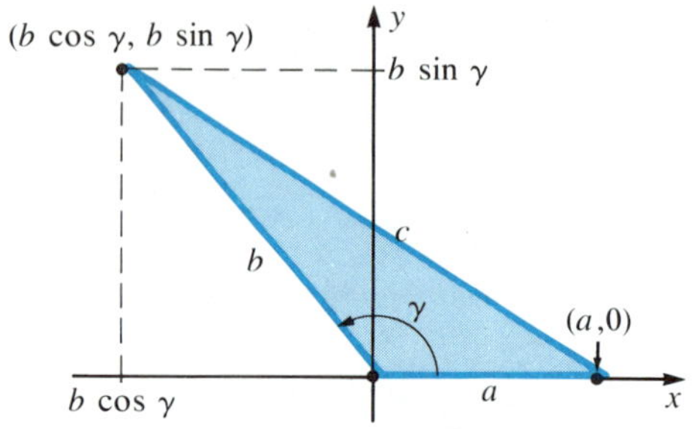

Fig. 2 Proof of the law of cosines

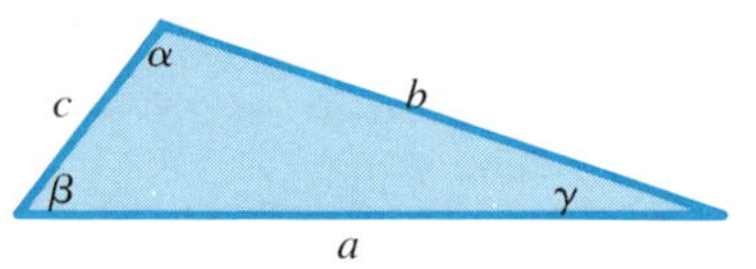

Fig. 3 Given $a = 7.135$, $b = 6.002$, $c = 2.508$

Given: Three Sides (SSS)

Given three sides of a triangle, its three angles can be found by the law of cosines.

EXAMPLE 1 Given (Fig. 3)

$$a = 7.135 \qquad b = 6.002 \qquad c = 2.508,$$

find α, β, and γ.

Solution By the law of cosines

$$a^2 = b^2 + c^2 - 2bc \cos \alpha.$$

Therefore

$$\cos \alpha = \frac{b^2 + c^2 - a^2}{2bc} = \frac{(6.002)^2 + (2.508)^2 - (7.135)^2}{(2)(6.002)(2.508)} \approx -0.2855,$$

$$\alpha \approx \text{arc cos}(-0.2855) \approx 106.59°$$

Similarly

$$\cos \beta = \frac{c^2 + a^2 - b^2}{2ca} = \frac{(2.508)^2 + (7.135)^2 - (6.002)^2}{(2)(2.508)(7.135)} \approx 0.5916$$

$$\beta \approx \text{arc cos}(0.5916) \approx 53.73°$$

$$\cos \gamma = \frac{a^2 + b^2 - c^2}{2ab} = \frac{(7.135)^2 + (6.002)^2 - (2.508)^2}{(2)(7.135)(6.002)} \approx 0.9415$$

$$\gamma \approx \text{arc cos}(0.9415) \approx 19.69°.$$

Check $106.59 + 53.73 + 19.69 = 180.01$

This is convincing since $\alpha + \beta + \gamma$ should be 180.00.

Answer $\alpha \approx 106.59° \qquad \beta \approx 53.73° \qquad \gamma \approx 19.69°$

EXAMPLE 2 Given

$$a = 4.102 \qquad b = 3.997 \qquad c = 8.165,$$

find α, β, and γ.

Solution By the law of cosines

$$\cos\alpha = \frac{b^2 + c^2 - a^2}{2bc} = \frac{(3.997)^2 + (8.165)^2 - (4.102)^2}{(2)(3.997)(8.165)} \approx 1.008.$$

This is impossible, since $-1 \le \cos\alpha \le 1$, so there is no such triangle.

Why? Because the sides of a triangle must satisfy the triangle inequalities

$$a < b + c \qquad b < c + a \qquad c < a + b$$

In this case

$$a + b = 4.102 + 3.997 = 8.099 < 8.165 = c.$$

Answer There is no such triangle.

EXAMPLE 3 A pilot intends to fly from A to B, a straight line distance of 200.0 miles. His flight plan calls for a straight leg of 97.0 miles from A to a navigation station C, then another straight leg of 108.0 miles from C to B. By what angle should he change course at C?

Solution The desired angle is θ in Fig. 4. But $\theta = 180 - \gamma$, and we can find γ by the law of cosines:

$$\cos\gamma = \frac{108^2 + 97^2 - 200^2}{(2)(108)(97)} \approx -0.9034$$

$$\gamma \approx \text{arc}\cos(-0.9034) \approx 154.6°$$

Therefore

$$\theta = 180 - \gamma \approx 180 - 154.6 = 25.4°$$

Answer 25.4°

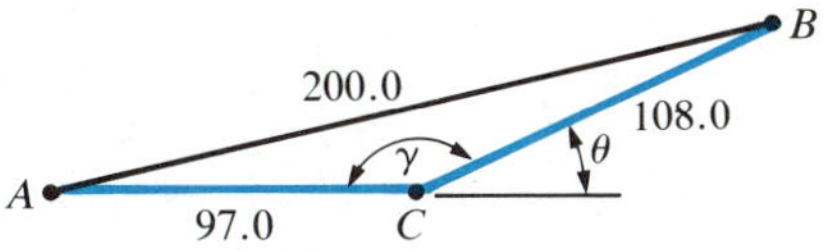

Fig. 4

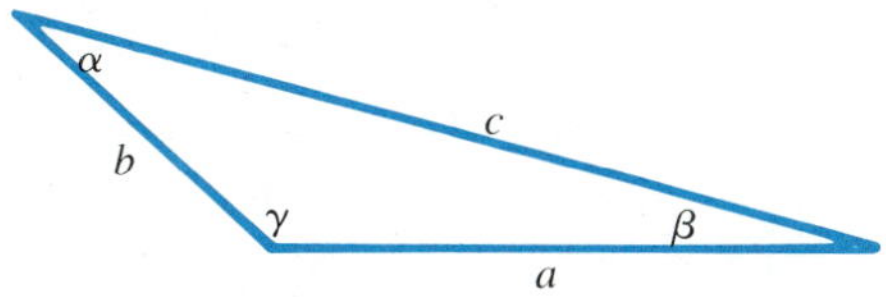

Fig. 5 Given $a = 14.722$, $b = 8.624$, $\gamma = 137.41°$

Given: Two Sides and Their Included Angle (SAS)

EXAMPLE 4 Given in Fig. 5:

$$a = 14.722 \qquad b = 8.624 \qquad \gamma = 137.41°,$$

find c, α, and β.

Solution By the law of cosines,

$$\begin{aligned} c^2 &= a^2 + b^2 - 2ab\cos\gamma \\ &= (14.722)^2 + (8.624)^2 - (2)(14.722)(8.624)(\cos 137.41°) \\ &\approx 478.054 \\ c &\approx 21.864 \end{aligned}$$

Now we have the three sides, so we can proceed as in Example 1 to find α and β:

$$\cos\alpha = \frac{b^2 + c^2 - a^2}{2bc} = \frac{(8.624)^2 + (21.864)^2 - (14.722)^2}{(2)(8.624)(21.864)} \approx 0.8901$$

$$\alpha \approx \text{arc cos } 0.8901 \approx 27.11°$$

$$\cos\beta = \frac{c^2 + a^2 - b^2}{2ca} = \frac{(21.864)^2 + (14.722)^2 - (8.624)^2}{(2)(21.864)(14.722)} \approx 0.9637$$

$$\beta \approx \text{arc cos } 0.9637 \approx 15.48°$$

Check $\alpha + \beta + \gamma \approx 27.11 + 15.48 + 137.41 = 180°$

Answer $c \approx 21.864 \quad \alpha \approx 27.11° \quad \beta \approx 15.48°$

EXAMPLE 5 Two planes at the same altitude are approaching an airport. Radar shows plane A at 18.0 miles due east of the tower and plane B at 22.5 miles from the tower on a line 10° north of east. How far apart are the planes?

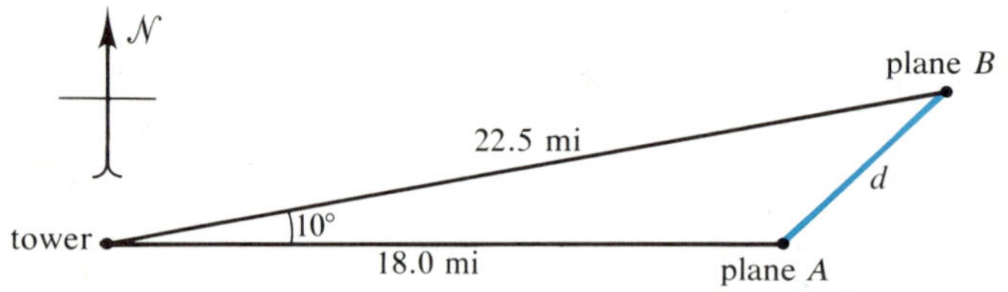

Fig. 6

Solution Draw Fig. 6. To find the distance d between the planes, use the law of cosines:

$$d^2 = (18.0)^2 + (22.5)^2 - (2)(18.0)(22.5)(\cos 10°) \approx 32.56$$

$$d \approx 5.7$$

Answer 5.7 miles

Further applications of the law of cosines are given in Section 5.

Exercises

Find α, β, and γ to 2 places

1 $a = 2.00 \quad b = 3.00 \quad c = 4.00$

2 $a = 7.00 \quad b = 6.00 \quad c = 5.00$

3 $a = 2.80 \quad b = 1.90 \quad c = 3.60$

4 $a = 10.20 \quad b = 13.60 \quad c = 8.300$

5 $a = 1.043 \quad b = 0.986 \quad c = 1.527$

6 $a = 200.8 \quad b = 157.7 \quad c = 166.5$

7 $a = 16.38 \quad b = 24.43 \quad c = 9.02$

8 $a = 7.382 \quad b = 7.889 \quad c = 14.001$

9 $a = 3048 \quad b = 3692 \quad c = 6480$

10 $a = 0.039 \quad b = 0.846 \quad c = 0.862$

11 $a = 0.416 \quad b = 0.752 \quad c = 0.329$

12 $a = 20.97 \quad b = 16.82 \quad c = 3.970.$

Find the other side and two angles

13 $a = 5.000 \quad b = 3.000 \quad \gamma = 39.00°$

14 $a = 4.000 \quad b = 7.000 \quad \gamma = 126.00°$

15 $a = 2.788 \quad b = 9.303 \quad \gamma = 11.14°$

16 $a = 7.980 \quad b = 8.003 \quad \gamma = 9.368°$

17 $b = 130.2 \quad c = 16.24 \quad \alpha = 152.37°$

18 $b = 0.01348 \quad c = 0.5172 \quad \alpha = 83.47°$

19 $b = 30348 \quad c = 52802 \quad \alpha = 39°17'12''$

20 $b = 17.634 \quad c = 62.115 \quad \alpha = 108°46'53''$

21 $c = 99.44 \quad a = 10.66 \quad \beta = 3.06°$

22 $c = 14.92 \quad a = 17.89 \quad \beta = 19.84°$

23 $c = 10{,}000 \quad a = 10{,}003 \quad \beta = 34''$

24 $c = 100.0 \quad a = 0.1362 \quad \beta = 179°21'.$

25 The diagonals of a parallelogram are 10 cm and 15 cm, and they intersect at an angle of 28.7°. Compute the sides of the parallelogram.

26 A point A is 10.00 km due north of here. Another point B is 12.36 km from here on a line 17.4° east of north. What is the distance from A to B?

27 A plane travels 50 miles due east after take-off, then adjusts its course 10° northward and flies 100 miles. How far is it from the point of departure?

28 Two boats, one traveling 19 km/hr and the other 23 km/hr, leave a point on straight paths 65° from each other. How far apart are they after 2.5 minutes?

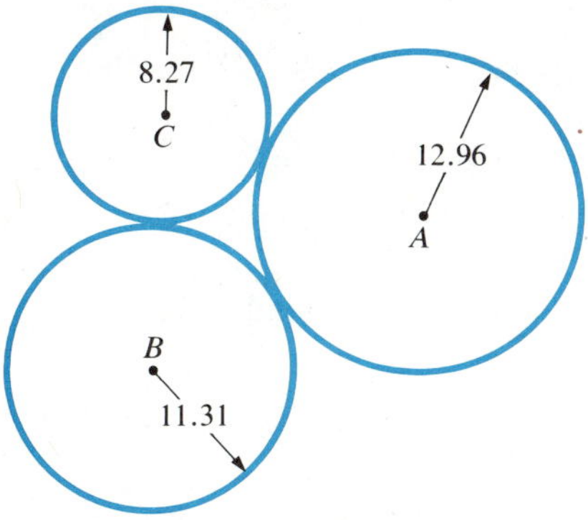

Fig. 7

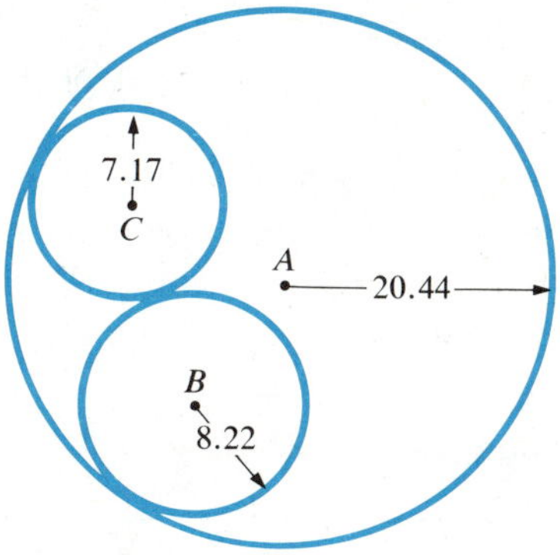

Fig. 8

29 The circles in Fig. 7 are mutually tangent. Find the angles of the triangle ABC.

30 The circles in Fig. 8 are mutually tangent. Find the angles of the triangle ABC.

31 Use the law of cosines to prove this assertion about any parallelogram: the sum of the squares of the two diagonals equals the sum of the squares of the four sides.

32 (cont.) How does it follow that the diagonals of a rhombus (parallelogram with equal sides) are perpendicular?

33 The centerfielder stands 340 ft from home plate in straight centerfield. How far is he from third base? (The distance between bases is 90 ft.)

34 The leftfielder stands 330 ft from home plate. A straight line from him to home plate bisects the line from second to third base. How far is he from second base? (The distance between bases is 90 ft.)

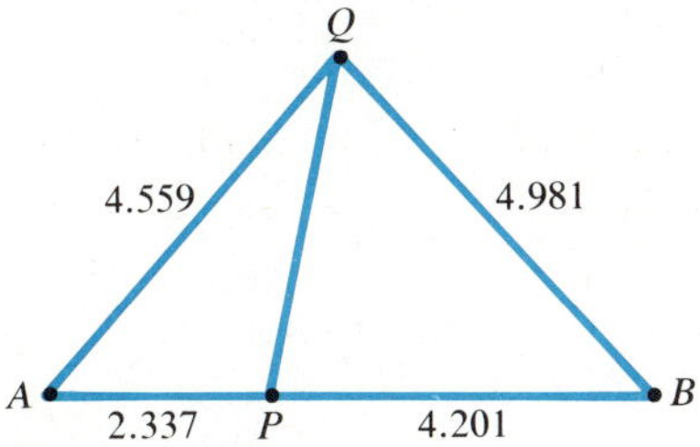

Fig. 9

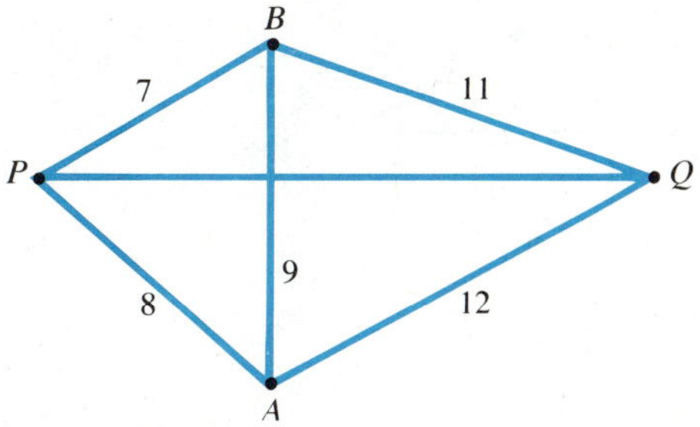

Fig. 10 ($\overline{AB} = 9$)

35* Find $\overline{PQ}$ in Fig. 9. [*Hint* Two steps are needed.]

36* Find $\overline{PQ}$ in Fig. 10. [*Hint* Three steps are needed.]

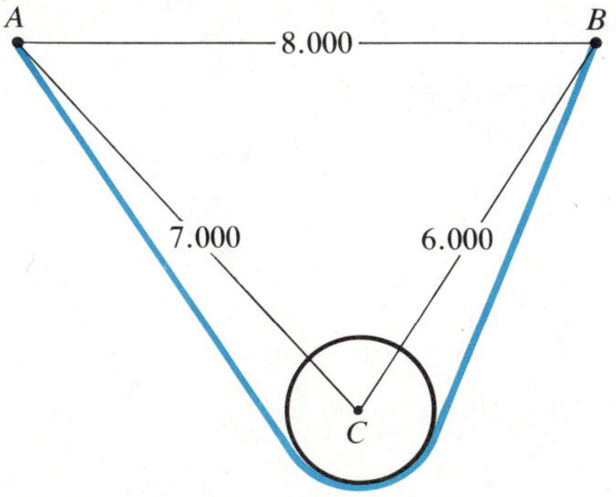

Fig. 11 (pulley radius = 1.000)

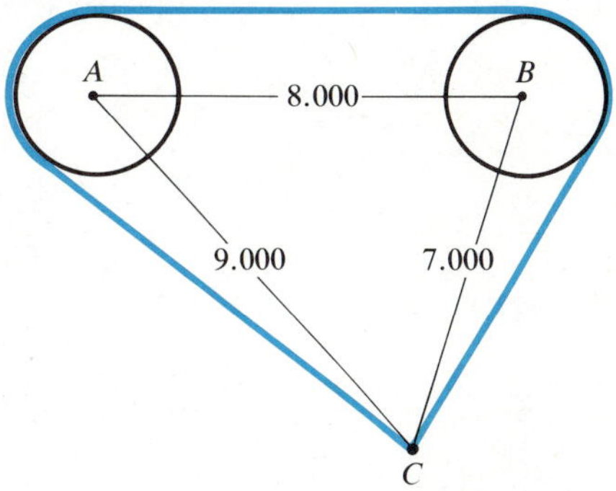

Fig. 12 (pulley radii = 1.500)

37 Find the length of the rope in Fig. 11, previous page. It is attached at A and at B, and taut around the pulley centered at C.

38* Find the length of the rope in Fig. 12, previous page. The two pulleys centered at A and at B have equal radii 1.500, and the rope is pulled taut at C.

4 LAW OF SINES

As we saw in Section 3, given SSS (three sides) or SAS (two sides and their included angle), the remaining parts of a triangle can be found using the law of cosines. The cases SAA, ASA, and SSA cannot be handled by the law of cosines; they require the law of sines. (Note that AAA is simply not enough information.)

Law of Sines $\dfrac{a}{\sin\alpha} = \dfrac{b}{\sin\beta} = \dfrac{c}{\sin\gamma}$

We'll postpone the proof until Section 5 and get right to its use in solving triangles.

Given: Two Angles and a Side (SAA or ASA)

Given any two angles of a triangle, we have the third angle automatically, because $\alpha + \beta + \gamma = 180°$. Therefore the cases SAA (two angles and an excluded side) and ASA (two angles and their included side) are really equivalent problems.

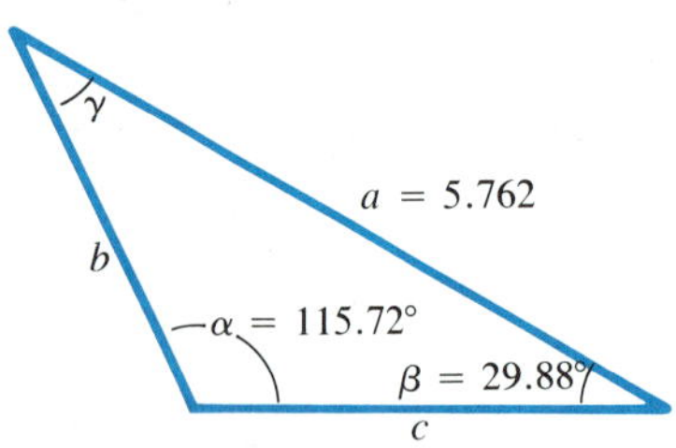

Fig. 1 (SAA)

EXAMPLE 1 Given

$$\alpha = 115.72° \qquad \beta = 29.88° \qquad a = 5.762,$$

find γ, b, and c.

Solution First,

$$\gamma = 180° - \alpha - \beta = 180° - 115.72° - 29.88° = 34.40°$$

To determine b we use part of the law of sines:

$$\frac{b}{\sin\beta} = \frac{a}{\sin\alpha}$$

$$b = \frac{a\sin\beta}{\sin\alpha} = \frac{(5.762)\sin 29.88°}{\sin 115.72°} \approx 3.186.$$

To determine c, we use the law of sines again:

$$\frac{c}{\sin\gamma} = \frac{a}{\sin\alpha}$$

$$c = \frac{a \sin \gamma}{\sin \alpha} = \frac{(5.762) \sin 34.40°}{\sin 115.72°} \approx 3.613 .$$

Check $b^2 + c^2 - 2bc \cos \alpha$

$$\approx (3.186)^2 + (3.613)^2 - (2)(3.186)(3.613)(\cos 115.72°)$$

$$\approx 33.195$$

$$a^2 \approx (5.762)^2 \approx 33.201 .$$

The law of cosines, $a^2 \approx b^2 + c^2 - 2bc \cos \alpha$, checks.

Answer $\gamma = 34.40°$ $\quad b \approx 3.186$ $\quad c \approx 3.613$

Given: Two Sides and an Excluded Angle (SSA)

Suppose sides a and c and angle γ are given. Then several possibilities can occur. Let us study them by drawing a few figures. To make life a little easier, we'll assume temporarily that γ is an acute angle.

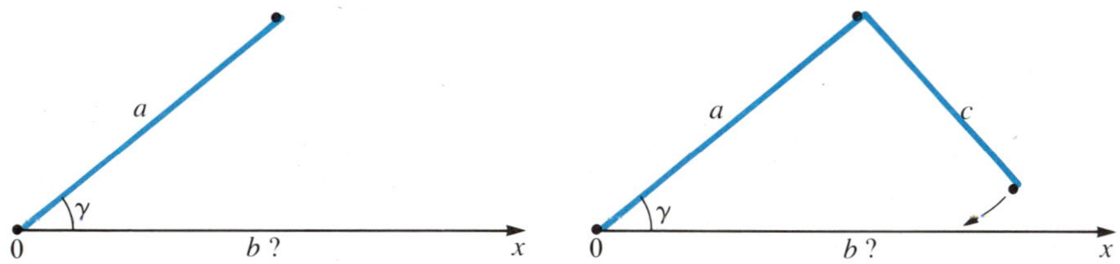

(a) Take b along the x-axis. (b) Swing c from the top of a.

Fig. 2 (SSA) Given: $a \quad c \quad \gamma$

Think of side b as the base of the triangle lying on the x-axis with its left end-point at the origin (Fig. 2a). Then side a will make angle γ with the positive x-axis. Side c will dangle from the top end-point of a. Imagine it swinging like a pendulum, the lower end tracing an arc (Fig. 2b). This arc may meet the x-axis twice, may just touch it (tangent), or may miss the axis altogether.

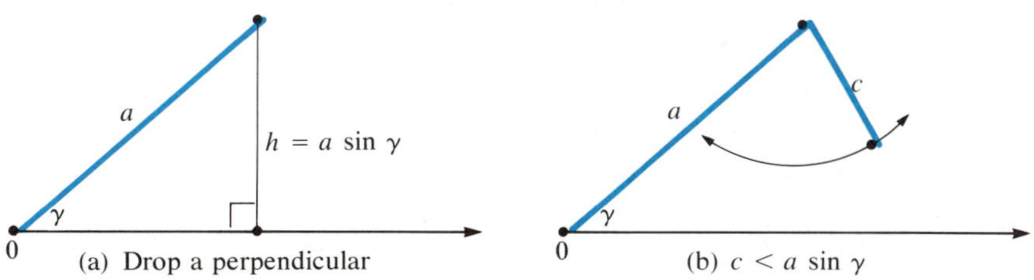

(a) Drop a perpendicular (b) $c < a \sin \gamma$

Fig. 3

From Fig. 3a the distance h from the top end of side a to the x-axis is $a \sin \gamma$. This is a critical number, for if side c is too short, that is, if $c < a \sin \gamma$, then the pendulum will never reach the x-axis (Fig. 3b). Hence there is no triangle with the given parts a, c, and γ.

If $c = a \sin \gamma$, then the pendulum just reaches the x-axis. Hence a, c, γ determine one triangle, a right triangle (Fig. 4a, next page).

If $c > a \sin \gamma$, then the pendulum is long enough to hit the x-axis twice, apparently forming two triangles (Fig. 4b) with the same parts a, c, and γ. But

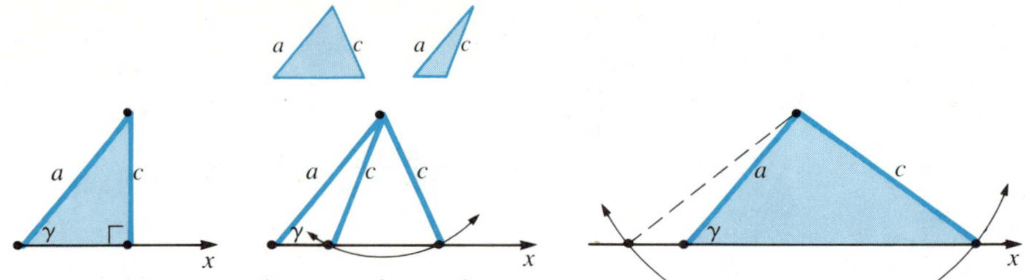

(a) $c = a \sin \gamma$ (b) $c > a \sin \gamma$ and $c < a$ (c) $c > a \sin \gamma$ and $c \geq a$

Fig. 4 $c \geq a \sin \gamma$

let's be careful: there really are *two* triangles formed *provided* c is not too large, that is, provided $c < a$. If $c \geq a$, the swinging pendulum hits the x-axis once on the right side of γ, and once (Fig. 4c) on the wrong (left) side; there is just one triangle with the given parts a, c, γ.

Remark The case $a \sin \gamma < c < a$ in which the same data determines two triangles is known as the **ambiguous case.**

A flowchart (Fig. 5) helps us work systematically.

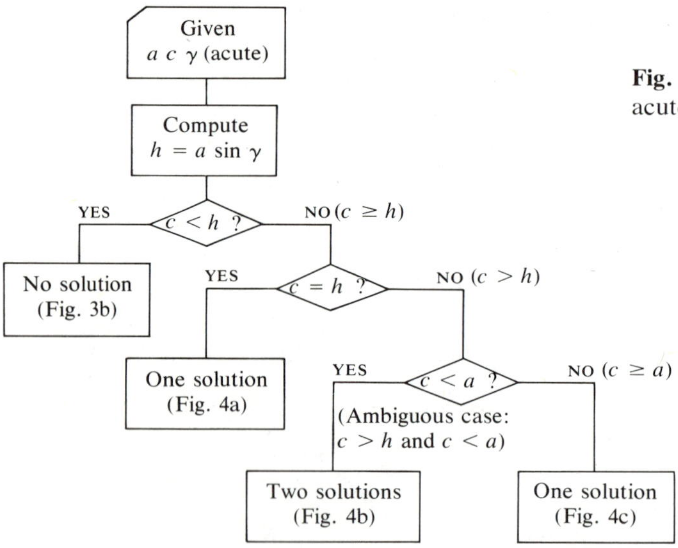

Fig. 5 Flowchart for SSA, acute case

Example 2 Find α, β, b for a triangle with

(a) $a = 5 \quad c = 2 \quad \gamma = 30°$

(b) $a = 5 \quad c = 2.5 \quad \gamma = 30°$.

Solution (a) γ is acute, so we compare c with $h = a \sin \gamma$. Now $c = 2$ and

$$h = a \sin \gamma = 5 \sin 30° = 2.5 \quad \text{so} \quad c < h.$$

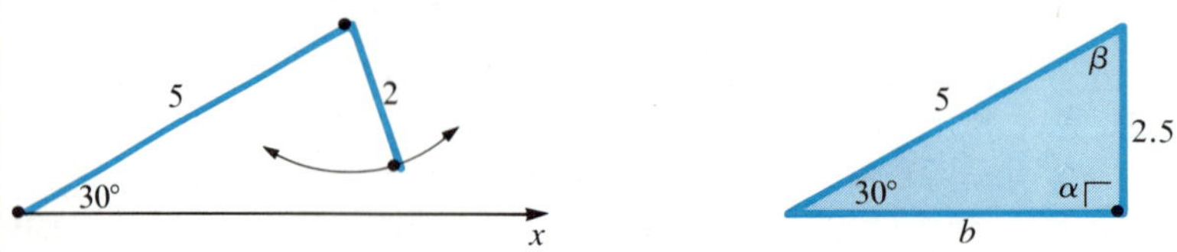

(a) (b)

Fig. 6

There is no such triangle (Fig. 6a).

(b) In this case $c = h$, so one triangle is formed, a right triangle (Fig. 6b). The right angle is α, and β is the complement of γ, that is, $\beta = 60°$. Finally,

$$b = 5 \cos \gamma = 5 \cos 30° = 2.5\sqrt{3} \approx 4.3301 .$$

Answer (a) no such triangle

(b) one possibility: $\alpha = 90°$ $\beta = 60°$ $b = 2.5\sqrt{3} \approx 4.3301$

Example 3 Given a triangle with

$$a = 3.802 \qquad c = 2.766 \qquad \gamma = 41.09°$$

Find α, β, b.

Solution Since γ is acute, we compute $h = a \sin \gamma$:

$$h = a \sin \gamma = (3.802) \sin 41.09° \approx 2.499 .$$

Since $c = 2.766$ and $a = 3.802$,

$$c > h \quad \text{and also} \quad a > c .$$

This is the situation in Fig. 4b (ambiguous case). There are *two* triangles that fit the data. In each,

$$\frac{a}{\sin \alpha} = \frac{c}{\sin \gamma}$$

so

$$\sin \alpha = \frac{a \sin \gamma}{c} = \frac{(3.802) \sin 41.09°}{2.766} \approx 0.9034 .$$

Expecting two answers, we are extra alert and realize that there are *two* angles α in the domain $0 < \alpha < 180°$ such that $\sin \alpha = 0.9034$. One is $\arcsin 0.9034$ and the other is its supplement, $180° - \arcsin 0.9034$. Thus we have two cases.

Case 1 $\alpha \approx \arcsin 0.9034 \approx 64.61°$

Then

$$\beta = 180° - \gamma - \alpha \approx 180° - 41.09° - 64.61° = 74.30° .$$

By the law of sines again

$$\frac{b}{\sin \beta} = \frac{c}{\sin \gamma} \qquad b = \frac{c \sin \beta}{\sin \gamma} = \frac{(2.766)(\sin 74.30°)}{\sin 41.09°} \approx 4.051 .$$

See Fig. 7a.

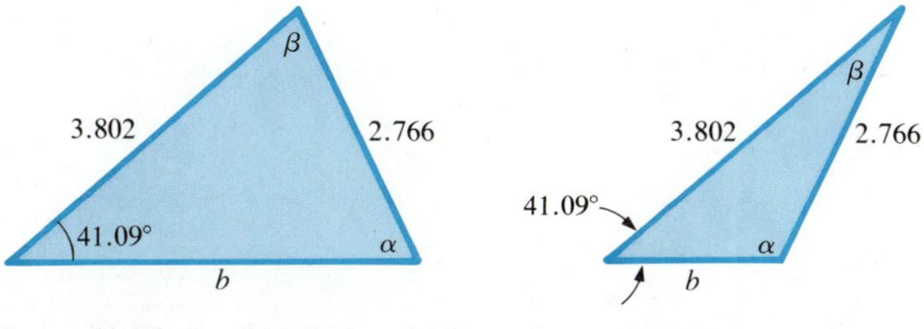

(a) $\alpha \approx 64.61°$ $\beta \approx 74.30°$ $b \approx 4.051$ (b) $\alpha \approx 115.39°$ $\beta \approx 23.52°$ $b \approx 1.679$

Fig. 7 Example of the ambiguous case

Check (law of sines)

$$\frac{a}{\sin\alpha} \approx \frac{3.802}{\sin 64.61°} \approx 4.208 \qquad \frac{b}{\sin\beta} \approx \frac{4.051}{\sin 74.30°} \approx 4.208$$

$$\frac{c}{\sin\gamma} \approx \frac{2.766}{\sin 41.09°} \approx 4.208$$

Check (law of cosines) Use the last angle found, β:

$$a^2 + c^2 - 2ac\cos\beta \approx (3.802)^2 + (2.766)^2 - (2)(3.802)(2.766)(\cos 74.30°) \approx 16.41$$

$$b^2 \approx (4.051)^2 \approx 16.41$$

Case 2 $\alpha \approx 180° - \arcsin 0.9034 \approx 115.39°$

$$\beta = 180° - \gamma - \alpha \approx 180° - 41.09° - 115.39° = 23.52°.$$

As in Case 1,

$$b = \frac{c\sin\beta}{\sin\gamma} \approx \frac{(2.766)(\sin 23.52°)}{\sin 41.09°} \approx 1.679.$$

See Fig. 7b, previous page.

Check (law of cosines)

$$a^2 + c^2 - 2ac\cos\beta \approx (3.802)^2 + (2.766)^2 - (2)(3.802)(2.766)(\cos 23.52°) \approx 2.821$$

$$b^2 \approx (1.679)^2 \approx 2.819$$

Check (law of sines) Note that $c/\sin\gamma$ and $a/\sin\alpha$ are exactly the same as in Case 1, approximately 4.208. This time

$$\frac{b}{\sin\beta} \approx \frac{1.679}{\sin 23.52°} \approx 4.207.$$

Answer Two answers: $\begin{cases} \alpha_1 \approx 64.61° & \beta_1 \approx 74.30° & b_1 \approx 4.051 \\ \alpha_2 \approx 115.39° & \beta_2 \approx 23.52° & b_2 \approx 1.679 \end{cases}$

The ratio $c/\sin\gamma$ was used several times in the solution of Example 3. The computation could have been shortened by first computing and storing this ratio. Let's flowchart how the computation should be organized (Fig. 8), then try another problem.

Fig. 8 Organized computation for the ambiguous case

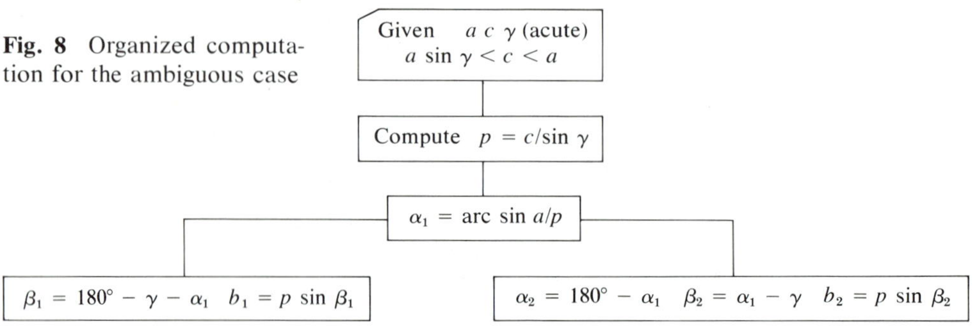

Explanation From the law of sines

$$\frac{a}{\sin\alpha} = \frac{c}{\sin\gamma} = p \quad \text{follows} \quad \sin\alpha = \frac{a}{p};$$

hence the third box in the flowchart. Similarly from

$$\frac{b}{\sin\beta} = \frac{c}{\sin\gamma} = p \quad \text{follows} \quad b_1 = p\sin\beta_1 \quad \text{and} \quad b_2 = p\sin\beta_2.$$

Finally, $\beta_2 = 180° - \gamma - \alpha_2 = 180° - \gamma - (180° - \alpha_1) = \alpha_1 - \gamma$.

EXAMPLE 4 Given a triangle with

$$a = 7002 \qquad c = 4468 \qquad \gamma = 19.91°$$

Find α, β, b.

Solution Clearly $c < a$. Also $a\sin\gamma \approx 2384 < c$, so this is the ambiguous case. Compute

$$p = \frac{c}{\sin\gamma} = \frac{4468}{\sin 19.91°} \approx 13120.$$

Then

$$\alpha_1 = \text{arc sin}\frac{a}{p} \approx \text{arc sin}\frac{7002}{13120} \approx 32.25°$$

$$\beta_1 = 180° - \gamma - \alpha_1 \approx 127.84° \qquad b_1 = p\sin\beta_1 \approx 10360$$

This completes the first case. Now

$$\alpha_2 = 180° - \alpha_1 \approx 147.75° \qquad \beta_2 = \alpha_1 - \gamma \approx 12.34°$$

$$b_2 = p\sin\beta_2 \approx 2804,$$

which completes the second case.

Check $b_1^2 \approx 1.073 \times 10^8 \qquad a^2 + c^2 - 2ac\cos\beta_1 \approx 1.074 \times 10^8$

$b_2^2 \approx 7.862 \times 10^6 \qquad a^2 + c^2 - 2ac\cos\beta_2 \approx 7.867 \times 10^6$

Answer Two answers: $\begin{cases} \alpha_1 \approx 32.25° & \beta_1 \approx 127.84° & b_1 \approx 10360 \\ \alpha_2 \approx 147.75° & \beta_2 \approx 12.34° & b_2 \approx 2804 \end{cases}$

EXAMPLE 5 Given a triangle with

$$a = 75.22 \qquad c = 84.37 \qquad \gamma = 62.17°$$

Find α, β, b.

Solution Since $c > a$, this is the case of Fig. 4c. No testing is required, only computing; there is one solution. Compute

$$p = \frac{c}{\sin\gamma} = \frac{84.37}{\sin 62.17°} \approx 95.40$$

By the law of sines,

$$\sin\alpha = a\frac{\sin\gamma}{c} = \frac{a}{p}$$

$$\alpha = \arcsin \frac{a}{p} \approx \arcsin \frac{75.22}{95.40} \approx 52.04°$$

$$\beta = 180° - \gamma - \alpha = 65.79°.$$

Again by the law of sines,

$$b = \frac{c}{\sin \gamma} \sin \beta = p \sin \beta \approx 95.40 \sin 65.79° \approx 87.01.$$

Check $b^2 \approx 7571 \qquad c^2 + a^2 - 2ac \cos \beta \approx 7571$

Answer $\alpha \approx 52.04° \qquad \beta \approx 65.79° \qquad b \approx 87.01$

SSA: The Obtuse Case

We have discussed the problem SSA assuming that γ is an acute angle. We know how to handle the triangle if γ is a right angle. Suppose γ is an obtuse angle, $90° < \gamma < 180°$. If $c < a \sin \gamma$, then just as before the pendulum will not reach the x-axis (Fig. 9a) and there is no solution. If $a \leq c < a \sin \gamma$, the pendulum will meet the x-axis twice (Fig. 9b), but on the wrong side of γ. In this case too, there is no solution. If $c > a$, the pendulum will meet the x-axis once on the right side of γ, hence there is one solution (Fig. 9c). We shall not give a worked example illustrating Fig. 9c; step-by-step, the calculation is exactly like that in Example 5.

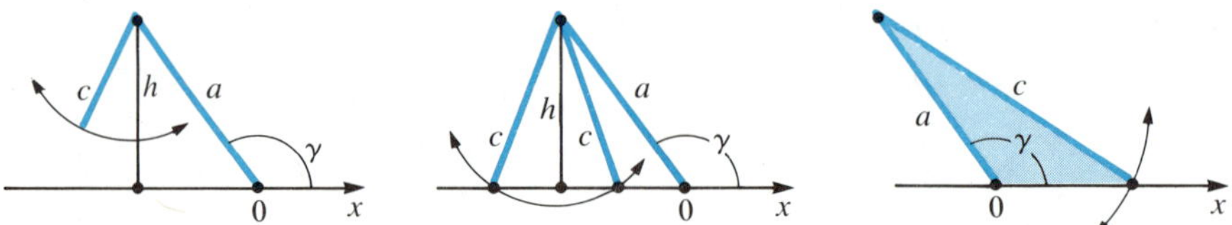

(a) $c < h$ no solution (b) $c > h$ $c \leq a$ no solution (c) $c > a$ one solution

Fig. 9 γ obtuse ($h = a \sin \gamma$)

Remark 1 In the worked examples of SSA, the given sides were always a and c, and the given angle was always γ. Of course, two sides and an excluded angle might be given with different names, such as b, c, α, etc. Therefore you must know the method rather than formulas involving particular letters (names). Perhaps it is best to remember the law of sines in words: *in any triangle, the three quotients* (side) ÷ sin(opp. angle) *are equal.* Always sketch the triangle(s)!

Remark 2 In the examples we have generally shown all numbers rounded to four significant figures. Of course, when doing this work on a calculator, you should not do any intermediate rounding. Only your answers should be rounded—to the accuracy of the data (or as you are instructed).

Exercises

Find the remaining angle and sides; check

1 $\alpha = 45.38° \quad \beta = 62.45° \quad a = 3.077$

2 $\alpha = 71.32° \quad \beta = 21.01° \quad a = 5.000$

3 $\alpha = 103.05° \quad \beta = 13.11° \quad c = 12.97$

4 $\alpha = 39.00° \quad \beta = 93.04° \quad c = 701.8$

5 $\beta = 24.17° \quad \gamma = 19.92° \quad b = 0.6523$

6 $\beta = 138.22° \quad \gamma = 27.72° \quad b = 0.1628$

7 $\beta = 24.17° \quad \gamma = 19.92° \quad a = 0.6523$

8 $\beta = 138.22° \quad \gamma = 27.72° \quad a = 0.1628$

9 $\beta = 24.17° \quad \gamma = 19.92° \quad c = 0.6523$

10 $\beta = 138.22° \quad \gamma = 27.72° \quad c = 0.1628.$

Find the remaining parts; give all solutions and check

11 $a = 3.602 \quad c = 3.403 \quad \gamma = 75.64°$

12 $a = 1042 \quad c = 445.82 \quad \gamma = 25.88°$

13 $a = 14.23 \quad c = 21.38 \quad \alpha = 47.08°$

14 $a = 89.22 \quad b = 103.4 \quad \alpha = 63.30°$

15 $a = 3\sqrt{2}$ $c = 3$ $\gamma = 45°$
16 $a = 7$ $b = 3$ $\beta = \text{arc sin } \frac{3}{7}$
17 $a = 4.990$ $c = 3.861$ $\gamma = 31.24°$
18 $a = 12.63$ $c = 9.241$ $\gamma = 43.24°$
19 $a = 3111$ $c = 3048$ $\gamma = 73.94°$
20 $a = 406.2$ $c = 378.7$ $\gamma = 63.78°$
21 $b = 0.6742$ $c = 0.5109$ $\gamma = 27.31°$
22 $b = 10.43$ $c = 9.687$ $\gamma = 57.44°$
23 $a = 41.53$ $b = 47.22$ $\alpha = 54.99°$
24 $a = 6.903$ $b = 8.211$ $\alpha = 31.31°$
25 $a = 40830$ $b = 20780$ $\beta = 18.35°$
26 $a = 766.2$ $b = 683.1$ $\beta = 23.94°$
27 $a = 15.21$ $c = 38.97$ $\gamma = 83.12°$
28 $a = 6.129$ $c = 9.204$ $\gamma = 52.51°$
29 $a = 524.6$ $b = 719.3$ $\beta = 9.38°$
30 $a = 6048$ $b = 9121$ $\beta = 43.15°$
31 $a = 67.11$ $c = 72.38$ $\gamma = 138.61°$
32 $a = 7.801$ $c = 9.316$ $\gamma = 100.41°$
33 $a = 3048$ $c = 2176$ $\alpha = 98.17°$
34 $a = 607.0$ $c = 431.3$ $\alpha = 161.23°$
35 $a = 7.213$ $b = 7.014$ $\alpha = 118.30°$
36 $b = 52.61$ $c = 61.52$ $\gamma = 126.34°$

5 APPLICATIONS

In this section, we discuss some applications of the law of sines and some further applications of the law of cosines.

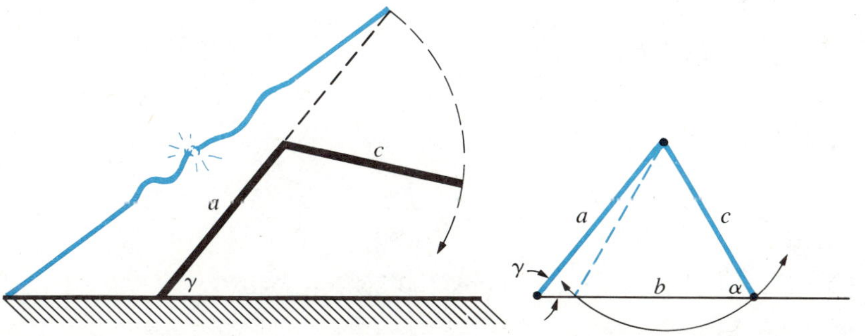

(a) $a = 18.4$ m $c = 16.7$ m $\gamma = 51.3°$

(b) Ambiguous case, but only α acute is physically possible.

Fig. 1 Falling crane

EXAMPLE 1 A cable supporting a large hinged crane snaps (Fig. 1a). How far from the base of the crane does the top hit the ground?

Solution This is a case of SSA. Since

$$a \sin \gamma = 18.4 \sin 51.3° \approx 14.4,$$

we have

$$a \sin \gamma < c < a,$$

so it is the ambiguous case; there are two triangles that fit the data. Obviously we want the triangle with α acute (Fig. 1b). As usual, we compute

$$p = \frac{c}{\sin \gamma} = \frac{16.7}{\sin 51.3°} \approx 21.4$$

and use the law of sines:

$$\sin \alpha = \frac{a}{p} \approx \frac{18.4}{21.4} \approx 0.860 \qquad \alpha \approx 59.3°$$

Then

$$\beta = 180° - \gamma - \alpha \approx 69.4° \qquad b = p \sin \beta \approx 20.0.$$

Answer 20.0 m

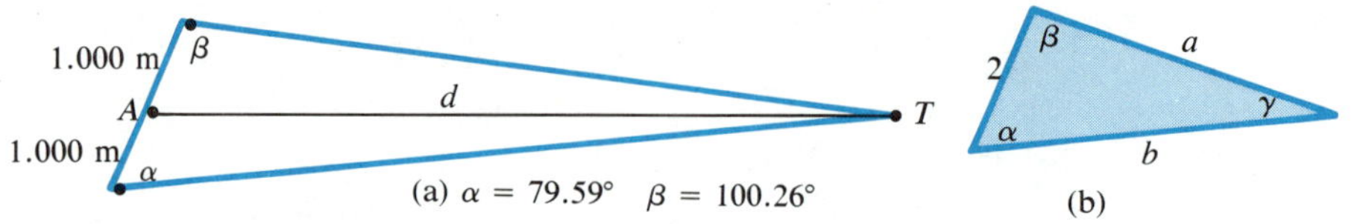

Fig. 2 Rangefinder (inaccurate figures)

EXAMPLE 2 A rangefinder (Fig. 2) sights a target at T. Find the distance d from A to T.

Solution The angle at T in the big triangle is

$$\gamma = 180° - \alpha - \beta = 0.15°$$

By the law of sines

$$b = \frac{2 \sin \beta}{\sin \gamma} = \frac{2 \sin 100.26°}{\sin 0.15°} \approx 751.73$$

By the law of cosines

$$d^2 = b^2 + 1^2 - 2 \cdot 1 \cdot b \cos \alpha$$

$$\approx (751.73)^2 + 1 - 2(751.73) \cos 79.59° \approx 564827$$

$$d \approx 751.5$$

Check $a = \dfrac{2 \sin \alpha}{\sin \gamma} = \dfrac{2 \sin 79.59°}{\sin 0.15°} \approx 751.37$

$$d^2 \approx (751.37)^2 + 1 - 2(751.37) \cos 100.26° \approx 564826$$

$$d \approx 751.5$$

Answer $d \approx 751.5$ m

Indirect Trigonometry Problems

Some geometric problems can be solved by trigonometry, but require several steps and, often, auxiliary constructions.

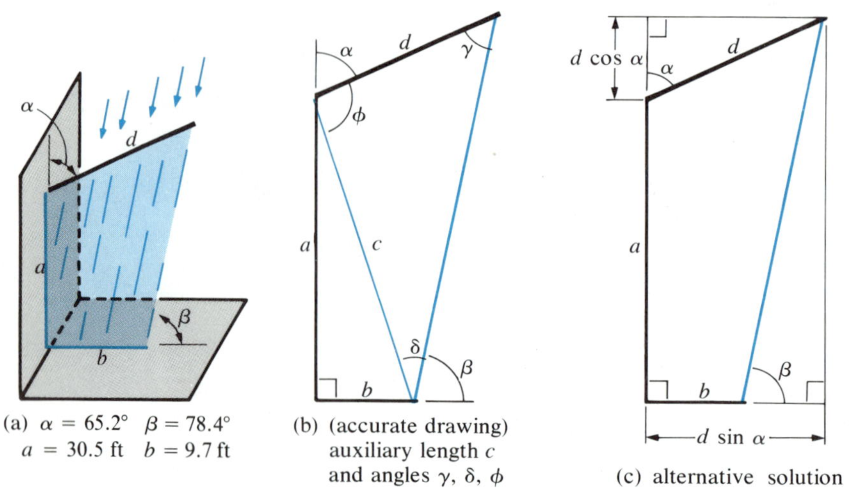

Fig. 3 Shadow of flagstaff

EXAMPLE 3 A flagstaff on the south wall of a building casts a shadow at noon as shown in Fig. 3a. Find the length d of the flagstaff.

Solution Since we would like d as the side of a triangle, we introduce c as in Fig. 3b, and mark the angles ϕ, δ, γ. From the shaded right triangle

$$c = \sqrt{a^2 + b^2} = \sqrt{(30.5)^2 + (9.7)^2} \approx 32.0$$

We could find d by the law of sines if we knew the angles δ and γ. Let's find them. First,

$$\beta + \delta + \arctan\frac{a}{b} = 180^\circ$$

so

$$\delta = 180^\circ - \beta - \arctan\frac{a}{b} = 180^\circ - 78.4^\circ - \arctan\frac{30.5}{9.7} \approx 29.2^\circ$$

Next we find γ by first finding ϕ:

$$\phi = 180^\circ - \alpha - \arctan\frac{b}{a} = 180^\circ - 65.2^\circ - \arctan\frac{9.7}{30.5} \approx 97.2^\circ$$

Therefore, since $\delta + \phi + \gamma = 180^\circ$,

$$\gamma = 180^\circ - \delta - \phi \approx 180^\circ - 29.2^\circ - 97.2^\circ = 53.6^\circ.$$

Now we are ready for the law of sines:

$$\frac{d}{\sin\delta} = \frac{c}{\sin\gamma}$$

$$d = \frac{c\sin\delta}{\sin\gamma} \approx \frac{32.0\sin 29.2^\circ}{\sin 53.6^\circ} \approx 19.4$$

Alternative Solution Complete the rectangle (Fig. 3c). The horizontal projection of d is $d\sin\alpha$, and the vertical projection of d is $d\cos\alpha$. Thus the shaded right triangle has base $d\sin\alpha - b$ and height $d\cos\alpha + a$. Therefore

$$\frac{d\cos\alpha + a}{d\sin\alpha - b} = \tan\beta$$

Solve for d:

$$d = \frac{a + b\tan\beta}{\sin\alpha\tan\beta - \cos\alpha} = \frac{30.5 + 9.7\tan 78.4^\circ}{\sin 65.2^\circ\tan 78.4^\circ - \cos 65.2^\circ} \approx 19.4$$

These are not the only ways to solve the problem!

Answer 19.4 ft

Area and the Law of Sines

In Section 2, we found a formula for the area of a triangle (Fig. 4, next page) given SAS. The area is

$$A = \tfrac{1}{2}ab\sin\gamma.$$

Similarly

$$A = \tfrac{1}{2}bc\sin\alpha \quad\text{and}\quad A = \tfrac{1}{2}ac\sin\beta.$$

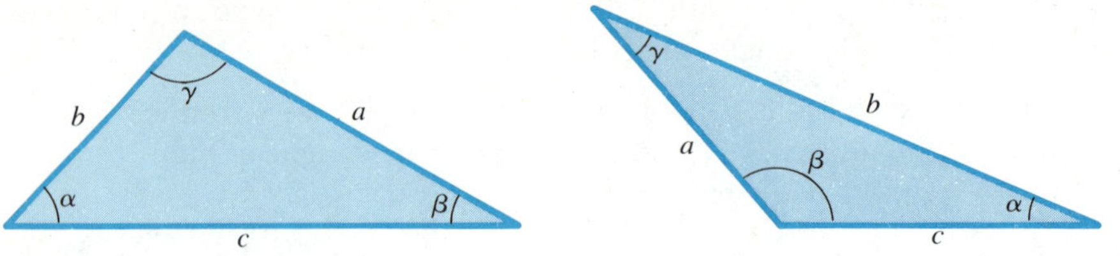

Fig. 4

> **Area** The area of any triangle is
>
> $$A = \tfrac{1}{2}ab \sin \gamma = \tfrac{1}{2}bc \sin \alpha = \tfrac{1}{2}ac \sin \beta.$$

A proof of the law of sines (postponed in Section 4 until now) drops out of these area formulas. For example, from

$$\tfrac{1}{2}bc \sin \alpha = \tfrac{1}{2}ac \sin \beta$$

follows

$$b \sin \alpha = a \sin \beta \qquad \frac{b}{\sin \beta} = \frac{a}{\sin \alpha}.$$

Similarly, from $\tfrac{1}{2}ab \sin \gamma = \tfrac{1}{2}ac \sin \beta$ follows

$$b \sin \gamma = c \sin \beta \qquad \frac{b}{\sin \beta} = \frac{c}{\sin \gamma}.$$

That's all there is to it!

The area formulas apply to the case SAS. Given SAA or SSS, however, we can compute the area of a triangle by combining the area formulas with the law of sines or the law of cosines.

Example 4 Find the area of a triangle given

(a) $a = 5.107 \quad \beta = 48.57° \quad \gamma = 26.19°$

(b) $a = 3.804 \quad b = 7.112 \quad c = 9.065$

Solution (a) The third angle is

$$\alpha = 180° - \beta - \gamma = 180° - 48.57° - 26.19° = 105.24°.$$

To apply one of the area formulas, we need another side, which we can find by the law of sines. For instance, let's find b:

$$\frac{b}{\sin \beta} = \frac{a}{\sin \alpha}$$

$$b = \frac{a \sin \beta}{\sin \alpha} = \frac{(5.107) \sin 48.57°}{\sin 105.24°} \approx 3.969.$$

Now it follows that

$$A = \tfrac{1}{2}ab \sin \gamma \approx \tfrac{1}{2}(5.107)(3.969) \sin 26.19° \approx 4.473.$$

(b) To apply one of the area formulas, this time we need an angle (actually the sine of an angle). Let's find $\sin \gamma$. From the law of cosines,

$$\cos \gamma = \frac{a^2 + b^2 - c^2}{2ab} = \frac{(3.804)^2 + (7.112)^2 - (9.065)^2}{2(3.804)(7.112)} \approx -0.3165$$

$$\sin \gamma = \sqrt{1 - \cos^2 \gamma} \approx 0.9486.$$

Therefore

$$A = \tfrac{1}{2}ab \sin\gamma \approx \tfrac{1}{2}(3.804)(7.112)(0.9486) \approx 12.83$$

Answer (a) 4.473 (b) 12.83

Heron's Formula

An ancient Greek geometer (probably Archimedes) found a beautiful formula for the area of a triangle in terms of its three sides.

Heron's Formula Set $s = \tfrac{1}{2}(a + b + c)$. Then

$$A = \sqrt{s(s-a)(s-b)(s-c)}.$$

Let's sketch a proof. We start with $A = \tfrac{1}{2}ab \sin\gamma$ and square:

$$4A^2 = a^2b^2 \sin^2\gamma = a^2b^2(1 - \cos^2\gamma) = a^2b^2 - (ab\cos\gamma)^2$$

We multiply by 4:

$$16A^2 = 4a^2b^2 - (2ab\cos\gamma)^2,$$

and use the law of cosines,

$$2ab\cos\gamma = a^2 + b^2 - c^2.$$

We find

$$16A^2 = 4a^2b^2 - (a^2 + b^2 - c^2)^2.$$

Factoring the right-hand side yields

$$16A^2 = (a + b + c)(a + b - c)(b + c - a)(c + a - b).$$

Now

$$a + b + c = 2s, \quad a + b - c = 2s - 2c, \quad \text{etc.},$$

and we have

$$16A^2 = 16s(s-a)(s-b)(s-c),$$

from which Heron's formula follows.

Example $a = 3.804 \quad b = 7.112 \quad c = 9.065$

$$s = \tfrac{1}{2}(3.804 + 7.112 + 9.065) = 9.9905$$

$$s - a = 6.1865 \qquad s - b = 2.8785 \qquad s - c = 0.9255$$

$$s(s-a)(s-b)(s-c) \approx 164.655$$

$$A \approx \sqrt{164.655} \approx 12.832$$

Compare Example 4b.

Exercises

1 Two forest rangers 10.0 miles apart at points A and B observe a fire (Fig. 5, next page). The ranger at A finds $\alpha = 52.6°$. The ranger at B finds $\beta = 31.5°$. How far is the fire from each ranger? From the line through A and B?

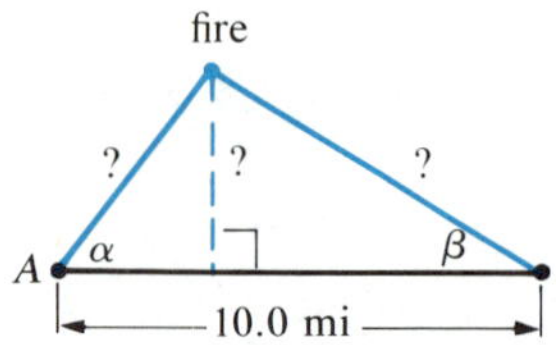

Fig. 5 $\alpha = 52.6°$ $\beta = 31.5°$

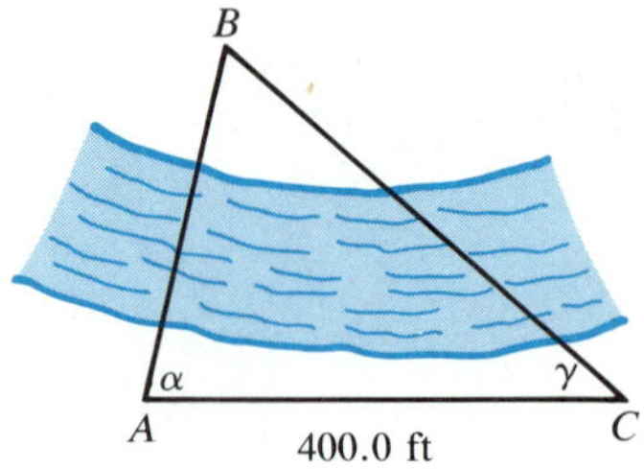

Fig. 6 $\alpha = 77.3°$ $\gamma = 41.8°$

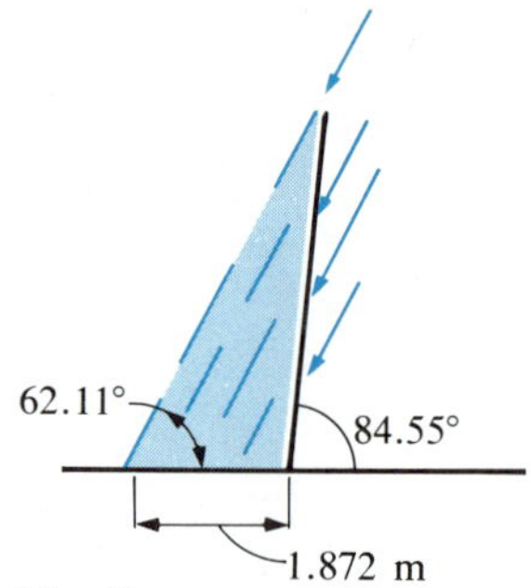

Fig. 7

2 Two points A and B are on opposite sides of a river (Fig. 6). A third point C is 400.0 ft from A. If $\alpha = 77.3°$ and $\gamma = 41.8°$, find the distance from A to B.

3 Find the length of the tilted pole (Fig. 7).

4 Find $\overline{CD}$ in Fig. 8.

5 Find $\overline{CD}$ in Fig. 9.

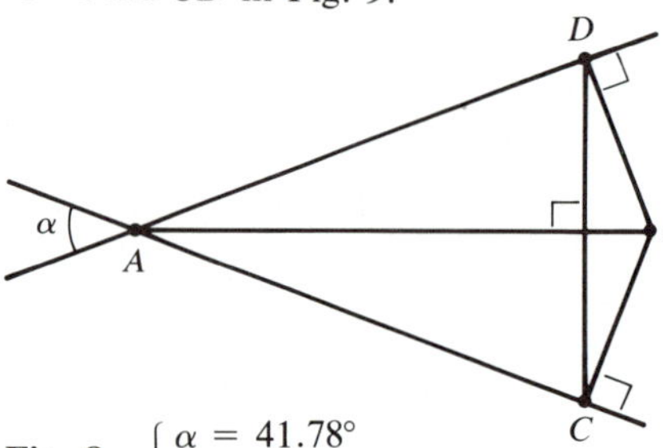

Fig. 8 $\begin{cases} \alpha = 41.78° \\ \overline{AB} = 12.72 \end{cases}$

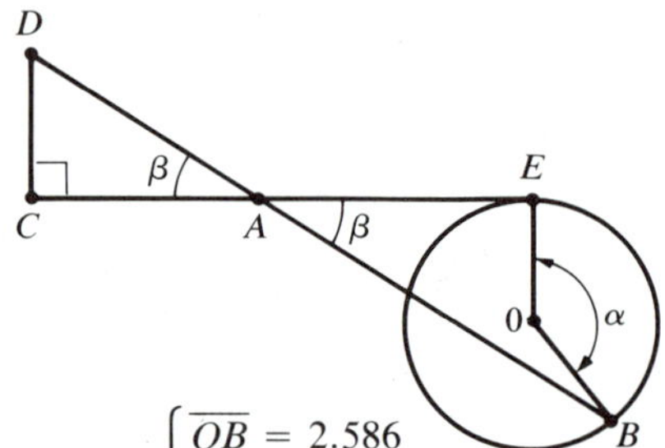

Fig. 9 $\begin{cases} \overline{OB} = 2.586 \\ \overline{AE} = 5.662 \\ \overline{AC} = 4.713 \\ \alpha = 141.63° \end{cases}$

6 Refer to Example 2. Suppose $\alpha = 83.16°$ and $\beta = 96.73°$. Now what is the distance to the target?

7 A vertical telephone pole stands by the side of a road that slopes at an angle of 8.5° with the horizontal. When the angle of elevation of the sun is 61.3°, the pole casts a shadow parallel to the road, downhill, and 13.2 ft long. How high is the pole?

8 (cont.) Suppose instead that the shadow is cast uphill. Then how high is the pole?

9 Find $\overline{BD}$ in Fig. 10.

10 Find $\overline{AB}$ in Fig. 11 to 4 significant figures.

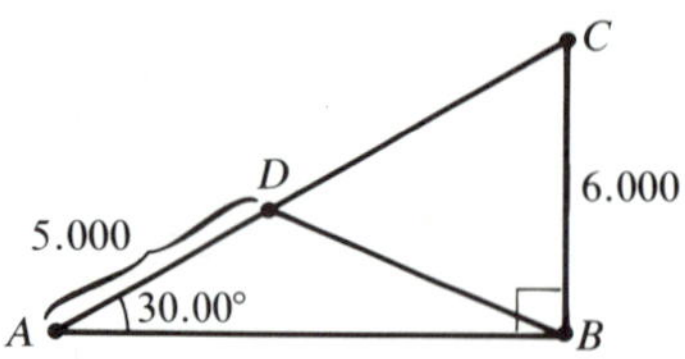

Fig. 10

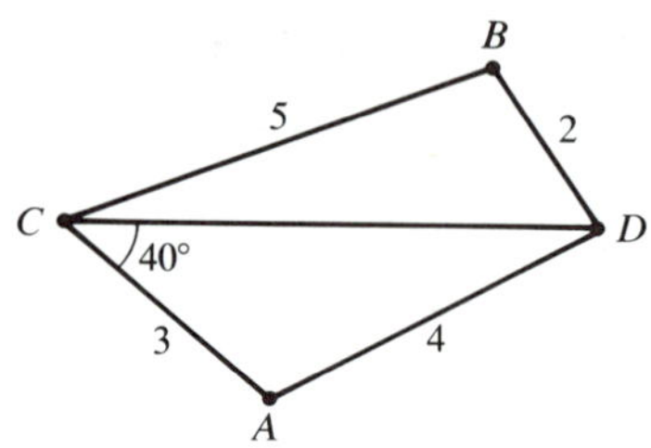

Fig. 11

11 Find $\overline{AB}$ and $\overline{BC}$ in Fig. 12 to 5 significant figures.

12* Find $\theta°$ in Fig. 13 to 4 significant figures.

13 Find $\overline{DC}$ and $\overline{EC}$ in Fig. 14 to 6 places.

14* (cont.) Now find $\overline{DE}$, $\angle$DEC, and θ. Surprised?

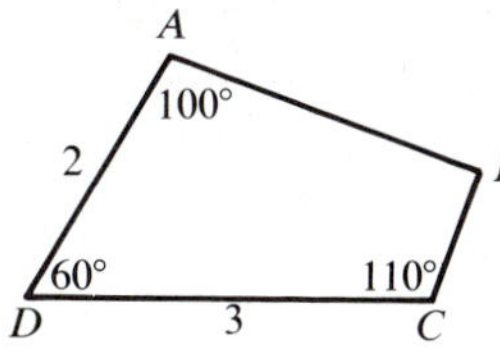

Fig. 12

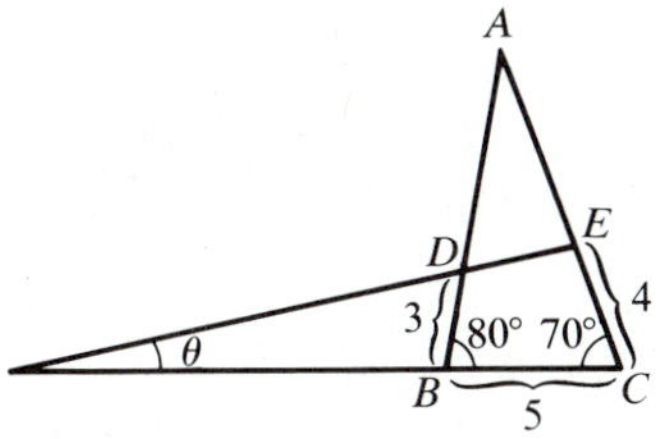

Fig. 13

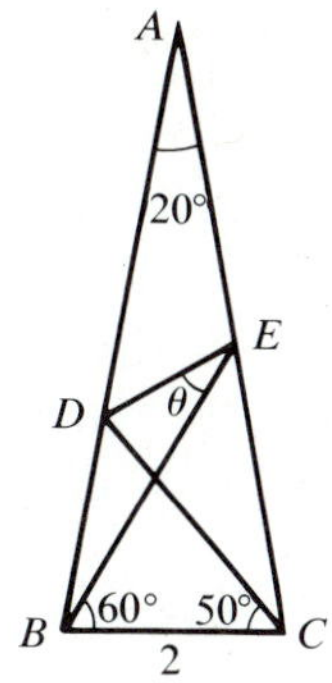

Fig. 14 (ABC isosceles)

Find the area of the triangle, given

15 $b = 3 \quad c = 5 \quad \alpha = \frac{1}{6}\pi$

16 $b = 4 \quad c = 1 \quad \alpha = \frac{1}{4}\pi$

17 $c = 2.807 \quad a = 1.442 \quad \beta = 27.91°$

18 $c = 12.66 \quad a = 17.41 \quad \beta = 119.28°$

19 $a = 108.6 \quad b = 349.7 \quad \gamma = 156.34°$

20 $a = 7.803 \quad b = 4.441 \quad \gamma = 81.19°$

21 $a = 12.66 \quad \beta = 15.29° \quad \gamma = 100.31°$

22 $a = 1069 \quad \beta = 11.93° \quad \gamma = 16.01°$

23 $c = 5.131 \quad \alpha = 63.88° \quad \beta = 41.70°$

24 $c = 15.31 \quad \alpha = 63.88° \quad \beta = 41.70°$

25 $a = 16.52 \quad b = 12.77 \quad c = 19.51$

26 $a = 8.413 \quad b = 2.718 \quad c = 10.145$

27 $a = 506.1 \quad b = 682.3 \quad c = 415.5$

28 $a = 3004 \quad b = 5172 \quad c = 6131$

29 $a = 19.33 \quad c = 17.61 \quad \gamma = 55.22°$

30 $b = 4.883 \quad c = 6.107 \quad \beta = 19.66°$.

Find the area by Heron's formula

31 $a = 11.26 \quad b = 18.81 \quad c = 23.44$

32 $a = 4.800 \quad b = 7.343 \quad c = 5.103$

33 $a = 703.1 \quad b = 698.9 \quad c = 721.5$

34 $a = 69.96 \quad b = 39.74 \quad c = 28.77$

35 The 8×8 square in Fig. 15a is cut into 4 pieces. The pieces are fitted together in Fig. 15b. But the square has area 64 and the rectangle 65. Explain this paradox.

36 Find the area to 3 places of triangle ABC in Fig. 16. Its vertices are on 3 vertical rods.

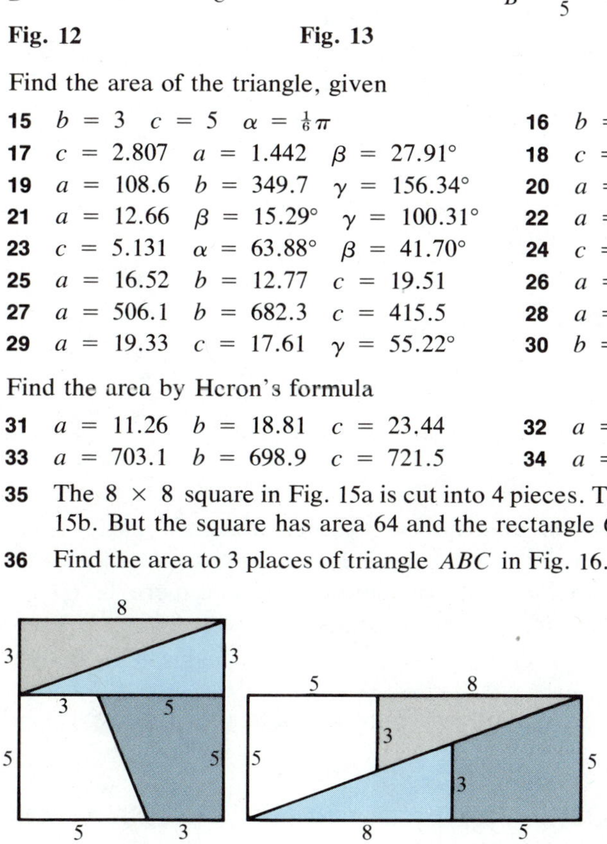

(a) area 64 (b) area 65

Fig. 15

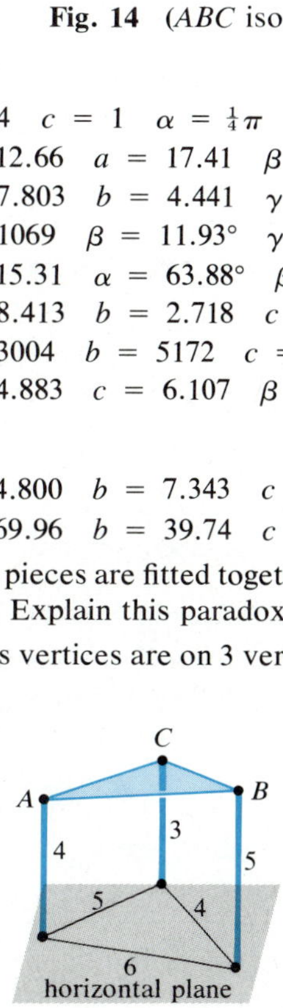

Fig. 16

6 MEASUREMENT AND ACCURACY [Optional]

We first describe briefly the tools currently used to measure angles and distances, and their accuracy. Then we look at some examples of how sensitive quantities computed by trigonometry are to small errors in the measured data.

Divided Circle

One of the most basic tools of angle measurement is the protractor, or equally divided circumference. If we assume that we can distinguish clearly rulings that are

$\frac{1}{2}$ mm apart, then a divided circumference that distinguishes degrees must have diameter at least

$$\frac{360 \times 0.05}{\pi} \approx 5.73 \text{ cm} \approx 2\tfrac{1}{4} \text{ in}.$$

But if we want the division marks to represent minutes, then we need a diameter of 60 times this, that is, about 60×5.73 cm ≈ 3.44 m $\approx 11\frac{1}{4}$ ft. It is unrealistic to make an accurate divided circle this large.

Astronomical Measurements

Astronomers are very dependent on highly accurate angle measurements. These measurements are used indirectly to determine distances. Telescopes commonly have divided circles attached to read the azimuth (horizontal) angle and the elevation (vertical) angle of rotation. These divided circles are read with microscopes, and the usual accuracy is to $0.03''$ with a maximum error of $\pm 0.003''$. The most accurate of these devices, used for elevation angles only, is called a **meridian circle.** It is generally about 1 m in diameter and can be read to $0.01''$. The reading is indirect and painstaking to avoid errors.

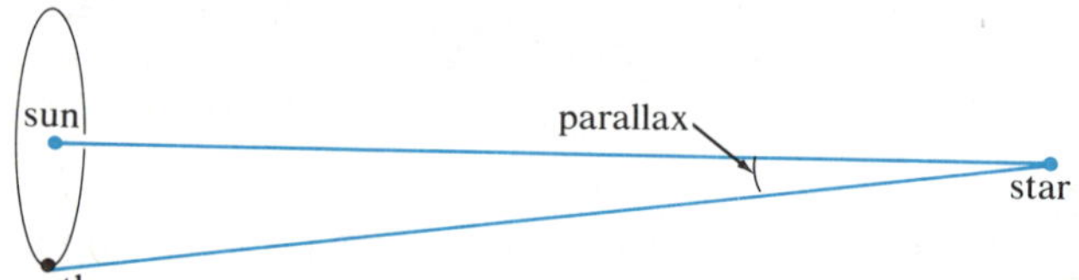

Fig. 1 Measurement of parallax

The distance to nearby stars is determined by measuring an angle called **parallax** (Fig. 1). It is the angle at the star subtended by the average radius of the Earth's orbit. For the *nearest* star, α-Centauri, the parallax is a mere $0.760''$.

The mean radius of the Earth's orbit is called an **astronomical unit** (A.U.), and

$$1 \text{ A.U.} \approx 1.495979 \times 10^8 \text{ km}.$$

Other important units of distance in astronomy are the **light year** (ly), where

$$1 \text{ ly} \approx 9.460530 \times 10^{12} \text{ km} \approx 6.32397 \times 10^4 \text{ A.U.}$$

and the **parsec.** One parsec (pc) is the distance from the Sun to a point in space with parallax $1''$. Thus

$$1 \text{ pc} \approx \frac{1 \text{ A.U.}}{\tan 1''} \approx 3.085678 \times 10^{13} \text{ km} \approx 3.26163 \text{ ly}.$$

By measuring parallax, distances up to 30 parsecs can be estimated within a 10% error.

Distances to the Sun and the nearby planets can be measured more accurately by using reflected radar signals. The most accurate distance measurement in space is the reflection of a laser beam off a reflector on the Moon. This measurement of about 3.8×10^5 km is accurate to ± 6 cm.

Machine Shop Measurements

Let us start with distance measurements. Here are three common measuring devices with their ranges and accuracies (for high quality models):

	range	error
Vernier caliper gauge	0–100 mm	± 0.025 mm
Dial gauge	0–25 mm	± 0.0025 mm
Micrometer gauge	0–25 mm	± 0.004 mm

For very accurate distance measurements, **gauge blocks** are used. These are metal rectangular solids, two of whose faces are ground parallel at a measured distance to high accuracy. They must be used at 20 °C to avoid errors of thermal expansion. For "grade 0" gauge blocks the range and errors are

range	error
0—20 mm	$\pm 10^{-4}$ mm
. . . .	. . . .
80—100 mm	$\pm 2.5 \times 10^{-4}$ mm

To be able to measure any distance from 0 to 10 cm by increments of, say, 0.002 mm would seem to require $100/0.002 = 50{,}000$ gauge blocks! Actually one simply "wrings" together (stacks up) pieces of standard sizes to make desired heights. A standard 112-piece set consists of the following:

Lengths (in mm)	Number of pieces
1.0005	1
1.001, 1.002, 1.003, · · · , 1.009	9
1.01, 1.02, 1.03, · · · , 1.49	49
0.5, 1.0, 1.5, 2.0, · · · , 24.5	49
25, 50, 75, 100	4
	total: 112

For example,

$$34.3755 = 1.0005 + 1.005 + 1.37 + 6 + 25.$$

Of course, there are many other devices used in precision machining. One device, called a coordinate cathetometer, combines a microscope and micrometer adjustments in two directions. It is used for two-dimensional measurements in a vertical plane. An instrument with a range of 75 mm $\times$ 75 mm is typically accurate to ± 0.001 mm.

Angles are measured in various ways. A precision protractor that is about 23 cm in diameter will have circumference ≈ 720 mm, so $\frac{1}{2}°$ will be indicated by division marks 1 mm apart. By adding a micrometer or vernier for fine adjustment, you can easily interpolate to $1'$ (or less) with about $\pm\frac{1}{10}'$ error.

For highly accurate angle measurements, a set of **angle gauge blocks** are used. Since they can be wrung together either to add or *to subtract* angles, very few are needed in a set. A standard 13 (thirteen!) piece set consists of

$$1° \quad 3° \quad 9° \quad 27° \quad 41° \quad 90° \quad 1' \quad 3' \quad 9' \quad 27' \quad 3'' \quad 9'' \quad 27''$$

The error is $\pm 0.5''$ for $1°$–$90°$ and $\pm 0.3''$ for $3''$–$27'$. Any angle can be "wrung" in increments of $3''$, with interpolation to $1\frac{1}{2}''$. See Fig. 2, next page, for an example of "wringing."

A 16-block set can produce all 324,000 angles from 0 to 90° in $1''$ increments:

1° 3° 5° 15° 30° 45° 1′ 3′ 5′ 20′ 30′ 1″ 3″ 5″ 20″ 30″

We mentioned the **sine bar** in the exercises on p. 319. Used with a set of (distance) gauge blocks, a 250 mm sine bar measures angles in the range 0°–30° with an accuracy of ±1″.

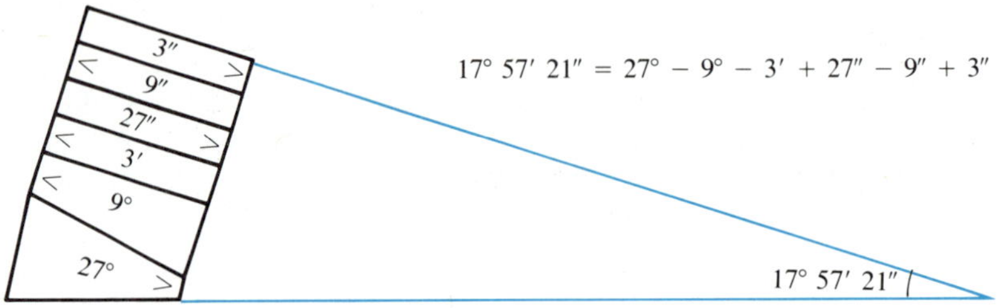

Fig. 2 Example of angle gauge blocks wrung together

Laboratory Distance Measurements

The ultimate distance measurement instrument is the **interferometer** with a laser light source. It measures the interference pattern of a split light beam, one half of which is reflected off the target. Accuracies of 3 parts in 10^8 are possible. That means that in a measurement of 1 m, the error is $\pm 3 \times 10^{-8}$ m.

Angle Measurement in Surveying

A surveyor's **transit** measures horizontal angles at 20″ intervals; the error is ±10″. The (generally) more accurate **theodolite** measures both horizontal and vertical angles at 1″ intervals, so the accuracy is about $\pm\frac{1}{2}''$. By averaging several readings, accuracies to 0.1″ are possible. We shouldn't forget the mariner's **sextant,** which measures vertical angles with an accuracy of ±12″, usually from the horizon to a star (*sextant* because its scale is $\frac{1}{6}$ of a circle). These instruments are entirely optical.

There is a new generation of electronic instruments using infrared beams, laser beams, and microwave radar. They are not more accurate than the best theodolites, but are much easier to use as they have digital readouts and data recording possibilities.

Distance Measurement in Surveying

Most basic is the good old steel tape, in lengths of 30 to 300 m, calibrated in cm or mm. At standard temperature and tension, the tape is accurate to about 1 part in 10,000. But thermal expansion, sag, uneven ground, etc. contribute to a possible large error.

Surveyors commonly use the angle subtended by a measured length at the target to estimate distance. Exs. 50–56 on p. 320 illustrate the principles. For accurate distance measurements, the standard is a 2-meter **subtense bar,** so called because you measure the angle it subtends. The main property of a subtense bar is that its length is nearly constant over a wide range of temperatures. This is accomplished by a clever design using metals with different coefficients of thermal expansion. A good subtense bar has length 2 ± 0.003 m over the temperature range from −10 °C to 50 °C.

The new generation electronic measuring instruments are remarkably precise, easy to use, and expensive! We list in Table 1 some instrument types, their ranges and accuracies.

Table 1

Type	Beam source	Range	Accuracy
E.O.	Infrared	0–1.6 km	±3 mm
E.O.	Infrared	0–10 km	±5 mm ± 1 mm/km
E.O.	Laser	0–60 m	$\pm 10^{-5}$ mm
E.O.	Laser	0–45 km	±5 mm ± 1 mm/km
EDMI	Microwave	50–2000 m	±10 mm
EDMI	Microwave	2–30 km	Depends on weather

E.O. = electro-optical
EDMI = electronic distance measuring instrument

A model of the third instrument listed, also capable of accurate angle measurement, sells for about $50,000.

Examples of Sensitivity

EXAMPLE 1 A surveyor estimates a distance (Fig. 3) by a subtense measurement. He measures $1°13'15.6''$ for the angle subtended by the subtense bar.

(a) Find his distance estimate for $h = \overline{AB}$.

(b) Suppose his angle measurement has an error of $\pm 0.3''$. How does this affect his distance estimate?

Solution (a) By the upper right triangle,

$$\frac{1}{h} = \tan\tfrac{1}{2}\theta \qquad h = \frac{1}{\tan\frac{1}{2}\theta} = \frac{1}{\tan 36'37.8''} \approx 93.847 \text{ m}.$$

(b) If the reading of θ is too large, then θ should be $1°13'15.6'' - 0.3'' = 1°13'15.3''$, so

$$h \approx \frac{1}{\tan\frac{1}{2}(1°13'15.3'')} = \frac{1}{\tan 36'37.65''} \approx 93.853 \text{ m}.$$

If the reading of θ is too small, then $\theta = 1°13'15.6'' + 0.3'' = 1°13'15.9''$, so

$$h \approx \frac{1}{\tan\frac{1}{2}(1°13'15.9'')} = \frac{1}{\tan 36'37.95''} \approx 93.841 \text{ m}.$$

Now

$$93.853 - 93.847 = 0.006 \text{ m} \qquad 93.847 - 93.841 = 0.006 \text{ m}.$$

Answer (a) 93.847 m (b) by ±6 mm.

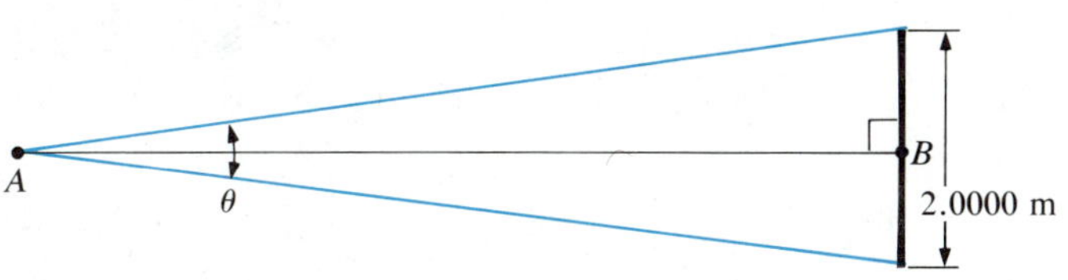

Fig. 3 $\theta = 1°\ 13'\ 15.6''$

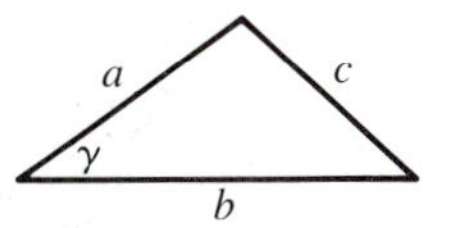

Fig. 4

EXAMPLE 2 In the triangle of Fig. 4, the measured parts are

$$a = 4 \qquad b = 6 \qquad \gamma = 35°.$$

(a) Compute c to 6 places.

(b) Suppose the measurements are accurate to 5 parts in 10,000. Find by how much c can differ from the value calculated in (a) in parts per 10,000.

Solution (a) By the law of cosines

$$c = \sqrt{4^2 + 6^2 - 2\cdot 4\cdot 6\cdot \cos 35°} \approx 3.560997$$

(b) It is clear from the figure that c increases if a decreases, if b increases, or if γ increases. Therefore the worst case for an increase in c occurs for

$$a = 4\cdot(1 - 5\times 10^{-4}) = 3.998 \qquad b = 6\cdot(1 + 5\times 10^{-4}) = 6.003$$

$$\gamma = 35\cdot(1 + 5\times 10^{-4}) = 35.0175°.$$

In this case the law of cosines yields

$$c = 3.564987.$$

The relative increase from the previous c is

$$\frac{3.564987 - 3.560997}{3.560997} \approx 0.00112 < 0.0012$$

Similarly, the worst case for a decrease in c occurs for

$$a = 4\cdot(1 + 5\times 10^{-4}) = 4.002 \qquad b = 6\cdot(1 - 5\times 10^{-4}) = 5.997$$

$$\gamma = 35\cdot(1 - 5\times 10^{-4}) = 34.9825°.$$

The corresponding c is 3.557010 and the relative decrease is

$$\frac{3.560997 - 3.557010}{3.560997} \approx 0.00112 < 0.0012$$

Answer (a) 3.560997 (b) 12 parts in 10,000

Exercises

1 Find the diameter of a circle graduated in seconds, if the division points are $\frac{1}{2}$ mm apart.

2 An eye with good vision can distinguish points that subtend 1′. From how far away can this eye distinguish two points the same distance from the eye and 1 cm apart? One meter apart?

3 How many A.U. are there in one parsec?

4 How many parsecs are there in one light year?

5 Show that $1/\alpha \approx d$, where α is the parallax in radians and d the distance in A.U. to a star.

6 Show that $1/\alpha \approx d$, where α is the parallax in seconds and d the distance in parsecs to a star.

Fill in the missing entries for the eight nearest stars to our Sun (use Ex. 6 and the conversion factors in the text).

	Star	Parallax (″)	pc	ly	km
7	α-Centauri	0.760″			
8	Barnard's	0.552″			
9	Wolf 359		2.32		
10	Lalande 21185		2.49		
11	Sirius			8.65	
12	Luyten 726-8			8.94	
13	Ross 154				8.94×10^{13}
14	Ross 248				9.73×10^{13}

Show how to make up the given length from the standard 112-piece gauge block set (p. 341)

15 153.976 mm

16 81.043 mm

Show how to make up the given angle from the standard 13-piece set of angle gauge blocks (p. 341)

17 82°

18 70°

19 14°38′51″

20 60°20′30″

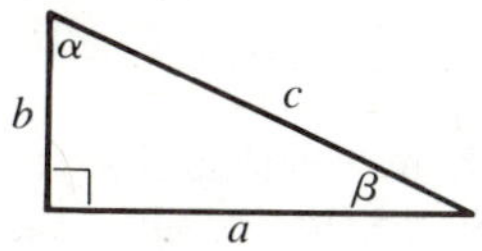

Fig. 5

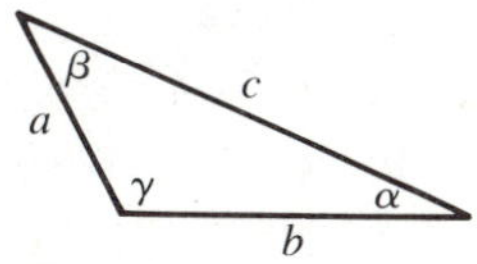

Fig. 6

21 In Fig. 5, suppose $a = 4$ and $b = 3$. Find c. Now suppose a and b can be in error by one part in 1000. Find the maximum error in c.

22 In Fig. 5, suppose $a = 4$ and $\beta = 25°$. Find c. Now suppose β can be in error by $\pm 10'$ and a by ± 0.003. Find the maximum error in c.

23 In Fig. 6, suppose $a = 3$, $b = 4$, and $\gamma = 118°$. Find c. Now suppose a, b, and γ can be in error by one part in 1000. Find the worst possible error in c in parts per 1000.

24 In Fig. 6, suppose $a = 4$, $b = 6$, and $\gamma = 123° \pm 5'$. Find the possible error in β in minutes.

25 In Fig. 6, suppose $b = 10$, $\alpha = 21°$, and $\gamma = 105°$. Find the worst possible error in c due to $\frac{1}{10}$ of 1% errors in α, b, and γ.

26 In Fig. 6, suppose $\alpha = 24°$, $c = 7$, $a = 3$ and γ is obtuse as drawn. If α has a possible error of 2 parts in 1000, find the worst possible error in b in parts per 1000.

27 The gunsights on a rifle are 25 in. apart, and the rifle is aimed at a target 170 ft from the near sight. Suppose the far sight is $\frac{1}{10}$ in. out of line. How much error will this cause?

28 (cont.) Suppose instead the near sight is $\frac{1}{10}$ in. out of line. Now how much error?

REVIEW EXERCISES

Find the remaining sides and angles of the right triangle (Fig. 1)

1 $a = 12.97 \quad \alpha = 21.14°$

2 $c = 9.38614 \quad \beta = 53°12'26''$

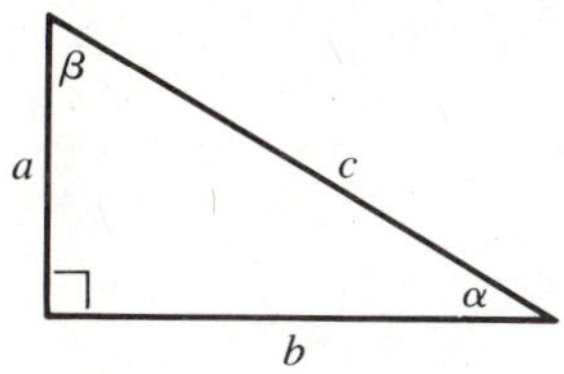

Fig. 1

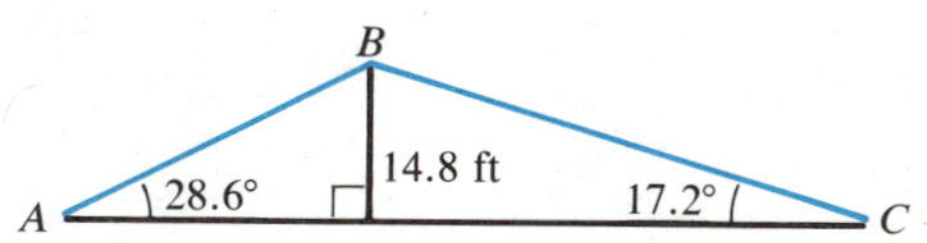

Fig. 2

3 The rope (Fig. 2) goes from A to C via the top of the pole at B. How long is the rope?

4 Find the perimeter and area of an isosceles triangle with base 17.62 cm and base angle 13.13°.

5 The hour hand of a clock is 3.7 cm long and the minute hand 5.2 cm. At a certain instant the ends of the hands are exactly 8 cm apart. What is the angle between the hands at that instant?

6 An oblique triangle has

$a = 4.308 \qquad b = 3.977 \qquad \gamma = 118.52°$

Find the length of the altitude on side c.

7 (cont.) Find the length of the median to side c.

Find the remaining parts; give all solutions and check

8 $\alpha = 42.78°$ $\qquad \beta = 97.71°$ $\qquad c = 112.4$

9 $a = 3.466$ $\qquad c = 4.991$ $\qquad \alpha = 14.20°$

10 $b = 146.2$ $\qquad c = 131.4$ $\qquad \beta = 127.82°$

11 A flagpole is tilted 15.0° to the west. When the sun is due south and making a 38.4° angle with the horizon, the shadow of the flagpole has length 4.60 m. How long is the flagpole? [*Hint* Draw a careful picture; express the shadow length in terms of the pole length.]

12* (cont.) Solve the same problem if the tilt is not due west, but is 21.0° south of west.

13 In Figure 3, the measurements are $a = 3 \quad b = 4 \quad \gamma = 25°$, accurate to 1%. Find c and the possible error in c.

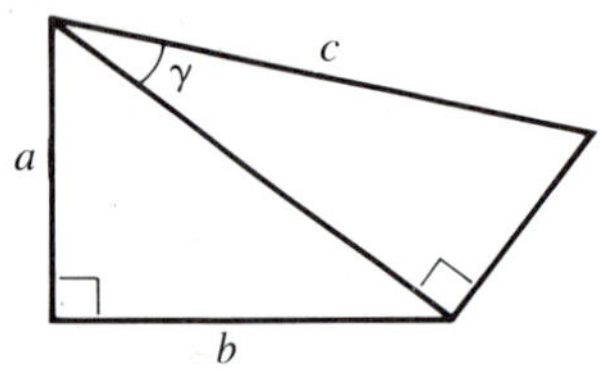

Fig. 3

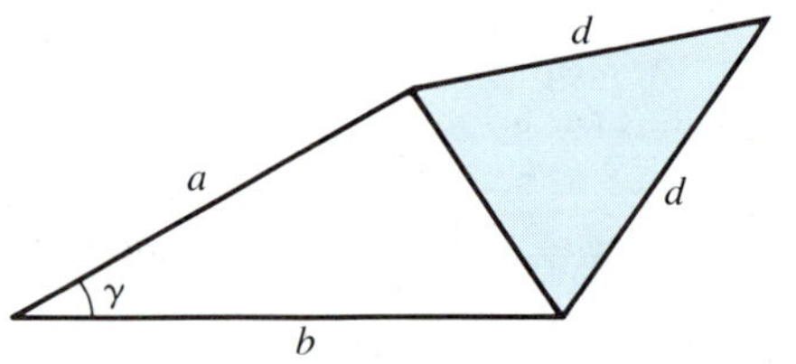

Fig. 4

14 In Figure 4, the measurements are $a = 5 \quad b = 6 \quad \gamma = 30° \quad d = 4$, accurate to one part in 50. Find the shaded area, and its possible error. (You may assume that the triangle is truly isosceles; both measurements of d give equal values.)

9 LINEAR AND POLYNOMIAL ALGEBRA

1 LINEAR SYSTEMS

In Chapter 2, Section 3, we discussed linear systems of two equations in two unknowns. You should review that material before continuing. Here we shall discuss the solution of linear systems of three equations in three unknowns.

We want to solve simultaneously three linear equations

$$\begin{cases} a_1x + b_1y + c_1z = d_1 \\ a_2x + b_2y + c_2z = d_2 \\ a_3x + b_3y + c_3z = d_3 \end{cases}$$

for x, y, z where $a_1, \cdots, d_3$ are real numbers. We shall usually write the solution as an ordered triple (x, y, z).

Gaussian Elimination

The general theory of linear systems of equations belongs to a subject called linear algebra. We shall discuss here a practical method of solution, the method of **elimination,** also called **Gaussian elimination.** The idea is to modify the given system by systematic steps to an equivalent system whose solution is obvious. It is just an extension of the method we used in the two-by-two case earlier.

Two systems

$$\begin{cases} a_1x + b_1y + c_1z = d_1 \\ a_2x + b_2y + c_2z = d_2 \\ a_3x + b_3y + c_3z = d_3 \end{cases} \quad \text{and} \quad \begin{cases} a_1'x + b_1'y + c_1'z = d_1' \\ a_2'x + b_2'y + c_2'z = d_2' \\ a_3'x + b_3'y + c_3'z = d_3' \end{cases}$$

are **equivalent** if each solution (x, y, z) of one is a solution of the other, that is, if they have the same solution set.

There are three basic operations on a system that produce an equivalent system, and hence do not change the solution set of the system:

(1) Replace the i-th equation by

$$ka_ix + kb_ix + kc_ix = kd_i$$

where $k \neq 0$.
(In other words, multiply the i-th equation by k. Since $k \neq 0$, you can divide it right out again, so *the solution set is unchanged.*)

(2) Interchange the i-th and j-th equations.
(The order in which the equations of the system are written down obviously can't make any difference. It's still the same system, hence *the solution set is unchanged.*)

(3) Replace the j-th equation by

$$(a_j + ka_i)x + (b_j + kb_i)y + (c_j + kc_i)z = d_j + kd_i$$

where k is any constant.
(In other words, add a multiple of the i-th equation to the j-th equation. Since you can subtract it right off again and get back to where you started, *the solution set is unchanged.*)

For instance, (3) says that

$$S_1 \begin{cases} a_1x + b_1y + c_1z = d_1 \\ a_2x + b_2y + c_2z = d_2 \\ a_3x + b_3y + c_3z = d_3 \end{cases}$$

and

$$S_2 \begin{cases} a_1x + b_1y + c_1z = d_1 \\ a_2x + b_2y + c_2z = d_2 \\ (a_3 + ka_1)x + (b_3 + kb_1)y + (c_3 + kc_1)z = d_3 + kd_1 \end{cases}$$

are equivalent, that is, have the same solution set. For let (x, y, z) be a solution of S_1. Then certainly (x, y, z) satisfies the first two equations in S_2 because they are in the system S_1. It also satisfies the third because

$$\begin{aligned} &(a_3 + ka_1)x + (b_3 + kb_1)y + (c_3 + kc_1)z \\ &\quad = (a_3x + b_3y + c_3z) + k(a_1x + b_1y + c_1z) \\ &\quad = d_3 + kd_1. \end{aligned}$$

Conversely, if (x, y, z) is a solution of S_2, then certainly (x, y, z) satisfies the first two equations of S_1 because they are in the system S_2. It also satisfies the third because

$$\begin{aligned} &a_3x + b_3y + c_3z \\ &\quad = (a_3 + ka_1)x + (b_3 + kb_1)y + (c_3 + kc_1)z - k(a_1x + b_1y + c_1z) \\ &\quad = (d_3 + kd_1) - kd_1 \\ &\quad = d_3. \end{aligned}$$

Not only do these basic operations take a system to an equivalent system, but any two equivalent systems can be gotten, one from the other, by a sequence of these basic operations. This is shown in linear algebra courses.

Examples

(1) $\begin{cases} 3x = 2 \\ y = 0 \\ z = 1 \end{cases}$ and $\begin{cases} x = \frac{2}{3} \\ y = 0 \\ z = 1 \end{cases}$

are equivalent. (Multiply the first equation by $\frac{1}{3}$.)

(2) $\begin{cases} x = 1 \\ z = 3 \\ y = 2 \end{cases}$ and $\begin{cases} x = 1 \\ y = 2 \\ z = 3 \end{cases}$

are equivalent. (Interchange the second and third equations.)

(3) $\begin{cases} 2x - 3y + z = 5 \\ 4x + 3y + 2z = -1 \\ y + z = 4 \end{cases}$ and $\begin{cases} 2x - 3y + z = 5 \\ 9y = -11 \\ y + z = 4 \end{cases}$

are equivalent. (Add -2 times the first to the second equation.)

Elimination

With the elimination method, we modify a given system by a sequence of basic operations to an equivalent system of the type shown on the right side of Example (3) above. Note that one equation involves only one unknown, and a second equation involves this and another unknown. Such a system can be solved by inspection.

Example 1 Solve

$$\begin{cases} 2x - y + z = 4 \\ 2x + 2y + 3z = 3 \\ 6x - 9y - 2z = 17 \end{cases}$$

Solution To eliminate x, subtract the first from the second; then subtract 3 times the first from the third (that is, add -3 times the first to the third). The result is

$$\begin{cases} 2x - y + z = 4 \\ 3y + 2z = -1 \\ -6y - 5z = 5 \end{cases}$$

Now to eliminate y, add twice the second to the third:

$$\begin{cases} 2x - y + z = 4 \\ 3y + 2z = -1 \\ -z = 3 \end{cases}$$

This is the desired equivalent system. The third equation involves only z; the second equation involves only y and z. In other words, we have eliminated x from two equations and y from one of these.

From the third equation, $z = -3$.

From the second equation,

$$3y = -1 - 2z = -1 + 6 = 5 \qquad y = \tfrac{5}{3}.$$

From the first equation,

$$2x = 4 + y - z = 4 + \tfrac{5}{3} - (-3) = \tfrac{26}{3} \qquad x = \tfrac{13}{3}.$$

Check $2x - y + z = 2(\tfrac{13}{3}) - \tfrac{5}{3} - 3 = \tfrac{21}{3} - 3 = 7 - 3 = 4$
$2x + 2y + 3z = 2(\tfrac{13}{3}) + 2(\tfrac{5}{3}) + 3(-3) = \tfrac{36}{3} - 9 = 12 - 9 = 3$
$6x - 9y - 2z = 6(\tfrac{13}{3}) - 9(\tfrac{5}{3}) - 2(-3) = 26 - 15 + 6 = 17.$

Answer $(\tfrac{13}{3}, \tfrac{5}{3}, -3)$

You may eliminate any variables you please, in any order you please. Try to make the arithmetic as easy as possible.

EXAMPLE 2 Solve

$$\begin{cases} 3x + y - 2z = 4 \\ -5x \phantom{{}+y} + 2z = 5 \\ -7x - y + 3z = -2 \end{cases}$$

Solution Since y is already missing from the second equation, we add the first to the third to eliminate y from the third also:

$$\begin{cases} 3x + y - 2z = 4 \\ -5x \phantom{{}+y} + 2z = 5 \\ -4x \phantom{{}+y} + z = 2 \end{cases}$$

Now we subtract twice the third from the second:

$$\begin{cases} 3x + y - 2z = 4 \\ 3x = 1 \\ -4x \phantom{{}+y} + z = 2 \end{cases}$$

Therefore

$$x = \tfrac{1}{3} \qquad z = 2 + 4x = 2 + \tfrac{4}{3} = \tfrac{10}{3}$$

$$y = 4 - 3x + 2z = 4 - 1 + \tfrac{20}{3} = \tfrac{29}{3}.$$

Answer $(\tfrac{1}{3}, \tfrac{29}{3}, \tfrac{10}{3})$

Please *check* this (and every) answer by substituting into the original equations.

An Application to Curve Fitting

The following example indicates a useful application of linear systems. This type of problem arises frequently in statistics and analysis of data.

EXAMPLE 3 Find a quadratic polynomial whose graph passes through

$$(1, -2) \qquad (3, 1) \qquad \text{and} \qquad (4, -1).$$

Solution Let the quadratic polynomial be $y = Ax^2 + Bx + C$, where the coefficients A, B, C must be determined. Since the graph passes through $(1, -2)$, substituting $x = 1$ and $y = -2$ yields

$$-2 = A + B + C \quad \text{that is} \quad A + B + C = -2.$$

Similarly if we substitute $x = 3$, $y = 1$ and then $x = 4$, $y = -1$,

$$9A + 3B + C = 1 \quad \text{and} \quad 16A + 4B + C = -1.$$

Altogether, we have a system for (A, B, C):

$$\begin{cases} A + B + C = -2 \\ 9A + 3B + C = 1 \\ 16A + 4B + C = -1 \end{cases}$$

We eliminate C from the second and third equation by subtracting the first from each:

$$\begin{cases} A + B + C = -2 \\ 8A + 2B = 3 \\ 15A + 3B = 1 \end{cases}$$

We eliminate B from the third; to avoid fractions we first multiply the third by 2, then subtract 3 times the second:

$$\begin{cases} A + B + C = -2 \\ 8A + 2B = 3 \\ 30A + 6B = 2 \end{cases} \qquad \begin{cases} A + B + C = -2 \\ 8A + 2B = 3 \\ 6A = -7. \end{cases}$$

Therefore $A = -\frac{7}{6}$. We substitute this into the second equation:

$$2B = 3 - 8A = 3 - 8(-\tfrac{7}{6}) = 3 + \tfrac{28}{3} = \tfrac{37}{3} \qquad B = \tfrac{37}{6}.$$

Finally we substitute the values of A and B we have found into the first equation:

$$C = -2 - A - B = -2 + \tfrac{7}{6} - \tfrac{37}{6} = -2 - 5 = -7.$$

Check

$$-\tfrac{7}{6} \cdot 1^2 + \tfrac{37}{6} \cdot 1 - 7 = 5 - 7 = -2$$
$$-\tfrac{7}{6} \cdot 3^2 + \tfrac{37}{6} \cdot 3 - 7 = -\tfrac{21}{2} + \tfrac{37}{2} - 7 = 8 - 7 = 1$$
$$-\tfrac{7}{6} \cdot 4^2 + \tfrac{37}{6} \cdot 4 - 7 = -\tfrac{56}{3} + \tfrac{74}{3} - 7 = 6 - 7 = -1$$

Answer $y = -\frac{7}{6}x^2 + \frac{37}{6}x - 7$

Remark 1 We avoided fractions as long as possible in the solution. There is nothing wrong with fractions, but their presence increases chances of an error.

Remark 2 To avoid errors, it is better to work with small integers than with large ones.

We could have done better at the first step of the solution of the system by subtracting the *second* from the third, then the first from the second, to obtain

$$\begin{cases} A + B + C = -2 \\ 8A + 2B = 3 \\ 7A + B = -2 \end{cases}$$

Inconsistent Systems

Sometimes the elimination process leads to an equation

$$0x + 0y + 0z = d,$$

where $d \neq 0$. Obviously no choice of x, y, and z will satisfy this equation. This means that the system simply has no solution, and it is called an **inconsistent system.** For example, look at the system

$$\begin{cases} x + 3y = -1 \\ 2x + 6y = 3. \end{cases}$$

Add -2 times the first equation to the second. The result is the system

$$\begin{cases} x + 3y = -1 \\ 0 = 5. \end{cases}$$

There is no solution.

What happened? Why is this system different from any other system? Notice that the left-hand side of the second equation is twice the left-hand side of the first, but the right-hand side is not twice the right-hand side of the first. This is a contradiction; if $x + 3y = -1$, then $2(x + 3y)$ should be -2, not 3.

The graphs of the two given equations are two parallel lines; they do not intersect. That explains geometrically why the system has no solution.

Underdetermined Systems

Some systems have more than one solution. This happens when one of the equations is a consequence of the other two, so that really there are only two (or fewer) equations. Such systems are called **underdetermined.**

EXAMPLE 4

$$\begin{cases} x + y + z = 1 \\ x - 2y + 2z = 4 \\ 2x - y + 3z = 5 \end{cases}$$

Solution Eliminate x from the second and third equations:

$$\begin{cases} x + y + z = 1 \\ -3y + z = 3 \\ -3y + z = 3 \end{cases}$$

The last two equations both say the same thing. Therefore the system is equivalent to the system

$$\begin{cases} x + y + z = 1 \\ \quad - 3y + z = 3 \end{cases}$$

This system has not just one solution, but a whole family of solutions. We can assign one of the unknowns an arbitrary value, and then solve the resulting system of two equations in the other two variables.

For instance, set $y = 0$. Then $z = 3$ and $x = 1 - 3 = -2$, so $(-2, 0, 3)$ is a solution. Now set $y = 2$. Then $z = 3 + 3y = 9$ and $x = 1 - y - z = -10$, so $(-10, 2, 9)$ is a solution. In general, if $y = t$, any real number, then $z = 3 + 3t$ and

$$x = 1 - y - z = 1 - t - (3 + 3t) = -4t - 2.$$

Therefore

$$(x, y, z) = (-4t - 2, t, 3t + 3).$$

This is the most general solution because in any solution y has *some* value—call it t—and then x and z are determined as indicated.

Example 5

$$\begin{cases} x + y + z = 1 \\ 2x + 2y + 2z = 2 \\ 3x + 3y + 3z = 3 \end{cases}$$

Solution This system is obviously equivalent to the single equation $x + y + z = 1$. Therefore we can assign two of the unknowns arbitrary values, and then solve the resulting single linear equation for the third. For instance, set $y = 0$ and $z = 0$. Then $x = 1$ so $(1, 0, 0)$ is a solution. Or set $y = 1$ and $z = 0$. Then $x = 0$, so $(0, 1, 0)$ is a solution. Or set $y = 3, z = -4$. Then $x = 2$, so $(2, 3, -4)$ is a solution. In general, if $y = s$ and $z = t$ are any real numbers, then $x = 1 - y - z = 1 - s - t$, so

$$(x, y, z) = (1 - s - t, s, t)$$

is the most general solution.

Remark We have discussed the elimination method for systems of three equations in three unknowns. The same method works for four equations in four unknowns, five equations in five unknowns, etc., and even when the number of equations is different from the number of unknowns.

Exercises

Solve

1 $\begin{cases} x + y + z = 0 \\ 2y - 3z = -1 \\ 3y + 5z = 2 \end{cases}$

2 $\begin{cases} 2x - y - z = 1 \\ 2y - 3z = -1 \\ -3y + 5z = 2 \end{cases}$

3 $\begin{cases} x + y - z = 0 \\ x - y + z = 0 \\ -x + y + z = 0 \end{cases}$

4 $\begin{cases} 2x + y + 3z = 1 \\ -x + 4y + 2z = 0 \\ 3x + y + z = -1 \end{cases}$

5 $\begin{cases} 2x - y - 3z = 1 \\ -x - 4y - 2z = 1 \\ 3x - y - z = 1 \end{cases}$

6 $\begin{cases} 4x + 2y - z = 0 \\ x + 3y + 2z = 0 \\ x + y + 3z = 4 \end{cases}$

7 $\begin{cases} x + 5y + 5z = 1 \\ 2x + 5y + 3z = 3 \\ x - 4y - 2z = 2 \end{cases}$ **8** $\begin{cases} x - y - z = 2 \\ -x + y - z = -2 \\ -x - y + z = 1 \end{cases}$ **9** $\begin{cases} x + y = -4 \\ y + z = 3 \\ z + x = 2 \end{cases}$

10 $\begin{cases} x + 2y = 3 \\ y + 2z = 4 \\ z + 2x = 5 \end{cases}$ **11** $\begin{cases} 3x + y - z = \frac{11}{6} \\ x + 2y + 2z = -\frac{17}{6} \\ 2x - 6y - 3z = 8 \end{cases}$ **12** $\begin{cases} 4x + 2y + 3z = 4 \\ 3x + 3y - 2z = \frac{8}{3} \\ -2x + 5y - 4z = \frac{19}{3} \end{cases}$

13 $\begin{cases} \frac{1}{2}x + \frac{1}{3}y - \frac{5}{6}z = 0 \\ \frac{1}{2}x - \frac{1}{10}y + \frac{2}{5}z = \frac{4}{5} \\ 3x - \frac{1}{3}y - \frac{5}{12}z = \frac{9}{4} \end{cases}$ **14** $\begin{cases} x + y + 3z = \frac{1}{4} \\ \frac{1}{3}x + \frac{1}{2}y + \frac{4}{3}z = 0 \\ 5x + 3y + 5z = \frac{3}{4} \end{cases}$

15 $\begin{cases} x + y + z = -0.1 \\ 3x - 2y + 2z = -7.6 \\ 5x + 3y + 4z = -1.5 \end{cases}$ **16** $\begin{cases} x + 2y + 3z = -1.9 \\ 3x - 2y - 4z = 1.4 \\ 4x + 5y + 3z = -0.4 \end{cases}$

17 $\begin{cases} 2.1x + 0.7y - 1.4z = 2.8 \\ -1.1x - 2.3y - 2.5z = -6.0 \\ 3.0x - 5.1y - 6.2z = 1.7 \end{cases}$ **18** $\begin{cases} -1.4x - 2.3y - 5.0z = 5.25 \\ 2.1x - 0.3y + 1.2z = 0.03 \\ 3.9x + 1.7y - 0.1z = -1.03 \end{cases}$

Find a quadratic function whose graph passes through the three given points

19 $(-1, 1)$, $(0, 0)$, $(1, 2)$ **20** $(-2, -5)$, $(0, 3)$, $(2, 0)$

21 $(-1, 1)$, $(1, 1)$, $(3, 5)$ **22** $(-3, 2)$, $(1, 1)$, $(2, -1)$.

Find all solutions

23 $\begin{cases} x + y + 2z = 1 \\ 3x + 5y + 7z = 2 \\ -x - y - 2z = 0 \end{cases}$ **24** $\begin{cases} x + y + 2z = 1 \\ -x + 2y + z = 3 \\ y + z = 1 \end{cases}$

25 $\begin{cases} 2x - y + z = 1 \\ 3x + y + z = 0 \\ 7x - y + 3z = 2 \end{cases}$ **26** $\begin{cases} 11x + 10y + 9z = 5 \\ x + 2y + 3z = 1 \\ 3x + 2y + z = 1 \end{cases}$

Solve by elimination

27 $\begin{cases} x + y + z + w = 10 \\ x + y + 2z = 11 \\ x - 3y + w = -14 \\ y + 3z - w = 7 \end{cases}$ **28** $\begin{cases} 2x + y + z = 3 \\ y + z + w = 5 \\ 4x + z + w = 0 \\ 3y - 2z - w = 6 \end{cases}$

29 $\begin{cases} 7x + 3y + 4z + 6w = -5 \\ x + y + 2z + 3w = -8 \\ 3x + 2y + 3z + 4w = -7 \\ 5x + 3y + 5z + 7w = -13 \end{cases}$ **30** $\begin{cases} 4x + 2y + 3z + 5w = 0 \\ 7x + y + 3z + 5w = 4 \\ 6x + 3y + 4z + 7w = 0 \\ 3x + y + 2z + 3w = 1 \end{cases}$

31 $\begin{cases} -2x + y - 5z - w = 1 \\ 5x + 5y + 7z + 2w = 0 \\ 3x + 4y + 2z + w = 1 \\ -3x - y - 9z - 2w = 1 \end{cases}$

32 $\begin{cases} 3x + 5y - 2z - 3w = 1 \\ 4x + 5y + z - w = 1 \\ 2x + 7y - 5z - 9w = 1 \\ x + 2y - z - 2w = 1 \end{cases}$

33 $\begin{cases} -4x + y - 5z - w = 0 \\ 8x + 5y + 7z + 2w = 3 \\ 3x + 4y + 2z + w = 0 \\ 8x + y + 9z + 2w = 0 \end{cases}$

34 $\begin{cases} x + 5y + 4z + w = 0 \\ -4x + 8y + 3z + 8w = 0 \\ -x + 2y + z + 2w = 1 \\ -5x + 7y + 2z + 9w = 3. \end{cases}$

Find the circle $(x^2 + y^2 + ax + by + c = 0)$ passing through

35 $(2,0)$ $(0,2)$ $(2,2)$

36 $(2,0)$ $(0,1)$ $(3,3)$

37 $(1,0)$ $(0,1)$ $(0,-3)$

38 $(5,-1)$ $(-1,-1)$ $(-1,7)$.

Find the cubic curve $y = ax^3 + bx^2 + cx + d$ passing through

39 $(1,-5)$ $(2,-5)$ $(3,9)$ $(-1,-11)$

40 $(1,0)$ $(2,1)$ $(3,0)$ $(4,1)$.

2 MATRICES

Matrix theory is a mathematical tool for processing two-dimensional discrete data. Matrices (plural of matrix) occur in a wide variety of mathematics and its applications: linear systems, differential equations, vibrations, stress analysis, quantum theory, electric circuit theory, statistics, numerical analysis, linear programming, graph theory, economic modelling, ecology, and mathematical psychology to mention a few. It is important and useful to acquire manipulative skill in the basic matrix techniques.

A **matrix** is a rectangular array of numbers. A matrix with m rows and n columns is called an $\boldsymbol{m \times n}$ **matrix.**

Examples

$$\begin{bmatrix} -3 & 5 & 7 \\ 2 & 0 & 4 \\ -1 & -10 & 3 \end{bmatrix} \qquad \begin{bmatrix} -2 \\ -2 \\ 7 \end{bmatrix} \qquad \begin{bmatrix} -2 & 3 & 4 & 5 \\ 1 & 0 & 7 & -2 \\ 4 & 3 & 0 & 0 \end{bmatrix}$$

3×3 matrix $\qquad 3 \times 1$ matrix $\qquad 3 \times 4$ matrix

The Matrix of a System

To each system

$$\begin{cases} a_1x + b_1y + c_1z = d_1 \\ a_2x + b_2y + c_2z = d_2 \\ a_3x + b_3y + c_3z = d_3 \end{cases}$$

corresponds a 3×4 matrix

$$\begin{bmatrix} a_1 & b_1 & c_1 & d_1 \\ a_2 & b_2 & c_2 & d_2 \\ a_3 & b_3 & c_3 & d_3 \end{bmatrix}.$$

Using this matrix, we can describe the system without writing the variables x, y, z and the equal signs. We can carry out Gaussian elimination in this shorthand, passing through a series of matrices of equivalent systems until we come to the desired form.

EXAMPLE 1 Solve the system

$$\begin{cases} 2x + 3y - z = -2 \\ -3x - y + z = -2 \\ x + y - 3z = -8 \end{cases}$$

by elimination, using matrix notation.

Solution The matrix of the system is

$$\begin{bmatrix} 2 & 3 & -1 & -2 \\ -3 & -1 & 1 & -2 \\ 1 & 1 & -3 & -8 \end{bmatrix}.$$

To eliminate z from the second and third equations, add the first row to the second and -3 times the first row to the third:

$$\begin{bmatrix} 2 & 3 & -1 & -2 \\ -1 & 2 & 0 & -4 \\ -5 & -8 & 0 & -2 \end{bmatrix}.$$

Now eliminate x from the third equation by adding -5 times the second row to the third row:

$$\begin{bmatrix} 2 & 3 & -1 & -2 \\ -1 & 2 & 0 & -4 \\ 0 & -18 & 0 & 18 \end{bmatrix}.$$

The third row of this matrix represents the equation $-18y = 18$. Thus $y = -1$. By the second row,

$$-x + 2y = -4 \qquad x = 4 + 2y = 4 - 2 = 2.$$

By the first row

$$2x + 3y - z = -2 \qquad z = 2 + 2x + 3y = 2 + 4 - 3 = 3.$$

Answer $(2, -1, 3)$

Matrix Multiplication

We shall define the product AB of two matrices A and B. This product is defined for matrices of all sizes with one restriction: the number of columns of A must equal the number of rows of B. If A is an $m \times n$ matrix and B is an $n \times p$ matrix, then their product AB is an $m \times p$ matrix.

We shall work up to the product of two 3×3 matrices in three steps. Once you understand these cases, then you will know how to multiply any two matrices, subject to the restriction just mentioned.

Before starting, let's note that a matrix with only one row is called a **row vector** and often is denoted with a small letter such as $\mathbf{a}$ or $\mathbf{x}$. Similarly a matrix with only one column is called a **column vector** and is also often denoted by a small letter.

Row Vector Times Column Vector

Let

$$\mathbf{a} = [a_1 \quad a_2 \quad a_3] \quad \text{and} \quad \mathbf{b} = \begin{bmatrix} b_1 \\ b_2 \\ b_3 \end{bmatrix}.$$

Then the definition of $\mathbf{ab}$ is

$$\mathbf{ab} = [a_1b_1 + a_2b_2 + a_3b_3].$$

Thus $\mathbf{ab}$ is a 1×1 matrix. Note carefully how it is formed: The first element of $\mathbf{a}$ is multiplied by the first element of $\mathbf{b}$, the second element of $\mathbf{a}$ by the second element of $\mathbf{b}$, and the third element of $\mathbf{a}$ by the third element of $\mathbf{b}$. Then these three products are summed to form the single element of $\mathbf{ab}$.

Examples

(1) $$[2 \quad 3 \quad 5]\begin{bmatrix} 3 \\ 4 \\ 7 \end{bmatrix} = [2\cdot 3 + 3\cdot 4 + 5\cdot 7] = [53]$$

(2) $$[-2 \quad 0 \quad 3]\begin{bmatrix} 5 \\ -1 \\ 4 \end{bmatrix} = [(-2)\cdot 5 + 0\cdot(-1) + 3\cdot 4] = [2]$$

(3) $$[4 \quad 3 \quad -2]\begin{bmatrix} x \\ y \\ z \end{bmatrix} = [4x + 3y - 2z]$$

Remark Matrices in real life seldom have integer entries, and a calculator is a useful aid in computing matrix products. For instance

$$[-1.32, 4.07, 2.91]\begin{bmatrix} 0.44 \\ -0.21 \\ 1.89 \end{bmatrix}$$

$$= [(-1.32)(0.44) + (4.07)(-0.21) + (2.91)(1.89)] = [4.0644].$$

The calculation is routine on an algebraic logic calculator or on a left-to-right logic calculator with [M+]. See p. 212 for details.

Square Matrix Times Column Vector

It is convenient first to introduce the **double subscript** notation for matrices. We write a 3 × 3 matrix as

$$A = \begin{bmatrix} a_{11} & a_{12} & a_{13} \\ a_{21} & a_{22} & a_{23} \\ a_{31} & a_{32} & a_{33} \end{bmatrix}$$

The first index i in a_{ij} is the **row index** and the second index j is the **column index.** Thus a_{ij} is the element of A in the i-th row and j-th column. The i-th row of A is

$$[a_{i1} \quad a_{i2} \quad a_{i3}] \quad \text{and the } j\text{-th column of } A \text{ is} \quad \begin{bmatrix} a_{1j} \\ a_{2j} \\ a_{3j} \end{bmatrix}.$$

Examples

$$\text{Third row: } [a_{31} \quad a_{32} \quad a_{33}] \qquad \text{Second column: } \begin{bmatrix} a_{12} \\ a_{22} \\ a_{32} \end{bmatrix}$$

Let

$$A = \begin{bmatrix} a_{11} & a_{12} & a_{13} \\ a_{21} & a_{22} & a_{23} \\ a_{31} & a_{32} & a_{33} \end{bmatrix} \quad \text{and} \quad \mathbf{b} = \begin{bmatrix} b_1 \\ b_2 \\ b_3 \end{bmatrix}$$

Then the definition of $A\mathbf{b}$ is

$$A\mathbf{b} = \begin{bmatrix} a_{11}b_1 + a_{12}b_2 + a_{13}b_3 \\ a_{21}b_1 + a_{22}b_2 + a_{23}b_3 \\ a_{31}b_1 + a_{32}b_2 + a_{33}b_3 \end{bmatrix}.$$

This looks like a lot to learn! But it really isn't. The *first element* of $A\mathbf{b}$ is the product of the *first row* of A by $\mathbf{b}$, the *second element* of $A\mathbf{b}$ is the product of the *second row* of A by $\mathbf{b}$, etc.

Examples

(1) $$\begin{bmatrix} 2 & 3 & -1 \\ 0 & 1 & 0 \\ -4 & 3 & -1 \end{bmatrix} \begin{bmatrix} 1 \\ 3 \\ 5 \end{bmatrix} = \begin{bmatrix} 2\cdot 1 + 3\cdot 3 + (-1)\cdot 5 \\ 0\cdot 1 + 1\cdot 3 + 0\cdot 5 \\ (-4)\cdot 1 + 3\cdot 3 + (-1)\cdot 5 \end{bmatrix} = \begin{bmatrix} 6 \\ 3 \\ 0 \end{bmatrix}$$

(2) $$\begin{bmatrix} a_1 & b_1 & c_1 \\ a_2 & b_2 & c_2 \\ a_3 & b_3 & c_3 \end{bmatrix} \begin{bmatrix} x \\ y \\ z \end{bmatrix} = \begin{bmatrix} a_1x + b_1y + c_1z \\ a_2x + b_2y + c_2z \\ a_3x + b_3y + c_3z \end{bmatrix}$$

Square Matrix Times Square Matrix

If A and B are 3×3 matrices, then the element in the i-th row and j-th column of AB is the product of the i-th row of A by the j-th column of B.

Let

$$A = \begin{bmatrix} a_{11} & a_{12} & a_{13} \\ a_{21} & a_{22} & a_{23} \\ a_{31} & a_{32} & a_{33} \end{bmatrix} \quad \text{and} \quad B = \begin{bmatrix} b_{11} & b_{12} & b_{13} \\ b_{21} & b_{22} & b_{23} \\ b_{31} & b_{32} & b_{33} \end{bmatrix}$$

Then by definition AB is a 3×3 matrix whose element in the i-th row and j-th column is

$$a_{i1}b_{1j} + a_{i2}b_{2j} + a_{i3}b_{3j}.$$

For instance, the element in the 2-nd row and 3-rd column of AB is

$$a_{21}b_{13} + a_{22}b_{23} + a_{23}b_{33}.$$

Example

$$\begin{bmatrix} 2 & 1 & 1 \\ -3 & 1 & -1 \\ 0 & 4 & 2 \end{bmatrix}\begin{bmatrix} -2 & 0 & 1 \\ 4 & -3 & -1 \\ -1 & 2 & 1 \end{bmatrix} = \begin{bmatrix} -1 & -1 & 2 \\ 11 & -5 & -5 \\ 14 & -8 & -2 \end{bmatrix}$$

Please check all 9 elements of the answer. For instance, in the 2-nd row and 1-st column

$$(-3)\cdot(-2) + 1\cdot 4 + (-1)\cdot(-1) = 11.$$

By now you should know how to form any matrix product AB where A has as many columns as B has rows.

Examples

(1) $$\begin{bmatrix} 2 & 3 & -1 \\ -1 & 0 & -3 \end{bmatrix}\begin{bmatrix} 1 & 1 \\ 2 & -2 \\ -3 & 2 \end{bmatrix} = \begin{bmatrix} 11 & -6 \\ 8 & -7 \end{bmatrix}$$

(2) $$\begin{bmatrix} 0 & 1 \\ 0 & 1 \end{bmatrix}\begin{bmatrix} 1 & 1 \\ 0 & 0 \end{bmatrix} = \begin{bmatrix} 0 & 0 \\ 0 & 0 \end{bmatrix}$$ (3) $$\begin{bmatrix} 1 & 1 \\ 0 & 0 \end{bmatrix}\begin{bmatrix} 0 & 1 \\ 0 & 1 \end{bmatrix} = \begin{bmatrix} 0 & 2 \\ 0 & 0 \end{bmatrix}$$

Note carefully the last two products. If

$$A = \begin{bmatrix} 0 & 1 \\ 0 & 1 \end{bmatrix} \quad \text{and} \quad B = \begin{bmatrix} 1 & 1 \\ 0 & 0 \end{bmatrix}$$

then $AB \neq BA$. In general *matrix multiplication is not commutative.*

Exercises

Solve the system whose matrix is

1 $$\begin{bmatrix} -1 & 2 & -1 & 1 \\ 2 & 0 & -3 & -1 \\ -3 & 0 & 5 & 2 \end{bmatrix}$$

2 $$\begin{bmatrix} 3 & 1 & -2 & -1 \\ 2 & 4 & 1 & -1 \\ 1 & 1 & -3 & -1 \end{bmatrix}$$

3 $$\begin{bmatrix} 3 & 2 & 1 & 1 \\ 2 & -1 & 4 & 0 \\ 1 & 3 & 1 & -1 \end{bmatrix}$$

4 $\begin{bmatrix} 5 & -5 & 1 & 1 \\ 3 & -5 & 2 & 3 \\ -2 & 4 & 1 & 2 \end{bmatrix}$ **5** $\begin{bmatrix} 1 & -2 & 1 & 0 \\ -3 & -4 & 0 & -1 \\ 5 & -6 & 0 & 2 \end{bmatrix}$ **6** $\begin{bmatrix} 12 & -1 & 2 & 0 \\ 3 & 2 & 3 & 0 \\ 3 & 3 & 1 & 4 \end{bmatrix}$

7 $\begin{bmatrix} 2 & 3 & 8 & 0 \\ 4 & 4 & 12 & 1 \\ 20 & 12 & 20 & 3 \end{bmatrix}$ **8** $\begin{bmatrix} 2 & -6 & -3 & 8 \\ 6 & 12 & 12 & -17 \\ 18 & 6 & -6 & 11 \end{bmatrix}$ **9** $\begin{bmatrix} 2 & 1 & 0 & 3 \\ 1 & 0 & -2 & 4 \\ 0 & 2 & -1 & 5 \end{bmatrix}$

10 $\begin{bmatrix} 2 & -2 & 2 & 4 \\ -1 & 1 & 1 & -2 \\ -1 & -1 & -1 & 1 \end{bmatrix}$

Compute the product

11 $[3 \quad 4 \quad 7]\begin{bmatrix} 2 \\ 3 \\ 5 \end{bmatrix}$ **12** $[5 \quad -1 \quad 4]\begin{bmatrix} -2 \\ 0 \\ 3 \end{bmatrix}$ **13** $[-1 \quad -1 \quad 2]\begin{bmatrix} 3 \\ 1 \\ -1 \end{bmatrix}$

14 $[3 \quad 2 \quad 1]\begin{bmatrix} 1 \\ 2 \\ 3 \end{bmatrix}$ **15** $\begin{bmatrix} 2 & 3 & -1 \\ 0 & -3 & 0 \\ 7 & -1 & 4 \end{bmatrix}\begin{bmatrix} 1 \\ 1 \\ 1 \end{bmatrix}$ **16** $\begin{bmatrix} -2 & -2 & -2 \\ 3 & 3 & 3 \\ -2 & -2 & -2 \end{bmatrix}\begin{bmatrix} 2 \\ 3 \\ -4 \end{bmatrix}$

17 $\begin{bmatrix} 1 & 0 & 0 \\ 2 & 1 & 0 \\ 3 & 2 & 1 \end{bmatrix}\begin{bmatrix} 2 \\ 4 \\ 1 \end{bmatrix}$ **18** $\begin{bmatrix} 3 & 2 & 1 \\ 0 & 3 & 2 \\ 0 & 0 & 3 \end{bmatrix}\begin{bmatrix} 1 \\ 0 \\ 8 \end{bmatrix}$ **19** $\begin{bmatrix} -7 & 8 & 0 \\ 3 & -5 & 0 \\ 6 & 3 & 0 \end{bmatrix}\begin{bmatrix} 0 \\ -1 \\ 1 \end{bmatrix}$

20 $\begin{bmatrix} 0 & 2 & 0 \\ -3 & 0 & 2 \\ 0 & -3 & 0 \end{bmatrix}\begin{bmatrix} -3 \\ 2 \\ -2 \end{bmatrix}$ **21** $\begin{bmatrix} 4 & 3 & -1 \\ 2 & -2 & 8 \\ 7 & 7 & 6 \end{bmatrix}\begin{bmatrix} 6 \\ 3 \\ -2 \end{bmatrix}$ **22** $\begin{bmatrix} -2 & -3 & -5 \\ -3 & -5 & -2 \\ -5 & -2 & -3 \end{bmatrix}\begin{bmatrix} 1 \\ -1 \\ 1 \end{bmatrix}$

23 $\begin{bmatrix} 3 & 5 & -2 \\ 4 & 0 & 7 \\ 0 & 0 & 2 \end{bmatrix}\begin{bmatrix} 3 & 5 & -2 \\ 4 & 0 & 7 \\ 0 & 0 & 2 \end{bmatrix}$ **24** $\begin{bmatrix} 2 & -1 & -2 \\ 3 & 4 & -2 \\ 6 & -6 & 3 \end{bmatrix}\begin{bmatrix} 1 & 0 & 1 \\ -2 & 3 & 1 \\ 7 & 0 & 2 \end{bmatrix}$

25 $\begin{bmatrix} 0 & 1 & 0 \\ 0 & 0 & 1 \\ -2 & -1 & 3 \end{bmatrix}\begin{bmatrix} 2 & 8 & 6 \\ 3 & 4 & 1 \\ 1 & 1 & 2 \end{bmatrix}$ **26** $\begin{bmatrix} 0 & 0 & 0 \\ 1 & 0 & 0 \\ 0 & 1 & 0 \end{bmatrix}\begin{bmatrix} 0 & 0 & 0 \\ 1 & 0 & 0 \\ 0 & 1 & 0 \end{bmatrix}$

27 $\begin{bmatrix} 1 & 2 & 3 \\ 2 & 3 & 1 \\ 3 & 1 & 2 \end{bmatrix}\begin{bmatrix} -1 & 0 & 1 \\ 0 & 1 & -1 \\ 1 & -1 & 0 \end{bmatrix}$ **28** $\begin{bmatrix} 4 & 0 & 4 \\ 0 & 2 & 0 \\ 4 & 0 & 4 \end{bmatrix}\begin{bmatrix} -1 & 1 & -1 \\ 1 & 4 & 1 \\ -1 & 1 & -1 \end{bmatrix}$

29 $\begin{bmatrix} 3 & 2 & 1 \\ 2 & 3 & 2 \\ 1 & 2 & 3 \end{bmatrix}\begin{bmatrix} 3 & 2 & 1 \\ 2 & 3 & 2 \\ 1 & 2 & 3 \end{bmatrix}$ **30** $\begin{bmatrix} 1 & 3 & 2 \\ 2 & 1 & 3 \\ 3 & 2 & 1 \end{bmatrix}\begin{bmatrix} -1 & 3 & 2 \\ 2 & -1 & 3 \\ 3 & 2 & -1 \end{bmatrix}$

31 $\begin{bmatrix} 0 & 2 & 1 \\ 3 & 0 & -3 \\ -2 & -1 & 0 \end{bmatrix}\begin{bmatrix} 0 & 3 & -2 \\ 2 & 0 & -1 \\ 1 & -3 & 0 \end{bmatrix}$

32 $\begin{bmatrix} 0 & 2 & -3 \\ 0 & 0 & 4 \\ 0 & 0 & 0 \end{bmatrix}\begin{bmatrix} 1 & 0 & 0 \\ -5 & 1 & 0 \\ 6 & -5 & 0 \end{bmatrix}$

33 $\begin{bmatrix} 2 & -3 \\ 4 & 5 \end{bmatrix}\begin{bmatrix} 5 & 3 \\ -4 & 2 \end{bmatrix}$

34 $\begin{bmatrix} 3 & 1 \\ 1 & 3 \end{bmatrix}\begin{bmatrix} 2 & -1 \\ -1 & 2 \end{bmatrix}$

35 $\begin{bmatrix} 2 & -1 \\ 3 & 1 \\ 4 & -1 \end{bmatrix}\begin{bmatrix} 1 & 0 \\ 0 & 1 \end{bmatrix}$

36 $\begin{bmatrix} -1 & 2 \\ 1 & 3 \\ -1 & 4 \end{bmatrix}\begin{bmatrix} 1 & -1 \\ -2 & 2 \end{bmatrix}$

37 $\begin{bmatrix} 1 & 0 & 4 \\ 4 & 2 & 0 \end{bmatrix}\begin{bmatrix} -1 & 3 \\ 3 & -1 \\ 1 & -3 \end{bmatrix}$

38 $\begin{bmatrix} -2 & 3 & 1 \\ 1 & 1 & 1 \end{bmatrix}\begin{bmatrix} 4 & 3 \\ 3 & -5 \\ 6 & -5 \end{bmatrix}.$

3 MATRIX INVERSE

Identity Matrix

The matrix $I = \begin{bmatrix} 1 & 0 & 0 \\ 0 & 1 & 0 \\ 0 & 0 & 1 \end{bmatrix}$ is called the 3×3 **identity matrix.**

The identity matrix is important because in matrix arithmetic it acts like 1 in ordinary arithmetic. Namely, if A is any 3×3 matrix, then

$AI = A$ and $IA = A$.

Example

$$\begin{bmatrix} 7 & -2 & -3 \\ 4 & 0 & 1 \\ -1 & -1 & 2 \end{bmatrix}\begin{bmatrix} 1 & 0 & 0 \\ 0 & 1 & 0 \\ 0 & 0 & 1 \end{bmatrix} = \begin{bmatrix} 7 & -2 & -3 \\ 4 & 0 & 1 \\ -1 & -1 & 2 \end{bmatrix}$$

Please check it.

The Inverse

Let A be a 3×3 matrix. A is called **invertible** if there exists a 3×3 matrix B such that

$AB = I$ and $BA = I$.

Then B is called the **inverse** of A, and we write $B = A^{-1}$.

Example

$$A = \begin{bmatrix} 3 & 0 & -1 \\ 2 & 1 & -2 \\ 4 & -3 & 1 \end{bmatrix} \qquad B = \begin{bmatrix} 1 & -\frac{3}{5} & -\frac{1}{5} \\ 2 & -\frac{7}{5} & -\frac{4}{5} \\ 2 & -\frac{9}{5} & -\frac{3}{5} \end{bmatrix}$$

Then $AB = I$ and $BA = I$ (please check!); hence A is invertible and $B = A^{-1}$.

Not every matrix is invertible. Obviously the zero matrix (all elements 0) has no inverse. But also many other matrices have no inverse. For instance,

$$A = \begin{bmatrix} 1 & 1 & 1 \\ 1 & 1 & 1 \\ 1 & 1 & 1 \end{bmatrix}$$

has no inverse. Indeed, if B is any matrix, then the first row of BA has the form $[b \quad b \quad b]$, where $b = b_{11} + b_{12} + b_{13}$, whereas the first row of I is $[1 \quad 0 \quad 0]$. Therefore $BA = I$ is impossible, else we would have $b = 1$ and $b = 0$ simultaneously.

Facts About Inverses

The following basic facts about matrix inverses are proved in courses on linear algebra.

(1) If A is an invertible matrix, then its inverse is unique.

(2) If either $AB = I$ or $BA = I$, then A is invertible and $B = A^{-1}$.

(3) If A is invertible, then A^{-1} is invertible and

$$(A^{-1})^{-1} = A$$

(4) If A and B are invertible, then AB is invertible and

$$(AB)^{-1} = B^{-1}A^{-1}.$$

Remark The hard statement to prove is (2). Statement (3) is pretty easy—try it. Statement (4) is straightforward provided you know the associative law for matrix multiplication:

$$(AB)C = A(BC).$$

(Its proof is an orgy of subscripts.) Indeed,

$$(AB)(B^{-1}A^{-1}) = A(BB^{-1})A^{-1} = AIA^{-1} = AA^{-1} = I.$$

Similarly $(B^{-1}A^{-1})(AB) = I$, so (4) holds. See if you can use the associative law to prove (1).

Calculating Inverses

Given a 3×3 matrix A, how can we tell whether A has an inverse? And if it does, how can we compute the inverse? We now discuss an algorithm (procedure) that either leads to A^{-1}, or shows us that A has no inverse. It is a natural extension of the method of Gaussian elimination for solving systems. We shall elaborate on the connection later.

Step 1 Write the 3×6 matrix

$$[A \quad I].$$

Its first 3 columns are the columns of A and its last 3 columns are the columns of I.

Step 2 Perform a sequence of basic row operations transforming $[A \quad I]$ into $[I \quad B]$. (Details later.)

Step 3 If you get stuck in step 2 with a column of 0's or a row of 0's in the left half (first three columns), then A is not invertible.

Step 4 If step 2 succeeds, then $B = A^{-1}$.

How we perform step 2 systematically is best seen through an example.

EXAMPLE 1 Find the inverse of $A = \begin{bmatrix} 0 & -3 & 2 \\ 2 & 4 & -1 \\ 1 & 3 & -1 \end{bmatrix}$

Solution We start with the 3×6 matrix

$$[A \quad I] = \begin{bmatrix} 0 & -3 & 2 & 1 & 0 & 0 \\ 2 & 4 & -1 & 0 & 1 & 0 \\ 1 & 3 & -1 & 0 & 0 & 1 \end{bmatrix}$$

We want to change the first three columns into those of the identity matrix by a sequence of basic row operations (p. 348). First we need a 1 in the 1-1 position, so we interchange rows 1 and 3:

$$\begin{bmatrix} 1 & 3 & -1 & 0 & 0 & 1 \\ 2 & 4 & -1 & 0 & 1 & 0 \\ 0 & -3 & 2 & 1 & 0 & 0 \end{bmatrix}$$

We want 0's elsewhere in the first column, so we subtract twice the first row from the second:

$$\begin{bmatrix} 1 & 3 & -1 & 0 & 0 & 1 \\ 0 & -2 & 1 & 0 & 1 & -2 \\ 0 & -3 & 2 & 1 & 0 & 0 \end{bmatrix}$$

Now we need a 1 in the 2-2 position, so we multiply the second row by $-\frac{1}{2}$:

$$\begin{bmatrix} 1 & 3 & -1 & 0 & 0 & 1 \\ 0 & 1 & -\frac{1}{2} & 0 & -\frac{1}{2} & 1 \\ 0 & -3 & 2 & 1 & 0 & 0 \end{bmatrix}$$

To clean up the rest of the second column, we add 3 times the second row to the third and subtract 3 times the second from the first:

$$\begin{bmatrix} 1 & 0 & \frac{1}{2} & 0 & \frac{3}{2} & -2 \\ 0 & 1 & -\frac{1}{2} & 0 & -\frac{1}{2} & 1 \\ 0 & 0 & \frac{1}{2} & 1 & -\frac{3}{2} & 3 \end{bmatrix}$$

Next we multiply the third row by 2 in order to have 1 in the 3-3 spot:

$$\begin{bmatrix} 1 & 0 & \frac{1}{2} & 0 & \frac{3}{2} & -2 \\ 0 & 1 & -\frac{1}{2} & 0 & -\frac{1}{2} & 1 \\ 0 & 0 & 1 & 2 & -3 & 6 \end{bmatrix}$$

Finally we subtract $\frac{1}{2}$ times the third row from the first and add $\frac{1}{2}$ times the third row to the second:

$$\begin{bmatrix} 1 & 0 & 0 & -1 & 3 & -5 \\ 0 & 1 & 0 & 1 & -2 & 4 \\ 0 & 0 & 1 & 2 & -3 & 6 \end{bmatrix}$$

We have reached the desired form $[I \quad B]$. Conclusion:

Answer $A^{-1} = \begin{bmatrix} -1 & 3 & -5 \\ 1 & -2 & 4 \\ 2 & -3 & 6 \end{bmatrix}$

Check

$$AA^{-1} = \begin{bmatrix} 0 & -3 & 2 \\ 2 & 4 & -1 \\ 1 & 3 & -1 \end{bmatrix}\begin{bmatrix} -1 & 3 & -5 \\ 1 & -2 & 4 \\ 2 & -3 & 6 \end{bmatrix} = \begin{bmatrix} 1 & 0 & 0 \\ 0 & 1 & 0 \\ 0 & 0 & 1 \end{bmatrix} = I$$

Warning It is *very important* to check your answer in matrix inverse problems.

Remark You can often use some ingenuity to simplify the arithmetic. For instance, starting in the solution above before we multiplied the second row by $-\frac{1}{2}$:

$$\begin{bmatrix} 1 & 3 & -1 & 0 & 0 & 1 \\ 0 & -2 & 1 & 0 & 1 & -2 \\ 0 & -3 & 2 & 1 & 0 & 0 \end{bmatrix} \xrightarrow{\text{(subtract 3-rd from 2-nd)}} \begin{bmatrix} 1 & 3 & -1 & 0 & 0 & 1 \\ 0 & 1 & -1 & -1 & 1 & -2 \\ 0 & -3 & 2 & 1 & 0 & 0 \end{bmatrix}$$

This puts a 1 in the 2-2 spot without introducing fractions. Now continue as before:

$$\longrightarrow \begin{bmatrix} 1 & 0 & 2 & 3 & -3 & 7 \\ 0 & 1 & -1 & -1 & 1 & -2 \\ 0 & 0 & -1 & -2 & 3 & -6 \end{bmatrix} \longrightarrow \begin{bmatrix} 1 & 0 & 2 & 3 & -3 & 7 \\ 0 & 1 & -1 & -1 & 1 & -2 \\ 0 & 0 & 1 & 2 & -3 & 6 \end{bmatrix}$$

$$\longrightarrow \begin{bmatrix} 1 & 0 & 0 & -1 & 3 & -5 \\ 0 & 1 & 0 & 1 & -2 & 4 \\ 0 & 0 & 1 & 2 & -3 & 6 \end{bmatrix}$$

EXAMPLE 2 Find the inverse of $A = \begin{bmatrix} 1 & -1 & 3 \\ 4 & 0 & 7 \\ 6 & -2 & 13 \end{bmatrix}$.

Solution Start with

$$[A \quad I] = \begin{bmatrix} 1 & -1 & 3 & 1 & 0 & 0 \\ 4 & 0 & 7 & 0 & 1 & 0 \\ 6 & -2 & 13 & 0 & 0 & 1 \end{bmatrix}$$

Add multiples of the first row to the second and third:

$$\begin{bmatrix} 1 & -1 & 3 & 1 & 0 & 0 \\ 0 & 4 & -5 & -4 & 1 & 0 \\ 0 & 4 & -5 & -6 & 0 & 1 \end{bmatrix}$$

Subtract the second row from the third:

$$\begin{bmatrix} 1 & -1 & 3 & 1 & 0 & 0 \\ 0 & 4 & -5 & -4 & 1 & 0 \\ 0 & 0 & 0 & -2 & -1 & 1 \end{bmatrix}$$

Since the left half contains a row of 0's, the process ends in failure.

Answer A is not invertible.

Linear Systems

A linear system

$$\begin{cases} a_1x + b_1y + c_1z = d_1 \\ a_2x + b_2y + c_2z = d_2 \\ a_3x + b_3y + c_3z = d_3 \end{cases}$$

can be written in matrix form:

$$\begin{bmatrix} a_1 & b_1 & c_1 \\ a_2 & b_2 & c_2 \\ a_3 & b_3 & c_3 \end{bmatrix} \begin{bmatrix} x \\ y \\ z \end{bmatrix} = \begin{bmatrix} d_1 \\ d_2 \\ d_3 \end{bmatrix}$$

Briefly, $A\mathbf{x} = \mathbf{d}$ where $A = \begin{bmatrix} a_1 & b_1 & c_1 \\ a_2 & b_2 & c_2 \\ a_3 & b_3 & c_3 \end{bmatrix} \quad \mathbf{x} = \begin{bmatrix} x \\ y \\ z \end{bmatrix} \quad \mathbf{d} = \begin{bmatrix} d_1 \\ d_2 \\ d_3 \end{bmatrix}.$

A is called the **coefficient matrix** of the system, $\mathbf{d}$ is called the **column of constants,** and $\mathbf{x}$ is the **column of unknowns.**

Now we state an important connection between systems and inverses.

> If A is invertible, then the solution of
>
> $$A\mathbf{x} = \mathbf{d}$$
>
> is
>
> $$\mathbf{x} = A^{-1}\mathbf{d}.$$

This result is true because if $\mathbf{x} = A^{-1}\mathbf{d}$, then

$$A\mathbf{x} = A(A^{-1}\mathbf{d}) = (AA^{-1})\mathbf{d} = I\mathbf{d} = \mathbf{d}.$$

(This proof uses the associative law for matrix multiplication, mentioned earlier.)

The result is useful in two instances: first, it may happen that you have A^{-1}; then you merely multiply it by $\mathbf{d}$ to solve $A\mathbf{x} = \mathbf{d}$. Second, you may have to solve several systems $A\mathbf{x} = \mathbf{d}$ in which the coefficient matrix A is fixed, but $\mathbf{d}$ takes on different values. An example is the problem of finding the stresses in the different trusses of a bridge under a variety of assumed loads.

EXAMPLE 3 Solve $A\mathbf{x} = \mathbf{d}$, where

$$A = \begin{bmatrix} 0 & -3 & 2 \\ 2 & 4 & -1 \\ 1 & 3 & -1 \end{bmatrix}$$

and

$$\text{(a)}\quad \mathbf{d} = \begin{bmatrix} 1 \\ 0 \\ 0 \end{bmatrix} \qquad \text{(b)}\quad \mathbf{d} = \begin{bmatrix} 1 \\ 1 \\ 0 \end{bmatrix} \qquad \text{(c)}\quad \mathbf{d} = \begin{bmatrix} 1 \\ 1 \\ 2 \end{bmatrix}$$

Solution Since A is the matrix of Example 1,

$$A^{-1} = \begin{bmatrix} -1 & 3 & -5 \\ 1 & -2 & 4 \\ 2 & -3 & 6 \end{bmatrix}$$

$$\text{(a)}\quad \mathbf{x} = A^{-1}\mathbf{d} = \begin{bmatrix} -1 & 3 & -5 \\ 1 & -2 & 4 \\ 2 & -3 & 6 \end{bmatrix} \begin{bmatrix} 1 \\ 0 \\ 0 \end{bmatrix} = \begin{bmatrix} -1 \\ 1 \\ 2 \end{bmatrix}$$

Check $$A\mathbf{x} = \begin{bmatrix} 0 & -3 & 2 \\ 2 & 4 & -1 \\ 1 & 3 & -1 \end{bmatrix} \begin{bmatrix} -1 \\ 1 \\ 2 \end{bmatrix} = \begin{bmatrix} 1 \\ 0 \\ 0 \end{bmatrix} = \mathbf{d}.$$

Similarly

$$\text{(b)}\quad \mathbf{x} = \begin{bmatrix} -1 & 3 & -5 \\ 1 & -2 & 4 \\ 2 & -3 & 6 \end{bmatrix} \begin{bmatrix} 1 \\ 1 \\ 0 \end{bmatrix} = \begin{bmatrix} 2 \\ -1 \\ -1 \end{bmatrix}$$

$$\text{(c)}\quad \mathbf{x} = \begin{bmatrix} -1 & 3 & -5 \\ 1 & -2 & 4 \\ 2 & -3 & 6 \end{bmatrix} \begin{bmatrix} 1 \\ 1 \\ 2 \end{bmatrix} = \begin{bmatrix} -8 \\ 7 \\ 11 \end{bmatrix}$$

and these check.

Answer $$\text{(a)}\quad \begin{bmatrix} -1 \\ 1 \\ 2 \end{bmatrix} \qquad \text{(b)}\quad \begin{bmatrix} 2 \\ -1 \\ -1 \end{bmatrix} \qquad \text{(c)}\quad \begin{bmatrix} -8 \\ 7 \\ 11 \end{bmatrix}$$

Note Now let's look at the connection between our elimination method of inverting A and linear systems. Let $\mathbf{e}_1, \mathbf{e}_2, \mathbf{e}_3$ denote the columns of the identity matrix I. Suppose we solve the three systems

(1) $A\mathbf{x}_1 = \mathbf{e}_1$ (2) $A\mathbf{x}_2 = \mathbf{e}_2$ (3) $A\mathbf{x}_3 = \mathbf{e}_3$.

Then the matrix B whose columns are $\mathbf{x}_1, \mathbf{x}_2, \mathbf{x}_3$ is the inverse of A. Indeed,

$$AB = A[\mathbf{x}_1 \;\; \mathbf{x}_2 \;\; \mathbf{x}_3] = [A\mathbf{x}_1 \;\; A\mathbf{x}_2 \;\; A\mathbf{x}_3] = [\mathbf{e}_1 \;\; \mathbf{e}_2 \;\; \mathbf{e}_3] = I.$$

To find $\mathbf{x}_1$, we can solve $A\mathbf{x}_1 = \mathbf{e}_1$ by Gaussian elimination; ditto for $A\mathbf{x}_2 = \mathbf{e}_2$ and for $A\mathbf{x}_3 = \mathbf{e}_3$. Our method for finding A^{-1} does these three elimination jobs *simultaneously*. That's really all there is to it.

Exercises

Find if the matrix is invertible; if so, calculate its inverse

1 $\begin{bmatrix} 1 & 0 & 0 \\ 1 & 1 & 0 \\ 1 & 1 & 1 \end{bmatrix}$ **2** $\begin{bmatrix} 1 & 2 & 2 \\ 0 & 1 & 2 \\ 0 & 0 & 1 \end{bmatrix}$ **3** $\begin{bmatrix} 1 & 0 & 0 \\ -1 & 2 & 0 \\ -1 & -1 & 3 \end{bmatrix}$ **4** $\begin{bmatrix} 2 & -1 & 2 \\ 0 & 2 & -1 \\ 0 & 0 & 1 \end{bmatrix}$

5 $\begin{bmatrix} 1 & 2 & 3 \\ -1 & -1 & -7 \\ 1 & 3 & 0 \end{bmatrix}$ **6** $\begin{bmatrix} 1 & -2 & 3 \\ 3 & -5 & 7 \\ 2 & -3 & 5 \end{bmatrix}$ **7** $\begin{bmatrix} 2 & -1 & -1 \\ 4 & -1 & -5 \\ 6 & -3 & -2 \end{bmatrix}$ **8** $\begin{bmatrix} -2 & -5 & -1 \\ -3 & -1 & -1 \\ 6 & 4 & 2 \end{bmatrix}$

9 $\begin{bmatrix} 1 & 1 & 0 \\ -3 & -6 & -3 \\ 3 & 7 & 4 \end{bmatrix}$ **10** $\begin{bmatrix} 2 & -1 & 3 \\ 3 & 2 & 8 \\ 5 & -3 & 7 \end{bmatrix}$ **11** $\begin{bmatrix} 2 & 1 & 5 \\ -4 & 1 & -1 \\ 3 & 1 & 6 \end{bmatrix}$ **12** $\begin{bmatrix} -5 & 2 & -3 \\ 3 & -2 & 1 \\ 1 & 2 & 3 \end{bmatrix}$

13 $\begin{bmatrix} 3 & -1 & 3 \\ -4 & 1 & -2 \\ -6 & 2 & -3 \end{bmatrix}$ **14** $\begin{bmatrix} 0 & 2 & 1 \\ -3 & 4 & 3 \\ 2 & -1 & -1 \end{bmatrix}$ **15** $\begin{bmatrix} 4 & 3 & 2 \\ 5 & 3 & 1 \\ 6 & 5 & 3 \end{bmatrix}$ **16** $\begin{bmatrix} 6 & -5 & 4 \\ 6 & -7 & 5 \\ 9 & -7 & 6 \end{bmatrix}$.

Calculate A^{-1}; then solve $A\mathbf{x} = \mathbf{d}$

17 $A = \begin{bmatrix} 0 & -1 & 1 \\ 1 & 0 & 0 \\ -1 & 1 & 0 \end{bmatrix}$ $\mathbf{d} = \begin{bmatrix} 1 \\ 0 \\ 2 \end{bmatrix}$ **18** $A = \begin{bmatrix} 1 & -4 & 1 \\ 2 & -1 & -1 \\ 1 & 1 & -1 \end{bmatrix}$ $\mathbf{d} = \begin{bmatrix} 0 \\ 2 \\ -1 \end{bmatrix}$

19 $A = \begin{bmatrix} -5 & 4 & 6 \\ 3 & -2 & -3 \\ -1 & 1 & 2 \end{bmatrix}$ $\mathbf{d} = \begin{bmatrix} -1 \\ 1 \\ 1 \end{bmatrix}$ **20** $A = \begin{bmatrix} 0 & -3 & 2 \\ 2 & 4 & -1 \\ 1 & 3 & -1 \end{bmatrix}$ $\mathbf{d} = \begin{bmatrix} 1 \\ 2 \\ -3 \end{bmatrix}$

21 $A = \begin{bmatrix} -\frac{13}{2} & -11 & -3 \\ \frac{1}{2} & 1 & 0 \\ 2 & 3 & 1 \end{bmatrix}$ $\mathbf{d} = \begin{bmatrix} -1 \\ 0 \\ 1 \end{bmatrix}$ **22** $A = \begin{bmatrix} 2 & 6 & 4 \\ 0 & 2 & 1 \\ -6 & -22 & -13 \end{bmatrix}$ $\mathbf{d} = \begin{bmatrix} -3 \\ -4 \\ 2 \end{bmatrix}$

23 $A = \begin{bmatrix} 4 & 5 & 6 \\ 3 & 3 & 5 \\ 2 & 1 & 3 \end{bmatrix}$ $\mathbf{d} = \begin{bmatrix} 1 \\ 1 \\ 1 \end{bmatrix}$ **24** $A = \begin{bmatrix} 4 & -5 & 6 \\ 5 & -7 & 6 \\ 6 & -7 & 9 \end{bmatrix}$ $\mathbf{d} = \begin{bmatrix} -3 \\ 4 \\ 0 \end{bmatrix}$

25 $A = \begin{bmatrix} 2 & -1 & 1 \\ -3 & 3 & -2 \\ 6 & -5 & 4 \end{bmatrix}$ $\mathbf{d} = \begin{bmatrix} 1 \\ 0 \\ 0 \end{bmatrix}$, $\mathbf{d} = \begin{bmatrix} 0 \\ 1 \\ 0 \end{bmatrix}$, $\mathbf{d} = \begin{bmatrix} 0 \\ 0 \\ 1 \end{bmatrix}$

26 $A = \begin{bmatrix} 4 & -1 & -1 \\ 1 & 1 & -2 \\ -1 & 1 & -1 \end{bmatrix}$ $\mathbf{d} = \begin{bmatrix} 1 \\ -1 \\ 0 \end{bmatrix}$, $\mathbf{d} = \begin{bmatrix} 0 \\ 1 \\ -1 \end{bmatrix}$, $\mathbf{d} = \begin{bmatrix} -1 \\ 0 \\ 1 \end{bmatrix}$.

4 DETERMINANTS

With each $n \times n$ (square) matrix A is associated a *number* called the determinant of A, and denoted $\det A$ or $|A|$. Determinants are important because they provide formulas in many situations involving linear algebra, and from these formulas many properties can be deduced. For instance, determinants are applied in finding the frequencies of vibrations, and in the stability theory of physical systems and of economic systems.

We shall limit our discussion to 2×2 and 3×3 determinants, since larger sizes introduce technicalities best left to more advanced courses.

Determinant of a 2 × 2 Matrix $\det \begin{bmatrix} a & b \\ c & d \end{bmatrix} = \begin{vmatrix} a & b \\ c & d \end{vmatrix} = ad - bc.$

Examples

$$\begin{vmatrix} 1 & 0 \\ 0 & 1 \end{vmatrix} = 1 \cdot 1 - 0 \cdot 0 = 1 \qquad \begin{vmatrix} 3 & 4 \\ -8 & -6 \end{vmatrix} = 3 \cdot (-6) - 4 \cdot (-8) = 14$$

Determinant of a 3 × 3 Matrix

$$\begin{vmatrix} a_1 & b_1 & c_1 \\ a_2 & b_2 & c_2 \\ a_3 & b_3 & c_3 \end{vmatrix} = a_1b_2c_3 + a_2b_3c_1 + a_3b_1c_2 - a_1b_3c_2 - a_2b_1c_3 - a_3b_2c_1.$$

There are six products; each has one factor from each column and one factor from each row. To remember the formula, follow the scheme in Fig. 1.

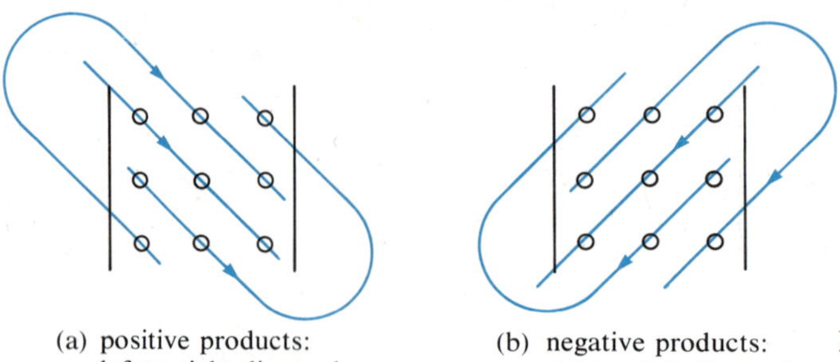

(a) positive products: left-to-right diagonals

(b) negative products: right-to-left diagonals

Fig. 1 Memory aid for 3×3 determinants

Example

$$A = \begin{bmatrix} -2 & -3 & 1 \\ 4 & 2 & -1 \\ 6 & 3 & 5 \end{bmatrix}$$

$$\begin{aligned} |A| &= [(-2)(2)(5) + (4)(3)(1) + (6)(-3)(-1)] \\ &\quad -[(-2)(3)(-1) + (4)(-3)(5) + (6)(2)(1)] \\ &= (-20 + 12 + 18) - (6 - 60 + 12) = (10) - (-42) = 52. \end{aligned}$$

Remark The 2×2 determinant formula has 2 terms; the 3×3 determinant formula has 6 terms. The 4×4 and 5×5 determinant formulas have 24 and 120 terms and follow a much more complicated scheme than that indicated in Fig. 1.

Minors

A 3×3 determinant can be evaluated in several ways by grouping terms suitably. For example,

$$\begin{vmatrix} a_1 & b_1 & c_1 \\ a_2 & b_2 & c_2 \\ a_3 & b_3 & c_3 \end{vmatrix} = a_1(b_2c_3 - b_3c_2) - b_1(a_2c_3 - a_3c_2) + c_1(a_2b_3 - a_3b_2).$$

Each quantity in parentheses is the determinant of a 2×2 matrix, so we can write

$$\begin{vmatrix} a_1 & b_1 & c_1 \\ a_2 & b_2 & c_2 \\ a_3 & b_3 & c_3 \end{vmatrix} = a_1\begin{vmatrix} b_2 & c_2 \\ b_3 & c_3 \end{vmatrix} - b_1\begin{vmatrix} a_2 & c_2 \\ a_3 & c_3 \end{vmatrix} + c_1\begin{vmatrix} a_2 & b_2 \\ a_3 & b_3 \end{vmatrix}.$$

The coefficients of the 2×2 determinants are the terms of the *first* row of the matrix, with suitable signs. There are corresponding formulas based on the second and third rows. Similarly, there are three formulas involving columns. For instance,

$$\begin{vmatrix} a_1 & b_1 & c_1 \\ a_2 & b_2 & c_2 \\ a_3 & b_3 & c_3 \end{vmatrix} = a_1\begin{vmatrix} b_2 & c_2 \\ b_3 & c_3 \end{vmatrix} - a_2\begin{vmatrix} b_1 & c_1 \\ b_3 & c_3 \end{vmatrix} + a_3\begin{vmatrix} b_1 & c_1 \\ b_2 & c_2 \end{vmatrix}.$$

All of these formulas can be described in a systematic way. First write the matrix in index notation,

$$\begin{vmatrix} a_{11} & a_{12} & a_{13} \\ a_{21} & a_{22} & a_{23} \\ a_{31} & a_{32} & a_{33} \end{vmatrix}.$$

Thus a_{32} denotes the element in the third row and second column, etc. Now we cross out the i-th row and j-th column. What remains is a 2×2 matrix, whose determinant m_{ij} is called the **minor** of a_{ij}. For instance,

$$m_{32} = \begin{vmatrix} a_{11} & a_{12} & a_{13} \\ a_{21} & a_{22} & a_{23} \\ a_{31} & a_{32} & a_{33} \end{vmatrix} = \begin{vmatrix} a_{11} & a_{13} \\ a_{21} & a_{23} \end{vmatrix}.$$

The six expansion formulas mentioned above are:

Expansion by Minors of Rows

$$\begin{vmatrix} a_{11} & a_{12} & a_{13} \\ a_{21} & a_{22} & a_{23} \\ a_{31} & a_{32} & a_{33} \end{vmatrix} = \begin{matrix} a_{11}m_{11} - a_{12}m_{12} + a_{13}m_{13} & \text{(by first row)} \\ -a_{21}m_{21} + a_{22}m_{22} - a_{23}m_{23} & \text{(by second row)} \\ a_{31}m_{31} - a_{32}m_{32} + a_{33}m_{33} & \text{(by third row)} \end{matrix}$$

Expansion by Minors of Columns

$$\begin{vmatrix} a_{11} & a_{12} & a_{13} \\ a_{21} & a_{22} & a_{23} \\ a_{31} & a_{32} & a_{33} \end{vmatrix} = \begin{matrix} a_{11}m_{11} - a_{21}m_{21} + a_{31}m_{31} & \text{(by first column)} \\ -a_{12}m_{12} + a_{22}m_{22} - a_{32}m_{32} & \text{(by second column)} \\ a_{13}m_{13} - a_{23}m_{23} + a_{33}m_{33} & \text{(by third column).} \end{matrix}$$

Note the pattern of the signs; the sign in front of the term $a_{ij}m_{ij}$ in any of these formulas is $(-1)^{i+j}$. It is the sign in the i-th row and j-th column of

$$\begin{bmatrix} + & - & + \\ - & + & - \\ + & - & + \end{bmatrix}.$$

EXAMPLE 1 Evaluate $d = \begin{vmatrix} 2 & -3 & 1 \\ 2 & 4 & -5 \\ 3 & 0 & 2 \end{vmatrix}$

(a) by the first row (b) by the second column.

Solution

(a) $d = 2\begin{vmatrix} 4 & -5 \\ 0 & 2 \end{vmatrix} - (-3)\begin{vmatrix} 2 & -5 \\ 3 & 2 \end{vmatrix} + 1\begin{vmatrix} 2 & 4 \\ 3 & 0 \end{vmatrix}$

$= 2(8) + 3(19) + (-12) = 61.$

(b) $d = -(-3)\begin{vmatrix} 2 & -5 \\ 3 & 2 \end{vmatrix} + 4\begin{vmatrix} 2 & 1 \\ 3 & 2 \end{vmatrix} - 0\begin{vmatrix} 2 & 1 \\ 2 & -5 \end{vmatrix}$

$= 3(19) + 4(1) - 0 = 61.$

Answer 61

Note The presence of a zero in the second column reduced the amount of computation. In general it pays to expand a determinant by a column that contains one or more zeros. For example,

$$\begin{vmatrix} 3 & 26 & -177 \\ 0 & 89 & 6 \\ 0 & 14 & 1 \end{vmatrix} = 3\begin{vmatrix} 89 & 6 \\ 14 & 1 \end{vmatrix} + 0 + 0 = 3(89 - 84) = 3(5) = 15$$

(expansion by the first column).

Remark The expansion formulas extend to higher order determinants. For instance, if A is a 4×4 matrix, then

$$\det A = a_{11}m_{11} - a_{12}m_{12} + a_{13}m_{13} - a_{14}m_{14}.$$

This formula reduces the evaluation of a 4×4 determinant to the evaluation of four 3×3 determinants.

Properties of Determinants

We state some basic properties of 2×2 and 3×3 determinants. (They are true as stated for $n \times$ n determinants, and can be used to evaluate them, even though we have not attempted a definition.)

Properties of Determinants

(1) If a row (column) has all zeros, then the determinant equals 0.

(2). If two rows (columns) are interchanged, the determinant changes sign.

(3) If two rows (columns) are equal term-by-term, the determinant equals 0.

(4) If the first row $\mathbf{r}_1$ is the sum term-by-term of two other row vectors $\mathbf{r}_1'$ and $\mathbf{r}_1''$, then the determinant is the sum of the two determinants that result when the first row $\mathbf{r}_1$ is replaced by $\mathbf{r}_1'$ and then by $\mathbf{r}_1''$. The same holds for the second row and for the third row. The same holds for the columns.

(5) If all terms of one row (column) are multiplied by a factor k, then the determinant is multiplied by k.

(6) $\det I = 1$ where I is the identity matrix.

In the 3×3 case, each of the properties follows from the definition

$$\begin{vmatrix} a_1 & b_1 & c_1 \\ a_2 & b_2 & c_2 \\ a_3 & b_3 & c_3 \end{vmatrix} = a_1b_2c_3 + a_2b_3c_1 + a_3b_1c_2 - a_1b_3c_2 - a_2b_1c_3 - a_3b_2c_1.$$

For instance, if a row contains all zeros, say $a_1 = b_1 = c_1 = 0$, then each product is 0, hence the determinant is 0. Similarly, if one column contains all zeros, say $b_1 = b_2 = b_3 = 0$, then the determinant is 0. This proves property (1).

If all terms of one row are multiplied by a factor k, say a_3, b_3, c_3 become ka_3, kb_3, kc_3, then each of the six products is multiplied by k, so the determinant itself is multiplied by k. This proves (5). We omit the verification of the other properties.

Note what (4) says. For instance

$$\begin{vmatrix} a_1 & b_1 & c_1 + d_1 \\ a_2 & b_2 & c_2 + d_2 \\ a_3 & b_3 & c_3 + d_3 \end{vmatrix} = \begin{vmatrix} a_1 & b_1 & c_1 \\ a_2 & b_2 & c_2 \\ a_3 & b_3 & c_3 \end{vmatrix} + \begin{vmatrix} a_1 & b_1 & d_1 \\ a_2 & b_2 & d_2 \\ a_3 & b_3 & d_3 \end{vmatrix}.$$

The first two columns stay fixed; the determinant splits according to its third column.

In practice, property (5) is most often used to factor out a common factor in any single row or column. A consequence of (4) and (5) is the following property, useful in practical computations.

> (7) If a multiple of one row (column) is added to another row (column) the determinant is unchanged.

For example, let us show that adding k times its second column to its third column does not change the value of a determinant. By property (4),

$$\begin{vmatrix} a_1 & b_1 & c_1 + kb_1 \\ a_2 & b_2 & c_2 + kb_2 \\ a_3 & b_3 & c_3 + kb_3 \end{vmatrix} = \begin{vmatrix} a_1 & b_1 & c_1 \\ a_2 & b_2 & c_2 \\ a_3 & b_3 & c_3 \end{vmatrix} + \begin{vmatrix} a_1 & b_1 & kb_1 \\ a_2 & b_2 & kb_2 \\ a_3 & b_3 & kb_3 \end{vmatrix}.$$

We need only show that the second determinant on the right is 0. But it is: we just factor out k from the third column, leaving a determinant with two identical columns, hence 0.

To simplify the computation of a determinant, we often use property (7) to introduce zeros in strategic places.

EXAMPLE 2 Evaluate

$$D = \begin{vmatrix} 4 & 3 & -3 \\ 2 & -2 & 3 \\ -6 & 2 & -5 \end{vmatrix}.$$

Solution Add -2 times the second row to the first, add 3 times the second row to the third, then expand by minors of the first column:

$$D = \begin{vmatrix} 0 & 7 & -9 \\ 2 & -2 & 3 \\ -6 & 2 & -5 \end{vmatrix} = \begin{vmatrix} 0 & 7 & -9 \\ 2 & -2 & 3 \\ 0 & -4 & 4 \end{vmatrix}$$

$$= -2\begin{vmatrix} 7 & -9 \\ -4 & 4 \end{vmatrix} = -(2)(4)\begin{vmatrix} 7 & -9 \\ -1 & 1 \end{vmatrix} = -8(7 - 9) = 16.$$

Answer 16

Example 3 Evaluate

$$f(x) = \begin{vmatrix} x & -1 & 0 \\ 0 & x & -1 \\ c & b & x+a \end{vmatrix}.$$

Solution Add x times the third column to the second column:

$$f(x) = \begin{vmatrix} x & -1 & 0 \\ 0 & 0 & -1 \\ c & x^2+ax+b & x+a \end{vmatrix}$$

Now expand the determinant using minors of the second row:

$$f(x) = -0 + 0 - (-1)\begin{vmatrix} x & -1 \\ c & x^2+ax+b \end{vmatrix}$$

$$= x(x^2+ax+b) + c = x^3 + ax^2 + bx + c.$$

Answer $f(x) = x^3 + ax^2 + bx + c$

We mention a further property of determinants:

(8) If the rows and columns of a matrix are interchanged with each other, the determinant is unchanged:

$$\begin{vmatrix} a_1 & a_2 & a_3 \\ b_1 & b_2 & b_3 \\ c_1 & c_2 & c_3 \end{vmatrix} = \begin{vmatrix} a_1 & b_1 & c_1 \\ a_2 & b_2 & c_2 \\ a_3 & b_3 & c_3 \end{vmatrix}.$$

This follows easily from the defining formula for 3×3 determinants. (It is also true for $n \times n$ determinants.)

Finally we mention what is perhaps the most important property of determinants, even though its proof is too advanced for this course:

Multiplicative Property of Determinants

(9) If A and B are $n \times n$ matrices, then

$$\det(AB) = (\det A)(\det B).$$

Example $A = \begin{bmatrix} 3 & -1 \\ 2 & 4 \end{bmatrix} \qquad B = \begin{bmatrix} 2 & -4 \\ 3 & 5 \end{bmatrix}$

$\det A = 3\cdot4 - (-1)\cdot2 = 14 \qquad \det B = 2\cdot5 - (-4)\cdot3 = 22$

$$AB = \begin{bmatrix} 3 & -1 \\ 2 & 4 \end{bmatrix}\begin{bmatrix} 2 & -4 \\ 3 & 5 \end{bmatrix} = \begin{bmatrix} 3 & -17 \\ 16 & 12 \end{bmatrix}$$

$\det(AB) = 3\cdot12 - (-17)\cdot16 = 36 + 272 = 308$

$(\det A)(\det B) = 14\cdot22 = 308$

Example $A = \begin{bmatrix} 3 & -4 & -6 \\ -1 & 1 & 2 \\ 3 & -2 & -3 \end{bmatrix} \qquad B = \begin{bmatrix} -2 & -5 & -1 \\ 3 & 1 & 1 \\ -6 & -4 & -2 \end{bmatrix}$

$$|A| = (-9 - 12 - 24) - (-12 - 12 - 18) = -3$$

$$|B| = (4 + 12 + 30) - (8 + 30 + 6) = 2$$

$$AB = \begin{bmatrix} 3 & -4 & -6 \\ -1 & 1 & 2 \\ 3 & -2 & -3 \end{bmatrix}\begin{bmatrix} -2 & -5 & -1 \\ 3 & 1 & 1 \\ -6 & -4 & -2 \end{bmatrix} = \begin{bmatrix} 18 & 5 & 5 \\ -7 & -2 & -2 \\ 6 & -5 & 1 \end{bmatrix}$$

$$\det(AB) = (-36 + 175 - 60) - (180 - 35 - 60) = -6$$

$$(\det A)(\det B) = (-3)(2) = -6$$

Cramer's Rule

One application of determinants is to the solution of linear systems. Using determinants, we can write formulas for the solution of the system

$$(*) \qquad \begin{cases} a_1x + b_1y + c_1z = d_1 \\ a_2x + b_2y + c_2z = d_2 \\ a_3x + b_3y + c_3z = d_3 . \end{cases}$$

The formulas can be proved by a lot of tedious computation, or systematically in a course on linear algebra.

> **Cramer's Rule** The system (*) has a unique solution if and only if
>
> $$\Delta = \begin{vmatrix} a_1 & b_1 & c_1 \\ a_2 & b_2 & c_2 \\ a_3 & b_3 & c_3 \end{vmatrix} \neq 0 .$$
>
> If $\Delta \neq 0$, then the solution (x, y, z) is given by
>
> $$x = \frac{1}{\Delta}\begin{vmatrix} d_1 & b_1 & c_1 \\ d_2 & b_2 & c_2 \\ d_3 & b_3 & c_3 \end{vmatrix} \qquad y = \frac{1}{\Delta}\begin{vmatrix} a_1 & d_1 & c_1 \\ a_2 & d_2 & c_2 \\ a_3 & d_3 & c_3 \end{vmatrix} \qquad z = \frac{1}{\Delta}\begin{vmatrix} a_1 & b_1 & d_1 \\ a_2 & b_2 & d_2 \\ a_3 & b_3 & d_3 \end{vmatrix}.$$

Note carefully what Cramer's rule says. We start with the coefficient matrix

$$\begin{bmatrix} a_1 & b_1 & c_1 \\ a_2 & b_2 & c_2 \\ a_3 & b_3 & c_3 \end{bmatrix}$$

Its determinant is Δ, and Cramer's rule covers only the case $\Delta \neq 0$. To find x we replace the x-column of the matrix by the column of constants in $(*)$, take the determinant and divide by Δ. Ditto for y and z. With obvious modifications, Cramer's rule applies to 2×2 systems and to $n \times n$ systems for any n.

EXAMPLE 4 Solve by Cramer's rule

$$\begin{cases} 3x \qquad\quad - z = 15 \\ 2x + y - 2z = 0 \\ 4x - 3y + z = 5. \end{cases}$$

Solution

$$\Delta = \begin{vmatrix} 3 & 0 & -1 \\ 2 & 1 & -2 \\ 4 & -3 & 1 \end{vmatrix} = -5 \neq 0 \qquad x = \frac{1}{-5}\begin{vmatrix} 15 & 0 & -1 \\ 0 & 1 & -2 \\ 5 & -3 & 1 \end{vmatrix} = \frac{-70}{-5} = 14$$

$$y = \frac{1}{-5}\begin{vmatrix} 3 & 15 & -1 \\ 2 & 0 & -2 \\ 4 & 5 & 1 \end{vmatrix} = \frac{-130}{-5} = 26$$

$$z = \frac{1}{-5}\begin{vmatrix} 3 & 0 & 15 \\ 2 & 1 & 0 \\ 4 & -3 & 5 \end{vmatrix} = \frac{-135}{-5} = 27.$$

Check $3x - z = 3 \cdot 14 - 27 = 15$

$$2x + y - 2z = 2 \cdot 14 + 26 - 2 \cdot 27 = 0$$

$$4x - 3y + z = 4 \cdot 14 - 3 \cdot 26 + 27 = 5$$

Answer $(14, 26, 27)$

Remark Cramer's rule provides explicit formulas for the solution of a linear system. Such formulas can be valuable for discovering properties of the solution, for example how the solution of $(*)$ varies when b_1 varies.

As noted, Cramer's rule also applies to $n \times n$ systems. But for large n, Cramer's rule is a disaster for the practical computation of solutions because its use involves an order of magnitude more arithmetic than does Gaussian elimination.

Exercises

Evaluate

1 $\begin{vmatrix} 2 & 1 \\ -1 & 3 \end{vmatrix}$ **2** $\begin{vmatrix} 1 & 3 \\ 1 & 1 \end{vmatrix}$ **3** $\begin{vmatrix} 4 & -3 \\ -3 & -2 \end{vmatrix}$ **4** $\begin{vmatrix} 2 & 3 \\ 3 & 5 \end{vmatrix}$

5 $\begin{vmatrix} a & 1 \\ 0 & a \end{vmatrix}$ **6** $\begin{vmatrix} 1 & b \\ 0 & 1 \end{vmatrix}$ **7** $\begin{vmatrix} 1 & 0 \\ b & 1 \end{vmatrix}$ **8** $\begin{vmatrix} a & 0 \\ 1 & a \end{vmatrix}$

Evaluate $\begin{vmatrix} -3 & 0 & 3 \\ 4 & -2 & 1 \\ 5 & 2 & -1 \end{vmatrix}$

9 by the defining formula
10 by minors of the second column
11 by minors of the third column
12 by minors of the first row
13 by minors of the second row
14 by minors of the first column

Evaluate in any way

15 $\begin{vmatrix} 1 & 2 & 3 \\ 3 & 2 & 1 \\ 2 & 1 & 3 \end{vmatrix}$ **16** $\begin{vmatrix} 4 & 5 & 6 \\ 5 & 6 & 4 \\ 6 & 4 & 5 \end{vmatrix}$ **17** $\begin{vmatrix} -1 & 1 & 2 \\ 3 & -1 & 3 \\ 4 & 0 & 2 \end{vmatrix}$ **18** $\begin{vmatrix} 1 & 2 & 3 \\ 1 & 4 & 9 \\ 1 & 8 & 27 \end{vmatrix}$

19 $\begin{vmatrix} 3 & 6 & -1 \\ 2 & 1 & -2 \\ 4 & -3 & 1 \end{vmatrix}$ **20** $\begin{vmatrix} 15 & 0 & -1 \\ 0 & 1 & -2 \\ 5 & -3 & 1 \end{vmatrix}$ **21** $\begin{vmatrix} 3 & 15 & -1 \\ 2 & 0 & -2 \\ 4 & 5 & 1 \end{vmatrix}$ **22** $\begin{vmatrix} 3 & 0 & 15 \\ 2 & 1 & 0 \\ 4 & -3 & 5 \end{vmatrix}$

23 $\begin{vmatrix} 1 & 1 & 1 \\ 1 & -1 & 2 \\ 1 & 1 & 4 \end{vmatrix}$ **24** $\begin{vmatrix} 2 & 0 & 0 \\ -1 & 3 & 0 \\ 1 & 2 & -1 \end{vmatrix}$ **25** $\begin{vmatrix} 2 & 3 & 6 \\ -1 & 2 & -9 \\ 1 & 3 & -2 \end{vmatrix}$ **26** $\begin{vmatrix} -1 & 2 & -9 \\ 1 & 3 & -2 \\ 2 & 2 & 6 \end{vmatrix}$

27 $\begin{vmatrix} 4 & -3 & 1 \\ 2 & 1 & -2 \\ 3 & 0 & -1 \end{vmatrix}$ **28** $\begin{vmatrix} 1 & 1 & 3 \\ 2 & 4 & 7 \\ 2 & 3 & 9 \end{vmatrix}$ **29** $\begin{vmatrix} x & 1 & 0 \\ -1 & x & 1 \\ 0 & -1 & x \end{vmatrix}$ **30** $\begin{vmatrix} x & y & y \\ y & x & y \\ y & y & x \end{vmatrix}$

31 $\begin{vmatrix} 2x & x+y & x+z \\ x+y & 2y & y+z \\ x+z & y+z & 2z \end{vmatrix}$ **32** $\begin{vmatrix} y+z & x & x \\ y & x+z & y \\ z & z & x+y \end{vmatrix}$

Verify that $\det(AB) = (\det A)(\det B)$

33 $A = \begin{bmatrix} 5 & -6 \\ 3 & -3 \end{bmatrix}$ $B = \begin{bmatrix} 13 & 7 \\ 3 & 2 \end{bmatrix}$ **34** $A = \begin{bmatrix} 5 & 3 \\ 7 & 4 \end{bmatrix}$ $B = \begin{bmatrix} 5 & 9 \\ 3 & 5 \end{bmatrix}$

35 $A = \begin{bmatrix} -2 & -1 & 1 \\ 21 & 9 & -11 \\ -7 & -3 & 4 \end{bmatrix}$ $B = \begin{bmatrix} 1 & 0 & 0 \\ -2 & 1 & 0 \\ 2 & -2 & 1 \end{bmatrix}$

36 $A = \begin{bmatrix} 4 & -1 & -1 \\ 1 & -1 & 1 \\ -1 & -1 & 2 \end{bmatrix}$ $B = \begin{bmatrix} 1 & -3 & 5 \\ -1 & 2 & -4 \\ -2 & 3 & -6 \end{bmatrix}$

Solve by Cramer's Rule

37 $\begin{cases} x + 2y = 1 \\ 3y = 2 \end{cases}$ **38** $\begin{cases} 2x = 3 \\ -x + y = 0 \end{cases}$ **39** $\begin{cases} x + 2y = 1 \\ x + 3y = 2 \end{cases}$

40 $\begin{cases} x + y = a \\ x - y = b \end{cases}$ **41** $\begin{cases} 2x - 3y = -1 \\ 3x + 5y = 2 \end{cases}$ **42** $\begin{cases} 2x - 3y = -1 \\ -3x + 5y = 2. \end{cases}$

43 $\begin{cases} x + y - z = 3 \\ x - y + z = 2 \\ -x + y + z = 1 \end{cases}$ **44** $\begin{cases} 2y - 3z = 0 \\ x + y + z = 1 \\ 3y + 5z = 0 \end{cases}$

45 $\begin{cases} 2x + y + 2z = 1 \\ -x + 4y + 2z = 0 \\ 3x + y + z = -1 \end{cases}$

46 $\begin{cases} 2x - y - 3z = 1 \\ -x + 4y - 2z = 1 \\ 3x - y - z = 1 \end{cases}$

47 $\begin{cases} 2x - y + z = 4 \\ 3y + 2z = -1 \\ -z = 3 \end{cases}$

48 $\begin{cases} x + z = -1 \\ 2x + 3y + z = 5 \\ 4x + 4y + 2z = -3 \end{cases}$

49 $\begin{cases} 3x - y - z = 1 \\ -x - 4y - 2z = 1 \\ 2x - y - 3z = 1 \end{cases}$

50 $\begin{cases} 2x + y + 3z = 1 \\ 3x + y + z = -1 \\ -x + 4y + 2z = 0. \end{cases}$

51 Show that

$$\begin{vmatrix} 3 & 5 & 1 \\ -1 & 2 & 1 \\ x & y & 1 \end{vmatrix} = 0$$

is the equation of the line through $(3, 5)$ and $(-1, 2)$.

52* (cont.) Let (a, b) and (c, d) be distinct points in the plane. Show that

$$\begin{vmatrix} a & b & 1 \\ c & d & 1 \\ x & y & 1 \end{vmatrix} = 0$$

is the equation for the line L joining them.

5 POLYNOMIAL DIVISION

Recall how we divide integers in arithmetic. For example,

```
   28 <--- quotient
 7|201
   14
    61
    56
     5 <--- remainder
```

(2 is chosen in the quotient because $20 - 7 \cdot 2 < 7$; then 8 is chosen because $61 - 7 \cdot 8 < 7$.) The result is

$$201 = 7 \cdot 28 + 5.$$

It can also be expressed in terms of fractions:

$$\tfrac{201}{7} = 28 + \tfrac{5}{7} = 28\tfrac{5}{7}.$$

When you divide by 7, *the remainder is less than* 7; that is the crucial feature of long division of integers.

There is a similar long division process for polynomials. When we divide a polynomial $f(x)$ by a polynomial $g(x)$, we obtain a quotient $q(x)$ and a remainder $r(x)$. The crucial feature of polynomial division is that the remainder $r(x)$, if not 0, has degree *less than* the degree of the divisor $g(x)$.

EXAMPLE 1 Divide

$$x^4 - 3x^3 + x^2 + 2x - 2 \quad \text{by} \quad x^2 + x + 3 .$$

Solution We first set up the usual form.

$$x^2 + x + 3 \overline{\big)\, x^4 - 3x^3 + x^2 + 2x - 2} .$$

For the first term of the quotient we choose a monomial so that it times the divisor will cancel the first term x^4 of the dividend when subtracted off. Clearly x^2 is this monomial, and the first step of the long division is

$$\begin{array}{rrrrrr}
 & x^2 & & & & \\
\cline{2-6}
x^2 + x + 3 \,\big) & x^4 & - 3x^3 & + x^2 & + 2x & - 2 \\
 & x^4 & + x^3 & + 3x^2 & & \\
\cline{2-4}
 & & -4x^3 & - 2x^2 & + 2x & \\
\end{array}$$

The new dividend is $-4x^3 - 2x^2 + 2x$, and $-4x$ times the divisor will cancel its first term when subtracted off. Thus the next step in the long division is

$$\begin{array}{rrrrrr}
 & x^2 & - 4x & & & \\
\cline{2-6}
x^2 + x + 3 \,\big) & x^4 & - 3x^3 & + x^2 & + 2x & - 2 \\
 & x^4 & + x^3 & + 3x^2 & & \\
\cline{2-4}
 & & -4x^3 & - 2x^2 & + 2x & \\
 & & -4x^3 & - 4x^2 & - 12x & \\
\cline{3-5}
 & & & 2x^2 & + 14x & - 2 \\
\end{array}$$

Next, 2 times the divisor will cancel the first term in $2x^2 + 14x - 2$, so the next step is

$$\begin{array}{rrrrrrl}
 & x^2 & - 4x & + 2 & & & \longleftarrow \text{quotient} \\
\cline{2-6}
x^2 + x + 3 \,\big) & x^4 & - 3x^3 & + x^2 & + 2x & - 2 & \\
 & x^4 & + x^3 & + 3x^2 & & & \\
\cline{2-6}
 & & - 4x^3 & - 2x^2 & + 2x & & \\
 & & - 4x^3 & - 4x^2 & - 12x & & \\
\cline{3-6}
 & & & 2x^2 & + 14x & - 2 & \\
 & & & 2x^2 & + 2x & + 6 & \\
\cline{4-6}
 & & & & 12x & - 8 & \longleftarrow \text{remainder} \\
\end{array}$$

We are done because $\deg(12x - 8) = 1$ is less than $\deg(x^2 + x + 3) = 2$.

Check

$$(x^2 + x + 3)(x^2 - 4x + 2) + (12x - 8)$$

$$= (x^4 - 3x^3 + x^2 - 10x + 6) + (12x - 8)$$

$$= x^4 - 3x^3 + x^2 + 2x - 2 .$$

Answer Quotient $x^2 - 4x + 2$ Remainder $12x - 8$

Remark The result of this example can be put in terms of rational functions:

$$\frac{x^4 - 3x^3 + x^2 + 2x - 2}{x^2 + x + 3} = x^2 - 4x + 2 + \frac{12x - 8}{x^2 + x + 3} .$$

The example illustrates the following basic property of polynomials.

> **Division Algorithm** Let $f(x)$ and $g(x)$ be polynomials, with $g(x) \neq 0$. Then there are unique polynomials $q(x)$ and $r(x)$ such that
>
> (1) $f(x) = g(x)q(x) + r(x)$
>
> (2) $r(x) = 0$ or $\deg r(x) < \deg g(x)$.

We call $q(x)$ the **quotient** and $r(x)$ the **remainder** resulting from the division of $f(x)$ by $g(x)$. (A complete proof of the division algorithm requires mathematical induction and is left as an exercise, p. 421.)

Examples

(1) $f(x) = x^3 + 1 \qquad g(x) = x^2 + 1$

$$x^3 + 1 = (x^2 + 1)(x) + (-x + 1) \qquad q(x) = x \qquad r(x) = -x + 1.$$

(2) $f(x) = 3x^4 - 1 \qquad g(x) = x^2 + 2$

$$3x^4 - 1 = (x^2 + 2)(3x^2 - 6) + 11 \qquad q(x) = 3x^2 - 6 \qquad r(x) = 11.$$

A Simplification of Notation

Here is a full-fledged long division problem in complete detail:

$$\begin{array}{r|rrrrrr}
 & 3x^3 & +\ x^2 & +\ 4x & +\ 11 & & \\ \hline
x^2 - 2x - 3 & 3x^5 & -\ 5x^4 & -\ 7x^3 & & -\ 25x & -\ 15 \\
 & 3x^5 & -\ 6x^4 & -\ 9x^3 & & & \\ \cline{2-4}
 & & x^4 & +\ 2x^3 & & & \\
 & & x^4 & -\ 2x^3 & -\ 3x^2 & & \\ \cline{3-5}
 & & & 4x^3 & +\ 3x^2 & -\ 25x & \\
 & & & 4x^3 & -\ 8x^2 & -\ 12x & \\ \cline{4-6}
 & & & & 11x^2 & -\ 13x & -\ 15 \\
 & & & & 11x^2 & -\ 22x & -\ 33 \\ \cline{5-7}
 & & & & & 9x & +\ 18.
\end{array}$$

Surely we can find some way to shorten this calculation, just to write less. The powers of x are dispensable, provided we keep their coefficients separated by spaces. Let's omit them:

$$\begin{array}{rrr|rrrrrr}
 & & & 3 & 1 & 4 & 11 & & \\ \hline
1 & -2 & -3 & 3 & -5 & -7 & 0 & -25 & -15 \\
 & & & 3 & -6 & -9 & & & \\ \cline{4-6}
 & & & & 1 & 2 & 0 & & \\
 & & & & 1 & -2 & -3 & & \\ \cline{5-7}
 & & & & & 4 & 3 & -25 & \\
 & & & & & 4 & -8 & -12 & \\ \cline{6-8}
 & & & & & & 11 & -13 & -15 \\
 & & & & & & 11 & -22 & -33 \\ \cline{7-9}
 & & & & & & & 9 & 18.
\end{array}$$

This is certainly easier to write and easier to read. We can simplify slightly more by not writing 0, -25, and -15 the second time, but that is hardly worth it.

The Remainder and Factor Theorems

Suppose a polynomial $f(x)$ is divided by a linear polynomial of the form $g(x) = x - a$. Then

$$f(x) = (x - a)q(x) + r(x),$$

where $r(x)$ is zero or a polynomial of degree zero, that is, a constant. Thus

$$f(x) = (x - a)q(x) + c.$$

What is the constant c? Just set $x = a$ in the formula;

$$f(a) = 0 + c,$$

hence $c = f(a)$. Thus the value of a polynomial $f(x)$ for $x = a$ is the remainder when $f(x)$ is divided by $x - a$. This fact is called the remainder theorem.

> **Remainder Theorem** If $f(x)$ is a polynomial and a is a real number, then $f(a)$ is the remainder when $f(x)$ is divided by $x - a$. Thus
>
> $$f(x) = (x - a)q(x) + f(a),$$
>
> where $q(x)$ is a polynomial.

EXAMPLE 2 Find the remainder when $x^{17} + 3x^9 - 2$ is divided by $x + 1$.

Solution In this example

$$f(x) = x^{17} + 3x^9 - 2 \quad \text{and} \quad a = -1,$$

so the remainder is $f(-1)$. But

$$f(-1) = (-1)^{17} + 3(-1)^9 - 2 = -1 - 3 - 2 = -6.$$

Answer -6

EXAMPLE 3 Use the remainder theorem to find $f(6)$, where

$$f(x) = 2x^3 - 3x^2 + x - 7.$$

Solution Divide $f(x)$ by $x - 6$:

$$\begin{array}{rr|rrrr}
 & & 2 & 9 & 55 & \\ \hline
1 & -6 & 2 & -3 & 1 & -7 \\
 & & 2 & -12 & & \\ \hline
 & & & 9 & 1 & \\
 & & & 9 & -54 & \\ \hline
 & & & & 55 & -7 \\
 & & & & 55 & -330 \\ \hline
 & & & & & 323
\end{array}$$

The remainder is $323 = f(6)$.

Answer 323

We deduce from the remainder theorem an important test for whether a is a root of the equation $f(x) = 0$. From the formula

$$f(x) = (x - a)q(x) + f(a)$$

we see that $f(a) = 0$ if and only if $f(x) = (x - a)q(x)$, that is, if and only if $x - a$ is a factor of $f(x)$.

Factor Theorem If $f(x)$ is a polynomial, then a real number a is a root of the equation $f(x) = 0$ if and only if $x - a$ is a factor of $f(x)$.

EXAMPLE 4 Show that -2 is a solution of

$$x^{18} + 2x^{17} - x^4 - 2x^3 + x + 2 = 0.$$

Solution Set

$$f(x) = x^{18} + 2x^{17} - x^4 - 2x^3 + x + 2$$

and show that $x + 2$ divides $f(x)$:

$$\begin{aligned} f(x) &= (x^{18} + 2x^{17}) - (x^4 + 2x^3) + (x + 2) \\ &= (x + 2)(x^{17} - x^3 + 1). \end{aligned}$$

Thus $x + 2$ is a factor of $f(x)$, so the factor theorem implies that $f(-2) = 0$.

Application to Oblique Asymptotes [Optional]

A line $y = ax + b$, where $a \neq 0$, is called an **oblique asymptote** of $y = f(x)$ if the graph of f comes nearer and nearer to the line as $x \longrightarrow \infty$ or $-\infty$, that is, if $f(x) - (ax + b) \longrightarrow 0$ as $x \longrightarrow \infty$ or $-\infty$. For example, if

$$f(x) = 3x + 4 + \frac{2}{x},$$

then $y = 3x + 4$ is an oblique asymptote because

$$f(x) - (3x + 4) = \frac{2}{x} \longrightarrow 0 \quad \text{as} \quad x \longrightarrow \infty \quad \text{or} \quad -\infty.$$

A rational function of the form

$$y = \frac{f(x)}{g(x)} \quad \text{where} \quad \deg f(x) = \deg g(x) + 1$$

always has an oblique asymptote. For if we divide $f(x)$ by $g(x)$:

$$f(x) = g(x)q(x) + r(x) \qquad \deg r(x) < \deg g(x),$$

then $\deg f(x) = \deg g(x) + \deg q(x)$, so $\deg q(x) = 1$. Therefore $q(x)$ is linear, $q(x) = ax + b$ with $a \neq 0$, so

$$y = \frac{f(x)}{g(x)} = q(x) + \frac{r(x)}{g(x)} = ax + b + \frac{r(x)}{q(x)}$$

and

$$y - (ax + b) = \frac{r(x)}{q(x)} \longrightarrow 0$$

as $x \longrightarrow +\infty$ or $-\infty$ since $\deg r(x) < \deg q(x)$. (Bottom heavy case; see p. 163.)

> **Oblique Asymptotes** If $R(x) = f(x)/g(x)$ is a quotient of two polynomials and $\deg f(x) = \deg g(x) + 1$, then the graph of $y = R(x)$ has an oblique asymptote.

EXAMPLE 5 Find the oblique asymptote of $y = \dfrac{3x^4 - 7x^3 + 1}{x^3 - x^2}$.

Solution First divide $3x^4 - 7x^3 + 1$ by $x^3 - x^2$:

$$\begin{array}{rrrr|rrrrr} & & & & 3 & -4 & & & \\ \hline 1 & -1 & 0 & 0 & 3 & -7 & 0 & 0 & 1 \\ & & & & 3 & -3 & 0 & 0 & \\ \hline & & & & & -4 & 0 & 0 & 1 \\ & & & & & -4 & 4 & 0 & 0 \\ \hline & & & & & & -4 & 0 & 1 \end{array}$$

Therefore

$$3x^4 - 7x^3 + 1 = (x^3 - x^2)(3x - 4) + (-4x^2 + 1)$$

so

$$y = \frac{3x^4 - 7x^3 + 1}{x^3 - x^2} = 3x - 4 + \frac{-4x^2 + 1}{x^3 - x^2}$$

We conclude that $y = 3x - 4$ is the oblique asymptote (Fig. 1).

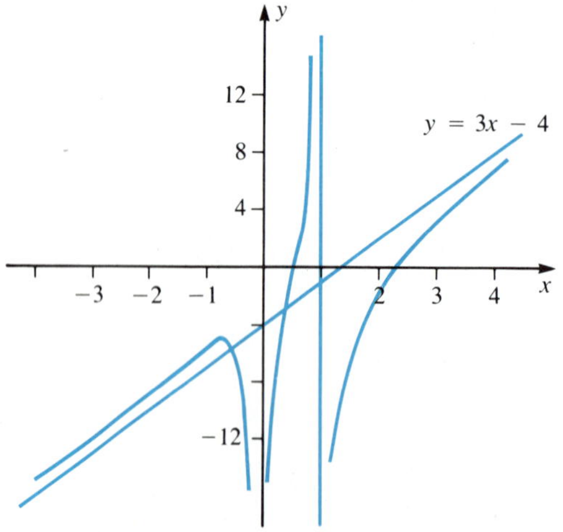

Fig. 1 $y = \dfrac{3x^4 - 7x^3 + 1}{x^3 - x^2}$
oblique asymptote: $y = 3x - 4$

Answer $y = 3x - 4$

Synthetic Division [Optional]

Long division can be drastically simplified if the divisor is a first degree polynomial of the special form

$$g(x) = x - a.$$

Let us observe how much repetition there is in a long division problem with such a divisor. For instance, let us divide

$$2x^3 + 4x^2 - 11x + 6 \quad \text{by} \quad x + 3:$$

$$\begin{array}{rr|rrrr} & & 2 & -2 & -5 & \\ \cline{3-6} 1 & 3 & 2 & 4 & -11 & 6 \\ & & 2 & 6 & & \\ \cline{3-4} & & & -2 & -11 & \\ & & & -2 & -6 & \\ \cline{4-5} & & & & -5 & 6 \\ & & & & -5 & -15 \\ \cline{5-6} & & & & & 21 \end{array}$$

The quotient is $q(x) = 2x^2 - 2x - 5$, and the remainder is $r(x) = 21$.

If we delete almost all of the repetition, what remains is

$$\begin{array}{rr|rrrr} & & 2 & -2 & -5 & \\ \cline{3-6} 1 & 3 & 2 & 4 & -11 & 6 \\ & & & \underline{6} & & \\ & & & & \underline{-6} & \\ & & & & & -15 \\ \cline{6-6} & & & & & 21 \end{array}$$

This can be compressed vertically into

$$\begin{array}{rr|rrrr} & & 2 & -2 & -5 & \\ \cline{3-6} 1 & 3 & 2 & 4 & -11 & 6 \\ & & & 6 & -6 & -15 \\ \cline{6-6} & & & & & 21 \end{array}$$

The top line is shorthand for the quotient $2x^2 - 2x - 5$. The left number 2 is just copied. The -2 is $4 - 6$; the -5 is $(-11) - (-6)$. Finally, the remainder 21 is $6 - (-15)$. It is customary to drop the 1 on the left and to put the top row underneath so that the results of subtractions occur below rather than above:

$$\begin{array}{r|rrrr} 3 & 2 & 4 & -11 & 6 \\ & & 6 & -6 & -15 \\ \hline & 2 & -2 & -5 & 21 \end{array}$$

This brief algorithm for dividing a polynomial by $x - a$ is called **synthetic division.**

EXAMPLE 6 Divide

$$3x^5 - 7x^4 + x^3 - 4x^2 + 1 \quad \text{by} \quad x - 2.$$

Solution The initial set-up is

$$\begin{array}{r|rrrrrr} -2 & 3 & -7 & 1 & -4 & 0 & 1 \\ & & & & & & \\ \hline & 3 & & & & & \end{array}$$

Multiply 3 by -2, write the answer under -7, and subtract:

$$\begin{array}{r|rrrrrr} -2 & 3 & -7 & 1 & -4 & 0 & 1 \\ & & -6 & & & & \\ \hline & 3 & -1 & & & & \end{array}$$

Now multiply -1 by -2, write the answer under 1, subtract, and continue in this way:

$$\begin{array}{r|rrrrrr} -2 & 3 & -7 & 1 & -4 & 0 & 1 \\ & & -6 & 2 & 2 & 12 & 24 \\ \hline & 3 & -1 & -1 & -6 & -12 & -23 \end{array}$$

Answer Quotient $3x^4 - x^3 - x^2 - 6x - 12$ Remainder -23

Exercises

Divide $f(x)$ by $g(x)$; give quotient and remainder

1 $f(x) = x + 3 \quad g(x) = x - 1$
2 $f(x) = 2x + 1 \quad g(x) = x + 2$
3 $f(x) = x + 1 \quad g(x) = 2x - 1$
4 $f(x) = x \quad g(x) = 3x + 2$
5 $f(x) = x^2 - 2x + 2 \quad g(x) = x - 3$
6 $f(x) = x^2 + x + 1 \quad g(x) = x + 2$
7 $f(x) = 2x^2 \quad g(x) = x^2 + 1$
8 $f(x) = 3x^2 + 2x + 5 \quad g(x) = x^2 - x + 1$
9 $f(x) = 2x^3 + 2x + 1 \quad g(x) = x - 1$
10 $f(x) = x^3 - 7x \quad g(x) = x + 2$
11 $f(x) = x^3 \quad g(x) = x^2 + x$
12 $f(x) = x^3 + 2x \quad g(x) = x^2 + x + 1$
13 $f(x) = x^3 + x^2 + x + 1 \quad g(x) = x^2 - 1$
14 $f(x) = 5x^3 - x^2 + 3x - 4 \quad g(x) = x^2 + 2$
15 $f(x) = x^3 \quad g(x) = 2x^3 - 4x + 1$
16 $f(x) = x^3 \quad g(x) = x^4$
17 $f(x) = x^4 + 1 \quad g(x) = x^2 + x + 1$
18 $f(x) = x^4 - 3x^3 + x^2 + 7x + 5 \quad g(x) = x^2 + 1$
19 $f(x) = x^5 \quad g(x) = x^2 + 1$
20 $f(x) = x^5 + x^4 - 7x + 2 \quad g(x) = x^2 - 3x - 3.$

Find the remainder when $f(x)$ is divided by $g(x)$

21 $f(x) = x^{10} + x^8 \quad g(x) = x - 1$
22 $f(x) = x^{12} + x^{10} - x^5 \quad g(x) = x + 1$
23 $f(x) = x^{20} + 4x^{10} + 1 \quad g(x) = x + 1$
24 $f(x) = x^{2n} + x^n + 1 \quad g(x) = x - 1$
25 $f(x) = x^{100} \quad g(x) = x^2 - 1$
26 $f(x) = x^{50} + x^{25} \quad g(x) = x^2 + 1$ [*Hint* $x^{25} = x(x^2)^{12}$].
27 Show that $x^{61} - x^{12} - x^4 + 1$ is divisible by $x - 1$.
28 Show that $x^6 - 4x^4 + 7x^3 - 10x^2 - x - 14$ is divisible by $x - 2$.

Find the oblique asymptote

29 $y = x - \dfrac{2}{x^2}$ **30** $y = 2x - 1 + \dfrac{3x}{x^3 - 1}$ **31** $y = \dfrac{x^3}{x^2 + 1}$

32 $y = \dfrac{x^3 - 2}{x^2 + 2}$ **33** $y = \dfrac{2x^4 - x - 1}{x^3}$ **34** $y = \dfrac{-x^4 + 5x^3 - x^2 + 1}{x^3}$

Graph

35 $y = \dfrac{x^2 + 1}{x}$ **36** $y = \dfrac{x^2}{x + 1}$ **37** $y = \dfrac{x^4 + 1}{x^3}$

38 $y = \dfrac{(x + 1)(x - 1)^2}{x^2}$ **39** $y = \dfrac{(x^2 - 1)(x^2 - 4)}{x^3}$ **40** $y = \dfrac{x^2(x - 1)^2}{(x + 1)^3}$

Find $f(a)$ by synthetic division

41 $f(x) = x^3 - x - 1 \quad a = 2$

42 $f(x) = 4x^3 - 2x^2 + 3 \quad a = -2$

43 $f(x) = x^4 + x^3 - 2x^2 + x + 1 \quad a = 3$

44 $f(x) = 2x^4 + 3x^3 + 4x^2 + 5x + 6 \quad a = -2$

45 $f(x) = x^3 - x^2 + 3x + 4 \quad a = 5$

46 $f(x) = x^5 + 5x + 1 \quad a = -4.$

Use synthetic division to decide whether -4 is a solution of the equation

47 $2x^3 + 14x^2 - 17x - 24 = 0$

48 $x^4 - 4x^3 - 31x^2 + 5x + 4 = 0$

49 $3x^5 + 11x^4 - 4x^3 + 2x^2 + 10x + 8 = 0$

50 $x^6 + 4x^5 + 2x^4 + 9x^3 + 4x^2 + x - 4 = 0.$

6 ZEROS OF POLYNOMIALS

A **zero** of a function $f(x)$ is a real number r such that $f(r) = 0$. As we know, the number r is also called a **root** of the equation $f(x) = 0$.

Graphically, a zero of any function $f(x)$ is simply a value of x where the graph $y = f(x)$ meets the x-axis. The function shown in Fig. 1a has three zeros, r_1, r_2, and r_3. A function need not have any real zeros, however. For example, $f(x) = x^2 + 1$ is positive for every real x, and hence $f(x)$ has no real zeros. Its graph never crosses the x-axis (Fig. 1b).

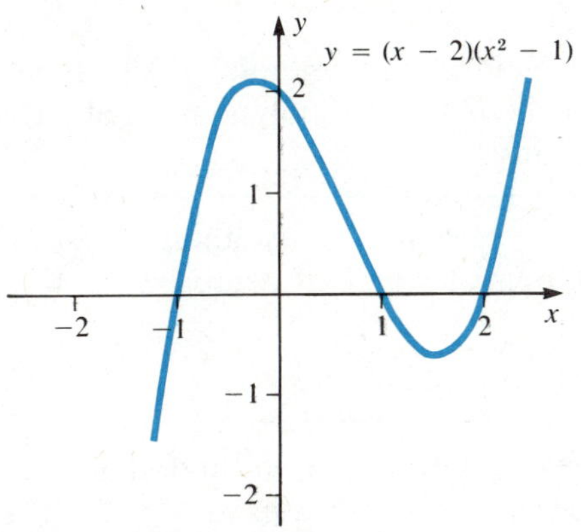

(a) a function with 3 zeros

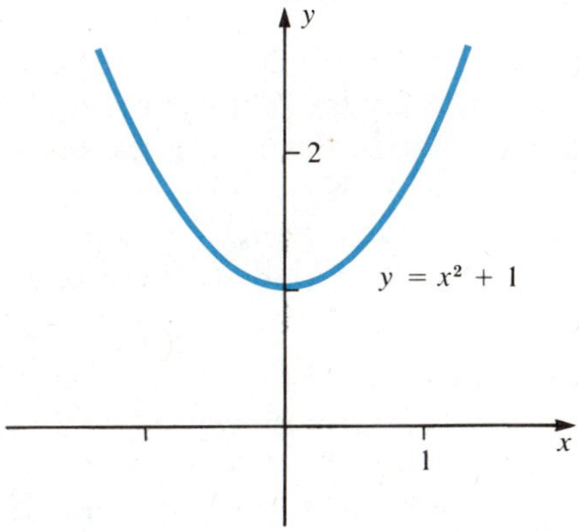

(b) a function without real zeros

Fig. 1

Linear and Quadratic Polynomials

In Chapter 2, we showed that a linear polynomial (Fig. 2a, next page)

$$f(x) = ax + b \qquad (a \neq 0)$$

has one zero $r = -b/a$, and

$$f(x) = a(x - r).$$

We also showed that a quadratic polynomial (Fig. 2b)

$$f(x) = ax^2 + bx + c \qquad (a \neq 0)$$

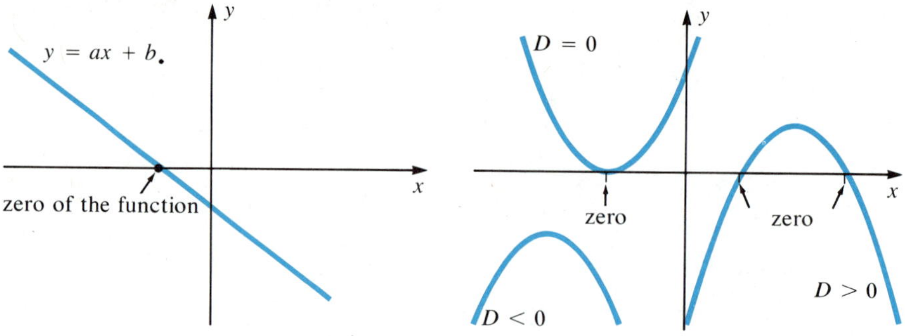

(a) linear polynomial: one real zero (b) quadratic polynomial: 3 cases

Fig. 2

has two, one, or no zeros, depending on its discriminant $D = b^2 - 4ac$:

(i) $D < 0$ no real zeros

(ii) $D = 0$ one real zero, $r = -b/2a$,
$f(x) = a(x - r)^2$

(iii) $D > 0$ two real zeros, $r_1, r_2 = (-b \pm \sqrt{D})/2a$,
$f(x) = a(x - r_1)(x - r_2)$.

The formulas for $f(x)$ in (ii) and (iii) are checked by direct substitution.

Higher Degree Polynomials

We have seen that if r is a zero of a linear or quadratic polynomial $f(x)$, then $x - r$ is a factor. By the factor theorem, the same is true for polynomials of any degree. Let us restate this fact in the present context.

> If r is a zero of $f(x) = a_n x^n + a_{n-1}x^{n-1} + \cdots + a_1 x + a_0$, then $x - r$ is a factor of $f(x)$. That is, there is a polynomial $g(x)$ of degree $n - 1$ such that
>
> $$f(x) = (x - r)g(x).$$

Now suppose that $f(x) = a_n x^n + \cdots + a_0$ with $a_n \neq 0$ and that $r_1, r_2, \cdots, r_m$ are *distinct* zeros of $f(x)$. Since r_1 is a zero we have

$$f(x) = (x - r_1)g(x) \qquad \deg g(x) = n - 1.$$

Since r_2 is a zero of $f(x)$ we have

$$f(r_2) = (r_2 - r_1)g(r_2) = 0.$$

But $r_2 - r_1 \neq 0$ since r_1 and r_2 are distinct. Therefore $g(r_2) = 0$. By the factor theorem again,

$$g(x) = (x - r_2)h(x) \qquad \deg h(x) = n - 2.$$

Hence

$$f(x) = (x - r_1)(x - r_2)h(x) \qquad \deg h(x) = n - 2.$$

By continuing the argument we eventually reach

$$f(x) = (x - r_1)(x - r_2)\cdots(x - r_m)k(x) \qquad \deg k(x) = n - m.$$

General Factor Theorem If $f(x)$ is a polynomial of degree n and if $r_1, \cdots, r_m$ are distinct zeros of $f(x)$, then

$$f(x) = (x - r_1)(x - r_2) \cdots (x - r_m)k(x)$$

where $k(x)$ is a polynomial of degree $n - m$.

Since $n - m \geq 0$ we have $m \leq n$. Thus we have an important corollary: a polynomial cannot have more zeros than its degree.

Bound on the Number of Zeros A polynomial of degree n can have at most n distinct zeros.

Remark This result has a consequence that is sometimes used to prove polynomials equal and that has applications in numerical analysis: Suppose $f(x)$ and $g(x)$ are two polynomials, each of degree at most n. Suppose that $f(x)$ and $g(x)$ agree at $n + 1$ distinct values of x, that is, $f(r_1) = g(r_1), \cdots, f(r_{n+1}) = g(r_{n+1})$ where $r_1, r_2, \cdots, r_{n+1}$ are distinct. Then $f(x) = g(x)$. To prove this, set $h(x) = f(x) - g(x)$. Then $h(x)$ is also a polynomial of degree at most n. But $h(r_1) = h(r_2) = \cdots = h(r_{n+1}) = 0$, so $h(x)$ has *more* zeros than its degree. This is possible only if $h(x) = 0$, that is, $f(x) = g(x)$.

Multiplicity

Consider

$$f(x) = x^5 - 2x^4 - 2x^3 + x^2 + 4x + 4$$

We easily check that $f(2) = 0$, so $x - 2$ is a factor. By long division,

$$f(x) = (x - 2)g(x) \quad \text{where} \quad g(x) = x^4 - 2x^2 - 3x - 2.$$

Again, we find $g(2) = 0$, so $x - 2$ is a factor of $g(x)$. By long division,

$$g(x) = (x - 2)h(x) \quad \text{where} \quad h(x) = x^3 + 2x^2 + 2x + 1.$$

Consequently

$$f(x) = (x - 2)^2 h(x).$$

It turns out that $h(-1) = 0$ so $x + 1$ is a factor of $h(x)$. After long division, the end result is

$$f(x) = (x - 2)^2(x + 1)(x^2 + x + 1).$$

The quadratic $x^2 + x + 1$ has no real zeros because its discriminant is negative, so we have factored $f(x)$ as far as we can. The factor $x + 1$, corresponding to the zero -1, occurs once in $f(x)$, and we say that -1 is a *simple* zero of $f(x)$. But the factor $x - 2$, corresponding to the zero 2, occurs twice in $f(x)$, and we say that 2 is a *double* zero of $f(x)$.

In general, let $r_1, r_2, \cdots, r_k$ be the distinct real zeros (if any) of a polynomial $f(x)$. Then

$$f(x) = (x - r_1)^{m_1}(x - r_2)^{m_2} \cdots (x - r_k)^{m_k} g(x),$$

where the factor $g(x)$ is a polynomial with no real zeros, possibly just a constant. In case $f(x)$ has no real zeros, then $f(x) = g(x)$. Otherwise each m_i is a positive integer called the **multiplicity** of the zero r_i. If $m_i = 1$, then r_i is called a

simple zero. If $m_i > 1$, then r_i is called a **multiple zero.** For example, if

$$f(x) = x^2(x + 1)^3(x - 2)(x^2 + 1),$$

then 2 is a simple zero, 0 is a zero of multiplicity 2 (a double zero), and -1 is a zero of multiplicity 3 (a triple zero).

Existence of Zeros

We have proved that a polynomial function of degree n has at most n real zeros, that is, the maximum possible number of zeros is n. But this does not assert that any real zeros exist; there may be none at all. For example, $f(x) = x^2 + 1$ and $f(x) = (x^2 + 1)(x^2 + 4)$ are positive for all x, hence they have no real zeros. The cubic $f(x) = x(x^2 + 1)$ has exactly one real zero, $r = 0$, less than the maximum possible number of real zeros of a cubic, three.

In our numerical work (p. 219) on estimating roots we noted that if $f(x)$ is negative at one point and positive at another,* then $f(x)$ has a zero between the points. This applies in particular to polynomials.

> **Existence of Zeros** Let $f(x)$ be a polynomial and suppose x_0 and x_1 are numbers for which $f(x_0)$ and $f(x_1)$ have opposite signs. Then there is at least one zero of $f(x)$ between x_0 and x_1.

Examples (1) $f(x) = x^{16} + 5x + 1$

$$f(-1) = -3 < 0 \qquad f(1) = 7 > 0.$$

Therefore $f(x)$ has a zero between -1 and 1.

(2) $f(x) = x^4 - 2x^3 + 3x^2 - 5x - 2$

$$f(-1) = 9 > 0 \qquad f(0) = -2 < 0.$$

Therefore $f(x)$ has a zero between -1 and 0.

Note We might need to estimate the zero in (2) by a calculator routine (p. 217). If we happen to notice in advance that $f(2) = 0$ and (factor theorem)

$$f(x) = (x - 2)g(x) \quad \text{where} \quad g(x) = x^3 + 3x + 1,$$

then it will be easier to calculate with $g(x)$ instead of $f(x)$. (Clearly $g(x)$ has a zero between -1 and 0 because $f(x)$ does.) This illustrates a practical use of the factor theorem for simplifying calculations.

Suppose $f(x)$ is a polynomial of *odd* degree:

$$f(x) = a_nx^n + \cdot \cdot \cdot + a_1x + a_0 \qquad (a_n \neq 0, n \text{ odd}).$$

We know (p. 146) that if $a_n > 0$, then $f(x) \longrightarrow \infty$ as $x \longrightarrow \infty$ and $f(x) \longrightarrow -\infty$ as $x \longrightarrow -\infty$. Therefore $f(x)$ takes *opposite signs* at x and $-x$ if x is sufficiently large. The same holds if $a_n < 0$. Conclusion:

> Each polynomial of odd degree has at least one real zero.

* Technically, $f(x)$ must be *continuous* on an interval including the two points where $f(x)$ has opposite signs. The concept of continuity is discussed in more advanced courses. Roughly, it means that the graph of $y = f(x)$ can be drawn without lifting your pencil. Polynomials are continuous.

Example $f(x) = x^{17} - 165x^{14} + 680x^{13} + 25x + 105$

has at least one real zero.

Rational Zeros

Sometimes it is necessary to find out if a polynomial with *integer* coefficients has any integer zeros or any rational zeros. There are two useful tests, important because they limit the search for rational zeros to a *finite* number of possibilities.

> **Integer Zeros** Let
>
> $$f(x) = x^n + \cdots + a_1x + a_0$$
>
> have integer coefficients and leading coefficient 1. Then each rational zero of $f(x)$ must be an integer, and this integer must divide a_0.

Proof This proof requires two properties of integers. First, if $b > 1$, then b has a prime divisor (itself if it is a prime). For instance, 119 has 7 as a prime divisor, also 17. But 19 has only 19 as a prime divisor. Second, if a prime p divides a power of an integer b, then it must divide b. For instance, if 13 divides b^{27}, then 13 divides b.

Now suppose r is a rational zero of $f(x)$. Then $r = b/c$ where b and c are integers, with $c > 0$, and this expansion of r is *in lowest terms,* that is, b and c have no common factor except 1. Since $f(r) = 0$, we have

$$\left(\frac{b}{c}\right)^n + a_{n-1}\left(\frac{b}{c}\right)^{n-1} + \cdots + a_1\frac{b}{c} + a_0 = 0.$$

We multiply by c^n:

$$b^n + a_{n-1}b^{n-1}c + \cdots + a_1bc^{n-1} + a_0c^n = 0.$$

Each term except b^n is c times something, so we deduce from this formula that $b^n = cu$, where u is an integer. Hence c divides b^n. If $c > 1$, then c has a prime divisor p. But then p divides b^n, so in turn divides b. Thus p divides *both* c *and* b. This contradicts the assumption that b/c is in lowest terms. Therefore $c = 1$, so $r = b/1 = b$ is an integer, and we have

$$b^n + a_{n-1}b^{n-1} + \cdots + a_1b + a_0 = 0.$$

We deduce that $bv = a_0$, where v is an integer. Therefore b divides a_0.

Example 1 Find all rational zeros of

$$f(x) = x^4 - 2x^3 - 8x^2 + 19x - 6.$$

Solution Let r be a rational zero. Then r is an integer and is a divisor of -6. The divisors of -6 are

$$1 \quad -1 \quad 2 \quad -2 \quad 3 \quad -3 \quad 6 \quad -6.$$

We start testing:

$$f(1) = 4 \qquad f(-1) = -30 \qquad f(2) = 0.$$

Therefore 2 is a zero but 1 and -1 are not.

Continuation 1 Continue testing:

$$f(-2) = -44 \qquad f(3) = 6 \qquad f(-3) = 0$$

$$f(6) = 684 \qquad f(-6) = 1320.$$

Therefore -3 is the only other rational zero.

Continuation 2 Divide out $x - 2$ and test the quotient:

$$\begin{array}{rrrrrrr}
 & & 1 & 0 & -8 & 3 & \\
\cline{3-7}
1 & -2 & \big) \; 1 & -2 & -8 & 19 & -6 \\
 & & 1 & -2 & & & \\
\cline{3-4}
 & & & 0 & -8 & 19 & \\
 & & & & -8 & 16 & \\
\cline{5-6}
 & & & & & 3 & -6 \\
 & & & & & 3 & -6 \\
\cline{6-7}
\end{array}$$

Thus $f(x) = (x - 2)(x^3 - 8x + 3)$. The only possible rational zeros of the quotient $g(x) = x^3 - 8x + 3$ are the integer divisors of 3, that is, 1, -1, 3, and -3. We have ruled out 1 and -1 (since $f(1) \neq 0$ and $f(-1) \neq 0$) so it remains to test 3 and -3:

$$g(3) = 6 \qquad g(-3) = 0.$$

Answer 2 -3

Next we consider polynomials with integer coefficients but leading coefficient not necessarily 1.

> **Rational Zeros** Let
>
> $$f(x) = a_n x^n + \cdots + a_1 x + a_0 \qquad (a_n \neq 0)$$
>
> have *integer* coefficients. Then each rational zero of $f(x)$ has the form $r = p/q$, where p and q are integers, p divides a_0, and q divides a_n.

EXAMPLE 2 Find all rational zeros of

$$f(x) = 3x^3 + 7x^2 + x - 4.$$

Solution If r is a rational zero, then $r = p/q$, where p divides -4 and q divides 3. The possibilities for p and q are

$$p = \pm 1 \quad \pm 2 \quad \pm 4 \qquad q = \pm 1 \quad \pm 3.$$

Hence the possible rational zeros are

$$r = \pm 1 \quad \pm 2 \quad \pm 4 \quad \pm \tfrac{1}{3} \quad \pm \tfrac{2}{3} \quad \pm \tfrac{4}{3}.$$

For each of these 12 values we must check whether $f(r) = 0$. Actually it is easy to eliminate a number of cases. For instance, in computing $f(1)$, we see not only that $f(1) > 0$ but that $f(x) > 0$ for all $x \geq 1$. Hence, we can rule out the values $\frac{4}{3}$, 2, and 4. By similar reasoning we can eliminate -2 and -4.

Now we must test the remaining possibilities. We find $f(-\frac{4}{3}) = 0$, but $f(x) \neq 0$ in all other cases.

If we are lucky enough to try $-\frac{4}{3}$ first, then we can reduce the computations considerably. For then $f(x)$ is divisible by $x + \frac{4}{3}$. By division we find

$$f(x) = 3(x + \tfrac{4}{3})(x^2 + x - 1).$$

The remaining zeros of $f(x)$ are zeros of the quadratic factor $x^2 + x - 1$. These are $\frac{1}{2}(-1 \pm \sqrt{5})$, not rational.

Answer $f(x)$ has one rational zero, $-\frac{4}{3}$.

Our final example illustrates most of the ideas of this section.

EXAMPLE 3 Locate the real zeros of

$$f(x) = 3x^5 + x^4 + x^3 - 8x^2 - 5x + 6.$$

Solution The polynomial has degree 5, odd, so we are sure of at least one real zero. Let us look for rational zeros. Each has the form $r = p/q$, where p divides 6 and q divides 3. The possibilities for p and q are

$$p = \pm 1 \quad \pm 2 \quad \pm 3 \quad \pm 6 \qquad q = \pm 1 \quad \pm 3,$$

so the possibilities for r are

$$r = \pm 1 \quad \pm 2 \quad \pm 3 \quad \pm 6 \quad \pm\tfrac{1}{3} \quad \pm\tfrac{2}{3}$$

First we test the integers:

x	1	2	3	6	-1	-2	-3	-6
$f(x)$	-2	84	L	L	0	$-$L	$-$L	$-$L

L = large

Thus $f(-1) = 0$ so $x + 1$ is a factor of $f(x)$. By long division,

$$f(x) = (x + 1)g(x) \quad \text{where} \quad g(x) = 3x^4 - 2x^3 + 3x^2 - 11x + 6.$$

Now we test $g(x)$ for rational zeros $r = p/q$. The possibilities are the same as for $f(x)$, except that we have already eliminated 1, ± 2, ± 3, and ± 6. We have not eliminated -1; it could be a multiple zero of $f(x)$, hence a zero of $g(x)$. Thus, we must test $-1 \quad \pm\frac{1}{3} \quad \pm\frac{2}{3}$:

x	-1	$\frac{1}{3}$	$\frac{2}{3}$	$-\frac{1}{3}$	$-\frac{2}{3}$
$g(x)$	25	$\frac{213}{81}$	0	$\frac{819}{81}$	$\frac{1284}{81}$

Thus $g(\frac{2}{3}) = 0$, so $x - \frac{2}{3}$ is a factor of $g(x)$. Actually, it is simpler to work with $3(x - \frac{2}{3}) = 3x - 2$; this must also be a factor of $g(x)$. By long division,

$$g(x) = (3x - 2)h(x) \quad \text{where} \quad h(x) = x^3 + x - 3.$$

The cubic polynomial $h(x)$ is strictly increasing because it is the sum of x^3 and $x - 3$, both strictly increasing. Therefore, $h(x)$ crosses the x-axis in exactly one point, that is, $h(x)$ has one real zero. By inspection

$$h(1) = -1 < 0 \quad \text{and} \quad h(2) = 7 > 0.$$

Therefore the zero of $h(x)$ is between 1 and 2. By the numerical methods of Chapter 5, p. 218, this zero is approximately 1.21341166.

Answer $f(x) = (x + 1)(3x - 2)(x^3 + x - 3)$ has the simple rational zeros $r_1 = -1$ and $r_2 = \frac{2}{3}$. It also has a real zero $r_3 \approx 1.21341166$.

Exercises

1 Find the most general linear function with $x = 3$ as a zero.

2 Prove that the graph of a linear function intersects each horizontal line exactly once.

3 Find the most general quadratic function with $x = 2$ as a zero.

4 Find a quadratic function $f(x)$ with zeros -1 and 3, and $f(0) = 6$.

5 Find the most general quadratic function whose graph is tangent to the x-axis at $(2, 0)$.

6 Let $f(x) = ax^2 + bx + c$ have real zeros. Prove that their sum is $-b/a$ and their product is c/a.

7 (cont.) Find the most general quadratic function, the sum of whose zeros is 5.

8 (cont.) Find the most general quadratic function, the product of whose zeros is 4.

9 Prove that the graph of a quadratic function intersects each horizontal line at most twice.

10 Let f be a linear function and g a quadratic function. Prove that their graphs intersect in at most two points.

11 Let f and g be quadratics. Prove that their graphs either coincide, or have at most two points in common.

12 Prove that if a and c have opposite signs, then $ax^2 + bx + c = 0$ has two real roots.

13 Prove that if $a > 0$, $b > 0$, and $c > 0$, then $ax^2 + bx + c$ has no positive zero.

14 Prove that if $a > 0$, $b < 0$, and $c > 0$, then $ax^2 + bx + c$ has no negative zero.

15 For what values of c does $x^2 - 3x + c$ have a real zero?

16 For what values of b does $x^2 + bx + b^2$ have a real zero?

Find the real zeros and their multiplicities

17 $(2x + 1)(x^2 + 1)$

18 $(x - 3)^2(x^2 + 2)$

19 $(x^2 - 4)^2(x^2 + 4)$

20 $(x^2 - 5x + 6)(x^2 + 1)^2$

21 $(x^2 + x - 2)(x^2 - 1)^2$

22 $x^3(x^2 - 4)^2(x^2 - 3)$.

23 Prove that the graph of a cubic function intersects each horizontal line at most three times.

24 Prove that the graph of a polynomial of degree 10 intersects each non-vertical line at most 10 times.

25 Let f be linear and g cubic. Prove that their graphs have at most three points of intersection.

26 Let f be quadratic and g cubic. Prove that their graphs have at most three points in common.

27 Find a polynomial $f(x)$ of degree 6 such that $f(x) \geq 0$ for all x and $f(1) = f(2) = f(3) = 0$.

28 Does there exist a real number that exceeds its cube by one?

29 Does there exist a real number that exceeds its 5-th power by 1000?

30 Is there a real number whose 9-th power is 4 less than its square?

31 Find a cubic polynomial $f(x)$ for which $f(0) = f(5) = f(8) = 0$ and $f(10) = 17$.

32* If $f(x)$ is a polynomial having only even powers of x and if $f(r) = 0$, prove that $f(x)$ is divisible by $x^2 - r^2$.

Find all rational zeros and their multiplicities

33 $x^3 + x^2 + 2x + 2$

34 $x^3 + 3x^2 + 2x + 6$

35 $x^4 - x^3 - x^2 - x - 2$

36 $x^6 + 2x^4 + x^2 + 2$

37 $3x^3 - 2x^2 + 3x - 2$

38 $2x^3 + 3x^2 + 4x + 6$

39 $2x^4 + 3x^3 + 8x^2 + 3x - 4$

40 $3x^4 + 5x^3 + x^2 + 5x - 2$.

Given that $f(x)$ has exactly one real zero, locate it between two integers

41 $f(x) = 12x^3 - 28x^2 - 7x - 10$

42 $f(x) = x^3 + 4x^2 + 10x + 15$.

43 Show that $f(x) = x^5 + x^2 - 7$ has a real zero between 1 and 2 but no zeros greater than 2.

44 If $f(x)$ is a polynomial with positive leading coefficient and if $f(0) < 0$, show that $f(x)$ has a real zero.

45 If $f(x)$ is a polynomial whose coefficients are all positive, show that $f(x)$ has no positive zeros.

46 If $f(x)$ is a polynomial whose coefficients alternate in sign, show that $f(x)$ has no negative zeros.

47* Let $f(x) = x^n + a_1x^{n-1} + \cdots + a_n$ have zeros $r_1, r_2, \cdots, r_n$. Prove that $r_1 + r_2 + \cdots + r_n = -a_1$.

48* (cont.) Prove that $r_1r_2 \cdots r_n = (-1)^n a_n$.

Locate the real zeros of

49 $x^4 - x^3 - x + 1$

50 $4x^4 + x^2 + 3x + 1$

51 $2x^4 + 5x^3 - x^2 - 3x + 1$

52 $3x^4 - 7x^3 - 12x^2 + 8x + 8$

53 $4x^5 - 12x^4 + 5x^3 + 20x^2 - 33x + 18$

54 $2x^5 + 5x^4 - 9x^3 - 13x^2 + 14x - 3$.

7 PARTIAL FRACTIONS [Optional]

In applications of rational functions, it is often useful to decompose a rational function into a sum of rational functions with simpler denominators. We shall describe how this is done in certain cases.

By long division (if necessary) we can reduce any rational function to the sum of a polynomial and a bottom-heavy rational function. For instance,

$$\frac{x^3 - 2x}{x^2 + x + 1} = x - 1 + \frac{-2x + 1}{x^2 + x + 1}.$$

Therefore, we can concentrate on bottom-heavy rational functions.

Quadratic Denominator

We begin with rational functions of the form

$$\frac{ax + b}{(x - r)(x - s)} \qquad (r \neq s).$$

Such a rational function can always be decomposed into two simpler functions, as we shall see.

Examples

(1) $$\frac{1}{(x + 1)(x - 1)} = \frac{-\frac{1}{2}}{x + 1} + \frac{\frac{1}{2}}{x - 1}$$

(2) $$\frac{x}{(x+1)(x+2)} = \frac{-1}{x+1} + \frac{2}{x+2}$$

(3) $$\frac{2x-3}{x(x+1)} = \frac{-3}{x} + \frac{5}{x+1}$$

In general,

$$\frac{ax+b}{(x-r)(x-s)} = \frac{A}{x-r} + \frac{B}{x-s}$$

for suitable constants A and B. The sum on the right-hand side is called the **partial fraction decomposition** of the rational function on the left-hand side. To find A and B, first clear of fractions:

$$ax + b = A(x-s) + B(x-r).$$

This relation between two *first degree* polynomials is an identity if it holds for *two distinct* values of x. (See the second box on p. 387.) We choose A and B to make it hold for $x = r$ and $x = s$:

$$ar + b = A(r-s) \qquad A = \frac{ar+b}{r-s}$$

$$as + b = B(s-r) \qquad B = \frac{as+b}{s-r}$$

Suppose $p(x)$ is a polynomial and $r \neq s$. Then there are constants A and B and a polynomial $f(x)$ such that

$$\frac{p(x)}{(x-r)(x-s)} = f(x) + \frac{A}{x-r} + \frac{B}{x-s}.$$

EXAMPLE 1 Decompose into partial fractions $\dfrac{3x+1}{(x+1)(x-2)}$.

Solution Write

$$\frac{3x+1}{(x+1)(x-2)} = \frac{A}{x+1} + \frac{B}{x-2}$$

and clear of fractions:

$$3x + 1 = A(x-2) + B(x+1).$$

Set $x = -1$:

$$-2 = -3A \qquad A = \tfrac{2}{3}.$$

Set $x = 2$:

$$7 = 3B \qquad B = \tfrac{7}{3}.$$

Check $$\frac{\frac{2}{3}}{x+1} + \frac{\frac{7}{3}}{x-2} = \frac{\frac{2}{3}(x-2) + \frac{7}{3}(x+1)}{(x+1)(x-2)} = \frac{3x+1}{(x+1)(x-2)}$$

Answer $$\frac{\frac{2}{3}}{x+1} + \frac{\frac{7}{3}}{x-2}$$

EXAMPLE 2 Decompose into partial fractions $\dfrac{2x^3 - 13x^2 + 22x - 14}{(x-2)(x-3)}$.

Solution First divide the numerator by the denominator $(x-2)(x-3) = x^2 - 5x + 6$:

$$\begin{array}{rrr|rrrr} & & & 2 & -3 & & \\ \hline 1 & -5 & 6 & 2 & -13 & 22 & -14 \\ & & & 2 & -10 & 12 & \\ \hline & & & & -3 & 10 & -14 \\ & & & & -3 & 15 & -18 \\ \hline & & & & & -5 & 4 \end{array}$$

Therefore

$$\frac{2x^3 - 13x^2 + 22x - 14}{(x-2)(x-3)} = 2x - 3 + \frac{-5x+4}{(x-2)(x-3)}.$$

Now write

$$\frac{-5x+4}{(x-2)(x-3)} = \frac{A}{x-2} + \frac{B}{x-3}$$

and clear of fractions:

$$-5x + 4 = A(x-3) + B(x-2).$$

Set $x = 2$: $\quad -6 = -A \qquad A = 6.$

Set $x = 3$: $\quad -11 = B \qquad B = -11.$

Answer $\quad 2x - 3 + \dfrac{6}{x-2} - \dfrac{11}{x-3}.$

Now we consider denominators of the form $(x-r)^2$, with a double root. The typical such bottom-heavy rational function

$$\frac{ax+b}{(x-r)^2} \quad \text{can be decomposed into} \quad \frac{A}{x-r} + \frac{B}{(x-r)^2}.$$

We simply write

$$ax + b = a(x-r) + (ar+b) = A(x-r) + B$$

and divide:

$$\frac{ax+b}{(x-r)^2} = \frac{A}{x-r} + \frac{B}{(x-r)^2}.$$

EXAMPLE 3 Decompose into partial fractions $\dfrac{2x^2 - 13x + 22}{(x-4)^2}$.

Solution First divide by $(x-4)^2 = x^2 - 8x + 16$:

$$\begin{array}{rrr|rrr} & & & 2 & & \\ \hline 1 & -8 & 16 & 2 & -13 & 22 \\ & & & 2 & -16 & 32 \\ \hline & & & & 3 & -10 \end{array}$$

Therefore

$$\frac{2x^2 - 13x + 22}{(x-4)^2} = 2 + \frac{3x - 10}{(x-4)^2}$$

Next,

$$3x - 10 = 3(x - 4) + 12 - 10 = 3(x - 4) + 2$$

$$\frac{3x - 10}{(x-4)^2} = \frac{3}{x-4} + \frac{2}{(x-4)^2}.$$

Answer $2 + \dfrac{3}{x-4} + \dfrac{2}{(x-4)^2}$

Cubic Denominator

We shall survey the partial fraction decompositions of rational functions with cubic denominators. As usual, we can make a preliminary long division to assure a bottom-heavy function. We shall also always assume leading coefficient 1 in the denominator. Thus we are considering

$$\frac{p(x)}{q(x)} = \frac{ax^2 + bx + c}{q(x)}$$

where $q(x) = x^3 + \cdots + d$. We know that $q(x)$ has at least one real zero.

Case 1 $q(x) = (x - r)(x - s)(x - t)$, three *distinct* real zeros. The partial fraction decomposition is

$$\frac{A}{x-r} + \frac{B}{x-s} + \frac{C}{x-t}.$$

EXAMPLE 4 Decompose into partial fractions $\dfrac{1}{(x-1)(x-2)(x-3)}$.

Solution Write

$$\frac{1}{(x-1)(x-2)(x-3)} = \frac{A}{x-1} + \frac{B}{x-2} + \frac{C}{x-3}.$$

Clear of fractions:

$$1 = A(x - 2)(x - 3) + B(x - 1)(x - 3) + C(x - 1)(x - 2).$$

Substitute the special values $x = 1$, $x = 2$, $x = 3$:

$$1 = 2A, \quad 1 = -B, \quad 1 = 2C.$$

Hence $A = \frac{1}{2}$, $B = -1$, $C = \frac{1}{2}$.

Check To avoid fractions, let us check that twice the answer equals twice the given rational function:

$$\frac{1}{x-1} - \frac{2}{x-2} + \frac{1}{x-3}$$

$$= \frac{(x-2)(x-3) - 2(x-1)(x-3) + (x-1)(x-2)}{(x-1)(x-2)(x-3)}$$

$$= \frac{(x^2 - 5x + 6) - 2(x^2 - 4x + 3) + (x^2 - 3x + 2)}{(x - 1)(x - 2)(x - 3)}$$

$$= \frac{2}{(x - 1)(x - 2)(x - 3)}.$$

Answer $\dfrac{\frac{1}{2}}{x - 1} + \dfrac{-1}{x - 2} + \dfrac{\frac{1}{2}}{x - 3}$

Case 2 $q(x) = (x - r)^2(x - s)$, a double zero and a distinct simple zero. The partial fraction decomposition is

$$\frac{A}{x - r} + \frac{B}{(x - r)^2} + \frac{C}{x - s}.$$

Example 5 Decompose into partial fractions $\dfrac{4x^2 - x - 9}{(x - 1)^2(x + 2)}$.

Solution Write

$$\frac{4x^2 - x - 9}{(x - 1)^2(x + 2)} = \frac{A}{x - 1} + \frac{B}{(x - 1)^2} + \frac{C}{x + 2}$$

and clear of fractions:

$$4x^2 - x - 9 = A(x - 1)(x + 2) + B(x + 2) + C(x - 1)^2.$$

Set $x = 1$: $\quad -6 = 3B \quad B = -2.$

Set $x = -2$: $\quad 9 = 9C \quad C = 1.$

We cannot find A by a substitution that wipes out the B and C terms; we must do something else. Actually, choosing *any* value for x different from 1 and -2 will do the trick. Try $x = 0$:

$$-9 = -2A + 2B + C = -2A - 4 + 1$$

$$-2A = -6 \quad A = 3.$$

Answer $\dfrac{3}{x - 1} - \dfrac{2}{(x - 1)^2} + \dfrac{1}{x + 2}.$

Case 3 $q(x) = (x - r)^3$, a triple zero. The partial fraction decomposition is

$$\frac{A}{x - r} + \frac{B}{(x - r)^2} + \frac{C}{(x - r)^3}.$$

Example 6 Decompose into partial fractions $\dfrac{-x^2 + 4x + 3}{(x - 1)^3}$.

Solution The most direct solution is to express the numerator as a polynomial in powers of $x - 1$ instead of powers of x. To do this, replace x by $(x - 1) + 1$:

$$\begin{aligned} -x^2 + 4x + 3 &= -[(x - 1) + 1]^2 + 4[(x - 1) + 1] + 3 \\ &= -[(x - 1)^2 + 2(x - 1) + 1] + [4(x - 1) + 4] + 3 \\ &= -(x - 1)^2 + 2(x - 1) + 6. \end{aligned}$$

Therefore

$$\frac{-x^2 + 4x + 3}{(x-1)^3} = \frac{-1}{x-1} + \frac{2}{(x-1)^2} + \frac{6}{(x-1)^3}.$$

Answer $\dfrac{-1}{x-1} + \dfrac{2}{(x-1)^2} + \dfrac{6}{(x-1)^3}$

Case 4 $q(x) = (x - r)(x^2 + bx + c)$, where the quadratic factor has no real zeros, that is, $b^2 - 4c < 0$. The partial fraction decomposition is

$$\frac{A}{x-r} + \frac{Bx + C}{x^2 + bx + c}.$$

Example 7 Decompose into partial fractions $\dfrac{x^2 - 5x - 2}{(x-3)(x^2+1)}$.

Solution Write

$$\frac{x^2 - 5x - 2}{(x-3)(x^2+1)} = \frac{A}{x-3} + \frac{Bx + C}{x^2+1}$$

and clear of fractions:

$$x^2 - 5x - 2 = A(x^2 + 1) + (Bx + C)(x - 3).$$

Set $x = 3$: $\quad -8 = 10A \qquad A = -\frac{4}{5}.$

We could now substitute $x = 0$ to determine C and then substitute, say, $x = 1$ to determine B. Another possibility is to replace A by its value and solve for $Bx + C$:

$$\begin{aligned} x^2 - 5x - 2 &= -\tfrac{4}{5}(x^2 + 1) + (Bx + C)(x - 3) \\ 5(Bx + C)(x - 3) &= 5(x^2 - 5x - 2) + 4(x^2 + 1) \\ &= 9x^2 - 25x - 6 \\ &= (9x + 2)(x - 3). \end{aligned}$$

Therefore

$$Bx + C = \tfrac{1}{5}(9x + 2).$$

Answer $\dfrac{-\frac{4}{5}}{x-3} + \dfrac{\frac{9}{5}x + \frac{2}{5}}{x^2+1}$

Higher Degree Denominators

The most general statement about partial fraction decompositions is beyond our scope. We can state the decomposition for a completely factored denominator with only simple zeros.

> If $r_1, r_2, \cdots, r_n$ are distinct real numbers and $\deg f(x) < n$, then there are constants $A_1, \cdots, A_n$ such that
>
> $$\frac{f(x)}{(x - r_1)\cdots(x - r_n)} = \frac{A_1}{x - r_1} + \cdots + \frac{A_n}{x - r_n}.$$

Examples 1 and 4 illustrate this for $n = 2$ and $n = 3$.

We shall mention one other type of partial fraction decomposition, that of

$$\frac{p(x)}{(x^2 + bx + c)^2}$$

where $\deg p(x) < 4$. The partial fraction decomposition is

$$\frac{p(x)}{(x^2 + bx + c)^2} = \frac{Ax + B}{x^2 + bx + c} + \frac{Cx + D}{(x^2 + bx + c)^2}.$$

EXAMPLE 8 Decompose into partial fractions $\dfrac{2x^3 + x^2 + 4x - 3}{(x^2 + x + 1)^2}$.

Solution Divide the numerator by $x^2 + x + 1$:

$$\begin{array}{ccc|cccc}
 & & & 2 & -1 & & \\
\hline
1 & 1 & 1 & 2 & 1 & 4 & -3 \\
 & & & 2 & 2 & 2 & \\
\hline
 & & & & -1 & 2 & -3 \\
 & & & & -1 & -1 & -1 \\
\hline
 & & & & & 3 & -2
\end{array}$$

Therefore

$$2x^3 + x^2 + 4x - 3 = (x^2 + x + 1)(2x - 1) + (3x - 2).$$

Divide by $(x^2 + x + 1)^2$:

$$\frac{2x^3 + x^2 + 4x - 3}{(x^2 + x + 1)^2} = \frac{2x - 1}{x^2 + x + 1} + \frac{3x - 2}{(x^2 + x + 1)^2}.$$

Answer $\dfrac{2x - 1}{x^2 + x + 1} + \dfrac{3x - 2}{(x^2 + x + 1)^2}$

Exercises

Decompose into partial fractions

1 $\dfrac{2}{x(x + 4)}$ **2** $\dfrac{x}{x^2 - 1}$ **3** $\dfrac{-x}{(x + 1)(x + 2)}$

4 $\dfrac{3}{x(x - 3)}$ **5** $\dfrac{x + 1}{x(x - 1)}$ **6** $\dfrac{x - 1}{x(x + 1)}$

7 $\dfrac{2x + 1}{(x - 1)(x + 2)}$ **8** $\dfrac{x + 3}{x(x - 3)}$ **9** $\dfrac{2x - 3}{(x + 2)(x + 3)}$

10 $\dfrac{3x - 3}{(x - 2)(x - 5)}$ **11** $\dfrac{x^2}{(2x + 1)(2x - 1)}$ **12** $\dfrac{x^4}{(3x - 2)(3x + 2)}$

13 $\dfrac{x + 1}{x^2}$ **14** $\dfrac{x}{(x + 1)^2}$ **15** $\dfrac{2x - 1}{(x + 2)^2}$

16 $\dfrac{4x - 3}{(x - 2)^2}$ **17** $\dfrac{x^2}{(x + 1)^2}$ **18** $\dfrac{x^3}{(x + 4)^2}$

19 $\dfrac{x^2}{x(x - 1)(x - 2)}$ **20** $\dfrac{x}{(x + 1)(x + 2)(x + 3)}$ **21** $\dfrac{x^2}{(x - 1)(x - 2)(x - 3)}$

22 $\dfrac{x^2 - 1}{x(x^2 - 4)}$ **23** $\dfrac{x + 1}{(x - 2)(x - 3)(x - 4)}$ **24** $\dfrac{x - 1}{(x + 2)(x + 3)(x + 4)}$

25 $\dfrac{2x^2 + 7x - 4}{x^3 - 4x}$ **26** $\dfrac{x^2 - x + 10}{(x - 1)(2x - 1)(3x - 1)}$ **27** $\dfrac{1}{x^2(x + 1)}$

28 $\dfrac{x}{(x-1)^2(x-2)}$ **29** $\dfrac{2x-1}{(x+1)^2(x-1)}$ **30** $\dfrac{x^2+3}{x(x-2)^2}$

31 $\dfrac{x^2-3}{(x-2)^2(x-3)}$ **32** $\dfrac{x^2+x+1}{(x-2)^2(x+1)}$ **33** $\dfrac{x}{(x+1)^3}$

34 $\dfrac{x^2}{(x-2)^3}$ **35** $\dfrac{x^2+3x+1}{x^3}$ **36** $\dfrac{3x^2-1}{(x-2)^3}$

37 $\dfrac{x^3}{(x+1)^3}$ **38** $\dfrac{x^3+1}{(x+1)^3}$ **39** $\dfrac{x^2}{(x+3)(x^2+1)}$

40 $\dfrac{x^2-1}{(x-2)(x^2+1)}$ **41** $\dfrac{x-2}{(x+2)(x^2+4)}$ **42** $\dfrac{x}{(x-1)(x^2+5)}$

43 $\dfrac{x^3}{(x^2+1)^2}$ **44** $\dfrac{-x^3+4}{(x^2+1)^2}$ **45** $\dfrac{x^3}{(x^2+2)^2}$

46 $\dfrac{x^3-x^2}{(x^2+3)^2}$ **47** $\dfrac{x^3-1}{(x^2+2x+3)^2}$ **48** $\dfrac{3x^2}{(x^2+x+2)^2}$

49 $\dfrac{6}{x(x-1)(x-2)(x-3)}$ **50** $\dfrac{x^3-2x+1}{(x+2)^4}.$

REVIEW EXERCISES

Solve by elimination

1 $\begin{cases} 2x+2y+3z=-1 \\ -5x-3y-8z=1 \\ 5x+8y+7z=2 \end{cases}$ **2** $\begin{cases} x+2y+3z=3 \\ 2x+3y+4z=2 \\ 3x+4y+6z=1. \end{cases}$

Solve the system whose matrix is

3 $\begin{bmatrix} 1 & 2 & 2 & 4 \\ -2 & -1 & 2 & 5 \\ 2 & -2 & 1 & -3 \end{bmatrix}$ **4** $\begin{bmatrix} 2 & 6 & 9 & 11 \\ 6 & 7 & -6 & -33 \\ 9 & -6 & 2 & 22 \end{bmatrix}$

Compute the product

5 $\begin{bmatrix} -1 & 3 & 0 \\ 4 & -2 & -1 \\ -1 & -1 & 3 \end{bmatrix}\begin{bmatrix} 5 \\ 6 \\ -4 \end{bmatrix}$ **6** $\begin{bmatrix} 6 & -2 & 1 \\ -1 & 4 & 5 \\ 2 & 0 & 7 \end{bmatrix}\begin{bmatrix} 2 & 2 & 2 \\ -1 & 3 & -1 \\ 0 & 4 & 3 \end{bmatrix}.$

Find the inverse of

7 $\begin{bmatrix} 10 & 8 & 3 \\ -6 & -3 & -2 \\ 3 & 2 & 1 \end{bmatrix}$ **8** $\begin{bmatrix} -7 & 4 & 4 \\ -4 & 1 & -8 \\ -4 & -8 & 1 \end{bmatrix}.$

Evaluate the determinant

9 $\begin{bmatrix} 2 & 5 & 8 \\ 4 & 5 & 13 \\ 1 & -6 & -1 \end{bmatrix}$ **10** $\begin{bmatrix} 1 & 2 & 2 \\ 2 & -2 & 1 \\ 2 & 1 & -2 \end{bmatrix}.$

Divide $f(x)$ by $g(x)$; give the quotient and remainder

11 $f(x) = -3x^5 + 2x^4 - x^3 - x + 2 \quad g(x) = x^2 - 2x + 2$

12 $f(x) = 2x^5 - 2x^4 + 2x^3 - 2x^2 + 2x + 2 \quad g(x) = (x + 1)^2.$

Find all real zeros and their multiplicities

13 $(x^2 + 2x - 1)^3(x^2 - 1)^2$

14 $(x^2 - 5x + 6)^4(x^2 + 4)^2(x - 1)^3.$

Find all rational zeros and their multiplicities

15 $4x^4 - 4x^3 - 35x^2 + 36x - 9$

16 $6x^4 + 25x^3 + 26x^2 - 4x - 8.$

Decompose into partial fractions

17 $\dfrac{x^2 + x + 1}{(x - 2)(2x + 1)}$

18 $\dfrac{3x - 4}{x^2(x - 1)}.$

10 DISCRETE ALGEBRA

1 SEQUENCES

The word **sequence** is the mathematical term for a list of objects. A sequence may be finite or infinite. For instance

bat , cat , fat , gat , hat , mat , pat , rat , sat , tat , vat

is a sequence (list) of 11 words. The first term of this sequence (list) is bat, the 11-th (last) is vat. We might abbreviate this sequence by

$$w_1, w_2, w_3, \cdots, w_{11}.$$

Thus $w_3 = \text{fat}$, $w_7 = \text{pat}$, etc.

The sequence

$$0, 1, 0, 1, 0, 1, \cdots \qquad \text{(forever)}$$

is an example of an infinite sequence. If we abbreviate it

$$a_1, a_2, a_3, \cdots$$

or simply

$$\{a_n\}_1^\infty$$

then $a_{2n-1} = 0$ and $a_{2n} = 1$ for $n = 1, 2, 3, \cdots$.

A sequence is generally numbered by a consecutive run of integers, usually starting with 0 or 1. Thus

$$b_0, b_1, b_2, b_3 \quad \text{and} \quad c_1, c_2, c_3, c_4$$

are both sequences of 4 terms.

In this section we shall study three useful types of sequences of numbers.

Arithmetic Progression

An **arithmetic progression** of n terms is a sequence of n equally spaced numbers. Such a sequence of numbers must have the form

$$a_1 = a, \quad a_2 = a + d, \quad a_3 = a + 2d, \cdots, \quad a_n = a + (n - 1)d.$$

Consecutive terms differ by the same number d, called the **common difference** of the arithmetic progression.

Examples

(1) $2, 7, 12, 17, 22, 27$ $\quad a = 2 \quad d = 5 \quad n = 6$

(2) $-1, -\frac{1}{2}, 0, \frac{1}{2}$ $\quad a = -1 \quad d = \frac{1}{2} \quad n = 4$

(3) $50, 48, 46, 44, \cdots, 2$ $\quad a = 50 \quad d = -2 \quad n = 25$

In many applications, we need the sum of (the terms of) an arithmetic progression. Let us find a formula for the sum. We observe in an arithmetic progression $a_1, \cdots, a_n$ of n terns (as written above) that

$$a_1 + a_n = a + [a + (n - 1)d] = 2a + (n - 1)d$$
$$a_2 + a_{n-1} = (a_1 + d) + (a_n - d) = a_1 + a_n = 2a + (n - 1)d$$
$$a_3 + a_{n-2} = (a_2 + d) + (a_{n-1} - d) = a_2 + a_{n-1} = 2a + (n - 1)d$$
$$\cdots\cdots\cdots\cdots\cdots\cdots\cdots\cdots\cdots\cdots\cdots\cdots$$
$$a_n + a_1 = (a_{n-1} + d) + (a_2 - d) = a_{n-1} + a_2 = 2a + (n - 1)d.$$

We add up these n equations. On the left we obtain

$$(a_1 + a_2 + \cdots + a_n) + (a_n + a_{n-1} + \cdots + a_1)$$
$$= 2(a_1 + \cdots + a_n) = 2S,$$

where S denotes the required sum. On the right we get simply $n[2a + (n - 1)d]$. Therefore

$$2S = n[2a + (n - 1)d] = n\{a + [a + (n - 1)d]\} = n(a_1 + a_n)$$
$$S = [\tfrac{1}{2}(a_1 + a_n)]n.$$

Sum of an Arithmetic Progression

$$S = a + (a + d) + (a + 2d) + \cdots + [a + (n - 1)d]$$
$$= \tfrac{1}{2}n[2a + (n - 1)d]$$
$$= [\tfrac{1}{2}(a_1 + a_n)]n.$$

The second expression for S says that the sum of an arithmetic progression is the average of its first and last terms times its number of terms.

Example 1 Compute the sum of the arithmetic progression

(a) all odd integers from 1 to 99.

(b) $-2.0 - 0.5 + 1.0 + 2.5 + \cdots + 31.0.$

Solution (a) The progression obviously has $\frac{1}{2}(100) = 50$ terms. By the formula for the sum, $S = [\frac{1}{2}(1 + 99)](50) = 2500$.

(b) In this case $a_1 = -2.0$, $d = 1.5$, and $a_n = 31.0$, so we can find n from

$$31.0 = -2.0 + (n - 1)(1.5)$$

$$(1.5)(n - 1) = 33 \qquad n - 1 = \frac{33}{1.5} = 22 \qquad n = 23.$$

Therefore $S = \frac{1}{2}(23)(-2.0 + 31.0) = \frac{1}{2}(23)(29) = 333.5$.

Answer (a) 2500 (b) 333.5

EXAMPLE 2 Find a formula for

$$(n + 1) + (n + 2) + \cdots + 3n.$$

Solution This arithmetic progression consists of the first $3n$ positive integers with the first n positive integers deleted, so it has $2n$ terms. By the formula for the sum,

$$S = [\tfrac{1}{2}(n + 1 + 3n)](2n) = n(4n + 1).$$

Answer $n(4n + 1)$

Geometric Progressions

In an arithmetic progression, we add a fixed number to each term to find the next term. Now we discuss sequences where we multiply by a fixed number. A **geometric progression** of n terms is a sequence of numbers

$$a_1 = a, \quad a_2 = ar, \quad a_3 = ar^2, \quad \cdots, \quad a_n = ar^{n-1}.$$

We call r the **common ratio** of the geometric progression.

Examples

(1)	$2, 6, 18, 54, 162, 486$	$a = 2$	$r = 3$	$n = 6$
(2)	$13, \frac{13}{2}, \frac{13}{4}, \frac{13}{8}$	$a = 13$	$r = \frac{1}{2}$	$n = 4$
(3)	$1.05, (1.05)^2, (1.05)^3, \cdots, (1.05)^{60}$	$a = 1.05$	$r = 1.05$	$n = 60$

Let us find a formula for the sum of a geometric progression. Let

$$S = a + ar + ar^2 + \cdots + ar^{n-1}.$$

If $r = 1$, then all n terms equal a, hence $S = na$. If $r \neq 1$, then

$$\begin{aligned} rS &= r(a + ar + ar^2 + \cdots + ar^{n-1}) \\ &= (ar + ar^2 + ar^3 + \cdots + ar^{n-1}) + ar^n \\ &= (S - a) + ar^n. \end{aligned}$$

Therefore

$$S - rS = a - ar^n \qquad (1 - r)S = a(1 - r^n)$$

$$S = a\frac{1 - r^n}{1 - r}.$$

Sum of a Geometric Progression

$$S = a + ar + \cdots + ar^{n-1} = \begin{cases} na & r = 1 \\ a\dfrac{1 - r^n}{1 - r} = a\dfrac{r^n - 1}{r - 1} & r \neq 1. \end{cases}$$

Remark The form $a(1 - r^n)/(1 - r)$ is better when $r < 1$ and the form $a(r^n - 1)/(r - 1)$ when $r > 1$, because the denominators are positive. Both versions of the sum formula are used in practice.

EXAMPLE 3 Compute the sum of the geometric progression

(a) $2 + 6 + 18 + 54 + 162 + 486$

(b) $3 - \frac{3}{2} + \frac{3}{4} - \frac{3}{8} + \frac{3}{16} - \frac{3}{32} + \frac{3}{64} - \frac{3}{128} + \frac{3}{256}$.

Solution (a) In this case $a = 2$, $r = 3$, and $n = 6$, so

$$S = a\frac{r^n - 1}{r - 1} = 2\frac{3^6 - 1}{3 - 1} = 2\frac{729 - 1}{2} = 728.$$

(b) In this case $a = 3$, $r = -\frac{1}{2}$, and $n = 9$, so

$$S = 3\frac{1 - (-\frac{1}{2})^9}{1 - (-\frac{1}{2})} = 3\frac{1 + \frac{1}{512}}{\frac{3}{2}} - 2(1 + \tfrac{1}{512}) = \tfrac{513}{256}.$$

Answer (a) 728 (b) $\frac{513}{256}$

EXAMPLE 4 A ball rebounds to half the height from which it starts to fall. If it is dropped from 10 ft, how far does it travel from the moment it is dropped until the moment of its eighth bounce?

Solution First it falls 10 ft to its first bounce. Then it rises 5 ft and falls 5 ft to its second bounce, 10 ft altogether from its first to its second bounce. From its second bounce to its third it travels half of this, that is, 5 ft, etc.

Let a_i be the distance the ball travels from its i-th bounce to its $(i + 1)$-th bounce. Thus $a_1 = 10$, and its total distance travelled is

$$D = 10 + a_1 + a_2 + \cdots + a_7.$$

Now $a_{i+1} = \frac{1}{2}a_i$, so (after the first term) we have a geometric progression with $a_1 = 10$, $r = \frac{1}{2}$, and $n = 7$. Hence

$$D = 10 + 10\frac{1 - (\frac{1}{2})^7}{1 - \frac{1}{2}} = 10 + 20 \times \tfrac{127}{128}$$

$$= 10 + 5 \times \tfrac{127}{32} = 29\tfrac{27}{32}.$$

Answer $29\frac{27}{32} \approx 29.84$ ft.

Factorials

We define an important sequence of numbers:

$$0! = 1 \qquad 1! = 1 \qquad 2! = 1 \cdot 2 = 2 \qquad 3! = 1 \cdot 2 \cdot 3 = 6$$

$$n! = 1 \cdot 2 \cdot 3 \cdots n = (n - 1)!n.$$

The number $n!$ is read ***n* factorial.** These numbers increase rapidly, as is shown in a short table:

j	0	1	2	3	4	5	6	7	8	9	10
$j!$	1	1	2	6	24	120	720	5040	40320	362880	3628800

Also $20! \approx 2.4329 \times 10^{18}$ and $100! \approx 9.3326 \times 10^{157}$.

Any product of consecutive integers can be expressed in terms of factorials. For example,

$$5 \cdot 6 \cdot 7 \cdot 8 \cdot 9 = \frac{1 \cdot 2 \cdot 3 \cdot 4 \cdot 5 \cdot 6 \cdot 7 \cdot 8 \cdot 9}{1 \cdot 2 \cdot 3 \cdot 4} = \frac{9!}{4!}.$$

In general, if $0 \le k < n$, then

$$(k + 1)(k + 2)(k + 3) \cdots n = \frac{n!}{k!}.$$

Remark Most scientific calculators have a key for $n!$ It is usually accurate up to $11! = 39{,}916{,}800$ on an 8-digit display calculator, and approximate up to $69! \approx 1.71 \times 10^{98}$. Larger factorials lead to overflow (error).

Example 28! 2 8 [n!] 3.0488834 $\times 10^{29}$

Exercises

Find the n-th term and the 12-th term of the arithmetic (or geometric) progression

1 $4, 11, 18, 25, 32, \cdots$

2 $8, 5, 2, -1, -4, \cdots$

3 $\frac{1}{8}, \frac{1}{4}, \frac{1}{2}, 1, 2, \cdots$

4 $ab^2c^5, ab^3c^7, ab^4c^9, ab^5c^{11}, \cdots$.

Compute the sum of the arithmetic (or geometric) progression

5 the first n positive integers

6 the first n even positive integers

7 $11 + 12 + 13 + \cdots + 35$

8 $8 + 11 + 14 + \cdots + 53$

9 $-15 - 13 - 11 + \cdots + 5$

10 $8.7 + 13.7 + 18.7 + \cdots + 108.7$

11 $x + (x + 2) + (x + 4) + \cdots + (x + 18)$

12 $(x + 3y) + (x + 5y) + (x + 7y) + \cdots + (x + 21y)$

13 $(2a + 3b) + (2a + 6b) + (2a + 9b) + \cdots + (2a + 3nb)$

14 $(2n + 1) + (2n + 2) + (2n + 3) + \cdots + 5n$

15 $3 + 6 + 12 + \cdots + 3 \cdot 2^9$

16 $3 - 6 + 12 - + \cdots + 3(-2)^9$

17 $8 + 4 + 2 + 1 + \cdots + \frac{1}{16}$

18 $1 + \dfrac{1}{3} + \dfrac{1}{3^2} + \cdots + \dfrac{1}{3^5}$

19 $2^4 + 2^5 + 2^6 + \cdots + 2^{10}$

20 $1 - \frac{2}{3} + (\frac{2}{3})^2 - (\frac{2}{3})^3 + (\frac{2}{3})^4$

21 $\dfrac{1}{x} + \dfrac{1}{x^2} + \dfrac{1}{x^3} + \cdots + \dfrac{1}{x^6}$

22 $x^{1/2} + x + x^{3/2} + \cdots + x^4$

23 $ab^2c^5 + ab^3c^7 + ab^4c^9 + \cdots + ab^8c^{17}$

24 $1 - y^2 + y^4 - + \cdots + y^{20}$.

25 Find all sequences a_1, a_2, a_3 that are both arithmetic and geometric progressions.

26 Find all arithmetic progressions a_1, a_2, a_3 such that a_1^2, a_2^2, a_3^2 is also an arithmetic progression.

27 Suppose $a_1, a_2, \cdots, a_n$ and $b_1, b_2, \cdots, b_n$ are arithmetic progressions. Prove that

$$a_1 + 3b_1, \; a_2 + 3b_2, \; \cdots, \; a_n + 3b_n$$

is also an arithmetic progression.

28 Suppose $a_1, a_2, \cdots, a_n$ and $b_1, b_2, \cdots, b_n$ are geometric progressions. Prove that

$$a_1b_1, a_2b_2, \cdots, a_nb_n$$

is also a geometric progression.

29 Show that each term of the sequence

$$1, 2, 4, 8, \cdots, 2^n$$

equals one plus the sum of all preceding terms.

30 Show that each term of the sequence

$$1, 2, 6, 18, \cdots, 2 \cdot 3^n$$

equals double the sum of all preceding terms.

31 Solve Example 4, assuming the ball rebounds to $\frac{9}{10}$ the height from which it is dropped.

32 A rajah once demanded of his subjects one grain of rice on the first square of his chessboard, two on the second, four on the third, and so on, doubling each time until the 64-th square. How much total rice did he demand?

Express by means of factorials

33 $18 \cdot 17 \cdot 16 \cdot 15 \cdot 14$

34 $7 \cdot 8 \cdot 9 \cdot 10 \cdot 11 \cdot 12 \cdot 13$

35 $2 \cdot 4 \cdot 6 \cdot 8 \cdot 10 \cdot 12$

36 $\dfrac{52 \cdot 51 \cdot 50 \cdot 49 \cdot 48}{120}$.

Compute

37 $\dfrac{9!}{7!}$ **38** $\dfrac{12!}{8!4!}$ **39** $\dfrac{10!}{(5!)^2}$

40 $\dfrac{14!}{8!4!2!}$ **41** $\dfrac{15!}{11!4!}$ **42** $\dfrac{25!}{3!22!}$.

Which is greater

43 2^n or $n!$ **44** $(n + 2)!n!$ or $[(n + 1)!]^2$?

Solve for n

45 $(n + 1)! = 10(n!)$

46 $(n + 2)! = (n + 1)! + 25(n!)$.

2 SUMMATION

In Section 1, we computed the sums of arithmetic and geometric sequences. In this section, we discuss techniques for computing other sums. We begin by introducing some standard notation.

Summation Notation

The symbol

$$\sum_{j=1}^{n} a_j$$

is used to abbreviate the sum

$$a_1 + a_2 + \cdots + a_n$$

of the terms of the sequence $a_1, a_2, \cdots, a_n$. The Greek sigma Σ stands for *sum,* and the **summation index** j labels the terms. For example, if $a_j = j^2$, the notation represents the sum of terms j^2, for $j = 1, 2, 3, \cdots, n$. Thus

$$\sum_{j=1}^{n} j^2 = 1^2 + 2^2 + \cdots + n^2.$$

There is nothing special about the letter j; any other letter will serve just as well for a summation index. For instance,

$$\sum_{j=1}^{n} j^2 = \sum_{k=1}^{n} k^2 = \sum_{i=1}^{n} i^2 = 1^2 + 2^2 + \cdots + n^2.$$

The index need not run from 1 to n. In general,

$$\sum_{j=s}^{t} a_j = a_s + a_{s+1} + \cdots a_t.$$

Here are a few examples of summation notation in action:

$$\sum_{j=6}^{10} j(j + 1) = 6 \cdot 7 + 7 \cdot 8 + 8 \cdot 9 + 9 \cdot 10 + 10 \cdot 11$$

$$\sum_{j=0}^{n} a_j x^j = a_0 + a_1 x + a_2 x^2 + \cdots + a_n x^n$$

$$\sum_{j=3}^{12} a_{j-1} x^{2j} = a_2 x^6 + a_3 x^8 + a_4 x^{10} + \cdots + a_{11} x^{24}.$$

A few more examples illustrate the versatility of the notation:

$$\sum_{j=1}^{n} a_{3j} = a_3 + a_6 + a_9 + \cdots + a_{3n}$$

$$\sum_{j=1}^{5} a_j b_{5+j} = a_1 b_6 + a_2 b_7 + a_3 b_8 + a_4 b_9 + a_5 b_{10}$$

$$\sum_{n=2}^{9} \frac{(-1)^n}{n!} x^{2n-1} = \frac{x^3}{1 \cdot 2} - \frac{x^5}{1 \cdot 2 \cdot 3} + \frac{x^7}{1 \cdot 2 \cdot 3 \cdot 4} - + \cdots - \frac{x^{17}}{1 \cdot 2 \cdot 3 \cdots 9}.$$

Let us express some previous results in summation notation.

Arithmetic Progression $$\sum_{k=1}^{n} [a + (k - 1)d] = \tfrac{1}{2} n[2a + (n - 1)d]$$

Geometric Progression $$\sum_{k=1}^{n} ar^{k-1} = a\frac{r^n - 1}{r - 1} \qquad (r \neq 1)$$

For later use, we note a special case of the first formula, which gives the sum of the first n positive integers. We just set $a = 1$ and $d = 1$:

$$1 + 2 + 3 + \cdots + n = \sum_{k=1}^{n} k = \tfrac{1}{2}n(n + 1).$$

Rules for Sums

There are two important rules concerning sums. The first is a generalized distributive law allowing you to take out a common factor; the second is a generalized commutative-associative law allowing you to rearrange certain sums in a natural way.

$$\sum_{j=s}^{t} ca_j = c\sum_{j=s}^{t} a_j$$

$$\sum_{j=s}^{t} (a_j + b_j) = \sum_{j=s}^{t} a_j + \sum_{j=s}^{t} b_j$$

Example $\sum_{j=1}^{n} j(j + 2) = \sum_{j=1}^{n} (j^2 + 2j) = \sum_{j=1}^{n} j^2 + 2\sum_{j=1}^{n} j.$

In a typical application, we might have formulas for the two sums on the right. Then we could obtain, free of charge, a formula for the sum on the left.

One sum is so simple that it sometimes causes confusion:

$$\sum_{j=1}^{n} c = nc.$$

The sum on the left is simply $c + c + \cdots + c$, with n summands.

The following example contains a useful formula from statistics.

EXAMPLE 1 The **mean** (or **average**) of a sequence $a_1, \cdots, a_n$ is

$$\bar{a} = \frac{1}{n}\sum_{1}^{n} a_i.$$

The **variance** of the sequence is $\sigma^2 = \dfrac{1}{n}\sum_{1}^{n} (a_i - \bar{a})^2.$

Show that $\sigma^2 = \dfrac{1}{n}\sum_{1}^{n} a_i^2 - \bar{a}^2.$

Solution Use $(a_i - \bar{a})^2 = a_i^2 - 2\bar{a}a_i + \bar{a}^2$ and the rules for sums:

$$\sigma^2 = \frac{1}{n}\sum_{1}^{n} a_i^2 - \frac{2\bar{a}}{n}\sum_{1}^{n} a_i + \frac{1}{n}\sum_{1}^{n} \bar{a}^2.$$

Now

$$\frac{1}{n}\sum_{1}^{n} a_i = \bar{a} \quad \text{and} \quad \frac{1}{n}\sum_{1}^{n} \bar{a}^2 = \frac{1}{n}\cdot n\bar{a}^2 = \bar{a}^2,$$

hence

$$\sigma^2 = \frac{1}{n}\sum_{1}^{n} a_i^2 - 2\bar{a}^2 + \bar{a}^2 = \frac{1}{n}\sum_{1}^{n} a_i^2 - \bar{a}^2.$$

Changing the Index

We can often simplify computations involving summation notation by changing the summation index. A typical change of index is based on the following observation. The quantity $j + 1$ runs from 1 to n as j runs from 0 to $n - 1$, and the quantity $j - 1$ runs from 1 to n as j runs from 2 to $n + 1$. Therefore we can describe the same sum in three ways:

$$\sum_{j=1}^{n} a_j = \sum_{j=0}^{n-1} a_{j+1} = \sum_{j=2}^{n+1} a_{j-1}.$$

Example

$$x^2 + 2x^4 + 3x^6 + 4x^8$$
$$= \sum_{j=1}^{4} jx^{2j} = \sum_{j=0}^{3} (j+1)x^{2(j+1)} = \sum_{j=2}^{5} (j-1)x^{2(j-1)}.$$

Another typical change: as j runs from 0 to n, the quantity $n - j$ runs from n to 0.

Example

$$a_0b_6 + a_1b_5 + a_2b_4 + \cdots + a_5b_1 + a_6b_0 = \sum_{j=0}^{6} a_jb_{6-j} = \sum_{j=0}^{6} a_{6-j}b_j.$$

The following example illustrates how changing the index can simplify expressions involving several sums.

EXAMPLE 2 Simplify $\displaystyle\sum_{j=2}^{10} \frac{1}{j} - \sum_{j=1}^{8} \frac{1}{j+2}$.

Solution Most terms are common to both sums and will cancel. In the second sum, let $k = j + 2$. Then as j runs from 1 to 8, the index k runs from 3 to 10. Therefore

$$\sum_{j=2}^{10} \frac{1}{j} - \sum_{j=1}^{8} \frac{1}{j+2} = \sum_{k=2}^{10} \frac{1}{k} - \sum_{k=3}^{10} \frac{1}{k} = \left(\frac{1}{2} + \sum_{k=3}^{10} \frac{1}{k}\right) - \sum_{k=3}^{10} \frac{1}{k} = \frac{1}{2}.$$

Answer $\frac{1}{2}$

Evaluation of Sums

Let us turn to the main business of this section, computing sums. We shall discuss a method, called telescoping,* that allows us to evaluate a variety of sums.

Suppose $b_1, b_2, \cdots, b_{n+1}$ is a sequence and $a_j = b_{j+1} - b_j$. Then

$$\sum_{j=1}^{n} a_j = \sum_{j=1}^{n} (b_{j+1} - b_j) = \sum_{j=1}^{n} b_{j+1} - \sum_{j=1}^{n} b_j = \sum_{k=2}^{n+1} b_k - \sum_{j=1}^{n} b_j$$

$$= \left(b_{n+1} + \sum_{k=2}^{n} b_k\right) - \left(b_1 + \sum_{j=2}^{n} b_j\right) = b_{n+1} - b_1.$$

Telescoping If $a_j = b_{j+1} - b_j$, then $\sum_{j=1}^{n} a_j = b_{n+1} - b_1$.

This statement seems simple, yet in practice it can be very powerful. Suppose we want to compute a sum

$$\sum_{j=1}^{n} a_j.$$

If we can find a sequence $b_1, b_2, \cdots$ such that $b_{j+1} - b_j = a_j$, then we can write down the answer immediately: $b_{n+1} - b_1$. For example, if $b_j = j^2$, then $b_{j+1} - b_j = 2j + 1$, hence

$$\sum_{j=1}^{n} (2j + 1) = b_{n+1} - b_1 = (n + 1)^2 - 1.$$

A corollary of this result is a familiar formula. Since

$$\sum_{j=1}^{n} (2j + 1) = 2\sum_{j=1}^{n} j + \sum_{j=1}^{n} 1 = 2\sum_{j=1}^{n} j + n$$

we have

$$2\sum_{j=1}^{n} j = (n + 1)^2 - 1 - n = n^2 + n = n(n + 1)$$

so we find again the sum of the first n positive integers

$$\sum_{j=1}^{n} j = \tfrac{1}{2}n(n + 1).$$

Example 3 Evaluate $\displaystyle\sum_{j=1}^{n} \frac{1}{j(j + 1)}$.

Solution We observe that

$$\frac{1}{j(j + 1)} = \frac{1}{j} - \frac{1}{j + 1} = b_{j+1} - b_j,$$

* Think of closing an old-fashioned mariner's telescope; the sections fit into each other.

where $b_j = -1/j$. Therefore

$$\sum_{j=1}^{n} \frac{1}{j(j+1)} = b_{n+1} - b_1 = -\frac{1}{n+1} + 1 = \frac{n}{n+1}.$$

Answer $\dfrac{n}{n+1}$

EXAMPLE 4 Evaluate $\sum_{j=1}^{n} j^2$.

Solution If $b_j = j^3$, then

$$b_{j+1} - b_j = (j+1)^3 - j^3 = (j^3 + 3j^2 + 3j + 1) - j^3 = 3j^2 + 3j + 1.$$

Hence

$$\begin{aligned}\sum_{j=1}^{n} (3j^2 + 3j + 1) &= b_{n+1} - b_1 \\ &= (n+1)^3 - 1^3\end{aligned}$$

$$3\sum_{j=1}^{n} j^2 + 3\sum_{j=1}^{n} j + \sum_{j=1}^{n} 1 = (n+1)^3 - 1$$

$$3\sum_{j=1}^{n} j^2 = (n+1)^3 - 1 - 3\sum_{j=1}^{n} j - \sum_{j=1}^{n} 1.$$

But we know the two sums on the right. The first is $\frac{1}{2}n(n+1)$ and the second is n. Therefore

$$\begin{aligned}3\sum_{j=1}^{n} j^2 &= (n+1)^3 - 1 - \tfrac{3}{2}n(n+1) - n \\ &= \tfrac{1}{2}[2(n+1)^3 - 2 - 3(n^2+n) - 2n] \\ &= \tfrac{1}{2}[2(n+1)^3 - 3n(n+1) - 2(n+1)] \\ &= \tfrac{1}{2}(n+1)[2(n+1)^2 - 3n - 2] \\ &= \tfrac{1}{2}(n+1)(2n^2 + n) \\ &= \tfrac{1}{2}n(n+1)(2n+1).\end{aligned}$$

Answer $\sum_{j=1}^{n} j^2 = \frac{1}{6}n(n+1)(2n+1)$

Remark In Example 4, we found $\sum_{1}^{n} j^2$ in terms of $\sum_{1}^{n} j$ and $\sum_{1}^{n} 1$, which were known. In a similar way, we can now find $\sum_{1}^{n} j^3$. See Ex. 29.

The method can also be applied to obtain certain estimates involving sums. In the following example, we are able to trap a complicated sum between successive integers. The solution, however, depends on a clever trick.

EXAMPLE 5 Prove

$$198 < 1 + \frac{1}{\sqrt{2}} + \frac{1}{\sqrt{3}} + \cdots + \frac{1}{\sqrt{10{,}000}} < 199.$$

Solution Start with

$$\sqrt{j+1} - \sqrt{j} = \frac{1}{\sqrt{j+1} + \sqrt{j}}.$$

Clearly $2\sqrt{j} < \sqrt{j} + \sqrt{j+1} < 2\sqrt{j+1}$; hence

$$\frac{1}{2\sqrt{j+1}} < \frac{1}{\sqrt{j} + \sqrt{j+1}} < \frac{1}{2\sqrt{j}},$$

so

$$\frac{1}{2\sqrt{j+1}} < \sqrt{j+1} - \sqrt{j} < \frac{1}{2\sqrt{j}}.$$

Now sum from $j = 1$ to $j = 10^4 - 1 = 9999$. The inner sum telescopes:

$$\sum_{j=1}^{9999} (\sqrt{j+1} - \sqrt{j}) = \sqrt{10{,}000} - \sqrt{1} = 100 - 1 = 99.$$

Therefore

$$\frac{1}{2}\sum_{j=1}^{9999} \frac{1}{\sqrt{j+1}} < 99 < \frac{1}{2}\sum_{j=1}^{9999} \frac{1}{\sqrt{j}}.$$

The left half of this inequality is equivalent to

$$\sum_{j=2}^{10000} \frac{1}{\sqrt{j}} < 2(99) = 198, \qquad \text{hence} \qquad \sum_{j=1}^{10000} \frac{1}{\sqrt{j}} < 198 + 1 = 199.$$

The right half is equivalent to

$$198 < \sum_{j=1}^{9999} \frac{1}{\sqrt{j}}, \qquad \text{hence} \qquad 198 < \sum_{j=1}^{10000} \frac{1}{\sqrt{j}}.$$

Exercises

Express in summation notation

1 $x_3 + x_4 + x_5 + \cdots + x_{20}$

2 $x_3 + x_5 + x_7 + \cdots + x_{19}$

3 the sum of the first 50 odd integers

4 the sum of the first n powers of 2 starting with $1 = 2^0$

5 $1^2 + 2^3 + 3^4 + \cdots + n^{n+1}$

6 $xy^{10} + x^2y^8 + x^3y^6 + x^4y^4 + x^5y^2 + x^6$.

Evaluate the sum

7 $\sum_{n=2}^{6} (-2)^n$

8 $\sum_{n=1}^{9} \frac{n}{10^n}$

9 $\sum_{k=1}^{30} (5k+2)$

10 $\sum_{j=10}^{50} (4-3j)$

11 $\sum_{n=-5}^{6} n^3$

12 $\sum_{k=-5}^{5} 3k^2$

13 $\sum_{n=1}^{99} \cos n\pi$

14 $\sum_{n=1}^{100} \sin \frac{1}{2}n\pi$

15 $\sum_{n=1}^{10} (10-n)10^n$

16 $\sum_{j=11}^{20} \frac{20-j}{100^j}$.

17 Write $\sum_{j=1}^{n} jx^{j-1}$ as a sum of multiples of x^j.

18 (cont.) Do the same for $\sum_{j=2}^{n} j(j-1)x^{j-2}$.

19 Prove that $\dfrac{1}{2}\displaystyle\sum_{j=1}^{n}\frac{1}{j} + \sum_{j=0}^{n-1}\frac{1}{2j+1} = \sum_{j=1}^{2n}\frac{1}{j}$.

20 Show that $\sum_{j=1}^{n} a_j b_j = \dfrac{1}{4}\left[\sum_{j=1}^{n}(a_j+b_j)^2 - \sum_{j=1}^{n}(a_j-b_j)^2\right]$.

21 Show that $\sum_{j=0}^{n} a_j b_{n-j} = \sum_{j=0}^{n} a_{n-j} b_j$.

Find the sum

22 $1\cdot 2 + 2\cdot 3 + 3\cdot 4 + \cdots + 99\cdot 100$

23 $\dfrac{1}{1\cdot 3} + \dfrac{1}{2\cdot 4} + \dfrac{1}{3\cdot 5} + \cdots + \dfrac{1}{49\cdot 51}$

24 $1^2 + 3^2 + 5^2 + 7^2 + \cdots + 99^2$.

25 Which is larger,

$$1+2+3+4+\cdots+100 \quad\text{or}\quad 1.1 + 1.1^2 + 1.1^3 + \cdots + 1.1^{65}\,?$$

26 Compute $\displaystyle\sum_{j=1}^{2n}\frac{1}{j(j+2)}$ using telescoping.

27 Compute $\sum_{j=1}^{n} j\cdot 2^j$. [*Hint* Consider $b_{j+1} - b_j$, where $b_j = j\cdot 2^j$.]

28 Compute $\sum_{j=1}^{n} j(j+1)(j+2)$.

[*Hint* Consider $b_j - b_{j-1}$, where $b_j = j(j+1)(j+2)(j+3)$.]

29 Compute $(j+1)^4 - j^4$. Use the result to derive the formula

$$1^3 + 2^3 + 3^3 + \cdots + n^3 = \frac{n^2(n+1)^2}{4} = (1+2+\cdots+n)^2.$$

30 Use the relation $\sum_{j=0}^{n} j^3 = \sum_{j=0}^{n}(n-j)^3$

to obtain the formula for $\sum_{j=1}^{n} j^3$ in another way.

31 Prove that no matter how large n is, $1 + \dfrac{1}{2^2} + \dfrac{1}{3^2} + \cdots + \dfrac{1}{n^2} < 2$.

$\left[\textit{Hint}\ \ \dfrac{1}{j^2} < \dfrac{1}{j(j-1)} \text{ for } j \ge 2.\right]$

32 Prove that no matter how large n is, $\dfrac{1}{1!} + \dfrac{1}{2!} + \cdots + \dfrac{1}{n!} < 2$.

[*Hint* Compare $1/j!$ with $1/2^{j-1}$.]

33* Let $A_i = a_1 + a_2 + \cdots + a_i$ and $B_i = b_1 + b_2 + \cdots + b_i$ for $i = 1, \cdots, n$. Prove

$$\sum_{i=1}^{n} A_i b_i = A_n B_n - \sum_{i=1}^{n-1} a_{i+1} B_i.$$

34* Find $\dfrac{3}{(1\cdot 2)^2} + \dfrac{5}{(2\cdot 3)^2} + \dfrac{7}{(3\cdot 4)^2} + \cdots + \dfrac{2n+1}{[n(n+1)]^2}$.

35 Find a telescoping principle for **products:** $\prod_{i=1}^{n} a_i = a_1 \cdot a_2 \cdot a_3 \cdots a_n$.
Apply it to find $\prod_{i=1}^{100} \left(1 - \dfrac{1}{i+1}\right)$.

36 Let $a_1, \cdots, a_n$ and $b_1, \cdots, b_n$ be sequences with means $\bar{a}$ and $\bar{b}$. (See Example 1.) Show that

$$\frac{1}{n}\sum_{i=1}^{n} (a_i - \bar{a})(b_i - \bar{b}) = \frac{1}{n}\sum_{i=1}^{n} a_i b_i - \bar{a}\bar{b}.$$

(The formula comes up in discussions of correlation coefficients in statistics.)

3 MATHEMATICAL INDUCTION

The word "induction" means discovery of a general principle from special cases. For instance, the special facts

$$1 = 1^2 \qquad 1 + 3 = 2^2 \qquad 1 + 3 + 5 = 3^2 \qquad 1 + 3 + 5 + 7 = 4^2$$

suggest that the general formula

$$1 + 3 + 5 + \cdots + (2n - 1) = n^2$$

may hold for *all* positive integers n. Similarly, the special inequalities

$$2^1 > 1 \qquad 2^2 > 2 \qquad 2^3 > 3 \qquad 2^4 > 4$$

suggest that the general inequality $2^n > n$ may hold for *all* positive integers n.

It's easy to jump to conclusions, however. For example, consider the formula

$$s_n = n^2 - n + 41.$$

For $n = 1, 2, 3, 4, 5$, it yields

$$s_1 = 41 \qquad s_2 = 43 \qquad s_3 = 47 \qquad s_4 = 53 \qquad s_5 = 61,$$

all primes (integers that cannot be factored into products of smaller integers). More arithmetic shows that $s_6, s_7, \cdots, s_{40}$ are also primes. The obvious "general principle" is that each s_n is a prime. But that is just wrong because

$$s_{41} = 41^2 - 41 + 41 = 41^2,$$

which is not a prime. After 40 correct cases, the "principle" fails.

As a more extreme example, take the statement "all positive integers n are less than 1,000,001." The statement is false, yet the first million cases are correct!

Thus no quantity of special cases can prove a general principle that contains infinitely many special cases. We need a method to prove conjectures based on experimental evidence without treating each case separately (life is too short). This is what mathematical induction does.

To understand the idea, imagine a row of dominoes standing on end (Fig. 1, next page). If they are spaced close to each other, and the first one is knocked down, they will all fall. We know this for two reasons:

(1) The first domino *will* fall.

(2) *If* a domino falls, then it will knock down the next one.

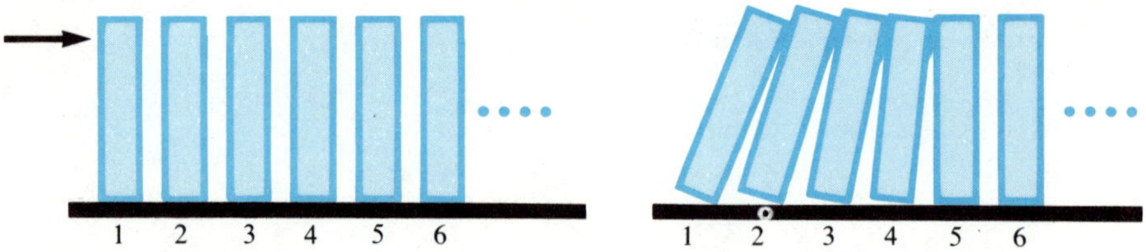

Fig. 1 Domino theory. If domino 1 is knocked down, all dominos fall, because domino k knocks down domino $k + 1$.

Be careful of (2). We do not assert a given domino *will* fall, only that *if* it does, it will knock down the next.

If the process starts by knocking down the k-th domino instead of the first, then all dominoes will fall from the k-th domino on. This idea is stated formally as an axiom for the positive integers.

> **Principle of Mathematical Induction** Let S be a set of positive integers such that
>
> (1) the integer k belongs to S
>
> (2) if an integer n belongs to S, then $n + 1$ also belongs to S.
>
> Then S contains all integers $n \geq k$.

Here is how we use the axiom. Suppose T is a theorem (statement) about all integers $n \geq k$. Then T consists of a sequence of statements T_k, T_{k+1}, T_{k+2}, $\cdots$. To prove all the statements we

(1) prove somehow that T_k is true

(2) prove that if T_n were true for an integer n, then T_{n+1} would also be true.

Once we carry out these two steps, we are finished. For let S be the set of positive integers n such that T_n is true. By (1), the set S contains k; by (2) if it contains n, then it contains $n + 1$. Hence the principle of mathematical induction guarantees that S contains all integers $n \geq k$. In other words, the theorem is true for all $n \geq k$. This two-step method of proof is called **proof by mathematical induction.**

Remark The critical step is going from n to $n + 1$. Remember, you do not prove the theorem for n; you merely show if it *were* true for n, then it *would be* true for $n + 1$. Thus you prove the theorem for $n + 1$, *assuming* it is true for n; in a sense, you lift yourself by your boot straps.

Example 1 Prove the formula

$$1 + 3 + 5 + \cdots + (2n - 1) = n^2 \qquad n \geq 1.$$

Solution We give a two-step proof by mathematical induction.

(1) For $n = 1$, the assertion is $1 = 1^2$, which is true.

(2) We assume the formula holds for an integer n and show that it must then hold for $n + 1$, that is,

$$1 + 3 + 5 + \cdots + (2n + 1) = (n + 1)^2.$$

We note that the left-hand side is the left-hand side of the preceding case, with one additional term:

$$\begin{aligned} &1 + 3 + 5 + \cdots + (2n + 1) \\ &\quad = [1 + 3 + 5 + \cdots + (2n - 1)] + (2n + 1) \\ &\quad = n^2 + (2n + 1) \qquad \text{[preceding case]} \\ &\quad = (n + 1)^2 \end{aligned}$$

as was to be proved. This completes the proof by mathematical induction; the formula holds for all $n \geq 1$.

EXAMPLE 2 Prove by mathematical induction that $2^n > n$ for all integers $n \geq 1$.

Solution (1) For $n = 1$, the statement is $2^1 > 1$, which is true.

(2) We show that if $2^n > n$, then it would follow that $2^{n+1} > n + 1$. We assume that $2^n > n$ and multiply both sides of the inequality by 2:

$$2^{n+1} > 2n.$$

But $2n \geq n + 1$ for all positive integers. Hence $2^{n+1} > n + 1$. This completes the proof by mathematical induction; the inequality holds for all $n \geq 1$.

EXAMPLE 3 Prove the general distributive law

$$a(b_1 + b_2 + \cdots + b_n) = ab_1 + ab_2 + \cdots + ab_n \qquad n \geq 2.$$

Solution Proof by mathematical induction.

(1) For $n = 2$, the assertion is $a(b_1 + b_2) = ab_1 + ab_2$, which is the distributive law.

(2) Assume the formula for n and show that it must then hold for $n + 1$:

$$\begin{aligned} &a(b_1 + b_2 + \cdots + b_{n+1}) \\ &\quad = a[(b_1 + b_2 + \cdots + b_n) + b_{n+1}] \\ &\quad = a(b_1 + b_2 + \cdots + b_n) + ab_{n+1} \qquad \text{[distr. law]} \\ &\quad = (ab_1 + ab_2 + \cdots + ab_n) + ab_{n+1} \qquad \text{[preceding case]} \\ &\quad = ab_1 + ab_2 + \cdots + ab_{n+1} \end{aligned}$$

which was to be proved. This completes the proof by mathematical induction.

Inductive Definitions

Suppose a sequence of numbers $a_1, a_2, a_3, \cdots$ is defined this way: $a_1 = 1$ and $a_{n+1} = 2a_n + 1$. The sequence is not given explicitly, only that it starts with 1 and that each number is one more than twice the preceding one. We say that the sequence is defined **inductively** or **recursively**. Another example is the sequence

$$a_1 = 0, \qquad a_2 = 1, \qquad a_{n+1} = \tfrac{1}{2}(a_n + a_{n-1});$$

the first two numbers are $0, 1$, and after that, each number is the average of the two preceding ones.

The method of mathematical induction is tailor-made for inductively defined sequences. Usually, we conjecture properties of such sequences and prove our conjectures by induction.

EXAMPLE 4 If $a_1 = 1$, and $a_{n+1} = 2a_n + 1$ for $n > 1$, find an explicit formula for a_n and prove it by mathematical induction.

Solution Trying a few cases, we see

$$a_1 = 1, \qquad a_2 = 3, \qquad a_3 = 7, \qquad a_4 = 15, \qquad a_5 = 31.$$

These numbers are 1 less than 2^1, 2^2, 2^3, 2^4, 2^5. We conjecture that $a_n = 2^n - 1$ for n ≥ 1.

Let us prove the conjecture by mathematical induction.

(1) Since $a_1 = 1 = 2^1 - 1$, the conjecture is true for $n = 1$.

(2) Now we assume $a_n = 2^n - 1$. If so, then

$$a_{n+1} = 2a_n + 1 = 2(2^n - 1) + 1 = 2^{n+1} - 2 + 1 = 2^{n+1} - 1.$$

Hence if the conjecture is true for n, it is true for $n + 1$. By mathematical induction, the conjecture is proved.

Remark Inductive definitions are important in a careful development of the rules of arithmetic for real and complex numbers. For example, sums and products are defined at first only for two numbers. Sums and products of n numbers, $n > 2$, are then defined inductively:

$$a_1 + a_2 + \cdots + a_{n+1} = (a_1 + \cdots + a_n) + a_{n+1}$$

$$a_1 a_2 \cdots a_{n+1} = (a_1 \cdots a_n)a_{n+1}.$$

Exercises

Prove by mathematical induction for $n \geq 1$

1 $2 + 5 + 8 + \cdots + (3n - 1) = \frac{1}{2}n(3n + 1)$

2 $2 + 4 + 6 + \cdots + 2n = n(n + 1)$

3 $\dfrac{1}{1\cdot 3} + \dfrac{1}{3\cdot 5} + \cdots + \dfrac{1}{(2n - 1)(2n + 1)} = \dfrac{n}{2n + 1}$

4 $\dfrac{1}{2} + \dfrac{2}{2^2} + \dfrac{3}{2^3} + \cdots + \dfrac{n}{2^n} = 2 - \dfrac{n + 2}{2^n}$

5 $1^2 + 2^2 + 3^2 + \cdots + n^2 = \frac{1}{6}n(n + 1)(2n + 1)$

6 $1^3 + 2^3 + 3^3 + \cdots . n^3 = [\frac{1}{2}n(n + 1)]^2$

7 $1\cdot 2 + 2\cdot 3 + 3\cdot 4 + \cdots + n(n + 1) = \frac{1}{3}n(n + 1)(n + 2)$

8 $1\cdot 2\cdot 3 + 2\cdot 3\cdot 4 + 3\cdot 4\cdot 5 + \cdots + n(n + 1)(n + 2) = \frac{1}{4}n(n + 1)(n + 2)(n + 3)$

Prove by mathematical induction

9 The sum of n even integers is even.

10 The sum of $2n - 1$ odd integers is odd.

11 $\log(a_1 a_2 a_3 \cdots a_n) = \log a_1 + \log a_2 + \log a_3 + \cdots \log a_n \qquad n \geq 2$

12 $\sqrt{a_1 a_2 a_3 \cdots a_n} = \sqrt{a_1}\cdot\sqrt{a_2}\cdot\sqrt{a_3} \cdots \sqrt{a_n} \qquad n \geq 2$

13 If n people stand in line at a ticket counter, and if the first person in line is a woman and the last person is a man, then somewhere in the line there is a man directly behind a woman.

14 An n-piece jigsaw puzzle, $n \geq 2$, requires $n - 1$ joinings (fittings of a piece to another piece, or a block of pieces to another block).

Prove the inequality by mathematical induction

15 $(\frac{3}{2})^n > n \quad n \geq 1$

16 $2^n > 7n \quad n \geq 6$

17 $n! > 10n \quad n \geq 5$

18 $3^n > 2^n + 10n \quad n \geq 4$

19 $(1 + a)^n > 1 + na \quad a > 0, n \geq 2$

20 $\dfrac{1}{n+1} + \dfrac{1}{n+2} + \dfrac{1}{n+3} + \cdots + \dfrac{1}{2n} > \dfrac{3}{5} \quad n \geq 3.$

Find a formula and prove it by induction for

21 $\left(1 - \frac{1}{4}\right)\left(1 - \frac{1}{9}\right)\left(1 - \frac{1}{16}\right) \cdots \left(1 - \frac{1}{n^2}\right) \quad n \geq 2$

22 $(1 - x)(1 + x)(1 + x^2)(1 + x^4) \cdots (1 + x^{2^n}) \quad n \geq 0$

23 $\sin\theta \cos\theta \cos 2\theta \cos 4\theta \cdots \cos 2^n\theta \quad n \geq 0$

24 $10 \cdot 10^2 \cdot 10^3 \cdots 10^n \quad n \geq 1$

25 the number of committees that can be formed from a group of n people (allowing committees of 1, 2, 3, $\cdots$, n members).

26 the number of committees that can be formed from a group of n people and having (a) an odd number of members, (b) an even number of members.

27 In the sequence of figures shown in Fig. 2, the first is an equilateral triangle of side 1. At each step an equilateral triangle is constructed on each side of the preceding figure with length $\frac{1}{3}$ of the side. Find the number of sides of the n-th figure. Prove your answer.

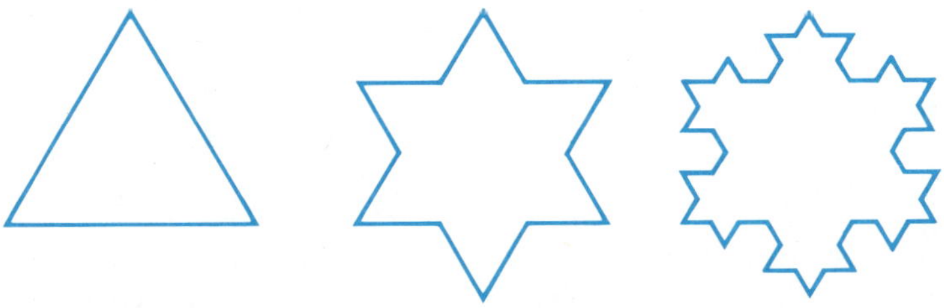

(a) step 1 (b) step 2 (c) step 3

Fig. 2

28 (cont.) Find the perimeter of the n-th figure.

29* (cont.) Show that the area of the n-th figure is $\frac{1}{20}\sqrt{3}[8 - 3(\frac{4}{9})^{n-1}]$.

30 (cont.) Deduce that the area of the n-th figure approaches $\frac{8}{5}$ of the area of the first figure as $n \longrightarrow \infty$.

Remark Exercises 28 and 30 show that the area of the figures is bounded but their perimeters grow without limit as $n \longrightarrow \infty$. (This is the same principle by which a fairly small state like Maine can have an extremely long coastline.)

31 Define a sequence by $a_0 = 0$, $a_1 = 1$, and $a_{n+2} = 2a_{n+1} - a_n$. Find a formula for a_n and prove it by induction.

32 Define a sequence by $x_0 = 1$, $x_1 = 0$, and $x_{n+2} = 2x_{n+1} - x_n$. Prove that $x_n = 1 - n$.

33 A pyramid of n decreasing rings is placed on one of the three pegs as shown in Figure 3, next page. The game (tower of Hanoi puzzle) is to transfer the pyramid to peg 2 in as few moves as possible. There are two rules: (1) only one ring can be moved at a time; (2) a ring may be moved from one peg to another but may not be placed on top of a ring smaller than itself. Find the least number of moves required and confirm your answer by induction.

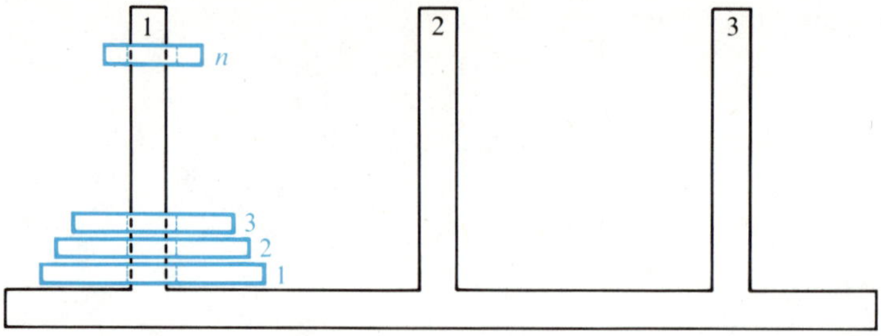

Fig. 3

34 Consider the sequence

$$x_1 = \sqrt{2}, \qquad x_2 = \sqrt{2 + \sqrt{2}}, \qquad \quad = \sqrt{2 + \sqrt{2 + \sqrt{2}}}, \qquad \text{etc.}$$

Prove that $x_n < 2$ for $n \geq 1$.

35 Define a sequence by $x_0 = 0$, $x_1 = 1$, and $x_{n+2} = x_{n+1} + x_n$. Prove that $x_3, x_6, x_9, \cdots$ are even integers and the others are odd.

36 (cont.) Compute $x_2, x_3, \cdots, x_{12}$. These numbers increase rapidly but it is not obvious at what rate. By induction prove for $n > 1$ the rough estimate

$$(\sqrt{2})^n < x_{n+2} < 2^n.$$

Find a formula for x_n and prove it by induction

37 $x_0 = 0$, $x_1 = 2$, and x_n is the average of $x_0, x_1, \cdots, x_{n-1}$ for $n \geq 2$

38 $x_0 = 1$, $x_1 = 1$, and $x_n = x_0 + x_1 + \cdots + x_{n-1}$ for n ≥ 2

39 $x_0 = 1$, $x_1 = 2$, and $x_n = 2(x_0 + x_1 + \cdots + x_{n-1})$ for $n \geq 2$

40 $x_1 = 5$ and $x_n = \sqrt{x_{n-1}}$ for $n \geq 2$.

41 Define a sequence as follows:

$$a_0 = 1 \qquad a_{n+1} = \frac{1}{2}\left(a_n + \frac{2}{a_n}\right).$$

Show that $1 \leq a_n \leq 2$ for $n \geq 0$.

42* Define a sequence by $x_0 = 0$, $x_1 = 1$, and $x_{n+2} = \frac{1}{2}(x_n + x_{n+1})$. Thus each number is the average of the two preceding. Find a formula for x_n and a number that x_n approaches as n increases.

43* Show that n lines in general position divide the plane into r_n regions, where $r_{n+1} = r_n + n + 1$. Conclude by induction that n such lines divide the plane into $\frac{1}{2}(n^2 + n + 2)$ regions. [*General position* means no two parallel and no three concurrent.]

44* Initially flask A contains 1 liter of wine and flask B contains 1 liter of water. A **transfer operation** consists of pouring $\frac{1}{2}$ liter from A into B, mixing, pouring $\frac{1}{2}$ liter from B into A, and mixing. How much wine is in A and how much in B after n transfer operations?

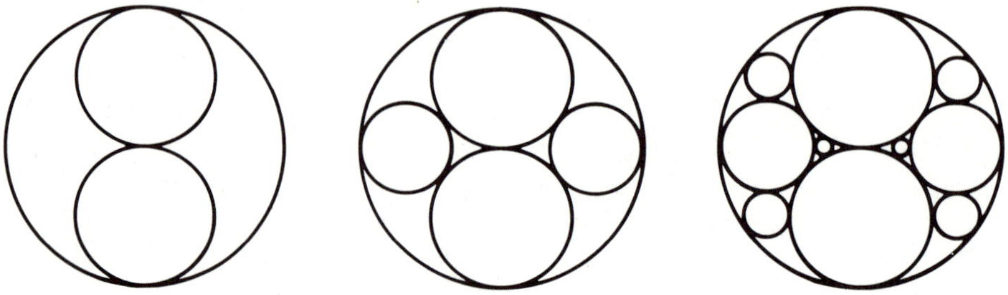

Case 0 Case 1 Case 2

Fig. 4

45 In Fig. 4, let s_r denote the number of spaces in case n into which circles will be put to reach case $n+1$. Thus $s_0 = 2, s_1 = 6, s_2 = 18$. Find and prove a formula for s_n.

46* (cont.) Find and prove a formula for the number c_n of circles in case n. Here $c_0 = 3, c_1 = 5, c_2 = 11$.

The following statement appears in a computer program subroutine for defining a function $f(n)$ for $n = 0, 1, 2, \cdots$ Describe $f(n)$ and prove what you claim

47 IF $n = 0$ THEN $f(n) = 1$ ELSE $f(n) = n * f(n-1)$
[The symbol $*$ means multiplication.]

48 IF $n = 0$ THEN $f(n) = 0$ ELSE $f(n) = n + f(n-1)$.

49 (division algorithm) Let a and b be integers with $a \geq 0$ and $b > 0$. Prove that there exist integers q and r such that $a = bq + r$ and $0 \leq r < b$. Use induction on a.

50 (division algorithm for polynomials; see p. 379) Let $f(x)$ and $g(x)$ be polynomials with real coefficients and $\deg g(x) > 0$. Prove that there exist polynomials $q(x)$ and $r(x)$ such that $f(x) = g(x)q(x) + r(x)$ and either $r(x) = 0$ or $\deg r(x) < \deg g(x)$. Use induction on $\deg f(x)$.

4 PERMUTATIONS AND COMBINATIONS

In this section, we discuss certain techniques of counting that are useful in many applications of mathematics.

Let us start with a basic principle. Suppose you must make a sequence of n independent decisions. For the first decision you have k_1 choices, for the second k_2 choices, etc. The total number of possible *sequences of decisions* is the product $k_1 \cdot k_2 \cdot k_3 \cdots k_n$.

For example, suppose there are three roads r_1, r_2, r_3 from A to B, two roads s_1, s_2 from B to C, and two roads t_1, t_2 from C to D. See Fig. 1. In how many ways can you choose a route from A to B to C to D? The answer is $3 \cdot 2 \cdot 2 = 12$. This is clear if you list the routes:

$r_1s_1t_1$	$r_1s_1t_2$	$r_1s_2t_1$	$r_1s_2t_2$
$r_2s_1t_1$	$r_2s_1t_2$	$r_2s_2t_1$	$r_2s_2t_2$
$r_3s_1t_1$	$r_3s_1t_2$	$r_3s_2t_1$	$r_3s_2t_2$

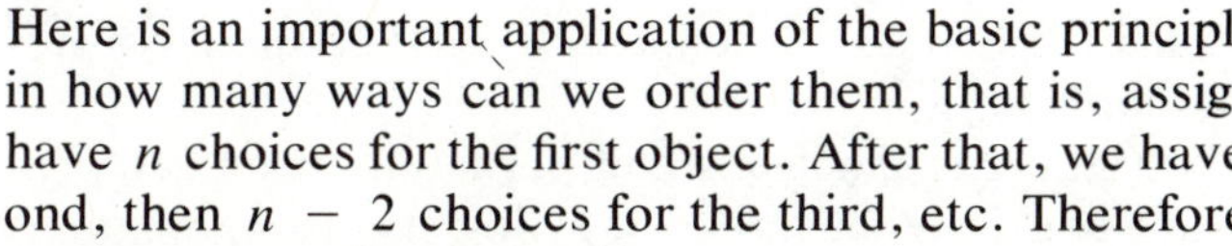

Fig. 1

Permutations

Here is an important application of the basic principle. Given n distinct objects, in how many ways can we order them, that is, assign a first, a second, etc? We have n choices for the first object. After that, we have $n-1$ choices for the second, then $n-2$ choices for the third, etc. Therefore, by the basic principle, the total number of orderings is

$$n(n-1)(n-2)\cdots 3\cdot 2\cdot 1 = n!$$

Each ordering of a set of objects is called a **permutation.**

> **Permutations** There are $n!$ permutations of n objects.

Examples

(1) There are $4! = 24$ ways to arrange 4 pictures on a wall from left to right.

(2) There are $9! = 362880$ batting orders for the 9 players on a baseball team.

(3) There are $52! \approx 8.066 \times 10^{67}$ arrangements of a deck of 52 cards.

Next, suppose we again have n objects and ask how many ways we can select k of them in order. (Last time we took all n of them.) Each such choice is called a **permutation of n objects k at a time.** We proceed as before, except that we stop after k selections. We have n choices for the first, $n - 1$ for the second, and $n - (k - 1) = n - k + 1$ choices for the k-th. Therefore, the total number of possible selections is

$$P_{n,k} = n(n-1)(n-2)\cdots(n-k+1).$$

(If the last factor is a little annoying, just remember that there are exactly k consecutive factors starting with n and going down.) The number $P_{n,k}$ can be expressed in terms of factorials:

> **Permutations of n Objects k at a Time**
>
> $$P_{n,k} = n(n-1)(n-2)\cdots(n-k+1) = \frac{n!}{(n-k)!}.$$

Examples (1) The number of choices in order for the first three batters from among 9 baseball players is

$$P_{9,3} = \frac{9!}{(9-3)!} = \frac{9!}{6!} = 9 \cdot 8 \cdot 7 = 504.$$

(2) The number of ways 5 cards can be dealt in order from a deck of 52 cards is

$$P_{52,5} = \frac{52!}{(52-5)!} = \frac{52!}{47!}$$
$$= 52 \cdot 51 \cdot 50 \cdot 49 \cdot 48 = 311{,}875{,}200.$$

Combinations

Sometimes we need the number of ways we can choose k objects from a set of n objects *without* regard to order. Each such choice is a **combination of n objects k at a time.** The number of such choices is denoted by

$$C_{n,k} \quad \text{or} \quad \binom{n}{k}$$

and is called a **binomial coefficient.** It is read "n choose k". (See the next section for the connection with binomials.)

There is a simple relation between combinations and permutations. Each *permutation* of n objects k at a time consists of a choice of k objects (a combination) followed by an ordering of these k objects. There are $\binom{n}{k}$ ways to choose k objects, then $k!$ ways to permute (order) them. Therefore by the general principle,

$$P_{n,k} = \binom{n}{k} \cdot k! \qquad \text{hence} \qquad \binom{n}{k} = \frac{P_{n,k}}{k!}.$$

From our earlier expression for $P_{n,k}$ we obtain the following useful formula:

Combinations of n Objects k at a Time

$$\binom{n}{k} = \frac{n!}{k!(n-k)!}.$$

Example 1 (a) In how many ways can an officer choose 5 out of his 18-man platoon for a patrol?
(b) In a race with 12 horses, how many outcomes are possible for the first three places (win, place, show)?

Solution (a) What counts is which five soldiers are chosen, not in what order they are chosen. So we want *combinations* of 18 soldiers 5 at a time:

$$\binom{18}{5} = \frac{18!}{5!13!} = \frac{18 \cdot 17 \cdot 16 \cdot 15 \cdot 14}{1 \cdot 2 \cdot 3 \cdot 4 \cdot 5} = 8568.$$

(b) Here the order counts, so we want the *permutations* of 12 horses 3 at a time:

$$P_{12,3} = \frac{12!}{9!} = 12 \cdot 11 \cdot 10 = 1320.$$

Answer (a) 8568 (b) 1320

Example 2 An urn contains 10 black balls and 10 white balls. How many possible choices are there of 5 balls containing at least one black ball?

Solution There are

$$\binom{20}{5} = \frac{20!}{5!\ 15!} = 15504$$

choices of 5 balls from the 20 balls. Of these, the ones that contain no black balls at all are the choices of 5 from the 10 white balls; there are

$$\binom{10}{5} = \frac{10!}{5!\ 5!} = 252$$

such choices. Each of the remaining choices must contain at least one black ball. There are

$$15504 - 252 = 15252$$

of these.

Answer $\binom{20}{5} - \binom{10}{5} = 15{,}252$

The binomial coefficients have a number of basic properties:

$$\binom{n}{0} = \binom{n}{n} = 1 \qquad \binom{n}{1} = \binom{n}{n-1} = n \qquad \binom{n}{k} = \binom{n}{n-k}$$

$$\binom{n+1}{k+1} = \binom{n}{k+1} + \binom{n}{k}, \qquad 0 \le k \le n-1.$$

These can be proved directly from the formula expressing $\binom{n}{k}$ in terms of factorials. They can also be proved by direct counting. For example, if you remove k objects from a box of n objects, then $n - k$ objects remain behind, so to each choice of k objects corresponds a complementary choice of $n - k$ objects. Hence

$$\binom{n}{k} = \binom{n}{n-k}.$$

Again, choosing $k + 1$ objects from a set $\{t_1, t_2, \cdots, t_n, t_{n+1}\}$ amounts either to choosing $k + 1$ objects from $\{t_1, t_2, \cdots, t_n\}$ or choosing the element t_{n+1} together with k objects from $\{t_1, t_2, \cdots, t_n\}$. Therefore

$$\binom{n+1}{k+1} = \binom{n}{k+1} + \binom{n}{k}.$$

Example $\binom{10}{4} = \binom{9}{4} + \binom{9}{3}$ $\quad 210 = 126 + 84$.

Exercises

1 There are 8 dogs in a race. How many orders of finishing are possible?

2 There are 8 dogs in a race. How many possibilities are there for win, place, show?

3 There are 100 joggers in a race. How many possibilities are there for the three first, disregarding order?

4 There are 100 joggers in a race. How many possibilities are there for the first three places, counting order?

5 How many 10-digit integers are there composed of all 10 digits $0, 1, 2, \cdots, 9$ if 0 is not allowed as the first digit?

6 How many 8-digit integers are there composed of 8 distinct digits, but 0 not allowed as the first digit?

7 How many 10-digit numbers are there composed of 6 ones and 4 fives?

8 How many 5-digit numbers are there composed only of 6's and 7's?

9 How many 11-digit binary integers (numbers composed only of 1's and 0's) are there if the first digit must be a one?

10 How many 8-digit ternary integers (digits 0, 1, 2) are there if the first digit is not zero?

11 From a committee of 11 people, how many subcommittees of 3 people are possible?

12 From a committee of 15 people, how many subcommittees of 4 people are possible?

13 From a committee of 12 people, how many ways can one choose a subcommittee of 2 people and another subcommittee of 3 people, with possible overlap?

14 From a committee of 15 people, how many ways can one appoint two subcommittees of 3 people, having exactly one common member?

15 How many ways can the 100-member U.S. Senate appoint a committee of 10 members?

16 (cont.) How many ways can the Senate appoint two 5-member committees with no overlap? Greater than, equal to, or less than the answer to Ex. 15?

17 The cabinet of a coalition government consists of 10 Conservatives and 13 Labors. In how many ways can the prime minister appoint a committee of 4 Conservatives and 4 Labors?

18 The cabinet of a coalition government consists of 7 Radicals, 6 Loyalists, and 5 Progressives. In how many ways can the prime minister appoint a committee of 3 Radicals, 2 Loyalists, and 2 Progressives?

19 In how many ways can a class of 8 boys and 8 girls be seated in a room with 16 chairs if the boys must take the odd-numbered seats?

20 (cont.) Suppose there are 20 seats; then how many ways are there?

21 From a deck of 52 cards, how many 5-card poker hands are possible?

22 (cont.) How many of these are flushes (all cards of the same suit)?

23 (cont.) How many are full houses (three of a kind plus a pair of another kind)?

24 (cont.) How many have four of a kind?

25 How many bridge hands (13 cards) are possible from a 52-card deck?

26* (cont.) How many of these contain two or more aces?

27 A woman wants to buy 4 dresses from a rack of 25. How many possibilities does she have?

28 A man wants to buy 2 sport jackets from a rack of 15 and 6 shirts from a selection of 12. How many choices does he have?

29* An urn contains 8 black balls and 6 white balls. How many possible choices are there of 4 balls containing at least 2 black balls?

30* An urn contains 10 balls each of colors white, black, and red. In how many ways is it possible to choose 5 balls so that at least one of each color is chosen?

31 Prove for $2 \leq k \leq n - 2$ that

$$\binom{n}{k} = \binom{n-2}{k-2} + 2\binom{n-2}{k-1} + \binom{n-2}{k}.$$

32 (cont.) Find a similar formula for $\binom{n}{k}$ in terms of the binomial coefficients $\binom{n-3}{j}$.

33 A committee of k is to be chosen from an assembly of n people, and then a subcommittee of r is to be chosen from that, where $1 \leq r < k < n$. By doing the selection in two different ways, prove that

$$\binom{n}{k}\binom{k}{r} = \binom{n}{r}\binom{n-r}{k-r}.$$

34* Prove that $\binom{2n}{n}$ is the largest of the binomial coefficients $\binom{2n}{j}$, where $0 \leq j \leq 2n$.

35 The grid lines on a piece of paper are one unit apart. Consider all paths that run along the grid lines from $(0, 0)$ to (n, n) moving either upwards or to the right. Prove that the number of such paths is $\binom{2n}{n}$.

36* (cont.) Each path in Ex. 35 passes through exactly one of the points $(k, n - k)$ where $0 \leq k \leq n$. Count the number of such paths for each k and derive the formula

$$\binom{2n}{n} = \binom{n}{0}^2 + \binom{n}{1}^2 + \cdots + \binom{n}{n}^2.$$

5 THE BINOMIAL THEOREM

The binomial theorem is a formula for the expanded product

$$(x + y)^n.$$

Let us compute a few cases:

$$(x+y)^1 = x + y$$
$$(x+y)^2 = x^2 + 2xy + y^2$$
$$(x+y)^3 = x^3 + 3x^2y + 3xy^2 + y^3$$
$$(x+y)^4 = x^4 + 4x^3y + 6x^2y^2 + 4xy^3 + y^4.$$

Obviously, the general formula should express $(x+y)^n$ as a sum of monomials $a_{n,k}x^{n-k}y^k$. The problem is, what are the correct coefficients? In multiplying out

$$(x+y)^n = (x+y)(x+y)(x+y)\cdots(x+y) \qquad (n \text{ factors})$$

we obtain terms that are products of n factors of x's and y's. We obtain $x^{n-k}y^k$ each way we can choose y from k of the factors $x + y$ and x from the remaining $n - k$ factors. The number of such ways is the number of combinations of n things taken k at a time. Therefore, the coefficient of $x^{n-k}y^k$ is $\binom{n}{k}$.

Binomial Theorem For each positive integer n,

$$(x+y)^n = x^n + \binom{n}{1}x^{n-1}y + \binom{n}{2}x^{n-2}y^2 + \binom{n}{3}x^{n-3}y^3$$
$$+ \cdots + \binom{n}{k}x^{n-k}y^k + \cdots + \binom{n}{n-1}xy^{n-1} + y^n.$$
$$= \sum_{k=0}^{n}\binom{n}{k}x^{n-k}y^k.$$

EXAMPLE 1 Expand (a) $(x+1)^5$ (b) $(2a-b)^6$.

Solution (a) Use the Binomial Theorem with $y = 1$ and $n = 5$. The coefficients needed are

$$\binom{5}{0} = 1 \quad \binom{5}{1} = 5 \quad \binom{5}{2} = 10 \quad \binom{5}{3} = 10 \quad \binom{5}{4} = 5 \quad \binom{5}{5} = 1.$$

By the formula,

$$(x+1)^5 = x^5 + 5x^4 + 10x^3 + 10x^2 + 5x + 1.$$

(b) Use the formula with $x = 2a$, $y = -b$, and $n = 6$. The coefficients are

$$\binom{6}{0} = \binom{6}{6} = 1 \quad \binom{6}{1} = \binom{6}{5} = 6 \quad \binom{6}{2} = \binom{6}{4} = 15 \quad \binom{6}{3} = 20.$$

The formula yields

$$(2a-b)^6 = (2a)^6 + 6(2a)^5(-b) + 15(2a)^4(-b)^2 + 20(2a)^3(-b)^3$$
$$+ 15(2a)^2(-b)^4 + 6(2a)(-b)^5 + (-b)^6$$
$$= 64a^6 - 192a^5b + 240a^4b^2 - 160a^3b^3$$
$$+ 60a^2b^4 - 12ab^5 + b^6.$$

EXAMPLE 2 Find the coefficient of x^9 in $\left(x^2 - \dfrac{2}{x}\right)^9$.

Solution The powers of x that occur in the expansion are

$$(x^2)^9 = x^{18} \qquad \frac{(x^2)^8}{x} = x^{15} \qquad \frac{(x^2)^7}{x^2} = x^{12} \qquad \frac{(x^2)^6}{x^3} = x^9 \qquad \frac{(x^2)^5}{x^4} = x^6$$

$$\frac{(x^2)^4}{x^5} = x^3 \qquad \frac{(x^2)^3}{x^6} = 1 \qquad \frac{(x^2)^2}{x^7} = x^{-3} \qquad \frac{(x^2)^1}{x^8} = x^{-6} \qquad \frac{(x^2)^0}{x^9} = x^{-9}.$$

The one we want, x^9, occurs in the term

$$\binom{9}{3}(x^2)^6\left(-\frac{2}{x}\right)^3 = \binom{9}{3}(-2)^3x^9.$$

The coefficient of x^9 is $\displaystyle\binom{9}{3}(-2)^3 = (84)(-8) = -672$.

Answer -672

EXAMPLE 3 Use the Binomial Theorem to prove that $(1.01)^{900} > 10$.

Solution

$$(1.01)^{900} = (1 + 0.01)^{900} = 1 + \binom{900}{1}(0.01) + \cdots$$
$$> 1 + (900)(0.01) = 1 + 9 = 10.$$

Pascal's Triangle

The binomial coefficients have many interesting properties, some which appear in an array called **Pascal's Triangle.** In the n-th row of the array (starting with $n = 0$) we write the coefficients

$$\binom{n}{k} \qquad k = 0, 1, 2, \cdots, n:$$

```
                  1
               1     1
            1     2     1
         1     3     3     1
      1     4     6     4     1
   1     5    10    10     5     1
1     6    15    20    15     6     1
.     .     .     .     .     .     .     .
```

Note that each interior entry is the sum of the two numbers above it. That is because

$$\binom{n+1}{k+1} = \binom{n}{k+1} + \binom{n}{k},$$

as proved in the last section. Thus the elements of the 7-th row are

$$1 \qquad 1+6=7 \qquad 6+15=21 \qquad 15+20=35$$
$$20+15=35 \qquad 15+6=21 \qquad 6+1=7 \qquad 1.$$

Note also that the sum of the numbers in the n-th row is 2^n. That is simply because

$$2^n = (1+1)^n = \binom{n}{0} + \binom{n}{1} + \cdots + \binom{n}{n}.$$

For instance

$$32 = 2^5 = 1+5+10+10+5+1.$$

Finally note that the alternating sum of each row is 0, that is,

$$\binom{n}{0} - \binom{n}{1} + \binom{n}{2} - + \cdots \pm \binom{n}{n} = 0.$$

True simply because $(1-1)^n = 0$. For instance $(n=6)$,

$$1-6+15-20+15-6+1=0.$$

Exercises

Expand

1 $(x+2)^3$ **2** $(x-y)^3$ **3** $(1+2x)^4$ **4** $(x-2y)^4$

5 $(x-y)^5$ **6** $(2+x)^5$ **7** $\left(x-\dfrac{1}{x}\right)^5$ **8** $\left(x+\dfrac{y}{x}\right)^5$

9 $\left(x+\dfrac{1}{x}\right)^6$ **10** $(2x-\frac{1}{2})^6$.

Find

11 $3^5 - \binom{5}{1}3^4\cdot 2 + \binom{5}{2}3^3\cdot 2^2 - \binom{5}{3}3^2\cdot 2^3 + \binom{5}{4}3\cdot 2^4 - 2^5$

12 $7^7 - \binom{7}{1}7^6\cdot 5 + \binom{7}{2}7^5\cdot 5^2 - \binom{7}{3}7^4\cdot 5^3 + \binom{7}{4}7^3\cdot 5^4 - \binom{7}{5}7^2\cdot 5^5 + \binom{7}{6}7\cdot 5^6 - 5^7.$

Find the coefficient of

13 x^5y^3 in $(x+2y)^8$ **14** x^8y^{12} in $(x^2-y^3)^8$

15 x^{-3} in $(2x+x^{-2})^6$ **16** $x^{10}y^4$ in $(x^2-2y)^9$

17 x^4y^{18} in $(3x-y^3)^{10}$ **18** 1 in $(x-\frac{1}{2}x^{-1})^{10}$.

Compute exactly by the binomial theorem

19 $(101)^5$ **20** $(1001)^4$ **21** $(99)^5$ **22** $(999)^4$.

Compute by the binomial theorem to 6-place accuracy

23 $(0.99)^{10}$ **24** $(1.01)^{12}$ **25** $(1.001)^{20}$ **26** $(0.999)^{25}$.

Prove by the binomial theorem

27 $(1.002)^{10} > 1.02$ **28** (cont.) $(1.002)^{10} > 1.02018$

29 $(1.0001)^{100} > 1.0100495$ **30** (cont.) $(1.0001)^{100} > 1.0100496617$

REVIEW EXERCISES

1 Let $b_n = n(n + 1)(n + 2)$ and $a_n = b_{n+2} - 2b_{n+1} + b_n$. Show that $\{a_n\}$ is an arithmetic progression.

2 Let $\{a_n\}$ be an arithmetic progression with common difference d and $b_n = 2^{a_n}$. Show that $\{b_n\}$ is a geometric progression and find its common ratio.

3 Find $\displaystyle\sum_{n=1}^{100} \log\left(\frac{n+1}{n}\right)$.

4 Evaluate $\displaystyle\sum_{k=-2}^{4} (k - 1)^2$.

5 Solve for n the equation $(n + 3)! = 720\, n!$.

6 Prove by mathematical induction

$$1 \cdot 1! + 2 \cdot 2! + 3 \cdot 3! + \cdots + n \cdot n! = (n + 1)! - 1.$$

7 Suppose $a_1 = \sqrt{3}$ and $a_{n+1} = \sqrt{3 + a_n}$ for $n = 1, 2, 3, \cdots$. Prove that $a_1 < a_2 < a_3 < \cdots$.

8 Suppose $a_1 = 1$ and $a_{n+1} = 1/(1 + a_n)$ for $n = 1, 2, 3, \cdots$. Prove that $a_{2n-1} > a_{2n}$ and $a_{2n} < a_{2n+1}$ for $n = 1, 2, 3, \cdots$.

9 Prove by induction that $x + 1$ divides $x^{2n-1} + 1$ for $n = 1, 2, 3, \cdots$.

10 How many 4 digit integers are there in which each digit is 1, 2, or 3, and the sum of the digits is even?

11 How many 5 digit integers are there with all distinct digits? Of course, 0 cannot be the first digit.

12 How many ways are there of appointing from 20 people a committee of 5 and another committee of 4, with no common members?

13 Expand $\left(\dfrac{x}{y} - \dfrac{y}{x}\right)^5$.

14 Find the constant term in $\left(2x + \dfrac{3}{x}\right)^{12}$.

11 Vectors in the Plane

1 VECTOR ALGEBRA

In this section we introduce and discuss addition of vectors and multiplication of vectors by numbers. In the next section, we apply this material to lines and segments. Later we introduce the inner product of vectors, which has many geometrical and physical applications.

The idea of vectors grew out of the study of forces in physics. A force applied at a point was represented by an arrow; it pointed in the direction of the force and its length represented the magnitude of the force. The concept of a quantity having direction and magnitude turned out to be a powerful tool not only in physics and engineering, but also in non-physical fields such as statistics, numerical analysis, and mathematical economics.

We choose a point **0** in the plane and call it the **origin.** A **vector** is a directed line segment from **0** to a point of the plane (Fig. 1a). The vector that goes from **0** to **0** (a degenerate line segment) we call the vector **0**.

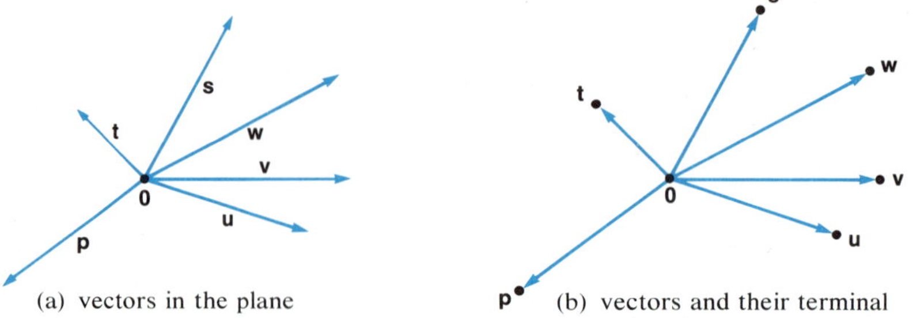

(a) vectors in the plane

(b) vectors and their terminal points identified

Fig. 1 Vectors

The vector from **0** to a point **z** is completely determined by its terminal point. For this reason, we shall often identify a vector and its terminal point as one and the same thing (Fig. 1b).

We have defined vectors without reference to coordinate systems. This is as it should be, for in real life a force is a force and a velocity is a velocity, regardless of how we choose to set up coordinate axes. Nevertheless, to compute with vectors we shall need coordinates. Given a cartesian coordinate system in the plane, we assign to each vector $\mathbf{v}$ the coordinates of its terminal point, and write

$$\mathbf{v} = (x, y).$$

The numbers x and y are called the **x-component** and the **y-component** of $\mathbf{v}$.

Addition

We now define the sum $\mathbf{v} + \mathbf{w}$ of two vectors $\mathbf{v}$ and $\mathbf{w}$. Draw a directed line segment parallel to $\mathbf{w}$, starting from the terminal point of $\mathbf{v}$. See Fig. 2a. Then the vector from $\mathbf{0}$ to the end point of this parallel segment is the vector $\mathbf{v} + \mathbf{w}$. See Fig. 2b.

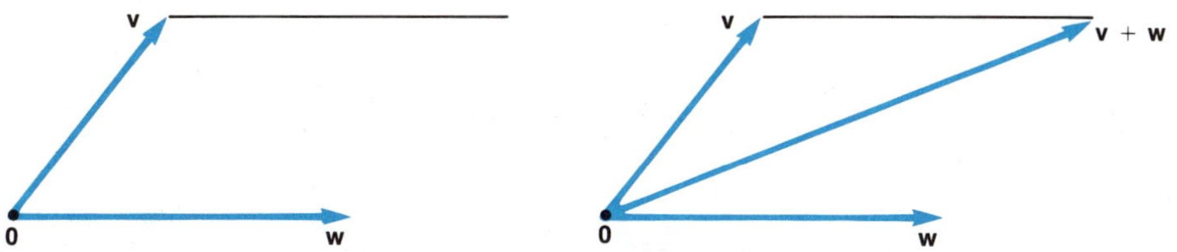

(a) Draw a segment from $\mathbf{v}$ parallel to $\mathbf{w}$. (b) Its endpoint determines $\mathbf{v} + \mathbf{w}$.

Fig. 2 Construction of $\mathbf{v} + \mathbf{w}$

By closing the parallelogram, we see that the roles of $\mathbf{v}$ and $\mathbf{w}$ can be interchanged in the definition (Fig. 3). The result is

$$\mathbf{v} + \mathbf{w} = \mathbf{w} + \mathbf{v}.$$

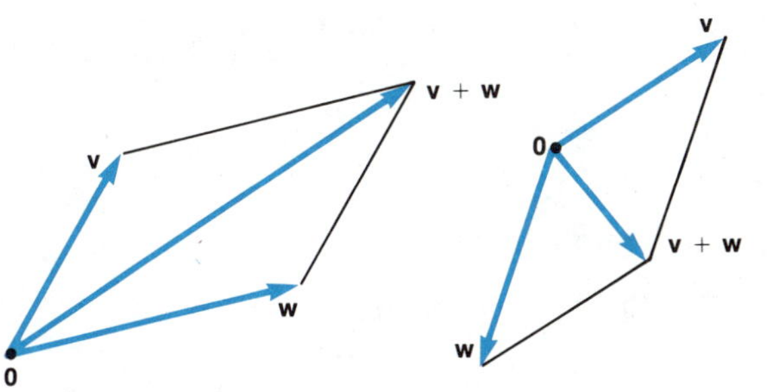

Fig. 3 $\mathbf{v} + \mathbf{w} = \mathbf{w} + \mathbf{v}$. Each figure can be interpreted as a "parallelogram of forces."

The simplest interpretation of addition of vectors is in terms of force vectors. If two forces $\mathbf{v}$ and $\mathbf{w}$ are applied at the same point, the resultant force is $\mathbf{v} + \mathbf{w}$; that is the law of "parallelogram of forces".

We need a practical way to compute $\mathbf{v} + \mathbf{w}$ when the coordinates of $\mathbf{v}$ and $\mathbf{w}$ are given. Suppose

$$\mathbf{v} = (a, b) \quad \text{and} \quad \mathbf{w} = (c, d).$$

The answer is

$$\mathbf{v} + \mathbf{w} = (a + c, b + d).$$

In words, vectors are added *componentwise;* the x-component of $\mathbf{v} + \mathbf{w}$ is the sum of the x-components of $\mathbf{v}$ and $\mathbf{w}$—ditto the y-component.

To see why, we go back to the geometric definition of $\mathbf{v} + \mathbf{w}$ and project everything onto the x-axis (Fig. 4, next page). The segment from $\mathbf{v}$ to $\mathbf{v} + \mathbf{w}$ is parallel to $\mathbf{w}$, so its projection on the x-axis has the same length and direction as the projection of $\mathbf{w}$, that is, c. Hence the x-component of $\mathbf{v} + \mathbf{w}$ is obtained by starting at a and adding c to it; that is, the x-component of $\mathbf{v} + \mathbf{w}$ is $a + c$. Similarly the y-component is $b + d$.

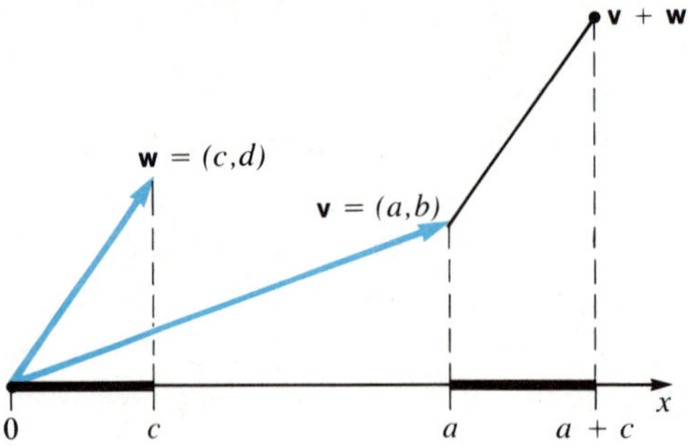

Fig. 4 The x-component of $\mathbf{v} + \mathbf{w}$ is $a + c$.

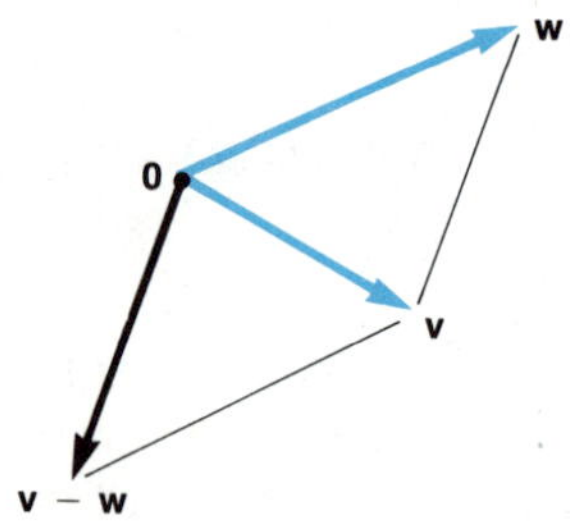

Fig. 5 Construction of $\mathbf{v} - \mathbf{w}$

> **Sum of Vectors** If $\mathbf{v} = (a, b)$ and $\mathbf{w} = (c, d)$, then
>
> $$\mathbf{v} + \mathbf{w} = (a + c, b + d).$$

The **difference** $\mathbf{v} - \mathbf{w}$ of two vectors $\mathbf{v}$ and $\mathbf{w}$ is that vector that satisfies

$$(\mathbf{v} - \mathbf{w}) + \mathbf{w} = \mathbf{v}.$$

Thus $\mathbf{v} - \mathbf{w}$ is the other side of a parallelogram, one side of which is $\mathbf{w}$ and the diagonal of which is $\mathbf{v}$. See Fig. 5. In components:

> If $\mathbf{v} = (a, b)$ and $\mathbf{w} = (c, d)$, then $\mathbf{v} - \mathbf{w} = (a - c, b - d)$.

Examples

$$(2, 7) + (1, 3) = (3, 10) \qquad (2, 7) - (1, 3) = (1, 4)$$
$$(2, 7) + (1, -3) = (3, 4) \qquad (2, 7) - (1, -3) = (1, 10)$$
$$(2, 7) + (-5, 0) = (-3, 7) \qquad (2, 7) - (-5, 0) = (7, 7).$$

Multiplication by a Scalar

In vector algebra, the word **scalar** means real number. We combine a scalar a and a vector $\mathbf{v}$ to form a new vector $a\mathbf{v}$ called the **scalar multiple** of a and $\mathbf{v}$.

> **Multiplication by a Scalar** If a is a scalar and $\mathbf{v}$ is a vector, then the scalar multiple $a\mathbf{v}$ is the vector defined by
>
> $$a\mathbf{v} = \begin{cases} \mathbf{0} & \text{if } a = 0. \\ \text{a vector in the same direction as } \mathbf{v} \text{ with length } a \text{ times the length of } \mathbf{v} & \text{if } a > 0. \\ \text{a vector in the opposite direction from } \mathbf{v} \text{ with length } |a| \text{ times the length of } \mathbf{v} & \text{if } a < 0. \end{cases}$$
>
> In components, if $\mathbf{v} = (x, y)$, then $a\mathbf{v} = (ax, ay)$.

The formula $a(x, y) = (ax, ay)$ can be verified by an argument using similar triangles (Fig. 6). We omit the details since the figure is pretty convincing.

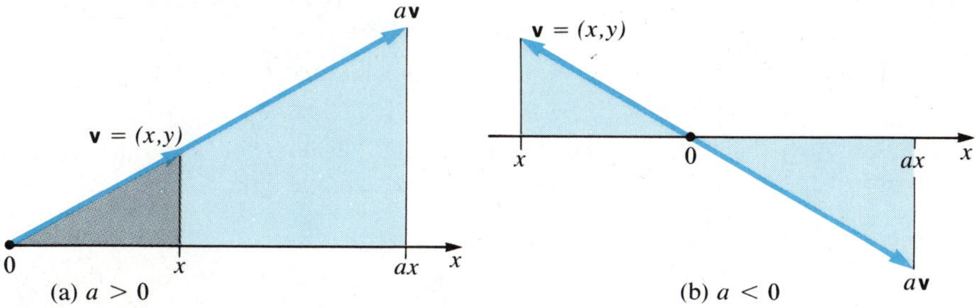

Fig. 6 Proof by similar triangles that $a(x, y) = (ax, ay)$.

One can interpret scalar multiplication physically in several ways. For example, a velocity vector $\mathbf{v}$ has direction and magnitude (speed). If the direction remains the same, but the speed triples, then the new velocity vector is $3\mathbf{v}$.

Examples

$$1(x, y) = (x, y) \qquad -3(5, -7) = (-15, 21)$$

$$3(0, 0) = (0, 0) = \mathbf{0} \qquad 0(5, -1) = (0, 0) = \mathbf{0}$$

$$-2(1, 0) = (-2, 0) \qquad 2(3, -2) = (6, -4).$$

We denote by $-\mathbf{v}$ the vector $(-1)\mathbf{v}$, opposite in direction but equal in magnitude to $\mathbf{v}$. Thus $-(3, -4) = (-3, 4)$. It is easily seen that

$$\mathbf{v} - \mathbf{w} = \mathbf{v} + (-\mathbf{w}).$$

For instance $(5, -7) - (3, -4) = (5, -7) + (-3, 4) = (2, -3)$.

The basic rules of vector algebra follow easily from the coordinate formulas for scalar multiplication and addition:

Rules of Vector Algebra

$$1 \cdot \mathbf{v} = \mathbf{v} \qquad \mathbf{v} + (-\mathbf{v}) = \mathbf{0}$$

$$\mathbf{v} + \mathbf{w} = \mathbf{w} + \mathbf{v} \qquad \text{(commutative law)}$$

$$\mathbf{u} + (\mathbf{v} + \mathbf{w}) = (\mathbf{u} + \mathbf{v}) + \mathbf{w} \qquad \text{(associative law)}$$

$$\begin{cases} a(\mathbf{v} + \mathbf{w}) = a\mathbf{v} + a\mathbf{w} \\ (a + b)\mathbf{v} = a\mathbf{v} + b\mathbf{v} \end{cases} \qquad \text{(distributive laws)}$$

$$a(b\mathbf{v}) = (ab)\mathbf{v}.$$

Exercises

Compute

1 $3(0, 1)$ **2** $-5(2, 0)$ **3** $2(1, -1)$
4 $7(3, -2)$ **5** $4(\frac{1}{2}, 3)$ **6** $-\frac{1}{3}(4, -2)$
7 $(3, -1) + (0, 1)$ **8** $(4, 2) + (3, 1)$ **9** $(-1, 1) + (1, -1)$
10 $(2, -2) + (3, 3)$ **11** $(4, -1) + (-3, 2)$ **12** $(7, 11) + (11, 7)$
13 $3(1, 1) + (7, 1)$ **14** $(2, -2) + 8(1, 2)$ **15** $-2(1, 3) + 4(5, 4)$
16 $3(2, 0) + 2(4, 5)$.

Plot the vectors and compute their sum graphically by the "parallelogram of forces." Check your answer:

17 $(4, 1) + (3, 4)$ **18** $(-3, 2) + (4, 4)$ **19** $(1, 5) + (-5, -1)$
20 $(2, -3) + (-1, -3)$.

21 Find a vector parallel to and having the same length as the directed segment from $(-2, 4)$ to $(-1, -6)$.

22 Find a point $\mathbf{x}$ such that the directed segment from $(2, 3)$ to $\mathbf{x}$ is parallel to the directed segment from $(1, 1)$ to $(-4, 0)$ and of the same length.

Find a and b; check your answer graphically:

23 $(-4, 5) = a(2, 0) + b(0, 1)$ **24** $(4, 2) = a(1, 0) + b(1, 1)$

2 LINES AND SEGMENTS

In this section we obtain a vector equation for a line and derive a number of useful facts from it. All we need are the basic ideas of vector addition and multiplication by a scalar.

Given a point $\mathbf{x}_0$ in the plane and a non-zero vector $\mathbf{v}$, we wish to describe the line that passes through $\mathbf{x}_0$ and is parallel to $\mathbf{v}$.

First we observe that the set of all real scalar multiples $t\mathbf{v}$ of $\mathbf{v}$ is the line through $\mathbf{0}$ and $\mathbf{v}$. See Fig. 1a. By the parallelogram rule for vector addition, the set of points $\mathbf{x} = \mathbf{x}_0 + t\mathbf{v}$ we get by adding $\mathbf{x}_0$ to these points $t\mathbf{v}$ is the line through $\mathbf{x}_0$ parallel to $\mathbf{v}$. See Fig. 1b.

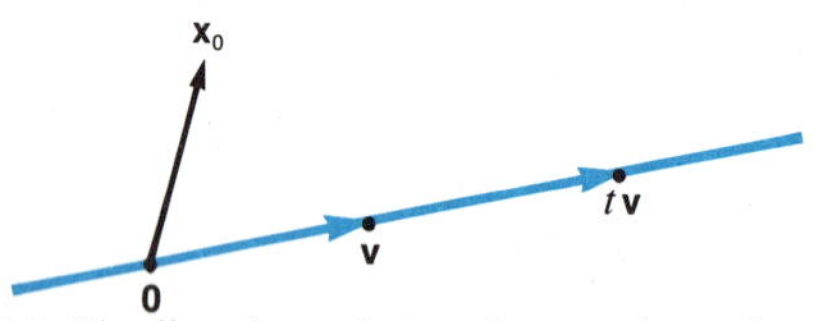

(a) The line through $\mathbf{0}$ and $\mathbf{v}$ consists of all $t\mathbf{v}$, where t is any real number.

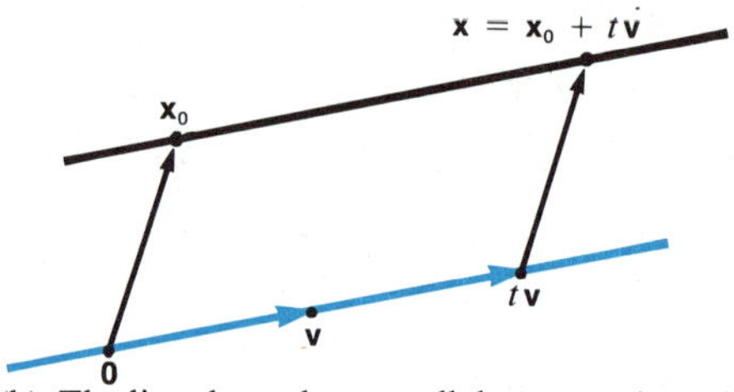

(b) The line through $\mathbf{x}_0$ parallel to $\mathbf{v}$ consists of all $\mathbf{x} = \mathbf{x}_0 + t\mathbf{v}$, where t is any real number.

Fig. 1 Parametric vector equation

> Given a point $\mathbf{x}_0$ in the plane, and a non-zero vector $\mathbf{v}$, the line through $\mathbf{x}_0$ parallel to $\mathbf{v}$ consists of all points
>
> $$\mathbf{x} = \mathbf{x}_0 + t\mathbf{v},$$
>
> where t ranges over all real numbers.

The equation $\mathbf{x} = \mathbf{x}_0 + t\mathbf{v}$ is called a **parametric vector equation** for the line, and the variable t is called the **parameter.** To each real number t corresponds the point $\mathbf{x} = \mathbf{x}_0 + t\mathbf{v}$ of the line, and to each point $\mathbf{x}$ of the line corresponds the unique value t of the parameter such that $\mathbf{x} = \mathbf{x}_0 + t\mathbf{v}$.

We can interpret the parametric vector equation $\mathbf{x} = \mathbf{x}_0 + t\mathbf{v}$ in terms of motion. Suppose at time $t = 0$ a particle starts at $\mathbf{x}_0$ and thereafter moves with constant velocity $\mathbf{v}$. Then it moves on the line and its position at time t is $\mathbf{x} = \mathbf{x}_0 + t\mathbf{v}$.

Scalar Equations

Let us write

$$\mathbf{x}_0 = (x_0, y_0) \qquad \mathbf{v} = (a, b) \qquad \mathbf{x} = (x, y)$$

Then the equation $\mathbf{x} = \mathbf{x}_0 + t\mathbf{v}$ is equivalent to

$$(x, y) = (x_0, y_0) + t(a, b) = (x_0 + ta, y_0 + tb).$$

By equating components, we conclude that the parametric vector equation is equivalent to the pair of **parametric scalar equations**

$$\begin{cases} x = x_0 + at \\ y = y_0 + bt. \end{cases}$$

To obtain a single **non-parametric scalar equation** for the line, we eliminate the parameter t. We multiply the first equation by b, the second by a, and subtract:

$$bx - ay = bx_0 - ay_0.$$

An alternative form of this equation is

$$a(y - y_0) = b(x - x_0).$$

EXAMPLE 1 Let $\mathbf{x}_0 = (2, 0)$ and $\mathbf{v} = (-3, 2)$. Find

(a) a parametric vector equation

(b) parametric scalar equations

(c) a non-parametric scalar equation

for the line through $\mathbf{x}_0$ parallel to $\mathbf{v}$.

Solution (a) $\mathbf{x} = \mathbf{x}_0 + t\mathbf{v} = (2, 0) + t(-3, 2) = (2 - 3t, 2t)$.

(b) Equate components in $\mathbf{x} = (x, y) = (2 - 3t, 2t)$:

$$x = 2 - 3t \qquad y = 2t$$

(c) Eliminate t. Multiply the first equation by 2, the second equation by 3 and add:

$$2x + 3y = 2(2 - 3t) + 3(2t) = 4.$$

Answer (a) $\mathbf{x} = (2 - 3t, 2t)$ (b) $\begin{cases} x = 2 - 3t \\ y = 2t \end{cases}$ (c) $2x + 3y = 4$

EXAMPLE 2 One line passes through $\mathbf{x}_0 = (2, 0)$ and is paralled to $\mathbf{u} = (3, -1)$. A second line passes through $\mathbf{y}_0 = (1, 1)$ and is parallel to $\mathbf{v} = (4, -2)$. Find their point of intersection.

Solution Let $\mathbf{x} = (x, y)$ be the point of intersection. Then $\mathbf{x}$ is on the first line so

$$\mathbf{x} = \mathbf{x}_0 + s\mathbf{u} \qquad \text{for some } s.$$

Similarly

$$\mathbf{x} = \mathbf{y}_0 + t\mathbf{v} \qquad \text{for some } t.$$

Equate these two expressions for $\mathbf{x}$ and substitute the given values of $\mathbf{x}_0$, $\mathbf{u}$, $\mathbf{y}_0$, and $\mathbf{v}$:

$$\mathbf{x}_0 + s\mathbf{u} = \mathbf{y}_0 + t\mathbf{v}$$

$$(2, 0) + s(3, -1) = (1, 1) + t(4, -2).$$

Equate components and rearrange:

$$\begin{cases} 2 + 3s = 1 + 4t \\ -s = 1 - 2t \end{cases} \qquad \begin{cases} 3s - 4t = -1 \\ -s + 2t = 1. \end{cases}$$

The resulting linear system can be solved, as usual, by elimination. Multiply the second equation by 2 and add to the first:

$$s = 1 \qquad 2t = 1 + s = 2 \qquad t = 1.$$

Then

$$\mathbf{x} = \mathbf{x}_0 + s\mathbf{u} = (2, 0) + (3, -1) = (5, -1).$$

Check $\mathbf{y}_0 + t\mathbf{v} = (1, 1) + (4, -2) = (5, -1).$

Answer $(5, -1)$

Division of a Segment

Suppose $\mathbf{v}$ and $\mathbf{w}$ are two points in the plane, and we want the point that is $\frac{3}{5}$ of the way from $\mathbf{v}$ to $\mathbf{w}$. By vectors, this is easy; we add $\frac{3}{5}$ of the vector $\mathbf{w} - \mathbf{v}$ to $\mathbf{v}$. The desired point is $\mathbf{v} + \frac{3}{5}(\mathbf{w} - \mathbf{v}) = \frac{2}{5}\mathbf{v} + \frac{3}{5}\mathbf{w}$. See Fig. 2a.

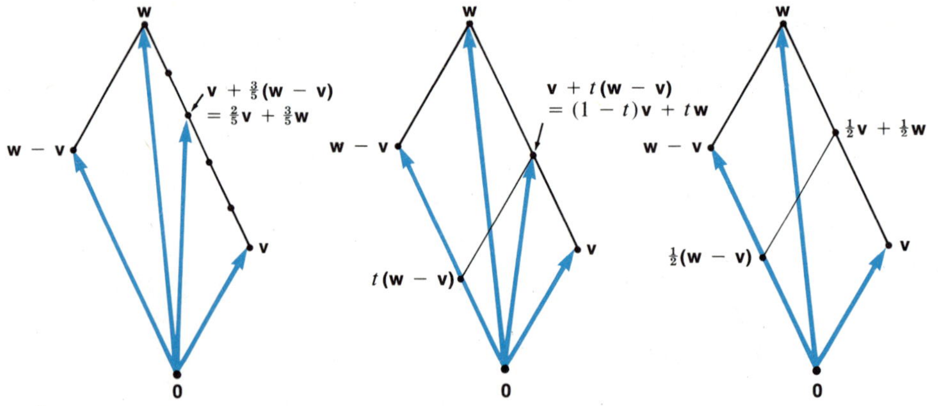

(a) $\frac{3}{5}$ of the way from $\mathbf{v}$ to $\mathbf{w}$ (b) t of the way from $\mathbf{v}$ to $\mathbf{w}$ (c) $t = \frac{1}{2}$: midpoint $= \frac{1}{2}\mathbf{v} + \frac{1}{2}\mathbf{w}$

Fig. 2 Division of a segment

The same reasoning applies in finding the point $\mathbf{x}$ that is a fractional part t of the way from $\mathbf{v}$ to $\mathbf{w}$:

$$\mathbf{x} = \mathbf{v} + t(\mathbf{w} - \mathbf{v}) = (1 - t)\mathbf{v} + t\mathbf{w}.$$

This formula is a parametric representation of the line segment from $\mathbf{v}$ to $\mathbf{w}$. As t runs from 0 to 1, we see that $\mathbf{x}$ traces the segment from $\mathbf{v}$ to $\mathbf{w}$. See Fig. 2b.

An important special case is the midpoint of the segment. The point $\mathbf{x}$ reaches the midpoint when $t = \frac{1}{2}$. See Fig. 2c.

Midpoint Formula The midpoint of the line segment between $\mathbf{v}$ and $\mathbf{w}$ is

$$\tfrac{1}{2}\mathbf{v} + \tfrac{1}{2}\mathbf{w}.$$

Example 3 Find

(a) the midpoint of the segment joining $\mathbf{v} = (-3, -4)$ to $\mathbf{w} = (-1, 6)$

(b) the point $\frac{2}{5}$ of the way from $\mathbf{v} = (-4, 7)$ to $\mathbf{w} = (2, -3)$.

Solution (a) $\frac{1}{2}\mathbf{v} + \frac{1}{2}\mathbf{w} = \frac{1}{2}(-3, -4) + \frac{1}{2}(-1, 6)$

$$= (-\tfrac{3}{2} - \tfrac{1}{2}, -2 + 3) = (-2, 1).$$

(b) $$\mathbf{v} + \tfrac{2}{5}(\mathbf{w} - \mathbf{v}) = (-4, 7) + \tfrac{2}{5}[(2, -3) - (-4, 7)]$$
$$= (-4, 7) + \tfrac{2}{5}(6, -10)$$
$$= (-4 + \tfrac{12}{5}, 7 - 4) = (-\tfrac{8}{5}, 3).$$

Answer (a) $(-2, 1)$ (b) $(-\frac{8}{5}, 3)$

The methods of this section can be used to give surprisingly snappy solutions of many geometry problems.

Example 4 Let $\mathbf{u}$, $\mathbf{v}$, $\mathbf{w}$ be the vertices of a triangle. Show that $\frac{1}{3}(\mathbf{u} + \mathbf{v} + \mathbf{w})$ is the intersection of its medians.

Solution The midpoint of the segment $\overline{\mathbf{vw}}$ is $\frac{1}{2}(\mathbf{v} + \mathbf{w})$. We set $\mathbf{m} = \frac{1}{3}(\mathbf{u} + \mathbf{v} + \mathbf{w})$ and rewrite $\mathbf{m}$ in the form

$$\mathbf{m} = \tfrac{1}{3}\mathbf{u} + \tfrac{2}{3}[\tfrac{1}{2}(\mathbf{v} + \mathbf{w})].$$

Thus $\mathbf{m}$ is the point $\frac{2}{3}$ of the way from $\mathbf{u}$ to $\frac{1}{2}(\mathbf{v} + \mathbf{w})$, the midpoint of $\mathbf{vw}$. So $\mathbf{m}$ is on one of the three medians (Fig. 3). Similarly

$$\mathbf{m} = \tfrac{1}{3}\mathbf{v} + \tfrac{2}{3}[\tfrac{1}{2}(\mathbf{w} + \mathbf{u})] = \tfrac{1}{3}\mathbf{w} + \tfrac{2}{3}[\tfrac{1}{2}(\mathbf{u} + \mathbf{v})]$$

shows that $\mathbf{m}$ is also on the other two medians.

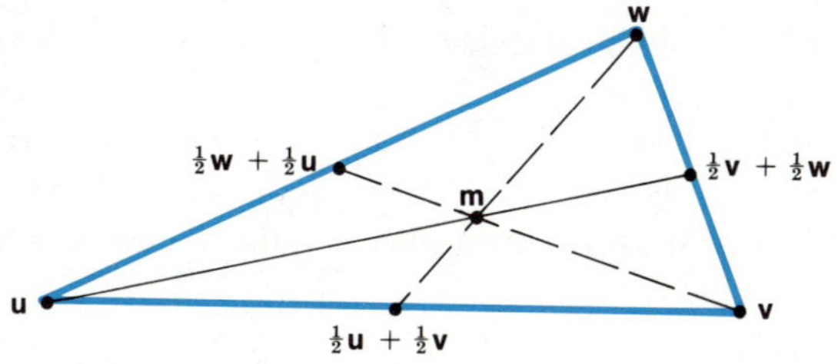

Fig. 3 $\mathbf{m} = \frac{1}{3}\mathbf{u} + \frac{2}{3}(\frac{1}{2}\mathbf{v} + \frac{1}{2}\mathbf{w})$
$= \frac{1}{3}\mathbf{u} + \frac{1}{3}\mathbf{v} + \frac{1}{3}\mathbf{w}$

Exercises

Find a parametric vector equation, parametric scalar equations, and a scalar equation for the line through $\mathbf{x}_0$ parallel to $\mathbf{v}$

1 $\mathbf{x}_0 = \mathbf{0}$ $\mathbf{v} = (1, 1)$

2 $\mathbf{x}_0 = \mathbf{0}$ $\mathbf{v} = (2, -1)$

3 $\mathbf{x}_0 = (1, 0)$ $\mathbf{v} = (0, 1)$

4 $\mathbf{x}_0 = (0, -1)$ $\mathbf{v} = (1, 0)$

5 $\mathbf{x}_0 = (2, 3)$ $\mathbf{v} = (2, -1)$

6 $\mathbf{x}_0 = (-1, 2)$ $\mathbf{v} = (-1, 2)$

7 $\mathbf{x}_0 = (3, 1)$ $\mathbf{v} = (4, 3)$

8 $\mathbf{x}_0 = (-1, -3)$ $\mathbf{v} = (-4, -3)$

9 $\mathbf{x}_0 = (5, 2)$ $\mathbf{v} = (-3, -1)$

10 $\mathbf{x}_0 = (-4, 7)$ $\mathbf{v} = (9, -2)$.

Find the point of intersection of the line through $\mathbf{x}_0$ parallel to $\mathbf{u}$ with the line through $\mathbf{y}_0$ parallel to $\mathbf{v}$

	$\mathbf{x}_0$	$\mathbf{u}$	$\mathbf{y}_0$	$\mathbf{v}$
11	$(0, 0)$	$(1, 1)$	$(1, 0)$	$(-1, 3)$
12	$(0, 1)$	$(1, -1)$	$(0, -1)$	$(-1, 0)$

	$\mathbf{x}_0$	$\mathbf{u}$	$\mathbf{y}_0$	$\mathbf{v}$
13	$(-1, 1)$	$(1, -2)$	$(2, 1)$	$(3, 2)$
14	$(3, -2)$	$(-1, 1)$	$(-2, 1)$	$(2, -1)$
15	$(-1, 3)$	$(1, -2)$	$(-2, -2)$	$(3, -1)$
16	$(-1, -2)$	$(3, -1)$	$(1, -1)$	$(-6, 2)$

Determine whether the two lines are the same or different

17 $\begin{cases} x = 2t + 1 \\ y = -3t + 2 \end{cases}$ $\begin{cases} x = 2t + 3 \\ y = -3t - 1 \end{cases}$ **18** $\begin{cases} x = t - 1 \\ y = 2t - 1 \end{cases}$ $\begin{cases} x = 3t - 1 \\ y = 6t - 1 \end{cases}$

19 $\begin{cases} x = t - 1 \\ y = -t + 1 \end{cases}$ $\begin{cases} x = 2t - 1 \\ y = -2t + 1 \end{cases}$ **20** $\begin{cases} x = t - 2 \\ y = 3t + 1 \end{cases}$ $\begin{cases} x = 2t - 4 \\ y = 6t - 5. \end{cases}$

Find the midpoint of the segment with given endpoints

21 $(2, 0)$ $(0, 4)$ **22** $(1, -2)$ $(3, 1)$
23 $(-1, 3)$ $(3, -1)$ **24** $(4, 1)$ $(-6, 2)$
25 $(3, 2)$ $(-1, -1)$ **26** $(-2, 5)$ $(3, 7)$
27 $(3, -7)$ $(3, 5)$ **28** $(2, -4)$ $(5, -4)$
29 $(-3, 5)$ $(3, -5)$ **30** $(15, 2)$ $(-6, -3)$.

Find the point t of the way from $\mathbf{v}$ to $\mathbf{w}$

	t	$\mathbf{v}$	$\mathbf{w}$
31	$\frac{1}{3}$	$(-1, 7)$	$(2, 4)$
32	$\frac{2}{3}$	$(3, -4)$	$(-1, 3)$
33	$\frac{1}{5}$	$(-2, 6)$	$(2, -1)$

	t	$\mathbf{v}$	$\mathbf{w}$
34	$\frac{4}{5}$	$(-1, 1)$	$(2, 6)$
35	$\frac{5}{6}$	$(0, 1)$	$(1, 0)$
36	$\frac{2}{7}$	$(3, -1)$	$(4, 0)$

37 Let $\mathbf{u}, \mathbf{v}, \mathbf{w}, \mathbf{z}$ be the vertices of a parallelogram, taken counterclockwise. Prove that $\frac{1}{2}(\mathbf{u} + \mathbf{w}) = \frac{1}{2}(\mathbf{v} + \mathbf{z})$ and interpret geometrically.

38 Prove that the midpoints of the sides of any quadrilateral are the vertices of a parallelogram.

39 Let L be a line and $c > 0$. Show that as $\mathbf{x}$ varies over L, the vector $c\mathbf{x}$ varies over another line.

40 Let L be a line and $\mathbf{c}$ a vector. Show that as $\mathbf{x}$ varies over L, the vector $\mathbf{c} + \mathbf{x}$ varies over another line.

3 LENGTH AND INNER PRODUCT

We start with some basic properties of the length of a vector. We denote the length of $\mathbf{v}$ by $\|\mathbf{v}\|$.

If $\mathbf{v} = (x, y)$, then

(1) $\|\mathbf{v}\|^2 = \|(x, y)\|^2 = x^2 + y^2$

(2) $\|\mathbf{0}\| = 0$, $\|\mathbf{v}\| > 0$ if $\mathbf{v} \neq \mathbf{0}$

(3) $\|a\mathbf{v}\| = |a| \cdot \|\mathbf{v}\|$

(4) $\|\mathbf{v} + \mathbf{w}\| \leq \|\mathbf{v}\| + \|\mathbf{w}\|$ (triangle inequality).

The first property comes directly from the distance formula, and the second follows from the first. The third is also proved by direct calculation:

$$\|a(x,y)\|^2 = \|(ax, ay)\|^2 = (ax)^2 + (ay)^2$$
$$= a^2(x^2 + y^2) = a^2\|(x,y)\|^2,$$

therefore

$$\|a\mathbf{v}\|^2 = a^2\|\mathbf{v}\|^2 \qquad \|a\mathbf{v}\| = |a|\cdot\|\mathbf{v}\|.$$

A useful fact: if $\mathbf{v}$ is any non-zero vector, then $\mathbf{v}/\|\mathbf{v}\|$ is a unit vector (length 1) in the same direction. Just apply (3) with $a = 1/\|\mathbf{v}\|$. [We used this device previously; we stretched (or shrank) a vector (a, b) to the unit circle by dividing its coordinates by $\sqrt{a^2 + b^2}$. This amounted to making the vector $\mathbf{v} = (a, b)$ into a unit vector by dividing it by $\|\mathbf{v}\| = \sqrt{a^2 + b^2}$.]

The triangle inequality (4) expresses a basic geometric fact: the length of each side of a triangle cannot exceed the sum of the lengths of the other two sides (Fig. 1).

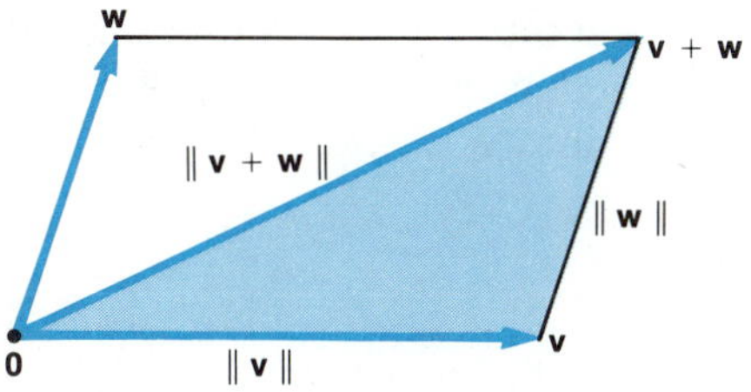

Fig. 1 Geometric proof that $\|\mathbf{v} + \mathbf{w}\| \leq \|\mathbf{v}\| + \|\mathbf{w}\|$

Inner Product

We now define another basic operation on vectors, the **inner product** (also called **dot product**). The inner product combines two vectors to produce a scalar. Suppose $\mathbf{v}$ and $\mathbf{w}$ are non-zero vectors, and $\cos\theta$ is the cosine of the angle between them. We define

$$\mathbf{v}\cdot\mathbf{w} = \|\mathbf{v}\|\cdot\|\mathbf{w}\|\cdot\cos\theta.$$

There are four possibilities for "the angle between $\mathbf{v}$ and $\mathbf{w}$." If one is θ, then the others are $-\theta$, $2\pi - \theta$, and $\theta - 2\pi$. But

$$\cos\theta = \cos(-\theta) = \cos(2\pi - \theta) = \cos(\theta - 2\pi),$$

so there is only one possibility for "the cosine of the angle between $\mathbf{v}$ and $\mathbf{w}$." If either $\mathbf{v} = \mathbf{0}$ or $\mathbf{w} = \mathbf{0}$, the angle between $\mathbf{v}$ and $\mathbf{w}$ is not defined. In this case we define

$$\mathbf{0}\cdot\mathbf{w} = \mathbf{v}\cdot\mathbf{0} = 0.$$

Notice that the inner product of two vectors is a real number, not a vector.

Three special cases that arise all the time follow directly from the definition of inner product:

(1) $\mathbf{v}\cdot\mathbf{v} = \|\mathbf{v}\|^2$.

(2) Vectors $\mathbf{v}$ and $\mathbf{w}$ are perpendicular if and only if $\mathbf{v}\cdot\mathbf{w} = 0$.

(3) If $\mathbf{v}$ and $\mathbf{w}$ are unit vectors ($\|\mathbf{v}\| = \|\mathbf{w}\| = 1$), then $\mathbf{v}\cdot\mathbf{w} = \cos\theta$.

Some examples of dot products are shown in Fig. 2.

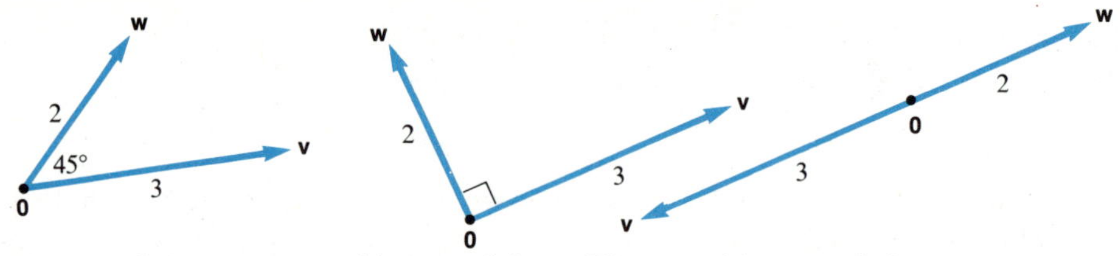

(a) $\mathbf{v}\cdot\mathbf{w} = 3\cdot 2\cdot \cos 45° = 3\sqrt{2}$ (b) $\mathbf{v}\cdot\mathbf{w} = 3\cdot 2\cdot \cos 90° = 0$ (c) $\mathbf{v}\cdot\mathbf{w} = 3\cdot 2\cdot \cos 180° = -6$

Fig. 2 Examples of inner products

There is another important geometric interpretation of the inner product. Suppose $\mathbf{v}$ and $\mathbf{w}$ are non-zero vectors and θ is the angle between them (we can always take θ so that $0 \le \theta \le \pi$). See Fig. 3. The quantity $\|\mathbf{v}\|\cdot\cos\theta$ is the length of the projection of $\mathbf{v}$ on $\mathbf{w}$ if $0 \le \theta \le \frac{1}{2}\pi$, and the negative of the length if $\frac{1}{2}\pi \le \theta \le \pi$. We call $\|\mathbf{v}\|\cdot\cos\theta$ the **signed projection** of $\mathbf{v}$ on $\mathbf{w}$. Its sign is + when $\mathbf{v}$ projects directly onto $\mathbf{w}$, and − when $\mathbf{v}$ projects onto $\mathbf{w}$ extended backwards. Since $\mathbf{v}\cdot\mathbf{w} = (\|\mathbf{v}\|\cos\theta)\cdot\|\mathbf{w}\| = (\|\mathbf{w}\|\cos\theta)\cdot\|\mathbf{v}\|$,

$$\begin{aligned}\mathbf{v}\cdot\mathbf{w} &= (\text{signed projection of } \mathbf{v} \text{ on } \mathbf{w})\cdot\|\mathbf{w}\| \\ &= (\text{signed projection of } \mathbf{w} \text{ on } \mathbf{v})\cdot\|\mathbf{v}\|.\end{aligned}$$

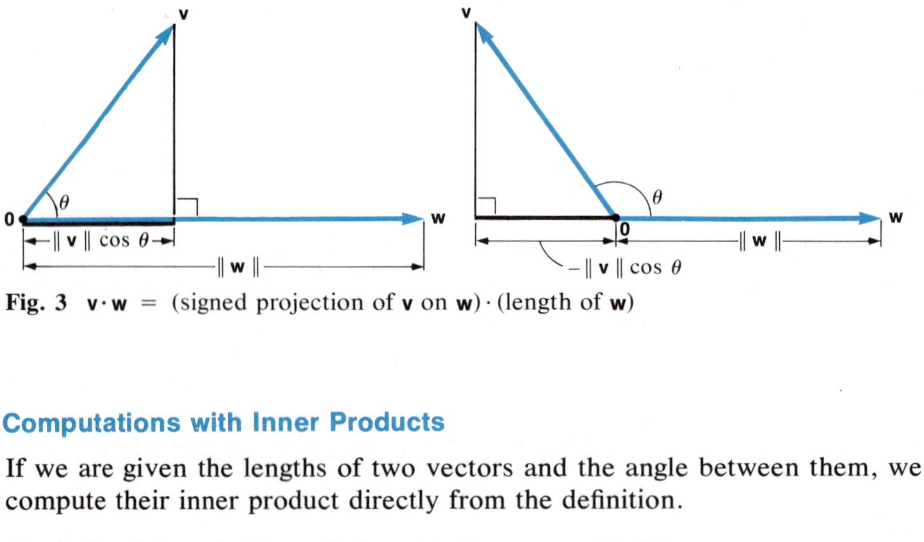

Fig. 3 $\mathbf{v}\cdot\mathbf{w}$ = (signed projection of $\mathbf{v}$ on $\mathbf{w}$) · (length of $\mathbf{w}$)

Computations with Inner Products

If we are given the lengths of two vectors and the angle between them, we can compute their inner product directly from the definition.

Example $\|\mathbf{v}\| = 8.000 \qquad \|\mathbf{w}\| = 11.00 \qquad \theta = 26.00°$

$$\mathbf{v}\cdot\mathbf{w} = 8\cdot 11\cdot\cos 26° \approx 79.09.$$

But what if we are given the components of $\mathbf{v}$ and $\mathbf{w}$? Then how do we find $\mathbf{v}\cdot\mathbf{w}$?

Inner Product If $\mathbf{v} = (a, b)$ and $\mathbf{w} = (c, d)$, then

$$\mathbf{v}\cdot\mathbf{w} = ac + bd.$$

Proof Consider the triangle in Fig. 4. By the law of cosines,

$$\|\mathbf{v} - \mathbf{w}\|^2 = \|\mathbf{v}\|^2 + \|\mathbf{w}\|^2 - 2\|\mathbf{v}\|\cdot\|\mathbf{w}\|\cdot\cos\theta = \|\mathbf{v}\|^2 + \|\mathbf{w}\|^2 - 2\mathbf{v}\cdot\mathbf{w}.$$

Therefore

$$\begin{aligned}2\mathbf{v}\cdot\mathbf{w} &= \|\mathbf{v}\|^2 + \|\mathbf{w}\|^2 - \|\mathbf{v} - \mathbf{w}\|^2 \\ &= (a^2 + b^2) + (c^2 + d^2) - [(a - c)^2 + (b - d)^2] \\ &= (a^2 + b^2 + c^2 + d^2) - (a^2 - 2ac + c^2) - (b^2 - 2bd + d^2) \\ &= 2(ac + bd).\end{aligned}$$

Consequently $\mathbf{v}\cdot\mathbf{w} = ac + bd$.

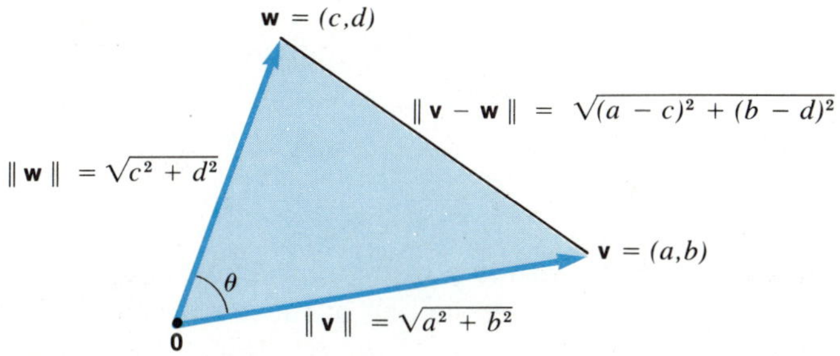

Fig. 4 Use of the law of cosines to obtain a formula for $\mathbf{v}\cdot\mathbf{w}$

Now let $\mathbf{v} = (a, b)$ and $\mathbf{w} = (c, d)$. Then

$$\mathbf{v}\cdot\mathbf{w} = \|\mathbf{v}\|\cdot\|\mathbf{w}\|\cdot\cos\theta = ac + bd$$

and

$$\|\mathbf{v}\| = \sqrt{a^2 + b^2} \qquad \|\mathbf{w}\| = \sqrt{c^2 + d^2}.$$

By putting these facts together we derive a formula for $\cos\theta$ in terms of the components of $\mathbf{v}$ and $\mathbf{w}$:

> If $\mathbf{v} = (a, b)$ and $\mathbf{w} = (c, d)$ are non-zero vectors, and θ is the angle between them, then
>
> $$\cos\theta = \frac{\mathbf{v}\cdot\mathbf{w}}{\|\mathbf{v}\|\cdot\|\mathbf{w}\|} = \frac{ac + bd}{\sqrt{a^2 + b^2}\sqrt{c^2 + d^2}}.$$

Example 1 Find the acute angle between $\mathbf{v} = (4, 3)$ and $\mathbf{w} = (1, 5)$.

Solution $\|\mathbf{v}\| = \sqrt{4^2 + 3^2} = 5$, $\|\mathbf{w}\| = \sqrt{1^2 + 5^2} = \sqrt{26}$, and $\mathbf{v}\cdot\mathbf{w} = 4\cdot1 + 3\cdot5 = 19$. Hence

$$\cos\theta = \frac{19}{5\sqrt{26}} \qquad \theta = \text{arc cos}\,\frac{19}{5\sqrt{26}}.$$

Answer $\text{arc cos}\,\dfrac{19}{5\sqrt{26}} \approx 41.82°$

The main properties of the inner product look just like the familiar rules of algebra.

Rules for Inner Products

(1) $\mathbf{v}\cdot\mathbf{w} = \mathbf{w}\cdot\mathbf{v}$

(2) $(a\mathbf{v})\cdot\mathbf{w} = \mathbf{v}\cdot(a\mathbf{w}) = a(\mathbf{v}\cdot\mathbf{w})$

(3) $(\mathbf{u} + \mathbf{v})\cdot\mathbf{w} = \mathbf{u}\cdot\mathbf{w} + \mathbf{v}\cdot\mathbf{w}$

(4) $\mathbf{u}\cdot(\mathbf{v} + \mathbf{w}) = \mathbf{u}\cdot\mathbf{v} + \mathbf{u}\cdot\mathbf{w}.$

These rules are easily proved by using components; the proofs are left as exercises.

Perpendicular Vectors

If $\mathbf{v} = (a, b)$ is a non-zero vector, then $(-b, a)$ is perpendicular to $\mathbf{v}$ and has the same length as $\mathbf{v}$:

$$(a, b)\cdot(-b, a) = -ab + ba = 0$$
$$\|(-b, a)\| = \sqrt{(-b)^2 + a^2} = \sqrt{a^2 + b^2} = \|(a, b)\|.$$

Therefore $(-b, a)$ is one of the two vectors shown in Fig. 5a. A direct check of the four quadrants shows that $(-b, a)$ is always the one of these two that is $\frac{1}{2}\pi$ counterclockwise from $\mathbf{v}$. See Fig. 5b.

The result of rotating a vector $\mathbf{v} = (a, b)$ counterclockwise $\frac{1}{2}\pi$ radians is $(-b, a)$.

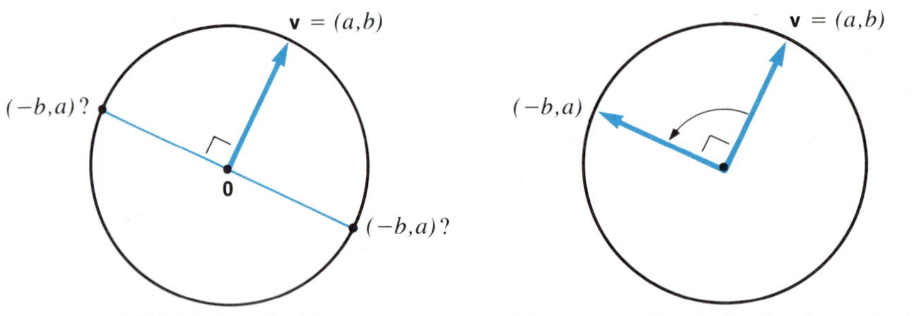

(a) Which is $(-b,a)$? (b) Answer: the choice $\frac{1}{2}\pi$ *forward* of $\mathbf{v}$

Fig. 5 Rotation

Applications

Physical quantities such as force, velocity, momentum, and magnetic field strength are vectors. Hence it is natural that vector algebra is useful in problems involving such quantities.

Example 2 A weight W hangs from a pulley that slides on a fixed rope (Fig. 6a). Find the tension T in the fixed rope for (a) $\theta = 10.0°$ and (b) $\theta = 5.0°$.

Solution Take the origin $\mathbf{0}$ at the pulley (Fig. 6b). Three forces act on the pulley: a downward force $\mathbf{w}$ with $\|\mathbf{w}\| = W$ and two forces along the fixed rope, $\mathbf{t}_1$ left and $\mathbf{t}_2$ right, with $\|\mathbf{t}_1\| = \|\mathbf{t}_2\| = T$, the tension. Since the pulley is in equilibrium, the three forces sum to zero:

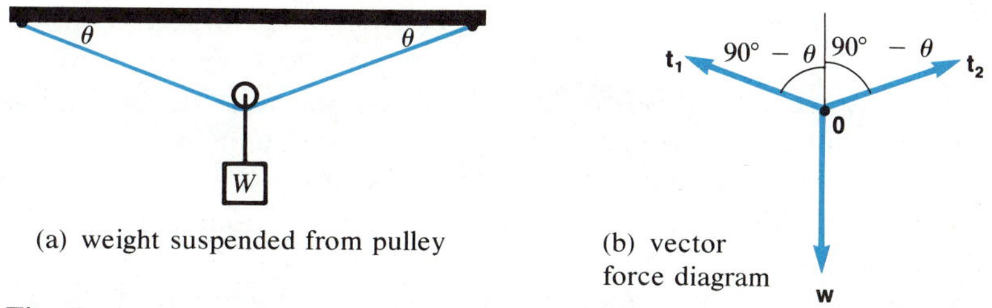

Fig. 6

$$\mathbf{w} + \mathbf{t}_1 + \mathbf{t}_2 = \mathbf{0}, \quad \text{that is,} \quad \mathbf{w} = -(\mathbf{t}_1 + \mathbf{t}_2).$$

The vertical components of $\mathbf{t}_1$ and $\mathbf{t}_2$ are both equal to

$$\|\mathbf{t}_1\| \cos(90° - \theta) = T \sin \theta .$$

Therefore, the condition for equilibrium is

$$2T \sin \theta = W, \quad \text{that is,} \quad T = \frac{W}{2 \sin \theta}.$$

$$\text{(a)} \quad \theta = 10.0° \qquad T = \frac{W}{2 \sin 10.0°} \approx 2.88\, W$$

$$\text{(b)} \quad \theta = 5.0° \qquad T = \frac{W}{2 \sin 5.0°} \approx 5.74\, W .$$

Answer (a) $T \approx 2.88\, W$ (b) $T \approx 5.74\, W$

The next example explains how a sailboat can sail "into the wind." The model is simplified by assuming (1) that the sail is straight (actually it is a more efficient airfoil) and (2) that the keel prevents any sideways motion.

Example 3 On a sailboat, the sail is set at 15° to the keel (Fig. 7a). The wind blows into the sail from forward, at an angle 45° to the keel, and exerts a total force of 78 kg against the sail. Find the effective wind force driving the boat forward.

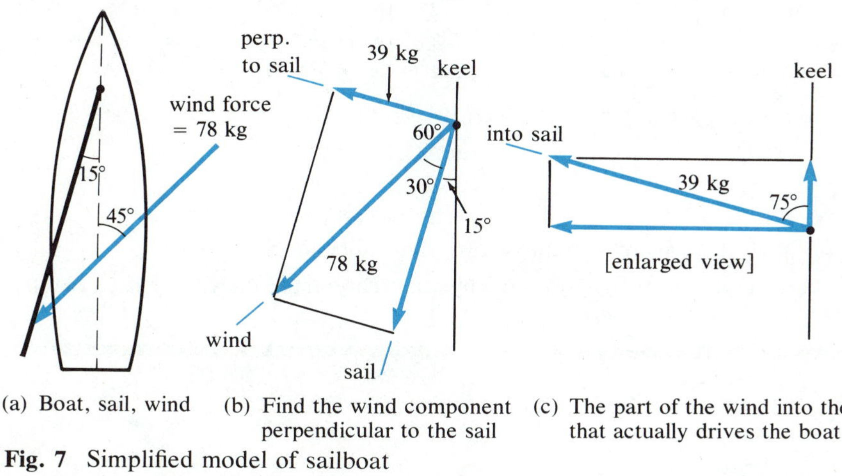

(a) Boat, sail, wind (b) Find the wind component perpendicular to the sail (c) The part of the wind into the sail that actually drives the boat forward

Fig. 7 Simplified model of sailboat

Solution The problem must be solved in two steps. First the wind force is resolved into two components, one parallel to the sail and the other perpendicular to the sail. The wind parallel to the sail slips past, giving no push to the boat.

Therefore, only the component of the wind perpendicular to the sail pushes the boat.

The second step is to resolve this (perpendicular) component further into two components, one parallel to the keel and the other perpendicular to the keel. The component perpendicular to the keel does nothing because the boat can move only in the direction of the keel. Therefore, the parallel component is the effective force pushing the boat forward.

By Figure 7b, the wind strikes the sail at $30°$, so the component of the wind perpendicular to the sail has magnitude

$$78 \sin 30° = 39 \text{ kg}.$$

By Figure 7c, this 39 kg force acts at angle $75°$ to the keel; hence its component in the direction of the keel has magnitude

$$39 \cos 75° \approx 10 \text{ kg}.$$

Answer 10 kg

Exercises

Compute

1 $\|(12, 5)\|$ **2** $\|(15, -8)\|$ **3** $\|(-5, 6)\|$
4 $\|(3, 11)\|$ **5** $(2, 0)\cdot(3, 1)$ **6** $(-1, -1)\cdot(1, 1)$
7 $(4, 1)\cdot(-1, 4)$ **8** $(2, 2)\cdot(-1, 2)$ **9** $(0, 3)\cdot(1, 1)$
10 $(2, 3)\cdot(-3, -1)$ **11** $(3.5, 4.1)\cdot(7.2, 1.5)$ **12** $(-5.8, 6.1)\cdot(3.0, -7.8)$.

Compute in two ways

13 $(2, 3)\cdot[(1, 4) + (2, 2)]$ **14** $(-1, 2)\cdot[(1, 3) + (1, 4)]$
15 $[(-1, 7) - (2, 3)]\cdot(4, 5)$ **16** $[(-2, -3) - (-1, 4)]\cdot(-2, 3)$
17 $[3(4, 5)]\cdot(-1, 2)$ **18** $[-2(3, -1)]\cdot(5, -6)$.

Prove

19 (1) on p. 442 **20** (2) on p. 442
21 (3) on p. 442 **22** (4) on p. 442
23 $\|\mathbf{v} + \mathbf{w}\|^2 = \|\mathbf{v}\|^2 + 2\mathbf{v}\cdot\mathbf{w} + \|\mathbf{w}\|^2$ **24** $\|\mathbf{v} + \mathbf{w}\|^2 + \|\mathbf{v} - \mathbf{w}\|^2 = 2\|\mathbf{v}\|^2 + 2\|\mathbf{w}\|^2$
25 $\|\mathbf{v} + \mathbf{w}\|^2 - \|\mathbf{v} - \mathbf{w}\|^2 = 4\mathbf{v}\cdot\mathbf{w}$ **26*** $|\mathbf{v}\cdot\mathbf{w}| \le \|\mathbf{v}\|\cdot\|\mathbf{w}\|$.

Compute the smallest positive angle between the vectors

27 $(3, 4)$ $(-4, 3)$ **28** $(1, -2)$ $(-2, 1)$
29 $(1, 3)$ $(3, 1)$ **30** $(1, -3)$ $(3, 6)$
31 $(1, 1)$ $(-1, 7)$ **32** $(3, 4)$ $(-5, -2)$.

33 Prove that the diagonals of a rhombus are perpendicular.

34 Prove that the midpoints of the sides of any quadrilateral are the vertices of a parallelogram.

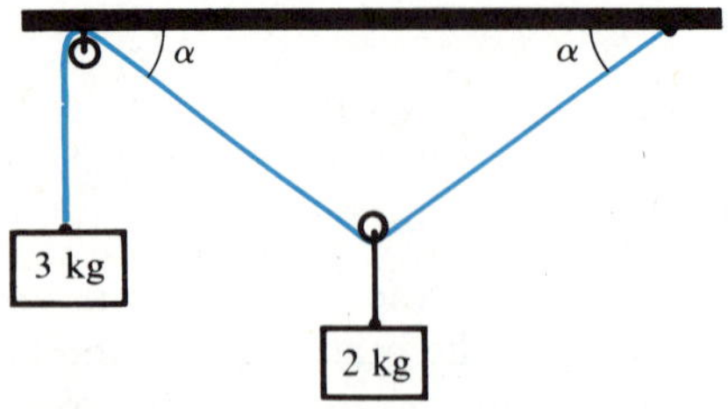

Fig. 8

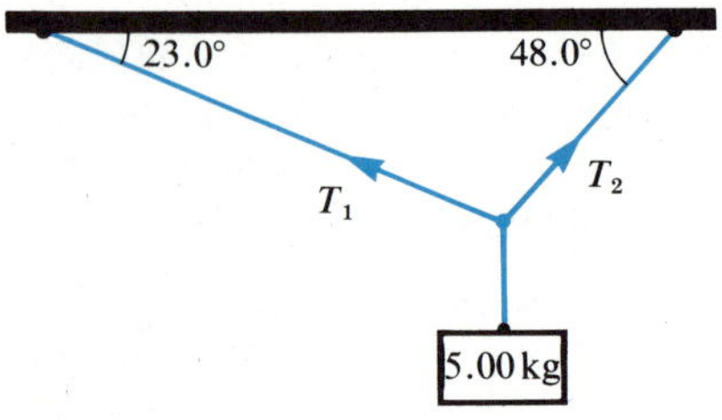

Fig. 9

35 Estimate $\alpha°$ to 2 places for the system of pulleys and weights in Fig. 8.

36 Find the tensions T_1 and T_2 in the ropes in Fig. 9. There are no pulleys, only knots.

37 Suppose the sail is set at 45° to the keel and the wind comes in directly from the side, that is, perpendicular to the keel, with total force W. Find the effective force pushing the sailboat forward.

38 Work Ex. 37 with the wind directly from behind.

39 A boat in open water is headed $26\frac{1}{2}°$ east of north at a water speed of 23 knots (nautical miles per hour). After 3 hours it is observed that the boat has actually moved 58 nautical miles 19° east of north. Find the current speed and direction.

40 A plane is headed due west at an air speed of 187 mph, hoping to reach a point 150 miles west. After $\frac{1}{2}$ hour the pilot finds he has actually gone 112 miles in a direction 9° south of west. Find the wind vector.

4 NORMAL FORM

Using inner products, we obtain a useful vector equation for a straight line, different from the parametric equations in Section 2.

Let L be any line (Fig. 1). Let $\mathbf{n}$ be either of the two unit vectors perpendicular to L. Let $p\mathbf{n}$ be the intersection of L and the line along $\mathbf{n}$. Clearly $p > 0$ if $\mathbf{n}$ points towards L. Also $p < 0$ if $\mathbf{n}$ points away from L, and $p = 0$ if L passes through $\mathbf{0}$.

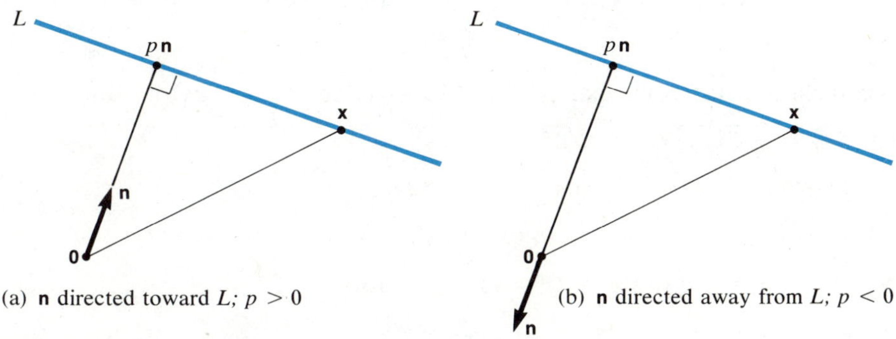

(a) $\mathbf{n}$ directed toward L; $p > 0$ (b) $\mathbf{n}$ directed away from L; $p < 0$

Fig. 1 Normal form of a line

Now let $\mathbf{x}$ be any point on L. Then the signed projection of $\mathbf{x}$ on $\mathbf{n}$ is $p\mathbf{n}$ because L and $\mathbf{n}$ are perpendicular. Hence

$$\mathbf{x}\cdot\mathbf{n} = p\mathbf{n}\cdot\mathbf{n} = p$$

since $\mathbf{n}\cdot\mathbf{n} = 1$. This simple relation, $\mathbf{x}\cdot\mathbf{n} = p$, is the equation for L that we want. To find its scalar form, set $\mathbf{n} = (a, b)$ and $\mathbf{x} = (x, y)$. Then

$$\mathbf{x}\cdot\mathbf{n} = (x, y)\cdot(a, b) = ax + by = p.$$

Normal Form Each line in the plane has an equation

$$\mathbf{x}\cdot\mathbf{n} = p,$$

where $\mathbf{n}$ is a unit vector and p is a scalar. This equation is called the **normal form** of the line. The scalar version of this normal form is

$$ax + by = p \qquad a^2 + b^2 = 1.$$

Remark Actually each line has two normal forms because there are two choices, $\mathbf{n}$ and $-\mathbf{n}$, for the unit perpendicular vector, hence two equivalent equations

$$\mathbf{x}\cdot\mathbf{n} = p \qquad \text{and} \qquad \mathbf{x}\cdot(-\mathbf{n}) = -p.$$

We try to ignore this ambiguity and we refer (incorrectly) to either as "the" normal form of the line.

EXAMPLE 1 Find the normal form of the line

(a) $3x + 4y = 10$ (b) $x - y = 1$.

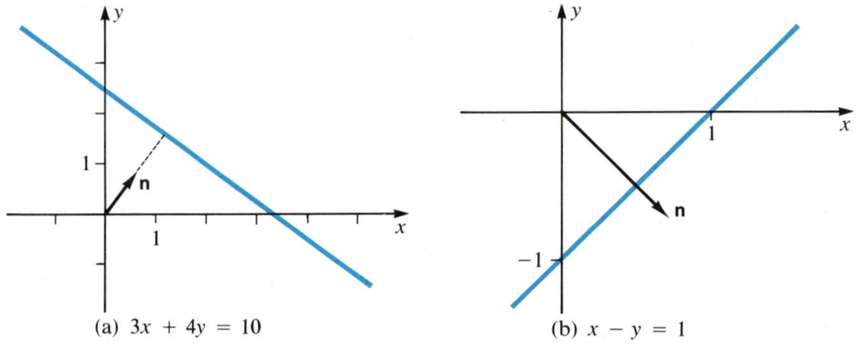

Fig. 2

Solution (a) See Fig. 2a. Write $3x + 4y = 10$ in the vector form

$$(3, 4)\cdot(x, y) = 10.$$

This is almost normal form, except that $(3, 4)$ is not a unit vector. But

$$\|(3, 4)\| = \sqrt{3^2 + 4^2} = \sqrt{25} = 5$$

so $\frac{1}{5}(3, 4)$ is a unit vector. Thus

$$\tfrac{1}{5}(3, 4)\cdot(x, y) = \tfrac{1}{5}\cdot 10 \qquad (\tfrac{3}{5}, \tfrac{4}{5})\cdot(x, y) = 2.$$

(b) See Fig. 2b. We write $x - y = 1$ in the form

$$(1, -1)\cdot(x, y) = 1$$

and divide both sides by

$$\|(1, -1)\| = \sqrt{1^2 + (-1)^2} = \sqrt{2}:$$

$$\frac{1}{\sqrt{2}}(1, -1)\cdot(x, y) = \frac{1}{\sqrt{2}} \qquad (\tfrac{1}{2}\sqrt{2}, -\tfrac{1}{2}\sqrt{2})\cdot(x, y) = \tfrac{1}{2}\sqrt{2}.$$

Answer (a) $(\frac{3}{5}, \frac{4}{5})\cdot\mathbf{x} = 2$ (b) $(\frac{1}{2}\sqrt{2}, -\frac{1}{2}\sqrt{2})\cdot\mathbf{x} = \frac{1}{2}\sqrt{2}$

Remark 1 Sometimes it is convenient to write the normal form of a line as

$$x\cos\alpha + y\sin\alpha = p.$$

The point is that the unit vector $\mathbf{n}$ in $\mathbf{x}\cdot\mathbf{n} = p$ can be written as $\mathbf{n} = (\cos\alpha, \sin\alpha)$, where α is determined up to $\alpha + 2\pi n$. See Fig. 3.

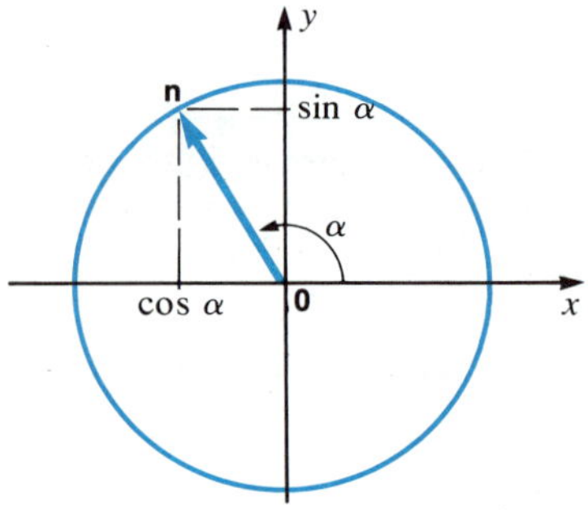

Fig. 3

Remark 2 The normal form is easy for horizontal lines and for vertical lines. The horizontal line $y = p$ is in scalar normal form. In vector form,

$$(0, 1)\cdot(x, y) = p \qquad \|(0, 1)\| = 1.$$

Similarly the vertical line $x = p$ is in scalar normal form, and its vector normal form is

$$(1, 0)\cdot(x, y) = p \qquad \|(1, 0)\| = 1.$$

EXAMPLE 2 Find the normal form of the line

$$\mathbf{x} = (3 + t, 2 - 4t),$$

given in parametric form.

Solution Write the line as

$$\mathbf{x} = (3, 2) + t(1, -4);$$

it is parallel to $(1, -4)$. We rotate this vector $\frac{1}{2}\pi$ to obtain $(4, 1)$, a vector perpendicular to the line. Then

$$\|(4, 1)\| = \sqrt{4^2 + 1^2} = \sqrt{17},$$

so

$$\mathbf{n} = \frac{1}{\sqrt{17}}(4, 1) = \tfrac{1}{17}\sqrt{17}(4, 1)$$

is a unit vector perpendicular to the line. We have

$$\begin{aligned}\mathbf{x}\cdot\mathbf{n} &= [(3, 2) + t(1, -4)]\cdot\mathbf{n} = (3, 2)\cdot\mathbf{n} + t(1, -4)\cdot\mathbf{n} \\ &= (3, 2)\cdot\mathbf{n} + 0 = (3, 2)\cdot\tfrac{1}{17}\sqrt{17}(4, 1) = \tfrac{14}{17}\sqrt{17}.\end{aligned}$$

Answer $(x, y)\cdot[\frac{1}{17}\sqrt{17}(4, 1)] = \frac{14}{17}\sqrt{17}$

Distance from a Point to a Line

As a first application of normal form, we derive a formula for the distance from a point $\mathbf{x}_0$ to a line L given in normal form $\mathbf{x}\cdot\mathbf{n} = p$.

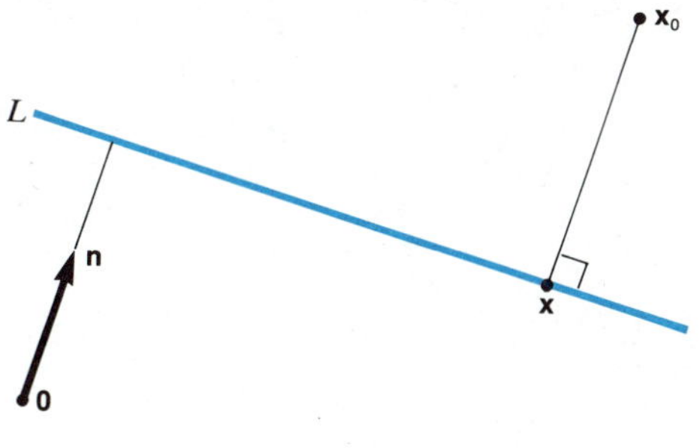

Fig. 4 $\mathbf{x}$ is the point of L closest to $\mathbf{x}_0$, so the vector $\mathbf{x} - \mathbf{x}_0$ is parallel to $\mathbf{n}$.

Let $\mathbf{x}$ be the point of L nearest to $\mathbf{x}_0$. Then $\mathbf{x} - \mathbf{x}_0$ is perpendicular to L, hence parallel to $\mathbf{n}$. This means that

$$\mathbf{x} - \mathbf{x}_0 = c\mathbf{n} \qquad c \text{ a scalar}.$$

The distance we want is

$$\|\mathbf{x} - \mathbf{x}_0\| = \|c\mathbf{n}\| = |c|.$$

But $\mathbf{x}$ is in L, so $\mathbf{x}\cdot\mathbf{n} = p$ and we have

$$(\mathbf{x} - \mathbf{x}_0)\cdot\mathbf{n} = c\mathbf{n}\cdot\mathbf{n} = c$$

$$\mathbf{x}\cdot\mathbf{n} - \mathbf{x}_0\cdot\mathbf{n} = c$$

$$p - \mathbf{x}_0\cdot\mathbf{n} = c.$$

Therefore the distance is

$$|c| = |p - \mathbf{x}_0\cdot\mathbf{n}|.$$

> Suppose $\mathbf{x}\cdot\mathbf{n} = p$ is the normal form of a line L, and $\mathbf{x}_0$ is a point of the plane. Then the distance from $\mathbf{x}_0$ to L is
>
> $$|p - \mathbf{x}_0\cdot\mathbf{n}|.$$
>
> If $ax + by = p$ is the normal form in coordinates, then the distance is
>
> $$|p - ax_0 - by_0|.$$

EXAMPLE 3 Find the distance from $(4, -1)$ to the line

(a) $\frac{3}{5}x - \frac{4}{5}y = 6$ (b) $y = 3x - 2$.

Solution (a) Since $\mathbf{n} = (\frac{3}{5}, -\frac{4}{5})$ is a unit vector, the equation is in normal form, with $p = 6$. Therefore the distance from $\mathbf{x}_0 = (4, -1)$ to the line is

$$|p - \mathbf{x}_0\cdot\mathbf{n}| = |6 - (4, -1)\cdot(\tfrac{3}{5}, -\tfrac{4}{5})|$$

$$= |6 - (\tfrac{12}{5} + \tfrac{4}{5})| = |6 - \tfrac{16}{5}| = \tfrac{14}{5}.$$

(b) We must first put $y = 3x - 2$ into normal form. We write it as

$$-3x + y = -2 \qquad (-3, 1)\cdot(x, y) = -2$$

and divide by $\|(-3, 1)\| = \sqrt{10}$:

$$\frac{1}{\sqrt{10}}(-3, 1)\cdot(x, y) = \frac{-2}{\sqrt{10}} \qquad \tfrac{1}{10}\sqrt{10}(-3, 1)\cdot(x, y) = -\tfrac{1}{5}\sqrt{10}.$$

Hence $\mathbf{n} = \frac{1}{10}\sqrt{10}(-3, 1)$ and $p = -\frac{1}{5}\sqrt{10}$. The distance from $\mathbf{x}_0 = (4, -1)$ to the line is

$$|p - \mathbf{x}_0\cdot\mathbf{n}| = |-\tfrac{1}{5}\sqrt{10} - (4, -1)\cdot[\tfrac{1}{10}\sqrt{10}(-3, 1)]|$$

$$= |-\tfrac{1}{5}\sqrt{10} - \tfrac{1}{10}\sqrt{10}(-12 - 1)|$$

$$= \sqrt{10}|-\tfrac{1}{5} + \tfrac{13}{10}| = \tfrac{11}{10}\sqrt{10}$$

Answer (a) $\frac{14}{5}$ (b) $\frac{11}{10}\sqrt{10}$

Angle Between Lines

Let M and N be intersecting lines (Fig. 5a). They determine two angles, θ and $\pi - \theta$. But $\cos\theta = \cos(\pi - \theta)$, so the "cosine of their angle of intersection" is a single number—no ambiguity. Let

$$M\colon \mathbf{x}\cdot\mathbf{m} = p \qquad N\colon \mathbf{x}\cdot\mathbf{n} = q$$

be the normal forms of the lines. Then the unit vector $\mathbf{m}$ is perpendicular to M and the unit vector $\mathbf{n}$ is perpendicular to N. It follows (Fig. 5b) that the angle

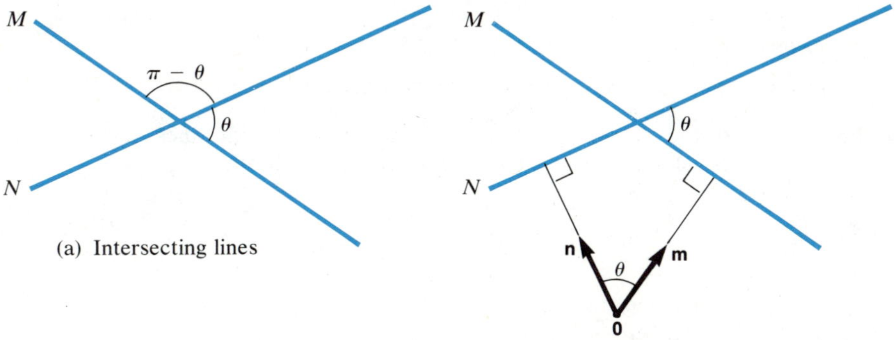

(a) Intersecting lines

(b) The angle between their normals equals one of the angles between the lines.

Fig. 5 Angle between lines

between **m** and **n** is one of the two angles between M and N. Therefore $\mathbf{m}\cdot\mathbf{n} = \cos\theta$.

> If $\mathbf{x}\cdot\mathbf{m} = p$ and $\mathbf{x}\cdot\mathbf{n} = q$ are two lines in normal form, then
> $$\mathbf{m}\cdot\mathbf{n} = \cos\theta,$$
> where θ is one of the angles of intersection of the lines.

EXAMPLE 4 Find the acute angle θ between the lines $3x - 4y = 2$ and $x + y = 1$.

Solution Write the lines in the form

$$(3, -4)\cdot(x, y) = 2 \qquad \text{and} \qquad (1, 1)\cdot(x, y) = 1.$$

Now

$$\|(3, -4)\| = 5 \quad \text{and} \quad \|(1, 1)\| = \sqrt{2}$$

so

$$\mathbf{m} = \tfrac{1}{5}(3, -4) \quad \text{and} \quad \mathbf{n} = \frac{1}{\sqrt{2}}(1, 1) = \tfrac{1}{2}\sqrt{2}(1, 1)$$

are corresponding normals to M and to N. Hence $\cos\theta = \mathbf{m}\cdot\mathbf{n}$ or $\cos\theta = -\mathbf{m}\cdot\mathbf{n}$, whichever is positive, since θ is acute. Clearly

$$\cos\theta = -\mathbf{m}\cdot\mathbf{n} = -\tfrac{1}{10}\sqrt{2}(3 - 4)\cdot(1, 1) = \tfrac{1}{10}\sqrt{2}$$
$$\theta = \arccos \tfrac{1}{10}\sqrt{2}.$$

Answer $\theta = \arccos \tfrac{1}{10}\sqrt{2} \approx 81.87°$

Remark 1 If two lines M and N are parallel (or coincide), then $\mathbf{m} = \pm\mathbf{n}$ so $\mathbf{m}\cdot\mathbf{n} = \pm 1$. But $\cos\theta = \pm 1$ means $\theta = 0$ or $\theta = \pi$. The result $\mathbf{m}\cdot\mathbf{n} = \cos\theta$ is thus correct provided we interpret $\theta = 0$ or $\theta = \pi$ to mean parallel (or coincident) lines.

Remark 2 Two lines M and N are perpendicular if $\mathbf{m}\cdot\mathbf{n} = 0$, that is, $\theta = \tfrac{1}{2}\pi$. Suppose the lines are given in the function forms

$$M: y = mx + b \qquad N: y = nx + c,$$

where m and n are their slopes. We may rewrite these as

$$(-m, 1)\cdot(x, y) = b \qquad \text{and} \qquad (-n, 1)\cdot(x, y) = c.$$

Clearly the vectors $(-m, 1)$ and $(-n, 1)$ are parallel to $\mathbf{m}$ and $\mathbf{n}$ respectively, so $\mathbf{m}\cdot\mathbf{n} = 0$ if and only if

$$(-m, 1)\cdot(-n, 1) = 0 \qquad mn + 1 = 0 \qquad mn = -1.$$

Thus we have derived the result (p. 126) that two non-vertical lines are perpendicular if and only if their slopes are negative reciprocals of each other.

Exercises

Find the normal form

1 $y = x$ **2** $y = -x$ **3** $y = \frac{3}{4}x$

4 $y = -\frac{12}{5}x$ **5** $3x + 4y = 2$ **6** $4x - 3y = 3$

7 $12x + 5y = 1$ **8** $-8x + 15y = 17$ **9** $7x - 24y = 25$

10 $24x + 7y = 1$ **11** $2x - 3y = 4$ **12** $3x + 2y = 2$

13 $y = 2x + 1$ **14** $y = -3x - 2$ **15** $y = ax + b$

16 $\dfrac{x}{a} + \dfrac{y}{b} = 1.$

Find the normal form of the line, given in parametric form

17 $\mathbf{x} = (1, -1) + t(\frac{3}{5}, -\frac{4}{5})$ **18** $\mathbf{x} = (1, 0) + t(\frac{12}{13}, \frac{5}{13})$

19 $\mathbf{x} = (1 + t, 1 + 2t)$ **20** $\mathbf{x} = (2 - t, 3 + t)$

21 $\begin{cases} x = 3 - 2t \\ y = 2 + 3t \end{cases}$ **22** $\begin{cases} x = 1 - \frac{3}{5}t \\ y = 2 - \frac{4}{5}t. \end{cases}$

Find the distance from $\mathbf{x}_0$ to L

	$\mathbf{x}_0$	L		$\mathbf{x}_0$	L
23	$(1, 0)$	$y = -2$	**24**	$(0, -1)$	$x = 3$
25	$(1, 1)$	$\frac{4}{5}x - \frac{3}{5}y = -2$	**26**	$(-2, 3)$	$\frac{3}{5}x + \frac{4}{5}y = 1$
27	$(0, -2)$	$4x + 3y = 7$	**28**	$(1, -1)$	$3x - 4y = 1$
29	$(10, 5)$	$12x + 5y = 26$	**30**	$(1, 3)$	$-5x + 12y = 13$
31	$(1, 0)$	$x + y = 2$	**32**	$(2, -1)$	$x = 4y - 3$

Find the acute angle between the lines M and N

33 $y = x$ $x = 0$ **34** $y = -x$ $y = 0$

35 $y = \frac{1}{2}x$ $y = 2x$ **36** $y = 3x$ $y = -x$

37 $3x + 4y = 1$ $5x + 12y = 1$ **38** $-3x + 4y = 1$ $12x - 5y = 1$

39 $x + y = 1$ $x + 2y = 2$ **40** $2x - y = 3$ $x + 3y = 2.$

41 Find the normal form of the line through $(1, 0)$ and perpendicular to $x - 2y = 6$.

42 Find the normal form of the line through $(-2, -3)$ and perpendicular to the line through $(1, 0)$ and $(5, 4)$.

43 Find the normal form of the perpendicular bisector of the line segment between $(0, 1)$ and $(5, -2)$.

44 Find the normal form of the perpendicular bisector of the line segment between $(1, 3)$ and $(-4, 4)$.

REVIEW EXERCISES

Compute

1 $-4(3, 1) + 7(2, 2)$ **2** $(8, -5) - (5, -8)$

3 $0(1, -1) - 3(-1, 1)$ **4** $2(3, 4) + 5(6, 7).$

5 Find a parametric vector equation and a scalar equation for the line through $(-3, 1)$ parallel to $(-2, -1)$.

6 Find the point $\frac{2}{5}$ of the way from $(4, -1)$ to $(-2, 3)$.

7 Find parametric scalar equations for the line through $(1, -1)$ parallel to $(x, y) = (3t - 1, -2t + 1)$.

8 Let $\mathbf{v} = (4, -3)$ and $\mathbf{w} = (-5, 12)$. Find $\mathbf{v} \cdot \mathbf{w}$ and $\|\mathbf{v}\|/\|\mathbf{w}\|$.

9 Compute the smallest positive angle between $\mathbf{v} = (3, -1)$ and $\mathbf{w} = (3 + \sqrt{3}, 3\sqrt{3} - 1)$.

10 If there were no current, the drifting float would be pushed due NE at 3 mph by the wind. But there is a 7 mph current moving due NW. Find the speed and direction of the float.

Find the normal form

11 $40x - 9y = 123$

12 $x = -5y + 5$.

13 Find the distance from $(1, -2)$ to $\frac{3}{5}x - \frac{4}{5}y = 10$.

14 Find the acute angle between the lines $x - y = 2$ and $\sqrt{3}\,x + y = 1$.

15 Find the normal form of $(x, y) = (3 - t, 1 + \frac{3}{2}t)$.

16 Find the acute angle between the lines $(x, y) = (1 + 2t, 2 - 3t)$ and $2x - 5y = 6$.

12 PLANE ANALYTIC GEOMETRY

1 TRANSLATION AND CIRCLES

In working problems, it is sometimes convenient to introduce new coordinate axes parallel to the given ones. This operation is called **shifting** or **translating** the axes. A translation is described by a single vector directed from the old origin to the new origin (Fig. 1).

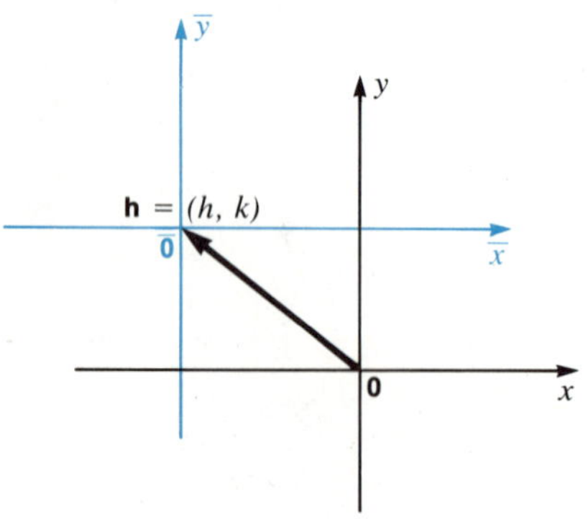

Fig. 1 New origin at $\mathbf{h} = (h, k)$

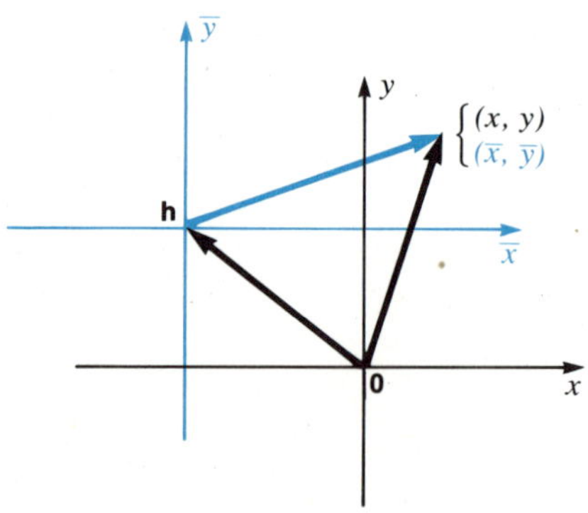

Fig. 2 Relation between old and new coordinates: $(x, y) = (\bar{x}, \bar{y}) + (h, k)$

When new axes are introduced, each point of the plane acquires new coordinates. What is the relation between the original coordinates (x, y) of a point and its new coordinates $(\bar{x}, \bar{y})$? See Fig. 2. Suppose the new origin is the point with old coordinates (h, k). From the triangle in Fig. 2, we see that

$$\mathbf{x} = \bar{\mathbf{x}} + \mathbf{h} \qquad \text{that is} \qquad (x, y) = (\bar{x}, \bar{y}) + (h, k) = (\bar{x} + h, \bar{y} + k).$$

Separate components:

$$x = \bar{x} + h \qquad y = \bar{y} + k.$$

If the coordinate axes are translated so that the origin is moved to (h, k), then the new coordinates $(\bar{x}, \bar{y})$ and the old coordinates (x, y) of a point are related by the equations

$$\begin{cases} x = \bar{x} + h \\ y = \bar{y} + k \end{cases} \qquad \begin{cases} \bar{x} = x - h \\ \bar{y} = y - k. \end{cases}$$

To get the signs right in this relation, just remember that $\bar{x} = 0$ and $\bar{y} = 0$ when $x = h$ and $y = k$.

Translation of axes is helpful in simplifying equations and computations. For example, take the equation $y + 3 = (x - 2)^2$. If we introduce a new coordinate system with origin at $(2, -3)$, then $x - 2 = \bar{x}$ and $y + 3 = \bar{y}$. Therefore the equation becomes $\bar{y} = \bar{x}^2$ in the new coordinates. Thus the graph of $y + 3 = (x - 2)^2$ is just the familiar quadratic curve $y = x^2$, but with its vertex shifted (Fig. 3).

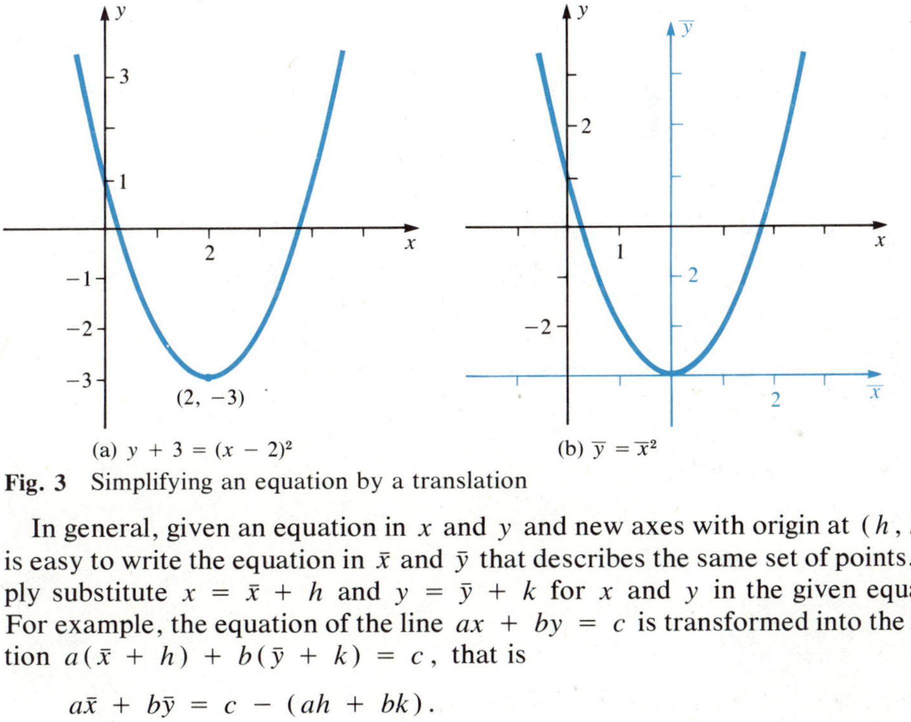

Fig. 3 Simplifying an equation by a translation

In general, given an equation in x and y and new axes with origin at (h, k), it is easy to write the equation in $\bar{x}$ and $\bar{y}$ that describes the same set of points. Simply substitute $x = \bar{x} + h$ and $y = \bar{y} + k$ for x and y in the given equation. For example, the equation of the line $ax + by = c$ is transformed into the equation $a(\bar{x} + h) + b(\bar{y} + k) = c$, that is

$$a\bar{x} + b\bar{y} = c - (ah + bk).$$

Note that the equation of the line is the same in either coordinate system except for the constant on the right side. Can you explain why geometrically?

The Circle

In vector notation, the circle with center $\mathbf{h} = (h, k)$ and radius $r > 0$ has the equation

$$\|\mathbf{x} - \mathbf{h}\| = r$$

in the x, y-coordinate system. Squared and expressed in components, this is equivalent to

$$(x - h)^2 + (y - k)^2 = r^2,$$

an equation already studied on p. 137.

If coordinates are translated so the $\bar{\mathbf{x}}$-origin is at $\mathbf{h}$, then the circle

$$\|\mathbf{x} - \mathbf{h}\| = r \quad \text{has equation} \quad \|\bar{\mathbf{x}}\| = r$$

in the new coordinate system.

Intersection of Line and Circle

Given a circle and a line, there are three possibilities: (1) they do not intersect, (2) they have exactly one common point—the line is a tangent to the circle, (3) they intersect in two distinct points.

From the equations of the circle and line, we can find explicitly the intersections (if any). The algebra is simplest if the circle has its center at the origin. Otherwise, we make a preliminary translation to move the origin to the center.

Example 1 Find the intersections (if any) of

$$x^2 + y^2 = 25 \quad \text{and} \quad x - y = 1.$$

Solution Solve the second equation for y in terms of x and substitute this expression into the first equation:

$$y = x - 1 \qquad x^2 + (x - 1)^2 = 25.$$

Expand, collect terms, and simplify:

$$x^2 + (x^2 - 2x + 1) = 25 \qquad 2x^2 - 2x - 24 = 0$$

$$x^2 - x - 12 = 0.$$

Solve the resulting quadratic for x; factoring is easiest:

$$(x - 4)(x + 3) = 0 \qquad x = -3, 4.$$

The corresponding values of y are

$$y = x - 1 = -3 - 1 = -4 \qquad y = x - 1 = 4 - 1 = 3$$

so there are two intersections,

$$(x, y) = (-3, -4) \quad \text{and} \quad (x, y) = (4, 3).$$

See Fig. 4.

Answer Two intersections: $(-3, -4)$ and $(4, 3)$

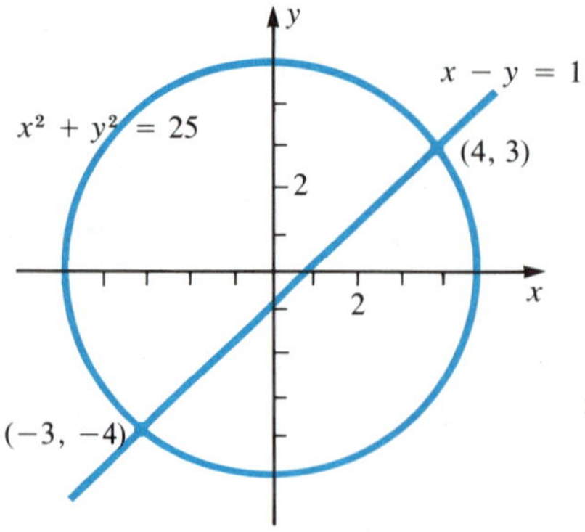

Fig. 4

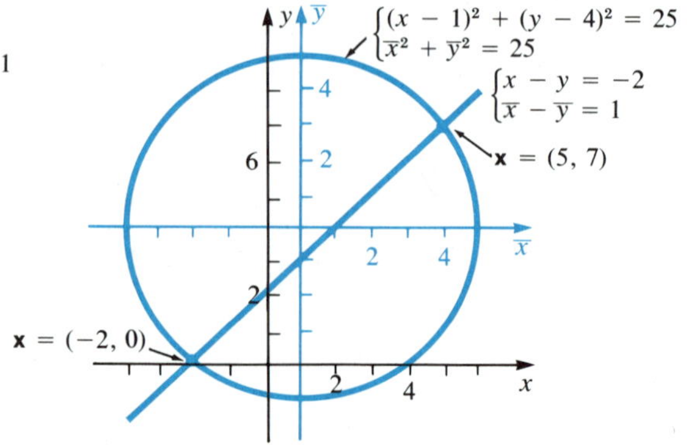

Fig. 5

EXAMPLE 2 Find the intersections (if any) of

$$(x - 1)^2 + (y - 4)^2 = 25 \quad \text{and} \quad x - y = -2$$

Solution Set $\bar{x} = x - 1$ and $\bar{y} = y - 4$. Then the new equation of the circle is $\bar{x}^2 + \bar{y}^2 = 25$, and the new equation of the line is

$$(\bar{x} + 1) - (\bar{y} + 4) = -2, \quad \text{that is,} \quad \bar{x} - \bar{y} = 1.$$

Thus we must solve simultaneously

$$\bar{x}^2 + \bar{y}^2 = 25 \quad \text{and} \quad \bar{x} - \bar{y} = 1.$$

This is the system of Example 1, but in $\bar{x}$ and $\bar{y}$ instead of x and y, so the solutions are $(\bar{x}, \bar{y}) = (-3, -4)$ and $(\bar{x}, \bar{y}) = (4, 3)$. But

$$\bar{x} = x - 1 \quad \text{so} \quad x = \bar{x} + 1, \qquad \text{and} \qquad \bar{y} = y - 4 \quad \text{so} \quad y = \bar{y} + 4.$$

Therefore

$$(x, y) = (\bar{x} + 1, \bar{y} + 4) = \begin{cases} (-3 + 1, -4 + 4) = (-2, 0) \\ (4 + 1, 3 + 4) = (5, 7). \end{cases}$$

See Fig. 5.

Answer Two intersections: $(-2, 0)$ and $(5, 7)$

Remark 1 Example 2 can be solved without shifting axes. For instance, x can be eliminated from the original pair of equations and the resulting quadratic solved for y. The computations involved are longer.

Remark 2 When one variable is eliminated from the equation of a line and the equation of a circle, the result is a *quadratic* equation for the other variable.

If this equation has negative discriminant (no real solution), then the line and circle do not intersect. If it has zero discriminant (only one solution), then the line is tangent to the circle, a situation we shall explore in Section 2.

Intersection of Two Circles

Given two non-concentric circles, how can we find their points of intersection, if any? We can reduce this problem to the previous problem of the intersection of a circle and a line. Indeed, let the equations of the circles be

(1) $x^2 + y^2 - 2a_1x - 2b_1y = c_1$

(2) $x^2 + y^2 - 2a_2x - 2b_2y = c_2$.

The centers are (a_1, b_1) and (a_2, b_2) respectively, as can be seen by completing the squares. Subtract the two equations:

(3) $2(a_1 - a_2)x + 2(b_1 - b_2)y = c_1 - c_2$.

Because the circles are not concentric, $(a_1, b_1) \neq (a_2, b_2)$; hence $a_1 - a_2$ and $b_1 - b_2$ are not both 0, so (3) is the equation of a line L.

This line L intersects either circle in precisely the same points where the circles intersect each other. Why? Because the intersections of the circles are the simultaneous solutions of the equations (1) and (2). The intersections of L and the first circle are the simultaneous solutions of equations (1) and (3). In either case the answers are the same. For if (x, y) satisfies both (1) and (2), then (x, y) also satisfies (3), the difference of (1) and (2). Likewise if (x, y) satisfies (1) and (3), then (x, y) also satisfies (2), the difference of (1) and (3). Similarly, the intersections of

L and the second circle are the same points. [In other words, the system (1) and (2) is equivalent to the system (1) and (3) and to the system (2) and (3).]

Thus finding the intersection of two circles is equivalent to finding the intersection of a circle and a line.

EXAMPLE 3 Find the intersections (if any) of

$$x^2 + y^2 = 25 \quad \text{and} \quad x^2 + y^2 - 4x + 4y = 21.$$

Solution First note that the second equation really defines a circle, because by completing squares it can be written in the form

$$(x - 2)^2 + (y + 2)^2 = 29.$$

This circle has center $(2, -2)$ and radius $\sqrt{29}$.

We subtract the equations of the two circles

$$\begin{array}{r} x^2 + y^2 \qquad\qquad\; = 25 \\ x^2 + y^2 - 4x + 4y = 21 \\ \hline 4x - 4y = \;\; 4 \end{array}, \quad \text{that is,} \quad x - y = 1.$$

Therefore the system

$$\begin{cases} x^2 + y^2 = 25 \\ x^2 + y^2 - 4x + 4y = 21 \end{cases} \quad \text{is equivalent to} \quad \begin{cases} x^2 + y^2 = 25 \\ x - y = 1. \end{cases}$$

This is precisely the problem of intersecting a circle and a line that we solved in Example 1. The solutions are $(-3, -4)$ and $(4, 3)$.

Answer Two intersections: $(-3, -4)$ and $(4, 3)$

Exercises

An $\bar{x}, \bar{y}$-coordinate system is introduced with its origin at $(-5, 2)$. Find the $\bar{x}, \bar{y}$-coordinates of the point (x, y)

1 $(1, 6)$ **2** $(5, 4)$ **3** $(0, 0)$

4 $(-8, 1)$ **5** $(-5, -6)$ **6** $(0, 2)$.

With the $\bar{x}, \bar{y}$-coordinate system as above, find the x, y-coordinates if

7 $(\bar{x}, \bar{y}) = (1, 3)$ **8** $(\bar{x}, \bar{y}) = (4, -2)$

9 $(\bar{x}, \bar{y}) = (-6, -7)$ **10** $(\bar{x}, \bar{y}) = (0, 3)$

11 Show that the distance formula gives the same distance between two points $\mathbf{x}_1$ and $\mathbf{x}_2$ regardless of whether the computation is done in x, y-coordinates or in $\bar{x}, \bar{y}$-coordinates.

12 Let $\mathbf{x}_1$ and $\mathbf{x}_2$ be vectors. Show that computing $\mathbf{x}_1 - \mathbf{x}_2$ in either x, y-coordinates or $\bar{x}, \bar{y}$-coordinates yields the same answer.

Describe a translation of axes that converts the first equation into the second:

13 $3x - 2y = 1 \qquad 3\bar{x} - 2\bar{y} = 0$ **14** $x + 2y = 5 \qquad \bar{x} + 2\bar{y} = 0$

15 $y = x^2 + 2x + 2 \qquad \bar{y} = \bar{x}^2 + 1$ **16** $y = 3 + \dfrac{x - 1}{1 + (x - 1)^2} \qquad \bar{y} = \dfrac{\bar{x}}{1 + \bar{x}^2}$

17 $y = \sin(x - \frac{1}{6}\pi) - 1 \qquad \bar{y} = \sin \bar{x}$

18 $y = \cos x + \cos(x + \frac{1}{3}\pi) + \cos(x + \frac{2}{3}\pi)$
$\bar{y} = \cos(\bar{x} - \frac{1}{3}\pi) + \cos \bar{x} + \cos(\bar{x} + \frac{1}{3}\pi)$.

Suppose the normal form of a line is $\mathbf{x} \cdot \mathbf{n} = p$.

19 If the origin is translated to the point $\mathbf{h}$, find the normal form in the new coordinates.

20 (cont.) Describe all translations which convert the normal form to $\bar{\mathbf{x}} \cdot \mathbf{n} = 0$. Interpret geometrically.

Find the intersections (if any) of the circle and the line

21 $x^2 + y^2 = 25 \quad y = 7x + 25$

22 $x^2 + y^2 = 25 \quad x - 7y = 25$

23 $x^2 + y^2 = 169 \quad x + y = 17$

24 $x^2 + y^2 = 169 \quad 7x + 17y = 169$

25 $x^2 + y^2 = 5 \quad x + 2y = 5$

26 $x^2 + y^2 = 10 \quad y = 10 - 3x$

27 $x^2 + y^2 = 10 \quad x + y + 2 = 0$

28 $x^2 + y^2 = 13 \quad y = x + 1$

29 $x^2 + y^2 = 4 \quad x + 2y = 1$

30 $x^2 + y^2 = 9 \quad y = x + 1$

31 $x^2 - 2x + y^2 + 4y = 20 \quad y = 7x + 16$

32 $x^2 + 6x + y^2 + 2y = 15 \quad x - 7y = 29$

33 $x^2 + 10x + y^2 + 24y = 0 \quad x + y = 0$

34 $x^2 - 6x + y^2 + 14y = 111 \quad 7x + 17y = 71$

35 $x^2 + y^2 + 6x + 6y + 13 = 0 \quad x + 2y + 4 = 0$

36 $x^2 + y^2 + 4x - 6y + 3 = 0 \quad 3x + y = 7$

37 $x^2 + y^2 - 10x + 10y + 40 = 0 \quad x + y + 2 = 0$

38 $x^2 + y^2 - 4x = 9 \quad y = x - 1$

39 $x^2 + y^2 = 6 \quad y = 2x - 7$

40 $x^2 + y^2 - 4x - 2y + 4 = 0 \quad 2x - 5y = 6$.

Find the points of intersection (if any) of the two circles

41 $x^2 + y^2 = 5 \quad x^2 + y^2 + 2x + 4y = 15$

42 $x^2 + y^2 = 13 \quad x^2 + y^2 - 4x + 4y = 17$

43 $x^2 + y^2 - 10 \quad x^2 + y^2 - 3x - y - 0$

44 $x^2 + y^2 = 169 \quad x^2 + y^2 - 34x - 14y = -169$

45 $x^2 + y^2 = 25 \quad x^2 + y^2 + 2x + 2y = 31$

46 $x^2 + y^2 = 4 \quad x^2 + y^2 - 4x - 8y = 0$

47 $x^2 + y^2 = 9 \quad x^2 + y^2 + 8x + 12 = 0$

48 $x^2 + y^2 - 2x - 2y = 0 \quad (x - 2)^2 + (y - 3)^2 = 4$

49 $x^2 + y^2 - 10x + 6y + 33 = 0 \quad x^2 + y^2 + 2x + 4y - 4 = 0$

50 $(x - 1)^2 + (y - 2)^2 = \frac{5}{4} \quad (x - 3)^2 + (y - 6)^2 = \frac{45}{4}$.

2 TANGENTS AND LOCI

Tangents

A line tangent to a circle is perpendicular to the radius at the point of contact. This simple fact makes it easy to find the equation of the tangent to a circle at one of its points.

EXAMPLE 1 Find the equation of the tangent to the circle

$$x^2 + y^2 = 25$$

at the point $(4, 3)$.

Solution The radius at the point of contact is the line segment from $(0, 0)$ to $(4, 3)$. Its slope is $\frac{3}{4}$; the slope of the tangent is the negative reciprocal, $-\frac{4}{3}$. Thus the tangent passes through $(4, 3)$ and has slope $-\frac{4}{3}$. By the point-slope form, its equation is $y - 3 = -\frac{4}{3}(x - 4)$. See Fig. 1, next page.

Answer $y = -\frac{4}{3}x + \frac{25}{3}$

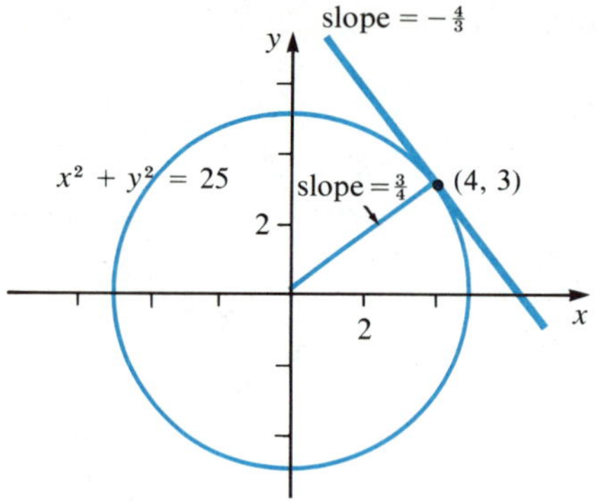

Fig. 1

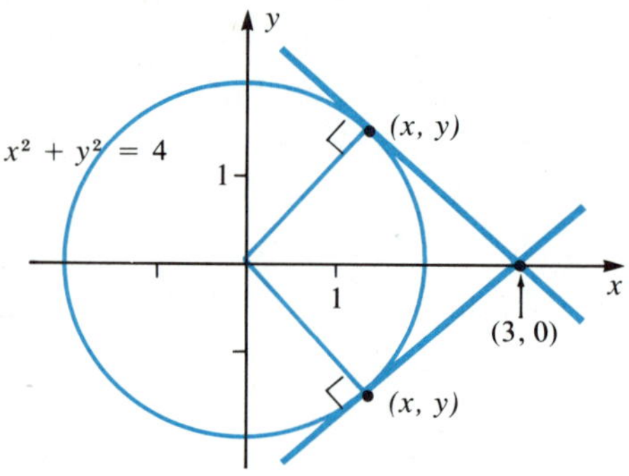

Fig. 2

The problem of finding the two tangent lines to a circle from an external point is slightly harder, but the basic idea is the same.

EXAMPLE 2 Find the two tangents to the circle

$$x^2 + y^2 = 4$$

from the point $(3, 0)$.

Solution Suppose a tangent meets the circle at (x, y). Then the vectors (x, y) and $(x, y) - (3, 0)$ are perpendicular (Fig. 2), so their dot product is 0:

$$(x, y)\cdot[(x, y) - (3, 0)] = (x, y)\cdot(x - 3, y) = 0$$

$$x^2 + y^2 = 3x.$$

Therefore the points of contact satisfy two equations:

$$x^2 + y^2 = 4 \qquad x^2 + y^2 = 3x.$$

Subtract: $3x = 4$, $x = \frac{4}{3}$. Hence

$$y^2 = 4 - (\tfrac{4}{3})^2 = \tfrac{36}{9} - \tfrac{16}{9} = \tfrac{20}{9} \qquad y = \pm\tfrac{2}{3}\sqrt{5}.$$

Each tangent passes through $(3, 0)$ and one of the two points $(\frac{4}{3}, \pm\frac{2}{3}\sqrt{5})$. By the two-point form, the two equations are

$$\frac{y - 0}{x - 3} = \frac{\pm\frac{2}{3}\sqrt{5}}{\frac{4}{3} - 3} = \frac{\pm\frac{2}{3}\sqrt{5}}{-\frac{5}{3}} = \mp\tfrac{2}{5}\sqrt{5}.$$

Answer $y = \pm\frac{2}{5}\sqrt{5}(x - 3)$.

EXAMPLE 3 Find the two tangents to the circle

$$x^2 + y^2 = 169$$

from the point $(-7, 17)$.

Solution Let (x, y) denote one of the points of tangency. Then $x^2 + y^2 = 169$, and the vectors $(x, y) - (-7, 17)$ and (x, y) are perpendicular. Therefore their dot product is 0:

$$(x, y)\cdot[(x, y) - (-7, 17)] = 0 \qquad x^2 + y^2 + 7x - 17y = 0.$$

Consequently (x, y) satisfies the system

$$\begin{cases} x^2 + y^2 + 7x - 17y = 0 \\ x^2 + y^2 = 169. \end{cases}$$

Subtract and solve for y in terms of x:

$$7x - 17y = -169 \qquad y = \tfrac{1}{17}(7x + 169).$$

Substitute this expression into $x^2 + y^2 = 169$ and simplify:

$$x^2 + \frac{1}{17^2}(7x + 169)^2 = 169$$

$$17^2x^2 + (49x^2 + 14\cdot 169x + 169^2) = 17^2\cdot 169$$

$$2\cdot 169x^2 + 14\cdot 169x + 169\cdot(169 - 17^2) = 0$$

$$2x^2 + 14x - 120 = 0 \qquad x^2 + 7x - 60 = 0.$$

Now solve for x:

$$(x + 12)(x - 5) = 0 \qquad x = -12,\ 5.$$

From $y = \frac{1}{17}(7x + 169)$ follows the corresponding values of y: $y = 5$ and $y = 12$.

Answer $(-12, 5)$ $(5, 12)$

Locus

Often a geometric figure is described as the set of all points that satisfy a certain condition. A curve specified in this way is called a **locus** (plural, **loci**). For example, the locus of all points at a distance 3 from a fixed point $\mathbf{p}$ is a circle of radius 3 centered at $\mathbf{p}$. The locus of all points equidistant from fixed points $\mathbf{p}$ and $\mathbf{q}$ is the perpendicular bisector of the segment $\mathbf{pq}$. The locus of all points (x, y) for which $y = f(x)$ is the graph of the function $f(x)$.

Example 4 Let $\mathbf{a}$ and $\mathbf{b}$ be distinct points in the plane. Show that the locus of all points that are twice as far from $\mathbf{a}$ as from $\mathbf{b}$ is a circle.

Solution Let $\mathbf{x}$ be any point of the plane. Then $\|\mathbf{x} - \mathbf{a}\|$ is the distance from $\mathbf{x}$ to $\mathbf{a}$, and $\|\mathbf{x} - \mathbf{b}\|$ is the distance from $\mathbf{x}$ to $\mathbf{b}$. Therefore the condition that $\mathbf{x}$ is twice as far from $\mathbf{a}$ as from $\mathbf{b}$ is expressed as the equation

$$\|\mathbf{x} - \mathbf{a}\| = 2\|\mathbf{x} - \mathbf{b}\|.$$

Now let us choose the coordinate system (Fig. 3) so that $\mathbf{a}$ and $\mathbf{b}$ are on the

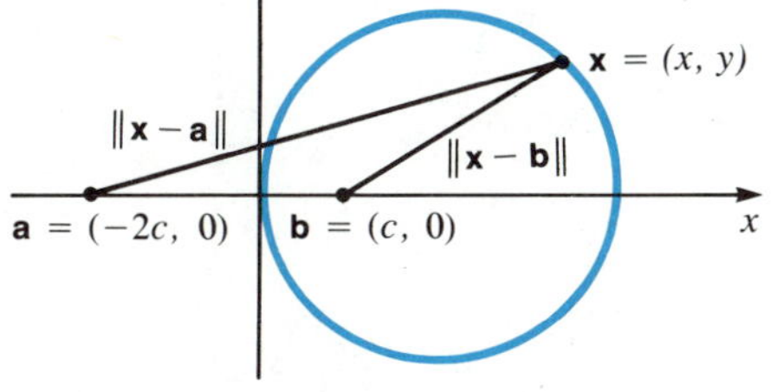

Fig. 3

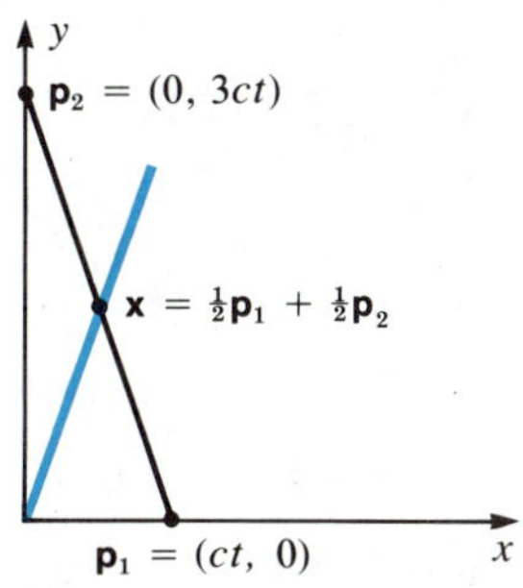

Fig. 4

x-axis and $(0, 0)$ is a point of the locus. We set $\mathbf{a} = (-2c, 0)$ and $\mathbf{b} = (c, 0)$, where c is a constant. (Actually then $c = \frac{1}{3}\|\mathbf{a} - \mathbf{b}\|$.) Then in terms of components, the condition that $\mathbf{x} = (x, y)$ is on the locus is

$$\sqrt{(x + 2c)^2 + y^2} = 2\sqrt{(x - c)^2 + y^2}.$$

Square and simplify:

$$(x + 2c)^2 + y^2 = 4[(x - c)^2 + y^2]$$

$$x^2 + 4cx + 4c^2 + y^2 = 4(x^2 - 2cx + c^2 + y^2)$$

$$12cx = 3(x^2 + y^2)$$

$$x^2 + y^2 - 4cx = 0.$$

Complete the square:

$$(x - 2c)^2 + y^2 = 4c^2.$$

This is the equation of the circle of radius $2c$ with center $(2c, 0)$.

EXAMPLE 5 A particle starts at the origin at time 0 and moves along the positive x-axis at uniform speed. A second particle also starts at the origin at time 0, but it moves along the positive y-axis, at three times the speed of the first particle. Find the locus of the midpoint of the segment joining the particles.

Solution (Fig. 4, previous page) Let c be the speed of the first particle. Then its position at time t is $\mathbf{p}_1 = (ct, 0)$. The speed of the second particle is $3c$ so its position at time t is $\mathbf{p}_2 = (0, 3ct)$. The midpoint of $\mathbf{p}_1\mathbf{p}_2$ is

$$\mathbf{x} = \tfrac{1}{2}(\mathbf{p}_1 + \mathbf{p}_2) = \tfrac{1}{2}[(ct, 0) + (0, 3ct)] = (\tfrac{1}{2}ct, \tfrac{3}{2}ct).$$

In components,

$$x = \tfrac{1}{2}ct \qquad y = \tfrac{3}{2}ct,$$

hence $y = 3x$.

Answer The ray $y = 3x$, $x \ge 0$

EXAMPLE 6 Let $\mathbf{z}$ trace the curve $y = x^2$. Find the locus of the point $\mathbf{x}$ that is $\frac{1}{3}$ of the way from $(-6, 0)$ to $\mathbf{z}$.

Solution $\mathbf{z}$ traces the curve $y = x^2$, so we can write $\mathbf{z} = (z, w)$, where $w = z^2$. We write $\mathbf{x} = (x, y)$ and seek the relation between x and y.

By the rule for division of a segment (p. 436)

$$\mathbf{x} = \tfrac{2}{3}(-6, 0) + \tfrac{1}{3}\mathbf{z}$$

hence

$$\mathbf{z} = 3\mathbf{x} - 2(-6, 0) \qquad (z, w) = 3(x, y) - 2(-6, 0),$$

$$z = 3x + 12 \qquad w = 3y.$$

Now substitute into $w = z^2$:

$$3y = (3x + 12)^2 = 9(x + 4)^2 \qquad y = 3(x + 4)^2.$$

Answer The quadratic curve $y = 3(x + 4)^2$

Remark Note carefully the change in notation. You cannot write *both* $\mathbf{x} = (x, y)$ *and* $\mathbf{z} = (x, y)$ because $\mathbf{x}$ and $\mathbf{z}$ are completely different points. So if you want the answer in x and y, you *must* give the coordinates of $\mathbf{z}$ some other names.

Exercises

Find the tangent to $x^2 + y^2 = 1$ at

1 $(-1, 0)$ **2** $(0, -1)$ **3** $(\frac{1}{2}\sqrt{2}, -\frac{1}{2}\sqrt{2})$
4 $(-\frac{1}{2}\sqrt{2}, -\frac{1}{2}\sqrt{2})$ **5** $(\frac{1}{2}\sqrt{3}, \frac{1}{2})$ **6** $(-\frac{1}{2}, \frac{1}{2}\sqrt{3})$.

Find the two tangents to the circle from the point

7 $x^2 + y^2 = 9$ from $(5, 0)$ **8** $x^2 + y^2 = 16$ from $(0, -5)$
9 $x^2 + y^2 = 1$ from $(2, 2)$ **10** $x^2 + y^2 = 4$ from $(2, 2)$
11 $x^2 + y^2 = 13$ from $(-5, 1)$ **12** $x^2 + y^2 = 13$ from $(5, -1)$
13 $(x - 1)^2 + y^2 = 1$ from $(3, 2)$ **14** $x^2 + (y + 3)^2 = 16$ from $(0, -8)$.

15 Show that the tangent to the circle $x^2 + y^2 = r^2$ at a point (x_0, y_0) on the circle is $x_0x + y_0y = r^2$.

16 (cont.) Show that the tangent to $(x - a)^2 + (y - b)^2 = r^2$ at a point (x_0, y_0) on the circle is $(x_0 - a)(x - a) + (y_0 - b)(y - b) = r^2$.

17 Let $\mathbf{a}$ and $\mathbf{b}$ be points and c a positive number. Show that the vector equation $(\mathbf{x} - \mathbf{a})\cdot(\mathbf{x} - \mathbf{b}) = c^2$ defines a circle. Find its center and radius.

18 Let $\mathbf{a}_1$, $\mathbf{a}_2$, $\mathbf{a}_3$, $\mathbf{a}_4$ be points and c a number. Show that

$$\|\mathbf{x} - \mathbf{a}_1\|^2 + \|\mathbf{x} - \mathbf{a}_2\|^2 + \|\mathbf{x} - \mathbf{a}_3\|^2 + \|\mathbf{x} - \mathbf{a}_4\|^2 = c^2$$

is the equation of a circle provided c is sufficiently large. Find the center of the circle.

19 A 10-ft ladder leans against a wall, and its foot slides along the floor as its top slides down the wall. Find the locus of its midpoint.

20 Find the locus of the midpoints of all chords of length 6 of a circle of radius 5.

21 At each point of a circle of radius 1 is drawn a segment of length 3, tangent at one end to the circle. Find the locus of the other ends of the segments.

22 Find the locus of all points whose distance from $(0, 0)$ is twice the distance from $(3, 4)$.

23 Find the locus of all midpoints of segments $\mathbf{pz}$, where $\mathbf{p} = (2, 0)$ and $\mathbf{z}$ varies over the circle $x^2 + y^2 = 1$.

24 Find the locus of all midpoints of segments $\mathbf{pz}$, where $\mathbf{p} = (1, 0)$ and $\mathbf{z}$ varies over the circle $x^2 + y^2 = 9$.

25 Find the locus of all points $\mathbf{x}$ such that the point $\frac{1}{3}$ of the way from $(-1, 0)$ to $\mathbf{x}$ lies on the circle $x^2 + y^2 = 25$.

26* Find the locus of all midpoints of chords of the circle $x^2 + y^2 = 4$ that pass through $(1, 0)$.

3 POLAR COORDINATES

Until now, we have always used rectangular coordinates to locate points in the plane. Other coordinate systems exist as well. The most important of these is the system of polar coordinates.

In a rectangular coordinate system (Fig. 1a, next page) two families of grid lines, $x =$ constant and $y =$ constant, fill the plane. Each point (a, b) is the intersection of two of these lines, $x = a$ and $y = b$.

In a polar coordinate system (Fig. 1b), the grid lines are (1) all circles centered at $(0, 0)$, and (2) all rays from $(0, 0)$. Each point (x, y) different from $(0, 0)$ is the intersection of one circle and one ray. The circle is identified by a positive number r, its radius, and the ray is identified by a real number θ, its angle in radians from the positive x-axis. Thus (x, y) is assigned the polar coordinates

$\{r, \theta\}$. Since θ is determined only up to a multiple of 2π, we agree that

$$\{r, \theta + 2\pi n\} = \{r, \theta\} \qquad (n \text{ any integer}).$$

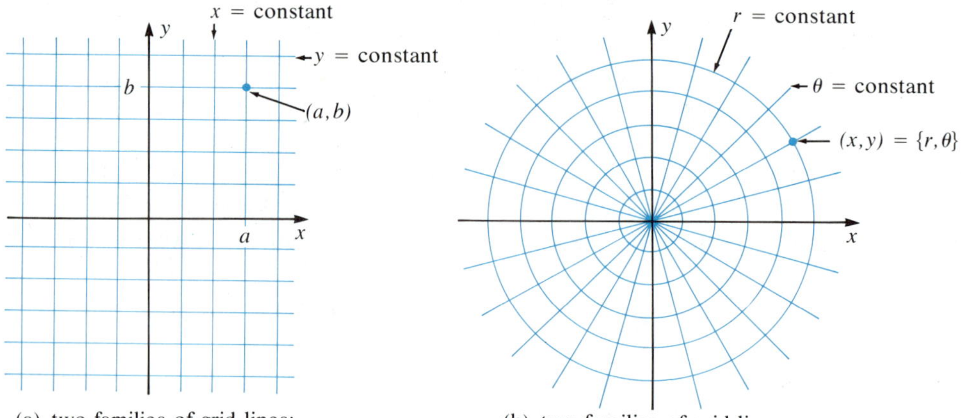

(a) two families of grid lines: x = constant and y = constant

(b) two families of grid lines: r = constant and θ = constant

Fig. 1 Rectangular and polar coordinates

The point $(0, 0)$ does not determine an angle θ. Nonetheless, it is customary to say that any pair $\{0, \theta\}$ represents $(0, 0)$.

The idea of polar coordinates is quite natural. You identify a point by telling how far it is from the origin and in what direction. (This is the principle of the radar screen, where the origin is your current position.) In Figure 2, we plot a few points in polar coordinates.

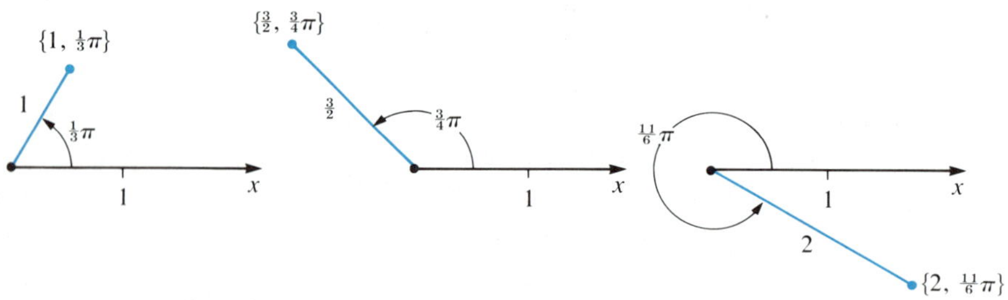

Fig. 2 Examples of polar coordinates

Given the polar coordinates $\{r, \theta\}$ of a point, what are its rectangular coordinates? The point is r units from the origin in the direction θ. Hence $x = r\cos\theta$, $y = r\sin\theta$. See Fig. 3a. Conversely, given the rectangular coordinates (x, y),

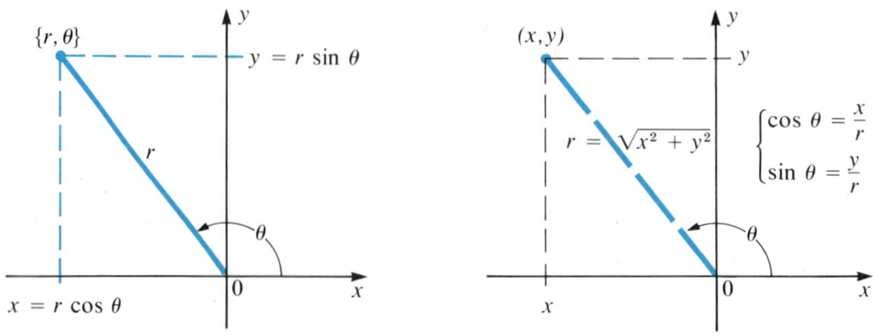

(a) $\{r, \theta\}$ determines x and y

(b) (x, y) determines r and θ

Fig. 3

what are the polar coordinates? Figure 3b shows that $r = \sqrt{x^2 + y^2}$, and that θ is determined by $\cos\theta = x/r$ and $\sin\theta = y/r$.

Polar to Rectangular	Rectangular to Polar
$x = r\cos\theta$	$r = \sqrt{x^2 + y^2}$
$y = r\sin\theta$	$\cos\theta = \dfrac{x}{\sqrt{x^2 + y^2}} = \dfrac{x}{r}$
	$\sin\theta = \dfrac{y}{\sqrt{x^2 + y^2}} = \dfrac{y}{r}$

EXAMPLE 1 (a) Convert $(2, -2\sqrt{3})$ to polar coordinates.
(b) Convert $\{3, \frac{1}{6}\pi\}$ to rectangular coordinates.

Solution (a) $r^2 = 4 + 12 = 16$, $r = 4$. Also $\cos\theta = \frac{2}{4} = \frac{1}{2}$ and $\sin\theta = \frac{1}{4}(-2\sqrt{3}) = -\frac{1}{2}\sqrt{3}$, so $\theta = \frac{5}{3}\pi$.

(b) $x = r\cos\theta = 3\cos\frac{1}{6}\pi = \frac{3}{2}\sqrt{3}$, and $y = r\sin\theta = 3\sin\frac{1}{6}\pi = \frac{3}{2}$.

Answer (a) $\{4, \frac{5}{3}\pi\}$ (b) $(\frac{3}{2}\sqrt{3}, \frac{3}{2})$

Remark Some calculators have built-in keys for rectangular-polar conversions.

Negative r

In applications it is convenient to allow points $\{r, \theta\}$ with $r < 0$. For example, consider a ray and a point $\{r, \theta\}$ on the ray (Fig. 4a). Suppose the point moves towards $(0, 0)$, through $(0, 0)$, and keeps on going! Then r decreases, becomes 0, but then what? So that θ won't jump abruptly to $\theta + \pi$, we agree that θ remains constant, but r becomes negative (Fig. 4b). This amounts to agreeing that

$$\{-r, \theta\} = \{r, \theta + \pi\}.$$

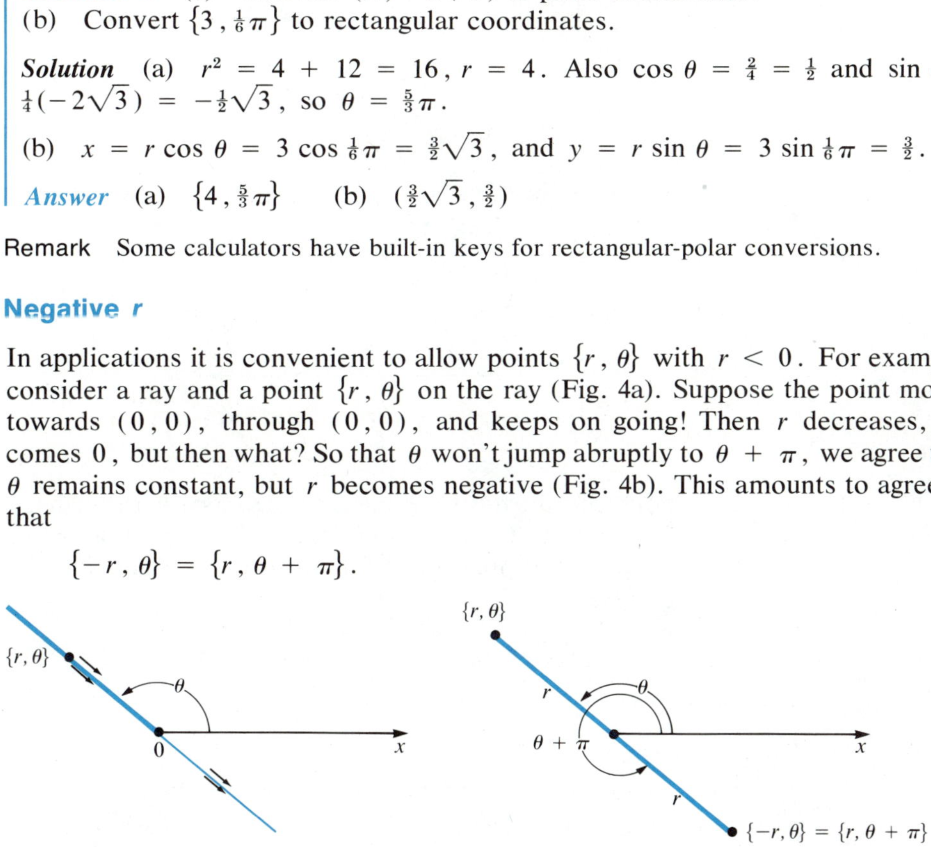

(a) θ fixed; $\{r, \theta\}$ moves through 0. (b) $\{-r, \theta\}$ and $\{r, \theta + \pi\}$ are the same point.

Fig. 4 Negative r

Just for practice, let us plot a few specific points with r negative (Fig. 5).

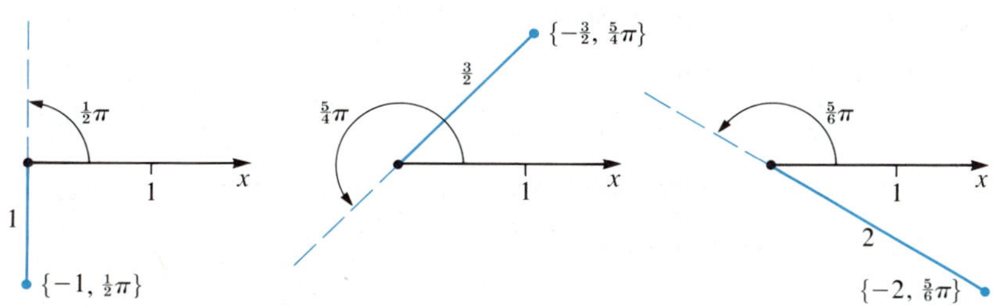

Fig. 5 Examples with $r < 0$

EXAMPLE 2 Convert to rectangular coordinates

(a) $\{-2, \frac{3}{4}\pi\}$ (b) $\{-5, -\frac{1}{3}\pi\}$.

Solution (a) $\{-2, \frac{3}{4}\pi\} = \{2, \frac{3}{4}\pi + \pi\} = \{2, \frac{7}{4}\pi\}$. Hence

$$x = 2\cos\tfrac{7}{4}\pi = \sqrt{2} \qquad y = 2\sin\tfrac{7}{4}\pi = -\sqrt{2}.$$

(b) $\{-5, -\frac{1}{3}\pi\} = \{5, -\frac{1}{3}\pi + \pi\} = \{5, \frac{2}{3}\pi\}$,

$$x = 5\cos\tfrac{2}{3}\pi = -\tfrac{5}{2} \qquad y = 5\sin\tfrac{2}{3}\pi = \tfrac{5}{2}\sqrt{3}$$

Answer (a) $(\sqrt{2}, -\sqrt{2})$ (b) $(-\frac{5}{2}, \frac{5}{2}\sqrt{3})$

Lines

Each curve $r = r_0$ is a circle with center $(0, 0)$, except for $r_0 = 0$. Each curve $\theta = \theta_0$ is a line through the origin (Fig. 6a).

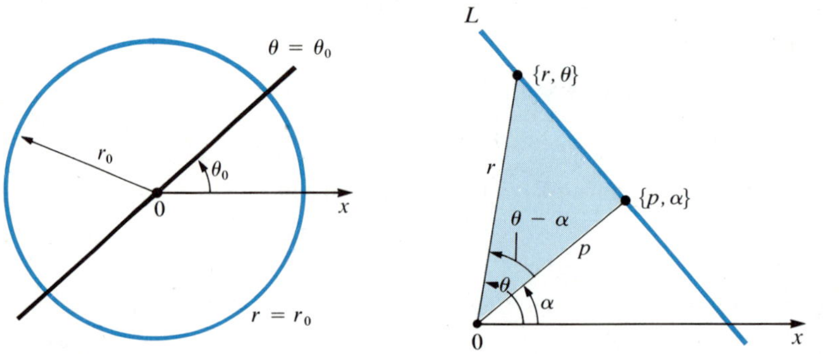

(a) $r = r_0$ and $\theta = \theta_0$ (b) equation of a line: $r\cos(\theta - \alpha) = p$

Fig. 6

What is the equation in polar coordinates for a line L not necessarily through the origin? Drop a perpendicular from $(0, 0)$ to L; let its foot have polar coordinates $\{p, \alpha\}$. See Fig. 6b. If $\{r, \theta\}$ is any point on L, then from the right triangle, $\cos(\theta - \alpha) = p/r$. Therefore

> The equation in polar coordinates of the general straight line is
>
> $$r\cos(\theta - \alpha) = p,$$
>
> where p and α are constants.

EXAMPLE 3 (a) Express in rectangular coordinates $r\cos(\theta - \frac{1}{6}\pi) = 3$.

(b) Express in polar coordinates $x + y = -5$.

Solution (a) Use the addition law for cosines:

$$r\cos(\theta - \tfrac{1}{6}\pi) = 3 \qquad r(\cos\theta\cos\tfrac{1}{6}\pi + \sin\theta\sin\tfrac{1}{6}\pi) = 3.$$

Replace $r\cos\theta$ by x and $r\sin\theta$ by y:

$$x\cos\tfrac{1}{6}\pi + y\sin\tfrac{1}{6}\pi = 3 \qquad \tfrac{1}{2}\sqrt{3}x + \tfrac{1}{2}y = 3.$$

(b) Substitute $x = r\cos\theta$ and $y = r\sin\theta$:

$$x + y = -5 \qquad r\cos\theta + r\sin\theta = -5.$$

To express this relation in polar form, we must have the right-hand side positive and the left-hand side in the form $r\cos(\theta - \alpha)$. So we bring the minus sign to the left and express the result as a constant times a cosine (review p. 283):

$$r(-\cos\theta - \sin\theta) = 5$$

$$r(-\tfrac{1}{2}\sqrt{2}\cos\theta - \tfrac{1}{2}\sqrt{2}\sin\theta) = \tfrac{5}{2}\sqrt{2}$$

$$r(\cos\tfrac{5}{4}\pi\cos\theta + \sin\tfrac{5}{4}\pi\sin\theta) = \tfrac{5}{2}\sqrt{2}$$

$$r\cos(\theta - \tfrac{5}{4}\pi) = \tfrac{5}{2}\sqrt{2}.$$

Answer (a) $\tfrac{1}{2}x\sqrt{3} + \tfrac{1}{2}y = 3$ (b) $r\cos(\theta - \tfrac{5}{4}\pi) = \tfrac{5}{2}\sqrt{2}$.

Remark The polar form $r\cos(\theta - \alpha) = p$ is equivalent to the normal form

$$x\cos\alpha + y\sin\alpha = p.$$

For substitute $x = r\cos\theta$ and $y = r\sin\theta$ into the normal form and use the addition formula for cosine:

$$r\cos\theta\cos\alpha + r\sin\theta\sin\alpha = p \qquad r\cos(\theta - \alpha) = p.$$

Exercises

Plot the points

1 $\{3, \tfrac{1}{6}\pi\}$ **2** $\{2, \pi\}$ **3** $\{2, \tfrac{5}{4}\pi\}$
4 $\{1, \tfrac{9}{4}\pi\}$ **5** $\{-3, \tfrac{2}{3}\pi\}$ **6** $\{-2, \tfrac{5}{3}\pi\}$.

Sketch the region in the plane

7 $r > 1$ **8** $\tfrac{1}{2} \le r < \tfrac{3}{2}$
9 $\tfrac{1}{4}\pi \le \theta \le \tfrac{1}{2}\pi$ **10** $1 \le r \le 2,\ \tfrac{4}{3}\pi \le \theta \le \tfrac{3}{2}\pi$.

Express in rectangular coordinates

11 $\{1, \tfrac{1}{2}\pi\}$ **12** $\{1, -\tfrac{1}{2}\pi\}$ **13** $\{1, -\tfrac{1}{6}\pi\}$
14 $\{1, \tfrac{1}{3}\pi\}$ **15** $\{2, -\tfrac{3}{4}\pi\}$ **16** $\{2, \tfrac{5}{4}\pi\}$.

Express in polar coordinates

17 $(1, 1)$ **18** $(0, -1)$ **19** $(-1, 1)$
20 $(-\tfrac{1}{2}, \tfrac{1}{2}\sqrt{3})$ **21** $(\sqrt{3}, -1)$ **22** $(\sqrt{2}, -\sqrt{2})$.

Find the equation in polar coordinates

23 line through $(0, 0)$ and $\{3, \tfrac{1}{4}\pi\}$ **24** circle, center $(0, 0)$, radius 5
25 line through $\{1, 0\}$ and $\{1, \tfrac{1}{2}\pi\}$ **26** line $ax + by = c$.

4 POLAR GRAPHS

We now find the equation in polar coordinates for a circle tangent to the y-axis at the origin (Fig. 1a, next page). Let the radius of the circle be a, so its center is $(a, 0)$ (provided the circle lies to the right of the y-axis). As seen in Fig. 1b, the point $\{r, \theta\}$ on the circle and the two endpoints of the horizontal diameter determine a right triangle. (Recall that an angle inscribed in a semicircle is a right angle.) Its hypotenuse is $2a$ and the side adjacent to θ is r, hence

$$r = 2a\cos\theta.$$

This is the polar equation of the circle.

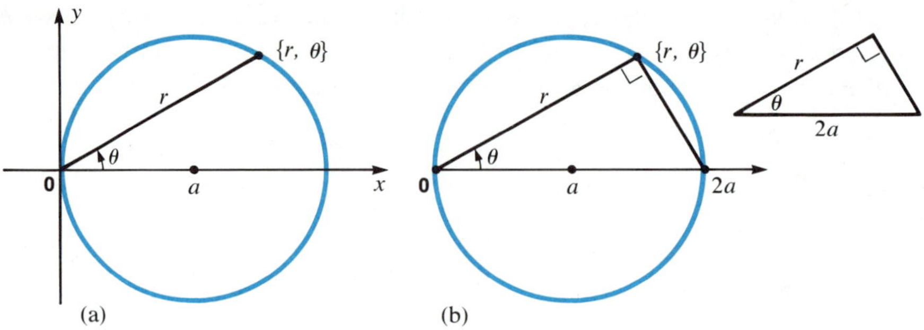

Fig. 1 Circle of radius a tangent to the y-axis at **0**

How does $\{r, \theta\}$ move on the circle $r = 2a \cos\theta$, as θ makes a complete revolution? If θ starts at 0, then r starts at $2a$. If θ increases to $\frac{1}{2}\pi$, then r decreases to 0. (Think of an arm rotating counterclockwise and shrinking.) Hence $\{r, \theta\}$ traces the upper half of the circle (Fig. 2a).

If θ increases from $\frac{1}{2}\pi$ to π, then r decreases from 0 to $-2a$. Since r is negative, the point $\{r, \theta\}$ is measured "backwards" and moves through the fourth quadrant, tracing the lower half of the circle (Fig. 2b).

Thus the full circle is described as θ runs from 0 to π. As θ runs from π to 2π, the same circle is traced again. For when θ is in the third quadrant, $r < 0$, so $\{r, \theta\}$ describes the semicircle in the first quadrant; when θ is in the fourth quadrant $r > 0$, so $\{r, \theta\}$ describes the semicircle in the fourth quadrant. Therefore, in one complete revolution of θ, the circle is traced *twice*.

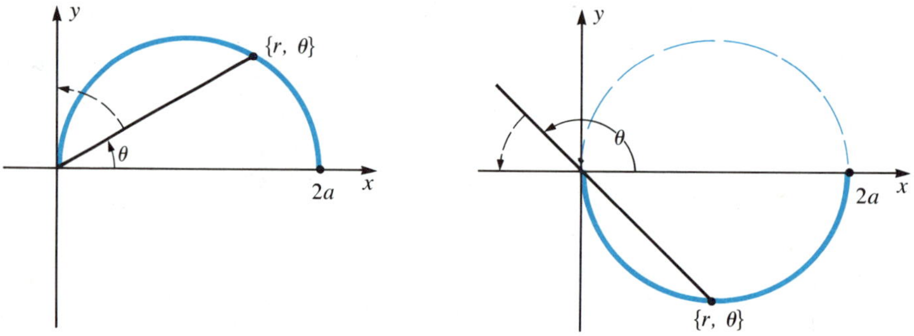

(a) As θ increases from 0 to $\frac{1}{2}\pi$, r decreases from $2a$ to 0.

(b) As θ increases from $\frac{1}{2}\pi$ to π, r decreases from 0 to $-2a$.

Fig. 2 The circle is traced by $\{r, \theta\}$ for $0 \leq \theta \leq \pi$

> The graph of the equation
>
> $$r = 2a \cos\theta$$
>
> is a circle of radius a and center $\{a, 0\}$. The circle is traced twice as θ makes a complete revolution.

Circles passing through the origin, but in other positions, can be handled similarly in polar coordinates.

Graphs of Functions

Graphing a function $r = f(\theta)$ in polar coordinates is tricky at first, because you must change your point of view. For $y = f(x)$, you think of x running along the horizontal axis, with the corresponding point (x, y) measured above or below. Basically your mental set is "left-right" and "up-down."

In polar coordinates, however, you must think of the angle θ swinging around (like a radar scope) and repeating after 2π. For each θ, you must measure forward from the origin a distance $f(\theta)$, or backward if $f(\theta) < 0$. Your mental set must be "round and round," "in and out," "forward or backward."

Because of the special nature of points $\{r, \theta\}$ where $r < 0$, pay close attention to the sign of $f(\theta)$ and be sure to plot points "backwards" if $f(\theta) < 0$.

Look for symmetries and periodicity. For example, if $f(\theta + 2\pi) = f(\theta)$, the polar graph $r = f(\theta)$ will repeat after 2π. There are many symmetries possible; we mention only two, $f(\theta)$ even and $f(\theta)$ odd. If $f(\theta)$ is even, that is, $f(-\theta) = f(\theta)$, then the point $\{r, -\theta\}$ is on the graph whenever $\{r, \theta\}$ is. The curve is symmetric in the x-axis (Fig. 3a). If $f(\theta)$ is odd, $f(-\theta) = -f(\theta)$, then the point $\{-r, -\theta\}$ is on the graph whenever $\{r, \theta\}$ is. The curve is symmetric in the vertical axis (Fig. 3b).

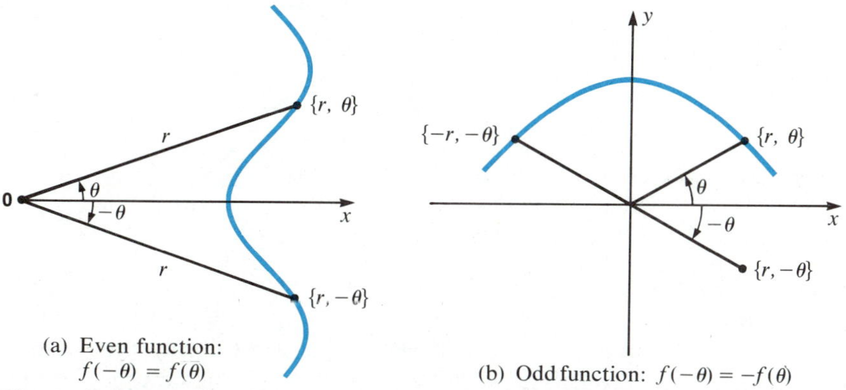

Fig. 3 Polar graphs of $r = f(\theta)$, where $f(\theta)$ is even or odd

EXAMPLE 1 Graph the **spiral of Archimedes,** $r = \theta$.

Solution As θ increases from 0 we see that r increases steadily. Hence the locus goes round and round, its distance from **0** greater and greater. The result is a spiral (Fig. 4a). Since $f(\theta) = \theta$ is an odd function, we obtain the locus for $\theta < 0$ by reflection in the vertical axis (Fig. 4b).

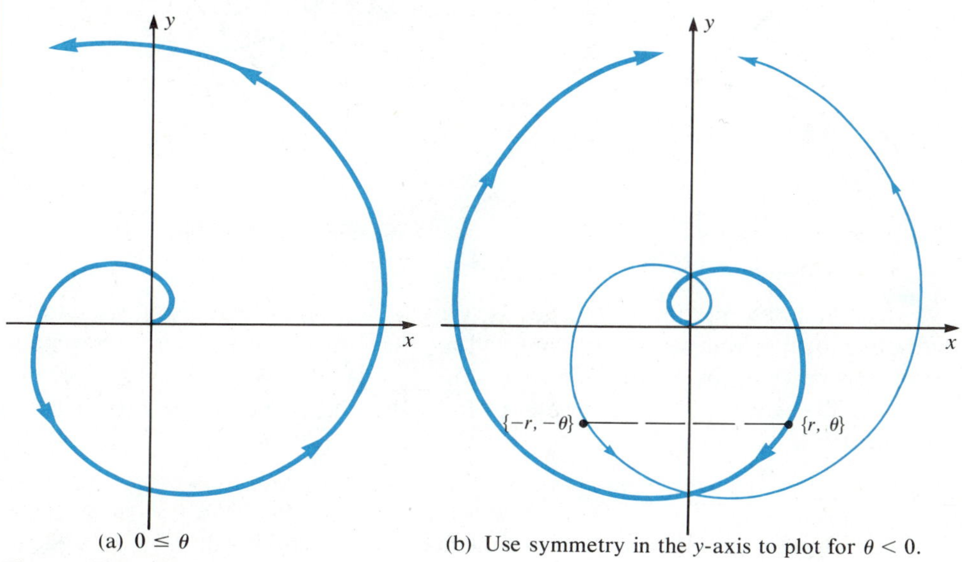

Fig. 4 The spiral $r = \theta$

EXAMPLE 2 Graph the **rose** $r = a \cos 2\theta$ where $a > 0$.

Solution Since $\cos 2(\theta + 2\pi) = \cos 2\theta$, the curve repeats every 2π, so we need plot it only for $0 \le \theta \le 2\pi$.

Since the sign of $\cos 2\theta$ fluctuates, it is advisable to make a preliminary sketch showing the proper sign (Fig. 5a). As θ starts at 0 and increases to $\frac{1}{4}\pi$, we see that $\cos 2\theta$ starts at 1 and decreases to 0. Since $\cos 2\theta$ is an even function, this part of the graph is repeated below, forming a loop (Fig. 5b).

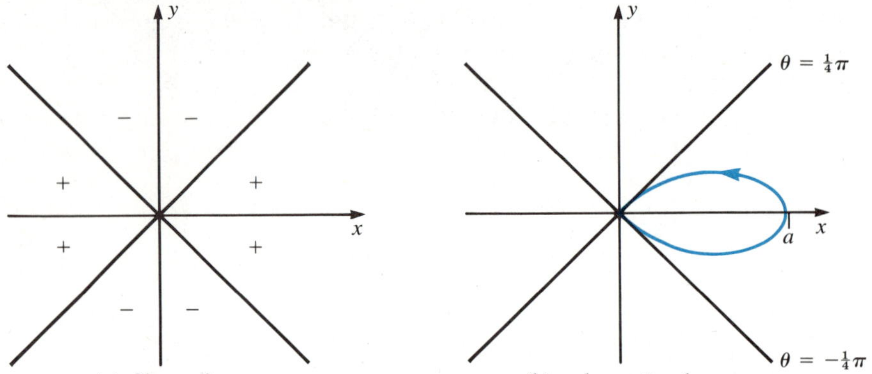

(a) Signs of r (b) $-\frac{1}{4}\pi \le \theta \le \frac{1}{4}\pi$

Fig. 5 Partial graph of $r = a \cos 2\theta$

As θ increases from $\frac{1}{4}\pi$ to $\frac{1}{2}\pi$ to $\frac{3}{4}\pi$, we see that $\cos 2\theta$ is negative and goes from 0 to -1 and back to 0. Thus we get another loop, but between $\frac{5}{4}\pi$ and $\frac{7}{4}\pi$. See Fig. 6a. For θ going from $\frac{3}{4}\pi$ to $\frac{5}{4}\pi$, we get a third loop plotted forward, and from $\frac{5}{4}\pi$ to $\frac{7}{4}\pi$ a fourth loop plotted backwards. The complete graph is shown in Fig. 6b.

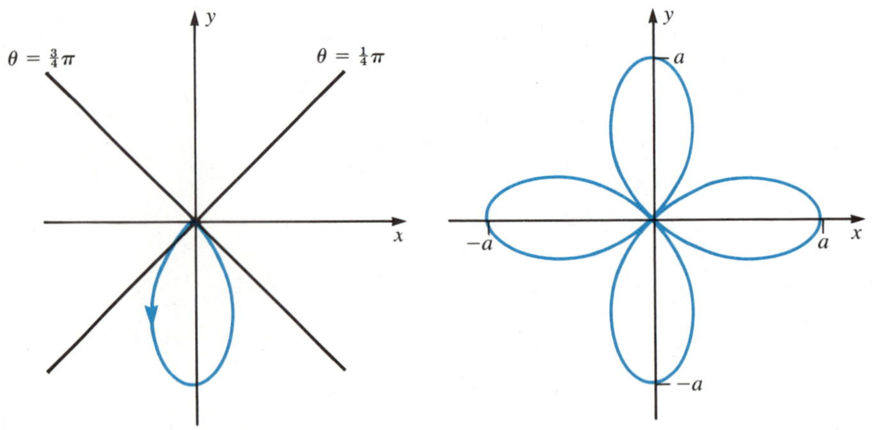

(a) $\frac{1}{4}\pi \le \theta \le \frac{3}{4}\pi$ (b) Complete graph

Fig. 6 Continuation of $r = a \cos 2\theta$

For an accurate picture of the petals, plot some points. One thing we can say without plotting: the petals are rounded at their ends, not pointed. That stems from a property of the cosine: for small angles, $\cos 2\theta$ is very close to 1. Hence for θ small (near the tip of the petal to the right) the curve $r = a \cos 2\theta$ looks like the circle $r = a$.

Hindsight It is necessary to plot only one of the petals in Fig. 6b. Since $\cos(\theta + \pi) = -\cos\theta$, we have $\cos 2(\theta + \frac{1}{2}\pi) = -\cos 2\theta$. Thus $\cos 2\theta$ repeats itself negatively after $\frac{1}{2}\pi$. We conclude that the first loop we plotted (for $-\frac{1}{4}\pi \le \theta \le \frac{1}{4}\pi$)

is repeated negatively as θ continues from $\frac{1}{4}\pi$ to $\frac{3}{4}\pi$. In other words, rotate the first loop backwards by $\frac{1}{2}\pi$; the result is another loop of the curve. Rotate again, and once again, and you have generated the whole curve.

Exercises

Find the equation in polar coordinates of the circle with

1 radius 2 center $(2, 0)$
2 radius 3 center $(-3, 0)$
3 radius 1 center $(0, 1)$
4 radius 1 center $(0, -1)$.

Graph

5 $r = 2\theta$
6 $r = -\theta$
7 $r = \theta^2$
8 $r = e^\theta$
9 $r = \theta + \pi$
10 $r = \ln(\theta/\pi)$
11 $r = \sin 2\theta$
12 $r = \cos 3\theta$
13 $r = \cos 4\theta$
14 $r = \sin 4\theta$
15 $r = \cos 5\theta$
16 $r = \sin 5\theta$
17 the **cardioid** $r = 1 - \cos\theta$
18 the **lemniscate** $r^2 = \cos 2\theta$
19 the **limaçon** $r = 2 + \cos\theta$
20 the **limaçon** $r = 1 + 2\cos\theta$
21 the **bifolium** $r = \sin\theta\cos^2\theta$
22 the **strophoid** $r = \cos 2\theta\sec\theta$
23 the **conchoid** $r = \csc\theta - 2$
24 the **cissoid** $r = \sin\theta\tan\theta$.

5 CONICS; THE PARABOLA

The ancient Greek geometers discovered that, on cutting a right circular cone by various planes, they obtained three types of remarkable curves called **conic sections,** or simply **conics** (Fig. 1). It is possible to define the conic sections as certain geometric loci. We shall do so because this is less complicated than starting from plane sections of a cone.

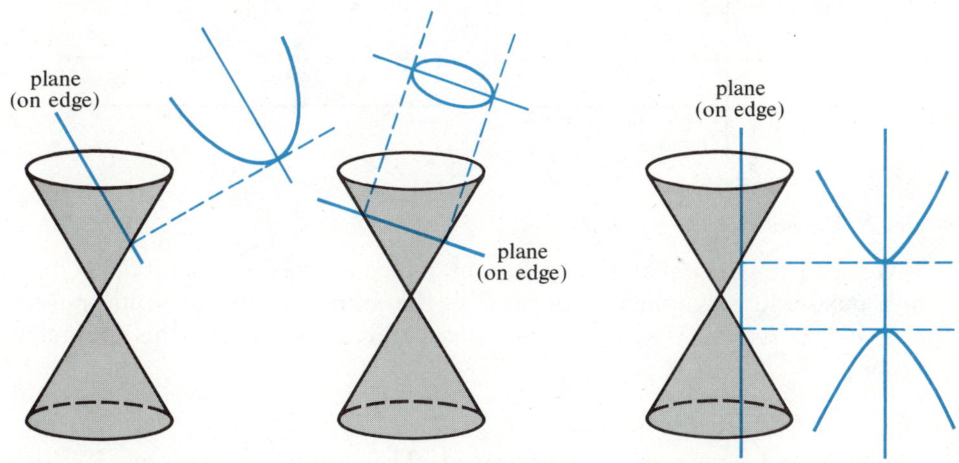

(a) Parabola: plane parallel to generator (b) Ellipse: plane cuts one nappe, not parallel to generator (c) Hyperbola: plane cuts both nappes

Fig. 1 Conic sections (conics)

The Parabola

A **parabola** is the locus of all points equidistant from a fixed line D and a fixed point $\mathbf{p}$ not on D. We call D the **directrix** and $\mathbf{p}$ the **focus** of the parabola. To find the equation of a parabola, we choose the coordinate system (Fig. 2a) so that $\mathbf{p} = (0, p)$ and D is the line $y = -p$. Then a point (x, y) is on the parabola if and only if the distance

$$\sqrt{(x - 0)^2 + (y - p)^2}$$

from (x, y) to $(0, p)$ equals the distance $|y + p|$ from (x, y) to D. Thus the condition is

$$\sqrt{x^2 + (y - p)^2} = |y + p|.$$

Square and simplify:

$$x^2 + (y - p)^2 = (y + p)^2$$

$$x^2 + y^2 - 2py + p^2 = y^2 + 2py + p^2$$

$$x^2 = 4py$$

The locus is shown in Fig. 2b.

> The equation of the parabola with focus $(0, p)$ and directrix $y = -p$ is
>
> $$y = \frac{1}{4p}x^2.$$
>
> The distance from the focus to the directrix equals $2p$.

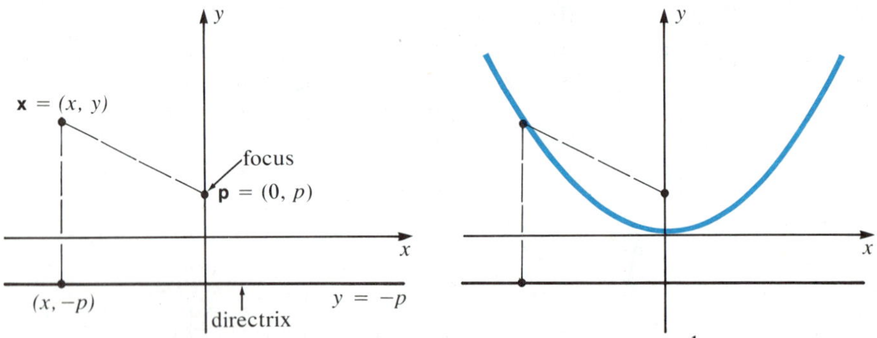

(a) Choose convenient coordinate axes (b) $y = \frac{1}{4p}x^2$

Fig. 2 Equation of the parabola

We see from the equation that for our choice of axes, the parabola is the graph of a quadratic polynomial. Conversely, the graph of any quadratic polynomial $y = ax^2 + bx + c$ (with $a \neq 0$) is a parabola. To prove this, complete the square:

$$y = a(x - h)^2 + k,$$

where h and k are easily determined. Then translate axes by $\bar{x} = x - h$, $\bar{y} = y - k$. The result is $\bar{y} = a\bar{x}^2$, a parabola with $4p = 1/a$.

The line through the focus perpendicular to the directrix is called the **axis** of the parabola. The point of intersection of the axis with the parabola is the **vertex** of the parabola (Fig. 3). Note that the vertex is halfway between the focus and the directrix. Hence the distance from the vertex to the focus equals p.

By interchanging the roles of x and y, we see that a parabola whose axis is parallel to the x-axis is the locus of an equation $x = ay^2 + by + c$. Some examples are shown in Fig. 4.

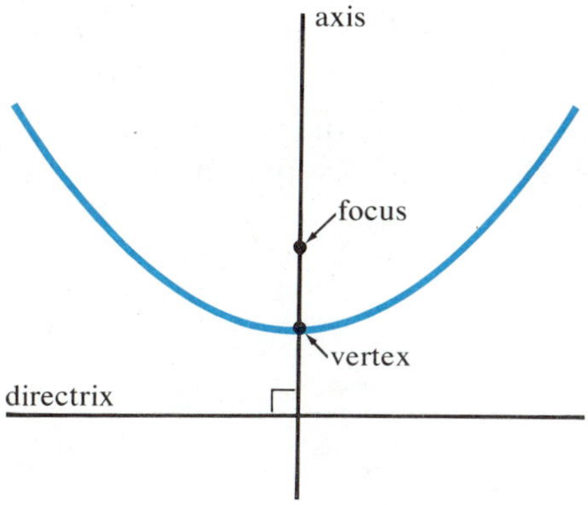

Fig. 3 Geometry of the parabola

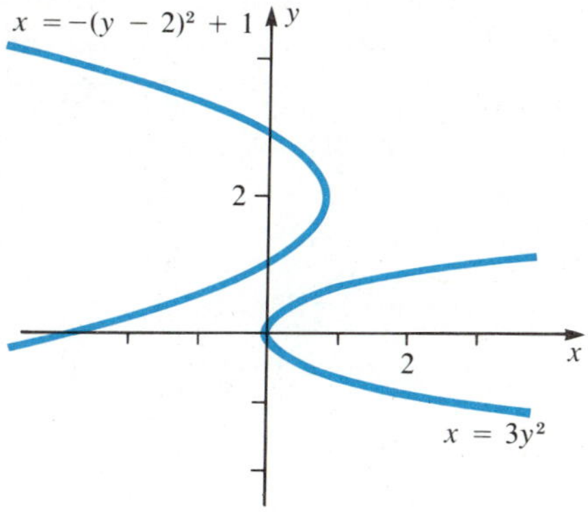

Fig. 4 Parabolas with horizontal axes

EXAMPLE 1 Sketch the parabola

$$8x = -y^2 + 2y + 15$$

Locate its focus, directrix, axis, and vertex.

Solution Rewrite the relation as

$$8x = -(y^2 - 2y) + 15$$

and complete the square:

$$8x = -(y^2 - 2y + 1) + 1 + 15$$

$$8x = -(y-1)^2 + 16$$

Therefore

$$8(x-2) = -(y-1)^2.$$

It follows that the vertex is $(2, 1)$, the axis is the horizontal line $y = 1$, and the parabola opens in the negative y-direction.

The focus is p units to the left of the vertex $(2, 1)$ and the directrix is p units to the right of $(2, 1)$. But what is p? Make the translation $\bar{x} = x - 2$ and $\bar{y} = y - 1$. In the new coordinates, the equation of the parabola is $8\bar{x} = -\bar{y}^2$. Clearly $4p = 8$, so $p = 2$. It follows that the focus is $(2-2, 1) = (0, 1)$, and the directrix is $x = 2 + 2$, that is, $x = 4$. The parabola is sketched in Fig. 5.

Answer focus $(0, 1)$ directrix $x = 4$

axis $y = 1$ vertex $(2, 1)$

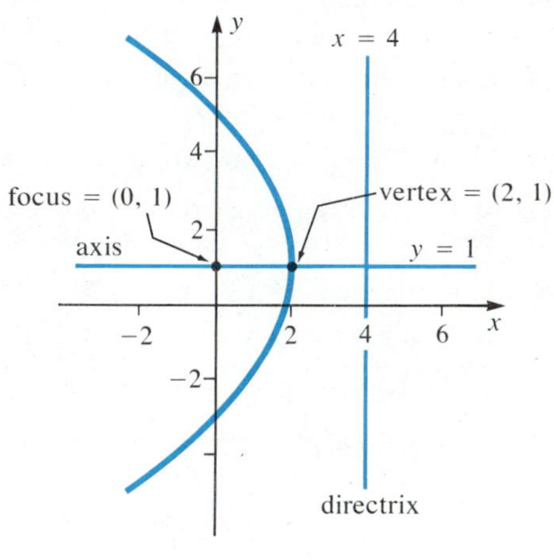

Fig. 5

Example 2 Find the equation of the parabola with focus $(-1, 1)$ and vertex $(-1, 3)$.

Solution The axis, which is the line through the focus and the vertex, is the vertical line $x = -1$. The parabola opens downward, and its highest point is its vertex $(-1, 3)$. Hence its equation is

$$y - 3 = -\frac{1}{4p}(x + 1)^2.$$

Since p is the distance from the vertex to the focus, $p = 2$.

Answer $y - 3 = -\frac{1}{8}(x + 1)^2$

Remark Parabolas have many applications. The path of a falling body, such as a bullet, with an initial non-zero horizontal component of velocity is a parabola with vertical axis. (Occasionally a mathematician testifies in court as to whether someone fell out of a window or was pushed.)

Suppose a parabola is rotated about its axis, forming a surface of revolution, and the inside of this surface is a mirror. Then each ray from the focus will be reflected to a ray parallel to the axis. This important focal property, proved by calculus, explains why headlight reflectors, telescope mirrors, and directional antennas are parabolic.

Exercises

Sketch the parabola and locate the focus, directrix, axis, and vertex

1 $y = x^2$
2 $x = 2y^2$
3 $x = -3y^2$
4 $y = -\frac{1}{2}x^2$
5 $x - 3 = 3y^2$
6 $x + 2 = -y^2$
7 $x = 2(y + 1)^2$
8 $2y = -3(x - 2)^2$
9 $x^2 + 4x - 6y = 0$
10 $2y^2 - 4y + x + 2 = 0$.

Find the equation of the parabola

11 vertex $(0, 0)$ through the point $(1, 3)$ axis vertical

12 vertex $(0, 0)$ through the point $(6, -1)$ axis vertical

13 vertex $(1, 2)$ through the point $(-3, 4)$ axis horizontal

14 vertex $(-5, 0)$ through the point $(0, 8)$ axis horizontal

15 vertex $(2, -3)$ focus $(2, 1)$

16 vertex $(2, -3)$ focus $(10, -3)$.

17 The line $y - 1 = -2(x - 1)$ intersects the parabola $y = x^2$ in two points. One of them is $(1, 1)$; find the other.

18 Does the point $(3, 7)$ lie inside or outside the parabola $y = x^2$? How can you tell without plotting the point?

19 Sketch the family of parabolas $4py = x^2 - 4p^2$ for $p > 0$. Locate their foci. Now sketch the family for $p < 0$.

20 (cont.) Show that no two of the curves for $p > 0$ intersect.

21 Let (x_0, y_0) be a point on the parabola $4ay = x^2$. Show that the line $2a(y + y_0) = x_0x$ meets the parabola only in (x_0, y_0). Interpret.

22 Find the locus of $(x, y) = (2at, at^2)$, where $-\infty < t < \infty$.

23 A shell fired with speed v_0 ft/sec at angle α with the ground will have coordinates

$$x = (v_0 \cos \alpha)t, \qquad y = (v_0 \sin \alpha)t - 16t^2$$

after t seconds (if air drag is ignored). Show that the trajectory is a parabolic arc.

24 (cont.) Find the highest point on the trajectory.

25 Show that the equation in polar coordinates of the parabola with focus at $\mathbf{0}$ and directrix $x = -p$ is $r(1 - \cos\theta) = p$.

26 Find the locus of all points $\mathbf{x}$ whose distance to the y-axis equals the length of the tangent(s) from $\mathbf{x}$ to the circle $(x - 1)^2 + y^2 = 1$.

27 For each m, the line $y - 1 = m(x - 1)$ intersects the parabola $y = x^2$ in $(1, 1)$ and another point. Find a value of m for which the second point coincides with $(1, 1)$. Interpret.

28* A circle C and a line L are given. Find the locus of all points $\mathbf{x}$ whose distance from L equals the length of the tangents from $\mathbf{x}$ to C. Discuss the case in which L passes through the center of C.

6 THE ELLIPSE

An **ellipse** is the locus of all points $\mathbf{x}$ such that the sum of the distances of $\mathbf{x}$ from two fixed points $\mathbf{p}$ and $\mathbf{q}$ is a constant greater than $\|\mathbf{p} - \mathbf{q}\|$. The points $\mathbf{p}$ and $\mathbf{q}$ are the **foci** (plural of **focus**) of the ellipse.

Our immediate goal is to derive the equation in rectangular coordinates of an ellipse whose foci lie on the x-axis. We shall do this in two steps; first we derive the equation in polar coordinates when one focus is at the origin, then we switch to rectangular coordinates.

Remark The orbit of a planet around a fixed star is an ellipse with the star at one focus. Because in astronomy one measures angles rather than distances, it is natural to study the polar equation of an ellipse with one focus at the origin.

Polar Equation of an Ellipse

Let $\mathbf{p}$ and $\mathbf{q}$ be the foci of an ellipse, and set

$$2c = \|\mathbf{p} - \mathbf{q}\|.$$

The ellipse, by definition, is the locus of all points $\mathbf{x}$ the sum of whose distances from $\mathbf{p}$ and $\mathbf{q}$ is a constant. We call this constant $2a$:

$$2a = \|\mathbf{x} - \mathbf{p}\| + \|\mathbf{x} - \mathbf{q}\|.$$

The ellipse is completely determined, except for *where* it is located, by these two constants a and c. We assume $0 < c < a$.

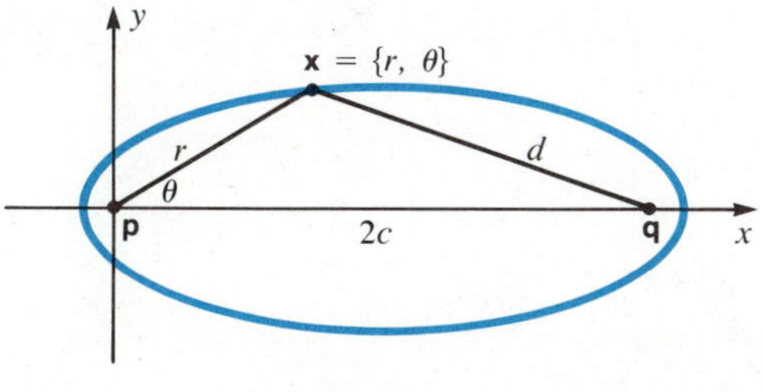

Fig. 1 Derivation of the polar equation of the ellipse

Now let us locate $\mathbf{p}$ at $\mathbf{0}$ and $\mathbf{q}$ on the positive x-axis (Fig. 1), so that $\mathbf{q} = (2c, 0)$. Let $\{r, \theta\}$ be the polar coordinates of a typical point $\mathbf{x}$ of the ellipse. Then $\|\mathbf{x} - \mathbf{p}\| = r$, so the ellipse is defined by

$$r + \|\mathbf{x} - \mathbf{q}\| = 2a, \quad \text{that is,} \quad \|\mathbf{x} - \mathbf{q}\| = 2a - r.$$

By the law of cosines,

$$\|\mathbf{x} - \mathbf{q}\|^2 = r^2 + (2c)^2 - 2r(2c)\cos\theta = r^2 + 4c^2 - 4rc\cos\theta.$$

Therefore the relation between r and θ is

$$(2a - r)^2 = r^2 + 4c^2 - 4rc\cos\theta,$$

that is,

$$4a^2 - 4ar + r^2 = r^2 + 4c^2 - 4rc\cos\theta.$$

Cancel r^2, divide by 4, and rearrange:

$$4a^2 - 4ar = 4c^2 - 4rc\cos\theta$$

$$a^2 - ar = c^2 - rc\cos\theta$$

$$r(a - c\cos\theta) = a^2 - c^2.$$

It is convenient to set $b^2 = a^2 - c^2$, $b > 0$.

Polar Equation The polar equation of the ellipse with foci $\mathbf{p} = (0, 0)$ and $\mathbf{q} = (2c, 0)$ and length sum $2a$ is

$$r(a - c\cos\theta) = b^2$$

where $a^2 = b^2 + c^2$.

Rectangular Equation of the Ellipse

We now convert the polar equation $r(a - c\cos\theta) = b^2$ to rectangular coordinates. Since $r\cos\theta = x$, we can write the equation as

$$ar - cx = b^2, \quad \text{that is,} \quad ar = cx + b^2.$$

Square both sides and replace r^2 by $x^2 + y^2$:

$$a^2(x^2 + y^2) = (cx + b^2)^2 = c^2x^2 + 2cb^2x + b^4$$

$$(a^2 - c^2)x^2 - 2cb^2x + a^2y^2 = b^4$$

$$b^2x^2 - 2cb^2x + a^2y^2 = b^4$$

$$b^2(x^2 - 2cx) + a^2y^2 = b^4.$$

Complete the square:

$$\begin{aligned} b^2(x - c)^2 + a^2y^2 &= b^4 + b^2c^2 \\ &= b^2(b^2 + c^2) \\ &= a^2b^2. \end{aligned}$$

Finally divide by a^2b^2:

$$\frac{(x - c)^2}{a^2} + \frac{y^2}{b^2} = 1.$$

This is the rectangular equation for the ellipse with foci $(0, 0)$ and $(2c, 0)$ and length sum $2a$. If we translate it to the left by c units, we have the ellipse with foci $(-c, 0)$ and $(c, 0)$ and length sum still $2a$.

Rectangular Equation The rectangular equation of the ellipse with foci $(-c, 0)$ and $(c, 0)$ and length sum $2a$ is

$$\frac{x^2}{a^2} + \frac{y^2}{b^2} = 1$$

where $a^2 = b^2 + c^2$.

Such an ellipse, whose foci are on the x-axis, symmetrically placed with respect to the origin, is said to be in **standard position** (Fig. 2).

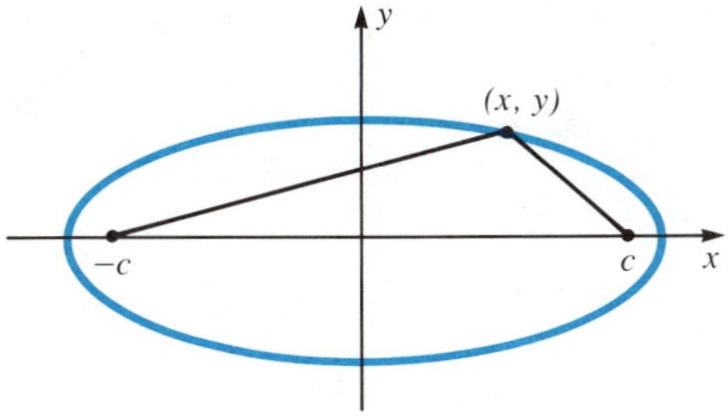

Fig. 2 Ellipse in standard position

Geometry of the Ellipse

We study the ellipse

$$\frac{x^2}{a^2} + \frac{y^2}{b^2} = 1 \qquad a^2 = b^2 + c^2$$

in standard position. Clearly for each point (x, y) on the ellipse

$$\frac{x^2}{a^2} \le \frac{x^2}{a^2} + \frac{y^2}{b^2} = 1, \quad \text{so} \quad x^2 \le a^2.$$

Therefore $-a \le x \le a$, and similarly $-b \le y \le b$; the ellipse is bounded by a rectangular box (Fig. 3a).

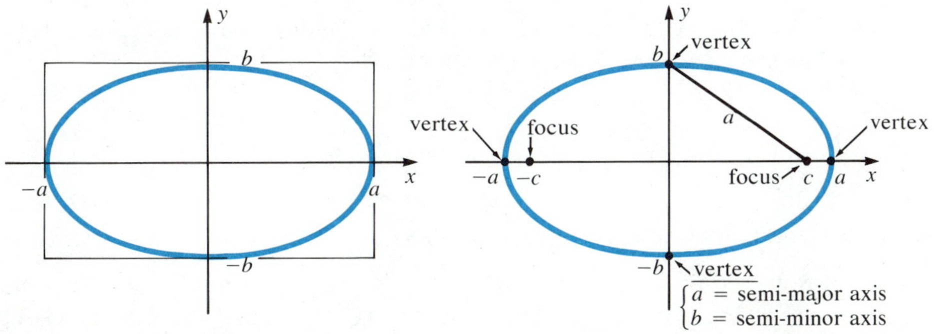

(a) The ellipse is bounded and symmetric. (b) Center, axes, and vertices

Fig. 3 Geometry of the ellipse

If (x, y) is on the ellipse, then so are $(-x, y)$, $(x, -y)$, and $(-x, -y)$, as follows readily from

$$\frac{(\pm x)^2}{a^2} + \frac{(\pm y)^2}{b^2} = \frac{x^2}{a^2} + \frac{y^2}{b^2} = 1.$$

Therefore the ellipse is symmetric with respect to both coordinate axes and with respect to the origin.

The points $(\pm a, 0)$ and $(0, \pm b)$ are on the ellipse, and are called its **vertices.**

The numbers a and b are known (historically) by the names **semi-major axis** and **semi-minor axis.** The midpoint of the foci is called the **center** of the ellipse. It is the origin for the ellipse in standard position.

Note from Fig. 3b that the distance from a focus to either of the points $(0, \pm b)$ is a; that is because $a^2 = b^2 + c^2$.

Remark 1 To construct an ellipse, tie a string of length $2a$ to two fixed pins $2c$ units apart $(a > c)$. Place your pencil against the string and move it so the string is taut. The locus generated is an ellipse. Why? If the pins are moved closer and closer together $(c \longrightarrow 0)$, the ellipse becomes more and more like a circle.

Remark 2 It is shown by calculus that on an elliptical billiard table, a ball shot in any direction from one focus will bounce once and pass through the other focus. (This accounts for the name focus, or focal point.) Whispering galleries are constructed in an elliptical shape because of this property.

Ellipses in Other Positions

By a translation of axes, we get

$$\frac{(x-h)^2}{a^2} + \frac{(y-k)^2}{b^2} = 1,$$

the equation of an ellipse centered at (h, k).

If $a > b$ then

$$\frac{x^2}{b^2} + \frac{y^2}{a^2} = 1$$

is an ellipse with major axis along the y-axis instead of the x-axis.

Remark If $a = b$, the equation $x^2/a^2 + y^2/b^2 = 1$ becomes

$$\frac{x^2}{a^2} + \frac{y^2}{a^2} = 1, \quad \text{that is,} \quad x^2 + y^2 = a^2$$

a circle of radius a. Thus a circle can be considered as a limiting case of an ellipse, where $c = 0$. (The foci come together at one point, the center.)

Example 1 Sketch the ellipse

$$\frac{(x-2)^2}{9} + \frac{(y+1)^2}{25} = 1$$

Locate the center, foci, and vertices.

Solution We have $a^2 = 25$ and $b^2 = 9$, so $a = 5$ and $b = 3$; and $c^2 = a^2 - b^2 = 16$, so $c = 4$. The center is $(2, -1)$ and the major axis is vertical. The foci are up and down $c = 4$ units from the center, so they are

$$(2, -1 + 4) = (2, 3) \quad \text{and} \quad (2, -1 - 4) = (2, -5).$$

The vertices on the major axis are up and down $a = 5$ units from the center, so they are

$$(2, -1 + 5) = (2, 4) \quad \text{and} \quad (2, -1 - 5) = (2, -6).$$

The vertices on the minor axis are left and right $b = 3$ units from the center, so they are

$$(2 + 3, -1) = (5, -1) \quad \text{and} \quad (2 - 3, -1) = (-1, -1).$$

Finally the curve is sketched in Fig. 4.

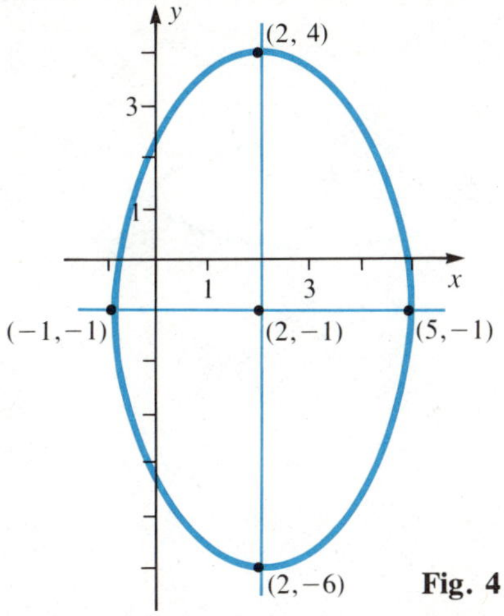

Fig. 4

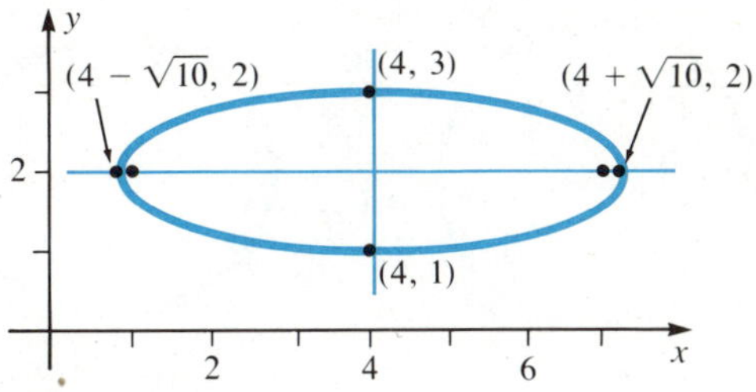

Fig. 5 Foci: $(1, 2)$, $(7, 2)$

EXAMPLE 2 Graph and find the equation of the ellipse with foci $(1, 2)$ and $(7, 2)$ and semi-minor axis 1.

Solution The center is $(4, 2)$, the midpoint of the foci. The distance from the center to either focus is 3, so $c = 3$. Since the major axis is horizontal, the semi-minor axis is $b = 1$, and

$$a^2 = b^2 + c^2 = 1 + 9 = 10.$$

Consequently the equation is

$$\frac{(x - 4)^2}{10} + \frac{(y - 2)^2}{1} = 1$$

and the graph is sketched in Fig. 5.

Answer $\frac{1}{10}(x - 4)^2 + (y - 2)^2 = 1$

Eccentricity

Let us return to the polar form

$$r(a - c \cos \theta) = b^2$$

of the ellipse with foci $(0, 0)$ and $(2c, 0)$ and length sum $2a$. As usual, $b^2 = a^2 - c^2$. We can rewrite the equation in the form

$$r\left(1 - \frac{c}{a} \cos \theta\right) = \frac{b^2}{a}$$

We define the **eccentricity** of the ellipse to be the number $e = c/a$. We also set $p = b^2/ae$. Then the polar equation of the ellipse is

$$\boxed{r(1 - e \cos \theta) = ep.}$$

Since $0 < c < a$, the eccentricity satisfies $0 < e < 1$. The eccentricity determines the shape of the ellipse. If e is near zero, then c is small compared to a.

That means that the foci are close together relative to the semi-major axis, hence the ellipse is circle-like. If e is near 1, the foci are relatively far apart and the ellipse is cigar-shaped. See Fig. 6. Once e is given, the scale factor p determines the size of the ellipse (as the radius does for a circle).

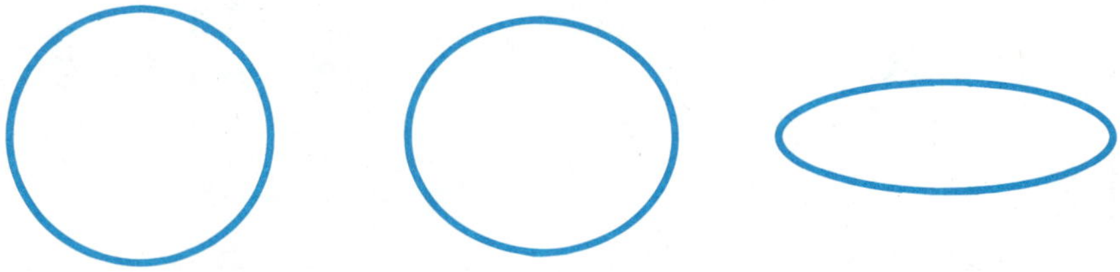

(a) $e = 0.1$ (b) $e = 0.5$ (c) $e = 0.95$

Fig. 6 Ellipses of various eccentricities

Consider again the ellipse

$$\frac{x^2}{a^2} + \frac{y^2}{b^2} = 1 \qquad (a > b)$$

in standard position. Since

$$\left(\frac{x}{a}\right)^2 + \left(\frac{y}{b}\right)^2 = 1$$

there is an angle θ such that $x/a = \cos\theta$ and $y/b = \sin\theta$, so

$$(x, y) = (a\cos\theta, b\sin\theta).$$

As θ makes a complete revolution, the point $(a\cos\theta, b\sin\theta)$ traverses the ellipse once. See Fig. 7. Note that θ is *not* the polar angle of (x, y).

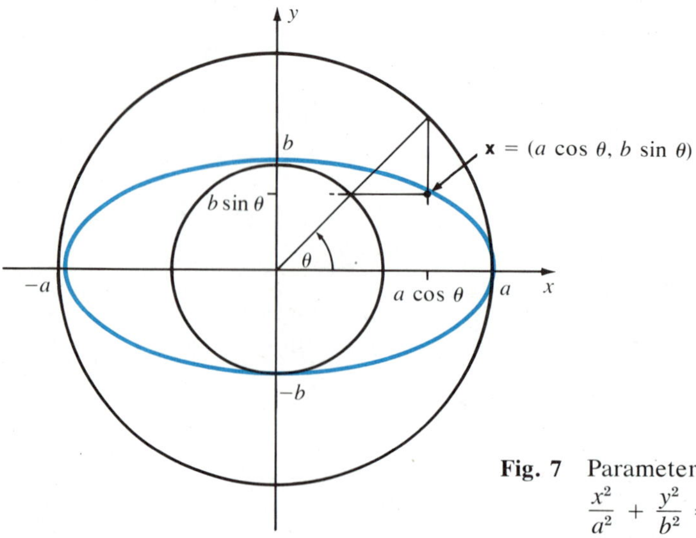

Fig. 7 Parameterization of the ellipse $\dfrac{x^2}{a^2} + \dfrac{y^2}{b^2} = 1$

Exercises

Give the center, foci, major and minor semi-axes (a and b), and vertices, and sketch

1 $\frac{1}{25}x^2 + \frac{1}{9}y^2 = 1$

2 $x^2 + 4y^2 = 4$

3 $2(x+1)^2 + (y-2)^2 = 2$

4 $4x^2 + y^2 - 2y = 0$

5 $2x^2 + y^2 - 12x - 4y = -21$

6 $x^2 + 2y^2 + 8y = 0$.

Write the equation of the ellipse

7 center at $(1, 4)$ vertices at $(10, 4)$ and $(1, 2)$

8 center at $(-2, -3)$ vertices at $(7, -3)$ and $(-2, -7)$

9 foci at $(2, 0)$ and $(8, 0)$ vertices at $(0, 0)$ and $(10, 0)$

10 foci at $(0, 3)$ and $(0, -3)$ semi-major axis $= 10$

11 foci at $(-1, 0)$ and $(3, 0)$ eccentricity $= \frac{1}{2}$

12 vertices at $(0, 2)$ and $(0, 6)$ eccentricity $= \frac{3}{4}$ major axis vertical.

13 Prove that the points on an ellipse farthest from its center are the two vertices on the major axis. [*Hint* Use parameterization.]

14 Prove that the points on an ellipse nearest to its center are the two vertices on the minor axis. [*Hint* Use parameterization.]

15 A 9-ft ladder slides along the floor as its top slides down a wall. If $\mathbf{x}$ is the point on the ladder $\frac{1}{3}$ of the way from its top, find the path of $\mathbf{x}$.

16 A rod moves with one end on the x-axis and the other on the y-axis. If P is a point on the rod, prove that the locus of P is an ellipse.

17* Let $\mathbf{p}$ be one end of the major axis of an ellipse. Find the locus of the midpoint of $\mathbf{pq}$ as $\mathbf{q}$ traces the ellipse.

18* Let E be an ellipse and L a line. Prove that the midpoints of all chords of E parallel to L are collinear.

19 Let an ellipse have eccentricity e, and write $e = \cos\theta°$ with $0 < \theta° < 90°$. Then we speak of a **$\vartheta°$-ellipse.** Interpret geometrically. Draw ellipses of $15°$, $30°$, $45°$, and $60°$.

20 Fix p and e with $0 < p$ and $0 < e < 1$. Find the locus of all points x whose distance from 0 is e times its distance from the line $x = -p$. [*Hint* Use polar coordinates.]

21 The orbit of the earth is approximately an ellipse with the sun at one focus and semi-major and semi-minor axes 9.3×10^7 and 9.1×10^7 miles respectively. Compute the eccentricity of the orbit.

22 (cont.) Find the distance from the sun to the other focus of the ellipse.

23 Show that

$$x = \frac{1 - t^2}{1 + t^2} \qquad y = \frac{2t}{1 + t^2}$$

parameterizes the unit circle by rational functions.

24* (cont.) Can you parameterize the ellipse $x^2/a^2 + y^2/b^2 = 1$ by rational functions? [*Hint* Intersect the ellipse with the line $y = t(x + a)$. One intersection is $(-a, 0)$; express the other in terms of t.]

7 THE HYPERBOLA

A **hyperbola** is the locus of all points $\mathbf{x}$ such that the absolute value of the difference of the distances of $\mathbf{x}$ from two fixed points $\mathbf{p}$ and $\mathbf{q}$, called the **foci,** is a constant, $2a$. Because $\mathbf{p}$, $\mathbf{q}$, and $\mathbf{x}$ form a triangle, $2a < \|\mathbf{p} - \mathbf{q}\|$. See Fig. 1, next page

Equation of the Hyperbola

Our derivation of the equation of a hyperbola parallels that of the ellipse, so we shall merely sketch the steps, and omit the details. First we work in polar coordinates with a hyperbola (Fig. 2, next page whose foci are $\mathbf{p} = (0, 0)$ and $\mathbf{q} = (2c, 0)$.

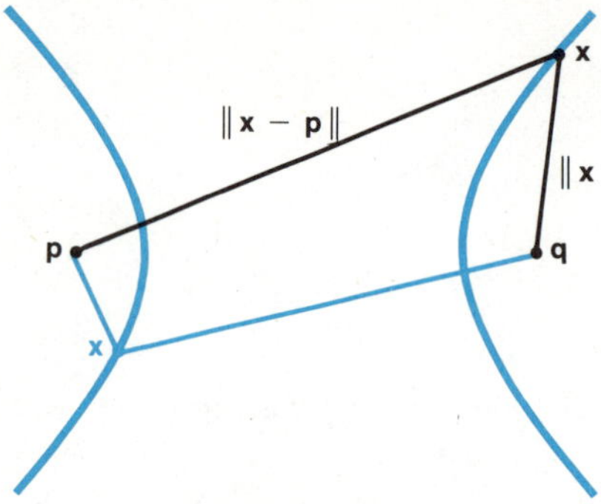

Fig. 1 The hyperbola:
$\big|\|\mathbf{x} - \mathbf{p}\| - \|\mathbf{x} - \mathbf{q}\|\big| = 2a$.
(Let the distance between the foci be $2c = \|\mathbf{p} - \mathbf{q}\|$. Then $2a > 2c$ by the triangle inequality, so $a > c$.)

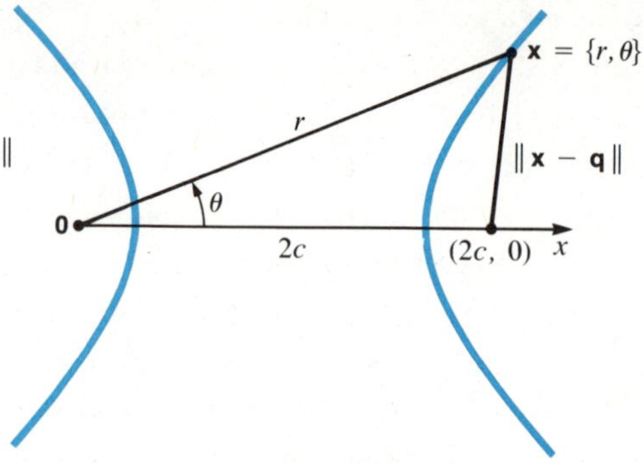

Fig. 2 Set-up for deriving the polar equation. Here the foci are $\mathbf{p} = \mathbf{0}$ and $\mathbf{q} = (2c, 0)$.

Then $\mathbf{x}$ is on the hyperbola provided

$$\|\mathbf{x} - \mathbf{q}\| = r \pm 2a.$$

By the law of cosines

$$\|\mathbf{x} - \mathbf{q}\|^2 = r^2 + 4c^2 - 4rc\cos\theta.$$

We eliminate $\|\mathbf{x} - \mathbf{q}\|$ and simplify; the result is:

Polar Equation The polar equation of the hyperbola with foci $\mathbf{p} = (0, 0)$ and $\mathbf{q} = (2c, 0)$ and absolute length difference $2a$ is

$$r(\pm a + c\cos\theta) = b^2$$

where $c^2 = a^2 + b^2$.

To derive the rectangular form, we first replace $r\cos\theta$ by x, transpose, square, and replace r^2 by $x^2 + y^2$:

$$\pm ar + cx = b^2 \qquad \pm ar = b^2 - cx$$

$$a^2(x^2 + y^2) = (b^2 - cx)^2.$$

After we simplify and complete the square, the result is

$$\frac{(x - c)^2}{a^2} - \frac{y^2}{b^2} = 1.$$

This is the rectangular equation for the hyperbola with foci $(0, 0)$ and $(2c, 0)$ and absolute length difference $2a$. If we translate this hyperbola to the left by c units, we obtain the hyperbola with foci $(-c, 0)$ and $(c, 0)$ and absolute length difference $2a$.

Rectangular Equation The rectangular equation of the hyperbola with foci $(-c, 0)$ and $(c, 0)$ and absolute length difference $2a$ is

$$\frac{x^2}{a^2} - \frac{y^2}{b^2} = 1$$

where $c^2 = a^2 + b^2$.

A hyperbola whose foci are on the x-axis, symmetrically placed with respect to the origin, is said to be in **standard position** (Fig. 3).

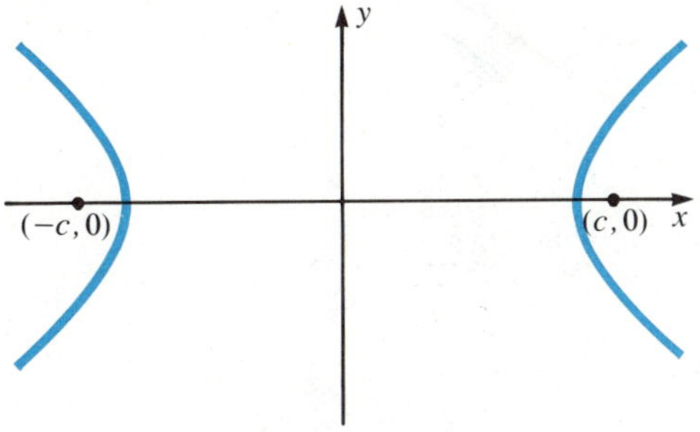

Fig. 3 Standard position of the hyperbola

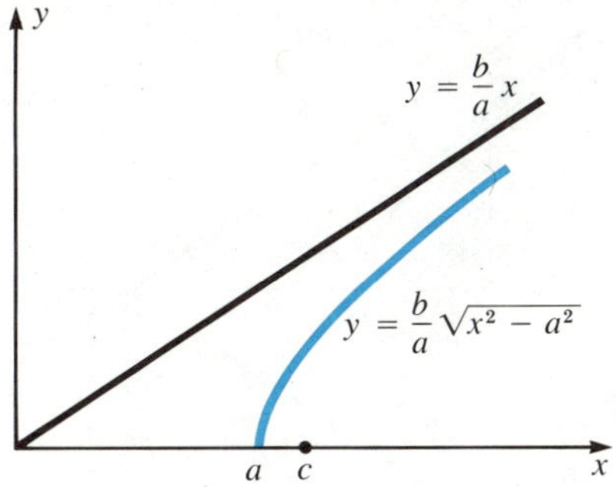

Fig. 4 First quadrant portion of $\dfrac{x^2}{a^2} - \dfrac{y^2}{b^2} = 1$

Note For the hyperbola in standard position, $a^2 = c^2 - b^2$, unlike for the ellipse in standard position, where $a^2 = b^2 + c^2$. Therefore both $a > b$ and $a \le b$ are possible for the hyperbola, not just $a > b$ as for the ellipse.

Geometry of the Hyperbola

Let us sketch the hyperbola

$$\frac{x^2}{a^2} - \frac{y^2}{b^2} = 1.$$

We first note symmetry; if (x, y) satisfies this equation, then so do $(-x, y)$, $(x, -y)$, and $(-x, -y)$. Therefore the curve is symmetric in both axes and in the origin; we need plot it only in the first quadrant, then extend the curve to the other quadrants by symmetry.

We solve for y:

$$y = \frac{b}{a}\sqrt{x^2 - a^2}.$$

(The positive square root applies for the first quadrant.) The locus is defined only for $x \ge a$. When x starts at a and increases, y starts at 0 and increases. When x is very large, we suspect that y is slightly less than bx/a. To confirm this suspicion, we write

$$\frac{b}{a}x - y = \frac{b}{a}(x - \sqrt{x^2 - a^2}\,) = \frac{b}{a}(x - \sqrt{x^2 - a^2}\,)\frac{x + \sqrt{x^2 - a^2}}{x + \sqrt{x^2 - a^2}}$$

$$= \left(\frac{b}{a}\right)\frac{x^2 - (x^2 - a^2)}{x + \sqrt{x^2 - a^2}} = \frac{ab}{x + \sqrt{x^2 - a^2}} < \frac{ab}{x}.$$

It follows that

$$\frac{b}{a}x - y \longrightarrow 0+ \quad \text{as} \quad x \longrightarrow \infty.$$

This means the curve approaches the line $y = bx/a$ (from below) as x increases; the line is an asymptote of the hyperbola. Now we can sketch the first quadrant portion (Fig. 4). We then extend the sketch to the other quadrants by symmetry (Fig. 5, next page).

In dealing with hyperbolas, the following terminology is customary. A hyperbola consists of two **branches:** one where $\|\mathbf{x} - \mathbf{p}\| - \|\mathbf{x} - \mathbf{q}\| = 2a$, and the other where $\|\mathbf{x} - \mathbf{q}\| - \|\mathbf{x} - \mathbf{p}\| = 2a$. The point half-way between the foci is

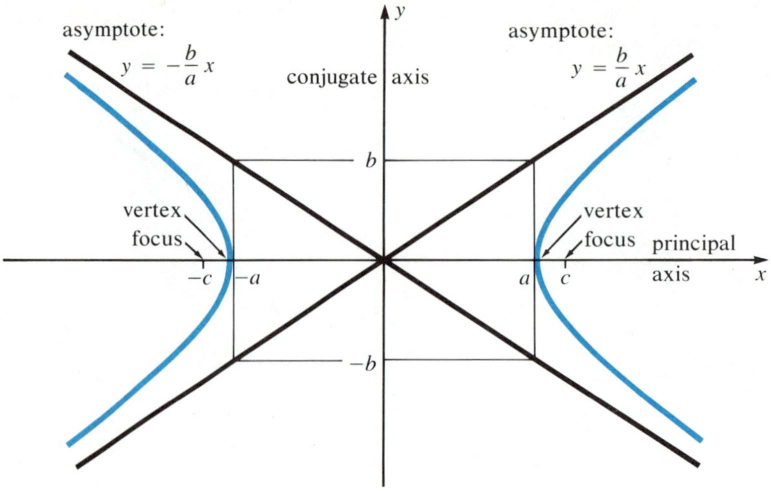

Fig. 5 Geometry of the hyperbola $\dfrac{x^2}{a^2} - \dfrac{y^2}{b^2} = 1$

the **center** of a hyperbola. The line through the foci is the **principal axis,** and the line through the center perpendicular to the principal axis is the **conjugate axis.** The points where the hyperbola meets its principal axis are its **vertices.** See Fig. 5 again.

The lines $y = \pm bx/a$ are the **asymptotes** of the hyperbola. A neat way to remember this is that the asymptotes of

$$\frac{x^2}{a^2} - \frac{y^2}{b^2} = 1 \quad \text{form the locus of} \quad \frac{x^2}{a^2} - \frac{y^2}{b^2} = 0.$$

This follows from the factorization

$$\frac{x^2}{a^2} - \frac{y^2}{b^2} = \left(\frac{x}{a} + \frac{y}{b}\right)\left(\frac{x}{a} - \frac{y}{b}\right).$$

The expression on the left is zero if and only if one of the factors on the right is zero, that is, if and only if $y = \pm bx/a$.

> The asymptotes of the hyperbola
>
> $$\frac{x^2}{a^2} - \frac{y^2}{b^2} = 1$$
>
> are the lines $y = \dfrac{b}{a}x$ and $y = -\dfrac{b}{a}x$,
>
> or equivalently, the locus of the equation $\dfrac{x^2}{a^2} - \dfrac{y^2}{b^2} = 0$.

Remark The equations

$$\frac{x^2}{a^2} + \frac{y^2}{b^2} = 1 \quad \text{and} \quad \frac{x^2}{a^2} - \frac{y^2}{b^2} = 1$$

differ by a little minus sign, but that makes all the difference in the world. The first equation, where the sign is plus, requires $x^2 \le a^2$ and $y^2 \le b^2$; the locus is confined. The second imposes no such restriction; x^2/a^2 and y^2/b^2 can both be enormous, yet differ by 1.

By translation,

$$\frac{(x-h)^2}{a^2} - \frac{(y-k)^2}{b^2} = 1$$

is the equation of a hyperbola with center at (h, k) and horizontal principal axis.

By interchanging the roles of x and y, we see that the equation

$$-\frac{x^2}{a^2} + \frac{y^2}{b^2} = 1$$

defines a hyperbola with center at the origin, vertical principal axis, and foci at $(0, \pm c)$, where $c^2 = a^2 + b^2$.

Example 1 Sketch the hyperbola

$$\frac{(x-1)^2}{9} - \frac{(y+2)^2}{16} = 1.$$

Locate the center, foci, vertices, and asymptotes.

Solution Clearly the center is $(1, -2)$, the principal axis is horizontal, and $a = 3$ and $b = 4$, so $c^2 = a^2 + b^2 = 25$, or $c = 5$. Therefore the foci are

$$(1 \pm 5, -2) = (-4, -2), \quad (6, -2).$$

The asymptotes are the locus of

$$\frac{(x-1)^2}{9} \quad \frac{(y+2)^2}{16} = 0,$$

that is,

$$\frac{y+2}{4} = \pm\frac{x-1}{3} \quad \text{that is} \quad y + 2 = \pm\tfrac{4}{3}(x-1).$$

They are the lines through $(1, -2)$ with slopes $\pm\frac{4}{3}$.

Since the principal axis is the horizontal line $y = -2$, the vertices are the points of the hyperbola for which $y = -2$, so

$$\frac{(x-1)^2}{9} = 1, \quad \text{that is,} \quad x - 1 = \pm 3.$$

Hence the vertices are $(-2, -2)$ and $(4, -2)$. All of this information and a sketch of the curve is shown in Fig. 6.

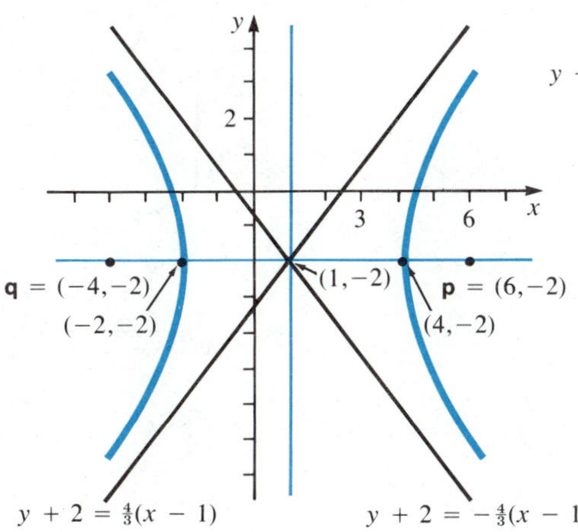

Fig. 6

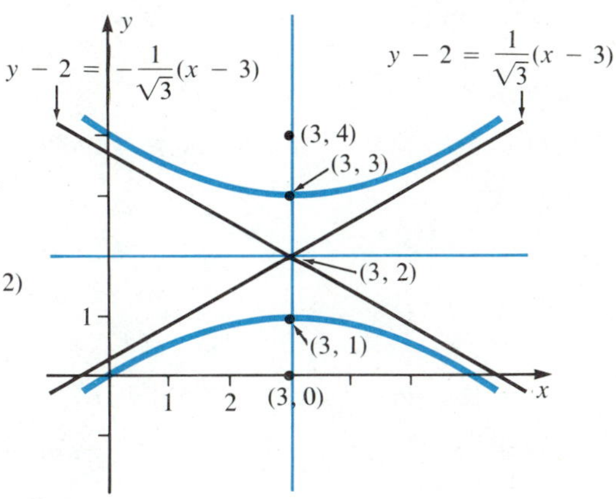

Fig. 7

EXAMPLE 2 Find the equation of the hyperbola with foci $(3, 0)$ and $(3, 4)$ and a vertex at $(3, 3)$. Sketch the hyperbola and its asymptotes.

Solution The center is the midpoint of the foci, so it is $(3, 2)$, and the equation of the hyperbola has the form

$$-\frac{(x-3)^2}{a^2}+\frac{(y-2)^2}{b^2}=1.$$

The distance between the foci is 4, so $2c = 4$ and $c = 2$. Therefore

$$a^2+b^2=c^2=4.$$

But $(3, 3)$ is a vertex, hence on the hyperbola. This implies

$$-\frac{0}{a^2}+\frac{(3-2)^2}{b^2}=1 \qquad b^2=1.$$

Hence $a^2 = 3$, so the equation of the hyperbola is

$$-\frac{(x-3)^2}{3}+\frac{(y-2)^2}{1}=1.$$

To find its asymptotes, we set

$$-\frac{(x-3)^2}{3}+\frac{(y-2)^2}{1}=0, \quad \text{that is,} \quad y-2=\pm\frac{1}{\sqrt{3}}(x-3).$$

Since $1/\sqrt{3} = \tan 30°$, we can make an accurate sketch (Fig. 7).

Answer $-\dfrac{(x-3)^2}{3}+(y-2)^2=1$

A hyperbola is called **rectangular** if its asymptotes are perpendicular. This happens when the slopes of the two asymptotes are negative reciprocals of each other.

$$\left(\frac{b}{a}\right)\left(-\frac{b}{a}\right)=-1 \qquad b^2=a^2 \qquad b=a.$$

Thus the locus of

$$x^2-y^2=a^2$$

is a rectangular hyperbola.

Exercises

Find the center, foci, and asymptotes, and sketch

1 $\frac{1}{4}x^2-\frac{1}{9}y^2=1$

2 $\frac{1}{9}x^2-\frac{1}{4}y^2=1$

3 $-\frac{1}{9}x^2+\frac{1}{4}y^2=1$

4 $-\frac{1}{4}x^2+\frac{1}{9}y^2=1$

5 $(x+1)^2-(y-1)^2=1$

6 $-(x-2)^2+4(y+1)^2=4$

7 $x^2-y^2=1$

8 $-2x^2+y^2=1$

9 $4x^2-y^2-24x-2y+31=0$

10 $3x^2-3y^2-3x-2y=\frac{31}{12}$.

Write the equation of the hyperbola with

11 foci $(0, \pm 5)$ vertices $(0, \pm 4)$

12 vertices $(\pm 3, 0)$ foci $(\pm 5, 0)$

13 asymptotes $y = \pm 2x$ vertices $(\pm 2, 0)$

14 foci $(1, 7), (1, -3)$ vertices $(1, 6), (1, -2)$

15 asymptotes $y = \pm(x - 1)$ curve passes through $(3, 1)$

16 asymptotes $y = \pm 2x$ curve passes through $(1, 1)$.

17 Show that the graph of $x^2 - y^2 + ax + by + c = 0$ is a rectangular hyperbola.

18 Show that the graph of $3x^2 - y^2 + ax + by + c = 0$ is a hyperbola whose asymptotes form a 60° angle.

19 Sketch the region $x^2/a^2 + y^2/b^2 < 1$ and the region $x^2/a^2 + y^2/b^2 > 1$.

20 Sketch the region $x^2/a^2 - y^2/b^2 < 1$ and the region $x^2/a^2 - y^2/b^2 > 1$.

21 Let C_1 and C_2 be circles external to each other. Prove that the locus of the center of a circle that touches C_1 and C_2 externally either is a branch of a hyperbola or a line.

22 A rifle at point $\mathbf{a}$ on level ground is shot at a target at point $\mathbf{b}$. Find the locus of all observers who hear the shot and the impact of the shell simultaneously.

8 ROTATION OF AXES

In Section 1 we studied how coordinates change under a translation of the rectangular axes. Now we shall consider rotations of the coordinate system about the origin, and some applications.

Rotation is much easier in polar coordinates than in rectangular coordinates. If the polar axis is rotated through an angle α, a point $\{r, \theta\}$ acquires new polar coordinates $\{\bar{r}, \bar{\theta}\}$. The relation between the new and old coordinates of a point are seen directly from Fig. 1. Since $\mathbf{0}$ is unchanged, $\bar{r} = r$. From the figure, $\bar{\theta} = \theta - \alpha$.

Rotation (Polar Coordinates) Suppose the polar axis is rotated by an angle α. Then the old polar coordinates $\{r, \theta\}$ and the new polar coordinates $\{\bar{r}, \bar{\theta}\}$ of a point are related by

$$\begin{cases} \bar{r} = r \\ \bar{\theta} = \theta - \alpha \end{cases} \quad \text{or equivalently} \quad \begin{cases} r = \bar{r} \\ \theta = \bar{\theta} + \alpha. \end{cases}$$

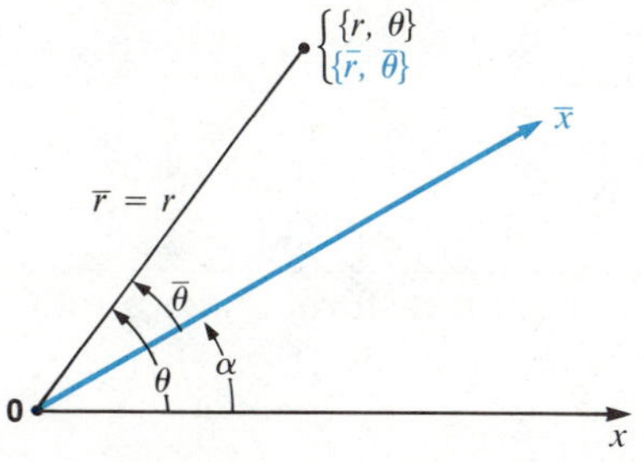

Fig. 1 Rotation in polar coordinates: $\bar{r} = r \quad \bar{\theta} = \theta - \alpha$

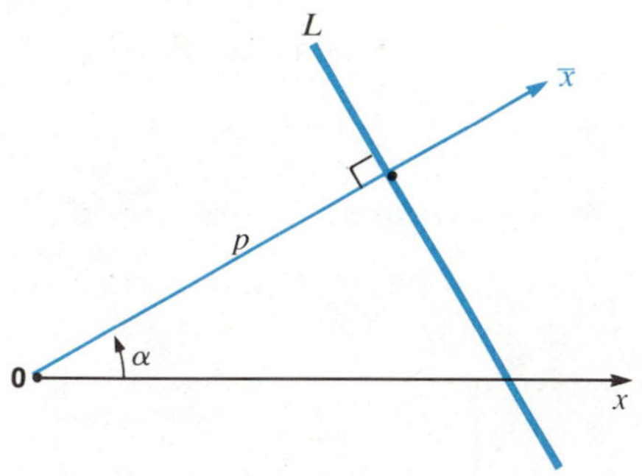

Fig. 2 Snappy derivation of the polar equation of a line

As an application, let us find the polar equation of the line L that is p units from the origin, perpendicular to the ray $\theta = \alpha$. See Fig. 2. Relative to the tilted axis, the line has equation $\bar{x} = p$ or $\bar{r}\cos\bar{\theta} = p$. Its r, θ-equation therefore is $r\cos(\theta - \alpha) = p$. This is a quick derivation of the result given on p. 464.

EXAMPLE 1 Find the equation in polar coordinates of the circle with center $(4, 3)$ and radius 5.

Solution Since $3^2 + 4^2 = 5^2$, the circle passes through the origin (Fig. 3a). Its diameter through $(0, 0)$ and its center $(4, 3)$ is inclined at angle α to the polar axis, where

$$\alpha = \arctan \tfrac{3}{4}.$$

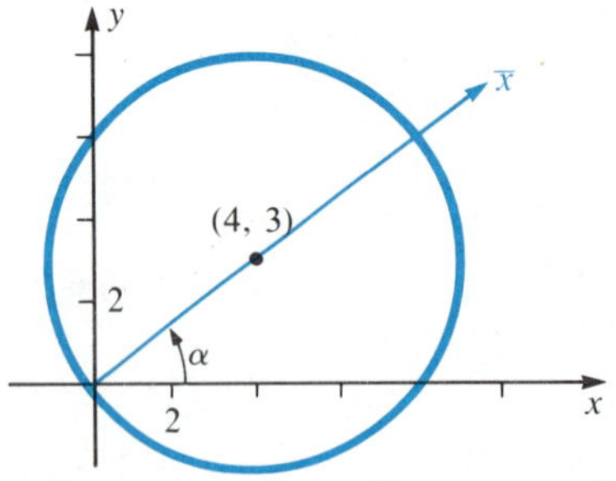

(a) Rotate through α, where $\tan \alpha = \frac{3}{4}$

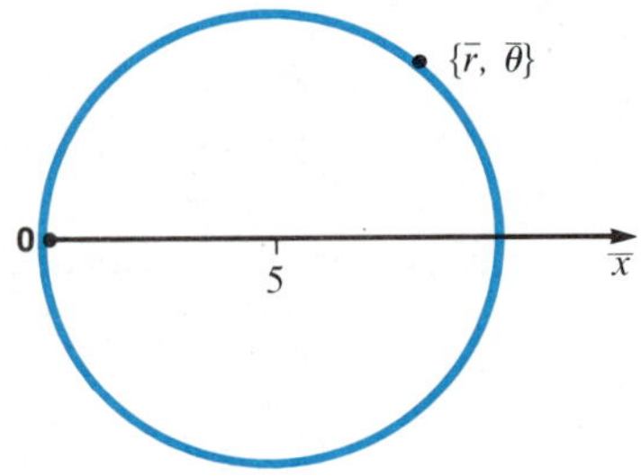

(b) Circle: $\bar{r} = 10 \cos \bar{\theta}$

Fig. 3

Now rotate the polar axis through angle α. See Fig. 3b. In the new $\bar{r}, \bar{\theta}$-coordinates the circle has center $(5, 0)$ and radius 5, a situation we studied on p. 466. Its equation in the new coordinates is therefore

$$\bar{r} = 10 \cos \bar{\theta},$$

hence in the old coordinates it is

$$r = 10 \cos(\theta - \alpha) = 10(\cos \alpha \cos \theta + \sin \alpha \sin \theta).$$

Clearly

$$\cos \alpha = \tfrac{4}{5} \quad \text{and} \quad \sin \alpha = \tfrac{3}{5},$$

so the equation is

$$r = 8 \cos \theta + 6 \sin \theta.$$

Check When $\theta = \alpha$, then $r = 8 \cdot \frac{4}{5} + 6 \cdot \frac{3}{5} = \frac{32}{5} + \frac{18}{5} = \frac{50}{5} = 10$.

When $\theta = \alpha + \frac{1}{2}\pi$, then $r = 8 \cdot (-\frac{3}{5}) + 6 \cdot (\frac{4}{5}) = 0$.

Answer $r = 8 \cos \theta + 6 \sin \theta$

Rotation in Rectangular Coordinates

Now let's find how the rectangular coordinates of a point transform under a rotation about the origin. As before, let α denote the rotation angle (Fig. 4).

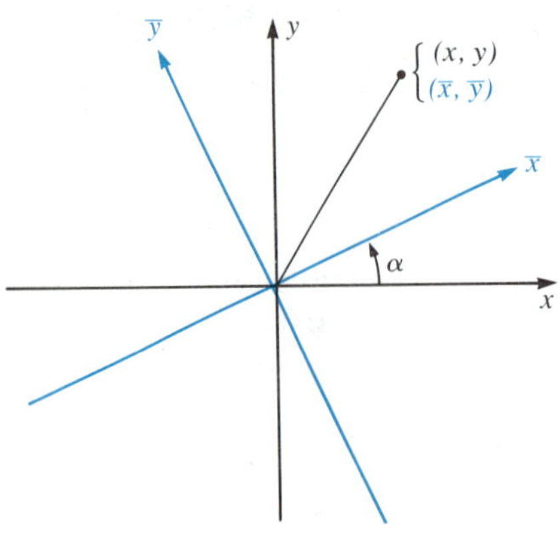

Fig. 4 Rotation in rectangular coordinates

A point has old coordinates (x, y) and new coordinates $(\bar{x}, \bar{y})$. We wish to express one set of coordinates in terms of the other and α. The simplest way to do this is to pass through polar coordinates, using

$$\begin{cases} x = r\cos\theta \\ y = r\sin\theta \end{cases} \qquad \begin{cases} \bar{x} = \bar{r}\cos\bar{\theta} \\ \bar{y} = \bar{r}\sin\bar{\theta} \end{cases}$$

and the relations between the old and new polar coordinates:

$$\bar{r} = r \qquad \bar{\theta} = \theta - \alpha.$$

We have

$$\begin{aligned} \bar{x} &= \bar{r}\cos\bar{\theta} = r\cos(\theta - \alpha) \\ &= r\cos\theta\cos\alpha + r\sin\theta\sin\alpha = x\cos\alpha + y\sin\alpha \end{aligned}$$

and

$$\begin{aligned} \bar{y} &= \bar{r}\sin\bar{\theta} = r\sin(\theta - \alpha) \\ &= r\sin\theta\cos\alpha - r\cos\theta\sin\alpha = y\cos\alpha - x\sin\alpha. \end{aligned}$$

We have proved that

$$\bar{x} = x\cos\alpha + y\sin\alpha \qquad \bar{y} = -x\sin\alpha + y\cos\alpha.$$

These formulas express $\bar{x}$ and $\bar{y}$ in terms of x and y (and of course α). To express x and y in terms of $\bar{x}$ and $\bar{y}$ we can solve the formulas as a linear system. Alternatively we can argue that a rotation of $-\alpha$ takes the $\bar{x},\bar{y}$-axes to the x,y-axes, hence

$$\begin{aligned} x &= \bar{x}\cos(-\alpha) + \bar{y}\sin(-\alpha) = \bar{x}\cos\alpha - \bar{y}\sin\alpha \\ y &= -\bar{x}\sin(-\alpha) + \bar{y}\cos(-\alpha) = \bar{x}\sin\alpha + \bar{y}\cos\alpha. \end{aligned}$$

Rotation (Rectangular Coordinates) Suppose the plane is rotated through angle α and the x- and y-axes, under this rotation, go to the $\bar{x}$- and $\bar{y}$-axes. Then the x,y-coordinates and $\bar{x},\bar{y}$-coordinates of any point are related by

$$\begin{cases} x = \bar{x}\cos\alpha - \bar{y}\sin\alpha \\ y = \bar{x}\sin\alpha + \bar{y}\cos\alpha \end{cases} \qquad \begin{cases} \bar{x} = x\cos\alpha + y\sin\alpha \\ \bar{y} = -x\sin\alpha + y\cos\alpha. \end{cases}$$

Examples

(1) $\alpha = \frac{1}{4}\pi$ $\qquad \begin{cases} x = \frac{1}{2}\sqrt{2}\,\bar{x} - \frac{1}{2}\sqrt{2}\,\bar{y} \\ y = \frac{1}{2}\sqrt{2}\,\bar{x} + \frac{1}{2}\sqrt{2}\,\bar{y} \end{cases}$

(2) $\alpha = -\frac{1}{6}\pi$ $\qquad \begin{cases} x = \frac{1}{2}\sqrt{3}\,\bar{x} + \frac{1}{2}\bar{y} \\ y = -\frac{1}{2}\bar{x} + \frac{1}{2}\sqrt{3}\,\bar{y} \end{cases}$

Conics

We have learned how to graph quadratic equations of the form

$$ax^2 + cy^2 + dx + ey + f = 0 \qquad (a^2 + c^2 > 0).$$

By completing squares, we generally obtain one of the conic sections. Now we tackle the most general quadratic equation

$$ax^2 + bxy + cy^2 + dx + ey + f = 0.$$

It is the term bxy that makes life difficult. Where does it come from and how can we get rid of it? We can learn a good deal from two experiments.

EXAMPLE 2 Find the equation of the ellipse

$$\frac{x^2}{9} + \frac{y^2}{4} = 1$$

in the $\bar{x},\bar{y}$-coordinate system that results from a $\frac{1}{4}\pi$ rotation of the x,y-coordinate system.

Solution The rotation formulas are

$$x = \tfrac{1}{2}\sqrt{2}(\bar{x} - \bar{y}) \qquad y = \tfrac{1}{2}\sqrt{2}(\bar{x} + \bar{y}).$$

Substitute:

$$\frac{x^2}{9} + \frac{y^2}{4} = \frac{1}{2}\left[\frac{(\bar{x} - \bar{y})^2}{9} + \frac{(\bar{x} + \bar{y})^2}{4}\right] = \frac{1}{2}\left[\frac{13}{36}\bar{x}^2 + \frac{10}{36}\bar{x}\bar{y} + \frac{13}{36}\bar{y}^2\right].$$

Answer $\frac{13}{72}\bar{x}^2 + \frac{10}{72}\bar{x}\bar{y} + \frac{13}{72}\bar{y}^2 = 1$ See Fig. 5.

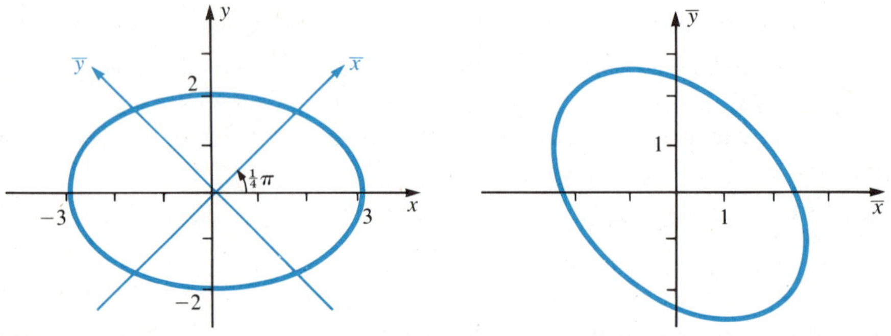

(a) $\frac{1}{9}x^2 + \frac{1}{4}y^2 = 1$ (b) $\frac{13}{72}\bar{x}^2 + \frac{10}{72}\bar{x}\bar{y} + \frac{13}{72}\bar{y}^2 = 1$

Fig. 5 An ellipse in rotated coordinates

The experiment suggests that the $\bar{x}\bar{y}$ term is due to the tilt of the coordinate axes relative to the axes of the ellipse. The same should be true for hyperbolas and parabolas. Let us try a parabola.

EXAMPLE 3 Apply the rotation of Example 2 to the parabola $y = x^2$.

Solution Substitute $x = \frac{1}{2}\sqrt{2}(\bar{x} - \bar{y})$ and $y = \frac{1}{2}\sqrt{2}(\bar{x} + \bar{y})$:

$$\tfrac{1}{2}\sqrt{2}(\bar{x} + \bar{y}) = [\tfrac{1}{2}\sqrt{2}(\bar{x} - \bar{y})]^2 = \tfrac{1}{2}(\bar{x}^2 - 2\bar{x}\bar{y} + \bar{y}^2),$$

$$\sqrt{2}(\bar{x} + \bar{y}) = \bar{x}^2 - 2\bar{x}\bar{y} + \bar{y}^2.$$

Answer $\bar{x}^2 - 2\bar{x}\bar{y} + \bar{y}^2 - \sqrt{2}\bar{x} - \sqrt{2}\bar{y} = 0$ See Fig. 6, next page.

Examples 8.2 and 8.3 suggest that an xy term occurs when the axes are placed obliquely. We suspect it can be eliminated by rotating the axes through a suitably chosen angle. Now the most general rotation of coordinates is

$$\begin{cases} x = \bar{x}\cos\alpha - \bar{y}\sin\alpha \\ y = \bar{x}\sin\alpha + y\cos\alpha. \end{cases}$$

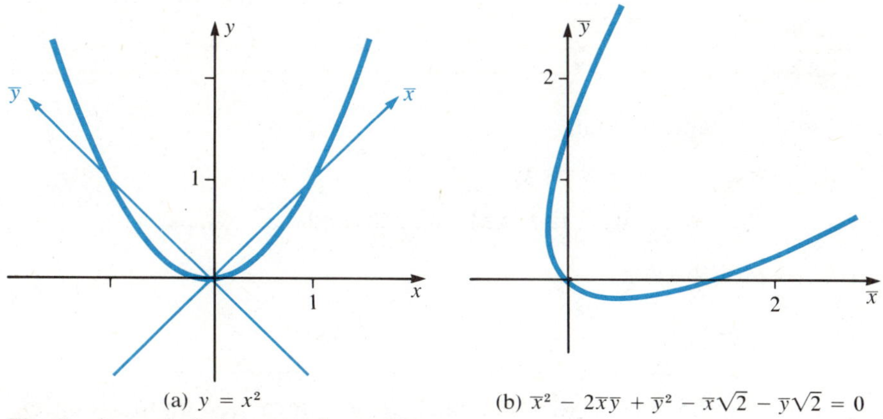

Fig. 6 A parabola in rotated coordinates

This rotation changes a linear polynomial $dx + ey + f$ in x and y into a linear polynomial in $\bar{x}$ and $\bar{y}$. Of more interest to us is what happens to a quadratic polynomial $ax^2 + bxy + cy^2$. Substitute:

$$ax^2 + bxy + cy^2 = a(\bar{x}\cos\alpha - \bar{y}\sin\alpha)^2$$
$$+ b(\bar{x}\cos\alpha - \bar{y}\sin\alpha)(\bar{x}\sin\alpha + \bar{y}\cos\alpha) + c(\bar{x}\sin\alpha + \bar{y}\cos\alpha)^2.$$

Multiply out and collect terms in $\bar{x}^2$, $\bar{x}\bar{y}$, and $\bar{y}^2$; the result is a quadratic polynomial:

Under a rotation through an angle α, the quadratic polynomial

$$ax^2 + bxy + cy^2 + dx + ey + f$$

is changed into

$$\bar{a}\bar{x}^2 + \bar{b}\bar{x}\bar{y} + \bar{c}\bar{y}^2 + \bar{d}\bar{x} + \bar{e}\bar{y} + \bar{f}$$

$$\text{where}\begin{cases} \bar{a} = a\cos^2\alpha + b\cos\alpha\sin\alpha + c\sin^2\alpha \\ \bar{b} = 2(c - a)\sin\alpha\cos\alpha + b(\cos^2\alpha - \sin^2\alpha) \\ \bar{c} = a\sin^2\alpha - b\sin\alpha\cos\alpha + c\cos^2\alpha. \end{cases}$$

For our purposes the most interesting of these formulas is the one for $\bar{b}$, which we can write in the form

$$\bar{b} = (c - a)\sin 2\alpha + b\cos 2\alpha.$$

From this formula we see that it is always possible to choose the rotation angle α so that $\bar{b} = 0$, that is, so that

$$(c - a)\sin 2\alpha + b\cos 2\alpha = 0.$$

For if $c = a$, then we simply choose $\alpha = \pm\frac{1}{4}\pi$; otherwise we choose α so that

$$\tan 2\alpha = \frac{b}{a - c}.$$

> A quadratic locus
>
> $$ax^2 + bxy + cy^2 + dx + ey + f = 0$$
>
> is changed into a quadratic locus
>
> $$\bar{a}\bar{x}^2 + \bar{c}\bar{y}^2 + \bar{d}\bar{x} + \bar{e}\bar{y} + \bar{f} = 0$$
>
> *without* an $\bar{x}\bar{y}$ term by rotating the axes through angle α, where
>
> $$\alpha = \pm\tfrac{1}{4}\pi \quad \text{if} \quad a = c,$$
>
> $$\tan 2\alpha = \frac{b}{a - c} \quad \text{if} \quad a \neq c.$$

Because the tangent has period π, the angle 2α is determined up to a multiple of π; hence α is determined only up to a multiple of $\frac{1}{2}\pi$. Therefore we can always choose α in the first quadrant.

In numerical examples, we must compute $\bar{a}$ and $\bar{c}$, having a, b, c, and $\tan 2\alpha$. We write the formulas for $\bar{a}$ and $\bar{c}$ in the form

$$\begin{cases} \bar{a} = a\cos^2\alpha + \frac{1}{2}b\sin 2\alpha + c\sin^2\alpha \\ \bar{c} = a\sin^2\alpha - \frac{1}{2}b\sin 2\alpha + c\cos^2\alpha. \end{cases}$$

From $\tan 2\alpha$ we can find $\sin 2\alpha$ and $\cos 2\alpha$:

$$\sin 2\alpha = \frac{\pm\tan 2\alpha}{\sqrt{1 + \tan^2 2\alpha}} \qquad \cos 2\alpha = \frac{\pm 1}{\sqrt{1 + \tan^2 2\alpha}}.$$

From $\cos 2\alpha$ we can find $\cos^2\alpha$ and $\sin^2\alpha$:

$$\cos^2\alpha = \tfrac{1}{2}(1 + \cos 2\alpha) \qquad \sin^2\alpha = \tfrac{1}{2}(1 - \cos 2\alpha).$$

EXAMPLE 4 Describe the locus of $xy = 1$.

Solution In this case $a = c = 0$, and $b = 1$. Therefore we choose $\alpha = \frac{1}{4}\pi$ to have $\bar{b} = 0$. The rotation is $x = \frac{1}{2}\sqrt{2}(\bar{x} - \bar{y})$ and $y = \frac{1}{2}\sqrt{2}(\bar{x} + \bar{y})$, so by direct computation.

$$xy = \tfrac{1}{2}(\bar{x} - \bar{y})(\bar{x} + \bar{y}) = \tfrac{1}{2}(\bar{x}^2 - \bar{y}^2).$$

The locus is $\frac{1}{2}(\bar{x}^2 - \bar{y}^2) = 1$.

Answer Rectangular hyperbola See Fig. 7.

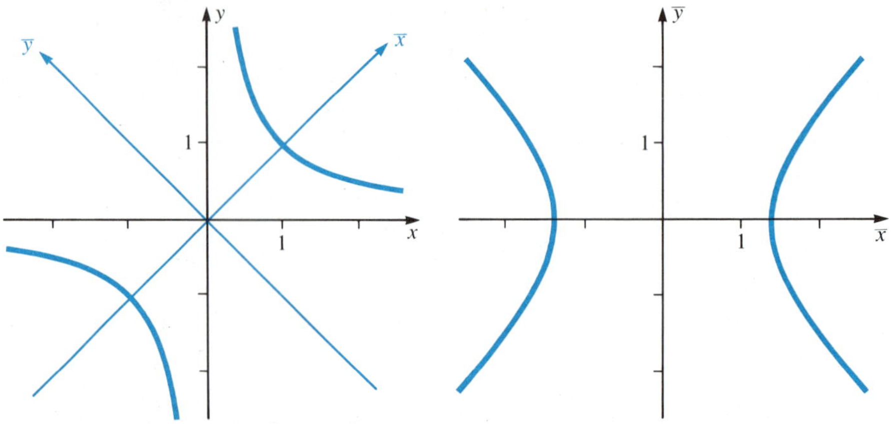

(a) Before rotation $[\alpha = \frac{1}{4}\pi]$ (b) In the $\bar{x},\bar{y}$-coordinates, $\frac{1}{2}\bar{x}^2 - \frac{1}{2}\bar{y}^2 = 1$

Fig. 7 Locus of $xy = 1$

EXAMPLE 5 Describe the locus of $x^2 - 2xy + 3y^2 = 1$.

Solution Rotate the axes through angle α where

$$\tan 2\alpha = \frac{b}{a-c} = \frac{-2}{-2} = 1 \qquad 2\alpha = \tfrac{1}{4}\pi \qquad \alpha = \tfrac{1}{8}\pi$$

In this case, $\sin 2\alpha = \cos 2\alpha = \frac{1}{2}\sqrt{2}$, so

$$\cos^2\alpha = \tfrac{1}{2}(1 + \tfrac{1}{2}\sqrt{2}) \qquad \sin^2\alpha = \tfrac{1}{2}(1 - \tfrac{1}{2}\sqrt{2}).$$

Substitute these values with $a = 1$, $b = -2$, $c = 3$ into the expressions preceding Example 4:

$$\begin{cases} \bar{a} = \frac{1}{2}(1 + \frac{1}{2}\sqrt{2}) - \frac{1}{2}\sqrt{2} + \frac{3}{2}(1 - \frac{1}{2}\sqrt{2}) = 2 - \sqrt{2}, \\ \bar{c} = \frac{1}{2}(1 - \frac{1}{2}\sqrt{2}) + \frac{1}{2}\sqrt{2} + \frac{3}{2}(1 + \frac{1}{2}\sqrt{2}) = 2 + \sqrt{2}. \end{cases}$$

Therefore, in the $\bar{x}, \bar{y}$-coordinate system, the locus is

$$(2 - \sqrt{2})\bar{x}^2 + (2 + \sqrt{2})\bar{y}^2 = 1.$$

Because $2 - \sqrt{2}$ and $2 + \sqrt{2}$ are both positive, this is an ellipse (Fig. 8). We may write

$$\frac{\bar{x}^2}{a^2} + \frac{\bar{y}^2}{b^2} = 1,$$

where $a^2 = 1/(2 - \sqrt{2})$ and $b^2 = 1/(2 + \sqrt{2})$. Clearly $a > b$.

Answer Ellipse with $a = \dfrac{1}{\sqrt{2 - \sqrt{2}}}$ and $b = \dfrac{1}{\sqrt{2 + \sqrt{2}}}$.

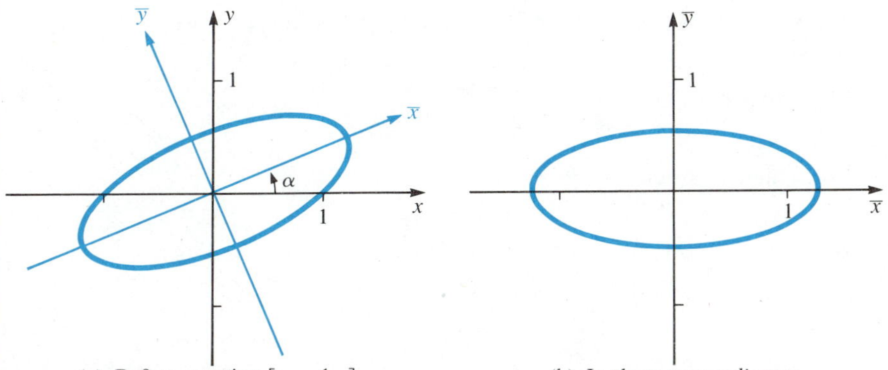

(a) Before rotation $[\alpha = \frac{1}{8}\pi]$

(b) In the $\bar{x}, \bar{y}$-*coordinates*, $(2 - \sqrt{2})\bar{x}^2 + (2 + \sqrt{2})\bar{y}^2 = 1$

Fig. 8 Locus of $x^2 - 2xy + 3y^2 = 1$

Exercises

Find the equation in polar coordinates of the circle with

1 center $(1, 1)$ radius $\sqrt{2}$

2 center $(-1, 1)$ radius $\sqrt{2}$

3 center $(1, \sqrt{3})$ radius 2

4 center $(\sqrt{3}, -1)$ radius 2

5 center $(5, 12)$ radius 13

6 center $(-12, -5)$ radius 13.

Make a suitable rotation and write the $\bar{x}, \bar{y}$-equation (without an $\bar{x}\bar{y}$ term)

7 $x^2 - xy = 1$

8 $xy - y^2 = 1$

9 $xy + y^2 = 1$

10 $2xy + y^2 = 1$.

Determine the type of the conic, the directions of its axes, and its equation in the rotated axes

11 $x^2 + xy + y^2 = 1$

12 $x^2 - xy + y^2 = 1$

13 $x^2 + xy - y^2 = 1$

14 $x^2 - xy - y^2 = 1$

15 $x^2 + xy + 2y^2 = 1$

16 $x^2 - xy + 2y^2 = 1$

17 $x^2 - 2xy + y^2 = 2y$

18 $x^2 - 4xy + 4y^2 = x$

19 $2x^2 - 6xy + y^2 = 1$

20 $x^2 + 3xy - y^2 = 1$.

21 Let $x = \bar{x}\cos\alpha - \bar{y}\sin\alpha$ and $y = \bar{x}\sin\alpha + \bar{y}\cos\alpha$. Compute $x^2 + y^2$. Explain your answer.

22 Let (x_1, y_1) and (x_2, y_2) be two points in the x, y-coordinate system. Let $(\bar{x}_1, \bar{y}_1)$ and $(\bar{x}_2, \bar{y}_2)$ be their coordinates in the $\bar{x}, \bar{y}$-coordinate system obtained by a rotation. Compute $x_1x_2 + y_1y_2$ in terms of $\bar{x}_1, \bar{x}_2, \bar{y}_1, \bar{y}_2$, and α, the angle of rotation. Explain your answer.

Suppose a rotation converts $ax^2 + bxy + cy^2$ into $\bar{a}\bar{x}^2 + \bar{b}\bar{x}\bar{y} + \bar{c}\bar{y}^2$. Prove

23 $a + c = \bar{a} + \bar{c}$

24* $4ac - b^2 = 4\bar{a}\bar{c} - \bar{b}^2$.

REVIEW EXERCISES

1 Find the translation of axes that moves the intersection of the lines

$$2x - y = 2 \qquad 5x - 3y = 3$$

to $(\bar{x}, \bar{y}) = (1, 1)$.

2 Find the intersections of the line $4x + 3y + 23 = 0$ and the circle $x^2 + y^2 + 4x + 10y + 4 = 0$.

3 Find the tangents to $x^2 + y^2 + 10x + 4y + 4 = 0$ from $(8, -2)$.

4 Find the locus of the midpoint of the segment **pz**, where $\mathbf{p} = (1, 1)$ and **z** varies over the curve $y = 1/x$.

Find the polar coordinates of

5 $(1, -1)$

6 $(1, -\sqrt{3})$.

Find the rectangular coordinates of

7 $\{2, \frac{5}{3}\pi\}$

8 $\{-2, \frac{5}{3}\pi\}$.

9 Find the polar equation for the circle with center $(3, 4)$ and radius 5.

Graph

10 $r = 2\sin 3\theta$

11 $r = 1 + \sin\theta$

12 $y + 2 = \frac{1}{4}(x - 3)^2$

13 $y^2 - 4x - 6y + 1 = 0$

14 $r\cos^2\theta = \sin\theta$

15 $\frac{1}{36}x^2 + \frac{1}{4}y^2 = 1$

16 $x^2 + 9y^2 - 2x + 36y = -28$.

Graph and find the asymptotes

17 $-\frac{1}{16}x^2 + \frac{1}{4}y^2 = 1$

18 $(x + 1)^2 - 9(y - 3)^2 = 9$.

19 Graph and find all intersections of the ellipse $\frac{1}{4}x^2 + y^2 = 1$ and the hyperbola $x^2 - y^2 = 1$.

20 Graph the conic $x^2 + 2xy + 2y^2 = 1$.

13 SOLID ANALYTIC GEOMETRY

1 COORDINATES AND VECTORS

The purpose of this chapter is to give a brief introduction to the analytic geometry of three-dimensional space. Perhaps the single most important skill you should get from this material is the ability to visualize simple geometric objects in space and to draw them on paper.

In order to study three-dimensional geometry, we introduce rectangular coordinate systems and vector algebra in space. Most of what we do is a direct extension of what we did previously in the plane (Chapter 11). Some of the figures will be more complicated because they are spatial, but the ideas are basically the same.

First we set up a rectangular coordinate system. We select an origin **0** and three mutually perpendicular real axes through the origin (Fig. 1a).

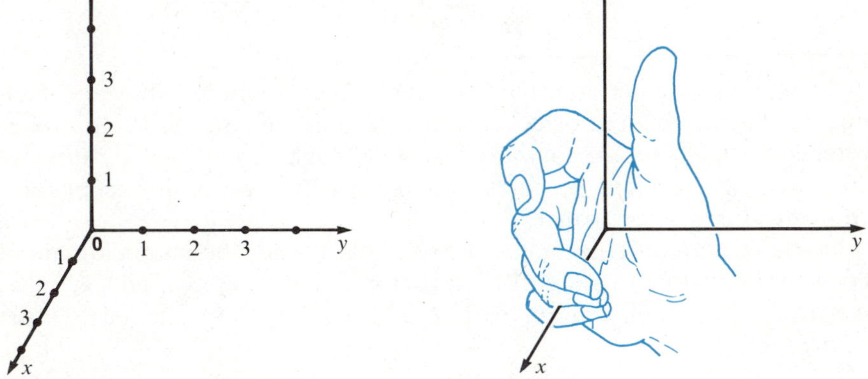

(a) Rectangular coordinates in space (b) Right-hand rule

Fig. 1

Once directions are fixed on the x- and y-axes, the direction on the z-axis is determined by the **right-hand rule:** curl the fingers of your right hand from the positive x-direction towards the positive y-direction; the thumb will point in the positive z-direction (Fig. 1b). (In drawings of 3-space, it is convenient to think of the x-axis as pointing straight up from the paper.)

We refer to the plane of the x-axis and y-axis as the ***x*,*y*-coordinate plane,** or simply the ***x*,*y*-plane,** etc. (Fig. 2a).

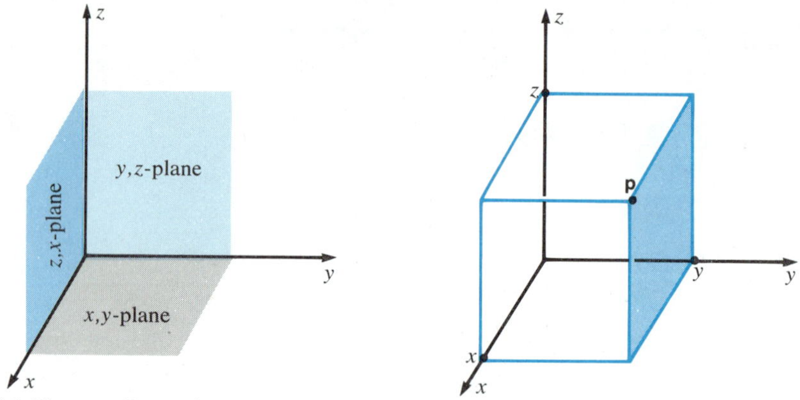

(a) The coordinate planes

(b) The planes through **p** parallel to the coordinate planes intersect the axes in the three coordinates of **p**.

Fig. 2

Now take any point $\mathbf{x}$ in space. Through $\mathbf{x}$ pass planes parallel to each coordinate plane. Their intersections with the coordinate axes determine three numbers x, y, z, called the **coordinates** or **components** of $\mathbf{x}$. See Fig. 2b. Conversely, each triple (x, y, z) of real numbers determines a unique point $\mathbf{x}$ in space. We shall write

$$\mathbf{x} = (x, y, z).$$

Vector Operations

The basic operations of multiplication by a scalar and addition of points (vectors) are defined componentwise, just as in the plane:

$$a\mathbf{x} = a(x, y, z) = (ax, ay, az)$$
$$\mathbf{x}_1 + \mathbf{x}_2 = (x_1, y_1, z_1) + (x_2, y_2, z_2)$$
$$= (x_1 + x_2, y_1 + y_2, z_1 + z_2).$$

Though defined in a particular coordinate system, these operations have geometric meanings quite independent of coordinates. Geometrically, the vector $a\mathbf{x}$ is just $\mathbf{x}$ stretched by the factor a, forward if $a > 0$, backward if $a < 0$. Proof of this fact requires an argument by similar triangles which we omit except for a suggestive figure (Fig. 3, next page).

The geometric interpretation of addition is our old friend, the parallelogram of forces. It can be shown that if $\mathbf{x}_1$ and $\mathbf{x}_2$ are vectors, then $\mathbf{0}, \mathbf{x}_1, \mathbf{x}_2$, and $\mathbf{x}_1 + \mathbf{x}_2$ are the vertices of a parallelogram with $\mathbf{0}$ and $\mathbf{x}_1 + \mathbf{x}_2$ opposite each other (Fig. 4, next page).

Examples (1) $(2, -1, 3) + (4, 0, -1) = (6, -1, 2)$

(2) $3(-7, 2, -1) = (-21, 6, -3)$

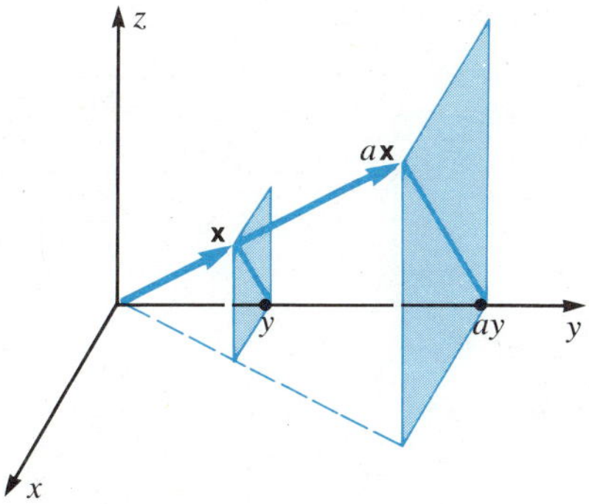

Fig. 3 Why $a\mathbf{x}$ is $\mathbf{x}$ stretched by the factor a.

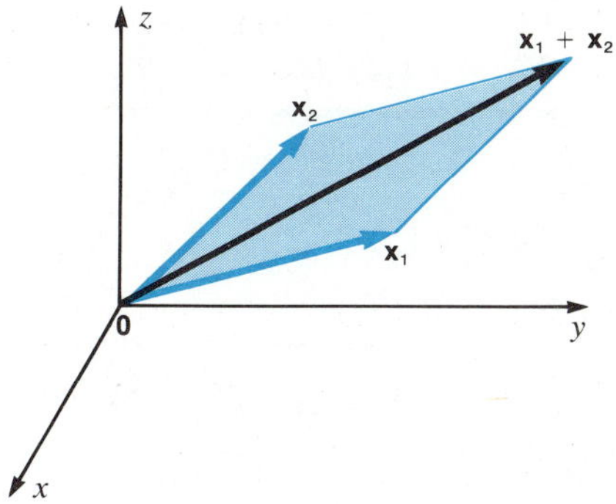

Fig. 4 Parallelogram rule for vector addition

The formal rules for multiplication by a scalar and addition are exactly the same as in the 2-dimensional case:

$$0\mathbf{x} = \mathbf{0} \qquad 1\mathbf{x} = \mathbf{x} \qquad a(b\mathbf{x}) = (ab)\mathbf{x}$$
$$(a+b)\mathbf{x} = a\mathbf{x} + b\mathbf{x} \qquad a(\mathbf{x}_1 + \mathbf{x}_2) = a\mathbf{x}_1 + a\mathbf{x}_2$$
$$\mathbf{x} + \mathbf{0} = \mathbf{0} + \mathbf{x} = \mathbf{0} \qquad \mathbf{x}_1 + \mathbf{x}_2 = \mathbf{x}_2 + \mathbf{x}_1$$
$$(\mathbf{x}_1 + \mathbf{x}_2) + \mathbf{x}_3 = \mathbf{x}_1 + (\mathbf{x}_2 + \mathbf{x}_3).$$

Length and Distance

The **length** $\|\mathbf{x}\|$ of a vector $\mathbf{x}$ in space is its distance from the origin. By regarding this distance as the length of the diagonal of a rectangular solid (Fig. 5), we see that

$$\|\mathbf{x}\|^2 = x^2 + y^2 + z^2.$$

This formula is the direct extension of the formula $\|\mathbf{x}\|^2 = x^2 + y^2$ for the length of a planar vector.

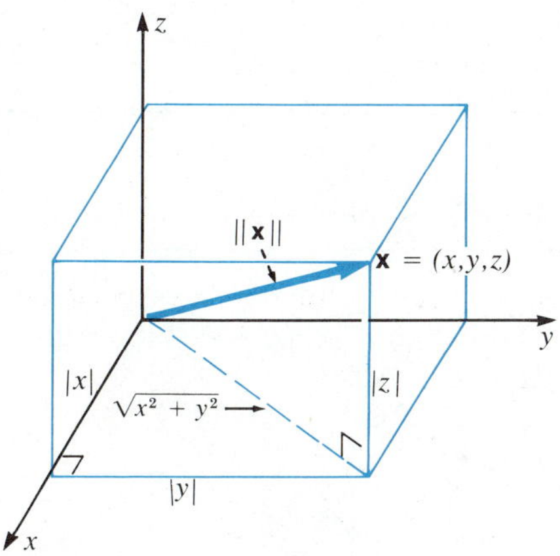

Fig. 5 $\|\mathbf{x}\|^2 = (\sqrt{x^2+y^2})^2 + z^2$
$= x^2 + y^2 + z^2$

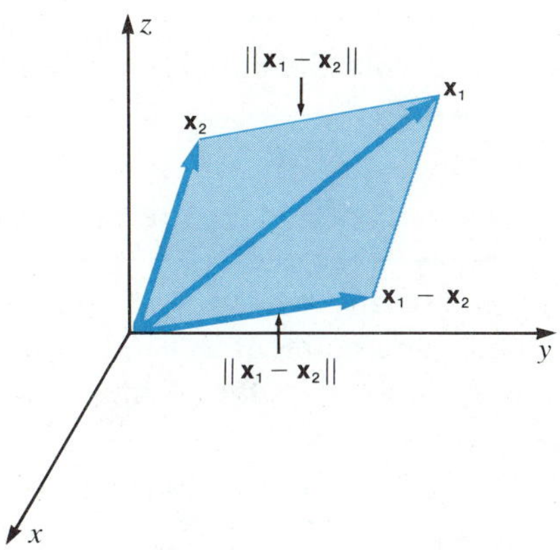

Fig. 6 (distance from $\mathbf{x}_1$ to $\mathbf{x}_2$)
$= \|\mathbf{x}_1 - \mathbf{x}_2\|$

Length of vectors in space has properties similar to the properties of length of vectors in the plane:

$$\|\mathbf{0}\| = 0, \qquad \|\mathbf{x}\| > 0 \text{ if } \mathbf{x} \neq \mathbf{0}.$$
$$\|a\mathbf{x}\| = |a| \cdot \|\mathbf{x}\|.$$
$$\|\mathbf{x}_1 + \mathbf{x}_2\| \leq \|\mathbf{x}_1\| + \|\mathbf{x}_2\| \quad \text{(triangle inequality).}$$

Examples (1) $\|(1, 2, -2)\| = \sqrt{1^2 + 2^2 + (-2)^2} = \sqrt{9} = 3$

(2) $\|(2, 5, -7)\| = \sqrt{9 + 25 + 49} = \sqrt{83}$

Since the distance between two points $\mathbf{x}_1$ and $\mathbf{x}_2$ is equal to $\|\mathbf{x}_1 - \mathbf{x}_2\|$, we have the distance formula in space (Fig. 6, previous page):

The distance between two points $\mathbf{x}_1 = (x_1, y_1, z_1)$ and $\mathbf{x}_2 = (x_2, y_2, z_2)$ is $\|\mathbf{x}_1 - \mathbf{x}_2\|$, where

$$\|\mathbf{x}_1 - \mathbf{x}_2\|^2 = (x_1 - x_2)^2 + (y_1 - y_2)^2 + (z_1 - z_2)^2.$$

Example $\mathbf{x}_1 = (1, 0, -2) \qquad \mathbf{x}_2 = (1, 1, 1)$

$$\|\mathbf{x}_1 - \mathbf{x}_2\|^2 = (1 - 1)^2 + (0 - 1)^2 + (-2 - 1)^2 = 10$$
$$\|\mathbf{x}_1 - \mathbf{x}_2\| = \sqrt{10}$$

Inner Product

We define the **inner product (dot product)** of two non-zero vectors exactly as in the plane:

$$\mathbf{x}_1 \cdot \mathbf{x}_1 = \|\mathbf{x}_1\| \cdot \|\mathbf{x}_2\| \cdot \cos\theta$$

where $\cos\theta$ is the cosine of the angle between $\mathbf{x}_1$ and $\mathbf{x}_2$. If either $\mathbf{x}_1 = \mathbf{0}$ or $\mathbf{x}_2 = \mathbf{0}$, then the angle between them is not defined. In this case we define

$$\mathbf{0} \cdot \mathbf{x} = \mathbf{x} \cdot \mathbf{0} = 0.$$

Two facts follow from the definition of inner product:

$$\mathbf{x} \cdot \mathbf{x} = \|\mathbf{x}\|^2.$$

Non-zero vectors $\mathbf{x}_1$ and $\mathbf{x}_2$ are perpendicular if and only if $\mathbf{x}_1 \cdot \mathbf{x}_2 = 0$.

If $\mathbf{x}_1$ and $\mathbf{x}_2$ are any two vectors, then they, together with $\mathbf{0}$, lie on a plane. Therefore the same argument as used on p. 440 shows that $\|\mathbf{x}_1\| \cos\theta$ is the signed projection of $\mathbf{x}_1$ on $\mathbf{x}_2$. See Fig. 7, next page. Hence

$$\mathbf{x}_1 \cdot \mathbf{x}_2 = \begin{cases} (\text{signed projection of } \mathbf{x}_1 \text{ on } \mathbf{x}_2) \cdot \|\mathbf{x}_2\| \\ (\text{signed projection of } \mathbf{x}_2 \text{ on } \mathbf{x}_1) \cdot \|\mathbf{x}_1\|. \end{cases}$$

There is a component formula for the inner product of two vectors in space, the

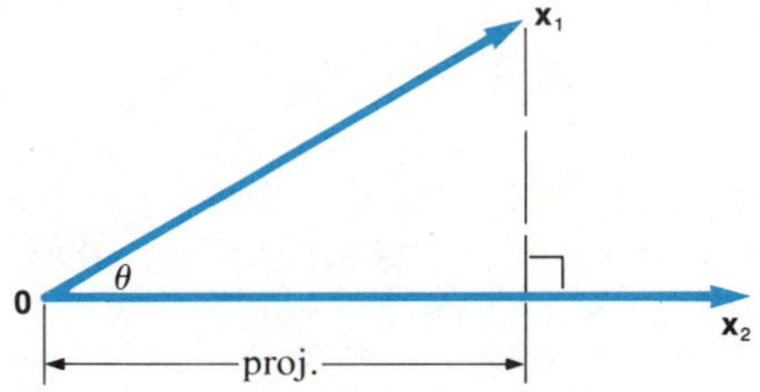

Fig. 7 $\mathbf{x}_1 \cdot \mathbf{x}_2$
$= (\text{proj. of } \mathbf{x}_1 \text{ on } \mathbf{x}_2) \cdot \|\mathbf{x}_2\|$
$= (\|\mathbf{x}_1\| \cos \theta) \cdot \|\mathbf{x}_2\|$

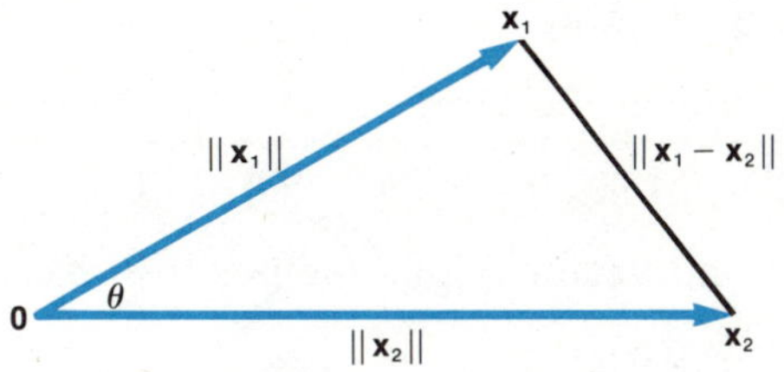

Fig. 8 $\|\mathbf{x}_1 - \mathbf{x}_2\|^2$
$= \|\mathbf{x}_1\|^2 + \|\mathbf{x}_2\|^2 - 2\|\mathbf{x}_1\| \cdot \|\mathbf{x}_2\| \cos \theta$.
The formula for $\mathbf{x}_1 \cdot \mathbf{x}_2$ drops out.

direct extension of the formula for vectors in the plane. It is proved (as on p. 441) by using the law of cosines (Fig. 8):

$$\mathbf{x}_1 \cdot \mathbf{x}_2 = x_1x_2 + y_1y_2 + z_1z_2.$$

Example $(3, 1, -2) \cdot (4, -7, -4) = 12 - 7 + 8 = 13$.

The formula for the inner product yields its main algebraic properties:

$$\mathbf{x}_1 \cdot \mathbf{x}_2 = \mathbf{x}_2 \cdot \mathbf{x}_1 \qquad (a\mathbf{x}_1) \cdot \mathbf{x}_2 = \mathbf{x}_1 \cdot (a\mathbf{x}_2) = a(\mathbf{x}_1 \cdot \mathbf{x}_2)$$
$$(\mathbf{x}_1 + \mathbf{x}_2) \cdot \mathbf{x}_3 = \mathbf{x}_1 \cdot \mathbf{x}_3 + \mathbf{x}_2 \cdot \mathbf{x}_3.$$

From the formula

$$\mathbf{x}_1 \cdot \mathbf{x}_2 = \|\mathbf{x}_1\| \cdot \|\mathbf{x}_2\| \cdot \cos \theta$$

follows

$$\cos \theta = \frac{\mathbf{x}_1 \cdot \mathbf{x}_2}{\|\mathbf{x}_1\| \cdot \|\mathbf{x}_2\|}$$

By expressing the dot product and lengths in terms of components, we obtain a useful formula for the cosine of the angle between two vectors:

If $\mathbf{x}_1 = (x_1, y_1, z_1)$ and $\mathbf{x}_2 = (x_2, y_2, z_2)$ are non-zero vectors, then the cosine of the angle between them is

$$\cos \theta = \frac{\mathbf{x}_1 \cdot \mathbf{x}_2}{\|\mathbf{x}_1\| \cdot \|\mathbf{x}_2\|} = \frac{x_1x_2 + y_1y_2 + z_1z_2}{\sqrt{x_1^2 + y_1^2 + z_1^2}\sqrt{x_2^2 + y_2^2 + z_2^2}}$$

EXAMPLE 1 Find the smallest positive angle between

$$\mathbf{x}_1 = (-2, 2, 1) \quad \text{and} \quad \mathbf{x}_2 = (2, 3, 6).$$

Solution $\mathbf{x}_1 \cdot \mathbf{x}_2 = -4 + 6 + 6 = 8$

$$\|\mathbf{x}_1\|^2 = 4 + 4 + 1 = 9 \quad \text{and} \quad \|\mathbf{x}_2\|^2 = 4 + 9 + 36 = 49.$$

Therefore

$$\cos \theta = \frac{8}{\sqrt{9}\sqrt{49}} = \frac{8}{21} \qquad \theta = \text{arc cos}\,\frac{8}{21}.$$

Answer $\theta = \text{arc cos}\,\frac{8}{21} \approx 67.61°$

Direction Cosines

It is customary to write

$$\mathbf{i} = (1, 0, 0), \quad \mathbf{j} = (0, 1, 0), \quad \mathbf{k} = (0, 0, 1)$$

for the unit vectors along the positive axes. If $\mathbf{x} = (x, y, z)$ is any vector, then

$$\mathbf{x} = x\mathbf{i} + y\mathbf{j} + z\mathbf{k}.$$

Note that

$$\mathbf{x}\cdot\mathbf{i} = (x, y, z)\cdot(1, 0, 0) = x,$$

and similarly $\mathbf{x}\cdot\mathbf{j} = y$ and $\mathbf{x}\cdot\mathbf{k} = z$.

> If $\mathbf{x}$ is any vector, then
>
> $$\mathbf{x} = (\mathbf{x}\cdot\mathbf{i})\mathbf{i} + (\mathbf{x}\cdot\mathbf{j})\mathbf{j} + (\mathbf{x}\cdot\mathbf{k})\mathbf{k}.$$

We apply this formula to a unit vector $\mathbf{u}$, a vector of length 1. Suppose $\mathbf{u}$ makes angles α, β, γ with the x-, y-, and z-axes respectively (Fig. 9). Then

$$\mathbf{u}\cdot\mathbf{i} = \|\mathbf{u}\|\cdot\|\mathbf{i}\|\cdot\cos\alpha = \cos\alpha, \qquad \mathbf{u}\cdot\mathbf{j} = \cos\beta, \qquad \mathbf{u}\cdot\mathbf{k} = \cos\gamma.$$

Therefore

$$\mathbf{u} = (\cos\alpha)\mathbf{i} + (\cos\beta)\mathbf{j} + (\cos\gamma)\mathbf{k} \qquad (\|\mathbf{u}\| = 1).$$

If $\mathbf{x}$ is any non-zero vector, then $\mathbf{u} = \mathbf{x}/\|\mathbf{x}\|$ is a unit vector. Therefore we may write $\mathbf{x} = \|\mathbf{x}\|\mathbf{u}$ and apply the last formula to $\mathbf{u}$.

> If $\mathbf{x}$ is any non-zero vector, then
>
> $$\mathbf{x} = \|\mathbf{x}\|[(\cos\alpha)\mathbf{i} + (\cos\beta)\mathbf{j} + (\cos\gamma)\mathbf{k}].$$
>
> The numbers $\cos\alpha$, $\cos\beta$, and $\cos\gamma$ are called the **direction cosines** of $\mathbf{x}$ and satisfy
>
> $$\cos^2\alpha + \cos^2\beta + \cos^2\gamma = 1.$$

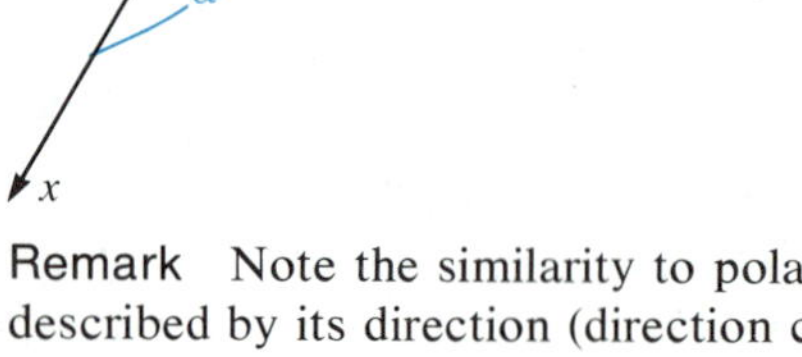

Fig. 9 The angles α, β, and γ determining the direction cosines of $\mathbf{u}$

Remark Note the similarity to polar coordinates in the plane. The non-zero vector $\mathbf{x}$ is described by its direction (direction cosines) and its length.

Example $\mathbf{x} = (6, -2, 3)$

$$\|\mathbf{x}\|^2 = 36 + 4 + 9 = 49 \qquad \|\mathbf{x}\| = 7.$$

Therefore

$$\mathbf{u} = \frac{\mathbf{x}}{\|\mathbf{x}\|} = (\tfrac{6}{7}, -\tfrac{2}{7}, \tfrac{3}{7})$$

is a unit vector, and

$$\cos\alpha = \tfrac{6}{7} \qquad \cos\beta = -\tfrac{2}{7} \qquad \cos\gamma = \tfrac{3}{7}$$

are the direction cosines of $\mathbf{x}$. It follows that

$$\alpha = \operatorname{arc\,cos} \tfrac{6}{7} \approx 31.00° \qquad \beta = \operatorname{arc\,cos}(-\tfrac{2}{7}) \approx 106.60°$$

$$\gamma = \operatorname{arc\,cos} \tfrac{3}{7} \approx 64.62°$$

Exercises

Draw coordinate axes as in Fig. 1 and locate each point accurately

1 $(3, 3, 3)$ **2** $(3, 0, -2)$ **3** $(-2, -2, 3)$ **4** $(-2, 2, -3)$

Set $\mathbf{x}_1 = (1, 2, -1)$ $\mathbf{x}_2 = (3, 0, 1)$ $\mathbf{x}_3 = (-3, 2, -1)$. Compute

5 $\mathbf{x}_1 + \mathbf{x}_2$ **6** $\mathbf{x}_2 - \mathbf{x}_3$ **7** $4\mathbf{x}_2$
8 $-2\mathbf{x}_1$ **9** $3\mathbf{x}_1 + \mathbf{x}_2$ **10** $2\mathbf{x}_2 - 3\mathbf{x}_3$
11 $\mathbf{x}_1 + \mathbf{x}_2 + \mathbf{x}_3$ **12** $2\mathbf{x}_1 - 3\mathbf{x}_2 + 5\mathbf{x}_3$ **13** $\mathbf{x}_1 \cdot \mathbf{x}_2$
14 $\mathbf{x}_2 \cdot \mathbf{x}_3$ **15** $(\mathbf{x}_1 + \mathbf{x}_2) \cdot \mathbf{x}_3$ **16** $(\mathbf{x}_1 - 2\mathbf{x}_2) \cdot \mathbf{x}_3$
17 $\|\mathbf{x}_1\|$ **18** $\|\mathbf{x}_3\|$.

Compute the distance between the points

19 $(0, 1, 2)$ $(5, -3, 1)$ **20** $(1, 1, 1)$ $(1, -1, 2)$
21 $(7, 0, 0)$ $(2, 3, 4)$ **22** $(8, 5, -1)$ $(7, 9, 3)$.

23 Describe geometrically all points $\mathbf{x} = (x, y, z)$ for which $x^2 + y^2 = z^2 = 1$.

24 If $\mathbf{x}_0 = (1, 2, 3)$, describe geometrically all points $\mathbf{x}$ for which $\|\mathbf{x} - \mathbf{x}_0\| = 5$.

Find the smallest positive angle between the vectors

25 $(4, 3, 0)$ $(-3, 0, 4)$ **26** $(1, 2, 2)$ $(-2, 1, -2)$
27 $(6, 1, 5)$ $(-2, -3, 3)$ **28** $(-5, 6, 1)$ $(2, 3, -8)$.

29 Find two non-collinear vectors perpendicular to $(1, -1, 2)$.

30 Let $\mathbf{u}_1 = (\cos\alpha_1, \cos\beta_1, \cos\gamma_1)$ and $\mathbf{u}_2 = (\cos\alpha_2, \cos\beta_2, \cos\gamma_2)$ be two unit vectors. Show that the angle θ between them satisfies

$$\cos\theta = \cos\alpha_1\cos\alpha_2 + \cos\beta_1\cos\beta_2 + \cos\gamma_1\cos\gamma_2.$$

31 Prove the inequality $|\mathbf{x}_1 \cdot \mathbf{x}_2| \le \|\mathbf{x}_1\| \cdot \|\mathbf{x}_2\|$.

32* (cont.) Now prove the triangle inequality: $\|\mathbf{x}_1 + \mathbf{x}_2\| \le \|\mathbf{x}_1\| + \|\mathbf{x}_2\|$.

[*Hint* $\|\mathbf{x}_1 + \mathbf{x}_2\|^2 = (\mathbf{x}_1 + \mathbf{x}_2)\cdot(\mathbf{x}_1 + \mathbf{x}_2) = (\mathbf{x}_1 + \mathbf{x}_2)\cdot\mathbf{x}_1 + (\mathbf{x}_1 + \mathbf{x}_2)\cdot\mathbf{x}_2$

$\le \|\mathbf{x}_1 + \mathbf{x}_2\| \cdot \|\mathbf{x}_1\| + \|\mathbf{x}_1 + \mathbf{x}_2\| \cdot \|\mathbf{x}_2\|$, etc.]

2 LINES AND PLANES

Parametric Equation of a Line

Take a non-zero vector $\mathbf{v}$. The set of all scalar multiples $\mathbf{x} = t\mathbf{v}$ is a line passing through the origin (Fig. 1a, next page). If $\mathbf{x}_0$ is any fixed point, the set of all points $\mathbf{x} = \mathbf{x}_0 + t\mathbf{v}$ is a line passing through $\mathbf{x}_0$ parallel to the first line (Fig. 1b).

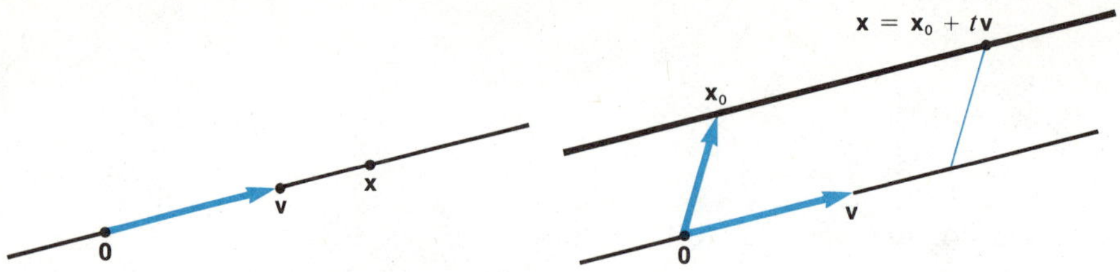

(a) Typical line through **0**: all $\mathbf{x} = t\mathbf{v}$ where $\mathbf{v}$ is a fixed non-zero vector and t is an arbitrary real number

(b) The line through $\mathbf{x}_0$ parallel to $\mathbf{v}$

Fig. 1 Parametric equation for a line in space

> If $\mathbf{x}_0$ is a point in space and $\mathbf{v}$ is a non-zero vector, then the line through $\mathbf{x}_0$ parallel to $\mathbf{v}$ consists of all points
>
> (1) $\mathbf{x} = \mathbf{x}_0 + t\mathbf{v} \qquad (t \text{ real}).$

The equation $\mathbf{x} = \mathbf{x}_0 + t\mathbf{v}$ is a **parametric vector equation** for the line. When written in coordinates, it is equivalent to three **parametric scalar equations.** If $\mathbf{x}_0 = (x_0, y_0, z_0)$ and $\mathbf{v} = (a, b, c)$, then

$$x = x_0 + ta \qquad y = y_0 + tb \qquad z = z_0 + tc.$$

Conversely, three equations of this type describe a line in space. For example, the scalar equations

$$x = 3 + 2t \qquad y = 5t \qquad z = 1 - t$$

are equivalent to the vector equation

$$(x, y, z) = (3, 0, 1) + t(2, 5, -1)$$

which describes the line through $(3, 0, 1)$ parallel to $(2, 5, -1)$.

Normal Form of a Plane

A plane can be described by an equation that is completely analogous to the normal form of a line (p. 445).

Let P be any plane in space. Draw the line L through **0** perpendicular to P and take one of the (two) unit vectors along L; call this vector $\mathbf{n}$. See Fig. 2. The line L meets P in a point $p\mathbf{n}$. If $\mathbf{x}$ is any point of P, then $\mathbf{x} - p\mathbf{n}$ is perpendicular to $\mathbf{n}$, hence

$$(\mathbf{x} - p\mathbf{n})\cdot\mathbf{n} = 0 \qquad \mathbf{x}\cdot\mathbf{n} = p\mathbf{n}\cdot\mathbf{n} = p.$$

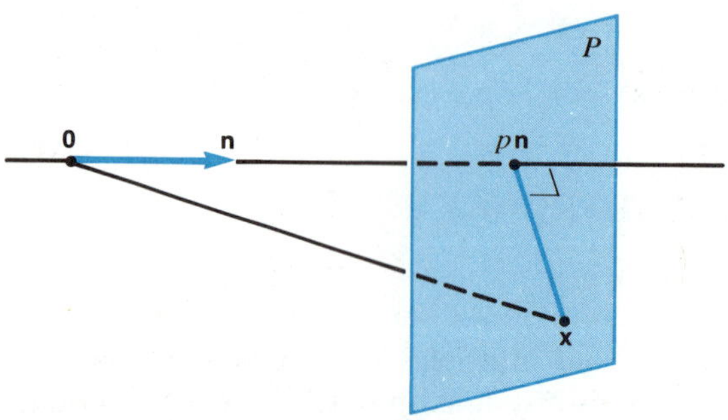

Fig. 2 The normal form of a plane: $\mathbf{x}\cdot\mathbf{n} = p$

Normal Form Each plane in space can be represented by an equation in normal form

$$\mathbf{x} \cdot \mathbf{n} = p$$

where $\mathbf{n}$ is a unit vector and p is a real number. The only other normal form of the same plane is

$$\mathbf{x} \cdot (-\mathbf{n}) = -p.$$

Conversely, each equation $\mathbf{x} \cdot \mathbf{n} = p$, where $\mathbf{n}$ is a unit vector, describes the plane perpendicular to $\mathbf{n}$ passing through the point $p\mathbf{n}$.

In components, if $\mathbf{n} = (a, b, c)$ and $\mathbf{x} = (x, y, z)$, then $\mathbf{x} \cdot \mathbf{n} = p$ becomes

$$ax + by + cz = p \qquad (a^2 + b^2 + c^2 = 1).$$

More generally, the locus of *any* linear equation

$$ax + by + cz = d \qquad (a^2 + b^2 + c^2 > 0)$$

is a plane. For if we write $\mathbf{a} = (a, b, c)$, then $\mathbf{a} \neq \mathbf{0}$ and the equation $ax + by + cz = d$ is equivalent to $\mathbf{a} \cdot \mathbf{x} = d$. This in turn is equivalent to

$$\frac{\mathbf{a}}{\|\mathbf{a}\|} \cdot \mathbf{x} = \frac{d}{\|\mathbf{a}\|}, \quad \text{that is, to} \quad \mathbf{x} \cdot \mathbf{n} = p$$

where $\mathbf{n} = \mathbf{a}/\|\mathbf{a}\|$, a unit vector, and $p = d/\|\mathbf{a}\|$. But $\mathbf{x} \cdot \mathbf{n} = p$ is the normal form of a planc.

Example $2x - 6y + 3z = 21$

Set $\mathbf{a} = (2, -6, 3)$ so

$$\|\mathbf{a}\|^2 = 4 + 36 + 9 = 49 \qquad \|\mathbf{a}\| = 7$$

Set $\mathbf{n} = \mathbf{a}/\|\mathbf{a}\| = (\frac{2}{7}, -\frac{6}{7}, \frac{3}{7})$, a unit vector. The given equation is equivalent to

$$\tfrac{2}{7}x - \tfrac{6}{7}y + \tfrac{3}{7}z = 3,$$

that is, to $\mathbf{x} \cdot \mathbf{n} = p = 3$, a normal form.

Parametric Form of a Plane

Let $\mathbf{u}$ and $\mathbf{v}$ be two non-zero, non-collinear vectors in space. They determine a plane Q, which consists precisely of all points

$$\mathbf{x} = s\mathbf{u} + t\mathbf{v} \qquad (s, t \text{ real}),$$

as is evident graphically (Fig. 3).

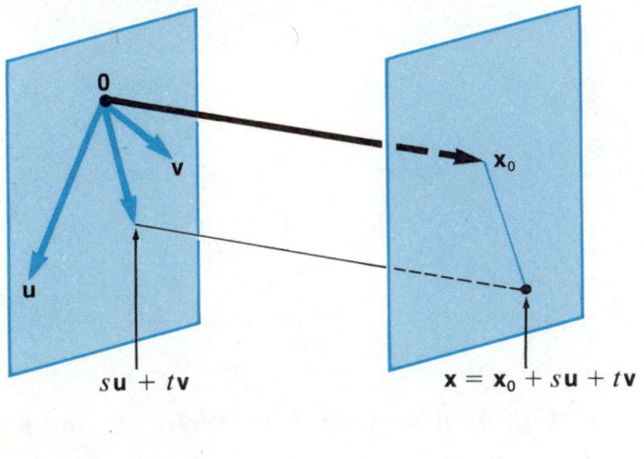

Fig. 3 The parametric form of a plane: $\mathbf{x} = \mathbf{x}_0 + s\mathbf{u} + t\mathbf{v}$, where $\mathbf{x}_0$, $\mathbf{u}$, and $\mathbf{v}$ are fixed vectors and s and t are arbitrary numbers.

Now take any point $\mathbf{x}_0$. The set of all points

$$\mathbf{x} = \mathbf{x}_0 + s\mathbf{u} + t\mathbf{v} \qquad (s, t \text{ real})$$

is the plane P through $\mathbf{x}_0$ parallel to Q. The representation $\mathbf{x} = \mathbf{x}_0 + s\mathbf{u} + t\mathbf{v}$ of the points of P, one for each pair (s, t) of real numbers, is called a **parametric form** of P. The numbers s and t are the **parameters.**

EXAMPLE 1 Give a parametric form and a normal form for the plane through $\mathbf{i}$, $\mathbf{j}$, and $\mathbf{k}$.

Solution Let P denote the desired plane. The vectors

$$\mathbf{u} = \mathbf{k} - \mathbf{i} = (-1, 0, 1) \quad \text{and} \quad \mathbf{v} = \mathbf{j} - \mathbf{i} = (-1, 1, 0)$$

determine a plane through $\mathbf{0}$ parallel to P. See Fig. 4. Since P passes through $\mathbf{x}_0 = \mathbf{i} = (1, 0, 0)$, a parametric form is

$$\mathbf{x} = (1, 0, 0) + s(-1, 0, 1) + t(-1, 1, 0),$$

that is,

$$x = 1 - s - t \qquad y = t \qquad z = s.$$

To obtain a normal form, we look for a non-zero vector perpendicular to P. It is enough to find a vector that is perpendicular to $\mathbf{u} = (-1, 0, 1)$ and to $\mathbf{v} = (-1, 1, 0)$. If $\mathbf{a} = (a, b, c)$ is such a vector, then $\mathbf{a} \cdot \mathbf{u} = 0$ and $\mathbf{a} \cdot \mathbf{v} = 0$, which means

$$-a + c = 0, \qquad -a + b = 0.$$

A natural choice is $\mathbf{a} = (1, 1, 1)$. Therefore a unit vector perpendicular to the plane (a unit normal) is

$$\mathbf{n} = \frac{\mathbf{a}}{\|\mathbf{a}\|} = \frac{1}{\sqrt{3}}(1, 1, 1),$$

and a normal form is $\mathbf{x} \cdot \mathbf{n} = p$. To find the constant p, we note that $\mathbf{i}$ is a point on the plane. Hence we must have $p = \mathbf{i} \cdot \mathbf{n} = 1/\sqrt{3}$, so a normal form is

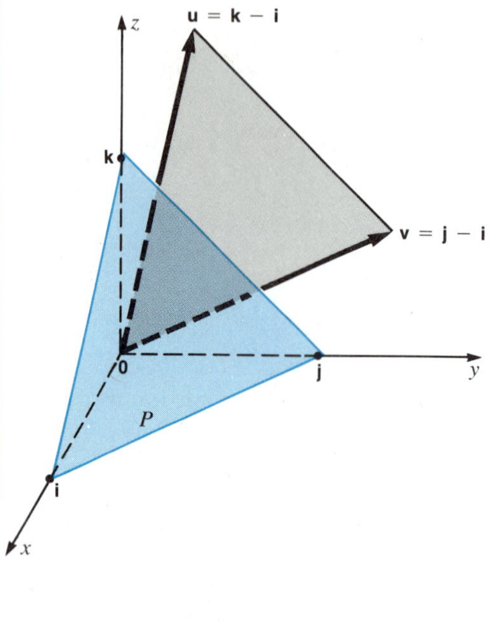

Fig. 4 The plane P through $\mathbf{i}$, $\mathbf{j}$, and $\mathbf{k}$

$$\frac{1}{\sqrt{3}}(x + y + z) = \frac{1}{\sqrt{3}}.$$

Answer $\mathbf{x} = (1 - s - t, t, s)$ $\quad \frac{1}{3}\sqrt{3}(x + y + z) = \frac{1}{3}\sqrt{3}$. See Fig. 4.

Remark The simplest equation of the plane is $x + y + z = 1$.

In Example 1, we went from a parametric form $\mathbf{x} = \mathbf{x}_0 + s\mathbf{u} + t\mathbf{v}$ of a plane to a normal form. The main step was finding a vector $\mathbf{a}$ perpendicular both to $\mathbf{u}$ and to $\mathbf{v}$, that is, a vector satisfying the pair of linear equations $\mathbf{a}\cdot\mathbf{u} = 0$ and $\mathbf{a}\cdot\mathbf{v} = 0$. In coordinates, this is a pair of linear equations of the form

$$\begin{cases} u_1x + u_2y + u_3z = 0 \\ v_1x + v_2y + v_3z = 0 \end{cases}$$

where $\mathbf{a} = (x, y, z)$. Such a system can be solved by elimination.

Example 2 Find a non-zero vector perpendicular to $(1, -1, 3)$ and to $(2, 3, -1)$.

Solution Suppose $\mathbf{x} = (x, y, z)$ satisfies

$$\mathbf{x}\cdot(1, -1, 3) = 0 \quad \text{and} \quad \mathbf{x}\cdot(2, 3, -1) = 0.$$

Then

$$\begin{cases} x - y + 3z = 0 \\ 2x + 3y - z = 0. \end{cases}$$

Subtract twice the first equation from the second:

$$5y - 7z = 0.$$

Hence the system is equivalent to

$$\begin{cases} x - y + 3z = 0 \\ \quad 5y - 7z = 0. \end{cases}$$

Now $y = 7$, $z = 5$ satisfies the second equation of this pair. Then

$$x = y - 3z = -8.$$

Answer $(-8, 7, 5)$ is one possibility.

Exercises

1 Let $\mathbf{a} \neq \mathbf{0}$. Show that a parametric equation for the line through $\mathbf{0}$ and $\mathbf{a}$ is $\mathbf{x} = t\mathbf{a}$.

2 Let $\mathbf{x}_0 \neq \mathbf{x}_1$. Show that the line through $\mathbf{x}_0$ and $\mathbf{x}_1$ has the parametric equation $\mathbf{x} = t\mathbf{x}_0 + (1 - t)\mathbf{x}_1$.

Find a parametric equation for the line through the two given points

3 $(1, 0, 1)$ $(0, 1, 1)$

4 $(-2, 1, 1)$ $(3, 2, 1)$

5 $(0, 1, 4)$ $(0, 2, 1)$

6 $(-1, 3, 0)$ $(2, 2, 2)$.

Express the plane in normal form

7 $x - 2y + 2z = 1$

8 $2x + 6y - 3z = 14$

9 $-8x + y - 4z = 27$

10 $3x - 2y - 6z = 4$

11 $x + y + z = 3$

12 $x - y + z = -12$.

Find a parametric form for the plane through the given points

13 $(0,0,0)$ $(1,1,1)$ $(1,1,-1)$

14 $(1,0,1)$ $(0,1,1)$ $(1,1,0)$

15 $(2,1,0)$ $(3,0,1)$ $(4,3,3)$

16 $(-1,1,-1)$ $(2,0,1)$ $(4,-1,2)$.

Find a normal form for the plane through the given points

17 the points in Ex. 13

18 the points in Ex. 14.

19 the points in Ex. 15

20 the points in Ex. 16.

21* Let $\mathbf{x}\cdot\mathbf{n} = p$ be a plane in normal form and let $\mathbf{x}_0$ be a point. Prove that the distance from $\mathbf{x}_0$ to the plane is $|\mathbf{x}_0\cdot\mathbf{n} - p|$.

22* (cont.) Prove that the point of the plane closest to $\mathbf{x}_0$ is $\mathbf{z} = \mathbf{x}_0 + (p - \mathbf{x}_0\cdot\mathbf{n})\mathbf{n}$.

23 Let $\mathbf{x} = \mathbf{x}_0 + t\mathbf{v}$ be a line in parametric form and let $\mathbf{x}\cdot\mathbf{n} = p$ be a plane in normal form. Prove that the line and plane are parallel if and only if $\mathbf{v}\cdot\mathbf{n} = 0$.

24 (cont.) Prove that the line is on the plane if and only if $\mathbf{v}\cdot\mathbf{n} = 0$ and $\mathbf{x}_0\cdot\mathbf{n} = p$.

25* (cont.) Suppose $\mathbf{v}\cdot\mathbf{n} \neq 0$. Prove that the point of intersection of the line and the plane is

$$\mathbf{x}_0 + [(p - \mathbf{x}_0\cdot\mathbf{n})/(\mathbf{v}\cdot\mathbf{n})]\mathbf{v}.$$

26 Let $\mathbf{x}\cdot\mathbf{m} = p$ and $\mathbf{x}\cdot\mathbf{n} = q$ be non-parallel planes in normal form. Let θ be one of their (dihedral) angles of intersection. Determine $\cos\theta$.

27* Let $\mathbf{x} = \mathbf{x}_0 + t\mathbf{u}$ be a line in parametric form, where $\mathbf{u}$ is a unit vector, and let $\mathbf{y}_0$ be any point. Find the point of the line closest to $\mathbf{y}_0$.

28 (cont.) Prove that the distance d from $\mathbf{y}_0$ to the line satisfies

$$d^2 = \|\mathbf{x}_0 - \mathbf{y}_0\|^2 - [(\mathbf{x}_0 - \mathbf{y}_0)\cdot\mathbf{u}]^2.$$

3 QUADRIC SURFACES

Suppose $f(x, y, z)$ is a polynomial of degree two (quadratic polynomial). The graph of the equation $f(x, y, z) = 0$ is called a **quadric surface.** Analysis of the most general quadric surface requires difficult tools from Linear Algebra analogous to the rotation of coordinates in the plane. We shall examine only a few of the most common quadratic polynomials.

We start with equations of the form

$$\pm\frac{x^2}{a^2} \pm \frac{y^2}{b^2} \pm \frac{z^2}{c^2} = 1 \qquad (a, b, c > 0).$$

The graph is symmetric in each of the three coordinate planes, because if (x, y, z) satisfies the equation, then so do all eight points $(\pm x, \pm y, \pm z)$. Therefore an accurate graph in just the first octant suffices to determine the whole graph.

Spheres

Consider the graph of the equation

$$x^2 + y^2 + z^2 = a^2 \qquad (a > 0).$$

The equation simply says $\|\mathbf{x}\| = a$, hence the graph consists of all points at distance a from $\mathbf{0}$. This is a sphere of radius a with center $\mathbf{0}$.

Ellipsoids

Consider the graph of

$$\frac{x^2}{a^2} + \frac{y^2}{b^2} + \frac{z^2}{c^2} = 1 \qquad (a, b, c > 0).$$

Since squares are non-negative, each point of the graph satisfies

$$\frac{x^2}{a^2} \le 1 \qquad \frac{y^2}{b^2} \le 1 \qquad \frac{z^2}{c^2} \le 1.$$

This means the graph is confined to the box

$$-a \le x \le a \qquad -b \le y \le b \qquad -c \le z \le c.$$

Suppose $-c < z_0 < c$. The intersection of the graph and the horizontal plane $z = z_0$ consists of all points (x, y, z_0) that satisfy

$$\frac{x^2}{a^2} + \frac{y^2}{b^2} = 1 - \frac{z_0^2}{c^2}.$$

This curve is an ellipse. It is as large as possible when $z_0 = 0$, and it becomes smaller and smaller as $z_0 \longrightarrow c$ or $z_0 \longrightarrow -c$. Thus each such cross-section by a horizontal plane is an ellipse, except at the extremes $z_0 = \pm c$, where it is a single point.

The same argument applies to plane sections parallel to the other coordinate planes. This gives us enough information for a sketch. The surface is called an **ellipsoid** (Fig. 1). In the special case $a = b = c$, it is a sphere.

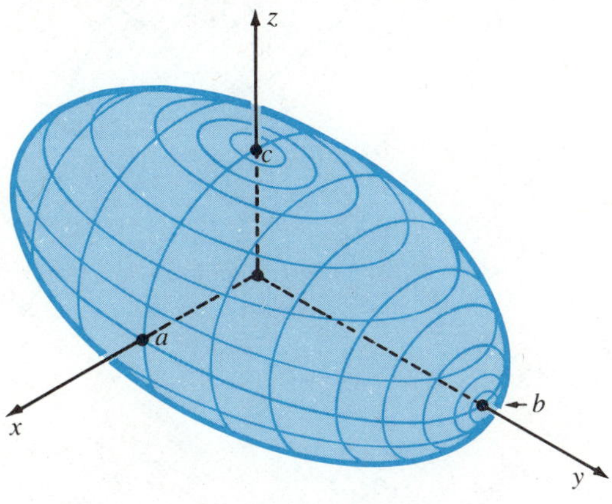

Fig. 1 Ellipsoid: $\frac{x^2}{a^2} + \frac{y^2}{b^2} + \frac{z^2}{c^2} = 1$

Hyperboloids of One Sheet

Consider the graph of

$$\frac{x^2}{a^2} + \frac{y^2}{b^2} - \frac{z^2}{c^2} = 1 \qquad (a, b, c > 0).$$

Each horizontal cross-section is an ellipse

$$\begin{cases} z = z_0 \\ \dfrac{x^2}{a^2} + \dfrac{y^2}{b^2} = 1 + \dfrac{z_0^2}{c^2}, \end{cases}$$

no matter what z_0 is. The ellipse is smallest for $z_0 = 0$. However, as $z_0 \longrightarrow \infty$ or $z_0 \longrightarrow -\infty$, the ellipse gets larger and larger.

The surface meets the y,z-plane in the hyperbola

$$\frac{y^2}{b^2} - \frac{z^2}{c^2} = 1,$$

and it meets the z,x-plane in the hyperbola

$$\frac{x^2}{a^2} - \frac{z^2}{c^2} = 1.$$

This information is enough to sketch the surface, called a **hyperboloid of one sheet** (Fig. 2a).

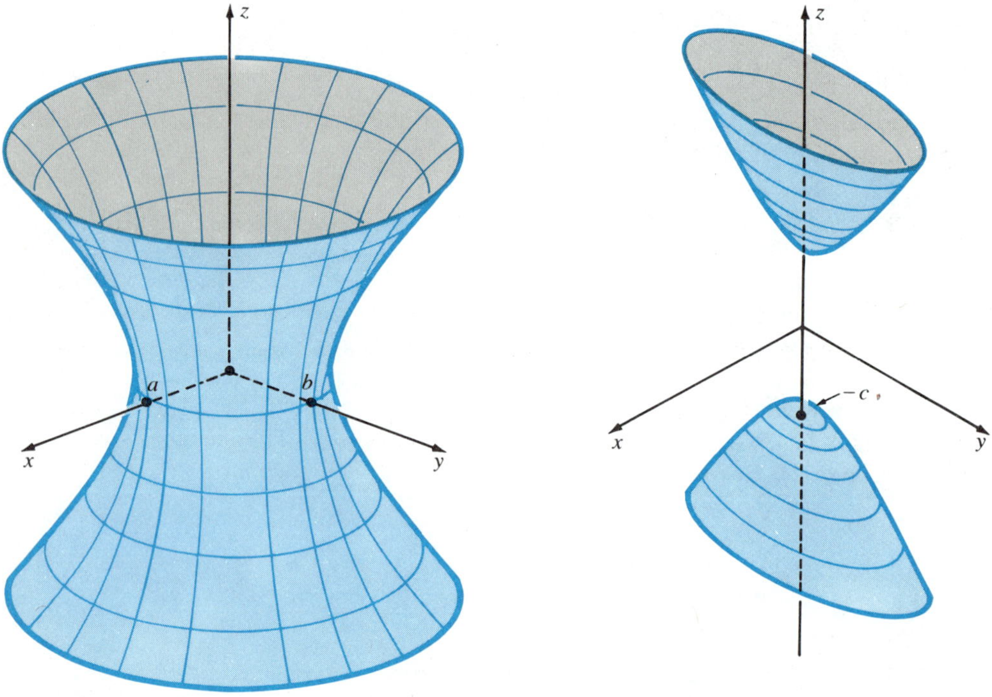

(a) Hyperboloid of one sheet: $\frac{x^2}{a^2} + \frac{y^2}{b^2} - \frac{z^2}{c^2} = 1$

(b) Hyperboloid of two sheets: $-\frac{x^2}{a^2} - \frac{y^2}{b^2} + \frac{z^2}{c^2} = 1$

Fig. 2 Hyperboloids

Hyperboloids of Two Sheets

Consider the equation

$$-\frac{x^2}{a^2} - \frac{y^2}{b^2} + \frac{z^2}{c^2} = 1 \qquad (a, b, c > 0).$$

If (x, y, z) is a point on the surface, then

$$\frac{z^2}{c^2} = 1 + \frac{x^2}{a^2} + \frac{y^2}{b^2} \geq 1,$$

hence $z^2 \geq c^2$. This means either $z \geq c$ or $z \leq -c$; that is, there are no points of the surface between the horizontal planes $z = c$ and $z = -c$.

If $z_0{}^2 > c^2$, the horizontal plane $z = z_0$ meets the surface in the curve

$$\begin{cases} z = z_0 \\ \dfrac{x^2}{a^2} + \dfrac{y^2}{b^2} = \dfrac{z_0{}^2}{c^2} - 1 > 0, \end{cases}$$

an ellipse. Also the surface meets the y,z-plane and the z,x-plane in the hyperbolas

$$-\frac{y^2}{b^2}+\frac{z^2}{c^2}=1 \quad \text{and} \quad -\frac{x^2}{a^2}+\frac{z^2}{c^2}=1$$

respectively. The surface breaks into two parts, and is called a **hyperboloid of two sheets** (Fig. 2b).

Paraboloids

The surface

$$z=\frac{x^2}{a^2}+\frac{y^2}{b^2} \qquad (a, b>0)$$

is called an **elliptic paraboloid.** Since $z \geq 0$, the surface lies above the x,y-plane, and it is symmetric in the y,z- and z,x-planes. Each horizontal cross-section

$$\frac{x^2}{a^2}+\frac{y^2}{b^2}=z_0>0$$

is an ellipse, and these ellipses grow larger as z_0 increases. The graph meets the y,z- and z,x-planes in parabolas (Fig. 3a).

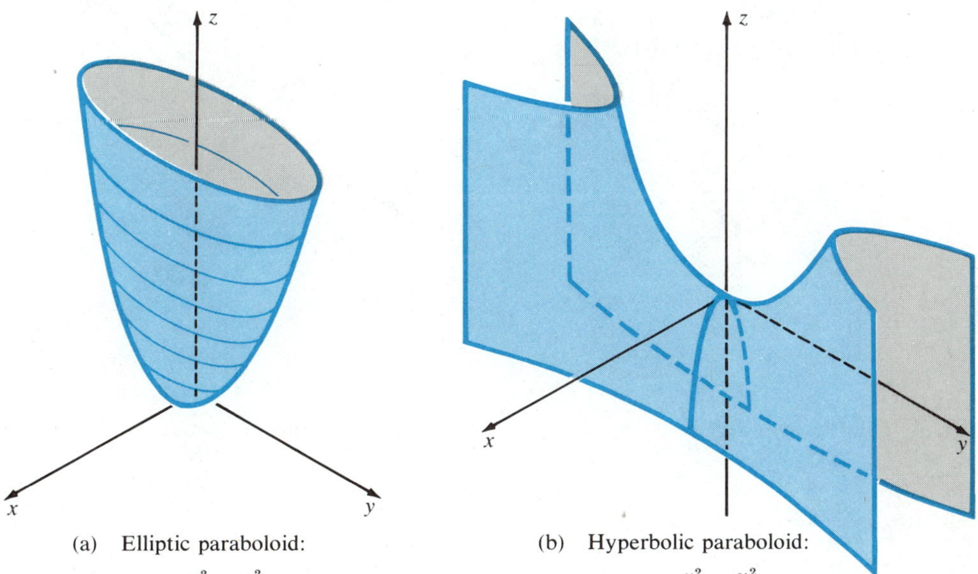

(a) Elliptic paraboloid: $z=\frac{x^2}{a^2}+\frac{y^2}{b^2}$

(b) Hyperbolic paraboloid: $z=-\frac{x^2}{a^2}+\frac{y^2}{b^2}$

Fig. 3 Paraboloids

Finally we consider the **hyperbolic paraboloid,** the locus of

$$z=-\frac{x^2}{a^2}+\frac{y^2}{b^2} \qquad (a, b>0).$$

It is symmetric in the y,z- and z,x-planes. The horizontal planes $z=z_0>0$ meet it in hyperbolas whose branches open out in the y-direction. The horizontal planes $z=z_0<0$ meet it in hyperbolas that open out in the x-direction. The y,z-plane meets the locus in the parabola $z=y^2/b^2$, which opens upwards, and the z,x-plane meets it in the parabola $z=-x^2/a^2$, which opens downwards. The best description is "saddle-shaped." See Fig. 3b.

Cylinders

Consider the locus of

$$\frac{x^2}{a^2} + \frac{y^2}{b^2} = 1 \qquad (a, b > 0)$$

in x,y,z-space. It meets each horizontal plane $z = z_0$ in the same ellipse. Therefore if (x_0, y_0, z_0) is any point of the surface, the whole vertical line (x_0, y_0, z) for $-\infty < z < \infty$ lies on the surface. The surface is an **elliptic cylinder,** and these vertical lines that lie on the surface are called **generators** of the cylinder (Fig. 4).

Any curve $f(x, y) = 0$ in the x,y-plane generates a cylinder in space consisting of all points (x_0, y_0, z) for which $f(x_0, y_0) = 0$. In particular, a circle generates a (right) circular cylinder, a hyperbola generates a hyperbolic cylinder, etc.

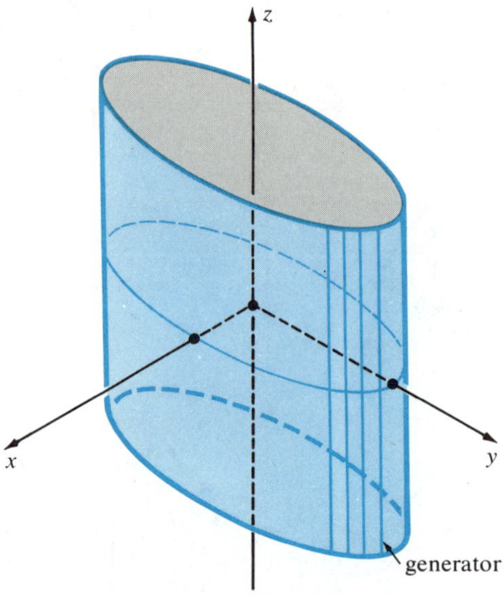

Fig. 4 Elliptic cylinder: $\frac{x^2}{a^2} + \frac{y^2}{b^2} = 1$

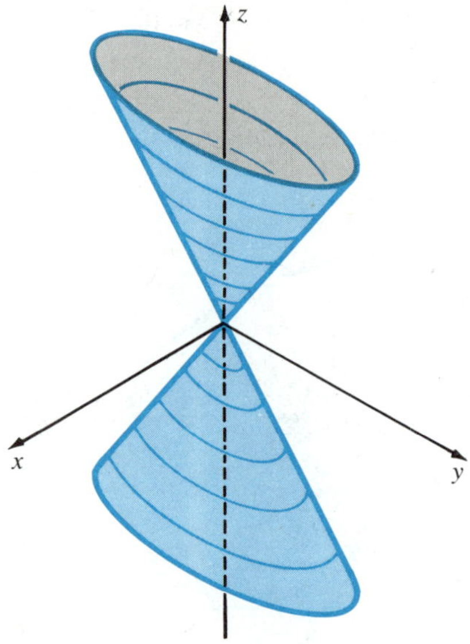

Fig. 5 Quadratic cone: $z^2 = \frac{x^2}{a^2} + \frac{y^2}{b^2}$

Cones

Consider the graph of

$$z^2 = \frac{x^2}{a^2} + \frac{y^2}{b^2} \qquad (a, b > 0).$$

It is a **cone,** a surface with the following property. For each point $\mathbf{x}_0$ on the surface, the entire line $\mathbf{x} = t\mathbf{x}_0$ lies on the surface. This line is called a **generator** of the cone.

To check that the graph of

$$z^2 = \frac{x^2}{a^2} + \frac{y^2}{b^2}$$

is a cone, we take any point $\mathbf{x}_0$ on the graph and we check that $t\mathbf{x}_0$ is also on the graph. If $\mathbf{x}_0 = (x_0, y_0, z_0)$, then $t\mathbf{x}_0 = (tx_0, ty_0, tz_0)$ and

$$(tz_0)^2 = t^2 z_0{}^2 = t^2\left(\frac{x_0{}^2}{a^2} + \frac{y_0{}^2}{b^2}\right) = \frac{(tx_0)^2}{a^2} + \frac{(ty_0)^2}{b^2}.$$

Thus $t\mathbf{x}_0$ is on the graph.

To sketch the cone, we first note that it meets the horizontal plane $z = 1$ in the ellipse

$$\frac{x^2}{a^2} + \frac{y^2}{b^2} = 1.$$

For each point on this ellipse, we draw the line through the point and $\mathbf{0}$. The surface is typical of quadratic cones (Fig. 5).

Exercises

Sketch the first octant portion

1 $\frac{1}{4}x^2 + y^2 + \frac{1}{4}z^2 = 1$

2 $x^2 + \frac{1}{9}y^2 + \frac{1}{4}z^2 = 1$

3 $x^2 + y^2 - z^2 = 1$

4 $-x^2 - y^2 + z^2 = 1$

5 $x^2 - y^2 + z^2 = 1$

6 $-x^2 + y^2 - z^2 = 1$

7 $z = x^2 + y^2$

8 $z = \frac{1}{4}x^2 + y^2$

9 $z = -x^2 + y^2$

10 $z = x^2 - y^2$.

11 Identify the quadric surface $z = x^2 + 2x + y^2$. [*Hint* Complete the square.]

12 Identify the quadric surface $z = xy$. [*Hint* Rotate 45° around the z-axis.]

Sketch the paraboloid

13 $x = y^2 + z^2$

14 $y = x^2 - z^2$.

Sketch the surface in x,y,z-space

15 $x - z = 1$

16 $y = x^2$

17 $xy = 1$

18 $-x^2 + y^2 = 1$

19 $x = z^2$

20 $y^2 + 4z^2 = 1$

21 $z^2 = x^2 + y^2$

22 $x^2 + 4z^2 = 1$

23 $y^2 = z^2 + 4x^2$

24 $z^2 = xy$.

25* Suppose $f(x, y) = 0$, $z = 1$ is a curve on the plane $z = 1$. Find an equation for the cone formed by all lines that pass through $\mathbf{0}$ and a point of the curve.

26* (cont.) Test your result on $x^2 + y^2 = 1$, $z = 1$.

27* Let $f(y, z) = 0$ be a curve in the y,z-plane. Find an equation for the surface of revolution obtained by revolving the curve around the y-axis.

28* (cont.) Test your result on the curve $y^2 + 9z^2 = 1$.

REVIEW EXERCISES

1 Show that $\mathbf{a} = (0, 0, 0)$, $\mathbf{b} = (1, 2, 2)$, and $\mathbf{c} = (-1, 4, 1)$ are the vertices of a right triangle.

2 Find the smallest positive angle θ between $\mathbf{x}_1 = (1, 4, 8)$ and $\mathbf{x}_2 = (-6, 2, 3)$.

3 Write the plane $4x - 8y + z = 27$ in normal form and find its point closest to $\mathbf{0}$.

4 Express the plane through $\mathbf{a} = (1, 1, 1)$, $\mathbf{b} = (2, 0, -5)$, and $\mathbf{c} = (3, -2, -2)$ in parametric form.

5 Show that the planes $2x - y + 2z = 0$ and $3x - 4y - 5z = 2$ intersect at right angles.

6 Find the smallest angle of intersection of the planes $x + y + z = 4$ and $2x - y - z = 6$.

7 Sketch the first octant portion of $x^2 + 9y^2 + 4z^2 = 4$.

8 Sketch $x^2 + 4y^2 - z^2 = 0$.

14 COMPLEX NUMBERS

1 COMPLEX ARITHMETIC

As we know, not every polynomial with real coefficients has a real zero. Even the simple quadratic polynomial $x^2 + 1$ has no real zero. For if r is any real number, then $r^2 \geq 0$ so $r^2 + 1 > 0$, not $r^2 + 1 = 0$.

The most general quadratic equation

$$ax^2 + bx + c = 0 \qquad (a \neq 0)$$

with real coefficients has no real solution if its discriminant is negative, that is, if $D = b^2 - 4ac < 0$. Imagine that there exists a number system containing the real number system, in which the special quadratic equation $x^2 + 1 = 0$ has a root, $\sqrt{-1}$. Then the general quadratic equation can be solved even when $D < 0$. Its roots are

$$\frac{-b \pm \sqrt{D}}{2a} = \frac{-b \pm \sqrt{-D}\sqrt{-1}}{2a}$$

provided the rules of ordinary arithmetic are valid in the extended number system.

We shall now construct such a new number system. It will contain the real number system as a part; it will contain a square root of -1; and it will obey the usual rules of arithmetic.

We start with the set of real numbers and a symbol i, which will play the role of $\sqrt{-1}$. The new number system, called the **complex number system,** consists of all formal expressions

$$a + bi,$$

where a and b are real numbers. We must now say how to operate with these formal symbols. First of all, since i is a new sort of object, it is natural to say that two complex numbers $a + bi$ and $c + di$ are equal if and only if $a = c$ and $b = d$.

Equality of Complex Numbers Two complex numbers $a + bi$ and $c + di$ are equal if and only if $a = c$ and $b = d$.

If ordinary rules of arithmetic are to hold, we must have

$$(a + bi) + (c + di) = (a + c) + (b + d)i.$$

This we take as the *definition* of addition in the complex number system.

Example $(3 - 7i) + (5 + 2i) = (3 + 5) + [(-7) + 2]i = 8 - 5i$

Since we want to have $i^2 = -1$, we agree to replace i^2 by -1. Thus it seems natural to compute the product of $a + bi$ and $c + di$ as follows:

$$\begin{aligned}(a + bi)(c + di) &= ac + a(di) + (bi)c + (bi)(di) \\ &= ac + (ad)i + (bc)i + (bd)i^2 \\ &= [ac + (bd)i^2] + [ad + bc]i \\ &= (ac - bd) + (ad + bc)i.\end{aligned}$$

This we take as the *definition* of multiplication in the complex number system.

Example $(-4 + 3i)(2 + 5i)$
$$= [(-4)(2) - (3)(5)] + [(-4)(5) + (3)(2)]i = -23 - 14i$$

Addition and Multiplication of Complex Numbers The **sum** of two complex numbers is defined by

$$(a + bi) + (c + di) = (a + c) + (b + d)i.$$

The **product** of two complex numbers is defined by

$$(a + bi)(c + di) = (ac - bd) + (ad + bc)i.$$

These definitions attempt to breathe life into the set of formal symbols $a + bi$. We shall show that they succeed; the complex number system obeys all the usual rules of arithmetic valid for the real numbers. The crucial definition is the product. It includes the relation $i^2 = -1$ that we wanted:

$$i^2 = (0 + 1i)(0 + 1i) = (0\cdot 0 - 1\cdot 1) + (0\cdot 1 + 1\cdot 0)i = -1.$$

We shall *identify* the complex number $a + 0i$ with the real number a. This is perfectly reasonable since complex numbers $a + 0i$ and $b + 0i$ add and multiply just as do the real numbers a and b:

$$(a + 0i) + (b + 0i) = (a + b) + (0 + 0)i = (a + b) + 0i,$$

$$(a + 0i)(b + 0i) = (ab - 0\cdot 0) + (a\cdot 0 + 0\cdot b)i = ab + 0i.$$

Thus, the complex number system contains a subsystem that we can identify with the real number system. In other words, the complex number system is an *extension* of the real number system, and its arithmetic is *consistent* with that of the real number system.

Notation Sometimes it is convenient to write $a + ib$ instead of $a + bi$. This is particularly useful with expressions such as $\cos\theta + i\sin\theta$ and $-1 + i\sqrt{3}$. In electrical engineering i is used for current and the symbol j is used for the complex number we are calling i.

Rules of Complex Arithmetic

We have referred to the "usual rules of arithmetic" being valid in the complex number system. Let us be precise now. There are two types of "rules": identities and existence of inverses.

Rules of Complex Arithmetic: Identities
Let α, β, γ be complex numbers. Then

Commutative Laws

$$\alpha + \beta = \beta + \alpha \qquad \alpha\beta = \beta\alpha$$

Associative Laws

$$\alpha + (\beta + \gamma) = (\alpha + \beta) + \gamma \qquad \alpha(\beta\gamma) = (\alpha\beta)\gamma$$

Distributive Law

$$\alpha(\beta + \gamma) = \alpha\beta + \alpha\gamma$$

Identity Laws

$$\alpha + 0 = \alpha \qquad \alpha \cdot 1 = \alpha$$

Each of these rules is a consequence of the rules of arithmetic for *real* numbers. For instance, let

$$\alpha = a + bi \quad \text{and} \quad \beta = c + di.$$

Then

$$\alpha\beta = (a + bi)(c + di) = (ac - bd) + (ad + bc)i$$

$$\beta\alpha = (c + di)(a + bi) = (ca - db) + (cb + da)i.$$

By the commutative laws for real numbers,

$$ac - bd = ca - db \quad \text{and} \quad ad + bc = cb + da,$$

hence $\alpha\beta = \beta\alpha$. This proves the commutative law for complex multiplication. Proofs of the remaining laws are left as exercises.

Example $\alpha = 1 - i \qquad \beta = 2 + 3i \qquad \gamma = -2 + i$

$$\alpha\beta = (1 - i)(2 + 3i) = 5 + i$$

$$\beta\gamma = (2 + 3i)(-2 + i) = -7 - 4i$$

$$(\alpha\beta)\gamma = (5 + i)(-2 + i) = -11 + 3i$$

$$\alpha(\beta\gamma) = (1 - i)(-7 - 4i) = -11 + 3i.$$

Hence $(\alpha\beta)\gamma = \alpha(\beta\gamma)$, as predicted by the associative law for multiplication.

Rules of Complex Arithmetic: Inverses

Additive Inverse If $\alpha = a + bi$, set $-\alpha = (-a) + (-b)i$. Then

$$\alpha + (-\alpha) = 0.$$

Multiplicative Inverse If $\alpha = a + bi \neq 0$, set

$$\alpha^{-1} = \left(\frac{a}{a^2 + b^2}\right) + \left(\frac{-b}{a^2 + b^2}\right)i = \frac{a - bi}{a^2 + b^2}.$$

Then $\alpha\alpha^{-1} = 1$.

The first is rather obvious; for instance

$$(-3 + 4i) + (3 - 4i) = 0 + 0i = 0.$$

The second says that each non-zero complex number has a reciprocal. It looks complicated, but it follows easily from the previous rules:

$$\alpha\alpha^{-1} = (a + bi)\left(\frac{a - bi}{a^2 + b^2}\right) = \frac{(a + bi)(a - bi)}{a^2 + b^2} = \frac{a^2 + b^2}{a^2 + b^2} = 1.$$

Just as for real numbers, division by a complex number β is defined as multiplication by β^{-1}.

$$\frac{\alpha}{\beta} = \alpha\beta^{-1} \qquad (\beta \neq 0).$$

Just as for real numbers $(\beta^{-1})^2 = (\beta^2)^{-1}$ since

$$(\beta^2)(\beta^{-1})^2 = (\beta\beta)(\beta^{-1}\beta^{-1}) = (\beta\beta^{-1})(\beta\beta^{-1}) = 1.$$

Thus we can define β^{-2}, and similarly $\beta^{-3}, \beta^{-4}, \cdots$ so that the usual rules of exponents hold for integer powers (positive and negative) of complex numbers.

Examples $\alpha = 1 - i \qquad \beta = 2 + 3i$

(1) $\beta^{-1} = (2 + 3i)^{-1} = \dfrac{2}{4 + 9} + \dfrac{-3}{4 + 9}i = \frac{2}{13} - \frac{3}{13}i.$

Check $\beta\beta^{-1} = (2 + 3i)(\frac{2}{13} - \frac{3}{13}i) = \frac{13}{13} + \frac{0}{13}i = 1.$

(2) $\dfrac{\alpha}{\beta} = \alpha\beta^{-1} = (1 - i)(2 + 3i)^{-1}$

$\quad = (1 - i)(\frac{2}{13} - \frac{3}{13}i) = -\frac{1}{13} - \frac{5}{13}i.$

Check $\beta(\alpha/\beta) = (2 + 3i)(-\frac{1}{13} - \frac{5}{13}i) = \frac{13}{13} - \frac{13}{13}i = \alpha.$

(3) $\beta^{-2} = (\beta^{-1})^2 = (\frac{2}{13} - \frac{3}{13}i)(\frac{2}{13} - \frac{3}{13}i) = -\frac{5}{169} - \frac{12}{169}i.$

Complex Conjugates and Absolute Values

Complex arithmetic is richer than real arithmetic. One feature is an operation, conjugation, that has no real counterpart.

Complex Conjugate Let $\alpha = a + bi$. Its **complex conjugate** is defined to be

$$\overline{\alpha} = a - bi.$$

The operation of complex conjugation satisfies the rules

$$\overline{\overline{\alpha}} = \alpha \qquad \overline{\alpha + \beta} = \overline{\alpha} + \overline{\beta} \qquad \overline{\alpha\beta} = \overline{\alpha}\,\overline{\beta}.$$

Examples

$$\overline{4} = 4 \qquad \overline{i} = -i \qquad \overline{-3i} = 3i \qquad \overline{4 - 3i} = 4 + 3i$$

The first two rules are left to prove as exercises. The third rule, $\overline{\alpha\beta} = \overline{\alpha}\,\overline{\beta}$, is by no means obvious; given the complicated definition of the complex product, it should come as a surprise. To prove it, set $\alpha = a + bi$ and $\beta = c + di$. Then

$$\overline{\alpha\beta} = \overline{(ac - bd) + (ad + bc)i} = (ac - bd) - (ad + bc)i$$

and

$$\overline{\alpha}\,\overline{\beta} = (a - bi)(c - di) = (ac - bd) - (ad + bc)i$$

So $\overline{\alpha\beta} = \overline{\alpha}\,\overline{\beta}$.

The rules extend to any number of summands or factors; for instance,

$$\overline{\alpha + \beta + \gamma} = \overline{\alpha} + \overline{\beta} + \overline{\gamma} \qquad \overline{\alpha\beta\gamma} = \overline{\alpha}\,\overline{\beta}\,\overline{\gamma} \qquad \text{etc.}$$

Each real number a has an absolute value, a nonnegative real $|a|$ that measures the magnitude of a. (Note that $|a| = \sqrt{a^2}$.) So does each complex number.

Absolute Value The **absolute value** (or **modulus**) of $\alpha = a + bi$ is

$$|\alpha| = \sqrt{a^2 + b^2}.$$

Examples

$$|3| = 3 \qquad |-5i| = 5 \qquad |-12 + 5i| = \sqrt{(-12)^2 + 5^2} = \sqrt{169} = 13$$

Rules for Absolute Values

Let α and β be complex numbers. Then

$$|\alpha| \geq 0; \qquad |\alpha| = 0 \quad \text{if and only if} \quad \alpha = 0.$$

$$|\alpha\beta| = |\alpha| \cdot |\beta| \qquad \left|\frac{\alpha}{\beta}\right| = \frac{|\alpha|}{|\beta|} \qquad |\overline{\alpha}| = |\alpha| \qquad |\alpha|^2 = \alpha\overline{\alpha}.$$

Triangle Inequality $|\alpha + \beta| \leq |\alpha| + |\beta|$.

Except for the triangle inequality, which will be discussed in the next section, the proofs of these properties are left as exercises.

Conjugates and absolute values help simplify division of complex numbers. Since $\alpha\overline{\alpha} = |\alpha|^2$, we can divide both sides by $|\alpha|^2$:

$$\alpha \cdot \frac{\overline{\alpha}}{|\alpha|^2} = 1.$$

Therefore

$$\alpha^{-1} = \frac{\overline{\alpha}}{|\alpha|^2} = \frac{a - bi}{a^2 + b^2} \qquad (\alpha \neq 0)$$

(This agrees with the formula for the multiplicative inverse given earlier.) It follows that

$$\frac{\alpha}{\beta} = \alpha\beta^{-1} = \frac{\alpha\overline{\beta}}{|\beta|^2} = \frac{\alpha\overline{\beta}}{\beta\overline{\beta}}.$$

Thus, to evaluate α/β, multiply both numerator and denominator by $\overline{\beta}$. For example,

$$\frac{2 + 3i}{1 + 2i} = \frac{2 + 3i}{1 + 2i}\,\frac{1 - 2i}{1 - 2i} = \frac{8 - i}{5} = \tfrac{8}{5} - \tfrac{1}{5}i$$

Remark The introduction of complex numbers was one of the greatest advances ever made in mathematics. It took hundreds of years to develop the idea once the need was felt, so don't expect to learn it perfectly in a few minutes. You should read and reread this section until you are confident you understand it.

Exercises

Express in the form $a + bi$

1 $(3 + 2i) + (6 - i)$ **2** $(1 - i) + (4 + 3i)$ **3** $2(1 + 4i) - 3(2 + i)$

4 $(-2 + 3i) + 5(1 - i)$ **5** $i(2 - 3i)$ **6** $i(8i + 5)$

7 $(1 + i)(3 - 4i)$ **8** $(1 - i)(2 + 7i)$ **9** $(5 + 4i)(3 + 2i)$

10 $(1 - 6i)(3 + i)$ **11** $(1 + i)^2$ **12** $(2 - i)(2 + i)$

13 $\dfrac{1}{3 + 4i}$ **14** $\dfrac{1}{6 - i}$ **15** $\dfrac{3 + i}{2 - i}$

16 $\dfrac{2 + 7i}{1 - 3i}$ **17** $\dfrac{1}{(2 + 3i)(5 - 4i)}$ **18** $\dfrac{1}{(1 - i)(8 + 3i)}$

19 $\dfrac{1 - 3i}{(2 + i)(2 + 5i)}$ **20** $\dfrac{(1 + i)(1 + 2i)}{(3 - 2i)(4 - 3i)}$ **21** $\dfrac{i}{2 + i} + \dfrac{3 + i}{4 + i}$

22 $\dfrac{1}{4 - 3i} + \dfrac{5 + 3i}{2 - i}$.

Compute $\overline{\alpha\beta}$ and $\overline{\alpha}\,\overline{\beta}$

23 $\alpha = 1 - i \quad \beta = 3i$ **24** $\alpha = 1 + i \quad \beta = 1 - i$

25 $\alpha = 2 - i \quad \beta = 3 + 2i$ **26** $\alpha = -1 - 2i \quad \beta = 3 - i$

27 $\alpha = (1 + i)^2 \quad \beta = 1 + 2i$ **28** $\alpha = (1 - i)^2 \quad \beta = 1 - 2i$.

Compute

29 $|3 + 4i|$ **30** $|3 - 4i|$ **31** $|(1 + i)(2 - i)|$

32 $|(2 - 3i)(6 + i)|$ **33** $|i(2 + i)| + |1 + 4i|$ **34** $|6(3 - 2i)| + |2 - i|$

35 $\left|\dfrac{1}{2 - i}\right|$ **36** $\left|\dfrac{2 + 5i}{3 + i}\right|$ **37** $\left|\dfrac{1 + i}{1 - i}\right|$ **38** $\left|\dfrac{3 + 4i}{4 - 3i}\right|$.

39 Simplify $i + i^2 + i^3 + \cdots + i^{11}$.

40 Compute $(1 + i)^2$ and use the result to compute $(1 + i)^{12}$.

Compute

41 $\left(\dfrac{-1 + i\sqrt{3}}{2}\right)^3$ **42** $\left(\dfrac{1 + i}{\sqrt{2}}\right)^8$.

Prove

43 $\alpha + \beta = \beta + \alpha$ **44** $\alpha + (\beta + \gamma) = (\alpha + \beta) + \gamma$

45 $\alpha(\beta + \gamma) = \alpha\beta + \alpha\gamma$ **46** $\alpha(\beta\gamma) = (\alpha\beta)\gamma$

47 $\bar{\bar{\alpha}} = \alpha$

48 $\overline{\alpha + \beta} = \bar{\alpha} + \bar{\beta}$

49 $|\alpha| \geq 0$

50 $|\alpha| = 0$ implies $\alpha = 0$

51 $|\bar{\alpha}| = |\alpha|$

52 $|\alpha|^2 = \alpha\bar{\alpha}$

53 $|\alpha\beta| = |\alpha||\beta|$

54 $|\alpha/\beta| = |\alpha|/|\beta|$.

2 THE COMPLEX PLANE

Each complex number $\alpha = a + bi$ is completely determined by the ordered pair (a, b) of real numbers. We call a the **real part** of α and b the **imaginary part** of α, and we write

$$a = \operatorname{Re}(\alpha) \quad \text{and} \quad b = \operatorname{Im}(\alpha).$$

If $\alpha = a + bi$, then $\bar{\alpha} = a - bi$. Therefore

$$\alpha + \bar{\alpha} = (a + bi) + (a - bi) = 2a$$

$$\alpha - \bar{\alpha} = (a + bi) - (a - bi) = 2bi.$$

From these equations follow the useful relations

$$\operatorname{Re}(\alpha) = \frac{\alpha + \bar{\alpha}}{2} \qquad \operatorname{Im}(\alpha) = \frac{\alpha - \bar{\alpha}}{2i}.$$

Example $\alpha = -2 + 5i$

$$\bar{\alpha} = -2 - 5i$$

$$\operatorname{Re}(\alpha) = -2 \qquad \operatorname{Im}(\alpha) = 5$$

$$\frac{\alpha + \bar{\alpha}}{2} = \frac{(-2 + 5i) + (-2 - 5i)}{2} = -2 = \operatorname{Re}(\alpha)$$

$$\frac{\alpha - \bar{\alpha}}{2i} = \frac{(-2 + 5i) - (-2 - 5i)}{2i} = \frac{10i}{2i} = 5 = \operatorname{Im}(\alpha)$$

Geometric Interpretation of Complex Numbers

The correspondence

$$\alpha = a + bi \longleftrightarrow (a, b)$$

between complex numbers and ordered pairs of real numbers strongly suggests a geometric representation for complex numbers. Indeed, we identify the *complex number* $\alpha = a + bi$ with the *point* (a, b) in the plane (Fig. 1).

This gives an entirely new interpretation to the cartesian (coordinate) plane: as the set of all complex numbers. When looked at this way, the plane is called the **complex plane** (also **Gaussian plane** after K.F. Gauss).

The definition of addition of complex numbers,

$$(a + bi) + (c + di) = (a + c) + (b + d)i,$$

has a useful geometric interpretation. To add $\beta = c + di$ to α, we must increase the x-coordinate of α by c and the y-coordinate of α by d. (See Fig. 2a.) Therefore we take the little rectangle built on β and move it lock, stock, and barrel, so that 0 is moved to α and β is moved to $\alpha + \beta$. Thus *the segment from α to $\alpha + \beta$ is the parallel displacement of the segment from* 0 *to* β.

This means that the three segments from 0 to β, from 0 to α, and from α to $\alpha + \beta$ make up three sides of a parallelogram. We complete the parallelogram

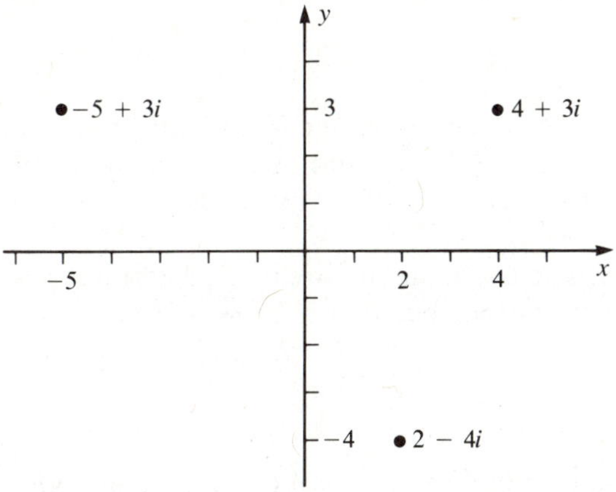

Fig. 1 Representation of complex numbers as points in the plane

(Fig. 2b), and we conclude that complex numbers can be added *graphically* by (what is called) the **parallelogram rule.**

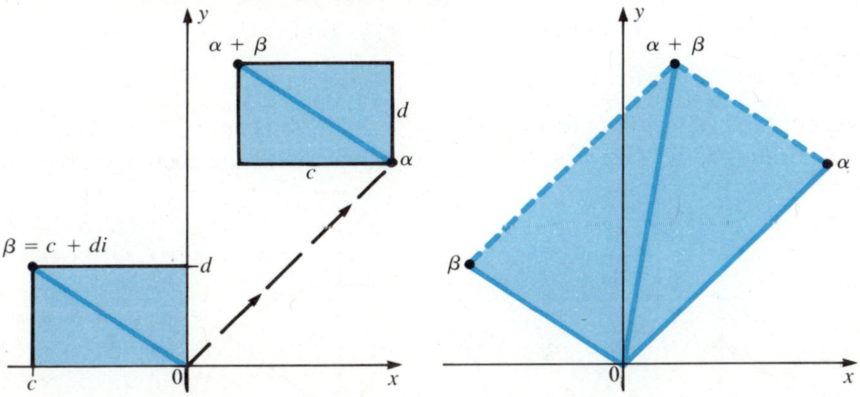

(a) to add β to α, displace the rectangle, corner 0 to α

(b) parallelogram rule

Fig. 2 Geometric meaning of addition of complex numbers

Remark The parallelogram rule of addition is the same as the rule for adding forces, or vectors in general. See p. 431.

Absolute values and complex conjugates also have geometric interpretations. The absolute value $|\alpha|$ is the distance of $\alpha = a + bi$ from 0, because $|\alpha|^2 = a^2 + b^2$. The complex conjugate $\overline{\alpha} = a - bi$ is the reflection (mirror image) of α in the x-axis (Fig. 3a).

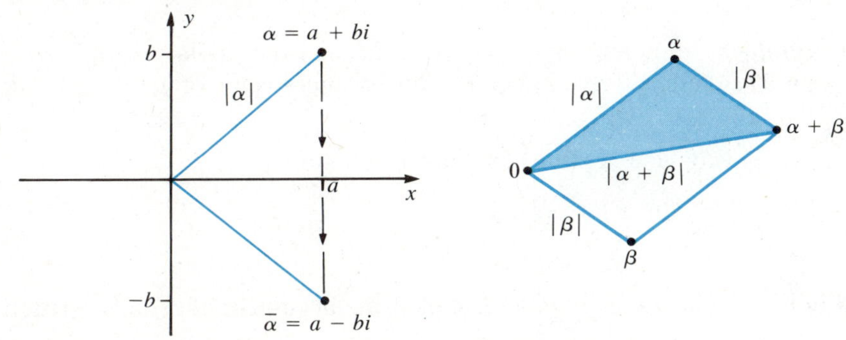

(a) geometric meaning of absolute value and complex conjugate

(b) proof of the triangle inequality $|\alpha + \beta| \leq |\alpha| + |\beta|$

Fig. 3

The triangle inequality

$$|\alpha + \beta| \leq |\alpha| + |\beta|$$

follows directly from the parallelogram rule for addition. We simply observe (Fig. 3b) that a triangle exists whose side lengths are $|\alpha|$, $|\beta|$, and $|\alpha + \beta|$. But in a triangle, each side is at most the sum of the other two sides (even if the triangle collapses), so in particular $|\alpha + \beta| \leq |\alpha| + |\beta|$.

Thus $\alpha + \beta$, $|\alpha|$, and $\overline{\alpha}$ have simple and useful geometric interpretations. The product $\alpha\beta$ also has a geometric interpretation, but that is more easily seen in polar coordinates.

Polar Form

Let $\alpha = a + bi$ be a non-zero complex number. The point (a, b) has polar coordinates $\{r, \theta\}$, where $r > 0$,

$$a = r\cos\theta \quad \text{and} \quad b = r\sin\theta.$$

Therefore $\alpha = a + bi = (r\cos\theta) + (r\sin\theta)i$, that is,

$$\alpha = r(\cos\theta + i\sin\theta).$$

The representation of α in the form $r(\cos\theta + i\sin\theta)$ is called the **polar form** of the complex number α. See Fig. 4a. (This contrasts with the **rectangular form** $\alpha = a + bi$.)

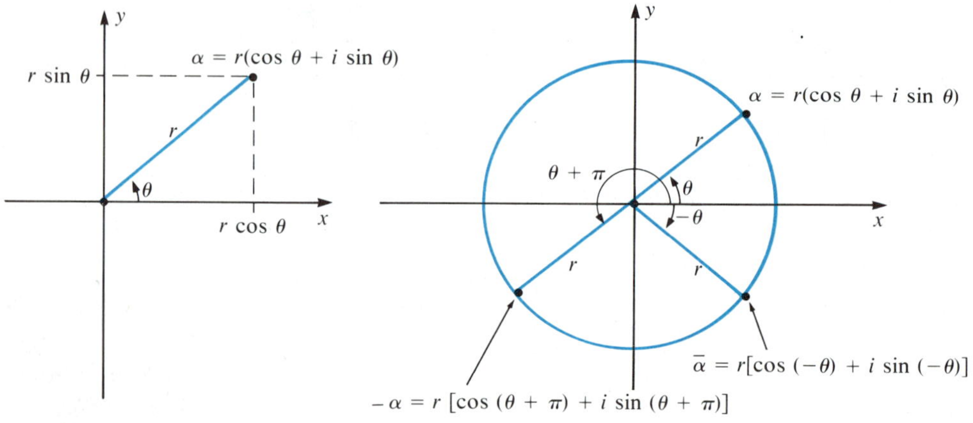

(a) polar form (b) polar forms of α, $\bar{\alpha}$, and $-\alpha$

Fig. 4

When a complex number α is expressed in the polar form $\alpha = r(\cos\theta + i\sin\theta)$, the positive number r is the absolute value of α. Just check it:

$$\begin{aligned} |\alpha|^2 &= \alpha\overline{\alpha} = [r(\cos\theta + i\sin\theta)][r(\cos\theta - i\sin\theta)] \\ &= r^2(\cos^2\theta + \sin^2\theta) = r^2 \\ |\alpha| &= r. \end{aligned}$$

The angle θ in $\alpha = r(\cos\theta + i\sin\theta)$ is called the **argument** of α, and written

$$\theta = \arg\alpha.$$

Warning The angle θ is only determined up to an integral multiple of 2π. Therefore, to be accurate, we should write $\arg \alpha = \theta + 2\pi n$ and thus let the symbol $\arg \alpha$ denote the whole family of possible angles θ such that $\alpha = r(\cos\theta + i\sin\theta)$. No one does this; the inaccurate notation $\theta = \arg\alpha$ is commonplace. But formulas involving $\arg\alpha$ must always be read with a grain of salt; you can always tack on $2\pi n$.

Polar Form Each non-zero complex number can be written in the polar form

$$\alpha = r(\cos\theta + i\sin\theta)$$

where $r = |\alpha|$ is the modulus (absolute value) of α and the angle $\theta = \arg\alpha$ is the argument of α.

How do we convert back and forth from rectangular to polar form? One way is easy. If

$$\alpha = r(\cos\theta + i\sin\theta)$$

is given in polar form, then

$$\alpha = (r\cos\theta) + i(r\sin\theta)$$

is rectangular form.

Example $\alpha = 5(\cos 30° + i\sin 30°) = (5\cos 30°) + (5\sin 30°)i$

$$= \tfrac{5}{2}\sqrt{3} + \tfrac{5}{2}i$$

If

$$\alpha = a + bi \neq 0$$

is given in rectangular form, then from $\alpha = r(\cos\theta + i\sin\theta)$ we have

$$r\cos\theta = a \qquad r\sin\theta = b$$

$$\cos\theta = \frac{a}{r} \qquad \sin\theta = \frac{b}{r}$$

We know $r = |\alpha| = \sqrt{a^2 + b^2}$, so we can solve for θ. It's the old story all over again of rectangular to polar coordinates.

Example 1 Convert into polar form

(a) $1 + i$ (b) $1 + i\sqrt{3}$.

Solution (a) $|1 + i| = \sqrt{1^2 + 1^2} = \sqrt{2}$.

Therefore

$$\cos\theta = \frac{a}{r} = \frac{1}{\sqrt{2}} \qquad \sin\theta = \frac{b}{r} = \frac{1}{\sqrt{2}}.$$

One choice for θ is $\theta = \frac{1}{4}\pi$, so $1 + i = \sqrt{2}(\cos\frac{1}{4}\pi + i\sin\frac{1}{4}\pi)$.

(b) $|1 + i\sqrt{3}| = \sqrt{1 + 3} = 2$

$$\cos\theta = \frac{a}{r} = \tfrac{1}{2} \qquad \sin\theta = \frac{b}{r} = \tfrac{1}{2}\sqrt{3}.$$

One choice for θ is $\theta = \frac{1}{3}\pi$, so $1 + i\sqrt{3} = 2(\cos \frac{1}{3}\pi + i \sin \frac{1}{3}\pi)$.

Answer (a) $\sqrt{2}(\cos \frac{1}{4}\pi + i \sin \frac{1}{4}\pi)$ (b) $2(\cos \frac{1}{3}\pi + i \sin \frac{1}{3}\pi)$

We next derive the polar forms for $\overline{\alpha}$ and $-\alpha$. Suppose we start with $\alpha = r(\cos \theta + i \sin \theta)$. Then

$$|\overline{\alpha}| = |-\alpha| = |\alpha| = r$$

so $\overline{\alpha}$ and $-\alpha$ have the same modulus as α. By Fig. 4b,

$$\arg \overline{\alpha} = -\theta \quad \text{and} \quad \arg(-\alpha) = \theta + \pi.$$

If $\alpha = r(\cos \theta + i \sin \theta)$, then

$$\overline{\alpha} = r[\cos(-\theta) + i \sin(-\theta)] \qquad -\alpha = r[\cos(\theta + \pi) + i \sin(\theta + \pi)].$$

Multiplication and Division

Suppose

$$\alpha_1 = r_1(\cos \theta_1 + i \sin \theta_1) \quad \text{and} \quad \alpha_2 = r_2(\cos \theta_2 + i \sin \theta_2)$$

are non-zero complex numbers in polar form. Then

$$\begin{aligned}\alpha_1\alpha_2 &\\ &= (r_1r_2)(\cos \theta_1 + i \sin \theta_1)(\cos \theta_2 + i \sin \theta_2)\\ &= (r_1r_2)[(\cos \theta_1 \cos \theta_2 - \sin \theta_1 \sin \theta_2) + i(\cos \theta_1 \sin \theta_2 + \sin \theta_1 \cos \theta_2)].\end{aligned}$$

But by the addition laws for sine and cosine,

$$\cos \theta_1 \cos \theta_2 - \sin \theta_1 \sin \theta_2 = \cos(\theta_1 + \theta_2)$$

and

$$\cos \theta_1 \sin \theta_2 + \sin \theta_1 \cos \theta_2 = \sin(\theta_1 + \theta_2).$$

Therefore

$$\alpha_1\alpha_2 = (r_1r_2)[\cos(\theta_1 + \theta_2) + i \sin(\theta_1 + \theta_2)].$$

Polar Form of Complex Multiplication

Given non-zero complex numbers

$$\alpha_1 = r_1(\cos \theta_1 + i \sin \theta_1) \quad \text{and} \quad \alpha_2 = r_2(\cos \theta_2 + i \sin \theta_2)$$

in polar form, the polar form of their product is

$$\alpha_1\alpha_2 = r_1r_2[\cos(\theta_1 + \theta_2) + i \sin(\theta_1 + \theta_2)].$$

Therefore

$$|\alpha_1\alpha_2| = |\alpha_1| \cdot |\alpha_2| \qquad \arg(\alpha_1\alpha_2) = \arg \alpha_1 + \arg \alpha_2.$$

In words, to multiply two complex numbers, multiply their moduli and add their arguments.

Examples (1) $\alpha = 1 + i \qquad \beta = 1 + i\sqrt{3}$

By Example 1,

$$\alpha = \sqrt{2}(\cos \tfrac{1}{4}\pi + i \sin \tfrac{1}{4}\pi) \qquad \beta = 2(\cos \tfrac{1}{3}\pi + i \sin \tfrac{1}{3}\pi).$$

Therefore $|\alpha\beta| = 2\sqrt{2} \qquad \arg(\alpha\beta) = \tfrac{1}{4}\pi + \tfrac{1}{3}\pi = \tfrac{7}{12}\pi,$

$$\alpha\beta = 2\sqrt{2}(\cos \tfrac{7}{12}\pi + i \sin \tfrac{7}{12}\pi).$$

(2) $\alpha = i \qquad \beta = r(\cos\theta + i\sin\theta)$

Then

$$|i| = 1 \qquad \arg i = \tfrac{1}{2}\pi \qquad i = \cos\tfrac{1}{2}\pi + i\sin\tfrac{1}{2}\pi$$

Therefore

$$i\beta = r[\cos(\theta + \tfrac{1}{2}\pi) + i\sin(\theta + \tfrac{1}{2}\pi)].$$

Thus multiplication by i simply rotates a complex number $\tfrac{1}{2}\pi$ radians counterclockwise about the origin (Fig. 5). Likewise, multiplication by $-i$ rotates it clockwise through a right angle.

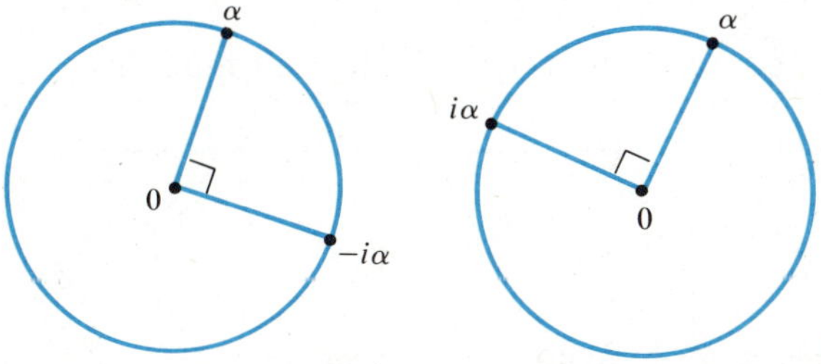

Fig. 5 Multiplication by i and by $-i$

Remark The factor 0 can be included in our recipes provided we agree that

$$|0| = 0 \qquad \arg 0 = \text{any } \theta \qquad 0 = 0(\cos\theta + i\sin\theta)$$

From the rule for multiplication of complex numbers in polar form follows a companion rule for division. First we note the form of the reciprocal of a complex number:

$$\begin{aligned}\frac{1}{r(\cos\theta + i\sin\theta)} &= \frac{1}{r(\cos\theta + i\sin\theta)}\,\frac{\cos\theta - i\sin\theta}{\cos\theta - i\sin\theta}\\ &= \frac{\cos\theta - i\sin\theta}{r(\cos^2\theta + \sin^2\theta)} = \frac{1}{r}(\cos\theta - i\sin\theta)\\ &= \frac{1}{r}[\cos(-\theta) + i\sin(-\theta)].\end{aligned}$$

> **Reciprocal in Polar Form** Let $\alpha = r(\cos\theta + i\sin\theta)$ Then
>
> $$\frac{1}{\alpha} = \frac{1}{r}[\cos(-\theta) + i\sin(-\theta)].$$
>
> Therefore $\left|\dfrac{1}{\alpha}\right| = \dfrac{1}{|\alpha|}$ and $\arg\dfrac{1}{\alpha} = -\arg\alpha.$

We can now derive the polar form of a quotient α_1/α_2. As before, we start with the polar forms of α_1 and α_2:

$$\alpha_1 = r_1(\cos\theta_1 + i\sin\theta_1) \qquad \alpha_2 = r_2(\cos\theta_2 + i\sin\theta_2).$$

Then

$$\frac{\alpha_1}{\alpha_2} = \alpha_1\alpha_2^{-1} \qquad \left|\frac{\alpha_1}{\alpha_2}\right| = |\alpha_1||\alpha_2^{-1}| = r_1\cdot\frac{1}{r_2} = \frac{r_1}{r_2}$$

$$\arg\frac{\alpha_1}{\alpha_2} = \arg(\alpha_1\alpha_2^{-1}) = \arg\alpha_1 + \arg\alpha_2^{-1}$$

$$= \arg\alpha_1 - \arg\alpha_2 = \theta_1 - \theta_2.$$

We fit the pieces together:

Quotient in Polar Form Let

$$\alpha_1 = r_1(\cos\theta_1 + i\sin\theta_2) \quad\text{and}\quad \alpha_2 = r_2(\cos\theta_2 + i\sin\theta_2) \neq 0.$$

Then $\dfrac{\alpha_1}{\alpha_2} = \dfrac{r_1}{r_2}[\cos(\theta_1 - \theta_2) + i\sin(\theta_1 - \theta_2)].$

Therefore $\left|\dfrac{\alpha_1}{\alpha_2}\right| = \dfrac{|\alpha_1|}{|\alpha_2|} \qquad \arg\dfrac{\alpha_1}{\alpha_2} = \arg\alpha_1 - \arg\alpha_2.$

In words, to divide two non-zero complex numbers, divide their moduli and subtract their arguments.

Example $\alpha = 1 + i \qquad \beta = 1 + i\sqrt{3}$

By Example 1,

$$\alpha = \sqrt{2}(\cos\tfrac{1}{4}\pi + i\sin\tfrac{1}{4}\pi) \qquad \beta = 2(\cos\tfrac{1}{3}\pi + i\sin\tfrac{1}{3}\pi).$$

Therefore

$$\left|\frac{\alpha}{\beta}\right| = \frac{\sqrt{2}}{2} = \tfrac{1}{2}\sqrt{2} \qquad \arg\frac{\alpha}{\beta} = \tfrac{1}{4}\pi - \tfrac{1}{3}\pi = -\tfrac{1}{12}\pi,$$

$$\frac{\alpha}{\beta} = \tfrac{1}{2}\sqrt{2}[\cos(-\tfrac{1}{12}\pi) + i\sin(-\tfrac{1}{12}\pi)]$$

Squares and Square Roots

If $\alpha = r(\cos\theta + i\sin\theta) \neq 0$, then the multiplication rule implies

$$\alpha^2 = r^2(\cos 2\theta + i\sin 2\theta).$$

Therefore, to square a complex number, you square its modulus and double its argument.

It stands to reason that to find a complex square root of α, you go backwards, *i.e.*, you take the square root of the modulus and half of the argument. But there are *two* geometric angles whose doubles are θ (plus a multiple of 2π), namely $\frac{1}{2}\theta$ and $\frac{1}{2}(\theta + 2\pi) = \frac{1}{2}\theta + \pi$.

Each non-zero complex number has two complex square roots:

$$\sqrt{r(\cos\theta + i\sin\theta)} = \begin{cases}\sqrt{r}(\cos\frac{1}{2}\theta + i\sin\frac{1}{2}\theta)\\ \sqrt{r}[\cos(\frac{1}{2}\theta + \pi) + i\sin(\frac{1}{2}\theta + \pi)].\end{cases}$$

Each square root is the negative of the other.

Example 2 Find the square roots of $9i$.

Solution In polar form,

$$9i = 9(\cos \tfrac{1}{2}\pi + i \sin \tfrac{1}{2}\pi).$$

According to the preceding formulas, one square root is

$$3(\cos \tfrac{1}{4}\pi + i \sin \tfrac{1}{4}\pi) = 3(\tfrac{1}{2}\sqrt{2} + \tfrac{1}{2}i\sqrt{2})$$

and the other square root is the negative of this number.

Check $[\tfrac{3}{2}\sqrt{2}(1 + i)]^2 = (\tfrac{3}{2}\sqrt{2})^2(1 + i)^2$
$= \tfrac{9}{2}(1 + 2i + i^2) = \tfrac{9}{2}(1 + 2i - 1)$
$= 9i.$

Answer $\pm\tfrac{3}{2}\sqrt{2}(1 + i)$.

Warning If r is a positive *real* number, the symbol $\sqrt{r}$ denotes the unique positive square root of r. But if α is a non-zero *complex* number, the symbol $\sqrt{\alpha}$ denotes *two* complex numbers, each a square root of α. We really should always write $\pm\sqrt{\alpha}$ to remember this ambiguity.

Be very careful of "rules" like $\sqrt{\alpha_1}\sqrt{\alpha_2} = \sqrt{\alpha_1\alpha_2}$. This is true only up to sign. For instance,

$$-1 = i \cdot i = \sqrt{-1}\sqrt{-1} = \sqrt{(-1)(-1)} = \sqrt{1} = 1$$

is nonsense.

Exercises

Express in polar form

1 $-3i$ **2** -5 **3** $1 - i$ **4** $3 + 3i$
5 $-\sqrt{3} - i$ **6** $-2 + 2i\sqrt{3}$ **7** $3\sqrt{3} + 3i$ **8** $4 - 4i$.

Express the factors and the products (quotients) in polar form

9 $(1 + i)(1 - i)$ **10** $(1 - i)(\sqrt{3} + i)$ **11** $(1 + i\sqrt{3})^2$
12 $(2 + 2i)(4\sqrt{3} - 4i)$ **13** $\dfrac{\sqrt{3} - i}{1 + i}$ **14** $\dfrac{-\sqrt{3} + i}{3 + 3i}$
15 $\dfrac{1 - i}{2 + 2i\sqrt{3}}$ **16** $\dfrac{3 - 3i}{\sqrt{3} - i}$.

Find Re α, Im α, $|\alpha|$, and arg α

17 $\alpha = \sqrt{3} + 3i$ **18** $\alpha = 1 - i$ **19** $\alpha = \dfrac{\sqrt{3} - i}{1 + i}$ **20** $\alpha = \dfrac{4 - 4i}{2\sqrt{3} - 2i}$.

Compute both $|\alpha| \cdot |\beta|$ and $|\alpha\beta|$ and compare

21 $\alpha = 1 - i \quad \beta = 3i$ **22** $\alpha = 1 + i \quad \beta = -1 + i$
23 $\alpha = 2 - i \quad \beta = 3 + 2i$ **24** $\alpha = -1 + 2i \quad \beta = 3 - i$
25 $\alpha = (1 + i)^2 \quad \beta = 1 + 2i$ **26** $\alpha = (1 - i)^2 \quad \beta = 1 - 2i$.

Compute $|\alpha|/|\beta|$ and $|\alpha/\beta|$

27 $\alpha = 1 + i \quad \beta = 2i$ **28** $\alpha = -i \quad \beta = 2 - i$
29 $\alpha = 4 \quad \beta = 3 - i$ **30** $\alpha = 1 + 2i \quad \beta = 3i$
31 $\alpha = 3 - i \quad \beta = 2 + 5i$ **32** $\alpha = 2 - 5i \quad \beta = -3 - i$.

Solve

33 $z^2 = -i$

34 $z^2 = -9$

35 $z^2 = 1 + i$

36 $z^2 = 2 - 2i$

37 $z^2 = \sqrt{3} - i$

38 $z^2 = 1 + i\sqrt{3}$

39 $z^2 = (1 + i)^3$

40 $z^2 = (1 - i)^3$.

41* Use complex numbers to prove the identity

$$(x^2 + y^2)(u^2 + v^2) = (xu - yv)^2 + (xv + yu)^2.$$

42 (cont.) From $13 = 2^2 + 3^2$ and $37 = 1^2 + 6^2$, express 481 as a sum of two perfect squares.

43 Justify geometrically the formula $\mathrm{Re}(\alpha) = \frac{1}{2}(\alpha + \bar{\alpha})$.

44 (cont.) Do the same for $\mathrm{Im}(\alpha) = (\alpha - \bar{\alpha})/2i$.

45 Calculate $(1 + i)(1 + i\sqrt{3})$ in two different ways to prove

$$\cos \tfrac{7}{12}\pi = \tfrac{1}{4}(\sqrt{2} - \sqrt{6}) \qquad \sin \tfrac{7}{12}\pi = \tfrac{1}{4}(\sqrt{2} + \sqrt{6})$$

Compare Example 1(a), page 283.

46 Calculate $(1 + i\sqrt{3})/(1 + i)$ in two different ways to prove

$$\cos \tfrac{1}{12}\pi = \tfrac{1}{4}(\sqrt{6} + \sqrt{2}) \qquad \sin \tfrac{1}{12}\pi = \tfrac{1}{4}(\sqrt{6} - \sqrt{2}).$$

3 ZEROS OF POLYNOMIALS

Let $ax^2 + bx + c$ be any quadratic polynomial with real coefficients and $a \neq 0$. The quadratic formula

$$r = \frac{-b \pm \sqrt{b^2 - 4ac}}{2a}$$

gives the zeros of the polynomial. If $b^2 - 4ac \geq 0$, the zeros are real; if $b^2 - 4ac < 0$, they are complex.

Example $2x^2 - x + 1$

The discriminant is

$$b^2 - 4ac = 1 - 8 = -7,$$

so the zeros are

$$\frac{1 + i\sqrt{7}}{4} \quad \text{and} \quad \frac{1 - i\sqrt{7}}{4}.$$

To check, set $r = \frac{1}{4}(1 \pm i\sqrt{7})$. Then

$$4r - 1 = \pm i\sqrt{7} \qquad (4r - 1)^2 = -7 \qquad 16r^2 - 8r + 1 = -7$$

$$16r^2 - 8r + 8 = 0 \qquad 2r^2 - r + 1 = 0.$$

At this point we can assert at least this much: we have succeeded in enlarging the real number system to the complex number system, and in this new system each *quadratic* polynomial with *real* coefficients has zeros.

The emphasis on *quadratic* and *real* suggests two very natural questions.

(1) Does each quadratic polynomial with *complex* coefficients have complex zeros?

(2) Does each real polynomial (real coefficients)

$$a_n x^n + a_{n-1} x^{n-1} + \cdots + a_1 x + a_0$$

of any degree have complex zeros?

If the answer to either question were no, we would be faced with enlarging the complex number system (perhaps over and over again) to handle more and more complicated polynomials. Fortunately, the answer to both questions is yes. The complex number system is big enough; it is the "right" system for polynomials.

The answer to question (1) is easy because of the quadratic formula. Suppose α, β, and γ are complex. Then the equation

$$\alpha z^2 + \beta z + \gamma = 0 \qquad (\alpha \neq 0)$$

has solutions

$$z = \frac{-\beta \pm \sqrt{\beta^2 - 4\alpha\gamma}}{2\alpha}.$$

This formula is meaningful since, as we have seen, complex numbers have complex square roots.

Example $z^2 + \sqrt{3}z + i = 0$

$$\alpha = 1 \qquad \beta = \sqrt{3} \qquad \gamma = i$$

$$\beta^2 \quad 4\alpha\gamma = 3 \quad 4i = (2 \quad i)^2$$

$$z = \frac{-\sqrt{3} \pm (2 - i)}{2} = \begin{cases} (1 - \frac{1}{2}\sqrt{3}) - \frac{1}{2}i \\ (-1 - \frac{1}{2}\sqrt{3}) + \frac{1}{2}i. \end{cases}$$

The proof that question (2) has a positive answer is very deep. It was given by K.F. Gauss and constitutes one of the most remarkable achievements in mathematics. Here is a precise statement of the result, in a generality that goes beyond questions (1) and (2).

Fundamental Theorem of Algebra Let

$$f(z) = \alpha_n z^n + \alpha_{n-1} z^{n-1} + \cdots + \alpha_1 z + \alpha_0 \qquad (\alpha_n \neq 0)$$

be any polynomial with complex coefficients and degree $n \geq 1$. Then there is a complex number β such that $f(\beta) = 0$.

There are many proofs, all beyond the scope of this course. Note that the theorem only guarantees the *existence* of a zero; it does not say a word about *how to find* one. However, there are numerical methods for approximating zeros to any degree of accuracy, some well suited for computers.

The same reasoning we used in Chapter 9 for real polynomials (p. 381) establishes the following result.

Factor Theorem Let $f(z)$ be a polynomial of degree $n \geq 1$ with complex coefficients. Suppose β is a complex zero of $f(z)$. Then $z - \beta$ divides $f(z)$, that is, $f(z) = (z - \beta)g(z)$, where $g(z)$ is a polynomial of degree $n - 1$.

Complete Factorization

Let $f(z)$ be a complex polynomial of positive degree. By the fundamental theorem, $f(z)$ has a complex zero β_1, and by the factor theorem

$$f(z) = (z - \beta_1)g(z),$$

where $\deg g = \deg f - 1$. If $g(z)$ has positive degree, then by the same reasoning

$$g(z) = (z - \beta_2)h(z),$$

where $\deg h = \deg g - 1 = \deg f - 2$. Clearly

$$f(z) = (z - \beta_1)(z - \beta_2)h(z).$$

The process can be repeated:

$$f(z) = (z - \beta_1)(z - \beta_2)(z - \beta_3)k(z) \qquad \deg k = \deg f - 3$$

etc., until we reach a constant factor, n steps in all.

Complete Factorization Let $f(z) = \alpha z^n + \cdots$ be a complex polynomial of degree $n \geq 1$. Then there are complex numbers $\beta_1, \cdots, \beta_n$ such that

$$f(z) = \alpha(z - \beta_1)\cdots(z - \beta_n).$$

Multiply out the last expression; αz^n really is the leading term of $f(z)$. Since $\deg f \geq 1$ is given, it is tacitly assumed that $\alpha \neq 0$.

Note The numbers $\beta_1, \cdots, \beta_n$ are the **zeros** of $f(z)$, and there can be repetitions among them. We could write the complete factorization of $f(z)$ in the form

$$f(z) = \alpha(z - \beta_1)^{m_1}\cdots(z - \beta_k)^{m_k},$$

where $\beta_1, \cdots, \beta_k$ are distinct from each other. The exponents $m_1, \cdots, m_k$ are positive integers with $m_1 + m_2 + \cdots + m_k = \deg f(z)$. We call m_j the **multiplicity** of the zero β_j. If $m_j = 1$, we call β_j a **simple zero** of $f(z)$.

Complex Zeros of Real Polynomials

Suppose $f(z) = az^2 + bz + c$ has *real* coefficients, $a \neq 0$. By the quadratic formula, its zeros are

$$\beta = -\frac{b}{2a} + \frac{1}{2a}\sqrt{D} \quad \text{and} \quad \gamma = -\frac{b}{2a} - \frac{1}{2a}\sqrt{D},$$

where $D = b^2 - 4ac$. The complete factorization is

$$f(z) = a(z - \beta)(z - \gamma).$$

(It can be checked directly in this case.) If $D \geq 0$, then β and γ are both real. But if $D < 0$, then $D = -d^2$, where d is real, $\sqrt{D} = \sqrt{-d^2} = di$, and

$$\beta = -\frac{b}{2a} + \frac{d}{2a}i \qquad \gamma = -\frac{b}{2a} - \frac{d}{2a}i.$$

By inspection, $\gamma = \overline{\beta}$ so

$$f(z) = a(z - \beta)(z - \overline{\beta}).$$

Thus the zeros of a real *quadratic* polynomial either are both real, or are a pair of non-real complex conjugates. What is remarkable is that this result can be generalized to real polynomials of any degree.

> Let $f(z) = a_n z^n + \cdot \cdot \cdot + a_0$ be a polynomial with *real coefficients*. If β is a complex zero of $f(z)$ *that is not real*, then $\overline{\beta}$ also is a zero of $f(z)$.

In simple words, non-real zeros of real polynomials occur in conjugate pairs. How can we prove this? In the quadratic case we used a formula for the zeros to make a proof. Of course, for polynomials of higher degree, there are no formulas for zeros in general. So we need an entirely different idea.

The idea is to use the rules

$$\overline{\alpha + \beta} = \overline{\alpha} + \overline{\beta} \qquad \overline{\alpha\beta} = \overline{\alpha}\,\overline{\beta} \qquad \overline{\alpha^n} = \overline{\alpha}^n .$$

Now suppose $f(\beta) = 0$. Then

$$a_n\beta^n + a_{n-1}\beta^{n-1} + \cdot \cdot \cdot + a_0 = 0 .$$

Take conjugates on both sides:

$$\overline{a_n\beta^n + a_{n-1}\beta^{n-1} + \cdot \cdot \cdot + a_0} = \overline{0}$$

$$\overline{a_n\beta^n} + \overline{a_{n-1}\beta^{n-1}} + \cdot \cdot \cdot + \overline{a_0} = 0$$

But a_j is real, so $\overline{a_j} = a_j$. Therefore $\overline{a_j\beta^j} = \overline{a_j}\,\overline{\beta^j} = a_j\overline{\beta}^j$, hence

$$a_n\overline{\beta}^n + a_{n-1}\overline{\beta}^{n-1} + \cdot \cdot \cdot + a_0 = \overline{0} = 0 .$$

This says $f(\overline{\beta}) = 0$. Done!

An immediate consequence is the factorization

$$f(z) = (z - \beta)(z - \overline{\beta})h(z),$$

where $\deg h(z) = n - 2$. Now observe that

$$(z - \beta)(z - \overline{\beta}) = z^2 - (\beta + \overline{\beta})z + \beta\overline{\beta} = z^2 - 2[\mathrm{Re}(\beta)]z + |\beta|^2 .$$

Therefore $g(z) = (z - \beta)(z - \overline{\beta})$ is a *real* quadratic. Since $f(z) = g(z)h(z)$, the polynomial $h(z) = f(z)/g(z)$ is the quotient of two real polynomials, hence real itself, since the long division of real polynomials can only lead to real polynomials. The same reduction can now be repeated on $h(z)$, etc.

Now let us fit all the pieces together. Start with any non-constant real polynomial $f(x)$. It has some (maybe no) real zeros and some (maybe no) conjugate pairs of non-real zeros. Each real zero yields a real linear factor $x - r$; each pair of non-real conjugate zeros yields a real quadratic factor.

> **Complete Factorization of Real Polynomials**
>
> Let $f(x)$ be a real polynomial of degree $n \geq 1$. Then
>
> $$f(x) = a_n(x - r_1) \cdot \cdot \cdot (x - r_k)g_1(x) \cdot \cdot \cdot g_s(x),$$
>
> where $r_1, \cdot \cdot \cdot, r_k$ are real and
>
> $$g_j(x) = x^2 + b_j x + c_j$$
>
> with b_j and c_j real and $b_j^2 - 4c_j < 0$. Here $k \geq 0$, $s \geq 0$, and $k + 2s = n$.

The quadratic factors $g_1(x), \cdots, g_s(x)$ cannot be split into real linear factors; we call such factors **irreducible.** Now we can restate the result above as follows:

> Each real polynomial is the product of real irreducible linear and quadratic factors.

EXAMPLE 1 Express $f(x) = x^5 - 3x^4 + 2x^3 - 6x^2 + x - 3$ as the product of real irreducible factors.

Solution By trial and error we find $f(3) = 0$. Hence $x - 3$ is a factor, and by division

$$f(x) = (x - 3)(x^4 + 2x^2 + 1).$$

But $x^4 + 2x^2 + 1 = (x^2 + 1)^2$, and $x^2 + 1$ is irreducible.

Answer $(x - 3)(x^2 + 1)^2$

Remark The irreducible factor $x^2 + 1$ corresponds to the complex zeros $\pm i$ since $x^2 + 1 = (x - i)(x + i)$. The factorization of $f(x)$ into complex linear factors is

$$f(x) = (x - 3)(x - i)^2(x + i)^2.$$

Exercises

Solve

1 $z^2 - 2z + 5 = 0$ **2** $z^2 + 4z + 13 = 0$ **3** $z^2 + z + 6 = 0$

4 $2z^2 - 3z + 10 = 0$ **5** $z^4 + 5z^2 + 4 = 0$ **6** $z^4 + 4z^2 + 29 = 0$

7 $z^3 - 1 = 0$ **8** $z^3 + 8 = 0$ **9** $z^4 - 1 = 0$

10 $z^3 + z - 2 = 0$.

11 Factor $x^3 + 1$ into real irreducible factors.

12 Factor $x^4 - 1$ into real irreducible factors.

13 Show from the factorization

$$z^5 + 1 = (z + 1)(z^4 - z^3 + z^2 - z + 1)$$

that -1 is a simple zero of $z^5 + 1$.

14 Show from the factorization

$$z^n - 1 = (z - 1)(z^{n-1} + z^{n-2} + \cdots + z + 1)$$

that $+1$ is a simple zero of $z^n - 1$.

15 Use the methods of this section to prove that a real polynomial of odd degree has a real zero.

16 If a polynomial $f(z)$ has zeros $\pm i$, prove that it is divisible by $z^2 + 1$.

Write down the most general real polynomial satisfying the given conditions

17 degree 4 and zeros ± 2, $1 \pm i$

18 degree 5 and a real zero of multiplicity 4

19 degree 4 and no real zeros

20 degree 6 and zeros $\pm i$, $\pm 2i$, and 0 with multiplicity 2.

The next four exercises deal with the set **T** of all *real* numbers of the form $\alpha = a + b\sqrt{2}$, where a and b are *rational* numbers.

21 Show that the sum, difference, and product of numbers in **T** are in **T**.

22 For $\alpha = a + b\sqrt{2}$ in $\mathbf{T}$, set $\alpha' = a - b\sqrt{2}$. Show that $(\alpha \pm \beta)' = \alpha' \pm \beta'$ and $(\alpha\beta)' = \alpha'\beta'$.

23* If $\alpha = a + b\sqrt{2}$ is in $\mathbf{T}$ and $\alpha \neq 0$, show that $\alpha\alpha'$ is a non-zero rational number and

$$\alpha^{-1} = \frac{a}{\alpha\alpha'} - \frac{b}{\alpha\alpha'}\sqrt{2} = \frac{\alpha'}{\alpha\alpha'} \quad \text{is in } \mathbf{T}.$$

24* Suppose

$$f(x) = a_n x^n + a_{n-1}x^{n-1} + \cdots + a_0$$

has rational coefficients and $f(\alpha) = 0$, where α belongs to $\mathbf{T}$. Show that $f(\alpha') = 0$. [*Hint* Proceed by analogy with the corresponding statement for complex zeros of real polynomials.]

4 DE MOIVRE'S THEOREM

De Moivre's Theorem provides an important formula for the integer powers of a complex number.

De Moivre's Theorem Let

$$\alpha = r(\cos\theta + i\sin\theta) \neq 0$$

be a complex number in polar form, and let n be an integer. Then

$$\alpha^n = r^n(\cos n\theta + i\sin n\theta).$$

In other words,

$$|\alpha^n| = |\alpha|^n \qquad \arg(\alpha^n) = n\arg\alpha.$$

Proof If $n = 0$, then $\alpha^n = 1$ by definition and $r^0(\cos 0 + i\sin 0) = 1$, so the formula is true. Clearly it is true for $n = 1$. We now prove it for $n \geq 1$ by mathematical induction. Suppose it is true for some $n \geq 1$. Then

$$\begin{aligned}\alpha^{n+1} &= \alpha\cdot\alpha^n = [r(\cos\theta + i\sin\theta)][r^n(\cos n\theta + i\sin n\theta)]\\ &= r^{n+1}[\cos(n+1)\theta + i\sin(n+1)\theta].\end{aligned}$$

Therefore the formula also holds for $n + 1$. By the principle of mathematical induction, it holds for all $n \geq 1$.

When $n < 0$, the formula follows from the positive case and the expression of the reciprocal in polar form. Set $n = -m$ with $m > 0$. Then

$$\begin{aligned}\alpha^n &= [r(\cos\theta + i\sin\theta)]^n\\ &= \{[r(\cos\theta + i\sin\theta)]^{-1}\}^m = \{r^{-1}[\cos(-\theta) + i\sin(-\theta)]\}^m\\ &= (r^{-1})^m[\cos m(-\theta) + i\sin m(-\theta)]\\ &= r^{-m}[\cos(-m\theta) + i\sin(-m\theta)] = r^n(\cos n\theta + i\sin n\theta).\end{aligned}$$

EXAMPLE 1 Find $(1 + i)^{15}$ in rectangular form.

Solution In polar form,

$$1 + i = \sqrt{2}(\cos\tfrac{1}{4}\pi + i\sin\tfrac{1}{4}\pi).$$

By De Moivre's Theorem,

$$(1 + i)^{15} = (\sqrt{2})^{15}(\cos \tfrac{15}{4}\pi + i \sin \tfrac{15}{4}\pi).$$

But

$$(\sqrt{2})^{15} = (\sqrt{2})^{2\times 7}(\sqrt{2}) = 2^7 \cdot \sqrt{2} = 128\sqrt{2}$$

and $\frac{15}{4}\pi = \frac{16}{4}\pi - \frac{1}{4}\pi = 4\pi - \frac{1}{4}\pi$. Therefore

$$\begin{aligned}(1 + i)^{15} &= 128\sqrt{2}[\cos(-\tfrac{1}{4}\pi) + i \sin(-\tfrac{1}{4}\pi)] \\ &= 128\sqrt{2}(\tfrac{1}{2}\sqrt{2} - \tfrac{1}{2}i\sqrt{2}) = 128 - 128i.\end{aligned}$$

Answer $128 - 128i$

Roots of Unity

We now discuss the zeros of the polynomial

$$f(z) = z^n - 1.$$

They are called **n-th roots of unity** because they are the roots of

$$z^n - 1 = 0, \quad \text{that is,} \quad z^n = 1.$$

Clearly the only *real* n-th roots of unity are $+1$ and -1 if n is even, only $+1$ if n is odd. So most n-th roots of unity are non-real; how do we find them?

Suppose $\zeta^n = 1$. Then $|\zeta|^n = |\zeta^n| = 1$, so $|\zeta| = 1$ and the polar form of ζ is

$$\zeta = \cos\theta + i \sin\theta.$$

By De Moivre's Theorem

$$\zeta^n = \cos n\theta + i \sin n\theta.$$

Therefore, to solve the equation $\zeta^n = 1$, we must find all possible angles θ such that $n\theta$ equals the argument of 1. But clearly $\arg 1 = 0$. Not so fast! Arg is determined up to multiples of 2π; any angle $2\pi k$, where k is an integer, is an argument of 1. That is,

$$\cos 2\pi k + i \sin 2\pi k = 1,$$

and these angles $2\pi k$ are the only arguments of 1. Hence $\zeta^n = 1$ if and only if

$$n\theta = 2\pi k \qquad (k \text{ an integer}).$$

Therefore θ must take one of the values

$$\theta = \frac{2\pi k}{n} \qquad k = 0, \pm 1, \pm 2, \pm 3, \cdots.$$

While these look like a lot of values of θ, they represent only n distinct angles. After n consecutive integers, the same values appear increased by 2π:

$$\frac{2\pi(k + n)}{n} = \frac{2\pi k}{n} + 2\pi.$$

Thus we can restrict θ to the values

$$\theta = \frac{2\pi k}{n} \qquad k = 0, 1, 2, \cdots, n - 1.$$

Other choices of k will produce a value of $2\pi k/n$ that differs from one of the above numbers by a multiple of 2π.

It follows that the n-th roots of unity are the n complex numbers

$$\zeta_k = \cos\frac{2\pi k}{n} + i\sin\frac{2\pi k}{n} \qquad k = 0, 1, 2, \cdots, n-1.$$

Geometrically, they are n equally spaced points on the unit circle; hence they form the vertices of a regular n-gon (Fig. 1).

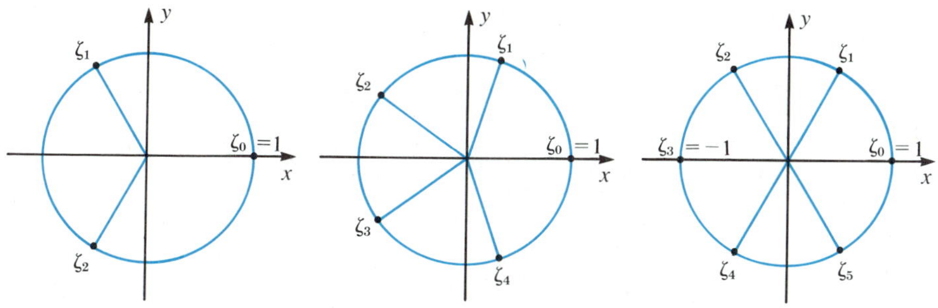

(a) cube roots of unity (b) fifth roots of unity (c) sixth roots of unity

Fig. 1 Roots of unity

Somewhat more is true. If we set

$$\zeta = \zeta_1 = \cos\frac{2\pi}{n} + i\sin\frac{2\pi}{n}$$

then by De Moivre's Theorem

$$\zeta^k = \left(\cos\frac{2\pi}{n} + i\sin\frac{2\pi}{n}\right)^k = \cos\frac{2\pi k}{n} + i\sin\frac{2\pi k}{n} = \zeta_k.$$

Therefore the n-th roots of unity are the powers of ζ:

$$1, \zeta, \zeta^2, \zeta^3, \cdots, \zeta^{n-1}.$$

These are the zeros of $z^n - 1$, so

$$z^n - 1 = (z-1)(z-\zeta)(z-\zeta^2)\cdots(z-\zeta^{n-1}).$$

n-th Roots of Unity The zeros of $z^n - 1$ are the complex numbers

$$\zeta_k = \cos\frac{2\pi k}{n} + i\sin\frac{2\pi k}{n} \qquad k = 0, 1, 2, \cdots, n-1.$$

If $\zeta = \zeta_1 = \cos(2\pi/n) + i\sin(2\pi/n)$, then $\zeta_k = \zeta^k$, and

$$z^n - 1 = (z-1)(z-\zeta)(z-\zeta^2)\cdots(z-\zeta^{n-1}).$$

EXAMPLE 2 Compute the cube roots of unity and express each in the form $a + bi$.

Solution Use the formula with $n = 3$ and $k = 0, 1, 2$:

$$k = 0 \qquad z_0 = \cos 0 + i\sin 0 = 1$$

$$k = 1 \qquad z_1 = \cos\frac{2\pi}{3} + i\sin\frac{2\pi}{3} = -\tfrac{1}{2} + \tfrac{1}{2}i\sqrt{3}$$

$$k = 2 \qquad z_2 = \cos\frac{4\pi}{3} + i\sin\frac{4\pi}{3} = -\tfrac{1}{2} - \tfrac{1}{2}i\sqrt{3}.$$

Answer $1 \qquad -\tfrac{1}{2} + \tfrac{1}{2}i\sqrt{3} \qquad -\tfrac{1}{2} - \tfrac{1}{2}i\sqrt{3}$

Remark Note that there is a conjugate pair of non-real roots. We could have predicted that, since the polynomial $z^3 - 1$ has real coefficients. Also note that it is not obvious that the cube of $-\tfrac{1}{2} \pm \tfrac{1}{2}i\sqrt{3}$ is 1; you should check it.

General n-th Roots

The solution of the equation

$$z^n = \alpha \qquad (\alpha \neq 0)$$

is similar to that of the special case $z^n = 1$. Write α in polar form:

$$\alpha = r_0(\cos\theta_0 + i\sin\theta_0) \qquad (r_0 > 0),$$

and write the unknown z as

$$z = r(\cos\theta + i\sin\theta) \qquad (r > 0).$$

Then $z^n = r^n(\cos n\theta + i\sin n\theta)$ by De Moivre's Theorem, so we must have

$$r^n(\cos n\theta + i\sin n\theta) = r_0(\cos\theta_0 + i\sin\theta_0).$$

This equation requires

$$r^n = r_0 \qquad n\theta = \theta_0 + 2\pi k \qquad k \text{ an integer.}$$

Therefore we choose

$$r = r_0^{1/n} \qquad \theta = \frac{1}{n}\theta_0 + \frac{2\pi k}{n} \qquad k = 0, 1, \cdots, n-1.$$

(As with roots of unity, there is no use taking other values of k; they yield no further angles.)

The n roots of the equation

$$z^n = r_0(\cos\theta_0 + i\sin\theta_0) \qquad (r_0 > 0)$$

are

$$\beta_k = r_0^{1/n}\left[\cos\left(\frac{1}{n}\theta_0 + \frac{2\pi k}{n}\right) + i\sin\left(\frac{1}{n}\theta_0 + \frac{2\pi k}{n}\right)\right]$$

$$k = 0, 1, \cdots, n-1.$$

They form the vertices of a regular n-gon centered at 0.

Example 3 Solve the equation $z^3 = 2i$.

Solution Write $z = r(\cos\theta + i\sin\theta)$ and $2i = 2(\cos\tfrac{1}{2}\pi + i\sin\tfrac{1}{2}\pi)$. Then $z^3 = 2i$ means

$$r^3(\cos 3\theta + i\sin 3\theta) = 2(\cos\tfrac{1}{2}\pi + i\sin\tfrac{1}{2}\pi).$$

Therefore

$$r^3 = 2 \quad \text{and} \quad 3\theta = \tfrac{1}{2}\pi + 2\pi k \qquad k = 0, 1, 2.$$

Consequently

$$r = \sqrt[3]{2} \quad \text{and} \quad \theta = \tfrac{1}{6}\pi + \tfrac{2}{3}\pi k \qquad k = 0, 1, 2.$$

The three values of θ are $\frac{1}{6}\pi$, $\frac{1}{6}\pi + \frac{2}{3}\pi = \frac{5}{6}\pi$, and $\frac{1}{6}\pi + \frac{4}{3}\pi = \frac{3}{2}\pi$. The corresponding roots of $z^3 = 2i$ are

$$\beta_0 = \sqrt[3]{2}(\cos\tfrac{1}{6}\pi + i\sin\tfrac{1}{6}\pi) = \sqrt[3]{2}(\tfrac{1}{2}\sqrt{3} + \tfrac{1}{2}i)$$
$$\beta_1 = \sqrt[3]{2}(\cos\tfrac{5}{6}\pi + i\sin\tfrac{5}{6}\pi) = \sqrt[3]{2}(-\tfrac{1}{2}\sqrt{3} + \tfrac{1}{2}i)$$
$$\beta_2 = \sqrt[3]{2}(\cos\tfrac{3}{2}\pi + i\sin\tfrac{3}{2}\pi) = -i\sqrt[3]{2}.$$

See Fig. 2.

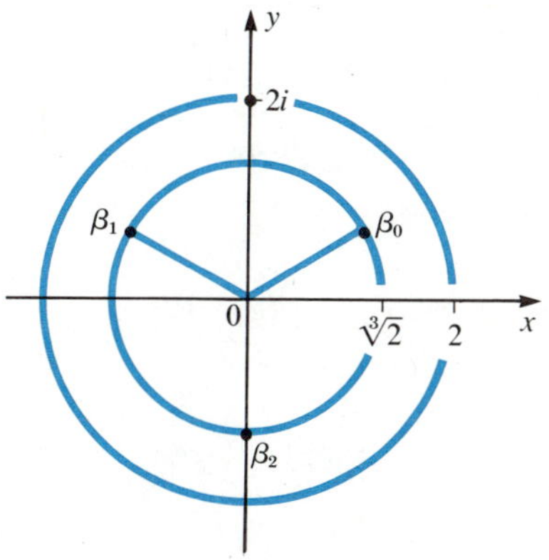

Fig. 2 Roots of $z^3 = 2i$ $\begin{cases} \beta_0 = \sqrt[3]{2}(\frac{1}{2}\sqrt{3} + \frac{1}{2}i) \\ \beta_1 = \sqrt[3]{2}(-\frac{1}{2}\sqrt{3} + \frac{1}{2}i) \\ \beta_2 = -i\sqrt[3]{2} \end{cases}$

Answer $\sqrt[3]{2}(\pm\frac{1}{2}\sqrt{3} + \frac{1}{2}i) \qquad -i\sqrt[3]{2}$

Exercises

Compute by De Moivre's theorem; give answer in rectangular form if possible

1 $(1 - i)^{11}$ **2** $(1 - i)^{27}$ **3** $(\sqrt{3} + i)^6$
4 $(1 + i\sqrt{3})^{-17}$ **5** $[\frac{1}{2}(-1 + i\sqrt{3})]^{13}$ **6** $(\cos\frac{1}{7}\pi + i\sin\frac{1}{7}\pi)^{50}$
7 $(\cos 15° + i\sin 15°)^{25}$ **8** $(\cos 12° + i\sin 12°)^{65}$ **9** $(\sqrt{3} - i)^{-10}$
10 $(\sqrt{3} - i)^{-28}$ **11** $(\cos\frac{1}{9}\pi - i\sin\frac{1}{9}\pi)^{-11}$ **12** $(\sqrt{3} + i)^{-20}$.

Solve completely

13 $z^4 = 1$ **14** $z^5 = 1$ **15** $z^6 = 1$
16 $z^8 = 1$ **17** $z^3 = 8$ **18** $z^4 = 16$
19 $z^4 = -25$ **20** $z^4 = i$ **21** $z^3 = i$
22 $z^4 = -i$ **23** $z^3 = -8i$ **24** $z^6 = -27$
25 $z^3 = 1 + i$ **26** $z^4 = -1 + i$ **27** $z^3 = \frac{1}{2} + \frac{1}{2}i\sqrt{3}$
28 $z^3 = -\frac{1}{2}\sqrt{3} + \frac{1}{2}i$ **29** $z^4 = 4\sqrt{3} - 4i$ **30** $z^5 = 3 - 3i$.

31 Let α be an n-th root of unity and $\alpha \neq 1$. Show that

$$\alpha^{n-1} + \alpha^{n-2} + \cdots + \alpha + 1 = 0.$$

32 (cont.) Solve $z^4 + z^3 + z^2 + z + 1 = 0$.

33 Let $\alpha = \cos\frac{2}{5}\pi + i\sin\frac{2}{5}\pi$. Set $\beta = \alpha + \alpha^{-1}$. Prove that $\beta^2 + \beta - 1 = 0$.

34 (cont.) Prove that $\beta = \frac{1}{2}(-1 + \sqrt{5})$.

35 (cont.) Prove that $\alpha^2 - \beta\alpha + 1 = 0$.

36 (cont.) Prove that $\alpha = \frac{1}{2}(\beta + i\sqrt{4 - \beta^2})$. Conclude that

$$\cos 72° = \frac{-1 + \sqrt{5}}{4} \qquad \sin 72° = \frac{\sqrt{10 + 2\sqrt{5}}}{4}.$$

Find all solutions (z, w) of the system

37* $\begin{cases} z^2 = w \\ w^2 = z \end{cases}$

38* $\begin{cases} z^2 = w \\ w^3 = z \end{cases}$

39 Verify that $\alpha = 2 - i$ is a root of the equation $z^4 = -7 - 24i$. Show that all roots are α, $i\alpha$, $-\alpha$, and $-i\alpha$, that is, α multiplied by each of the 4-th roots of unity.

40 Prove $1 + \cos\frac{2\pi}{n} + \cos\frac{4\pi}{n} + \cdots + \cos\frac{2(n-1)\pi}{n} = 0$,

$$\sin\frac{2\pi}{n} + \sin\frac{4\pi}{n} + \cdots + \sin\frac{2(n-1)\pi}{n} = 0.$$

[*Hint* Use Ex. 31.]

A number α is a **primitive** n-th root of unity if $\alpha^n = 1$, but $\alpha^m \neq 1$ for $0 < m < n$. Find the primitive n-th roots of unity for

41 $n = 4$ **42** $n = 5$ **43** $n = 6$

44 $n = 7$ **45** $n = 8$ **46** $n = 9$

47 $n = 10$ **48** $n = 12$

49* Show that each primitive 6-th root of unity satisfies $z^2 - z + 1 = 0$.

50* Show that each primitive 8-th root of unity satisfies $z^4 + 1 = 0$.

51* Show that each primitive 9-th root of unity satisfies $z^6 + z^3 + 1 = 0$.

52* Show that each primitive 12-th root of unity satisfies $z^4 - z^2 + 1 = 0$.

REVIEW EXERCISES

Express in the form $a + bi$

1 $\dfrac{2 - 5i}{1 + 3i}$ **2** $\dfrac{(4 - i)(2 + i)}{3 + 4i}$ **3** $\overline{(1 + i)(3 - 2i)}$

4 $(1 + i)^2\overline{(3 + 2i)}$ **5** i^{73}.

Express in polar form

6 $\dfrac{2 + 2i}{\sqrt{3} + i}$ **7** $-5i$ **8** $\cos\frac{1}{7}\pi - i\sin\frac{1}{7}\pi$.

9 Show that $z^3 + 1$ has three simple zeros by finding them.

10 Solve $z^4 + z^2 + 1 = 0$.

11 Compute $(\sqrt{3} + i)^{14}$.

12 Find the square roots of $1 - i\sqrt{3}$.

13 Plot in the complex plane the number $\alpha = 1 + i$ and all solutions of the equation $z^3 = \alpha$.

14 Find all 6-th roots of -1.

15 The point α moves counterclockwise around the circle $|z| = 2$ in the complex plane. Describe the corresponding motion of α^{-1}.

16 Let d_1, d_2, d_3, d_4 be the distances from a point z in the complex plane to the points $2, -2, 2i, -2i$. Show that $d_1 \cdot d_2 \cdot d_3 \cdot d_4 = |z^4 - 16|$.

17 If α, β, γ are the vertices of an equilateral triangle in the complex plane, prove that $\gamma - \alpha = \omega(\beta - \alpha)$, where ω is a 6-th root of unity.

18 Factor $z^4 + 4$ into

(a) linear factors (b) irreducible real quadratic factors.

ANSWERS TO ODD-NUMBERED EXERCISES

CHAPTER 1

Section 1 page 5

1 16,412 **3** 4009 **5** 8426

7 $(5x)(6y) - 10u = (5x)[(2\cdot 3)y] - 10u =_{(\text{assoc})} (5x)[2(3y)] - 10u =_{(\text{assoc})} [(5x)\cdot 2](3y) - 10u =_{(\text{comm})} [2(5x)](3y) - 10u =_{(\text{assoc})} [(2\cdot 5)x](3y) - 10u =_{(\text{assoc})} 10[x(3y)] - 10u =_{(\text{dist})} 10[x(3y) - u] = 10(3xy - u)$, since $x(3y) = (x\cdot 3)y = (3x)y = 3xy$ by assoc, comm, assoc.

9 $6xyz + x^2y^2z =_{(\text{comm})} x^2y^2z + 6xyz$. Now $x^2y^2z = x(xy)(yz) = x(yz)(xy) = (xyz)(xy) = (xyz)(yx)$ by assoc, comm, assoc, comm. Also $6xyz = 6(xyz) = (xyz)\cdot 6$ by assoc, comm. Hence LHS $= (xyz)(yz) + (xyz)\cdot 6 =_{(\text{dist})} xyz(yx + 6)$.

11 By the unity law, $a + b + c + 1 = 1\cdot(a + b + c + 1)$, so LHS $=$ $(a + b + c + 1)^2 - 1\cdot(a + b + c + 1) =$ $_{(\text{dist})}[(a + b + c + 1) - 1](a + b + c + 1) =$ $_{(\text{assoc})}[(a + b + c) + (1 - 1)](a + b + c + 1) =$ $_{(\text{inverse})}[(a + b + c) + 0](a + b + c + 1) =_{(\text{zero})} (a + b + c)(a + b + c + 1)$.

13 $(x + y)(x - y) =_{(\text{dist})} x(x - y) + y(x - y) =_{(\text{dist})} (x^2 - xy) + (yx - y^2) =$ $_{(\text{assoc})} x^2 + [-xy + (yx - y^2)] =_{(\text{comm})} x^2 + [-xy + (xy - y^2)] =$ $_{(\text{assoc})} x^2 + [(-xy + xy) - y^2] =_{(\text{inverse})} x^2 + (0 - y^2) =$ $_{(\text{zero})} x^2 + (-y^2) = x^2 - y^2$ by definition.

15 no; $2 \div 1 \neq 1 \div 2$ **17** no; $2 \div (2 \div 2) \neq (2 \div 2) \div 2$ **29** $3^4 > 4^3$

31 neither; they are equal **33** $3^2 > (-3)^3$ **35** 3.572 left **37** -0.019 left

39 $\frac{47}{15}$ **41** $\frac{2}{5}$ **43** $\frac{5}{36}$ **45** $\frac{7}{12}$ **47** $-\frac{2}{5}$ **49** $\frac{13}{14}$ **51** $\frac{3}{2}$

Section 2 page 9

1 $x + \frac{1}{2}y$ **3** $2(a + b)$ **5** $0.03x$ **7** $\dfrac{350}{r}$ hr **9** $\frac{1}{6}b^2$ **11** $\pi(r + 1)^2$

13 $(x + y)(x - y)$ **15** $G = \frac{1}{2}B + 17$ **17** $1, -\frac{1}{7}$ **19** 7, 7 **21** 10, -6

23 $\frac{13}{3}, \frac{5}{3}$ **25** $x = 1, y = 0$ yields 0 and 1, for example **27** $x = 1, y = -1$ yields 0, 2 for example **29** $x = y = 0$ yields 4 and 0, for example **31** 12, 0 **33** 4, 5

35 $\dfrac{ab - c}{d} + e,\ ab - \dfrac{c}{d} + e$ **37** $x \boxed{+} 3 \boxed{\div} y \boxed{=},\ x \boxed{+} 3 \boxed{=} \boxed{\div} y \boxed{=}$

39 $x \boxed{\div} y \boxed{\div} z \boxed{=}$, either logic **41** $a \boxed{\div} b \boxed{+} c \boxed{\div} 2 \boxed{=},\ a \boxed{\div} b \boxed{+} c \boxed{=} \boxed{\div} 2 \boxed{=}$

43 -2 **45** 945 **47** 1

Section 3 page 13

1 27 **3** $\frac{1}{36}$ **5** 4 **7** $-\frac{32}{243}$ **9** $\frac{1}{72}$ **11** $\frac{1}{2}$ **13** 2^4 **15** 2^{-7} **17** 2^3

19 2^4 **21** x^5y^4 **23** $a + \dfrac{1}{a}$ **25** $\dfrac{1}{b^{16}}$ **27** $\dfrac{-1}{16x^5y^2}$ **29** $\dfrac{1}{4^5x^{15}y^{10}z^5}$ **31** $\dfrac{2y^2}{x^2}$

33 $\dfrac{b^2}{a^2}$ **35** 4×10^{-1} **37** 2.37 **39** 7.6×10^{-5} **41** 1.7×10

43 1.24×10^4 **45** 3.832×10^6 **47** 6.48×10^5 **49** 9.6×10^{-9} **51** 6.4×10^{31}

53 6.4×10^{-29} **55** 6.336×10^6 in. **57** 4.8×10^{-18} grams

59 275,854.74 **61** 2.599273 **63** 1.7176×10^{-65}

Section 4 page 18

1 9 **3** $\frac{1}{3}$ **5** $3a^3$ **7** $5\sqrt{2}$ **9** $1/(2\sqrt{3}) = \frac{1}{6}\sqrt{3}$ **11** $3x\sqrt{2x}/(y + z)^2$

13 10 **15** $\frac{3}{4}$ **17** $\frac{1}{2}$ **19** $2\sqrt[4]{2}$ **21** $2ab^3$ **23** $2uv\sqrt[5]{2v}$ **25** $\frac{1}{6}\sqrt{6}$

27 $(x + 2\sqrt{xy} + y)/(x - y)$ **29** $(19 - 8\sqrt{3})/13$ **31** $(7\sqrt{2} + 2\sqrt{3})/86$

33 $1/[4(\sqrt{3} - \sqrt{2})]$ **35** $1/(\sqrt{x + h} + \sqrt{x - h})$ **37** $a^2 - 2b^2$

39 x^2 **41** $3x^2\sqrt{5y}$ **43** x^2 **45** $2y\sqrt{5xz}$

47 $\sqrt{a^2 + x^2} = \sqrt{a^2\left(1 + \frac{x^2}{a^2}\right)} = a\sqrt{1 + \frac{x^2}{a^2}} = a\sqrt{1 + \left(\frac{x}{a}\right)^2}$

49 $\frac{2}{\sqrt{3} + 1} - \frac{1}{\sqrt{3} - 1} = \frac{2(\sqrt{3} - 1) - (\sqrt{3} + 1)}{(\sqrt{3} + 1)(\sqrt{3} - 1)} = \frac{\sqrt{3} - 3}{2}$ **51** This is just the definition of the n-th root. **53** Let $\sqrt[n]{a} = r$ and $\sqrt[n]{b} = s$. Then $a = r^n$ and $b = s^n$, hence $ab = (rs)^n$. Therefore $\sqrt[n]{ab} = rs = \sqrt[n]{a}\sqrt[n]{b}$. **55** 2.4 **57** 0.77 **59** 5.2 **61** 6.2 **63** 54.690036 **65** 1.25892 54 **67** 2236.5045 **69** 0.04423 9617 **71** 3.58338 94 **73** 1.96157 06 **75** both $\approx$ 3.14626 44 **77** both $\approx$ 1.23153 77 **79** $(\sqrt{2} + \sqrt{3})^2 = 2 + 2\sqrt{6} + 3 = 5 + 2\sqrt{6}$. Take square roots: $\sqrt{2} + \sqrt{3} = \sqrt{5 + 2\sqrt{6}}$.

Section 5 page 22

1 9 **3** $\frac{1}{5}$ **5** 100,000 **7** $\frac{343}{8}$ **9** $2^{7/3}$ **11** $2^{7/2}$ **13** $2^{3/2}$ **15** $2^{-9/4}$ **17** $1/125x^6$ **19** u^3/v^9 **21** $x^{10}y^{15}/z^{20}$ **23** x^4/y^2 **25** $32x^2\sqrt{x}$ **27** $2u - u^{1/6}$ **29** $\sqrt[6]{32}$ **31** 1 **33** $3/2\sqrt[3]{6a^2}$ **35** x^3 **37** $\sqrt[8]{a}$ **39** $\sqrt{3^5 \cdot 2^2}$ **41** $1/x^{1/5}y^{2/5}$ **43** $y^{9/20}/x^{3/2}$ **45** 1 **47** 4.75682 85 **49** 1.79734 30 **51** $1.07326\,70 \times 10^8$

Section 6 page 25

1 $6x + 1$ **3** $8x^2 + 8x + 2$ **5** $-6x^2 + 6x - 3$ **7** $\frac{5}{3}x^2 - 2x - \frac{13}{2}$ **9** $x^3 + 8x^2 - 16x - 4$ **11** $-x^5 - 2x^4 - 2x^3 - 2x^2 - x + 7$ **13** $5x^8 + 36x^6 - 30x^5 + 20x^4 - 24x^3 + 10x^2$ **15** $x^2 - x - 6$ **17** $6x^2 + 17x + 5$ **19** $x^3 + 7x^2 + 5x - 6$ **21** $x^5 + 2x^3 + x$ **23** $x^7 - x^6 - \frac{1}{2}x^5 + \frac{1}{2}x^4 - \frac{1}{4}x^3 - \frac{1}{4}x^2$ **25** $x^4 + 2x^3 + 2x^2 + x$ **27** $x^5 - 1$ **29** $1 + x + x^2 + x^3$ **31** $(1 + x)(1 + x^2)\cdots(1 + x^{2^n}) = 1 + x + x^2 + \cdots + x^{2^{n+1}-1}$ **33** $1 - x^4$ **35** $(1 - x)(1 + x)(1 + x^2)(1 + x^{2^2})\cdots(1 + x^{2^n}) = 1 - x^{2^{n+1}}$ **37** $x^4 + x^2 + 1$ **39** $2x^3 + 4x^2 - x + 5$ **41** $x^2 + (a + b)x + ab$ **43** $x^4 + 2x^3 + 2x^2 + 2x + 1$ **45** $abx^2 + (a^2 + b^2)x + ab$ **47** $18x^2 + 6x + 25$ **49** $x^4 + 8x^3 + 11x^2 - 5x - 6$ **51** 2 **53** 7 **55** -20 **57** 1 **59** 3

Section 7 page 28

1 $5x + y + 2$ **3** $-5x + 6y + 5z - 2$ **5** $2x^2 + 4y^2 - 3y$ **7** $-2x^2 + yz + zx - x - y$ **9** $2x^2 + 5xy + 3y^2$ **11** $25x^2 - 4y^2$ **13** $16u^2 - 24uv + 9v^2$ **15** $x^3 + 3x^2y + 3xy^2 + y^3$ **17** $2x^2 - y^2 + xy + 2xz - yz$ **19** $x^3 + x^2y - xy^2 - y^3 + y^2 + xy - x$ **21** $x^4 - 2x^2y^2 - 3y^4$ **23** $x^2 + y^2 + z^2 + 2xy + 2yz + 2zx$ **25** $rst + rs + st + tr + r + s + t + 1$ **27** $a^3 + 6a^2b + 11ab^2 + 6b^3$ **29** $x^4 - 4x^2y^2 + 8xy^3 - 4y^4$ **31** $x^2 + 4xy + 4y^2$ **33** $x^2 - 8xy + 16y^2$ **35** $9a^2 + 12ab + 4b^2$ **37** $u^4 + 2u^2v^2 + v^4$ **39** $(100 + 2)^2 = 10{,}000 + 400 + 4 = 10{,}404$ **41** yes, of $x + 8$ **43** yes, of $z^3 + 2$ **45** no **47** yes, of $(x^2 + 1) + 1$, that is, $x^2 + 2$ **49** yes, of $ab^4 - 2c^2$ **51** $[(x + 1)^2 - (x - 1)^2]/x = 4x/x = 4$ **53** $3n^2 + 6n + 5 = n^2 + (n^2 + 2n + 1) + (n^2 + 4n + 4) = n^2 + (n + 1)^2 + (n + 2)^2$ **55** 16 **57** 12

Section 8 page 33

1 $(x + 1)(x + 3)$ **3** $(x + 3)(x - 2)$ **5** $(x - 2)(x - 7)$ **7** $(x - 3)^2$ **9** $x^2(x + 5)(x - 4)$ **11** $(x + 4)(2x + 3)$ **13** $(2x + 1)^2$ **15** $(x^2 - 2)(x^2 - 3)$ **17** $(2x + 9)(2x - 9)$ **19** $(x + y + z)(x + y - z)$ **21** $(a^2 + 4b)(a^2 - 4b)$ **23** $(v^2 + 9)(v^2 + v - 9)$ **25** $(2c + 3)(4c^2 - 6c + 9)$ **27** $(5x + 2y)(7x^2 + 2xy + 4y^2)$ **29** $(x - 1)(x^4 + x^3 + x^2 + x + 1)$ **31** $x^2(x - 1)(x^2 + x + 1)$ **33** $(x - 1)(x + 1)(x^2 + 1)(x^4 + 1)$ **35** $8(xy^2 + 5)(x^2y^4 - 5xy^2 + 25)$ **37** $(x + y^2z^3)(x - y^2z^3)$ **39** $a^2(a + 3b)^2(a - 3b)^2$ **41** $(r^2 + 2s + 2t)(r^2 - 2s - 2t)$ **43** $-(a + b + c)(a + b + c + 10)$ **45** $(2x + 1)(3y + 2)$ **47** $(2q + 5)(p - 4)$ **49** $(u + v)(u - v)(u^2 + v^2 + 1)$ **51** $(x + y)(x^2 - xy + y^2) = x^3 + (x^2y - x^2y) + (xy^2 - xy^2) + y^3 = x^3 + y^3$

53 $x^5 - y^5 = (x-y)(x^4 + x^3y + x^2y^2 + xy^3 + y^4)$. The right-hand side is $x(x^4 + x^3y + x^2y^2 + xy^3 + y^4) - y(x^4 + x^3y + x^2y^2 + xy^3 + y^4) = x^5 + (x^4y - x^4y) + (x^3y^2 - x^3y^2) + (x^2y^3 - x^2y^3) + (xy^4 - xy^4) - y^5 = x^5 - y^5$
55 $1027 = 10^3 + 3^3$, divisible by $10 + 3$.

Section 9 page 35

1 $\dfrac{1}{x+4}$ **3** $\dfrac{1}{x^2 - x + 1}$ **5** $\dfrac{x+y}{x^2+xy+y^2}$ **7** $\dfrac{1}{x+y+2z}$ **9** $\dfrac{1}{x(x-1)}$
11 $\dfrac{2}{(2x+3)(2x+5)}$ **13** $\dfrac{5x+4}{x^2-4}$ **15** $\dfrac{x^2+y^2}{xy}$ **17** $\dfrac{x^2+3xy+y^2}{(x+2y)(x+3y)}$
19 $-\dfrac{x+1}{x}$ **21** $\dfrac{x^2+x}{y^2+y}$ **23** $\dfrac{1}{xy}$ **25** $-3xy + 2zx + yz$ **27** $\dfrac{x+3}{2x}$
29 $\dfrac{a-1}{a}$ **31** $\dfrac{u^2-v^2}{uv}$ **33** $\dfrac{u^2v^2}{u^4-v^4}$ **35** u^2+v^2 **37** $\dfrac{5r+2}{r}$
39 $\dfrac{4x^2-8x+6}{(2x-1)^3}$ **41** $\dfrac{x(3x+2)}{(x+1)^2(2x+1)^2}$

Section 10 page 38

1 $(x/5)^2 = x^2/25$ **3** $\sqrt{4x+4} = 2\sqrt{x+1}$ **5** $\sqrt{2^9} = 2^{9/2} = 16\sqrt{2}$
7 $\sqrt[3]{\sqrt[3]{x}} = \sqrt[9]{x}$ **9** Leave as $\sqrt{x^2+1}$. **11** $2a(a^2+4a+1) = 2a^3 + 8a^2 + 2a$
13 $a^3 + b^3 = (a+b)(a^2 - ab + b^2)$ **15** $a^{-2}a^{-2} = a^{-4}$
17 $\sqrt{x^3 + 2x^2 + x} = (x+1)\sqrt{x}$ **19** $(25x)(4x) = 100x^2$
21 $(x+y)/(x+z) = x/(x+z) + y/(x+z)$
23 $(x+1)(y+1)(z+1) = xyz + xy + yz + zx + x + y + z + 1$
25 $(4x^3 - 7x^2 + 1)/x = 4x^2 - 7x + 1/x$ **27** $(x+y)^3 = x^3 + 3x^2y + 3xy^2 + y^3$
29 Leave as $x^2/(x+y)$. **31** Leave as -2^4. Note that $-2^4 = -16$ and $2^{-4} = 1/16$.
33 Leave as $(3+x)/(x+2)$.

Review Exercises page 38

1 $x^2 - y^2 = (x^2 + 0) - y^2 = [x^2 + (xy - xy)] - y^2 = [(x^2 + xy) - yx] - y^2 = (x^2 + xy) - (yx + y^2) = x(x+y) - y(x+y) = (x-y)(x+y)$ by zero, inverse, assoc. and comm., assoc., dist., and dist. Hence $3x + 3y + 4x^2 - 4y^2 = (3x + 3y) + (4x^2 - 4y^2) = 3(x+y) + 4(x^2 - y^2) = 3(x+y) + 4[(x-y)(x+y)] = 3(x+y) + [4(x-y)](x+y) = 3(x+y) + (4x-4y)(x+y) = [3 + (4x - 4y)](x+y) = (3 + 4x - 4y)(x+y)$ by assoc., dist., substitution, assoc., dist., dist., assoc. **3** $\frac{25}{17}$ **5** $\frac{3}{2}$ **7** 3^7
9 $1.17073\,53 \times 10^{83}$ **11** $1.64789\,81$ **13** $\dfrac{1}{x^{41/18}y^{7/3}}$ **15** 14
17 $(2y-1)(16y^4 + 8y^3 + 4y^2 + 2y + 1)$ **19** $\dfrac{(x-y)^2(x^2+y^2)}{x^2y^2}$

CHAPTER 2

Section 1 page 43

1 cond. eq. **3** identity **5** cond. eq. **7** identity **9** cond. eq. **11** 6
13 $1, -5$ **15** $0, 2, 7$ **17** 3 **19** $2, -2, 3, -3$ **21** $1, 2$ **23** 1
25 no real solutions **27** no real solutions **29** $\frac{2}{3}, -6$ **31** $0, 1, -1$
33 $3, -3, \frac{15}{2}$ **35** no real solutions

Section 2 page 48

1 $32 + \frac{9}{5}C = C + 20$ **3** $x(x+1)(x+2) = 15600$
5 $48 + 6 \cdot 18 + 11(w - 7) = 14.95w$
7 $32\left(1 + \dfrac{r}{100}\right) - (0.9)(32)\left(1 + \dfrac{r}{100}\right) = 3.36$ **9** $0.8B + 3500 = B + 500$
11 $\dfrac{N+14}{3} + 10 = N$ **13** $20 + \frac{1}{12}M = M$
15 $[(N^2+3)^2 + 3]^2 + 3 = 619{,}372$ **17** $\frac{1}{2}\{\frac{1}{2}[\frac{1}{2}(A - 50) - 50] - 50\} = 0.025A$
19 $\dfrac{30000}{N} = \dfrac{30000}{N-70} - 26.75$ **21** $\dfrac{5280 \cdot 12}{L} = \dfrac{5280 \cdot 12}{L-2} - 132$ L in inches
23 $P + N = 56$ $P + 5N = 100$ **25** $\dfrac{N}{N+3} + \dfrac{1}{10} = \dfrac{N+1}{N+4}$

27 $b + 2\sqrt{(\frac{1}{2}b)^2 + 3^2} = 12$ **29** y, where $\frac{1}{2}xy = 100$ and $x = \frac{1}{4}\sqrt{x^2 + y^2}$
31 $2.5k + 3m + 3.5n = 19$, where k, m, n are positive integers
33 $x + y = 30 \quad 2x = 3y$ **35** $0.8S$ where $\begin{cases} S = W + 99 \\ 0.8S = S - 2W \end{cases}$
37 $\begin{cases} J - \frac{1}{2}S = S - (J + 2) \\ J - \frac{1}{3}S = S - (J - 2) \end{cases}$

Section 3 page 54

1 6 **3** 2 **5** $-\frac{45}{4}$ **7** $\frac{5}{6}$ **9** 1 **11** -4 **13** $\dfrac{d - b}{a - c}$ **15** $\frac{7}{5}$ **17** $-\frac{19}{2}$
19 $\dfrac{1 - a^2 - 2b^2 - 3c^2}{1 + a^2 + b^2 + c^2}$ **21** 2 **23** $x = 2 - \frac{3}{2}y$ **25** $x = \dfrac{y + 1}{3y + 1}$
27 $x = \dfrac{-5y + 1}{3y - 2}$ **29** $x = a\left(1 - \dfrac{y}{b}\right)$ **31** $x = \dfrac{1 + 3y - y^3}{1 + y + 7y^2}$
33 $h = \dfrac{50}{\pi r(r + 2)}$ **35** $r^2 = \dfrac{s^4}{1 - s}$ **37** $\frac{1}{5}$ **39** $x = 1 \quad y = 1$
41 $x = 4 \quad y = 3$ **43** $x = 1 \quad y = 3$ **45** $x = \frac{10}{29} \quad y = \frac{7}{29}$ **47** no solutions
49 four solutions $x = \pm 2 \quad y = \pm 1$

Section 4 page 61

1 \$88.24 **3** $21\frac{9}{11}$ min **5** \$5000 **7** $\frac{1}{9}$ **9** $\frac{41}{333}$ **11** $\frac{9}{20} + \frac{61}{9900} = \frac{1129}{2475}$
13 \$909.09 **15** 80 77.5 **17** $\frac{18}{19}$ hr **19** $33\frac{1}{3}$ min **21** 12 hr **23** 5.071%
25 $15\frac{1}{6}$% **27** $66\frac{2}{3}$ g **29** 375 g **31** 8 mph **33** 7 **35** \$6250 in X and \$3750 in B **37** $P \approx \$12.27 \quad Q \approx 363{,}500$ **39** 700 cm³ gold, 250 cm³ silver

Section 5 page 67

1 0 $\frac{4}{3}$ **3** 7 -2 **5** $\frac{1}{2}$ $\frac{1}{3}$ **7** $\frac{3}{5}$ -4 **9** $(x - 4)^2 - 16$
11 $(x + 3)^2 - 8$ **13** $(x + k)^2 - k^2$ **15** $4(x + \frac{3}{2})^2 - 9$ **17** $2(x - \frac{1}{8})^2 - \frac{1}{32}$
19 $-(x - 2)^2 + 4$ **21** $3(x - \frac{5}{6})^2 - \frac{13}{12}$ **23** $-3(x - \frac{1}{6})^2 + \frac{61}{12}$ **25** $-2 \pm \sqrt{2}$
27 $-1 \pm \frac{1}{2}\sqrt{2}$ **29** no real solution **31** $\frac{1}{6}(-5 \pm \sqrt{19})$ **33** no real solution
35 9 -1

Section 6 page 72

1 none **3** one **5** two **7** 2 3 **9** $1 \pm \sqrt{5}$ **11** $-5 \pm \sqrt{23}$
13 $\frac{1}{4}(-1 \pm \sqrt{33})$ **15** 9 -11 **17** no real solutions **19** $\frac{3}{4}$ -2 **21** 3 $-\frac{2}{5}$
23 $\frac{1}{10}(4 \pm \sqrt{6})$ **25** no real solutions **27** $\frac{1}{3}(1 \pm \sqrt{5})$ **29** $\frac{1}{5}(1 \pm \sqrt{2})$
31 $-\dfrac{b}{a}$ **33** 1 **35** $\frac{1}{5}(2 \pm \sqrt{39})$ **37** $-3y^2 \pm \sqrt{9y^4 + y}$ **39** $\frac{1}{4}y(-5 \pm \sqrt{17})$
41 $\frac{1}{2}y$ $\frac{1}{3}y$ **43** 0.08082 606 0.83449 175 **45** $-1.41179\,576$ $-0.32478\,960$
47 0.50845 484 62.20893 6 **49** $-0.90228\,575$ 0.39250 010
51 no real solutions

Section 7 page 78

1 ± 1 $\pm\sqrt{6}$ **3** $\pm\frac{1}{2}\sqrt{2}$ $\pm\frac{1}{5}\sqrt{5}$ **5** 4 9 **7** -2 $\sqrt[3]{10}$ **9** $\pm\frac{1}{2}$ $\pm\frac{1}{2}\sqrt{2}$
11 $\pm\sqrt{\frac{1}{2}(1 + \sqrt{37})}$ **13** 0 -2 -3 **15** $-\frac{1}{2}$ $-\frac{1}{3}$ 2 **17** 2 $\frac{1}{2}(-3 \pm \sqrt{5})$
19 $\pm\frac{1}{2}$ ± 3 **21** -5 **23** ± 1 **25** 3 **27** 11 **29** 1 **31** 9 **33** 6
35 $(7, 9)$ $(9, 7)$ **37** $(2, -1)$ $(-1, 2)$ **39** $(\frac{1}{2}, 1)$ $(-\frac{1}{4}, -\frac{7}{2})$
41 $(3, 0)$ $(-5, \frac{20}{3})$ **43** $(5, 6)$ $(6, 5)$ **45** $(3, 2)$ $(-2, -3)$ **47** $(16, 9)$

Section 8 page 83

1 base $2/\sqrt[4]{15}$, slant sides $4/\sqrt[4]{15}$ **3** yes, $2 \pm \sqrt{5}$ **5** $\frac{9}{2}$ **7** 320 **7** $2\frac{13}{22}$ sec
11 $x + \dfrac{1}{x} = x - 2 + \dfrac{1}{x} + 2 = \left(\sqrt{x} - \dfrac{1}{\sqrt{x}}\right)^2 + 2 \geq 2$ **13** 17, 18 **15** $5\sqrt{5}$ cm
17 For $x = 1/r$, $cx^2 + bx + a = (c + br + ar^2)/r^2 = 0$ **19** $1 + \sqrt{5}, 3 + \sqrt{5}$ or $1 - \sqrt{5}, 3 - \sqrt{5}$ **21** If the number is $10x + y$, then you get $(10x + y)(10y + x) - 10(x^2 + y^2) = 101xy$, divisible by 101. **23** cL and $(1 - c)L$, where $c = \frac{1}{2}(\sqrt{3} - 1) \approx 0.366$ **25** The system $x + y = 30$, $xy = 240$ leads to the quadratic $x^2 - 30x + 240 = 0$, with discriminant $D = -60 < 0$, so no real solution.
27 \$17,500 **29** $D/(1 + \sqrt{S_2/S_1})$ from the first bulb **31** 2.58×10^5 km from the earth **33** $\sqrt{2} - 1$

Section 9 page 89

1 $4 < x < 9$ **3** $x \geq 0$ **5** $x < y \leq z$ **7** 

9 

11 

13 $5667 < 6093$, hence $5667 - 4128 < 6093 - 4128$. **15** $0 < \pi < 4$, hence $1/\pi > 1/4$ **17** $9 \cdot 11 \cdot 13 = 99 \cdot 13 < 100 \cdot 13 = 1300$
19 $\frac{1}{3}(8561 + 8774 + 8819) > \frac{1}{3}(8561 + 8561 + 8561) = 8561$
21 $\pi^2 < 10$, hence $\pi^6 = (\pi^2)^3 < 10^3$. [$\pi < 3.15$, so $\pi^2 < 9.9225$.]
23 $9 + 99 + 999 + 9999 < 10 + 100 + 1000 + 10000 = 11{,}110$
25 $\dfrac{1}{\sqrt{5}} + \dfrac{1}{\sqrt{6}} + \dfrac{1}{\sqrt{7}} + \dfrac{1}{\sqrt{8}} < \dfrac{1}{\sqrt{4}} + \dfrac{1}{\sqrt{4}} + \dfrac{1}{\sqrt{4}} + \dfrac{1}{\sqrt{4}} = 2$
$\dfrac{1}{\sqrt{5}} + \dfrac{1}{\sqrt{6}} + \dfrac{1}{\sqrt{7}} + \dfrac{1}{\sqrt{8}} > \dfrac{1}{\sqrt{9}} + \dfrac{1}{\sqrt{9}} + \dfrac{1}{\sqrt{9}} + \dfrac{1}{\sqrt{9}} = \dfrac{4}{3}$
27 $b^2 - a^2 = (b - a)(b + a) > 0$ since each factor is positive. **29** $a < 0$ and $b > 0$. Hence $1/a < 0$ and $1/b > 0$, so $1/a < 1/b$. **31** Not necessarily. For instance, $1 < 10$ and $-2 < -1$ but $(-2)(1) > (-1)(10)$.
33 $1/a - 1/b = (b - a)/ab > 0$ since both numerator and denominator are positive. Hence $1/b < 1/a$. **35** $(A + B) - (a + b) = (A - a) + (B - b) > 0$ since $A - a > 0$ and $B - b > 0$. Hence $a + b < A + B$. **37** $12 < ab < 13.65$
39 $60 < S < 65$ **41** 68^2: $(67)(69) = (68 - 1)(68 + 1) = 68^2 - 1 < 68^2$.
43 123^4: $(121)(122)(124)(125) = (123 - 2)(123 - 1)(123 + 1)(123 + 2) = (123^2 - 4)(123^2 - 1) < (123)^2(123)^2 = (123)^4$. **45** best, large economy; worst, family **47** best, mammoth; worst, super **49** decreases

Section 10 page 94

1 4 **3** 7 **5** 5 **7** 30 **9** $\frac{4}{9}$ **11** 36 **13** 0 **15** ± 3 **17** 3 5
19 $-\frac{1}{2} \leq x \leq \frac{1}{2}$ **21** $-4 < x < 4$ **23** $-2 < x < 4$ **25** $-3\frac{1}{2} < x < 4\frac{1}{2}$
27 $2 \leq x \leq 3$ **29** ± 3 $\pm\sqrt{5}$ **31** $-\frac{11}{3} < x < 3$ **33** $-\frac{2}{11} < x < \frac{2}{9}$, but $x \neq 0$
35 $1 \leq x \leq \sqrt{7}$ or $-\sqrt{7} \leq x \leq -1$ **37** 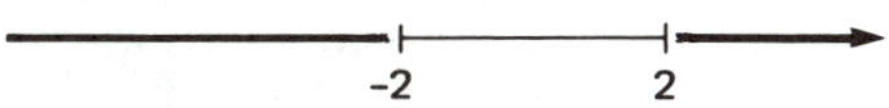

39 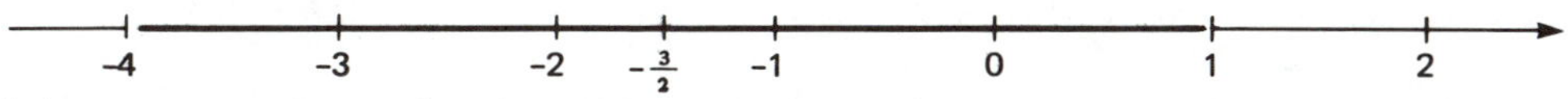

41 $|x| = 2$ **43** $|x - 1| < |x - 5|$ **45** $|x - 7| < 3$ **47** -1 2 **49** There is no point within 2 units of 1 and also within 3 units of 12. **51** If $a = b = c = 0$, then $|a| + |b| + |c| = 0$. Otherwise, at least one of the non-negative numbers $|a|$, $|b|$, $|c|$ is positive, hence $|a| + |b| + |c| > 0$. **53** $|7x - 7a| = 7|x - a| < 7 \cdot 10^{-6} < 10^{-5}$
55 First, $|x| \leq 5 + |x - 5| < 5 + \frac{1}{10} < 6$. Now $|xy - 35| = |x(y - 7) + 7(x - 5)| \leq |x(y - 7)| + |7(x - 5)| = |x| \cdot |y - 7| + 7 \cdot |x - 5| < 6 \cdot \frac{1}{10} + 7 \cdot \frac{1}{10} = 13 \cdot \frac{1}{10} = 1.3$ **57** First, $|x| \leq 3 + |x - 3| < 3 + 10^{-6} < 4$, so $x + 3 < 4 + 3 = 7$. Now $|x^2 - 9| = |(x + 3)(x - 3)| = |x + 3| \cdot |x - 3| < 7 \times 10^{-6} < 10^{-5}$.
59 By rule (9), $|a| - |b| \leq |a - b|$ and $|b| - |a| \leq |b - a| = |a - b|$. But $||a| - |b|| = |a| - |b|$ or $|b| - |a|$, so $||a| - |b|| \leq |a - b|$.
61 $(\operatorname{sgn} a)a = \begin{Bmatrix} a \text{ if } a \geq 0 \\ -a \text{ if } a < 0 \end{Bmatrix} = |a|$

Section 11 page 99

1 $x < \frac{2}{3}$ **3** $x < -\frac{3}{2}$ **5** $-\frac{13}{2} < x < \frac{7}{2}$ **7** $x \geq -\frac{1}{4}$ **9** $-5 < x < 3$
11 $x > 7$ or $x < -1$ **13** all $x \neq 0$ **15** all x **17** $-\frac{3}{2} < x < \frac{5}{2}$
19 $-2 < x < 2$ **21** $x > -\frac{5}{2}$ **23** $-2 < x < -1$ **25** $x < -\frac{3}{2}$ or $x > -1$
27 all x **29** $x < 0$ or $x > \frac{3}{8}$ **31** $x > 1$ or $x < \frac{1}{2}$ **33** $x \geq 2$ **35** $1 < x < 4$ or $x < 0$ **37** $5 < t < 15$ **39** $x(1 - x) = -(x^2 - x + \frac{1}{4}) + \frac{1}{4} = -(x - \frac{1}{2})^2 + \frac{1}{4} \leq \frac{1}{4}$ **41** $c \geq 20$ or $c \leq -20$ **43** $\dfrac{x^2 + y^2}{2} - \left(\dfrac{x + y}{2}\right)^2 = \dfrac{x^2 + y^2}{2} - \dfrac{x^2 + 2xy + y^2}{4} = \dfrac{x^2 - 2xy + y^2}{4} = \dfrac{(x - y)^2}{4} \geq 0$
45 $\frac{1}{4}(x + y + z + w) = \frac{1}{2}[\frac{1}{2}(x + y) + \frac{1}{2}(z + w)] \geq \frac{1}{2}(\sqrt{xy} + \sqrt{zw}) \geq \sqrt{\sqrt{xy}\sqrt{zw}} = \sqrt[4]{xyzw}$ **47** $x^4 + x^2 - 2x + 3 = x^4 + (x - 1)^2 + 2 \geq 2 > 0$
49 From $(x - a)^2 \geq 0$ follows $x^2 + a^2 \geq 2ax$. Now divide by x.

Review Exercises page 99

1 $\pm\sqrt{3}$ **3** $(N + 1) + (N + 2) + (N + 3) + (N + 4) + (N + 5) = (N + 3)^2$
5 4 **7** 44,444 bushels **9** $\frac{40}{3}$ hours **11** -5 6 **13** $\frac{1}{2}(5 \pm \sqrt{5})$
15 6.02701 9954 0.10673 8008 **17** 4 9 **19** $\frac{8}{3}$ 6 **21** $\frac{5}{8}(1 + \sqrt{17}) \approx 3.2$ sec

23

25
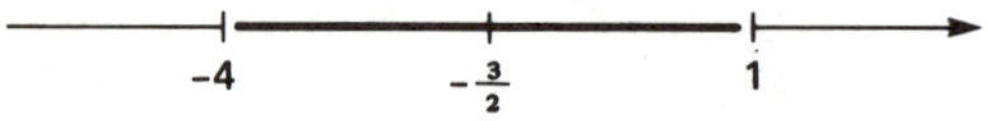

27 $x > -2$ **29** $x < 0$ $1 < x < 2$ $3 < x < 4$

CHAPTER 3

Section 2 page 105

1
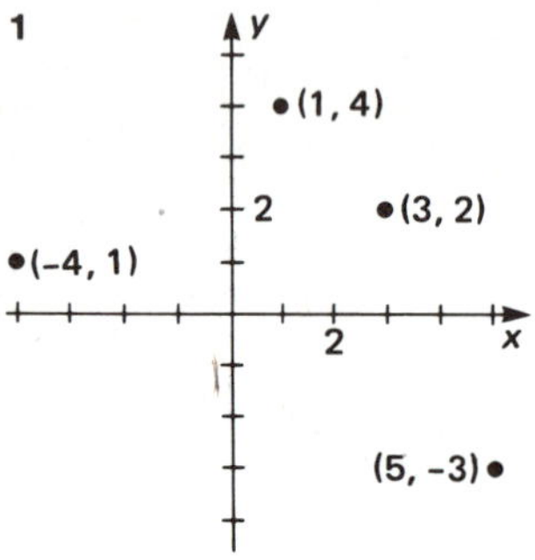

3
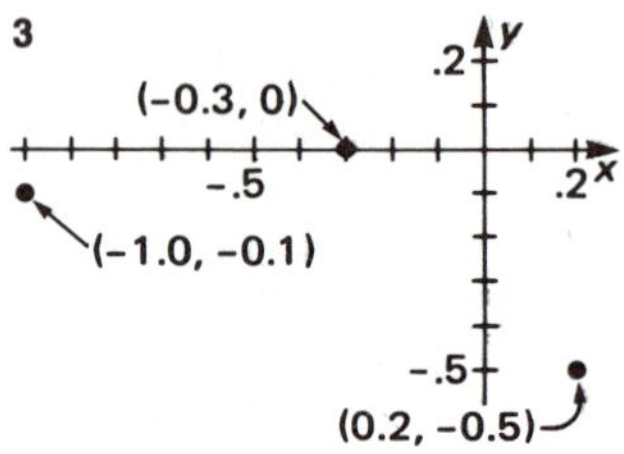

5
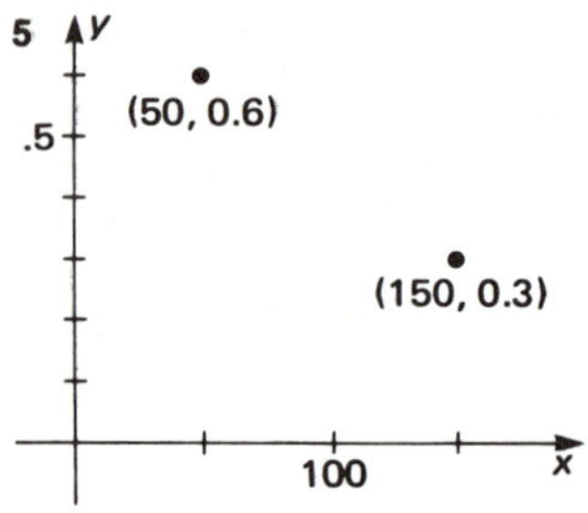

7
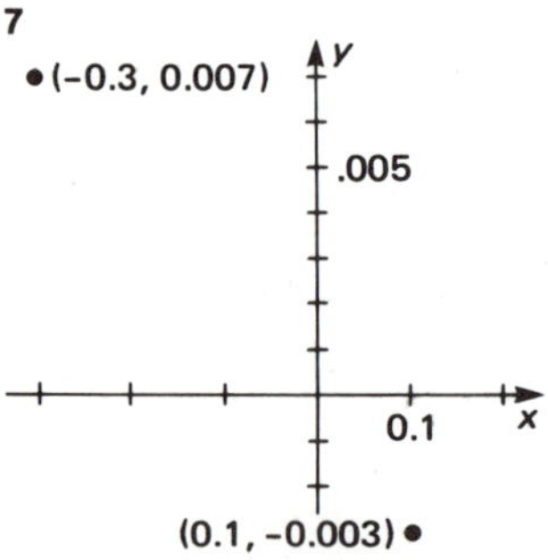

9
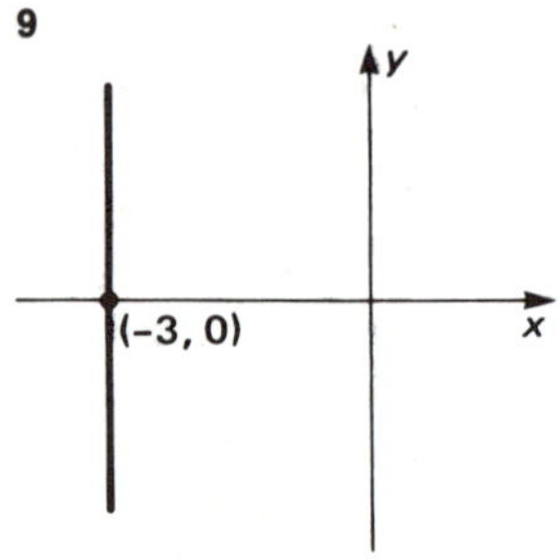

11
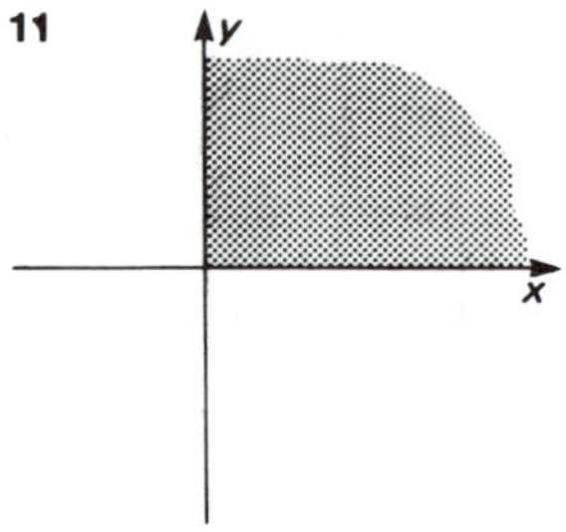

13
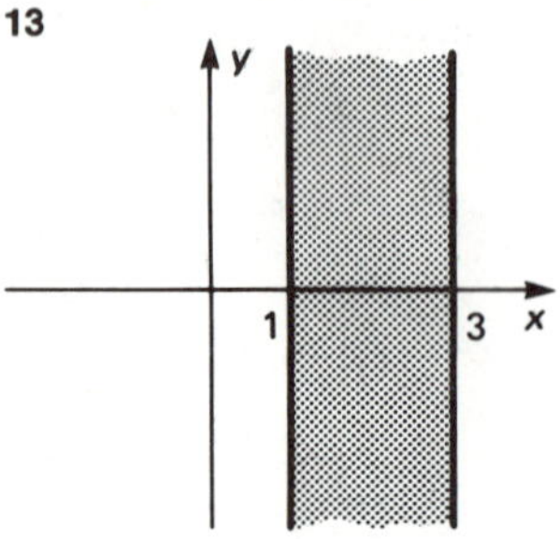

15
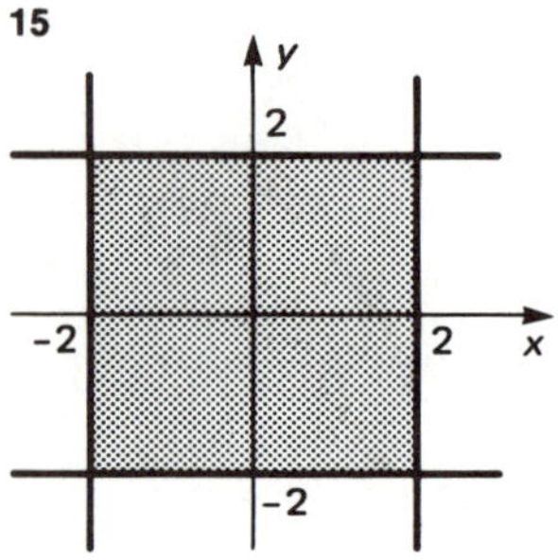

17
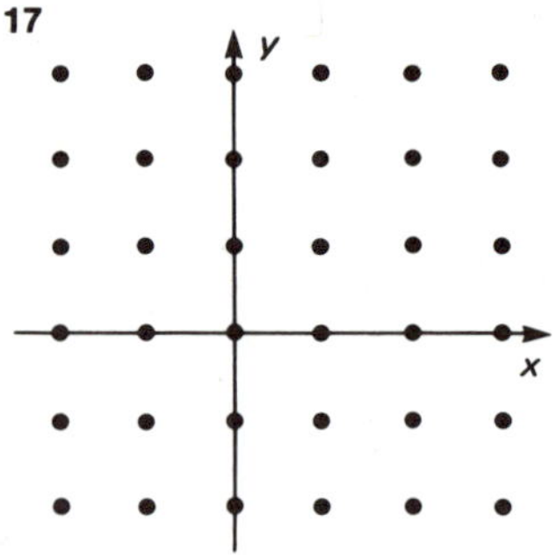

19
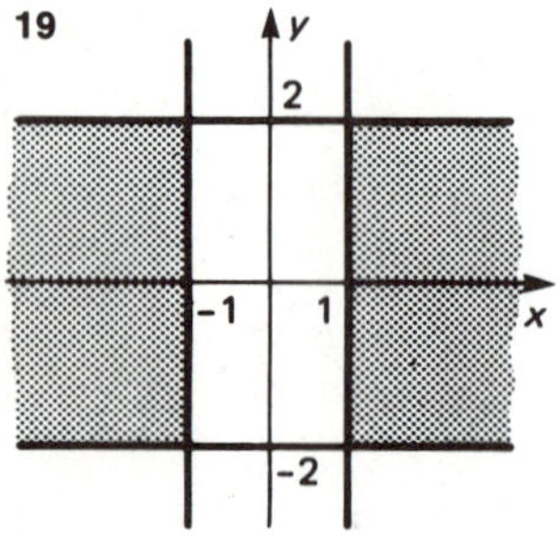

21
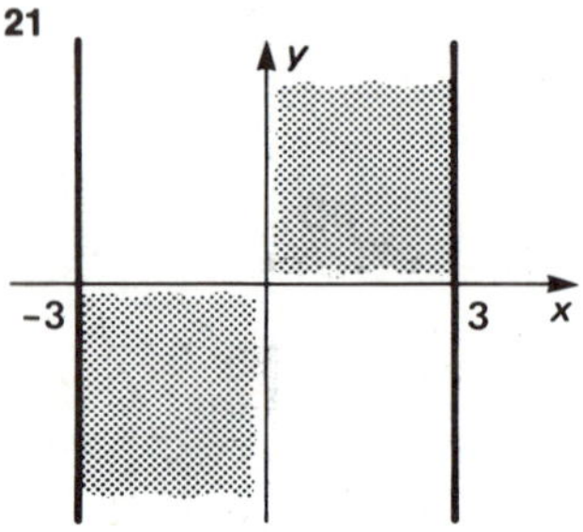

23
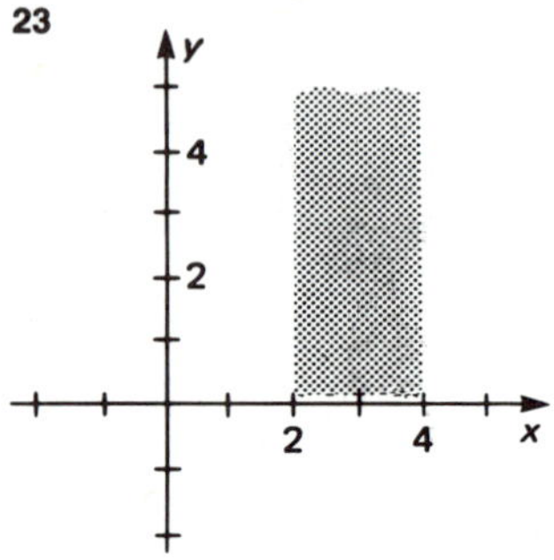

25

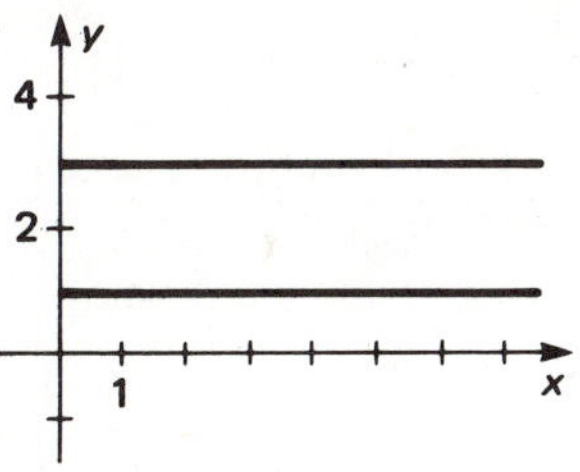

27 (1, 1) (1, −1) (−1, 1) (−1, −1) **29** (0, 0), (0, 9), and (12, 0) **31** 10
33 $2\sqrt{17}$ **35** $\sqrt{53}$ **37** Call the points A, B, C respectively. Then $\overline{AB}^2 = 130 = 32 + 98 = \overline{AC}^2 + \overline{BC}^2$. **39** Call the points A, B, C, D. Then $\overline{AB} = \overline{BC} = \overline{CD} = \overline{DA} = 5$. Also $AB \perp DA$ since $\overline{BD}^2 = 50 = 5^2 + 5^2 = \overline{AB}^2 + \overline{DA}^2$.
41 The distance between each pair of successive points is 2, and each of the points is 2 units from (0, 0). **43** No. The distance between their centers is 25, greater than the sum of their radii, 24. **45** $2x + 3y = 13$ **47** (−1, 2) only **49** $(\pm\sqrt{3}, 1)$ $(1, \pm\sqrt{3})$
51 The vertical distance from (6, 9) to $y = -1$ is $9 + 1 = 10$ = distance from (6, 9) to (0, 1).

Section 3 page 109

1 (a) 2 (b) −4 (c) −0.55 (d) −37 **3** (a) $\frac{1}{5}$ (b) 1000 (c) x (d) $\frac{2}{\sqrt{x}}$
5 (a) and (b) **7** (a) and (b) **9** $f(x_1x_2) = (x_1x_2)^n = x_1{}^n x_2{}^n = f(x_1)f(x_2)$. If $f(x) = x + 1$ for example, then $f(x_1x_2) = x_1x_2 + 1 \neq (x_1 + 1)(x_2 + 1) = f(x_1)f(x_2)$. **11** 0 for all x **13** $f(2x) = f(x)$ **15** $xf(x) = \begin{Bmatrix} x \text{ if } x \geq 0 \\ -x \text{ if } x < 0 \end{Bmatrix} = |x|$
17 all real x **19** all real x **21** $x \neq \frac{3}{2}$ **23** $x \neq \frac{5}{3}$ **25** $x \geq 6$ **27** $|x| \leq \frac{2}{3}$
29 $x \geq \frac{3}{2}$ **31** $|x| \leq \frac{1}{2}$ **33** $x \geq 4$ or $x \leq 1$ **35** $x \leq -3$ or $x \geq 4$
37 $f(x) = 3600x$ **39** $f(x) = (\sqrt{2})x$ **41** $f(x) = 1.06x + 1.09(10{,}000 - x)$

Section 4 page 115

1

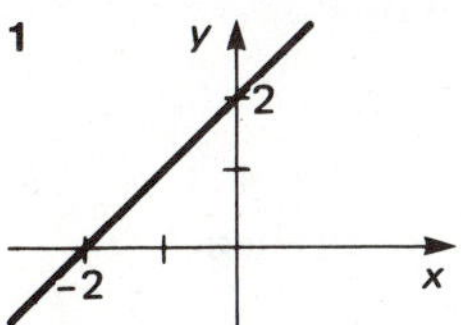

3

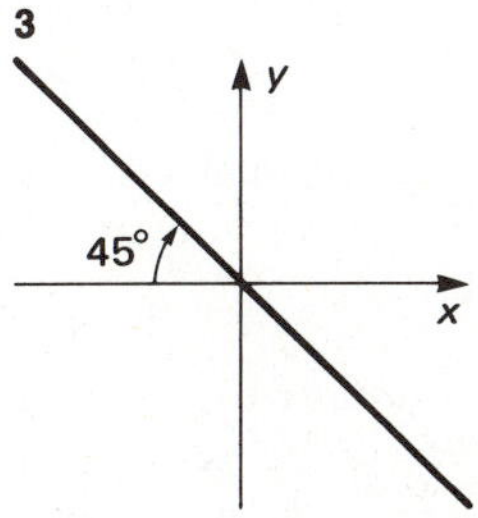

5

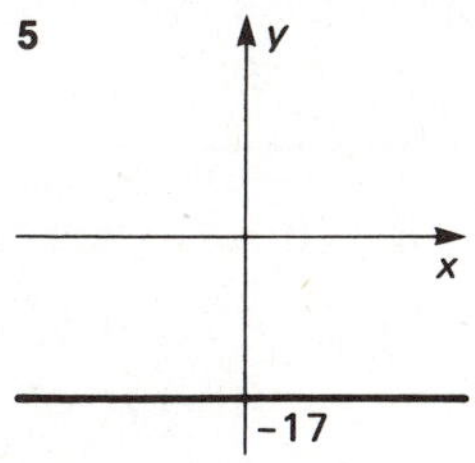

7

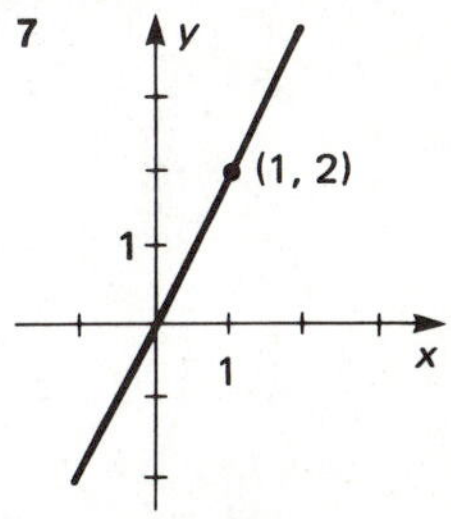

9

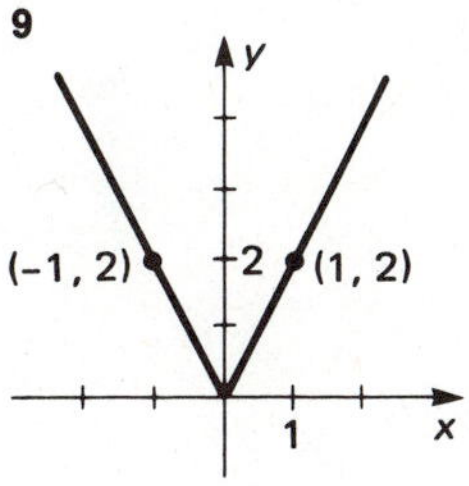

11

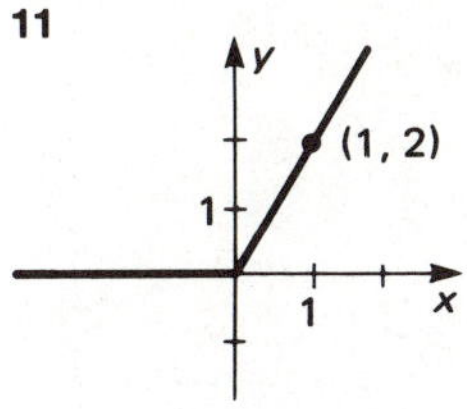

13

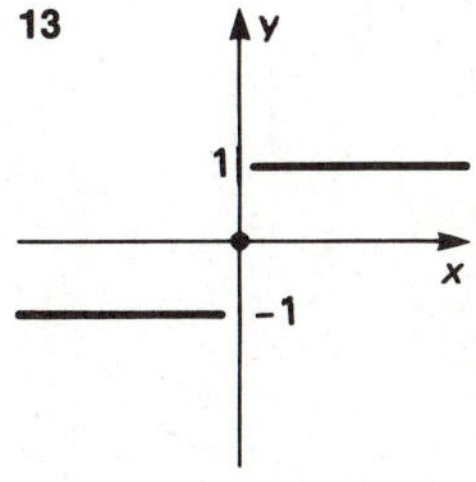

15

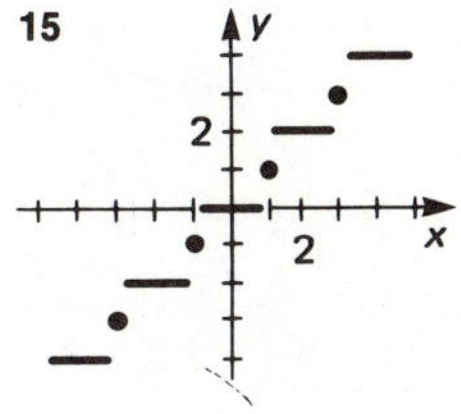

17

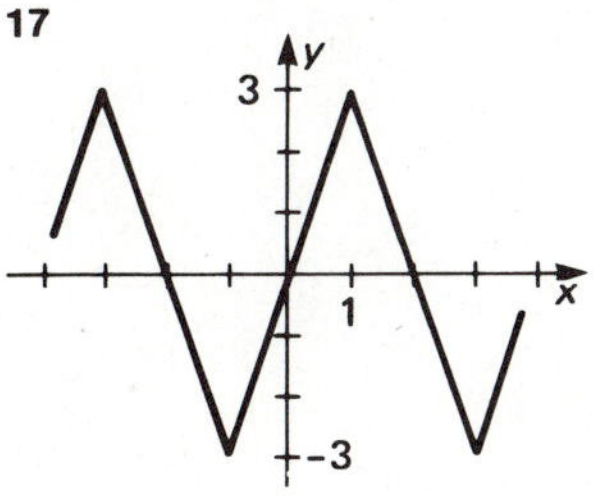

19

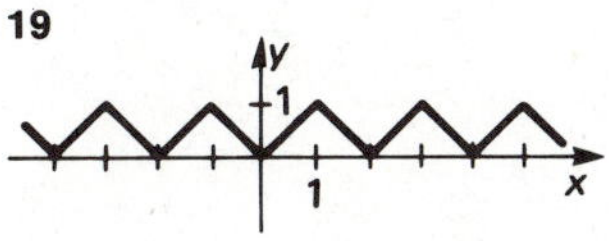

21

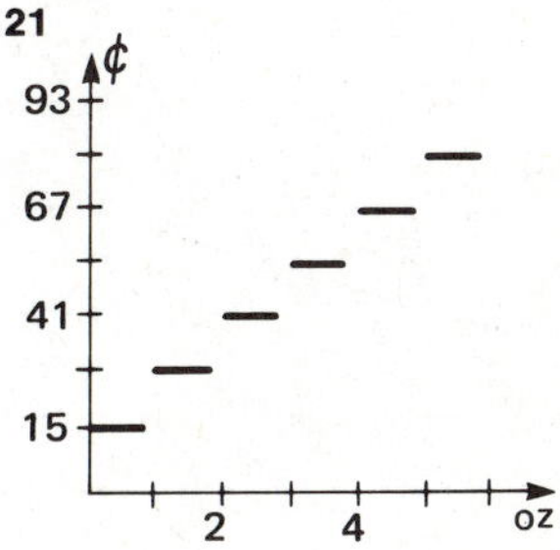

23

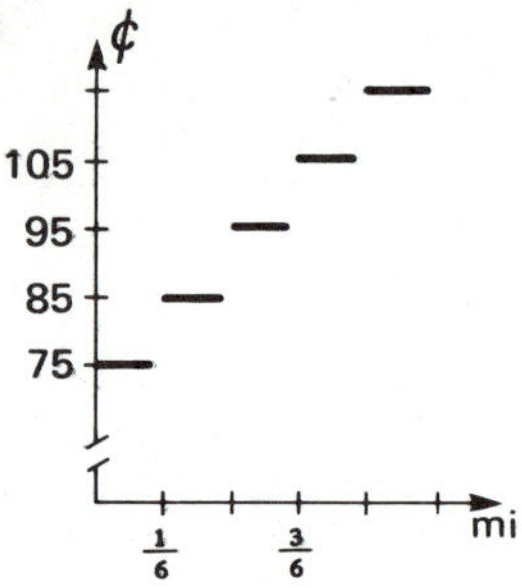

25

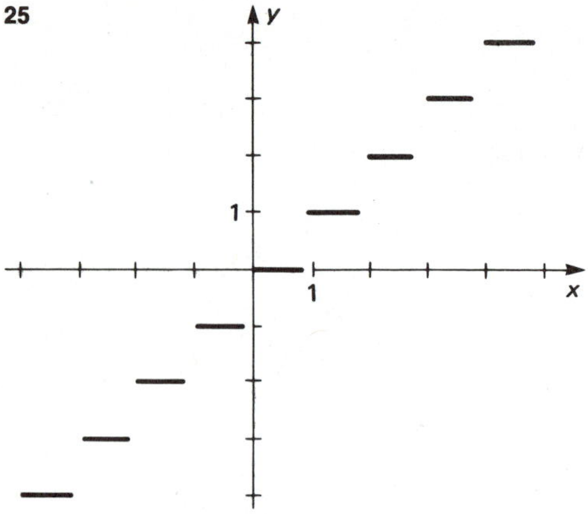

27

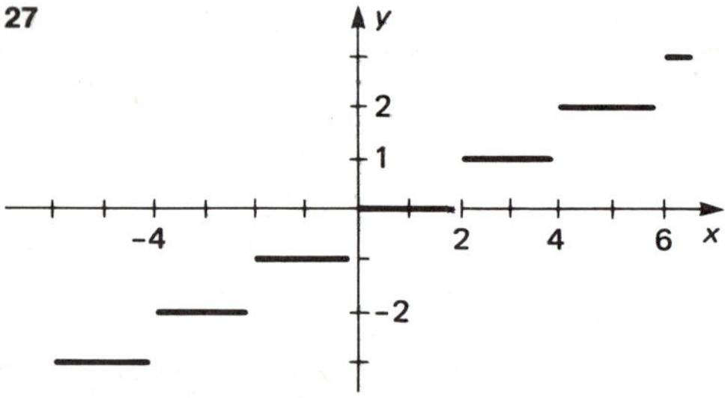

29

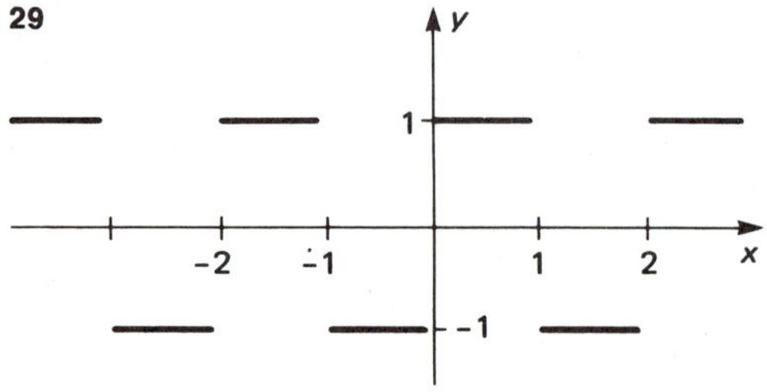

Section 5 page 118

1 $4x - 3 \quad 3x^2 - 11x - 4 \quad$ all x **3** $x^2 - 2x + 1 \quad -2x^3 + x^2 \quad$ all x
5 $x + \dfrac{1}{x} \quad 1 \quad x \neq 0$ **7** Both are undefined since the domains of f and g have no points in common. **9** $3x - 5 \quad 3x - 1 \quad$ all x **11** $2(x + 1)^2 \quad -2x^2 - 1 \quad$ all x
13 $-4x \quad -4x \quad$ all x **15** $9 \quad 3 \quad$ all x **17** $\sqrt{x^2 - 9} \quad x \le -3$ or $x \ge 3$; $x - 9 \quad x \ge 0$ **19** $\dfrac{1}{x^2 + 2}$ all x; $\dfrac{1}{x^2} + 2 \quad x \neq 0$ **21** $x \quad x \quad$ all x **23** $g(x)$
25 x **27** No; $\sqrt{-3 - 2x^2}$ is undefined for all real x. **29** $3 \le x \le 5$
31 $0 \le x \le 1$ **33** If $f(x_1) < f(x_2)$ and $g(x_1) < g(x_2)$ for $x_1 < x_2$, then

$$[f + g](x_1) = f(x_1) + g(x_1) < f(x_2) + g(x_2) = [f + g](x_2).$$

35 $g(x) = x - 2$ **37** Let $f(x) = \sqrt{x}$ and $g(x) = x^2$. The first sequence computes $[f \circ g](x) = \sqrt{x^2} = |x|$ for all x. The second computes $[g \circ f](x) = (\sqrt{x})^2 = x$, but only for $x \ge 0$. (They agree for $x \ge 0$.) **39** $y = \frac{80}{21}x$ **41** $y = 4x$

Section 6 page 126

1

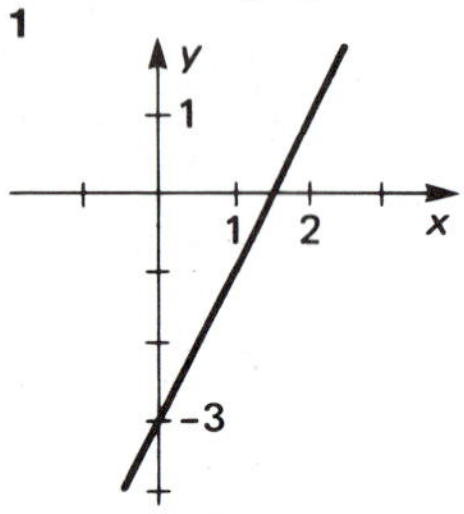

3

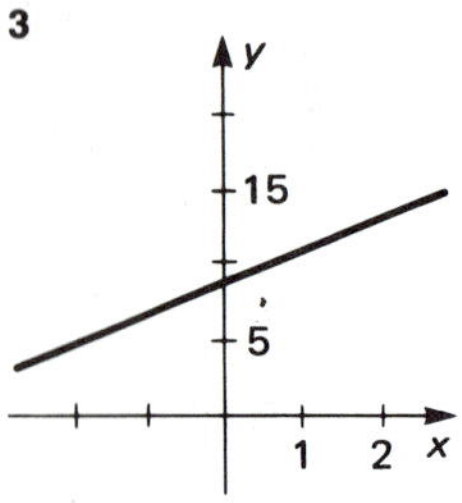

5

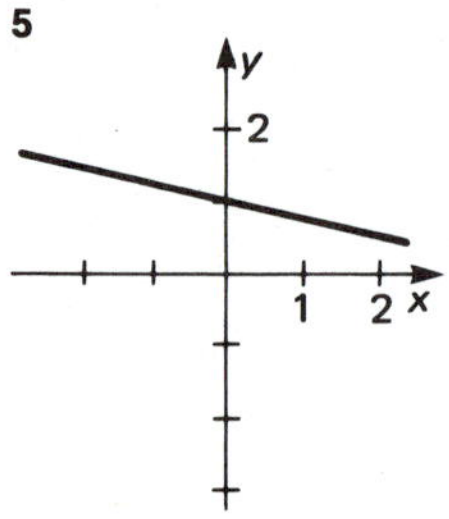

7

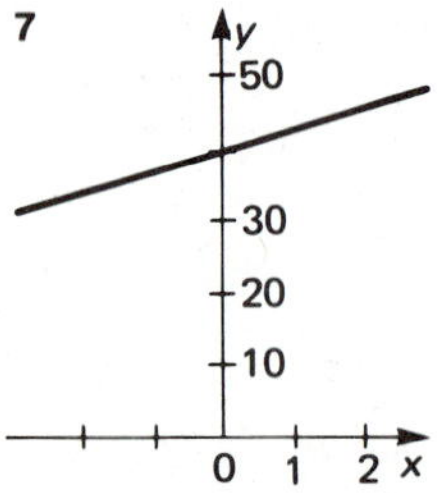

9

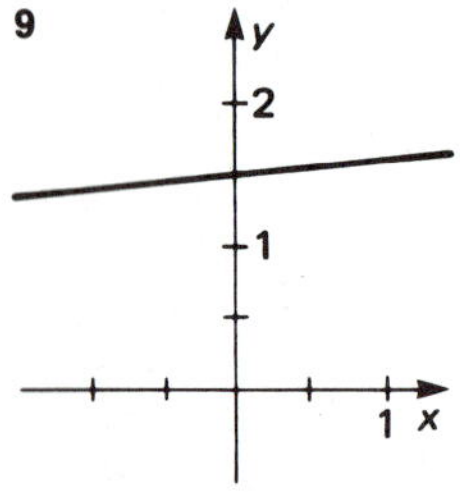

11 $\frac{4}{3}$ **13** 0 **15** $\frac{10}{3}$ **17** yes **19** yes **21** Two sides are vertical. The other two have equal slopes: $\dfrac{9-5}{3-0} = \dfrac{4}{3}$ and $\dfrac{4-0}{3-0} = \dfrac{4}{3}$. **23** yes **25** yes

27 $y = x - 2 \quad -2$ **29** $y = 3 \quad 3$ **31** $y = \frac{1}{2}x - 3 \quad -3$ **33** $y = 2x$
35 $y = \frac{4}{3}x + \frac{4}{3}$ **37** $y = x + \frac{1}{2}$ **39** $x = 0$ **41** $3 \quad -7$ **43** $-1 \quad 7$

45 2 3; as given **47** $\frac{1}{2}$ $\frac{1}{3}$; $\dfrac{x}{\frac{1}{2}} + \dfrac{y}{\frac{1}{3}} = 1$

49

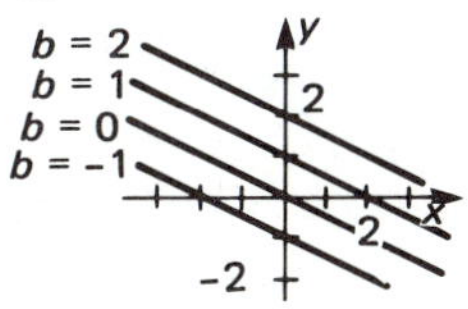

51

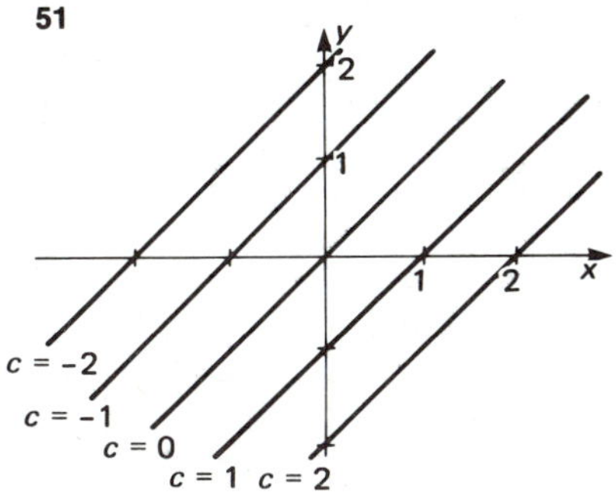

53

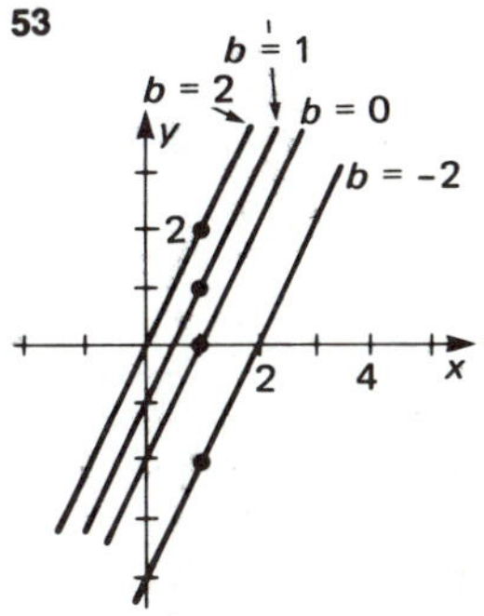

55 $y = 2x - 2$ **57** $y = -\frac{5}{9}x + \frac{59}{9}$ **59** $y = x \pm 1 \quad y = -x \pm 1$

61 Call the points A, B, C. Then $\overline{AB} \perp \overline{AC}$ because the slopes, respectively, are -1 and 1, hence have product -1. **63** Their slopes are $-\dfrac{a}{b}$ and $\dfrac{b}{a}$, with product -1.

Section 7 page 133

1

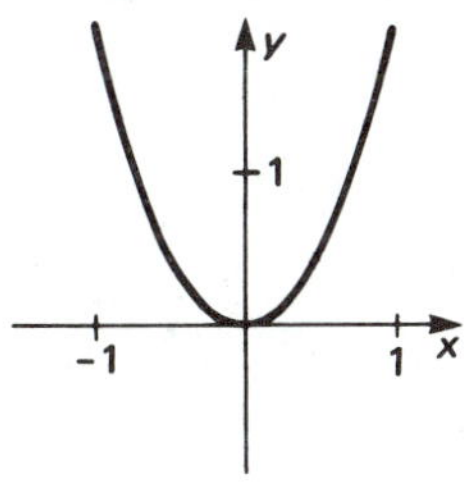

3

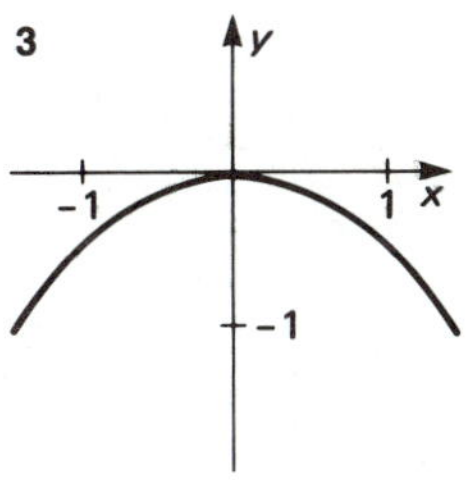

5

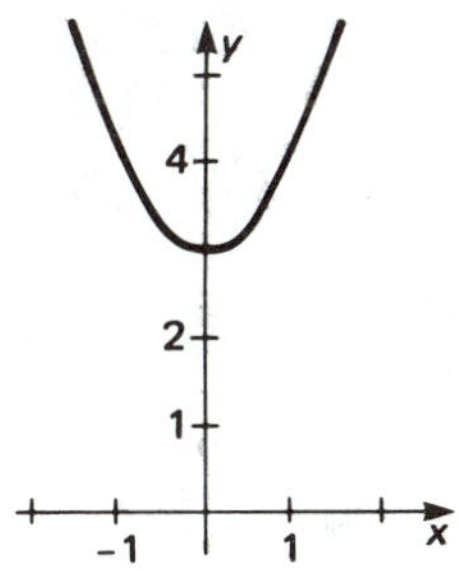

7

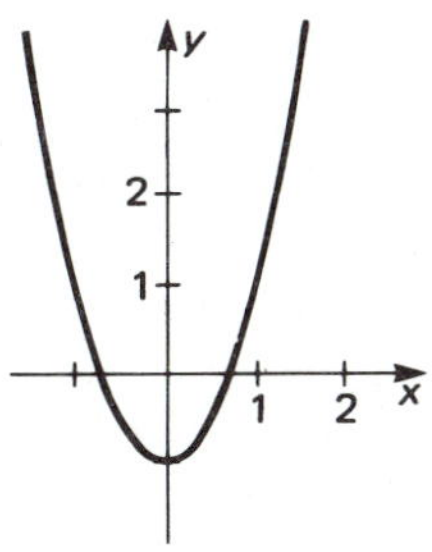

9

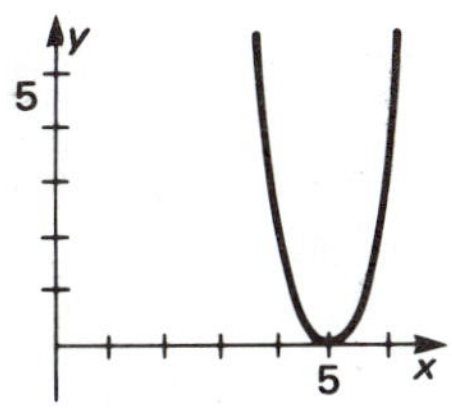

11

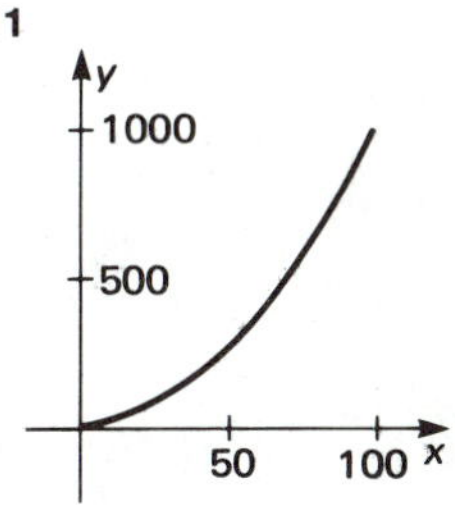

13 $y = (x-2)^2 - 3$

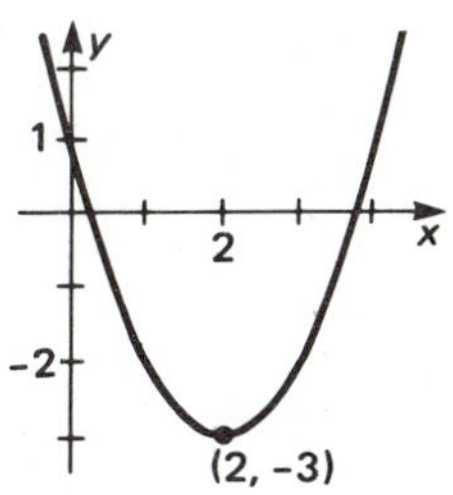

15 $y = (x + \frac{1}{2})^2 + \frac{3}{4}$

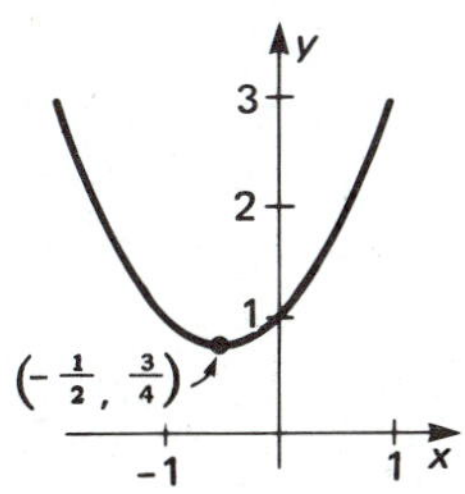

17 $y = -(x+1)^2 + 1$

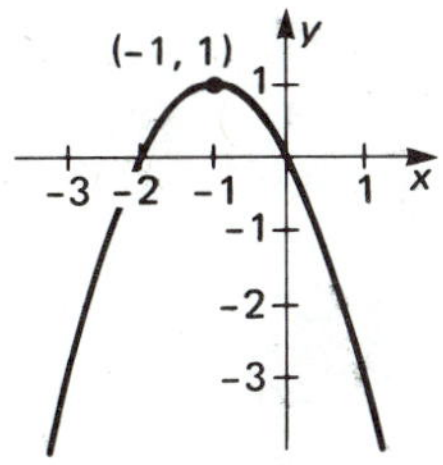

19 $y = -(x+2)^2 + 1$

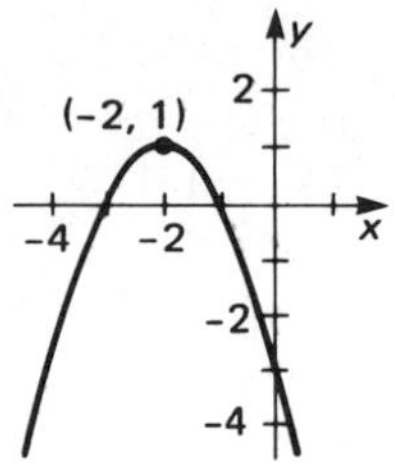

21 $y = 2(x - \frac{3}{2})^2 - \frac{7}{2}$

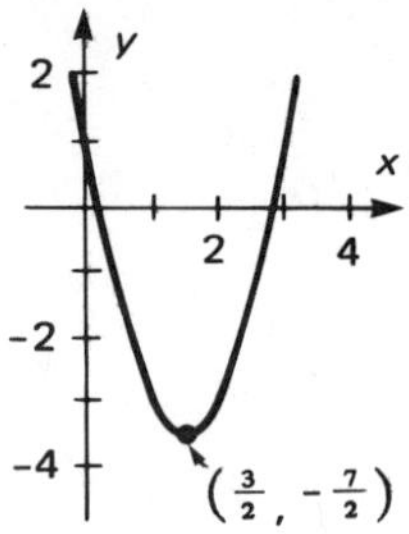

23 $y = 3(x+2)^2 - 20$

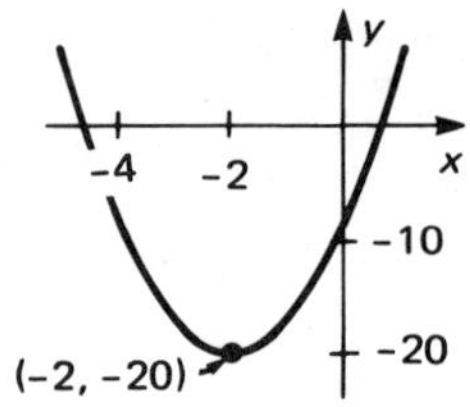

25 $y = -2(x-2)^2 - 2$

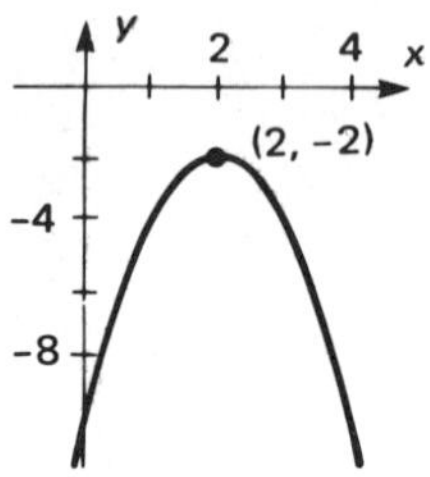

27 $y = -4(x - \frac{1}{8})^2 + \frac{1}{16}$

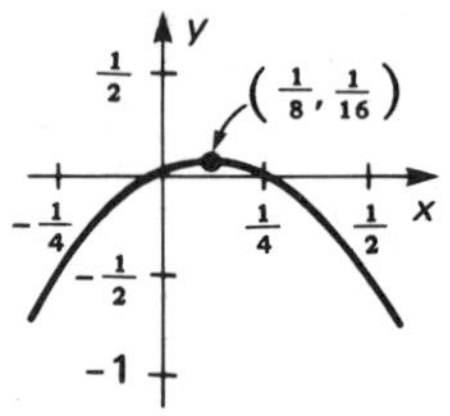

29 $y = 2(x - \frac{3}{4})^2 - \frac{9}{8}$

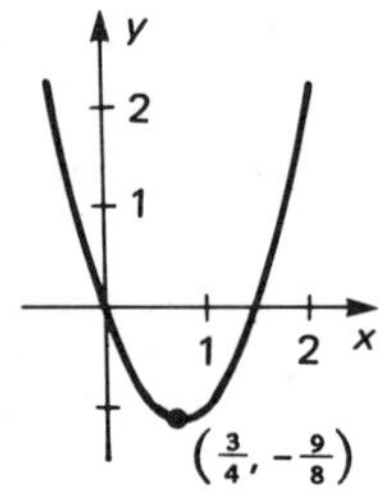

31 $y = (x + \frac{1}{2})^2 - \frac{17}{4}$

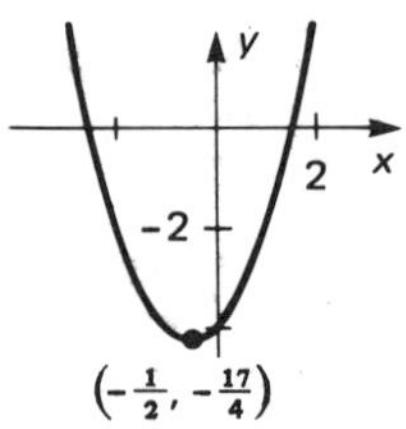

33 $y = -(x - \frac{1}{2})^2 - \frac{7}{4}$

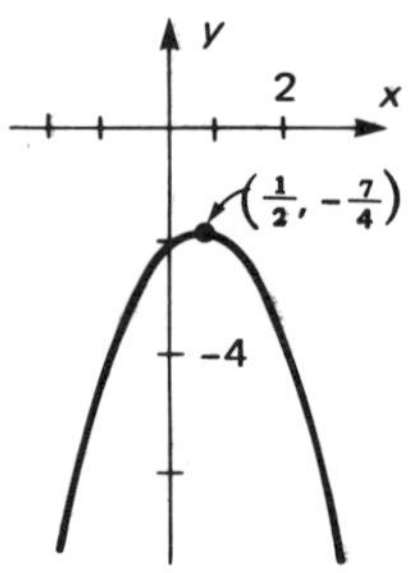

35 $y = 2(x+3)^2 - 8$

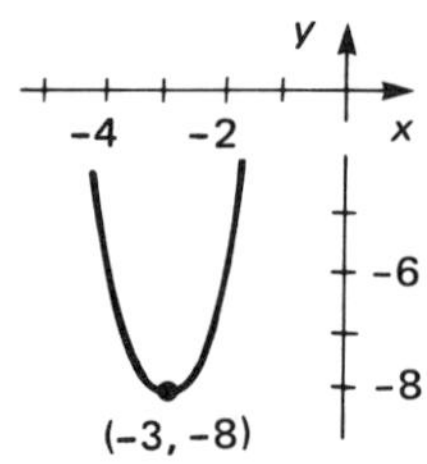

37 $y = 0$ when $x = 0$, so $(0, 0)$ is on the graph.

39 $b = 0$

41 $y = |x^2 - 4|$

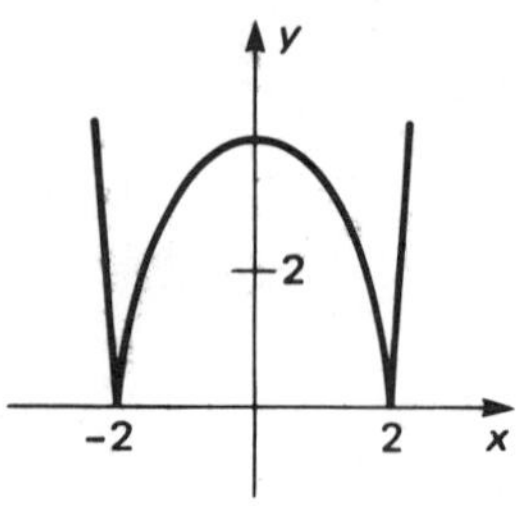

43 $x(1-x) = -x^2 + x = -(x - \frac{1}{2})^2 + \frac{1}{4} \le \frac{1}{4}$

45
$$\begin{aligned} A &= \tfrac{1}{2}x\sqrt{16 - x^2}, \\ A^2 &= \tfrac{1}{4}x^2(16 - x^2) = \tfrac{1}{4}(-x^4 + 16x^2) \\ &= -\tfrac{1}{4}(x^2 - 8)^2 + 16 \le 16 \\ A &\le 4 \end{aligned}$$

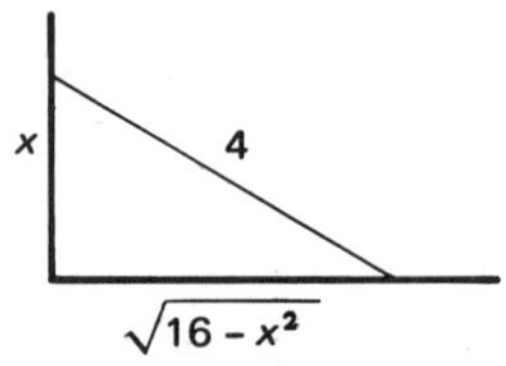

47 800 ft/sec **49** $x^2 + 6x + 10 = (x+3)^2 + 1 \ge 1 > 0$

51 $x^2 + 2x + y^2 - 4y + \frac{11}{2} = (x+1)^2 + (y-2)^2 + \frac{1}{2} \ge \frac{1}{2} > 0$

53 $ax + \dfrac{b}{x} = ax - 2\sqrt{ax}\sqrt{\dfrac{b}{x}} + \dfrac{b}{x} + 2\sqrt{ab} = \left(\sqrt{ax} - \sqrt{\dfrac{b}{x}}\right)^2 + 2\sqrt{ab} \le 2\sqrt{ab}$

Section 8 page 140

1

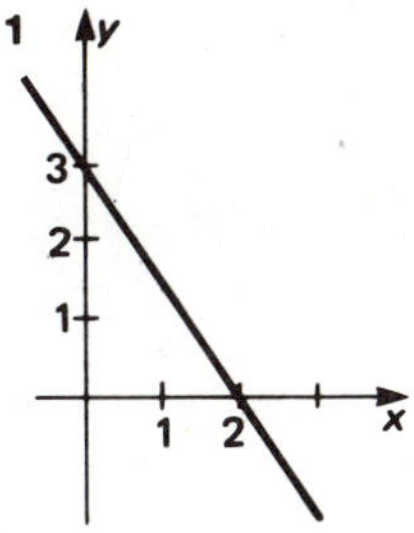

3

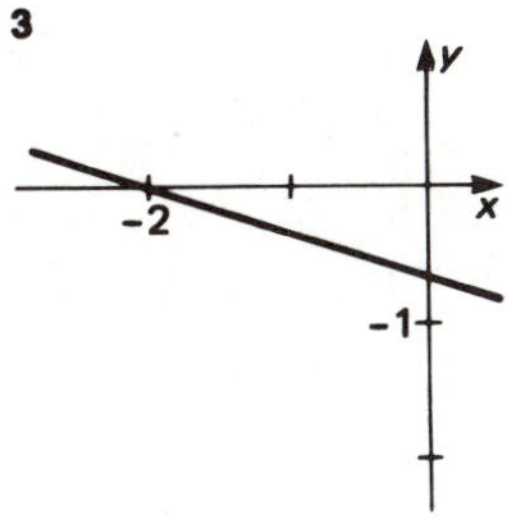

5

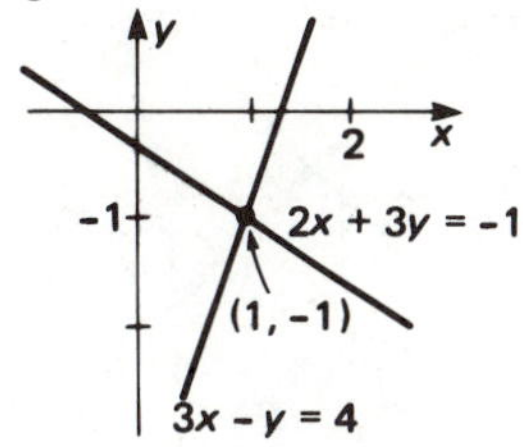

7

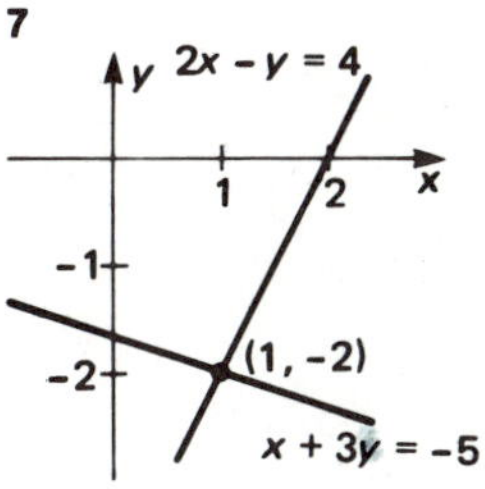

9

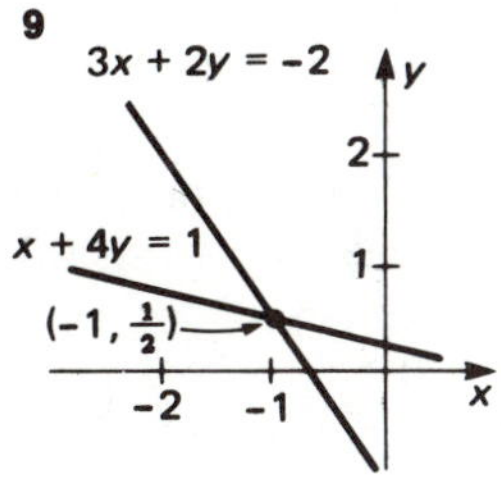

11 $(x - 2)^2 + (y - 2)^2 = 8$

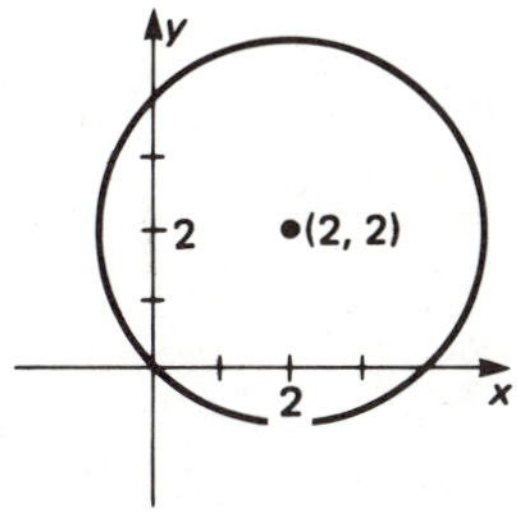

13 $(x + 1)^2 + (y + 3)^2 = 4$

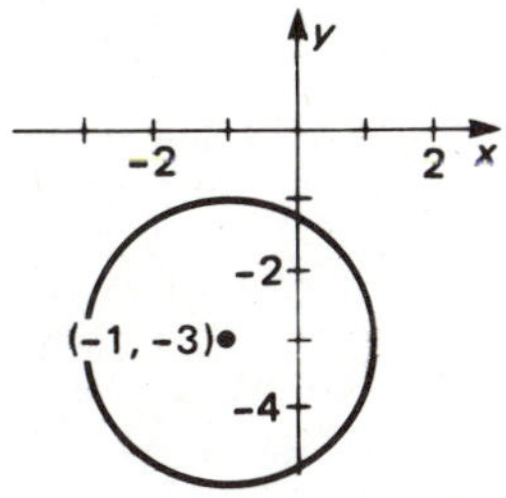

15 $(x - \frac{1}{2})^2 + (y + 1)^2 = \frac{5}{4}$

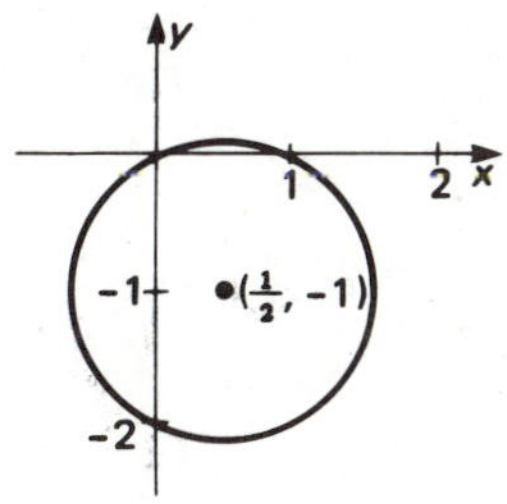

17 $(x - \frac{3}{4})^2 + (y - \frac{5}{4})^2 = 1$

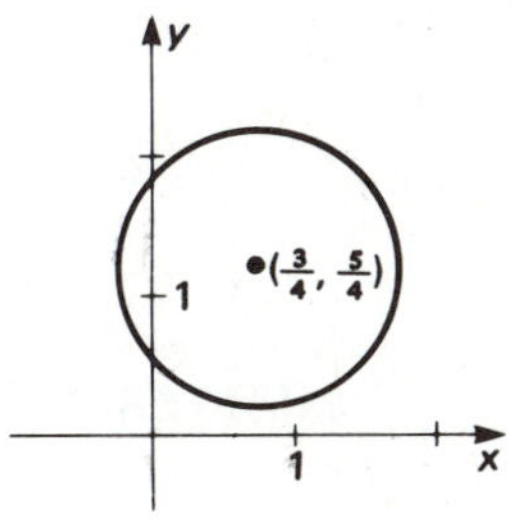

19 Rewrite as $(x - 4)^2 + (y - 3)^2 = 9$, a circle of center $(4, 3)$ and radius 3. The radius equals the distance from the center to the x-axis, so the x-axis is a tangent.

21

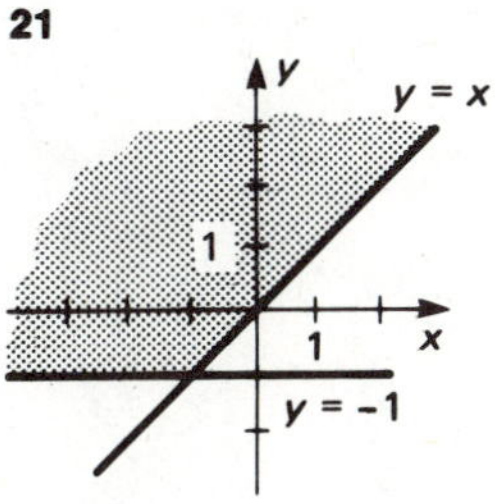

23

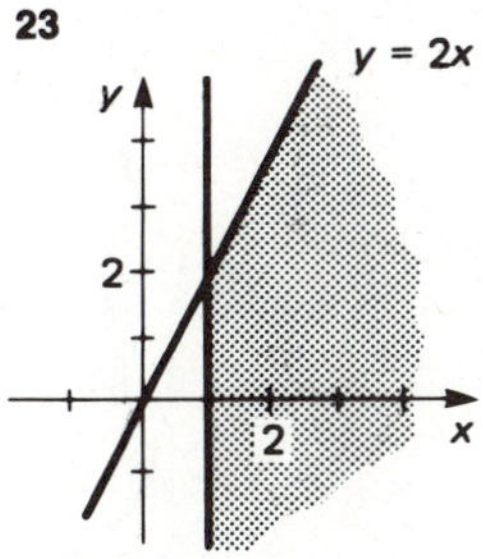

25

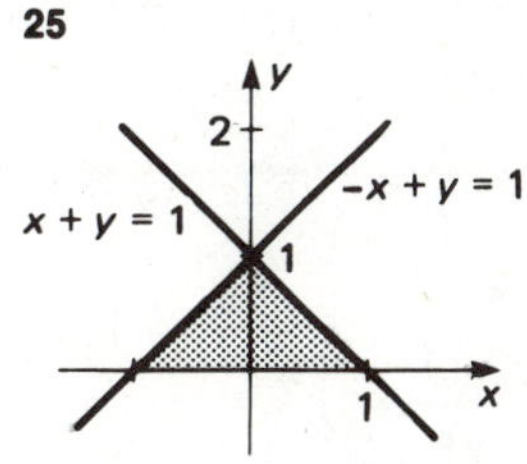

27

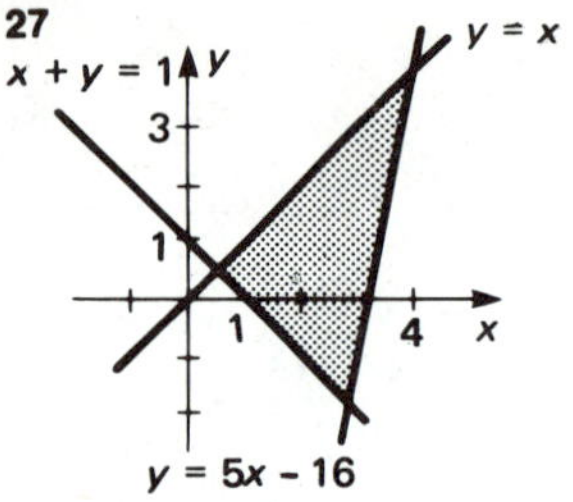

29

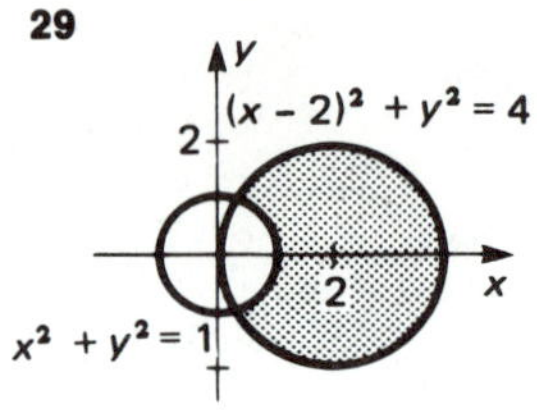

31

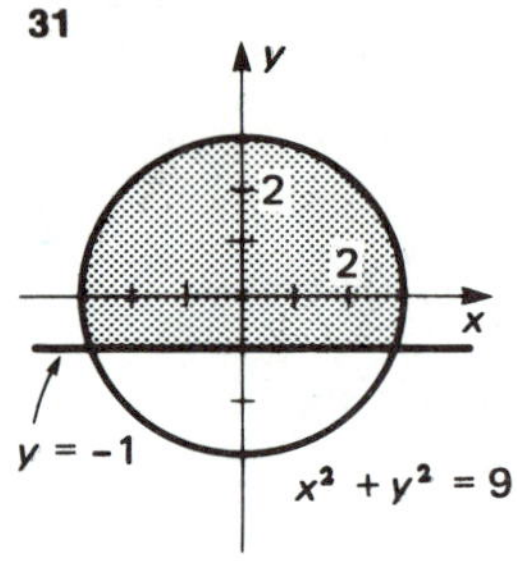

Review Exercises page 141

1

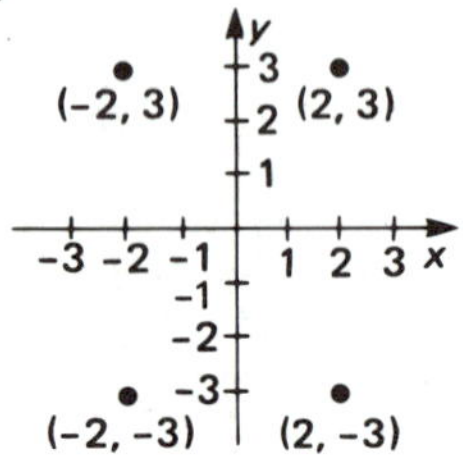

Two sides, $y = \pm 3$, are horizontal; two sides, $x = \pm 2$, are vertical; hence a rectangle.

3 (a) $f(0) = 3 \cdot 0 + 1 = 1 \quad f(-2) = 3 \cdot (-2) + 1 = -5$

$f[f(x)] = 3(3x + 1) + 1 = 9x + 4$

(b) $f(a + b) = 3(a + b) + 1 = 3a + 3b + 1 = (3a + 1) + (3b + 1) - 1 = f(a) + f(b) - 1$

5

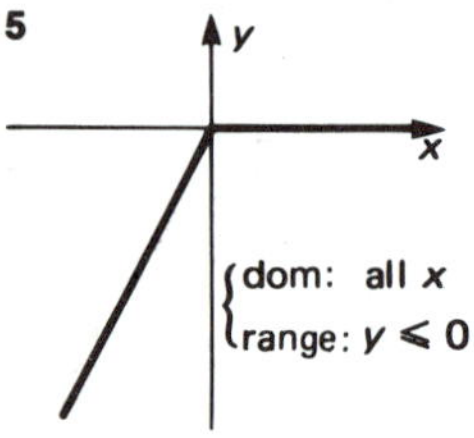

7 $h(x) = 6x^2 + 2$

9 No. The line through $(0, 0)$ and $(2, 5)$ has slope $\frac{5}{2}$; the line through $(0, 0)$ and $(3, 8)$ has slope $\frac{8}{3}$, different from $\frac{5}{2}$.

11 $y = -x + 6$

13 $y = 2(x - 3)^2 - 4$

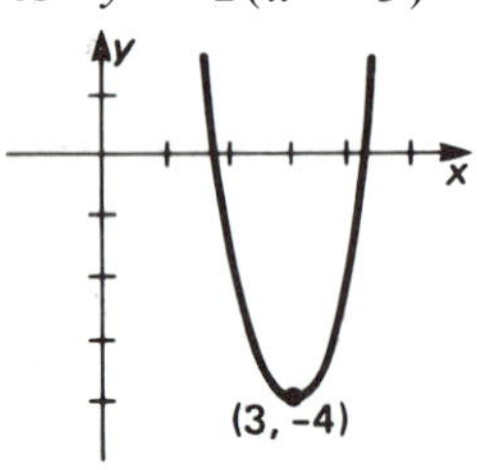

15 $-1 < b < 1$

17 $(x - 3)^2 + (y + 3)^2 = 7$

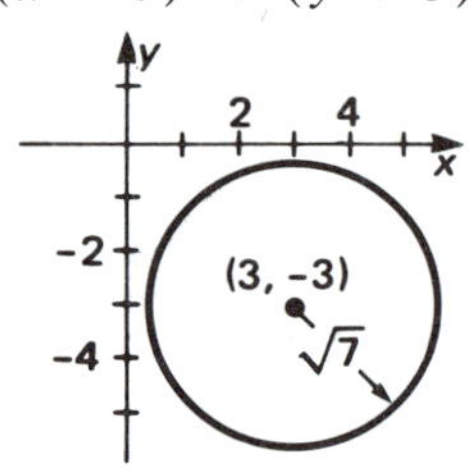

CHAPTER 4

Section 2 page 145

1

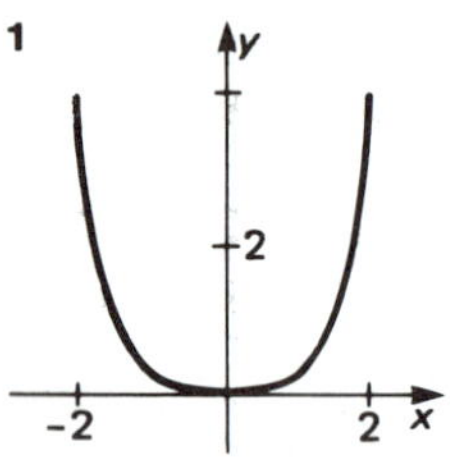

3

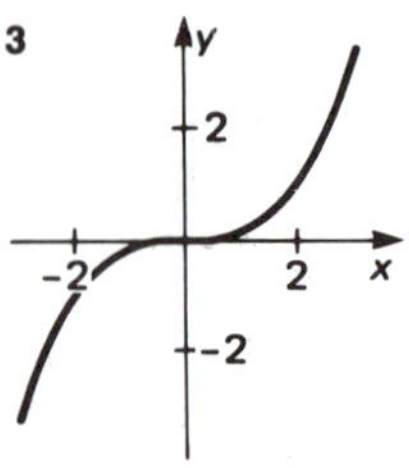

5

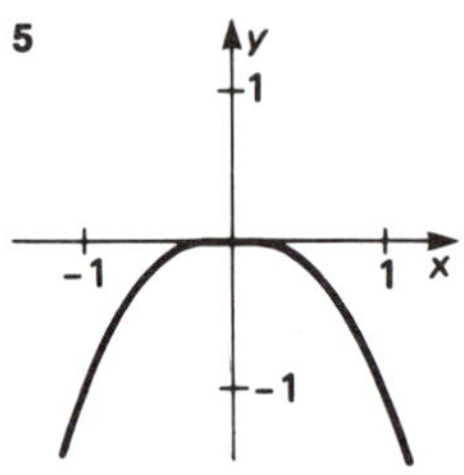

7

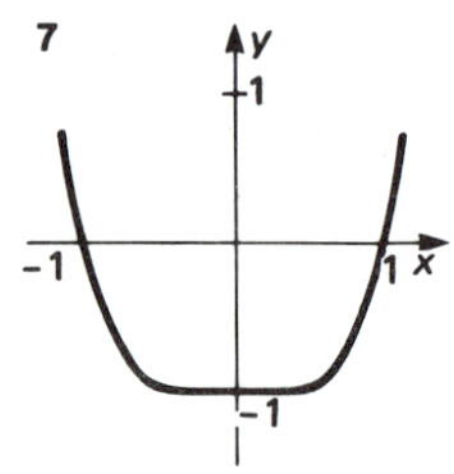

9

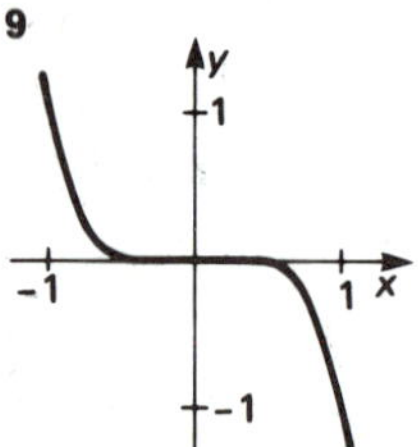

11

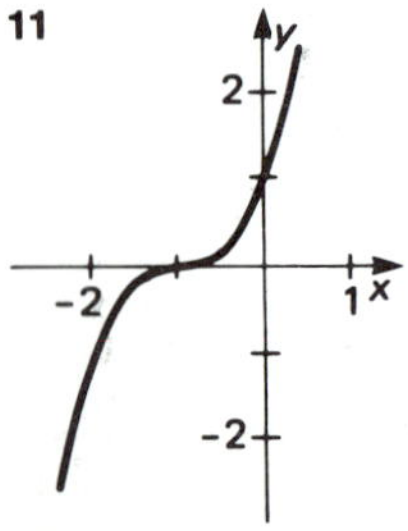

13

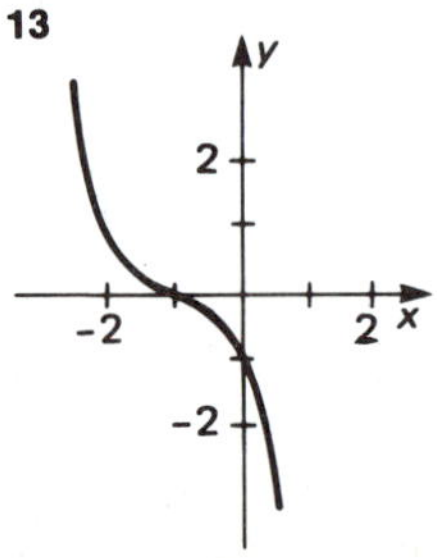

15

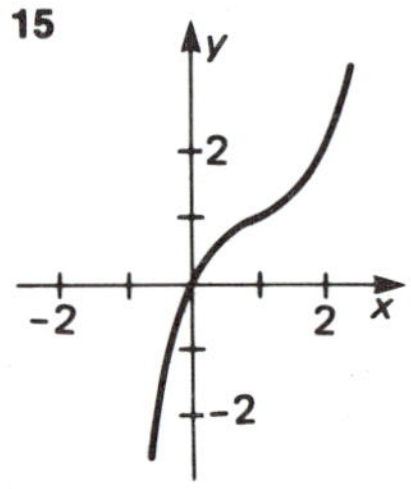

17

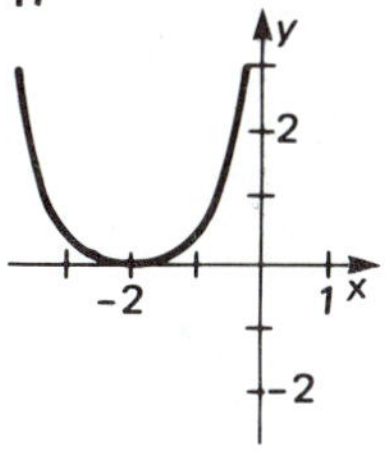

19

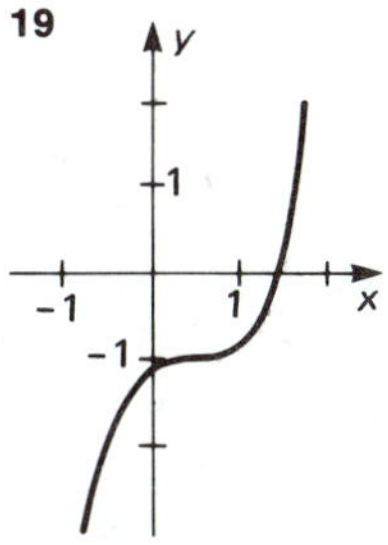

21

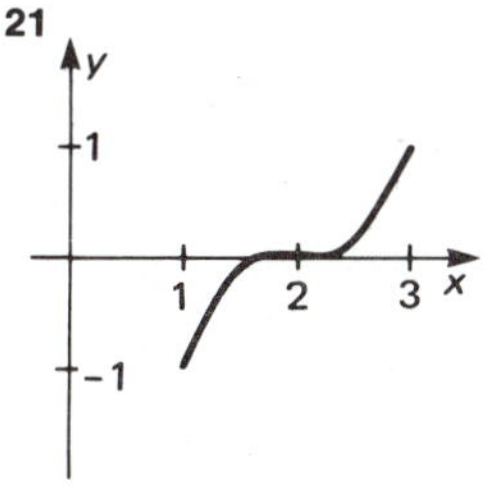

23

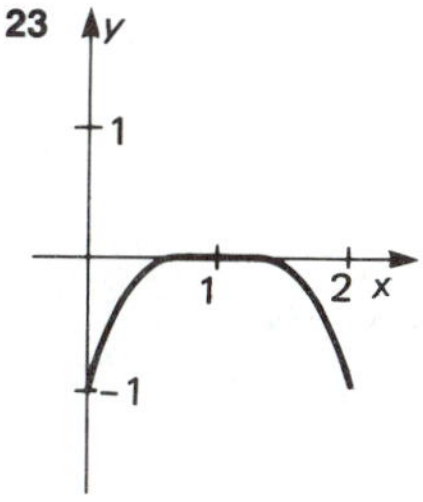

Section 3 page 151

1

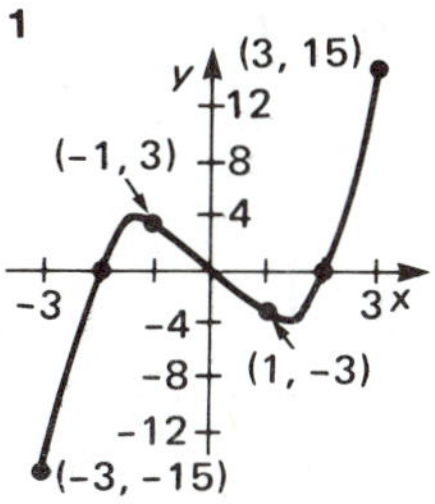

3

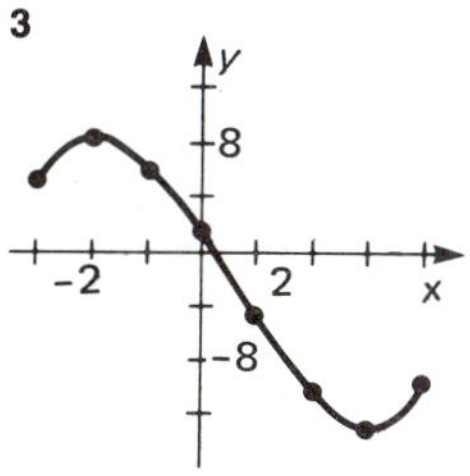

5

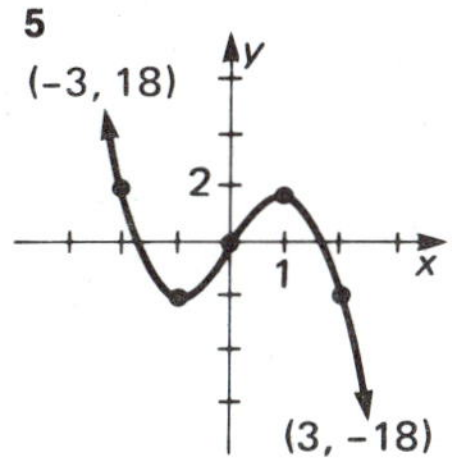

7

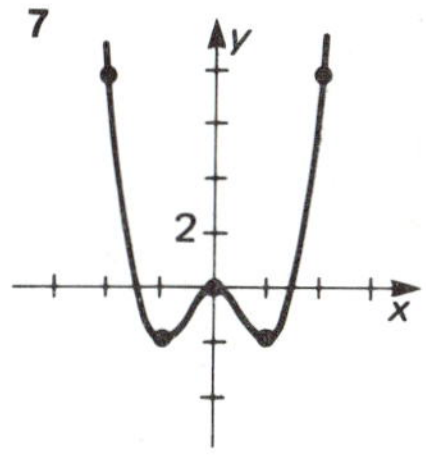

9

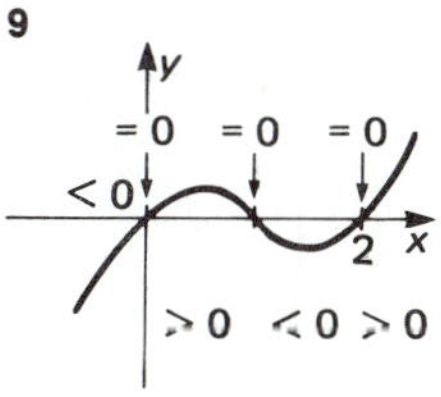

11

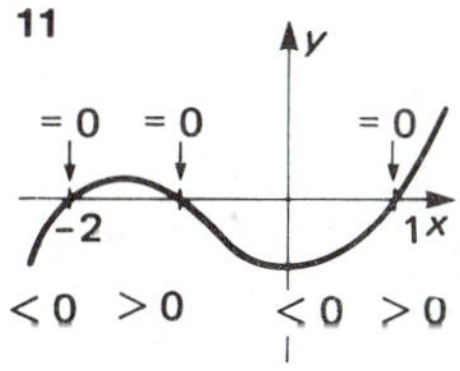

13

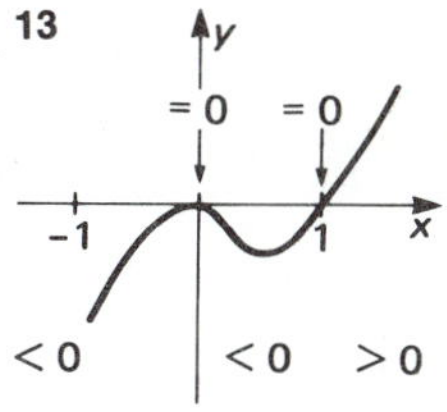

15

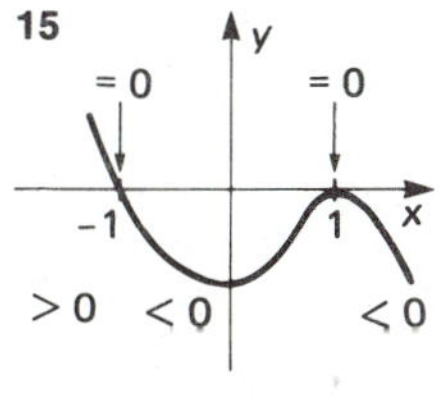

17

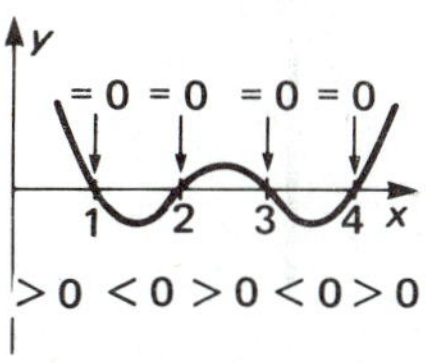

19

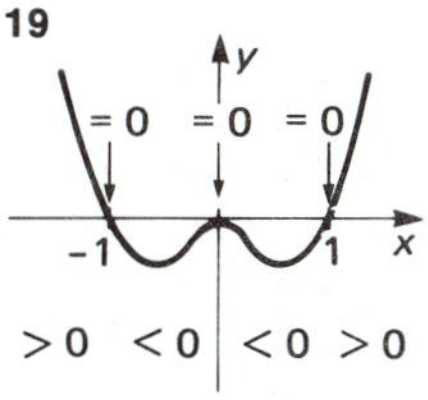

21

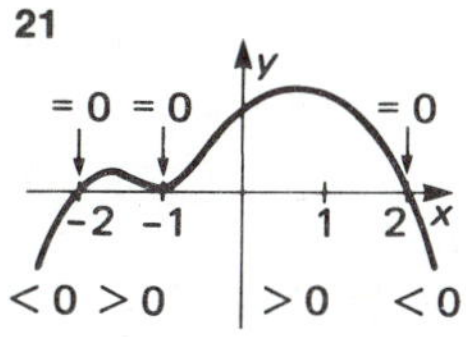

23

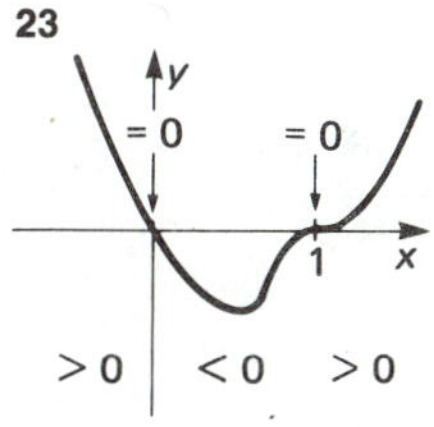

25 At the very least, the graph goes up, down, up, down, up, down, changing direction at least 5 times, so $\deg f(x) \geq 6$. An example of degree 6 is $y = -\frac{1}{9}x(x-2)^2(x-4)^2(x-6)$.

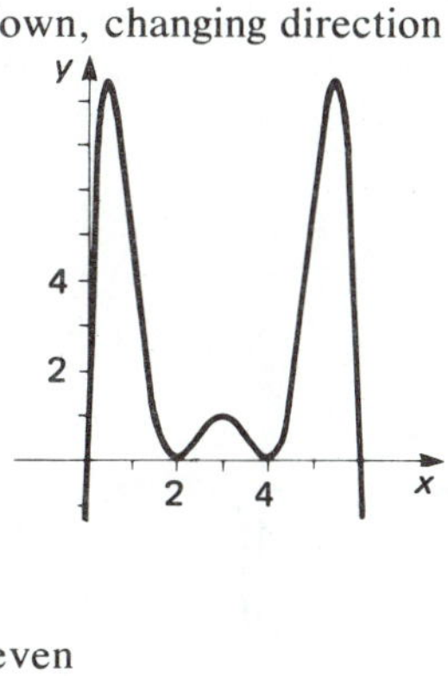

Section 4 page 156

1 odd **3** odd **5** odd **7** odd **9** neither **11** even
13 $g(x) = -5x^2 - 2$; even **15** $g(x) = 4x^2/(x^2 - 16)$; even
17 $h(x) = -5x^3 - 7x$; odd **19** $h(x) = \frac{1}{2}(\sqrt{1-x} - \sqrt{1+x})$, $h(-x) = \frac{1}{2}(\sqrt{1+x} - \sqrt{1-x}) = -h(x)$; odd
21 $g(-x) = \frac{1}{2}[f(-x) + f(x)] = g(x)$; even **23** $f(x) = (2x^4 + x^2 + 2) + (-7x^3 - 3x)$ **25** $f(x) = h_1(x) + h_2(x)$; $f(-x) = h_1(-x) + h_2(-x) = -h_1(x) - h_2(x) = -[h_1(x) + h_2(x)] = -f(x)$, so $f(x)$ is odd.
27 $f(x) = h(x)g(x)$; $f(-x) = h(-x)g(-x) = [-h(x)]g(x) = -h(x)g(x) = -f(x)$, so $f(x)$ is odd. **29** $[f\circ g](-x) = f[g(-x)] = f[-g(x)] = -f[g(x)] = -[f\circ g](x)$, so $f\circ g$ is also odd.

31

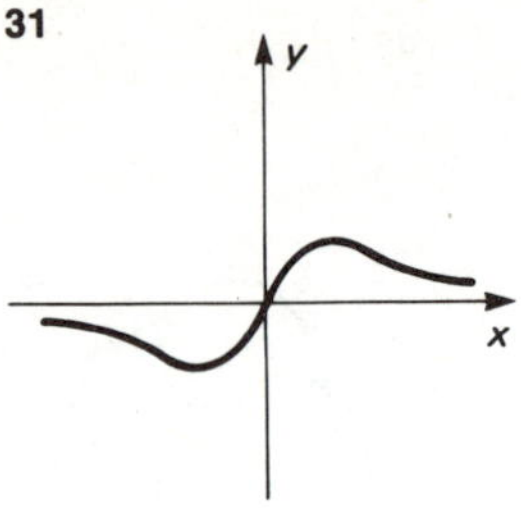

33

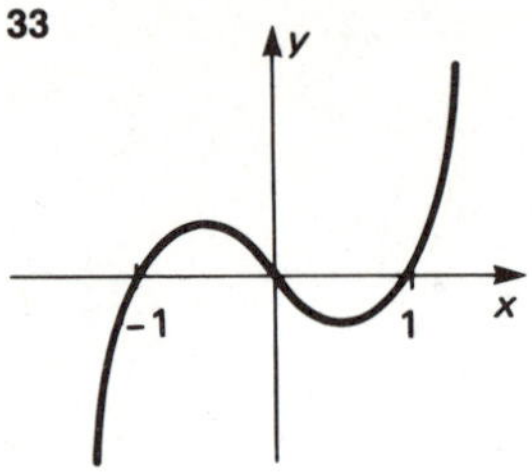

35

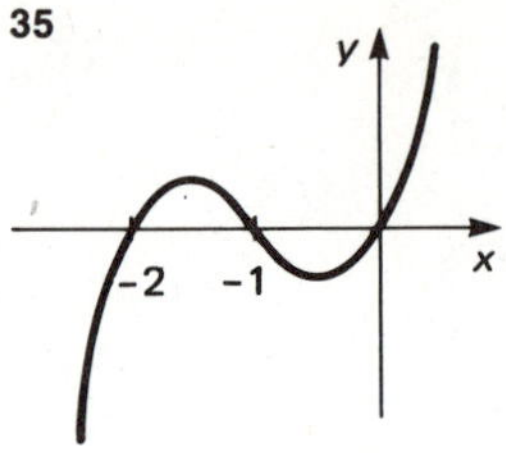

Section 5 page 160

1 $x \neq 4$ **3** $x \neq -1$ **5** $x \neq \pm 2$ **7** $x \neq 0, 2$ **9** $x \neq 1$ **11** $x \neq \pm 1$
13 $x \neq \frac{1}{6}(5 \pm \sqrt{61})$ **15** yes; all x **17** yes; $x \neq -1, \pm 2$ **19** no

21

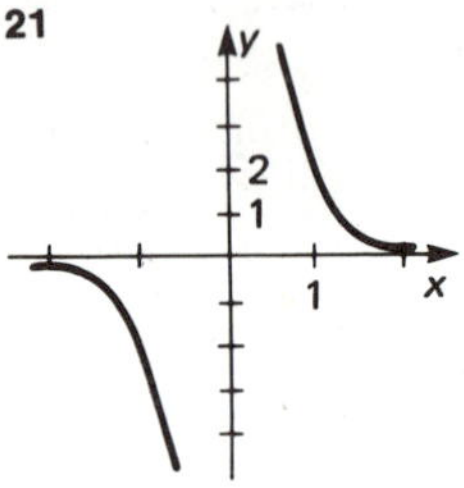

23

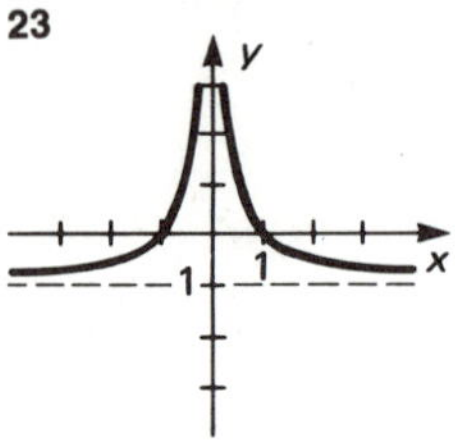

25

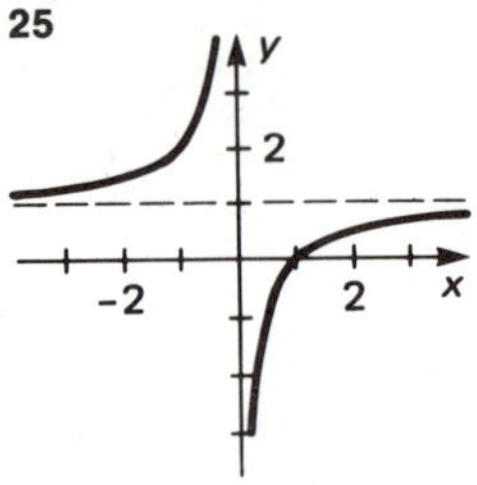

27

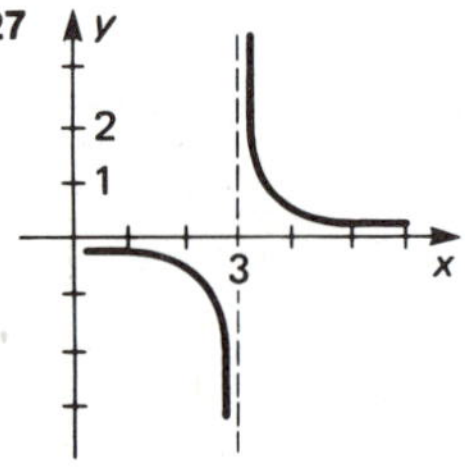

29

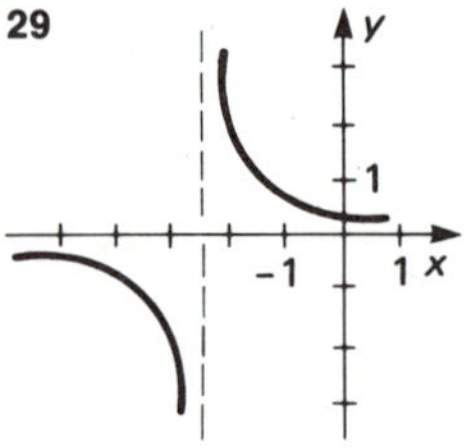

31

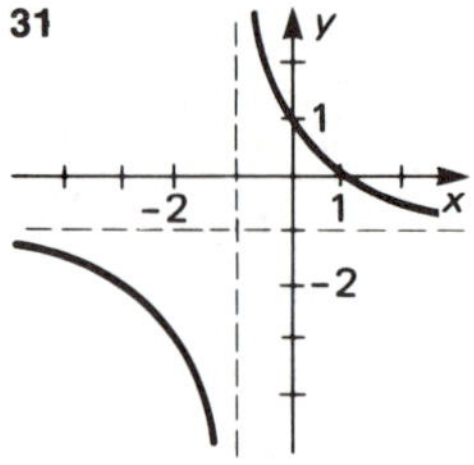

Section 6 page 165

1

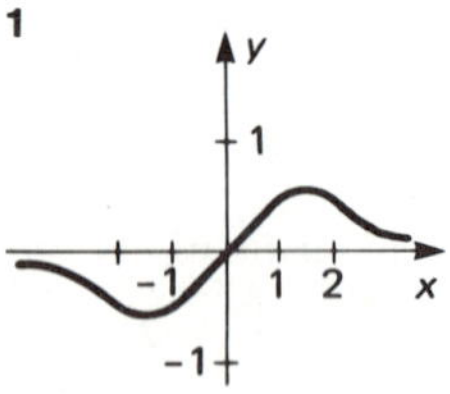

3

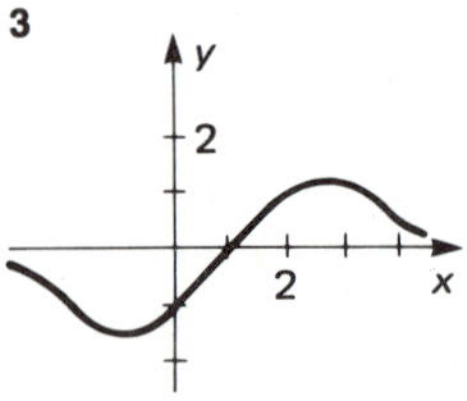

5

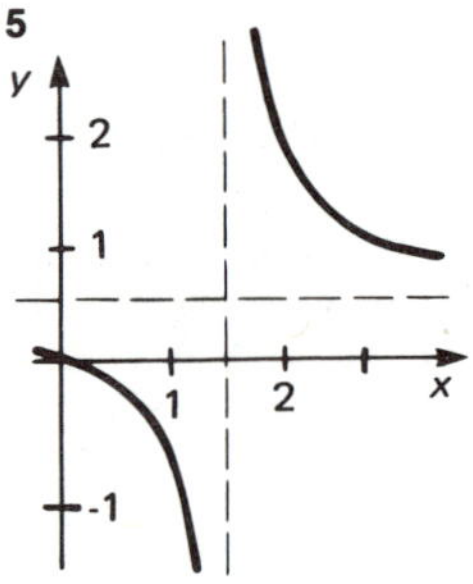

7

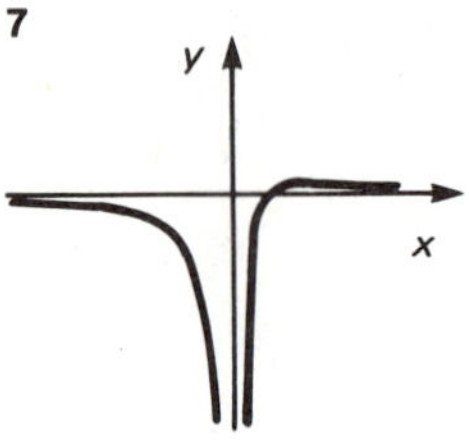

9

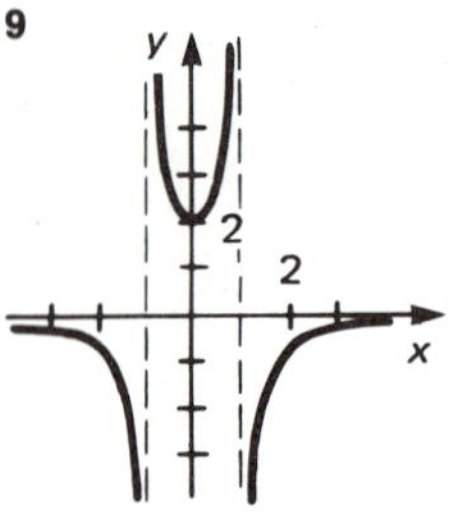

11

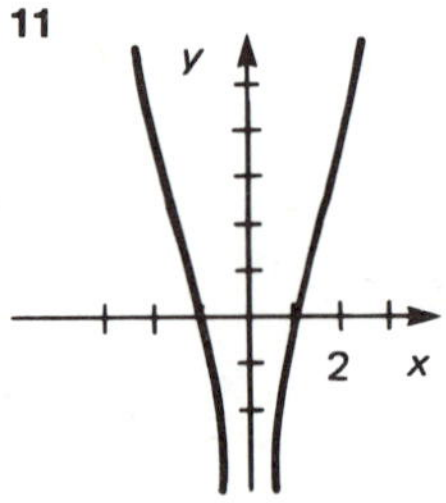

13 $r(x) \to \frac{1}{3}$ **15** $r(x) \to \pm\infty$ as $x \to \pm\infty$

Section 7 page 168

1

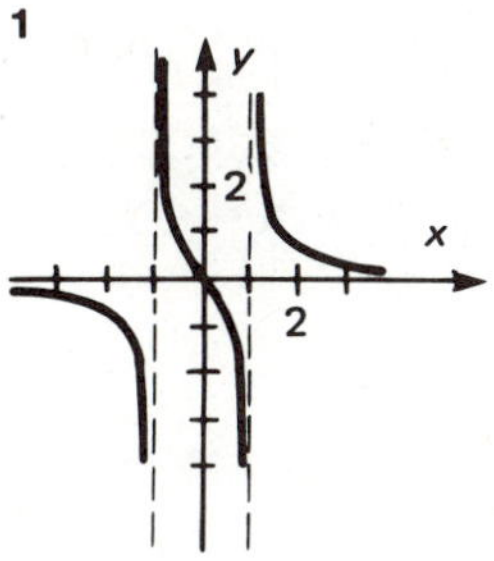

3

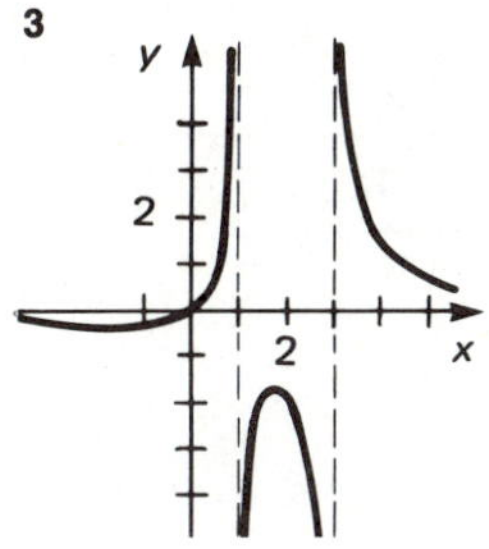

5

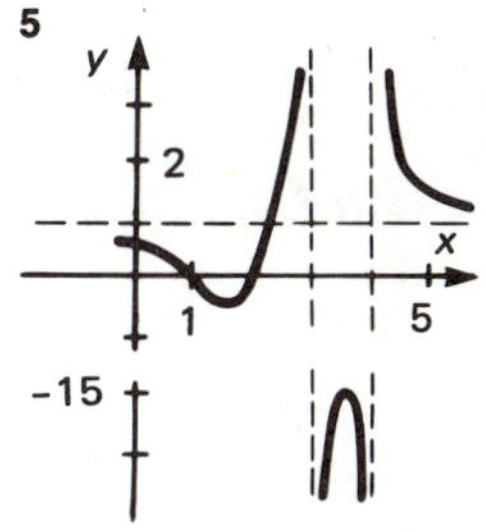

7

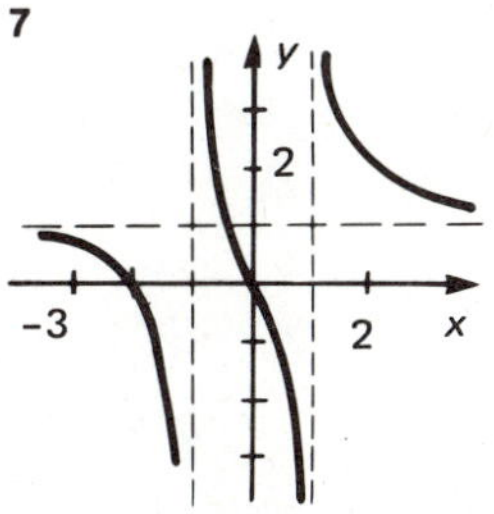

9

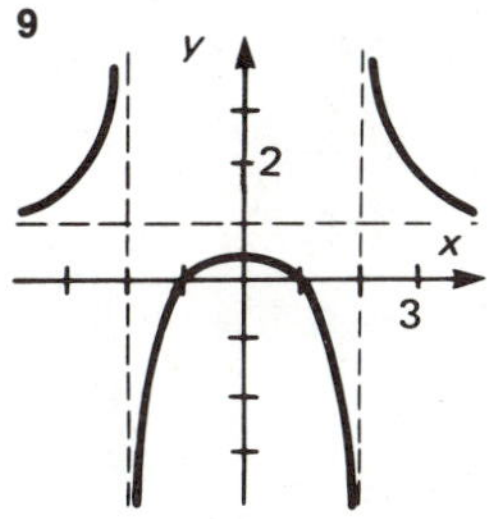

11

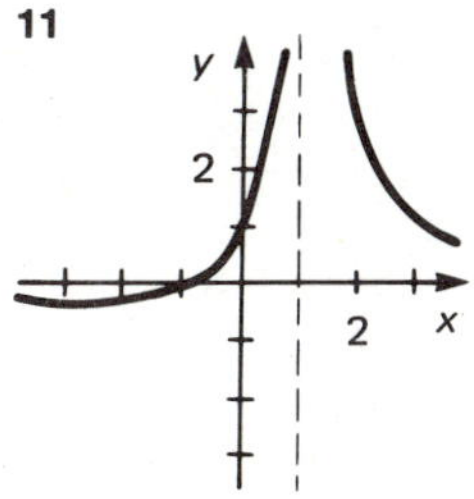

13

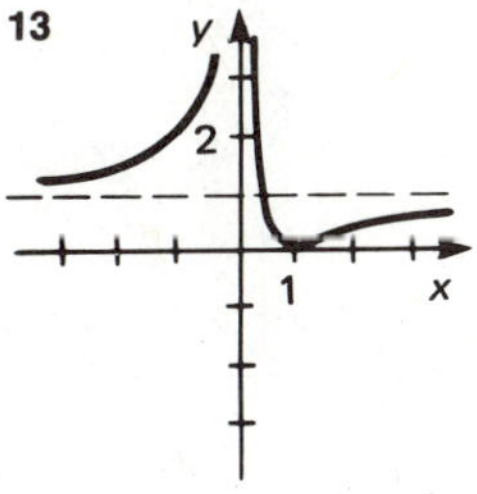

15

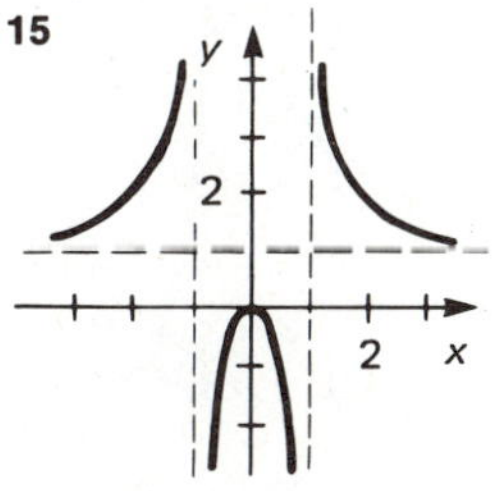

17

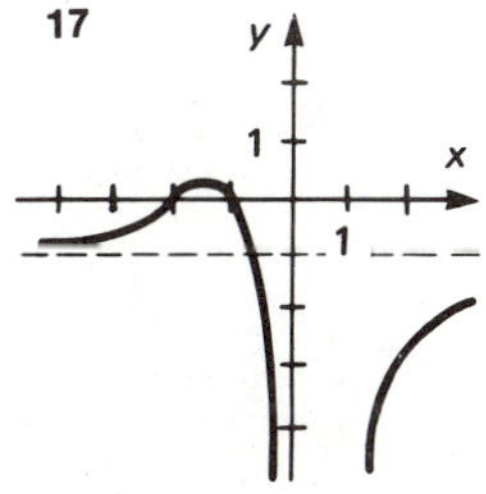

19

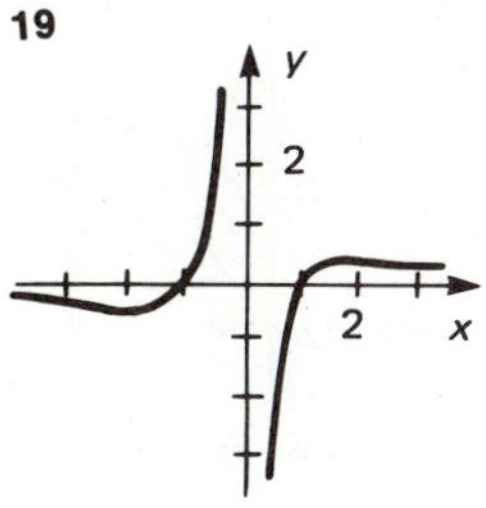

21

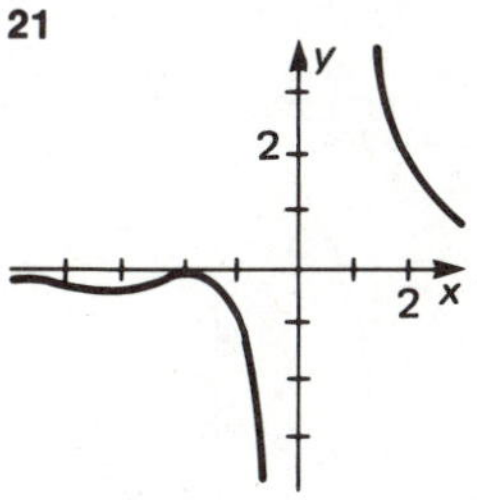

23

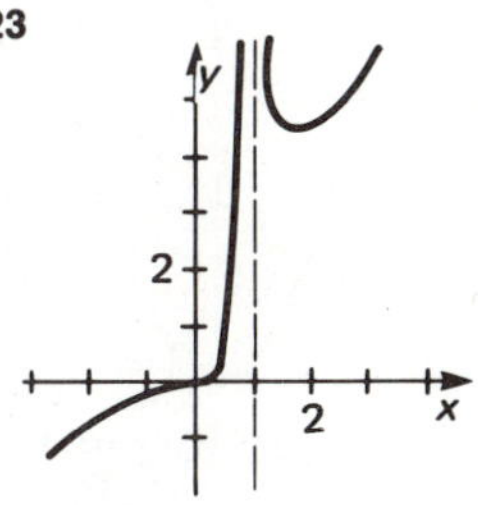

25

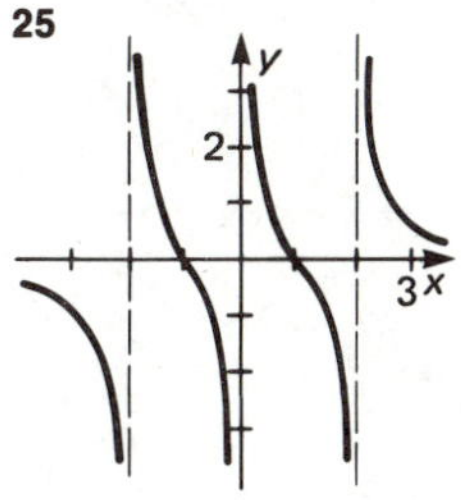

Review Exercises page 169

1

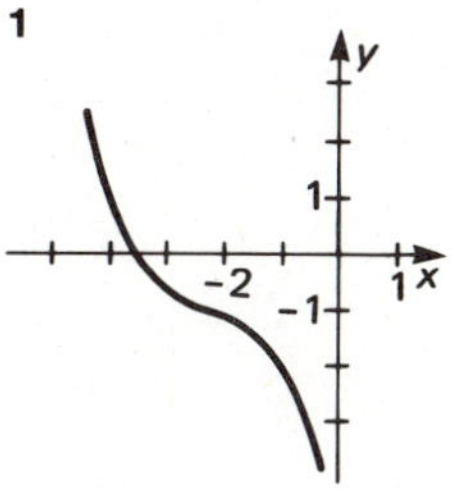

3

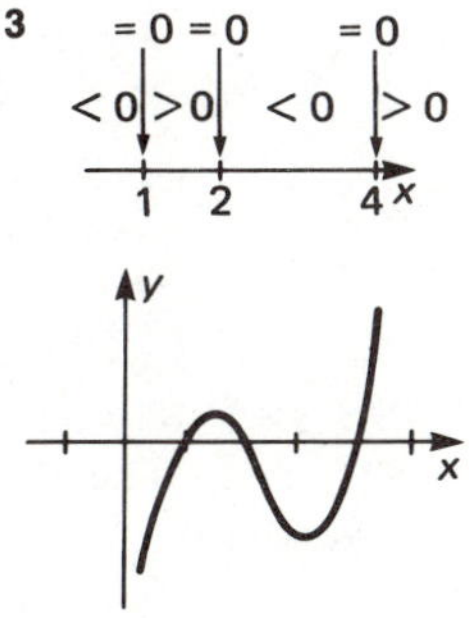

5

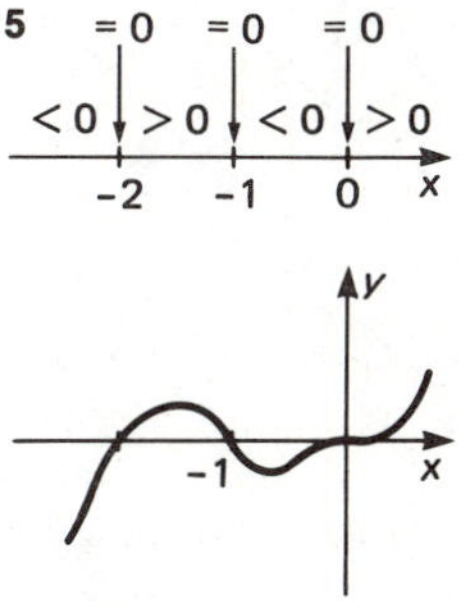

7
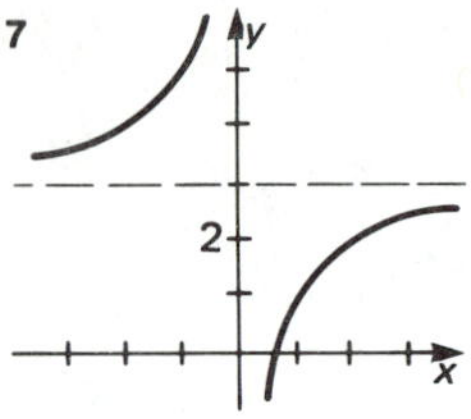

9
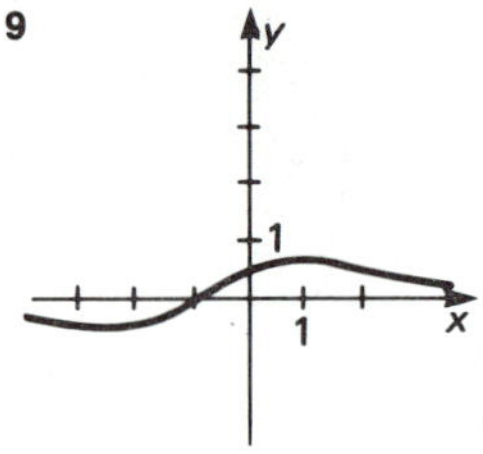

11
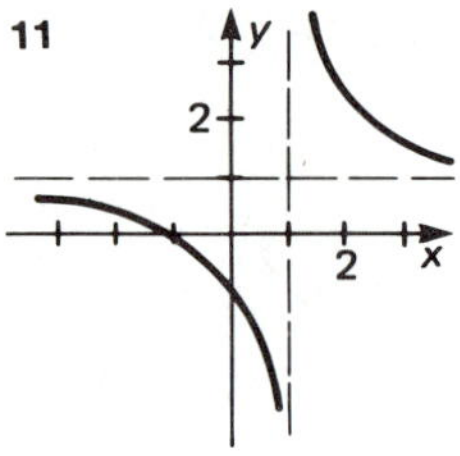

13
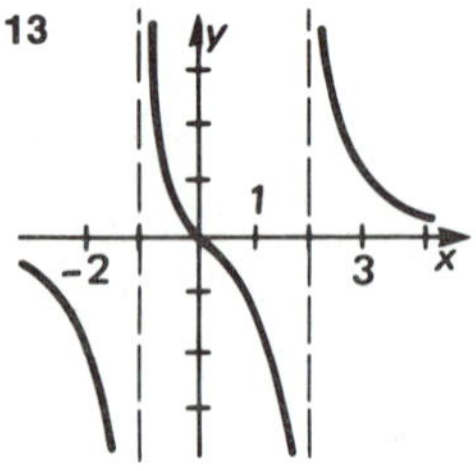

15
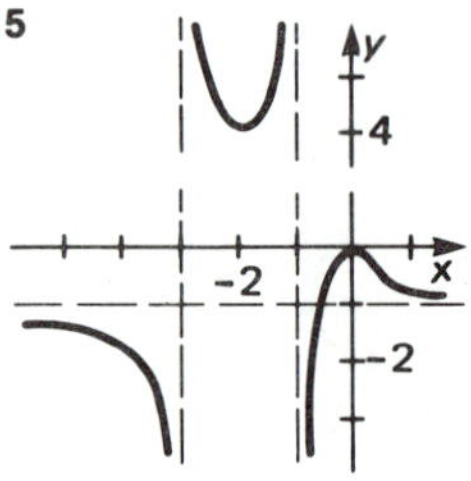

CHAPTER 5

Section 1 page 173

1
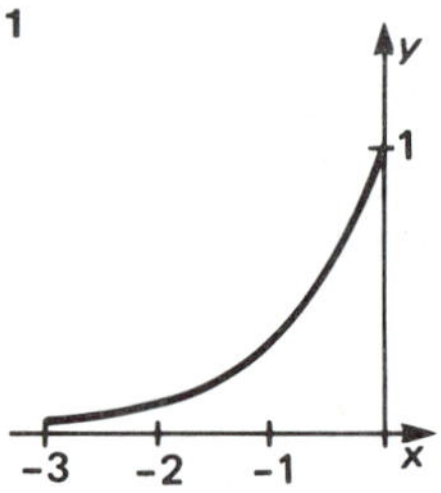

3
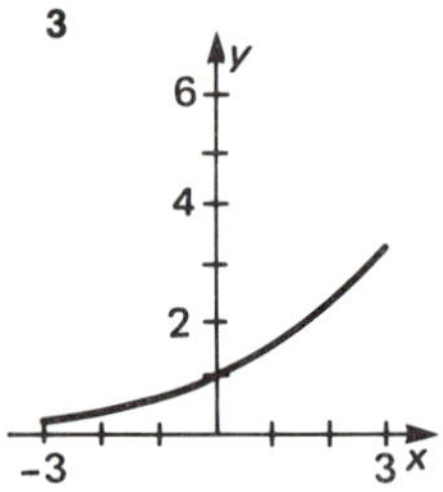

5
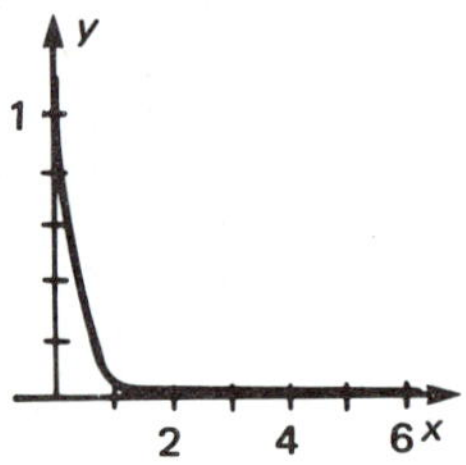

7
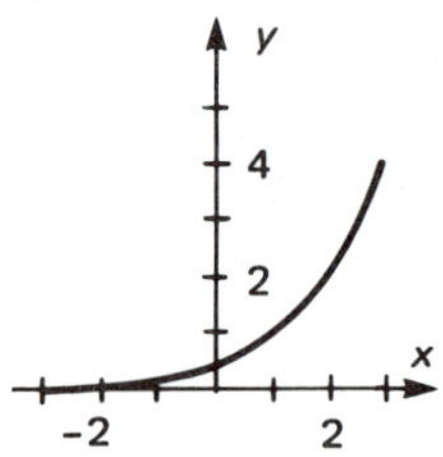

9
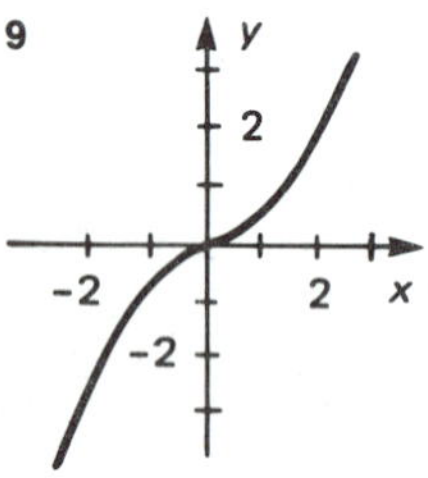

11
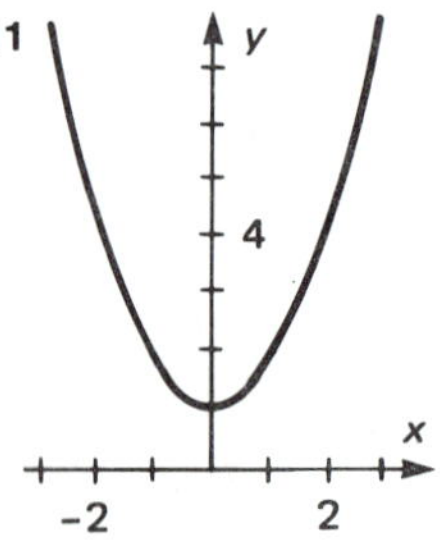

13
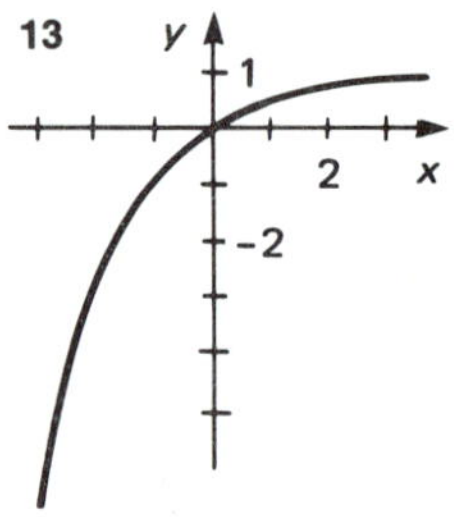

15
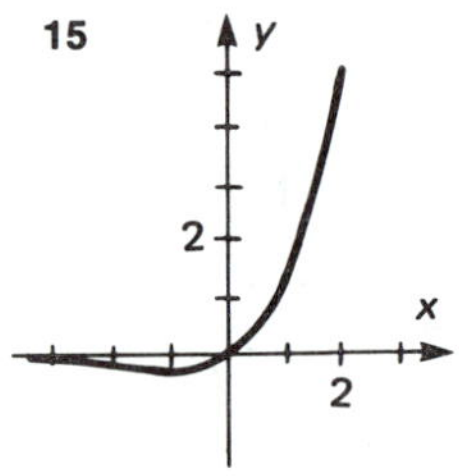

17 1 solution
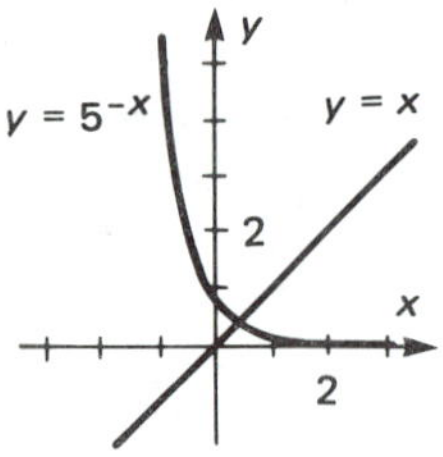

19 2 solutions
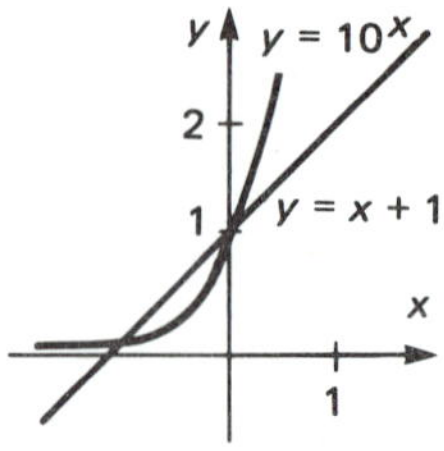

21 $f(x) = a^x$ by inspection **23** $2^{10/4.5} \approx 4.67$ times **25** (a) 155 °C (b) 25 °C (c) $\approx$38.46 °C **27** $(1.16)^{4+2/3} \approx 2.00$ **29** The 8% investment is worth \$1586.87, which is more than the 4% investment, worth only \$1539.45

Section 2 page 181

1 $f(x)$ is strictly increasing, and $\text{dom}(g) = \text{range}(f) = \{\text{all } y\}$ **3** If $0 < x_1 < x_2$, then $1/x_2 < 1/x_1$, so $-1/x_1 < -1/x_2$, that is, $f(x_1) < f(x_2)$. Therefore $f(x)$ is strictly increasing, hence has an inverse, and $\text{dom}(g) = \text{range}(f) = \{y < 0\}$. **5** $f(x)$ is strictly increasing. Since $3^2 = 9$ and $4^2 = 16$, $\text{dom}(g) = \text{range}(f) = \{9 < y < 16\}$. **7** Both x^5 and x^3 are strictly increasing; so is their sum, $f(x)$. It is clear from the graph of $f(x)$ and the facts that $f(x) \to \infty$ as $x \to \infty$ and $f(x) \to -\infty$ as $x \to -\infty$, that $\text{dom}(g) = \text{range}(f) = \{\text{all } y\}$. **9** The function x^3 is strictly increasing; hence $f(x)$ is strictly decreasing, so has an inverse, and $\text{dom}(g) = \text{range}(f) = \{y > 0\}$.
11 $f(x)$ is strictly decreasing, so has an inverse, and $\text{dom}(g) = \text{range}(f) = \{y \geq 0\}$.
13 $x = 2\sqrt[3]{y}$ $\{\text{all } y\}$ **15** $x = -1/y$ $\{y < 0\}$ **17** $x = \sqrt{5/y}$ $\{y > 0\}$
19 $x = y^2 - 1$ $\{y \geq 0\}$ **21** $x = \sqrt{y^2 - 4}$ $\{y \geq 2\}$
23 $x = -\sqrt[4]{1/y - 10}$ $\{0 < y \leq \frac{1}{10}\}$ **25** $x = -\sqrt{-1/y}$ $\{y < 0\}$ **27** 0 **29** $\frac{4}{5}$

31

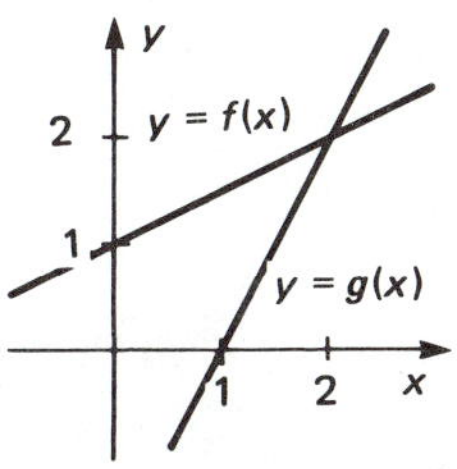

33

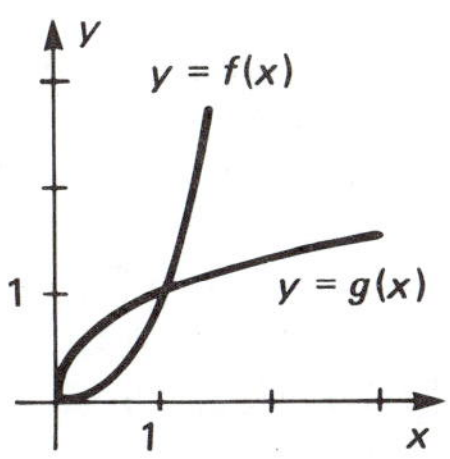

35

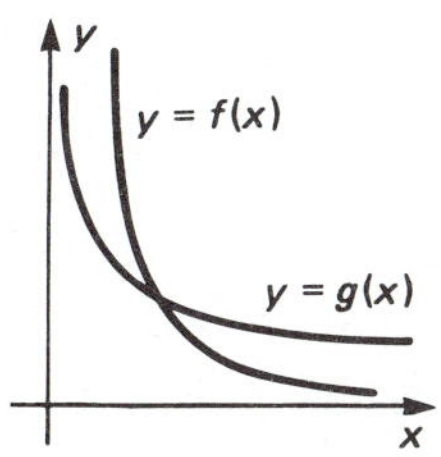

37

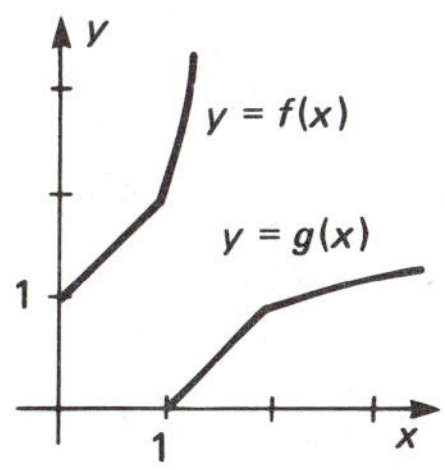

39 $x = -g(y)$ $\text{dom}(-g) = \text{range}(f)$

Section 3 page 186

1

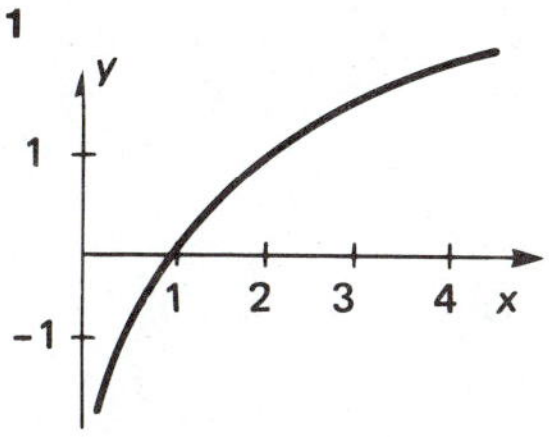

3

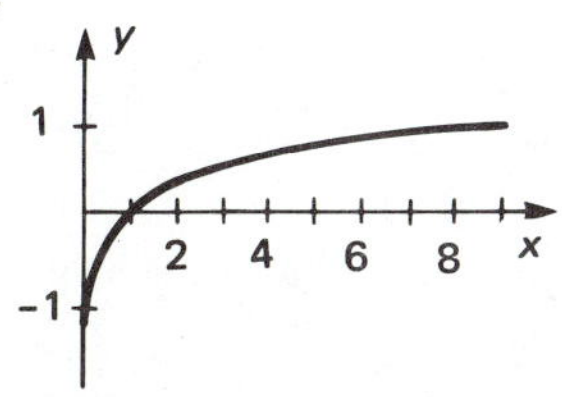

5

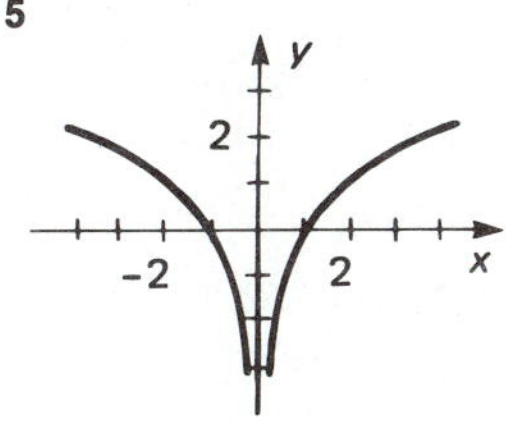

7 for $x = 10, 10^2, 10^{10}$ **9** $\log_2 1024 = 10$ **11** $\log_8 \frac{1}{4} = -\frac{2}{3}$ **13** 7 **15** $\frac{4}{5}$
17 $\frac{5}{2}$ **19** $\frac{1}{2}$ **21** 12 **23** -4 **25** 0 **27** -5 **29** 125 **31** 1 **33** 76
35 $-\frac{1}{2}$ **37** $\pm\sqrt{19}$ **39** 7 **41** 2 **43** 0.7473 **45** -0.2400 **47** 1.5876
49 -1.5681 **51** -0.8842 **53** $\log_a 48$ **55** $\log_a \frac{1}{3}$ **57** 3 **59** $\frac{1}{64}$
61 -2 8 **63** Set $y_1 = \log_a x_1$ and $y_2 = \log_a x_2$. Then $x_1 = a^{y_1}$ and $x_2 = a^{y_2}$, so $x_1/x_2 = a^{y_1}/a^{y_2} = a^{y_1 - y_2}$. Hence (2) follows. **65** $2 = 1 + 1$, but $\log_a 2 \neq 0 = \log_a 1 + \log_a 1$ **67** $\frac{1}{100} < x < \frac{1}{10}$

Section 4 page 190

1 0.377488 **3** 26.377488 **5** 1.144729 **7** 3.32193 **9** 5.65678 **11** 1.04822
13 1.10659 **15** 2303.74 **17** -0.990050 **19** 0.434294 **21** 2.01308×10^{10}
23 0.224207 **25** 0.827087 **27** 0.333333 **29** 2.011155 **31** 4.83946×10^8
33 1 200.717 **35** 21.54435

Section 5 page 198

1 6.97 hr 9.97 hr **3** $N = 10^5 \times 10^{t/5}$ **5** 24 March 2004 **7** N triples in k units of time **9** 6.58×10^9 4.76×10^{10} 3.45×10^{11}

11 $k = 1.9806 \times 10^{11}$

2000	2100	2200	ult.
5.74×10^9	9.49×10^9	9.84×10^9	9.86×10^9

13 12.09 days **15** 0.693 sec **17** 2.19 days **19** 33 min **21** 2.59 min
23 1.32×10^{10} times brighter **25** 6.3×10^5 times more intense
27 (a) $\log y = \log 3 + 2 \log x$ $v = 2u + \log 3$, a linear function (b) $y = cx^a$ $c > 0$
29 about 12,370 terms

Section 6 page 202

1

x	10	20	30	40	50	60	70	80	90	100
$2^x/x^{10}$	1.0×10^{-7}	1.0×10^{-7}	1.8×10^{-6}	1.0×10^{-4}	1.2×10^{-2}	1.9	4.2×10^2	1.1×10^5	3.6×10^7	1.3×10^{10}

3

x	1	2	3	4	5	6	7	8	9	10
$2^{-x}/x^{-2}$	0.5	1.0	1.1	1.0	.78	.56	.38	.25	.16	.098

5 167 **7** 10,000 **9** $2^x/x^{100} > 1$ for $x = 10^3$, $2^x/x^{100} > 10^{299,400}$ for $x = 10^6$
11 For $x = 10^{10}$, $(\log x)/x = 10^{-9}$. For $x = 10^{100}$, $(\log x)/x = 10^{-98}$
13 $x \log x \to 0-$ as $x \to 0+$ **15** $\log_2 x > \log_3 x$ for $x > 1$ **17** 1.0000053×10^{10} on a Sharp EL-5806S

Section 7 page 207

1 0.7419 **3** $0.7767 - 1 = -0.2233$ **5** 4.7686 **7** 3.7406 **9** 5.7517
11 0.1770 **13** 1.2660 **15** $0.2245 - 3 = -2.7755$ **17** 1.805 **19** 179,300
21 0.1780 **23** 0.44 0.31 0.11 0.26 **25** 0.000 0.000 16.244 3.786
27 (a) 5.102 5.10 (b) $1.26 + 0.40 + 2.12 + 1.34 = 5.12$ **29** 1.05×10^3 55.5 10.0
31 0.40 **33** 2.3

Section 8 page 214

1 $a + \sqrt{b} + \sqrt{c}$ **3** $\dfrac{a}{b} + \dfrac{1}{c}$ **5** $\left(\dfrac{ab}{c} + a\right)\dfrac{c}{b}$ **7** $(a + 10^b + c)^2 - a$

9 $\{[(a + 1)^{-1} + 1]^{-1} + 1\}^{-1}$ **11** $\dfrac{ab}{c} + \dfrac{ac}{b}$ **13** $a + b^c d$

15 $(a - b)c - (b - c)a$ **17** $(a + b^{a-b})(a - b)$ **19** $a + \dfrac{b}{a}$

21 a [STO] [+] 3 [=] [log] [×] [RCL] [=] **23** a [STO] [y^x] [RCL] [1/x] [−] [RCL] [=] [+/−]

25 1 [STO] a [+/−] [M+] [+/−] [+] 1 [÷] [RCL] [=] or 1 [−] a [÷] 2 [=] [1/x] [−] 1 [=].

Note $\dfrac{1 + a}{1 - a} = \dfrac{2}{1 - a} - 1.$

27 a [STO] [x^2] [x^2] [×] 5 [+] 1 [=] [√] [√] [−] [RCL] [=] [y^x] [RCL] [−] [RCL] [=]

29 3 [STO] a [+/−] [M+] [+/−] 2 [=] [÷] [RCL] [=] or 3 [−] a [=] [1/x] [×] 5 [−] 1 [=].

Note $\dfrac{2 + a}{3 - a} = \dfrac{5}{3 - a} - 1$

31 1 [+] a [STO] [y^x] [RCL] [1/x] [=] [×] [(] 1 [+] [RCL] [1/x] [y^x] [RCL] [)] [=]

33 a [STO] [y^x] 3 [−] 2 [=] [log] [+] [RCL] [x^2] [=] **35** 3.94482 37 **37** 0.30888 640

39 $\approx 4.32 \times 10^{6001}$ **41** 1.57×10^{-4365} **43** 1.73205 08075 68

45 [M+] [−] [RCL] [=] [M+] [+/−]:

dis.	a	a	$-b$	$-b$	b
mem.	b	$a + b$	$a + b$	a	a

Section 9 page 221

1 24.344 −27,870 **3** 28,940.24 −1928.26208 **5** 10.474993 2331.1309
7 3573.0717 885.67423 **9** −3414.0768 −693.62666 **11** 2 **13** −2 3
15 0.75487 767 **17** −0.86287 114 **19** 0.39901 298 **21** 0.63084 829
23 3.57715 21 **25** −0.82621 781

Section 10 page 228

1 (a) \$122.04 (b) \$122.14 **3** B is better, the equivalent of 7.19% simple annual interest against C's 6.98%. **5** 22.523% **7** \$3803.52 **9** \$142,710.57 **11** \$3029.06
13 \$28.31 **15** 12 yr 7 mo 1 day **17** \$22,580.33 **19** $7\frac{3}{8}\%$ **21** \$9565.33
23 \$815.56 **25** \$338.86 **27** 17.785% **29** \$154.10

Review Exercises page 229

1

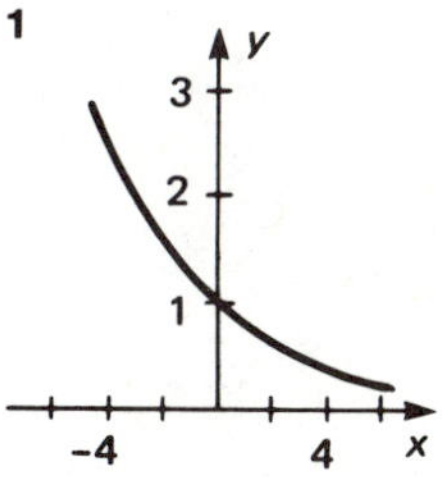

3 $x = \begin{cases} \frac{1}{3}y \\ \frac{1}{2}y \end{cases}$ for $\begin{cases} y \le 0 \\ y \ge 0 \end{cases}$
dom $= \{\text{all } y\}$ range $= \{\text{all } x\}$

5 $x = \log_2(y - 1)$ dom $= \{y > 1\}$ range $= \{\text{all } x\}$

7

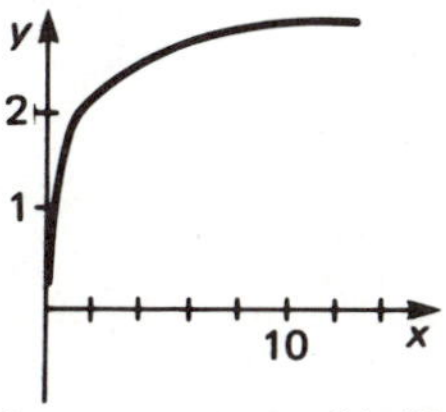

9

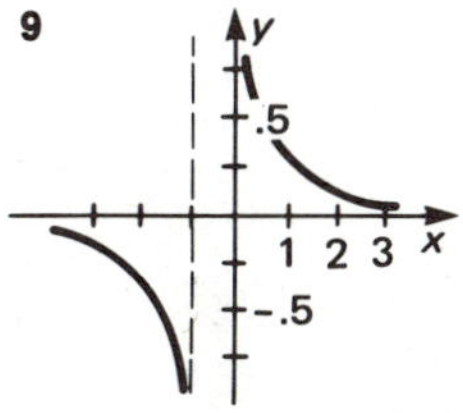

11 0 **13** $x \approx 21.434$ **15** to 4 places **17** 12 hr 1.2 min
19 3^x grows faster: $\dfrac{3^x}{2^{x+\log_2 x}} \to \infty$ as $x \to \infty$ **21** 3.60813 **23** 17.980186
25 1.394955 **27** \$71.66

CHAPTER 6

Section 1 page 237

1, 3, 5, 7

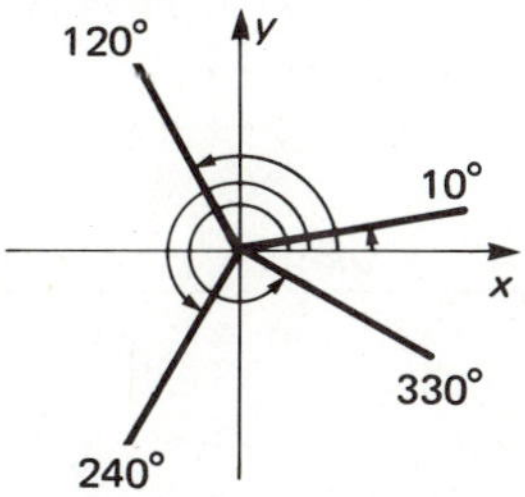

9 $\frac{3}{5}$ $\frac{4}{5}$ **11** $-\frac{12}{13}$ $-\frac{5}{13}$ **13** $\frac{24}{25}$ $\frac{7}{25}$
15 $-\frac{9}{41}$ $\frac{40}{41}$ **17** $-\frac{60}{61}$ $-\frac{11}{61}$ **19** $\sqrt{\frac{2}{3}}$ $\sqrt{\frac{1}{3}}$

21 $\frac{1}{3}\sqrt{7}$ $\frac{1}{3}\sqrt{2}$ **23** $-\sqrt{\frac{5}{8}}$ $\sqrt{\frac{3}{8}}$ **25** 0.3025 **27** 0.9959 **29** 0.9829
31 -0.0654 **33** -0.4410 **35** -0.9454 **37** -0.9876 **39** 0.1805

Section 2 page 243

1 $\frac{1}{2}\sqrt{3}$ **3** $-\frac{1}{2}$ **5** $-\frac{1}{2}\sqrt{3}$ **7** $-\frac{1}{2}\sqrt{3}$ **9** $-\frac{1}{2}$ **11** 0 **13** $\frac{1}{3}\sqrt{3}$
15 $-\sqrt{3}$ **17** $\frac{1}{3}\sqrt{3}$ **19** $\sqrt{3}$ **21** -1 **23** $\sqrt{1 - K^2}$ **25** $-K$
27 $\sqrt{1 - K^2}$ **29** $-\sqrt{1 - K^2}/K$ **31** $1/T$ **33** $1/\sqrt{1 + T^2}$ **35** $1/T$
37 $-T/\sqrt{1 + T^2}$ **39** 0.3174 **41** -5.3383 **43** 70.53° **45** 26.05°
47 101.54° **49** 102.59° **51** 19.47° 160.53° **53** 67.56° 112.44°

Section 3 page 250

1 18° **3** 108° **5** 40° **7** 195° **9** 285° **11** $\frac{1}{18}\pi$ **13** $\frac{1}{5}\pi$ **15** $\frac{3}{10}\pi$
17 $\frac{119}{60}\pi$ **19** $\frac{19}{15}\pi$ **21** 0.2243 **23** 3.0178 **25** 30.94° **27** 1.81° **29** 13.02
31 1.0167 **33** 0 -1 **35** 0 -1 **37** $\frac{1}{2}\sqrt{2}$ $-\frac{1}{2}\sqrt{2}$ **39** $-\frac{1}{2}\sqrt{2}$ $-\frac{1}{2}\sqrt{2}$
41 $-\frac{1}{2}\sqrt{3}$ $-\frac{1}{2}$ **43** $\frac{1}{2}\sqrt{3}$ $-\frac{1}{2}$ **45** $-\frac{1}{6}\pi$ $-\frac{5}{6}\pi$ **47** $\pm\frac{1}{4}\pi$ **49** $-\frac{1}{2}\pi$
51 $\pm\frac{1}{6}\pi$ **53** $\theta = n\pi$ **55** $\theta = (2n + \frac{1}{2})\pi$ **57** $0 \le \theta \le \frac{1}{2}\pi$ or $\frac{3}{2}\pi \le \theta \le 2\pi$
59 $\frac{1}{4}\pi \le \theta \le \frac{5}{4}\pi$ **61** $0 \le \theta \le \frac{1}{2}\pi$ **63** $-\sin\theta$ **65** $-\sin\theta$
67 $\cos\theta$ **69** $-y$ **71** $-\sqrt{1 - y^2}$ **73** $-x$ **75** $-x$
77 $\cos(\theta + \frac{1}{2}\pi) + \cos(\theta - \frac{1}{2}\pi) = -\sin\theta + \cos(\frac{1}{2}\pi - \theta) = -\sin\theta + \sin\theta = 0$
79 Because he calculated $\cos(\frac{1}{4}\pi)°$ rather than $\cos\frac{1}{4}\pi$. **81** radian mode
83 To prove $\sin(\frac{9}{2}\pi + \theta) = \sin(\frac{9}{2}\pi - \theta)$: $\sin(\frac{9}{2}\pi + \theta) = \sin(\theta - \frac{9}{2}\pi + 9\pi) = \sin(\theta - \frac{9}{2}\pi + \pi) = \sin[\pi - (\frac{9}{2}\pi - \theta)] = \sin(\frac{9}{2}\pi - \theta)$.

Section 4 page 257

1 0 **3** -1 **5** $-\sqrt{3}$ **7** $\sqrt{3}$ **9** $\sqrt{3}$ **11** $n\pi$
13 $\frac{5}{6}\pi + n\pi$ or $-\frac{1}{6}\pi + n\pi$ **15** $n\pi < \theta < (n + \frac{1}{2})\pi$ for any integer n
17 $\tan(\pi - \theta) = \dfrac{\sin(\pi - \theta)}{\cos(\pi - \theta)} = \dfrac{\sin\theta}{-\cos\theta} = -\tan\theta$
19 $(1 + \tan^2\theta)\cos^2\theta = \left(1 + \dfrac{\sin^2\theta}{\cos^2\theta}\right)\cos^2\theta = \left(\dfrac{\cos^2\theta + \sin^2\theta}{\cos^2\theta}\right)\cos^2\theta =$
$\cos^2\theta + \sin^2\theta = 1$ **21** $\pm\frac{3}{4}$ **23** $\pm\sqrt{3}$ **25** 1.15782 **27** 0.691831
29 -1581.67
31

$10^3\,\theta$	2	4	6	8	10
$10^3 \sin\theta$	1.9999987	3.9999893	5.9999640	7.9999147	9.9998333

33 ≈ 0.14 **35** $1° = \dfrac{\pi}{180}$, so $\sin 1° = \sin\dfrac{\pi}{180} \approx \dfrac{\pi}{180} \approx 0.01745$
37 $\tan(90° - \theta°) = \tan(\frac{1}{2}\pi - \dfrac{\pi}{180}\theta) \approx \dfrac{1}{\pi\theta/180} = \dfrac{180}{\pi\theta}$
39 $\tan(\theta - \frac{1}{2}\pi) = -\tan(\frac{1}{2}\pi - \theta) \approx -1/\theta$
41 $\frac{1}{2}(\sin\theta + \tan\theta) \approx \frac{1}{2}(\theta + \theta) = \theta$. Set $f(\theta) = \frac{1}{2}\sin\theta + \frac{1}{2}\tan\theta$. Then:

θ	0.1	0.05	0.01	0.005
$\sin\theta$	.0998334166	.0499791693	.0099998333	.0049999792
$\tan\theta$	.1003346721	.0500417084	.0100003333	.0050000417
$f(\theta)$	.1000840444	.0500104388	.0100000833	.0050000104

Section 5 page 261

1 1 $\sqrt{2}$ $\sqrt{2}$ **3** $-\frac{1}{3}\sqrt{3}$ 2 $-\frac{2}{3}\sqrt{3}$ **5** $-\frac{1}{3}\sqrt{3}$ 2 $-\frac{2}{3}\sqrt{3}$ **7** 1 $-\sqrt{2}$ $-\sqrt{2}$
9 $-\frac{1}{3}\sqrt{3}$ 2 $-\frac{2}{3}\sqrt{3}$ **11** $\pm\frac{1}{3}\pi + 2\pi n$ **13** $\frac{1}{6}\pi \pm \pi n$ **15** $(2n + 1)\pi$
17 3.10789 **19** -1.24822 **21** 3.73819 **23** $\cot\theta = 1/\tan\theta \approx 1/\theta$
25 $1/\theta - \cot\theta \approx \frac{1}{3}\theta$. Set $g(\theta) = 1/\theta - \cot\theta - \frac{1}{3}\theta$:

θ	0.10	0.08	0.06	0.04	0.02	0.01	0.005
$g(\theta)$	2.2×10^{-5}	1.1×10^{-5}	4.8×10^{-6}	1.4×10^{-6}	1.8×10^{-7}	2.2×10^{-8}	2.8×10^{-9}

27 $\csc\theta = 1/\sin\theta \approx 1/\theta$

θ	0.1	0.05	0.01	0.005	0.001
$\csc\theta$	10.0167	20.0083	100.0017	200.00083	1000.00017
$1/\theta$	10	20	100	200	1000

29 $(n + \frac{1}{4})\pi \le \theta < (n + \frac{1}{2})\pi$ or $(n + \frac{3}{4})\pi \le \theta < (n + 1)\pi$ **31** $\tan\theta$
33 $\sec\theta$ **35** $\cos\theta$ **37** θ

Section 6 page 268

1

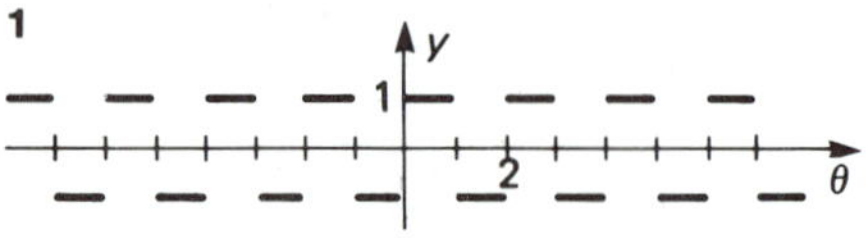

3

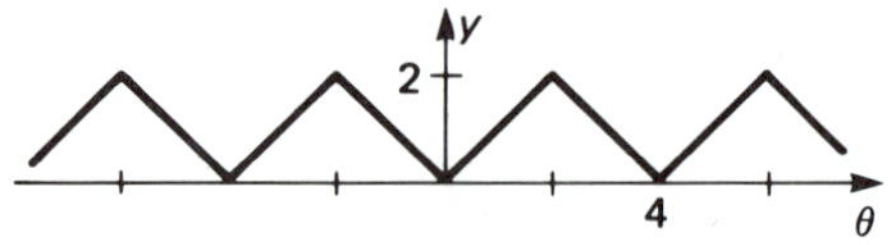

5 20π **7** 1 **9** π **11** π

13

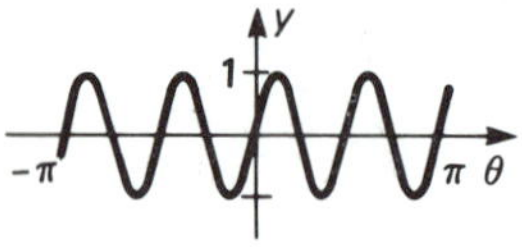

15

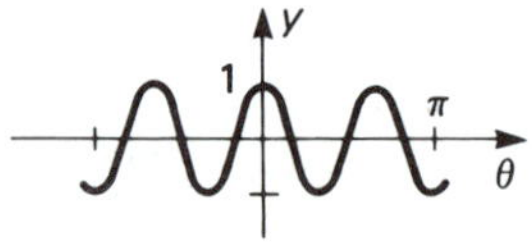

17

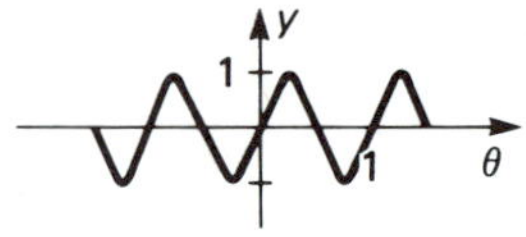

19

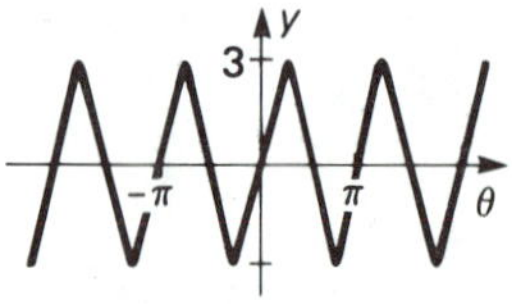

21

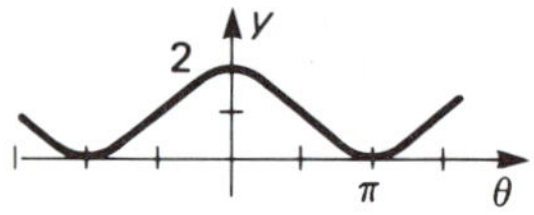

23

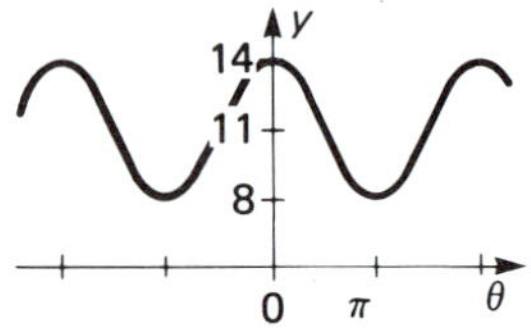

25

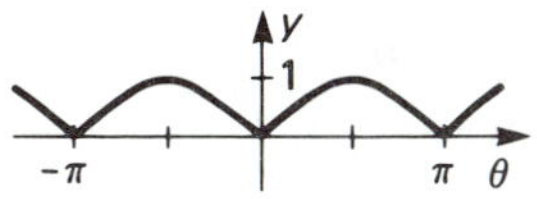

27

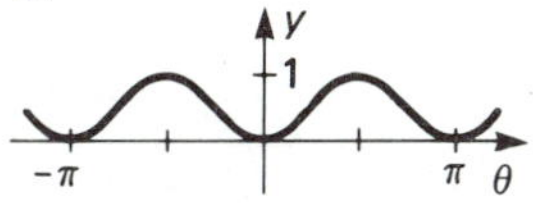

29

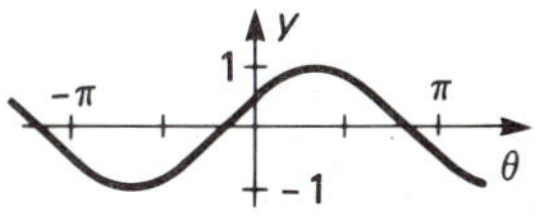

31

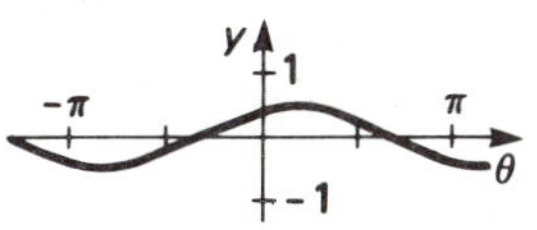

33

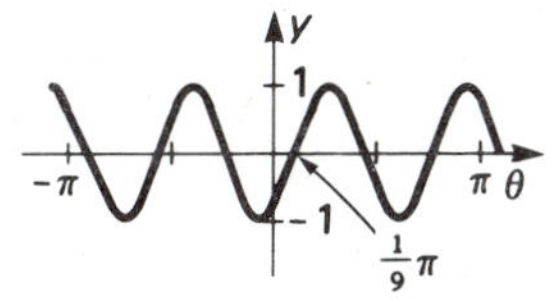

35

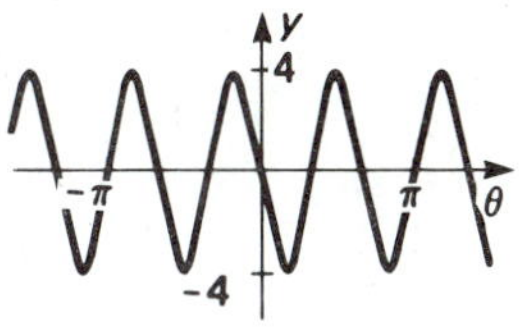

37 $x = 3\cos[2\pi(t + \frac{1}{2})]$
$= -3\cos(2\pi t)$

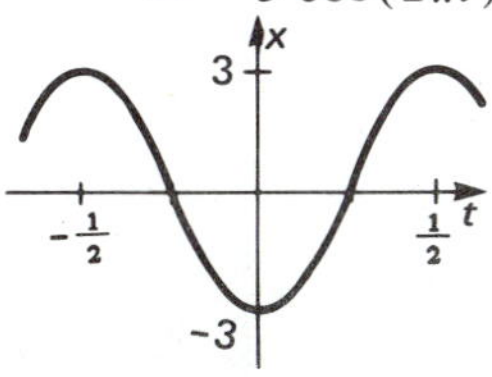

39 $x = 1 + \cos(20\pi t)$

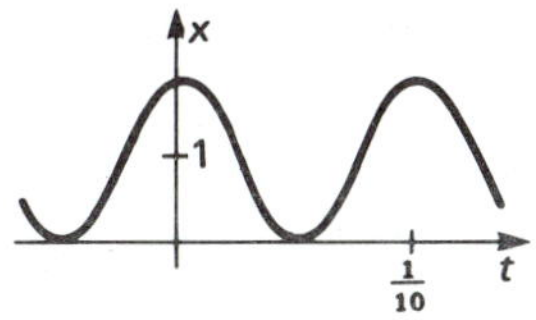

41 odd, per $= 2\pi$

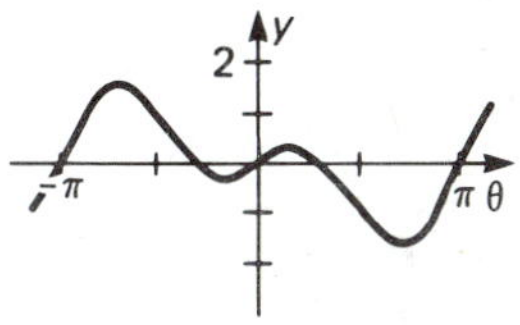

43 shifted sine, per $= 2\pi$

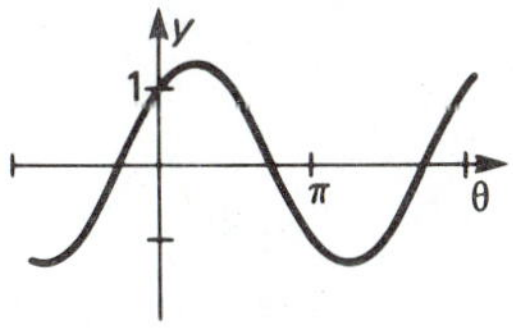

45 per $= \pi$, odd

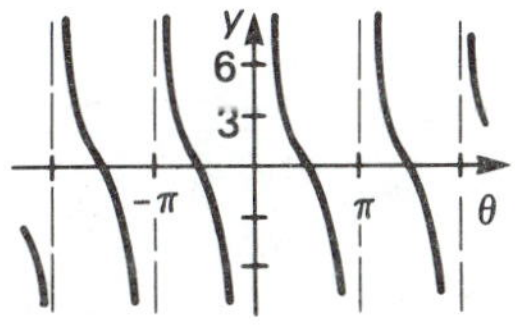

47 odd, per $= \pi$

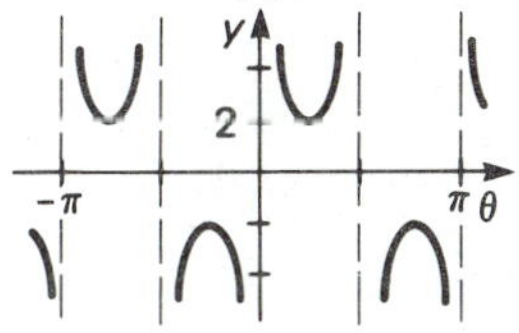

49 per $= 2\pi$

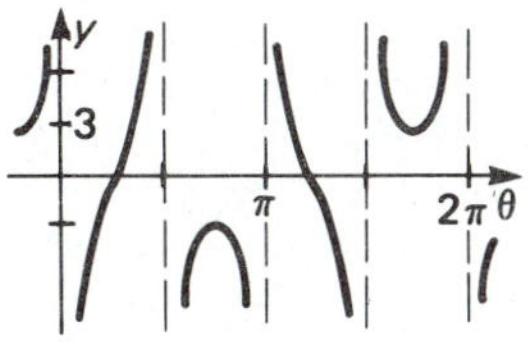

51 odd

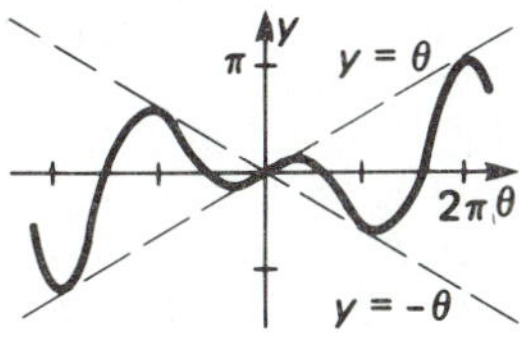

53

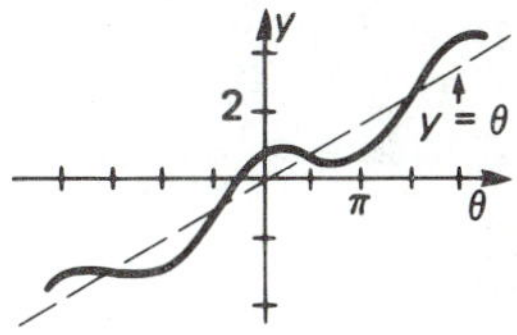

55 odd

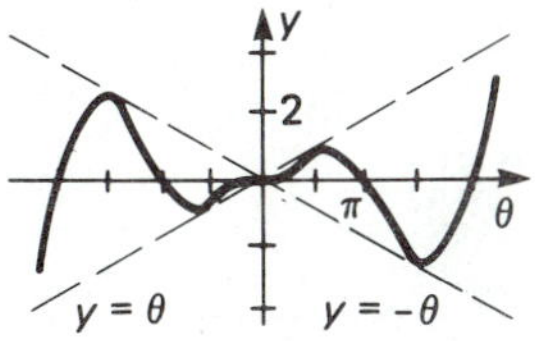

57

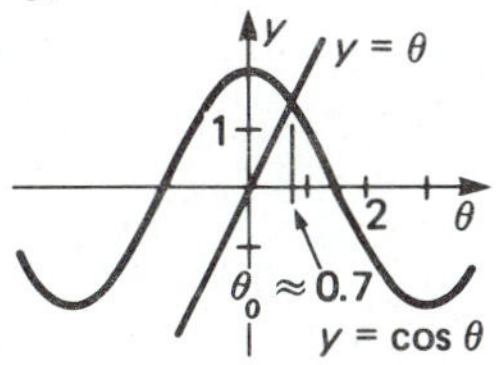

59

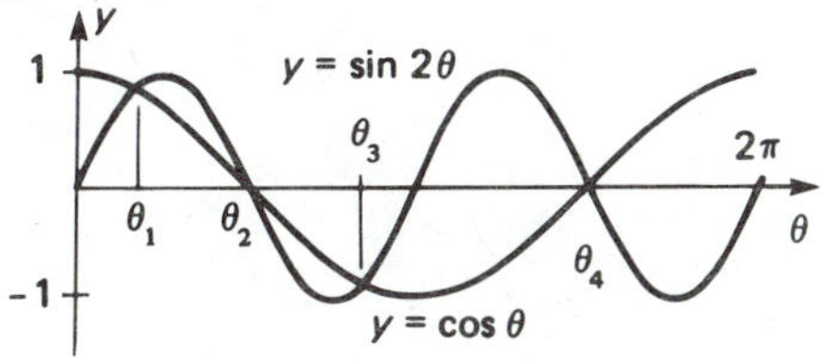

$\Theta_1 \approx 0.5$
$\Theta_2 = \frac{1}{2}\pi$
$\Theta_3 \approx 2.6$
$\Theta_4 = \frac{3}{2}\pi$

61 $y = 0$ for $\theta = 0.1n$. The points are too far apart; 0.1 is the period of y.

Review Exercises page 269

1 $\sqrt{\frac{7}{12}}$ $\sqrt{\frac{5}{12}}$ **3** 0.54106 **5** $-S$ **7** $77\frac{1}{7}°$
9 $\frac{1}{4}\pi$ $\frac{5}{4}\pi$ **11** 1 **13** 1.000000 **15** $\pm\frac{1}{6}\pi + 2\pi n$
19 $\approx 4.46095 \times 10^{-5}$ on a Sharp EL-5806S

17

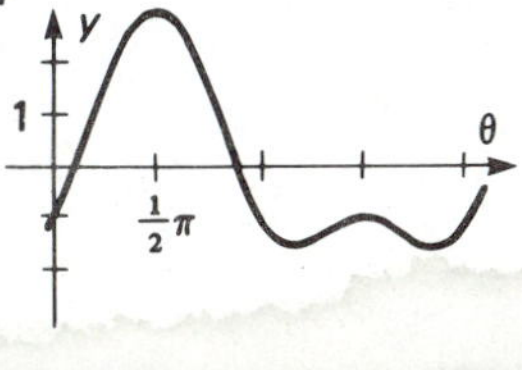

CHAPTER 7

Section 1 page 273

1 $\sin\theta\sec\theta = \sin\theta(1/\cos\theta) = \sin\theta/\cos\theta = \tan\theta$
3 $\tan\theta\csc\theta = (\sin\theta/\cos\theta)(1/\sin\theta) = 1/\cos\theta = \sec\theta$
5 $\sin^2\theta/(1 - \sin^2\theta) = \sin^2\theta/\cos^2\theta = (\sin\theta/\cos\theta)^2 = \tan^2\theta$
7 $\sec^2\theta/(\sec^2\theta - 1) = \sec^2\theta/\tan^2\theta = (1/\cos\theta)^2/(\sin\theta/\cos\theta)^2 = 1/\sin^2\theta$
9 $\sin\theta/(1 - \sin^2\theta) = \sin\theta/\cos^2\theta = (\sin\theta/\cos\theta)(1/\cos\theta) = \tan\theta\sec\theta$
11 $\sec^4\theta - \sec^2\theta = \sec^2\theta(\sec^2\theta - 1) = \sec^2\theta\tan^2\theta$
13 $1 - (\cos\theta - \sin\theta)^2 = 1 - (\cos^2\theta - 2\cos\theta\sin\theta + \sin^2\theta) =$
$1 - (1 - 2\sin\theta\cos\theta) = 2\sin\theta\cos\theta$ **15** $\cos^2\theta/(1 + \sin\theta) =$
$(1 - \sin^2\theta)/(1 + \sin\theta) = 1 - \sin\theta$

	$\sin\theta$	$\cos\theta$	$\tan\theta$	$\cot\theta$	$\sec\theta$	$\csc\theta$
17	$-\frac{3}{5}$		$\frac{3}{4}$	$\frac{4}{3}$	$-\frac{5}{4}$	$-\frac{5}{3}$
19	$3/\sqrt{10}$	$1/\sqrt{10}$		$\frac{1}{3}$	$\sqrt{10}$	$\frac{1}{3}\sqrt{10}$
21	$\frac{1}{2}\sqrt{3}$	$-\frac{1}{2}$	$-\sqrt{3}$	$-1/\sqrt{3}$		$2/\sqrt{3}$
23	$-2/\sqrt{13}$	$3/\sqrt{13}$	$-\frac{2}{3}$		$\frac{1}{3}\sqrt{13}$	$-\frac{1}{2}\sqrt{13}$

17
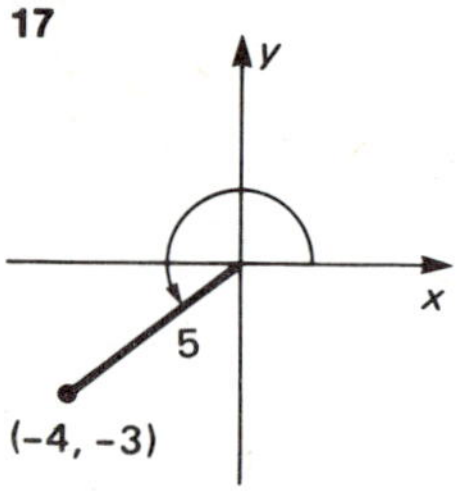

19
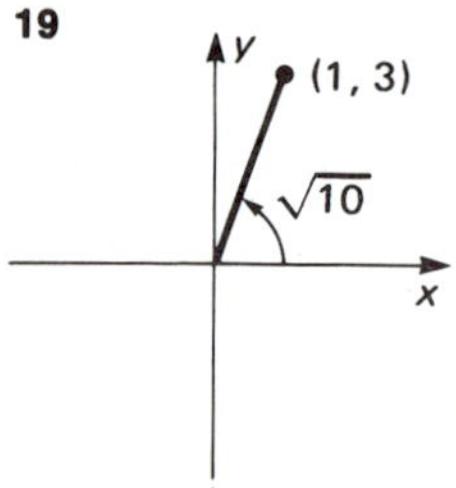

21
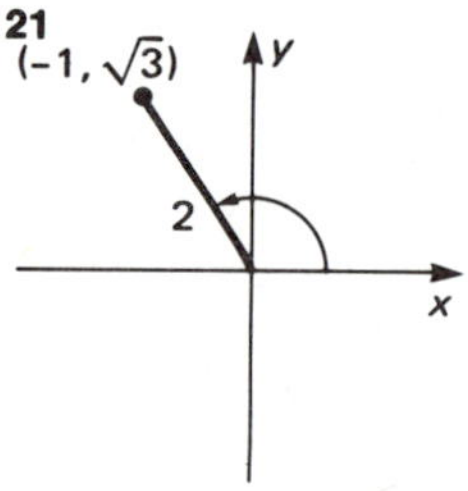

23
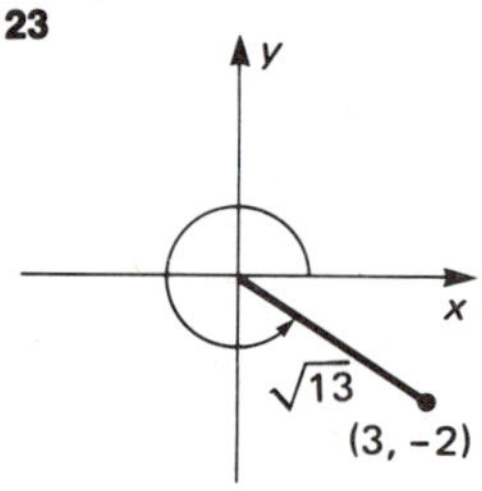

25 $1/\sqrt{1 + \cot^2\theta}$ **27** $\sqrt{1 + \cot^2\theta}$ **29** $-\sqrt{1 + \cot^2\theta}/\cot\theta$
31 $\sin\theta/\sqrt{1 - \sin^2\theta}$ **33** $1/\sqrt{1 - \cos^2\theta}$

Section 2 page 276

1 $s^4 - c^4 = (s^2 + c^2)(s^2 - c^2) = s^2 - c^2 = \sin^2\theta - \cos^2\theta$
3 $(1 - c^2)/s = s^2/s = s = \sin\theta$ **5** $(\sec^2\theta - \tan^2\theta)^3 = 1^3 = 1$
7 $\cos^2\theta + \cos^2(\frac{1}{2}\pi - \theta) = \cos^2\theta + \sin^2\theta = 1$ **9** $\tan\theta + \cot\theta = s/c + c/s =$
$(s^2 + c^2)/cs = 1/cs = \sec\theta\csc\theta$ **11** $(1 - \sin\theta)(1 + \csc\theta) =$
$(1 - s)(1 + 1/s) = 1 - s + 1/s - 1 = 1/s - s = (1 - s^2)/s = c^2/s =$
$c(c/s) = \cos\theta\cot\theta$ **13** $\sec\theta - \cos\theta = 1/c - c = (1 - c^2)/c = s^2/c =$
$s(s/c) = \sin\theta\tan\theta$ **15** $\sec^2\theta - \csc^2\theta = (1 + \tan^2\theta) - (1 + \cot^2\theta) =$
$\tan^2\theta - \cot^2\theta$ **17** $\sec^2\theta + \csc^2\theta = 1/c^2 + 1/s^2 = (s^2 + c^2)/s^2c^2 = 1/s^2c^2 =$
$\sec^2\theta\csc^2\theta$ **19** $\sec^4\theta - \tan^4\theta = (\sec^2\theta + \tan^2\theta)(\sec^2\theta - \tan^2\theta) =$
$\sec^2\theta + \tan^2\theta = (1 + \tan^2\theta) + \tan^2\theta = 1 + 2\tan^2\theta$ **21** $\cot^4\theta + \cot^2\theta =$
$\cot^2\theta(\cot^2\theta + 1) = \cot^2\theta\csc^2\theta = (\csc^2\theta - 1)\csc^2\theta = \csc^4\theta - \csc^2\theta$
23 $\text{LHS} = (c^2 - 2cs + s^2) + (c^2 + 2cs + s^2) = 2(c^2 + s^2) = 2$
25 $\text{LHS} = \dfrac{1 - s/c}{1 + s/c} = \dfrac{c - s}{c + s}$, $\text{RHS} = \dfrac{c/s - 1}{c/s + 1} = \dfrac{c - s}{c + s}$ **27** equivalent relation:
$\tan^2\theta = \sec\theta(\sec\theta - \cos\theta) = \sec^2\theta - 1$, which is an identity
29 $\text{RHS} = \left(\dfrac{s}{c} + \dfrac{c}{s}\right)(c + s) = \left(\dfrac{s^2 + c^2}{cs}\right)(c + s) = \dfrac{1}{cs}(c + s) = \dfrac{1}{s} + \dfrac{1}{c} = \text{LHS}$
31 $\text{LHS} = \dfrac{s/c - sc}{1/c} = s - sc^2 = s(1 - c^2) = ss^2 = s^3 = \text{RHS}$
33 $c^4 - c^2 = c^2(c^2 - 1) = (1 - s^2)(-s^2) = s^4 - s^2$ **35** $\text{LHS} = 2 - \tan\theta - \cot\theta =$
$2 - \dfrac{s}{c} - \dfrac{c}{s} = 2 - \dfrac{s^2 + c^2}{sc} = 2 - \dfrac{1}{sc} = \text{RHS}$ **37** $\text{LHS} = \dfrac{c}{1 - s/c} + \dfrac{s}{1 - c/s} =$
$\dfrac{c^2}{c - s} + \dfrac{s^2}{s - c} = \dfrac{c^2 - s^2}{c - s} = c + s = \text{RHS}$
39 Equivalent relations:

$$(\cos\alpha - \sin\beta)(\cos\alpha + \sin\beta) = (\cos\beta + \sin\alpha)(\cos\beta - \sin\alpha),$$
$$\cos^2\alpha - \sin^2\beta = \cos^2\beta - \sin^2\alpha,$$
$$\cos^2\alpha + \sin^2\alpha = \cos^2\beta + \sin^2\beta.$$

The last is an identity, hence so is the given relation.

Section 3 page 280

1 $\cos(\theta + 2\pi) = \cos\theta\cos 2\pi - \sin\theta\sin 2\pi = \cos\theta$
$\sin(\theta + 2\pi) = \sin\theta\cos 2\pi + \cos\theta\sin 2\pi = \sin\theta$
3 $\cos(\theta + \frac{1}{2}\pi) = \cos\theta\cos\frac{1}{2}\pi - \sin\theta\sin\frac{1}{2}\pi = -\sin\theta$
$\sin(\theta + \frac{1}{2}\pi) = \sin\theta\cos\frac{1}{2}\pi + \cos\theta\sin\frac{1}{2}\pi = \cos\theta$
5 $\frac{1}{2}\sqrt{2}(\cos\theta - \sin\theta)$ $\quad \frac{1}{2}\sqrt{2}(\cos\theta + \sin\theta)$
7 $(\cot\alpha\cot\beta - 1)/(\cot\alpha + \cot\beta)$ **9** $\frac{1}{4}(\sqrt{2} + \sqrt{6})$ **11** $\frac{1}{4}(\sqrt{2} + \sqrt{6})$
13 $\frac{1}{2}$ **15** $\frac{1}{2}\sqrt{2}$ **17** $\frac{5}{4}\pi$ **19** $\frac{1}{20} + \frac{3}{10}\sqrt{10}$ **21** $-\frac{4}{9}\sqrt{2}$ $\quad -\frac{7}{9}$ **23** $-\frac{15}{17}$ $\quad -\frac{8}{15}$
25 $4\cos^3\theta - 3\cos\theta$ **27** $\csc^3\theta/(3\csc^2\theta - 4)$ **29** $\sin 4\theta = 2\sin 2\theta\cos 2\theta = 2(2\sin\theta\cos\theta)(1 - 2\sin^2\theta) = 4\sin\theta\cos\theta - 8\sin^3\theta\cos\theta$ **31** $\cot 2\theta + \csc 2\theta = \dfrac{\cos 2\theta}{\sin 2\theta} + \dfrac{1}{\sin 2\theta} = \dfrac{\cos 2\theta + 1}{\sin 2\theta} = \dfrac{2\cos^2\theta}{2\sin\theta\cos\theta} = \dfrac{\cos\theta}{\sin\theta} = \cot\theta$ **33** LHS $= \dfrac{\sin\theta}{\cos\theta} + \dfrac{\cos\theta}{\sin\theta} = \dfrac{\sin^2\theta + \cos^2\theta}{\sin\theta\cos\theta} = \dfrac{1}{\sin\theta\cos\theta} = \dfrac{2}{2\sin\theta\cos\theta} = \dfrac{2}{\sin 2\theta} = 2\csc 2\theta$
35 $\tan 2\theta$ **37** $\cot 2\theta = \dfrac{1}{\tan 2\theta} = \dfrac{1 - \tan^2\theta}{2\tan\theta} = \dfrac{\cot^2\theta - 1}{2\cot\theta}$ **39** $\cot(\alpha + \beta) = \dfrac{1}{\tan(\alpha + \beta)} = \dfrac{1 - \tan\alpha\tan\beta}{\tan\alpha + \tan\beta} =$ RHS **41** $\dfrac{1 + \cos 2\theta}{1 - \cos 2\theta} = \dfrac{2\cos^2\theta}{2\sin^2\theta} = \cot^2\theta$
43 LHS $= \dfrac{1/c - 1/s}{1/c + 1/s} = \dfrac{s - c}{s + c} = \dfrac{(s - c)^2}{s^2 - c^2} = \dfrac{s^2 - 2sc + c^2}{-\cos 2\theta} = \dfrac{1 - \sin 2\theta}{-\cos 2\theta} = \tan 2\theta - \sec 2\theta$

Section 4 page 285

1 $\sin 22.5° = \sqrt{\frac{1}{2}(1 - \cos 45°)} = \sqrt{\frac{1}{2}(1 - \frac{1}{2}\sqrt{2})} = \frac{1}{2}\sqrt{2 - \sqrt{2}}$
3 $\tan 67.5° = \dfrac{\sin 135°}{1 + \cos 135°} = \dfrac{\frac{1}{2}\sqrt{2}}{1 - \frac{1}{2}\sqrt{2}} = \dfrac{\frac{1}{2}\sqrt{2}(1 + \frac{1}{2}\sqrt{2})}{1 - \frac{1}{2}} = \sqrt{2} + 1$
5 $\tan\frac{1}{2}\theta = \dfrac{\sin\theta}{1 + \cos\theta} = \dfrac{\sin\theta(1 - \cos\theta)}{1 - \cos^2\theta} = \dfrac{\sin\theta(1 - \cos\theta)}{\sin^2\theta} = \dfrac{1 - \cos\theta}{\sin\theta}$
7 $5\cos(\theta - \theta_0)$ $\quad \theta_0 = \arcsin\frac{3}{5} \approx 0.6435$ **9** $\sqrt{5}\cos(\theta + \theta_0)$ $\quad \theta_0 = \arcsin(2/\sqrt{5}) \approx 1.1071$ **11** $\sqrt{13}\cos(\theta - \theta_0)$ $\quad \theta_0 = \arcsin(3/\sqrt{13}) \approx 0.9828$
13 $\sqrt{10}\cos(\theta + \theta_0)$ $\quad \theta_0 = \arcsin(1/\sqrt{10}) \approx 0.3218$ **15** $|\sin\theta + \cos\theta| = |\sqrt{2}\cos(\theta - \frac{1}{4}\pi)| \le \sqrt{2}$ **17** $\sqrt{10}$ **19** $\frac{1}{2}(\cos\theta - \cos 3\theta)$
21 $\frac{1}{2}(\sin 7\theta - \sin\theta)$ **23** $2\sin\frac{3}{2}\theta\cos\frac{1}{2}\theta$ **25** $2\cos\frac{9}{2}\theta\sin\frac{1}{2}\theta$
27 $-2\sin\frac{11}{2}\theta\sin\frac{1}{2}\theta$ **29** LHS $= 2\sin 45°\cos 30° = 2(\frac{1}{2}\sqrt{2})(\frac{1}{2}\sqrt{3}) = \frac{1}{2}\sqrt{6}$
31 LHS $= 2\cos 60°\cos 45° = \frac{1}{2}\sqrt{2}$ **33** $\frac{1}{4}(\sqrt{6} + \sqrt{2})$ $\quad \frac{1}{4}(\sqrt{6} - \sqrt{2})$
35 $\sin(\alpha + \beta)/\cos\alpha\cos\beta = (\sin\alpha\cos\beta + \cos\alpha\sin\beta)/\cos\alpha\cos\beta = (\sin\alpha/\cos\alpha) + (\sin\beta/\cos\beta) = \tan\alpha + \tan\beta$ **37** $\sin^4\theta = [\frac{1}{2}(1 - \cos 2\theta)]^2 = \frac{1}{4}(1 - 2\cos 2\theta + \cos^2 2\theta) = \frac{1}{4}[1 - 2\cos 2\theta + \frac{1}{2}(1 + \cos 4\theta)] =$ RHS
39 $\cos(\alpha + \beta)\cos(\alpha - \beta) = (\cos\alpha\cos\beta - \sin\alpha\sin\beta)(\cos\alpha\cos\beta + \sin\alpha\sin\beta) = \cos^2\alpha\cos^2\beta - \sin^2\alpha\sin^2\beta = \cos^2\alpha(1 - \sin^2\beta) - (1 - \cos^2\alpha)\sin^2\beta = \cos^2\alpha - \sin^2\beta$ **41** LHS $= \sin\theta(\sin 2\theta\cos 2\theta) = \sin\theta(\frac{1}{2}\sin 4\theta) = \frac{1}{2}\sin\theta\sin 4\theta = \frac{1}{4}(\cos 3\theta - \cos 5\theta)$ **43** $4t(1 - t^2)/(1 + t^2)^2$
45 $\dfrac{\sin(\beta - \alpha)}{\sin\alpha\sin\beta} = \dfrac{\sin\beta\cos\alpha - \sin\alpha\cos\beta}{\sin\alpha\sin\beta} = \dfrac{\cos\alpha}{\sin\alpha} - \dfrac{\cos\beta}{\sin\beta} = \cot\alpha - \cot\beta$
47 LHS $= \dfrac{2\cos\frac{1}{2}(\alpha + \beta)\sin\frac{1}{2}(\alpha - \beta)}{2\cos\frac{1}{2}(\alpha + \beta)\cos\frac{1}{2}(\alpha - \beta)} = \dfrac{\sin\frac{1}{2}(\alpha - \beta)}{\cos\frac{1}{2}(\alpha - \beta)} = \tan\frac{1}{2}(\alpha - \beta)$
49 LHS $= \dfrac{2\sin\frac{1}{2}(\alpha + \beta)\cos\frac{1}{2}(\alpha - \beta)}{2\cos\frac{1}{2}(\alpha + \beta)\cos\frac{1}{2}(\alpha - \beta)} = \dfrac{\sin\frac{1}{2}(\alpha + \beta)}{\cos\frac{1}{2}(\alpha + \beta)} = \tan\frac{1}{2}(\alpha + \beta)$
51 LHS $\approx 0.642788 + 0.342020 = 0.984808$ $\quad$ RHS ≈ 0.984808
53 $\sin\frac{1}{6}\pi < \frac{1}{6}\pi < \tan\frac{1}{6}\pi$ $\quad \frac{1}{2} < \frac{1}{6}\pi < \frac{1}{3}\sqrt{3}$ $\quad 3 < \pi < 2\sqrt{3} < 3.47$
55 $3.132 < \pi < 3.160$

Section 5 page 292

1 $-\frac{1}{6}\pi$ **3** $-\frac{1}{4}\pi$ **5** $\frac{2}{3}\pi$ **7** $\frac{5}{6}\pi$ **9** $\frac{1}{6}\pi$ **11** $\frac{1}{3}\pi$ **13** 0.3 **15** 1
17 $\frac{1}{3}$ **19** Set $\theta = \arcsin x$ so $-\frac{1}{2}\pi \le \theta \le \frac{1}{2}\pi$. Then $-\frac{1}{2}\pi \le -\theta \le \frac{1}{2}\pi$ and $\sin(-\theta) = -\sin\theta = -x$, hence $\arcsin(-x) = -\theta = -\arcsin x$. **21** Same solution as Ex. 19 with "sin" replaced by "tan" **23** $\cot(\arctan x) = 1/\tan(\arctan x) = 1/x$

25 $-\frac{1}{2}\pi \le \frac{1}{2}\pi - \theta \le \frac{1}{2}\pi$ and $\sin(\frac{1}{2}\pi - \theta) = \cos\theta$, hence $\arcsin(\cos\theta) = \frac{1}{2}\pi - \theta$ **27** Set $\theta = \arccos x$ so $0 \le \theta \le \frac{1}{2}\pi$. Then $2x^2 - 1 = 2\cos^2\theta - 1 = \cos 2\theta$ and $0 \le 2\theta \le \pi$, hence $\arccos(2x^2 - 1) = 2\theta = 2\arccos x$. **29** Set $\alpha = \arctan\frac{1}{2}$ and $\beta = \arctan\frac{1}{3}$ so $0 < \alpha < \frac{1}{4}\pi$ and $0 < \beta < \frac{1}{4}\pi$. We have $0 < \alpha + \beta < \frac{1}{2}\pi$ and $\tan(\alpha + \beta) = (\frac{1}{2} + \frac{1}{3})/(1 - \frac{1}{2}\cdot\frac{1}{3}) = 1$, hence $\alpha + \beta = \arctan 1 = \frac{1}{4}\pi$. **31** $\arccos\sqrt{1 - x^2} = -\arcsin x$ **33** $55°$
35 $\sqrt{0.51}$ **37** $\frac{2}{9}\sqrt{8}$ **39** $-\frac{5}{12}$ **41** $-\frac{16}{65}$ **43** $\cot\theta$ is strictly decreasing on $\{0 < \theta < \pi\}$ with range $\{\text{all } x\}$, hence it has an inverse, also strictly decreasing, with domain $\{\text{all } x\}$ and range $\{0 < \theta < \pi\}$. **45** Let $\theta = \arctan x$, so $-\frac{1}{2}\pi < \theta < \frac{1}{2}\pi$ and hence $0 < \frac{1}{2}\pi - \theta < \pi$. Now $\cot(\frac{1}{2}\pi - \theta) = \tan\theta = x$, so $\frac{1}{2}\pi - \theta = \operatorname{arc\,cot} x$, that is, $\arctan x + \operatorname{arc\,cot} x = \frac{1}{2}\pi$.
47 The function is strictly increasing, with range $\{|x| \ge 1\}$, hence has an inverse function with domain $\{|x| \ge 1\}$ and range $\{0 \le \theta \le \pi, \theta \ne \frac{1}{2}\pi\}$

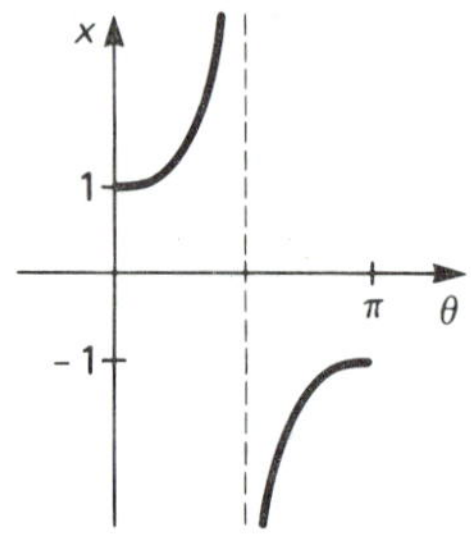

49 The function is strictly decreasing, with range $\{|x| \ge 1\}$, hence has an inverse function with domain $\{|x| \ge 1\}$ and range $\{-\frac{1}{2}\pi \le \theta \le \frac{1}{2}\pi, \theta \ne 0\}$

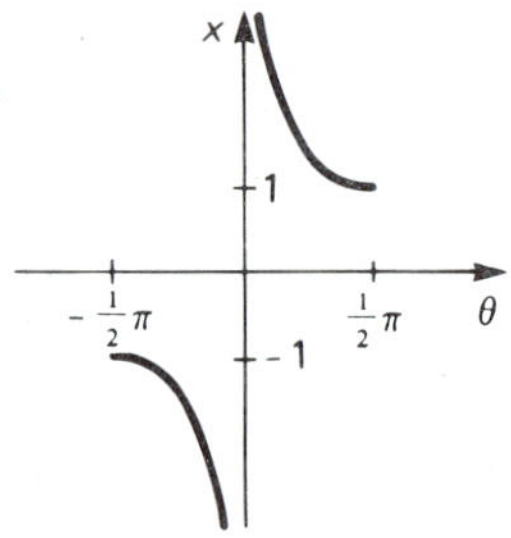

51 0.2527 **53** 1.2793 **55** −0.9582
57 6.91° **59** −89.81° **61** 90°
63

x	0.1	0.05	0.01
$\arcsin x$	0.1001674	0.050020857	0.0100001667
$x + \frac{1}{6}x^3$	0.1001667	0.050020833	0.0100001667

Conclusion: $\arcsin x \approx x + \frac{1}{6}x^3$ for $x \approx 0$
65 $\alpha + \beta = \arctan\frac{1}{3} + \arctan\frac{1}{2} = \frac{1}{4}\pi = \gamma$ by Ex. 29

Section 6 page 296

1 $\frac{1}{4}\pi$ $\frac{3}{4}\pi$ **3** ± 2.5185 **5** $\frac{1}{3}\pi$ $-\frac{2}{3}\pi$ **7** $\frac{1}{6}\pi$ $-\frac{5}{6}\pi$ $\frac{1}{3}\pi$ $-\frac{2}{3}\pi$
9 $\pm\frac{1}{12}\pi$ $\pm\frac{3}{4}\pi$ $\pm\frac{7}{12}\pi$ **11** $\frac{1}{6}\pi$ $\frac{5}{12}\pi$ $\frac{2}{3}\pi$ $\frac{11}{12}\pi$ $-\frac{1}{12}\pi$ $-\frac{1}{3}\pi$ $-\frac{7}{12}\pi$ $-\frac{5}{6}\pi$
13 ± 0.7782 ± 2.8726 ± 1.3162 **15** 0.4774 −2.6642 1.0934 −2.0482
17 210° 330° **19** 30° 210° **21** $\pm\frac{1}{4}\pi$ $\pm\frac{3}{4}\pi$ **23** ± 1.2310 **25** ± 1.9455
27 $-\frac{1}{4}\pi$ $\frac{3}{4}\pi$ 0.3218 −2.8198 **29** ± 1.2995 **31** no solutions **33** 0 π $\frac{1}{6}\pi$ $\frac{5}{6}\pi$
35 0 π **37** 0 π $\pm\frac{1}{2}\pi$ **39** $\frac{1}{4}\pi$ $-\frac{3}{4}\pi$ **41** $\pm\frac{2}{3}\pi$ **43** 0 $\pm\frac{2}{3}\pi$
45 all $\theta \ne n\pi$, $(\frac{1}{2} + n)\pi$ **47** 0 $\pm\frac{1}{2}\pi$ π **49** 0
51 0 $\pm\frac{1}{4}\pi$ $\pm\frac{1}{2}\pi$ $\pm\frac{3}{4}\pi$ π

Section 7 page 301

1 1.5023 **3** 1.9346 **5** 0.6533 **7** 2.8064 **9** −1.1541 **11** 1.4973
13 0.4502 **15** 4.5364

Review Exercises page 306

1 $-\frac{2}{3}\sqrt{2}$ **3** $\tan^2\theta - \sin^2\theta\tan^2\theta = \tan^2\theta(1 - \sin^2\theta) = \tan^2\theta\cos^2\theta = \sin^2\theta$, hence $\tan^2\theta - \sin^2\theta = \sin^2\theta\tan^2\theta$ **5** $\text{LHS} = \dfrac{1}{1/c + s/c} = \dfrac{c}{1 + s} = \dfrac{c(1 - s)}{1 - s^2} = \dfrac{c(1 - s)}{c^2} = \dfrac{1 - s}{c} = \text{RHS}$ **7** $4\cos\frac{1}{2}\theta\cos\frac{1}{4}\theta\sin\frac{1}{4}\theta = 2\cos\frac{1}{2}\theta(2\cos\frac{1}{4}\theta\sin\frac{1}{4}\theta) = 2\cos\frac{1}{2}\theta\sin\frac{1}{2}\theta = \sin\theta$ **9** $\sin 105° = \sin 60°\cos 45° + \cos 60°\sin 45° = \frac{1}{2}\sqrt{3}\cdot\frac{1}{2}\sqrt{2} + \frac{1}{2}\cdot\frac{1}{2}\sqrt{2} = \frac{1}{4}(\sqrt{6} + \sqrt{2})$ **11** $13\cos(\theta - \theta_0)$ where $\cos\theta_0 = -\frac{5}{13}$ and $\sin\theta_0 = \frac{12}{13}$ **13** $2\sin 2\theta\sin\theta$ **15** $-\frac{3}{14}\pi$

17

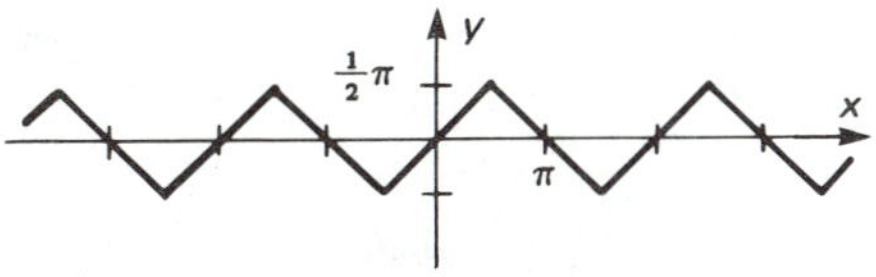

19 $-\frac{7}{9}\pi$ $-\frac{1}{9}\pi$ $\frac{5}{9}\pi$ $-\frac{8}{9}\pi$ $-\frac{2}{9}\pi$ $\frac{4}{9}\pi$ **21** $\frac{2}{3}\pi n$ or $(2n + 1)\pi$
23 1.3987 −0.7966

CHAPTER 8

Section 1 page 310

	a	b	c	α	β
1		5		67.4°	22.6°
3		138.9		37.69°	52.31°
5	16.1			47.0°	43.0°
7	48.56			46.67°	43.33°
9			6.4	38.7°	51.3°
11			28.68	71.04°	18.96°
13	7.51	29.0			75.5°
15		31.3	34.7		64.4°
17	194.8	132.4		55.8°	
19	58.22		80.80	46.1°	
21	0.22078	1.0017			77.57°
23		35.0636	35.2019		84.92°

25 0.000278° **27** 0.016389° **29** 0.203333° **31** 0.5° **33** 17.060833°
35 73.006389° **37** 71.964167° **39** 2.616667° **41** 19° 6′ 29.5″
43 72° 56′ 59.6″ **45** 1′ 12″ **47** 1.72″ **49** 20° 35′ 11″ **51** 69° 24′ 49″
53 72° 8′ 50″ **55** 1″ **57** 0.26228 **59** 0.27797 **61** 1.33495
63 4.8481×10^{-6} **65** −0.68924 **67** −0.96109 **69** about -2.5×10^{-11}
71 1.7×10^{-7} **73** 4.8×10^{-6} **75** $\frac{1}{30}(D + \frac{1}{60}M + \frac{1}{3600}S)$ hours
77 $\frac{1}{6}(D + \frac{1}{60}M + \frac{1}{3600}S)$ sec

Section 2 page 317

1 77.8 ft **3** 14.1 ft **5** 0.39 km **7** 2.2 km **9** 0.885
11 Refer to Fig. 4(b) of the text, with 6 replaced by R_0, 3 by $R_1 - R_0$, and 20 by D. Then $\theta = \text{arc cos}[(R_1 - R_0)/D]$, $L_0 = 2R_0\theta$, $L_1 = R_1(2\pi - 2\theta)$, and $L_2 = \sqrt{D^2 - (R_1 - R_0)^2}$. Finally $L = L_0 + L_1 + 2L_2$.
13 When $R_0 = R_1$, the formula yields $L = 2D + 2\pi R_1$, which checks with the figure.

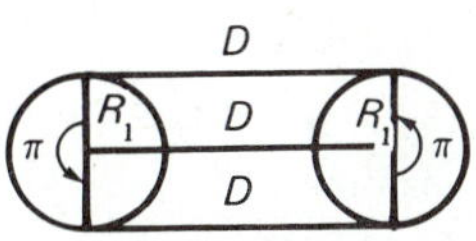

15 $\sqrt{D^2 - 4R_1^2} + 4R_1(\pi - \theta)$ **17** 50.41 m 5.53 m **19** 113.3 m 11.2 m
21 22.1 m 1.6 m **23** 84.52 m 3.96 m **25** 54.78° 43.87° 67.39° **27** $3r\sqrt{3}$
29 $8r\sqrt{2 - \sqrt{2}}$ **31** $s = 2a \text{ arc tan}(y/a)$ **33** $A = \frac{1}{4}b^2/\tan\frac{1}{2}\beta$ **35** $2r^2\sqrt{2}$
37 30° **39** $b = \sqrt{h^2 + \frac{1}{2}a^2}$ $\theta = 2 \text{ arc tan}(a/2\sqrt{h^2 + \frac{1}{4}a^2})$ **41** 5°26′19″
43 24°51′54″ **45** 2.4060 cm **47** 2.3196 cm **49** 34′22.6″ **51** 1°30′54.9″
53 12°42′50″ 87.592 **55** 39.37 m 136.93 m

Section 3 page 324

	α	β	γ
1	28.96°	46.57°	104.48°
5	42.61°	39.80°	97.59°
9	14.47°	17.62°	147.92°

	α	β	γ
3	50.35°	31.50°	98.16°
7	21.56°	146.76°	11.68°
11	no such triangle		

	c	α	β
13	3.269	105.72°	35.28°
15	6.590	4.69°	164.17°

	a	β	γ
17	144.8	24.65°	2.98°
19	35050	33° 14′ 50″	107° 27′ 58″

	b	γ	α
21	88.80	176.57°	0.37°

	b	γ	α
23	3.4231	28° 47′ 8″	151° 12′ 18″

25 3.9324 cm 12.1258 cm **27** 149.5 mi **29** 50.44° 56.71° 72.86°

31 $d_1^2 + d_2^2 = [a^2 + b^2 - 2ab\cos\theta] + [a^2 + b^2 - 2ab\cos(\pi - \theta)]$
$= 2(a^2 + b^2) - 2ab\cos\theta + 2ab\cos\theta = 2(a^2 + b^2)$

33 283.6 ft **35** 3.522 **37** 14.979

Section 4 page 332

1 $b \approx 3.833$ $c \approx 4.115$ $\gamma = 72.17°$ **3** $b \approx 3.278$ $a \approx 14.08$ $\gamma = 63.84°$
5 $c \approx 0.5428$ $a \approx 1.108$ $\alpha = 135.91°$ **7** $b \approx 0.3839$ $c \approx 0.3194$ $\alpha = 135.91°$
9 $b \approx 0.7839$ $a \approx 1.332$ $\alpha = 135.91°$ **11** no sol.: $a \sin\gamma > c$
13 no sol.: $c \sin\alpha > a$ **15** $b = 3$ $\beta = 45°$ $\alpha = 90°$
17 $\alpha \approx 42.09°$ $\beta \approx 106.67°$ $b \approx 7.132$; $\alpha \approx 137.91°$ $\beta \approx 10.85°$ $b \approx 1.401$
19 $\alpha \approx 78.76°$ $\beta \approx 27.30°$ $b \approx 1455$; $\alpha \approx 101.24°$ $\beta \approx 4.82°$ $b \approx 266.8$
21 $\alpha \approx 115.43°$ $\beta \approx 37.26°$ $a \approx 1.006$; $\alpha \approx 9.95°$ $\beta \approx 142.74°$ $a \approx 0.1924$
23 $\beta \approx 68.63°$ $\gamma \approx 56.38°$ $c \approx 42.22$; $\beta \approx 111.37°$ $\gamma \approx 13.64°$ $c \approx 11.96$
25 $\alpha \approx 38.21°$ $\gamma \approx 123.44°$ $c \approx 55080$; $\alpha \approx 141.79°$ $\gamma \approx 19.86°$ $c \approx 22430$
27 $\alpha \approx 22.80°$ $\beta \approx 74.08°$ $b \approx 37.75$ **29** $\alpha \approx 6.83°$ $\gamma \approx 163.79°$ $c \approx 1232$
31 $\alpha \approx 37.81°$ $\beta \approx 3.58°$ $b \approx 6.836$ **33** $\beta \approx 36.87°$ $\gamma \approx 44.96°$ $b \approx 1847$
35 $\beta \approx 58.89°$ $\gamma \approx 2.81°$ $c \approx 0.4015$

Section 5 page 337

1 5.3 mi from A, 8.0 mi from B, 4.2 mi from the line **3** 4.335 m **5** 2.992
7 21.9 ft **9** $\sqrt{43} \approx 6.557$ **11** 2.4718 0.94356 **13** 2.571150 2.694593
15 $\frac{15}{4}$ **17** 0.9473 **19** 7620. **21** 23.05 **23** 8.162 **25** 104.6
27 1047×10^2 **29** two triangles: 148.0 27.05 **31** 104.5 **33** 2168×10^2
35 The second figure is inaccurate; there is a skinny diamond missing from the center of the figure. For instance, the sum of the two angles at the lower left corner is $\theta = \arctan\frac{3}{8} + \arctan\frac{5}{2} < 90°$; in fact,

$$\tan\theta = \frac{\frac{3}{8} + \frac{5}{2}}{1 - \frac{3}{8}\cdot\frac{5}{2}} = \frac{\frac{23}{8}}{\frac{1}{16}} = 46 \qquad \theta \approx 88.75°$$

Section 6 page 344

1 $206\frac{1}{4}$ m **3** 2.063×10^5 **5** $1 \approx d\tan\alpha \approx d\alpha$ so $1/\alpha \approx d$

	parallax (″)	pc	ly	km
7		1.32	4.29	4.06×10^{13}
9	0.431		7.57	7.16×10^{13}
11	0.377	2.65		8.18×10^{13}
13	0.345	2.90	9.45	

15 $75 + 50 + 24.5 + 2.0 + 1.47 + 1.006 = 153.976$ mm **17** $90° - 9° + 1° = 82°$
19 $41° - 27° + 1° - 27' + 9' - 3' - 9'' = 14°38'51''$ **21** 5; one part in 1000
23 6.02224; 1.6 parts in 1000 **25** $\frac{2}{10}$ of 1% **27** 0.68 ft

Review Exercises page 345

1 $\beta = 68.86°\quad c \approx 35.96\quad b \approx 33.54$ **3** 81.0 ft **5** 127.2° **7** 2.122
9 $b \approx 8.081\quad \beta \approx 145.11°\quad \gamma \approx 20.69°;\quad b \approx 1.596\quad \beta \approx 6.49°\quad \gamma \approx 159.31°$
11 3.69 m **13** 5.5169; 1.2%

CHAPTER 9

Section 1 page 353

1 $(-\frac{8}{19}, \frac{1}{19}, \frac{7}{19})$ **3** $(0, 0, 0)$ **5** $(\frac{3}{14}, -\frac{1}{4}, -\frac{3}{28})$ **7** $(\frac{12}{7}, 0, -\frac{1}{7})$
9 $(-\frac{5}{2}, -\frac{3}{2}, \frac{9}{2})$ **11** $(\frac{1}{2}, -\frac{2}{3}, -1)$ **13** $(1, 1, 1)$ **15** $(1.0, 2.1, -3.2)$
17 $(3, -1, 2)$ **19** $y = \frac{3}{2}x^2 + \frac{1}{2}x$ **21** $y = \frac{1}{2}x^2 + \frac{1}{2}$ **23** inconsistent
25 $(x, y, z) = (t, -\frac{1}{2}t - \frac{1}{2}, -\frac{5}{2}t + \frac{1}{2})$, t arbitrary **27** $(5, 6, 0, -1)$
29 $(2, 1, -1, -3)$ **31** $(-\frac{1}{4}, 0, -\frac{3}{4}, \frac{13}{4})$ **33** $(6, -\frac{3}{2}, -\frac{9}{2}, -3)$
35 $x^2 + y^2 - 2x - 2y = 0$ **37** $x^2 + y^2 + 2x + 2y - 3 = 0$
39 $y = 2x^3 - 5x^2 + x - 3$

Section 2 page 359

1 $(1, \frac{3}{2}, 1)$ **3** $(\frac{11}{14}, -\frac{3}{7}, -\frac{1}{2})$ **5** $(\frac{7}{19}, -\frac{1}{38}, -\frac{8}{19})$ **7** $(\frac{1}{2}, -1, \frac{1}{4})$ **9** $(\frac{2}{3}, \frac{5}{3}, -\frac{5}{3})$

11 53 **13** -6 **15** $\begin{bmatrix} 4 \\ -3 \\ 10 \end{bmatrix}$ **17** $\begin{bmatrix} 2 \\ 8 \\ 15 \end{bmatrix}$ **19** $\begin{bmatrix} -8 \\ 5 \\ -3 \end{bmatrix}$ **21** $\begin{bmatrix} 35 \\ -10 \\ 51 \end{bmatrix}$

23 $\begin{bmatrix} 29 & 15 & 25 \\ 12 & 20 & 6 \\ 0 & 0 & 4 \end{bmatrix}$ **25** $\begin{bmatrix} 3 & 4 & 1 \\ 1 & 1 & 2 \\ -4 & -17 & -7 \end{bmatrix}$ **27** $\begin{bmatrix} 2 & -1 & -1 \\ -1 & 2 & -1 \\ -1 & -1 & 2 \end{bmatrix}$ **29** $\begin{bmatrix} 14 & 14 & 10 \\ 14 & 17 & 14 \\ 10 & 14 & 14 \end{bmatrix}$

31 $\begin{vmatrix} 5 & -3 & -2 \\ -3 & 18 & -6 \\ -2 & -6 & 5 \end{vmatrix}$ **33** $\begin{bmatrix} 22 & 0 \\ 0 & 22 \end{bmatrix}$ **35** $\begin{vmatrix} 2 & -1 \\ 3 & 1 \\ 4 & -1 \end{vmatrix}$ **37** $\begin{bmatrix} 3 & -9 \\ 2 & 10 \end{bmatrix}$

Section 3 page 367

1 $\begin{bmatrix} 1 & 0 & 0 \\ -1 & 1 & 0 \\ 0 & -1 & 1 \end{bmatrix}$ **3** $\begin{bmatrix} 1 & 0 & 0 \\ \frac{1}{2} & \frac{1}{2} & 0 \\ \frac{1}{2} & \frac{1}{6} & \frac{1}{3} \end{bmatrix}$ **5** $\begin{bmatrix} 21 & 9 & -11 \\ -7 & -3 & 4 \\ -2 & -1 & 1 \end{bmatrix}$ **7** $\begin{bmatrix} -\frac{13}{2} & \frac{1}{2} & 2 \\ -11 & 1 & 3 \\ -3 & 0 & 1 \end{bmatrix}$

9 no inverse **11** no inverse **13** $\begin{bmatrix} -\frac{1}{3} & -1 & \frac{1}{3} \\ 0 & -3 & 2 \\ \frac{2}{3} & 0 & \frac{1}{3} \end{bmatrix}$ **15** $\begin{bmatrix} \frac{4}{3} & \frac{1}{3} & -1 \\ -3 & 0 & 2 \\ \frac{7}{3} & -\frac{2}{3} & -1 \end{bmatrix}$

17 $\begin{bmatrix} 0 & 1 & 0 \\ 0 & 1 & 1 \\ 1 & 1 & 1 \end{bmatrix} \begin{bmatrix} 0 \\ 2 \\ 3 \end{bmatrix}$ **19** $\begin{bmatrix} 1 & 2 & 0 \\ 3 & 4 & -3 \\ -1 & -1 & 2 \end{bmatrix} \begin{bmatrix} 1 \\ -2 \\ 2 \end{bmatrix}$ **21** $\begin{bmatrix} 2 & 4 & 6 \\ -1 & -1 & -3 \\ -1 & -5 & -2 \end{bmatrix} \begin{bmatrix} 4 \\ -2 \\ -1 \end{bmatrix}$

23 $\begin{bmatrix} \frac{4}{3} & -3 & \frac{7}{3} \\ \frac{1}{3} & 0 & -\frac{2}{3} \\ -1 & 2 & -1 \end{bmatrix} \begin{bmatrix} \frac{2}{3} \\ -\frac{1}{3} \\ 0 \end{bmatrix}$ **25** $\begin{bmatrix} 2 & -1 & -1 \\ 0 & 2 & 1 \\ -3 & 4 & 3 \end{bmatrix} \begin{bmatrix} 2 \\ 0 \\ -3 \end{bmatrix} \begin{bmatrix} -1 \\ 2 \\ 4 \end{bmatrix} \begin{bmatrix} -1 \\ 1 \\ 3 \end{bmatrix}$

Section 4 page 375

1 7 **3** -17 **5** a^2 **7** 1
9 $(-3)(-2)(-1) + 0 \cdot 1 \cdot 5 + 3 \cdot 4 \cdot 2 - (-3) \cdot 1 \cdot 2 - 0 \cdot 4 \cdot (-1) - 3 \cdot (-2) \cdot 5 = 54$
11 $3\begin{vmatrix} 4 & -2 \\ 5 & 2 \end{vmatrix} - 1 \cdot \begin{vmatrix} -3 & 0 \\ 5 & 2 \end{vmatrix} + (-1)\begin{vmatrix} -3 & 0 \\ 4 & -2 \end{vmatrix} = 3(18) - (-6) - (6) = 54$
13 $-4\begin{vmatrix} 0 & 3 \\ 2 & -1 \end{vmatrix} + (-2)\begin{vmatrix} -3 & 3 \\ 5 & 1 \end{vmatrix} - 1 \cdot \begin{vmatrix} -3 & 0 \\ 5 & 2 \end{vmatrix} = -4(-6) - 2(-12) - (-6) = 54$
15 -12 **17** 16 **19** -65 **21** -130 **23** -6
25 -17 **27** 5 **29** $x^3 + 2x$ **31** 0
33 $AB = \begin{bmatrix} 47 & 23 \\ 30 & 15 \end{bmatrix}$ $\det(AB) = 15 \quad (\det A)(\det B) = 3 \cdot 5 = 15$
35 $AB = \begin{bmatrix} 2 & -3 & 1 \\ -19 & 31 & -11 \\ 7 & -11 & 4 \end{bmatrix}$ $\det(AB) = 1 \quad (\det A)(\det B) = 1 \cdot 1 = 1$
37 $(-\frac{1}{3}, \frac{2}{3})$ **39** $(-1, 1)$ **41** $(\frac{1}{19}, \frac{7}{19})$ **43** $(\frac{5}{2}, 2, \frac{3}{2})$ **45** $(-\frac{8}{15}, -\frac{13}{15}, \frac{22}{15})$
47 $(\frac{13}{3}, \frac{5}{3}, -3)$ **49** $(\frac{3}{14}, -\frac{1}{4}, -\frac{3}{28})$ **51** Expand by the last row: $3x - 4y + 11 = 0$, a linear equation satisfied by $(3, 5)$ and by $(-1, 2)$.

Section 5 page 384

1 $q(x) = 1 \quad r(x) = 4$ **3** $q(x) = \frac{1}{2} \quad r(x) = \frac{3}{2}$ **5** $q(x) = x + 1 \quad r(x) = 5$
7 $q(x) = 2 \quad r(x) = -2$ **9** $q(x) = 2x^2 + 2x + 4 \quad r(x) = 5$
11 $q(x) = x - 1 \quad r(x) = x$ **13** $q(x) = x + 1 \quad r(x) = 2x + 2$
15 $q(x) = \frac{1}{2} \quad r(x) = 2x - \frac{1}{2}$ **17** $q(x) = x^2 - x \quad r(x) = x + 1$
19 $q(x) = x^3 - x \quad r(x) = x$ **21** 2 **23** 6 **25** 1 **27** $1^{61} - 1^{12} - 1^4 + 1 = 0$, so $x - 1$ is a divisor by the factor theorem. **29** $y = x$ **31** $y = x$ **33** $y = 2x$

35

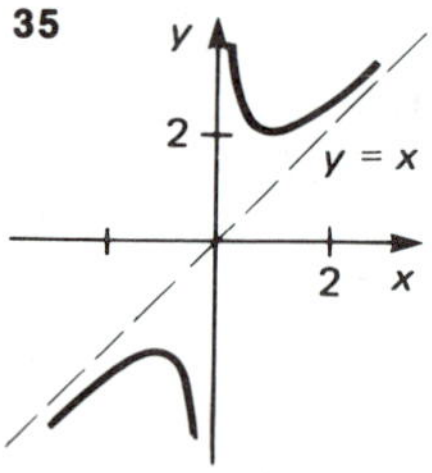

37

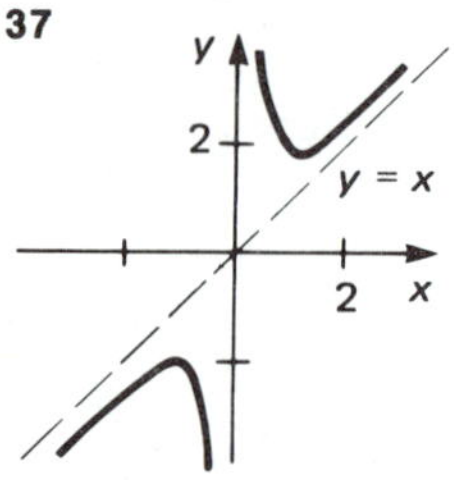

39

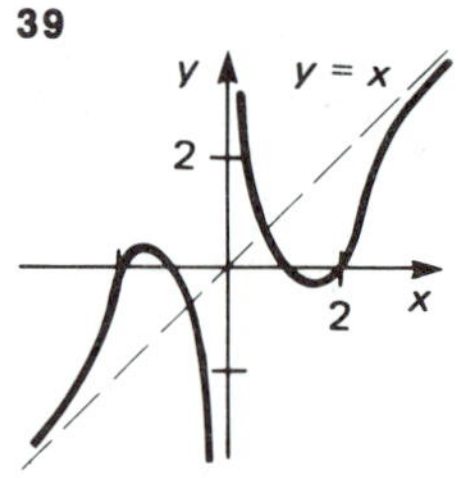

41 5 **43** 94 **45** 119 **47** no **49** yes

Section 6 page 392

1 $a(x - 3) \quad a \neq 0$ **3** $(x - 2)(ax + b) \quad a \neq 0$ **5** $a(x - 2)^2 \quad a \neq 0$
7 $ax^2 - 5ax + c \quad a \neq 0$ **9** The graph of $y = ax^2 + bx + c \quad (a \neq 0)$ meets $y = k$ if and only if $ax^2 + bx + c = k$, that is, $ax^2 + bx + (c - k) = 0$. This quadratic has at most two real solutions. **11** The quadratic equation $a_1x^2 + b_1x + c_1 = a_2x^2 + b_2x + c_2$ has at most two real roots, except when the two sides are identical; then the graphs coincide. **13** If $x > 0$, then $ax^2 > 0$, $bx > 0$, so $ax^2 + bx + c > 0$, not $= 0$. **15** $c \leq \frac{9}{4}$ **17** $-\frac{1}{2}$, mult. 1
19 -2, mult. 2; 2, mult. 2 **21** -2, mult. 1; -1, mult. 2; 1, mult. 3
23 $(ax^3 + bx^2 + cx + d) - k = 0 \quad (a \neq 0)$ is a cubic equation; at most 3 roots.
25 $(ax^3 + bx^2 + cx + d) - (mx + n) = 0 \quad (a \neq 0)$ is a cubic equation; at most 3 roots.
27 $a(x - 1)^2(x - 2)^2(x - 3)^2 \quad a > 0$ **29** Yes, because $x - x^5 - 1000$ is a polynomial of odd degree, hence has a real zero. **31** $\frac{17}{100}x(x - 5)(x - 8)$
33 -1, mult. 1 **35** -1 and 2 of mult. 1 **37** $\frac{2}{3}$, mult. 1 **39** -1 and $\frac{1}{2}$ of mult. 1
41 $2 < r < 3$ **43** $f(1) = -5 < 0$ and $f(2) = 29 > 0$, so there is a zero between 1 and 2. Clearly $f(x)$ is strictly increasing for $x > 0$, so $f(x) > 29$ for $x > 2$.
45 Each term $a_rx^r > 0$ for $x > 0$, hence $f(x) > 0$ for $x > 0$.
47 $x^n + a_1x^{n-1} + \cdots + a_n = (x - r_1)(x - r_2) \cdots (x - r_n)$. The coefficient of x^{n-1} on the RHS is $-(r_1 + \cdots + r_n)$. **49** 1, mult. 2 **51** $-1, \frac{1}{2}, -1 - \sqrt{2}$, $-1 + \sqrt{2}$, all of mult. 1 **53** $\frac{3}{2}$, mult. 2; an irrational zero r of mult. 1 in the interval $-2 < r < -1$

Section 7 page 399

1 $\dfrac{\frac{1}{2}}{x} - \dfrac{\frac{1}{2}}{x + 4}$ **3** $\dfrac{1}{x + 1} - \dfrac{2}{x + 2}$ **5** $-\dfrac{1}{x} + \dfrac{2}{x - 1}$ **7** $\dfrac{1}{x - 1} + \dfrac{1}{x + 2}$

9 $\dfrac{-7}{x + 2} + \dfrac{9}{x + 3}$ **11** $\frac{1}{4} + \dfrac{\frac{1}{8}}{2x - 1} - \dfrac{\frac{1}{8}}{2x + 1}$ **13** $\dfrac{1}{x} + \dfrac{1}{x^2}$

15 $\dfrac{2}{x + 2} - \dfrac{5}{(x + 2)^2}$ **17** $1 - \dfrac{2}{x + 1} + \dfrac{1}{(x + 1)^2}$ **19** $\dfrac{2}{x - 2} - \dfrac{1}{x - 1}$

21 $\dfrac{\frac{1}{2}}{x - 1} - \dfrac{4}{x - 2} + \dfrac{\frac{9}{2}}{x - 3}$ **23** $\dfrac{\frac{3}{2}}{x - 2} - \dfrac{4}{x - 3} + \dfrac{\frac{5}{2}}{x - 4}$

25 $\dfrac{1}{x} + \dfrac{\frac{9}{4}}{x - 2} - \dfrac{\frac{5}{4}}{x + 2}$ **27** $-\dfrac{1}{x} + \dfrac{1}{x^2} + \dfrac{1}{x + 1}$

29 $\dfrac{-\frac{1}{4}}{x + 1} + \dfrac{\frac{3}{2}}{(x + 1)^2} + \dfrac{\frac{1}{4}}{x - 1}$ **31** $\dfrac{-5}{x - 2} - \dfrac{1}{(x - 2)^2} + \dfrac{6}{x - 3}$

33 $\dfrac{1}{(x + 1)^2} - \dfrac{1}{(x + 1)^3}$ **35** $\dfrac{1}{x} + \dfrac{3}{x^2} + \dfrac{1}{x^3}$

37 $1 - \dfrac{3}{x + 1} + \dfrac{3}{(x + 1)^2} - \dfrac{1}{(x + 1)^3}$ **39** $\dfrac{\frac{9}{10}}{x + 3} + \dfrac{\frac{1}{10}x - \frac{3}{10}}{x^2 + 1}$

41 $\dfrac{-\frac{1}{2}}{x + 2} + \dfrac{\frac{1}{2}x}{x^2 + 4}$ **43** $\dfrac{x}{x^2 + 1} - \dfrac{x}{(x^2 + 1)^2}$ **45** $\dfrac{x}{x^2 + 2} - \dfrac{2x}{(x^2 + 2)^2}$

47 $\dfrac{x - 2}{x^2 + 2x + 3} + \dfrac{x + 5}{(x^2 + 2x + 3)^2}$ **49** $-\dfrac{1}{x} + \dfrac{3}{x - 1} - \dfrac{3}{x - 2} + \dfrac{1}{x - 3}$

Review Exercises page 400

1 $(-47, 6, 27)$ **3** $(-\frac{4}{3}, 1, \frac{5}{3})$ **5** $\begin{bmatrix} 13 \\ 12 \\ -23 \end{bmatrix}$ **7** $\begin{bmatrix} 1 & -2 & -7 \\ 0 & 1 & 2 \\ -3 & 4 & 18 \end{bmatrix}$ **9** -1

11 $q(x) = -3x^3 - 4x^2 - 3x + 2$ $r(x) = 9x - 2$ **13** $-1 - \sqrt{2}$ and $-1 + \sqrt{2}$ of mult. 3; -1 and 1 of mult. 2 **15** -3 and 3 of mult. 1; $\frac{1}{2}$ of mult. 2

17 $\frac{1}{2} + \dfrac{\frac{7}{5}}{x - 2} - \dfrac{\frac{3}{10}}{2x + 1}$

CHAPTER 10

Section 1 page 406

1 $7n - 3$ 81 **3** 2^{n-4} 256 **5** $\frac{1}{2}n(n + 1)$ **7** 575 **9** -55

11 $10x + 90$ **13** $2na + \frac{3}{2}n(n + 1)b$ **15** 3069 **17** $\frac{255}{16}$ **19** 2032

21 $\dfrac{x^6 - 1}{x^6(x - 1)}$ **23** $ab^2c^5\dfrac{b^7c^{14} - 1}{bc^2 - 1}$ **25** $a_1 = a_2 = a_3$

27 If $a_j = a_1 + (j - 1)d$ and $b_j = b_1 + (j - 1)e$, then $a_j + 3b_j = (a_1 + 3b_1) + (j - 1)(d + 3e)$, an arithmetic progression with common difference $d + 3e$.

29 $1 + (1 + 2 + \cdots + 2^{j-1}) = 1 + (2^j - 1)/(2 - 1) = 2^j$

31 total dist. $= 10 + 18\dfrac{1 - (\frac{9}{10})^7}{1 - \frac{9}{10}} \approx 103.9$ ft **33** $\dfrac{18!}{13!}$ **35** $2^6 \cdot 6!$ **37** 72

39 252 **41** 1365 **43** $n! > 2^n$ for $n \geq 4$; $2^n > n!$ for $n = 1, 2, 3$, $2^0 = 0!$

45 $n = 9$

Section 2 page 413

1 $\sum_{j=3}^{20} x_j$ **3** $\sum_{j=1}^{50} x_{2j-1}$ **5** $\sum_{j=1}^{n} j^{j+1}$ **7** 44 **9** 2385 **11** 216 **13** -1

15 1,234,567,890 **17** $\sum_{j=0}^{n-1} (j + 1)x^j$

19
$$\text{LHS} = \sum_{j=1}^{n} \frac{1}{2j} + \sum_{j=0}^{n-1} \frac{1}{2j + 1}$$
$$= \left(\frac{1}{2} + \frac{1}{4} + \cdots + \frac{1}{2n}\right) + \left(\frac{1}{1} + \frac{1}{3} + \cdots + \frac{1}{2n - 1}\right)$$
$$= \frac{1}{1} + \frac{1}{2} + \cdots + \frac{1}{2n} = \sum_{j=1}^{2n} \frac{1}{j}$$

21 Set $k = n - j$: $\sum_{j=0}^{n} a_j b_{n-j} = \sum_{k=0}^{n} a_{n-k} b_k = \sum_{j=0}^{n} a_{n-j} b_j$ **23** $\frac{931}{1275}$

25 The second sum is larger. The first $= 5050$; the second ≈ 5383. **27** $(n - 1)2^{n+1} + 2$

29 $(j + 1)^4 - j^4 = [(j + 1) - j][(j + 1)^3 + (j + 1)^2 j + (j + 1)j^2 + j^3]$ $= (j^3 + 3j^2 + 3j + 1) + (j^3 + 2j^2 + j) + (j^3 + j^2) + j^3 = 4j^3 + 6j^2 + 4j + 1$. Now sum (left side by telescoping):

$$(n + 1)^4 - 1 = 4\sum_{j=1}^{n} j^3 + 6[\tfrac{1}{6}n(n + 1)(2n + 1)] + 4[\tfrac{1}{2}n(n + 1)] + n$$

$$4\sum_{j=1}^{n} j^3 = (n + 1)[(n + 1)^3 - n(2n + 1) - 2n - 1]$$
$$= (n + 1)[(n + 1)^3 - (2n + 1)(n + 1)] = (n + 1)^2[(n + 1)^2 - (2n + 1)]$$

$= n^2(n + 1)^2$, hence $S = \frac{1}{4}n^2(n + 1)^2$. But $(1 + 2 + \cdots + n) = \frac{1}{2}n(n + 1)$, hence $S = (1 + 2 + \cdots + n)^2$ also.

31
$$\text{LHS} < 1 + \sum_{j=2}^{n} \frac{1}{j(j - 1)} = 1 + \sum_{j=2}^{n} \left(\frac{1}{j - 1} - \frac{1}{j}\right)$$
$$= 1 + \left(1 - \frac{1}{n}\right) = 2 - \frac{1}{n} < 2.$$

33 Set $A_0 = B_0 = 0$. Then $\sum_{i=1}^{n} A_i b_i = \sum_{i=1}^{n} A_i(B_i - B_{i-1}) = \sum_{i=1}^{n} A_i B_i - \sum_{i=1}^{n} A_i B_{i-1}$

$= \left(A_n B_n + \sum_{i=1}^{n-1} A_i B_i\right) - \left(\sum_{i=1}^{n-1} A_{i+1} B_i + A_1 B_0\right) = A_n B_n + \sum_{i=1}^{n-1} (A_i - A_{i+1}) B_i$

$= A_n B_n - \sum_{i=1}^{n-1} a_{i+1} B_i$.

35 If $a_i = \dfrac{b_{i+1}}{b_i}$, then $\prod_{i=1}^{n} a_i = \dfrac{b_{n+1}}{b_1}$; $\frac{1}{101}$

Section 3 page 418

1 True for $n = 1$: LHS $=$ RHS $= 2$. If true for some n, then
$2 + 5 + \cdots + [3(n+1) - 1]$
$= [2 + 5 + \cdots + (3n - 1)] + (3n + 2)$
$= \frac{1}{2}n(3n + 1) + (3n + 2) = \frac{1}{2}(3n^2 + 7n + 4)$
$= \frac{1}{2}(n + 1)(3n + 4) = \frac{1}{2}(n + 1)[3(n + 1) + 1]$,
so true for $n + 1$. Hence true for all $n \geq 1$.

3 True for $n = 1$: LHS $=$ RHS $= \frac{1}{3}$. If true for some n, then

$$\frac{1}{1\cdot 3} + \frac{1}{3\cdot 5} + \cdots + \frac{1}{(2n-1)(2n+1)} + \frac{1}{[2(n+1)-1][2(n+1)+1]}$$
$$= \frac{n}{2n+1} + \frac{1}{(2n+1)(2n+3)} = \frac{1}{2n+1}\left(n + \frac{1}{2n+3}\right)$$
$$= \frac{2n^2+3n+1}{(2n+1)(2n+3)} = \frac{(2n+1)(n+1)}{(2n+1)(2n+3)} = \frac{n+1}{2n+3} = \frac{(n+1)}{2(n+1)+1},$$

so true for $n + 1$. Hence true for all $n \geq 1$.

5 True for $n = 1$: LHS $=$ RHS $= 1$. If true for some n, then
$1^2 + 2^2 + \cdots + n^2 + (n+1)^2 = \frac{1}{6}n(n+1)(2n+1) + (n+1)^2$
$= \frac{1}{6}(n+1)[n(2n+1) + 6(n+1)] = \frac{1}{6}(n+1)(2n^2 + 7n + 6)$
$= \frac{1}{6}(n+1)(n+2)(2n+3) = \frac{1}{6}(n+1)[(n+1)+1][2(n+1)+1]$,
so true for $n + 1$. Hence true for all $n \geq 1$.

7 True for $n = 1$: LHS $=$ RHS $= 2$. If true for some n, then
$1\cdot 2 + 2\cdot 3 + \cdots + n(n+1) + (n+1)(n+2)$
$= \frac{1}{3}n(n+1)(n+2) + (n+1)(n+2)$
$= (n+1)(n+2)(\frac{1}{3}n + 1) = \frac{1}{3}(n+1)(n+2)(n+3)$,
so true for $n + 1$. Hence true for all $n \geq 1$.

9 True for $n = 1$. If true for some n, then

$$\begin{aligned} 2b_1 + \cdots + 2b_{n+1} &= (2b_1 + \cdots + 2b_n) + 2b_{n+1} = (\text{even}) + 2b_{n+1} \\ &= 2c + 2b_{n+1} = 2(c + b_{n+1}) = (\text{even}), \end{aligned}$$

so true for $n + 1$. Hence true for all $n \geq 1$.

11 True for $n = 2$. If true for some n, then

$$\begin{aligned} \log(a_1 \cdots a_{n+1}) &= \log[(a_1 \cdots a_n)a_{n+1}] = \log(a_1 \cdots a_n) + \log a_{n+1} \\ &= \log a_1 + \log a_2 + \cdots + \log a_n + \log a_{n+1}, \end{aligned}$$

so true for $n + 1$. Hence true for all $n \geq 2$.

13 Obvious for $n = 2$. Assume for some n and consider a line of length $n + 1$. If the next to last is a woman, she is directly in front of a man. If it is a man, we are back to the previous case.

15 True for $n = 1$ and $n = 2$. If true for some $n \geq 2$, then

$(\frac{3}{2})^{n+1} = \frac{3}{2}\cdot(\frac{3}{2})^n > \frac{3}{2}n \geq n + 1$ (for $n \geq 2$),

hence true for $n + 1$. Therefore true for all $n \geq 1$.

17 For $n = 5$, $5! = 120 > 10\cdot 5 = 50$. If true for some $n \geq 5$, then

$$(n+1)! = (n+1)\cdot n! > (n+1)\cdot 10n > 10(n+1),$$

so true for $n + 1$. Hence true for all $n \geq 5$.

19 True for $n = 2$: $(1 + a)^2 = 1 + 2a + a^2 > 1 + 2a$. If true for some $n \geq 2$, then

$$\begin{aligned} (1+a)^{n+1} &= (1+a)(1+a)^n > (1+a)(1+na) \\ &= 1 + (n+1)a + na^2 > 1 + (n+1)a, \end{aligned}$$

so true for $n + 1$. Hence true for all $n \geq 2$.

21 $\dfrac{n+1}{2n}$ **23** $\dfrac{\sin 2^{n+1}\theta}{2^{n+1}}$ **25** $2^n - 1$ **27** Let f_n be the number of sides of the n-th figure. By counting, $f_1 = 3$, $f_2 = 12$, $f_3 = 48$, so we guess $f_n = 3\cdot 4^{n-1}$. Suppose this is true for some $n \geq 1$. When we pass from the n-th to the $(n+1)$-th figure, we replace each old side by 4 new sides; hence

$$f_{n+1} = 4f_n = 4\cdot 3\cdot 4^{n-1} = 3\cdot 4^n,$$

so the guess is true for $n+1$. Hence true for all $n \geq 1$.
29 Let A_n be the area of the n-th figure. Then $A_1 = \frac{1}{4}\sqrt{3}$. To the n-th figure are added $3 \times 4^{n-1}$ equilateral triangles of side length $1/3^n$ (proved by induction). Hence

$$A_{n+1} = A_n + (3)(4^{n-1})\left(\frac{\sqrt{3}}{(4)(3^{2n})}\right) = A_n + \frac{3}{16}\left(\frac{4}{9}\right)^n\sqrt{3}.$$

Now set $B_n = \frac{1}{20}\sqrt{3}[8 - 3(\frac{4}{9})^{n-1}]$. Then $B_1 = \frac{1}{4}\sqrt{3} = A_1$ and $B_{n+1} - B_n = \frac{3}{16}(\frac{4}{9})^n\sqrt{3} = A_{n+1} - A_n$. By induction, $A_n = B_n$. **31** $a_n = n$
33 $2^n - 1$ moves **35** We try to prove simultaneously the three statements x_{3n-2} is odd, x_{3n-1} is odd, x_{3n} is even, for $n = 1, 2, \cdots$. This is true for $n = 1$. Assume it for some n. Then

$$\begin{aligned} x_{3n+1} &= x_{3n-1} + x_{3n} = \text{odd} + \text{even} = \text{odd},\\ x_{3n+2} &= x_{3n} + x_{3n+1} = \text{even} + \text{odd} = \text{odd},\\ x_{3n+3} &= x_{3n+1} + x_{3n+2} = \text{odd} + \text{odd} = \text{even}. \end{aligned}$$

37 $x_n = 1$ for $n \geq 2$ **39** $x_n = 2\cdot 3^{n-1}$ for $n \geq 1$ **41** True for $n = 0$. If true for some $n \geq 0$, then $1 \leq a_n \leq 2$, hence $1 \leq \dfrac{2}{a^n} \leq 2$, so

$$a_{n+1} = \frac{1}{2}\left(a_n + \frac{2}{a_n}\right) \leq \frac{1}{2}(2+2) = 2$$

and

$$a_{n+1} = \frac{1}{2}\left(a_n + \frac{2}{a_n}\right) \geq \frac{1}{2}(1+1) = 1;$$

the assertion is true for $n+1$. Therefore it is true for all $n \geq 0$. **43** Cut the plane into r_n regions by the first n lines. Now introduce the $(n+1)$-th line, not parallel to any of the others and not passing through any of the previous intersections. It intersects each of the first n lines in a distinct point. Think of drawing this new line in one stroke. As the pencil crosses one of the old lines, the new line leaves one of the old regions and enters another of the old regions, splitting it into two regions. Since there are n crossings, $n+1$ regions are split. Therefore $r_{n+1} = r_n + n + 1$. Assume $r_n = \frac{1}{2}(n^2 + n + 2)$ (valid for $n = 1$). Then $r_{n+1} = \frac{1}{2}(n^2 + n + 2) + n + 1 = \frac{1}{2}(n^2 + 3n + 4) = \frac{1}{2}[(n+1)^2 + (n+1) + 2]$, etc. **45** $s_n = 2\cdot 3^n$ **47** $f(n) = n!$
49 True if $0 \leq a < b$; take $q = 0$, $r = a$. Suppose $a \geq b$ and it is true for all smaller a's. Then $0 \leq a - b < a$, so $a - b = bq_1 + r$, where $0 \leq r < b$. Then $a = b(q_1 + 1) + r = bq + r$, so it is true up to a. Hence true for all $a \geq 0$.

Section 4 page 424

1 $8! = 40{,}320$ **3** $\binom{100}{3} = 161{,}700$ **5** $9P_{9,9} = 3{,}265{,}920$ **7** $\binom{10}{4} = 210$
9 $2^{10} = 1024$ **11** $\binom{11}{3} = 165$ **13** $\binom{12}{2}\binom{12}{3} = 14{,}520$
15 $\binom{100}{10} \approx 1.731 \times 10^{13}$ **17** $\binom{10}{4}\binom{13}{4} = 150{,}150$
19 $P_{8,8}\cdot P_{8,8} = 1{,}625{,}702{,}400$ **21** $\binom{52}{5} = 2{,}598{,}960$
23 $13\cdot\binom{4}{3}\cdot 12\cdot\binom{4}{2} = 3744$ **25** $\binom{52}{13} \approx 6.350 \times 10^{11}$
27 $\binom{25}{4} = 12{,}650$ **29** 826
31 $\binom{n}{k} = \binom{n-1}{k} + \binom{n-1}{k-1} = \left[\binom{n-2}{k} + \binom{n-2}{k-1}\right] + \left[\binom{n-2}{k-1} + \binom{n-2}{k-2}\right] = \binom{n-2}{k-2} + 2\binom{n-2}{k-1} + \binom{n-2}{k}$

33 First choose the committee in $C_{n,k}$ ways, then the subcommittee in $C_{k,r}$ ways. Alternatively, first choose the subcommittee in $C_{n,r}$ ways, then the remaining $k - r$ committee members in $C_{n-r,k-r}$ ways. Result: $C_{n,k}C_{k,r} = C_{n,r}C_{n-r,k-r}$. **35** $2n$ moves are required, n moves up and n moves right, and these may occur in any order. A path is completely determined by choosing the n up moves out of the $2n$ possible moves, hence the given number of paths.

Section 5 page 428

1 $x^3 + 6x^2 + 12x + 8$ **3** $1 + 8x + 24x^2 + 32x^3 + 16x^4$
5 $x^5 - 5x^4y + 10x^3y^2 - 10x^2y^3 + 5xy^4 - y^5$
7 $x^5 - 5x^3 + 10x - 10/x + 5/x^3 - 1/x^5$
9 $x^6 + 6x^4 + 15x^2 + 20 + 15x^{-2} + 6x^{-4} + x^{-6}$ **11** 1 **13** 448 **15** 160
17 17,010 **19** 10,510,100,501 **21** 9,509,900,499 **23** 0.904382 **25** 1.020191
27 $(1.002)^{10} = (1 + 0.002)^{10} > 1 + \binom{10}{1}\cdot 0.002 = 1.02$
29 $(1.0001)^{100} = (1 + 10^{-4})^{100} > 1 + \binom{100}{1}\cdot 10^{-4} + \binom{100}{2}\cdot 10^{-8} =$
$1 + 10^{-2} + 4.95 \times 10^{-5} = 1.0100495$

Review Exercises page 429

1 $a_n = 12 + 6n$, so $a_{n+1} = a_n + 6$ **3** log 101 **5** $n = 7$ **7** To prove: $a_{n-1} < a_n$ for $n \geq 2$. True for $n = 2$ since $a_2 = \sqrt{3 + \sqrt{3}} > \sqrt{3} = a_1$. If true for some $n \geq 2$, then $a_n > a_{n-1}$, hence

$$a_{n+1} = \sqrt{3 + a_n} > \sqrt{3 + a_{n-1}} = a_n,$$

so true for $n + 1$. Therefore true for all $n \geq 2$.
9 True for $n = 1$. Suppose true for some $n \geq 1$. Then

$$x^{2n+1} + 1 = x^2(x^{2n-1} + 1) - (x^2 - 1).$$

Both $x^2(x^{2n-1} + 1)$ and $x^2 - 1 = (x + 1)(x - 1)$ are divisible by $x + 1$; so is $x^{2n+1} + 1$. The claim is true for $n + 1$; hence it is true for all $n \geq 1$. **11** 27,216
13 $\dfrac{x^5}{y^5} - 5\dfrac{x^3}{y^3} + 10\dfrac{x}{y} - 10\dfrac{y}{x} + 5\dfrac{y^3}{x^3} - \dfrac{y^5}{x^5}$

CHAPTER 11

Section 1 page 433

1 $(0, 3)$ **3** $(2, -2)$ **5** $(2, 12)$ **7** $(3, 0)$ **9** $(0, 0)$ **11** $(1, 1)$
13 $(10, 4)$ **15** $(18, 10)$

17

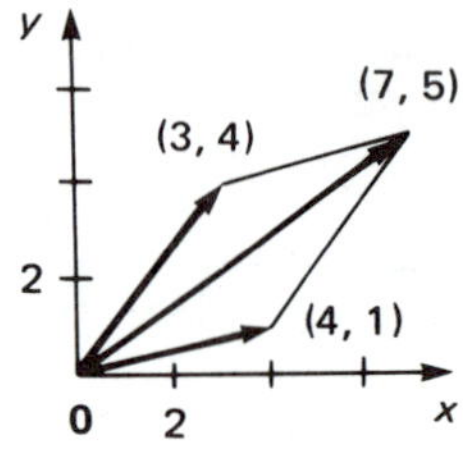

19

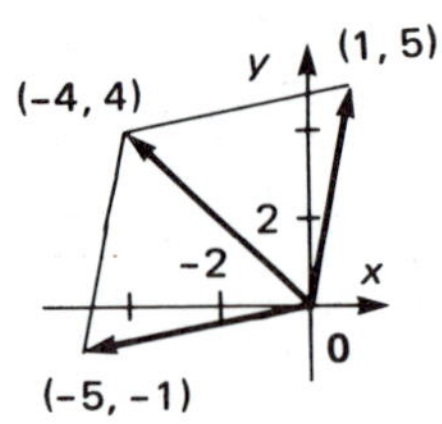

21 $(1, -10)$
23 $a = -2$ $b = 5$

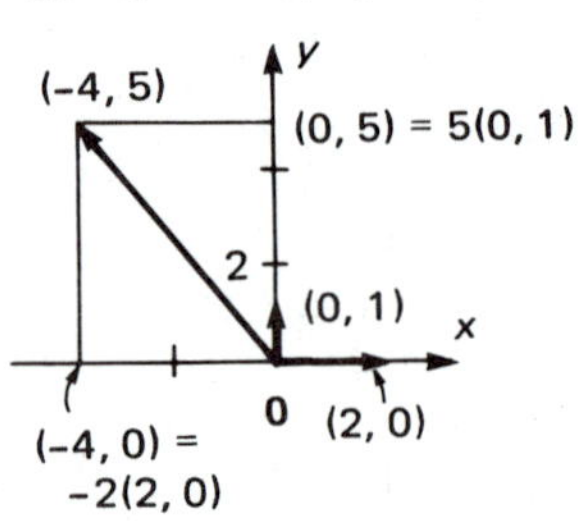

Section 2 page 437

1 $\mathbf{x} = (t, t)$; $x = t$, $y = t$; $x - y = 0$ **3** $\mathbf{x} = (1, t)$; $x = 1$, $y = t$, $x = 1$
5 $\mathbf{x} = (2 + 2t, 3 - t)$; $x = 2 + 2t$, $y = 3 - t$; $x + 2y = 8$
7 $\mathbf{x} = (3 + 4t, 1 + 3t)$; $x = 3 + 4t$, $y = 1 + 3t$; $3x - 4y = 5$
9 $\mathbf{x} = (5 - 3t, 2 - t)$; $x = 5 - 3t$, $y = 2 - t$; $x - 3y = -1$ **11** $(\frac{3}{4}, \frac{3}{4})$

13 $(-\frac{1}{4}, -\frac{1}{2})$ **15** $(\frac{11}{5}, -\frac{17}{5})$ **17** same **19** same **21** $(1, 2)$ **23** $(1, 1)$
25 $(1, \frac{1}{2})$ **27** $(3, -1)$ **29** $(0, 0)$ **31** $(0, 6)$ **33** $(-\frac{6}{5}, \frac{23}{5})$ **35** $(\frac{5}{6}, \frac{1}{6})$
37 $\mathbf{w} - \mathbf{v} = \mathbf{z} - \mathbf{u}$ because opposite sides are parallel and of equal length. Hence $\mathbf{u} + \mathbf{w} = \mathbf{v} + \mathbf{z}$, so $\frac{1}{2}(\mathbf{u} + \mathbf{w}) = \frac{1}{2}(\mathbf{v} + \mathbf{z})$. This relation means that the diagonals bisect each other. **39** Take L in parametric form: $\mathbf{x} = \mathbf{x}_0 + t\mathbf{v}$. Then $\mathbf{y} = c\mathbf{x} = c\mathbf{x}_0 + t(c\mathbf{v})$ is a line through $\mathbf{y}_0 = c\mathbf{x}_0$ and parallel to $c\mathbf{v}$, that is, to $\mathbf{v}$.

Section 3 page 444

1 13 **3** $\sqrt{61}$ **5** 6 **7** 0 **9** 3 **11** 31.35
13 $(2, 3)\cdot(3, 6) = 24 \quad (2, 3)\cdot(1, 4) + (2, 3)\cdot(2, 2) = 14 + 10 = 24$
15 $(-3, 4)\cdot(4, 5) = 8 \quad (-1, 7)\cdot(4, 5) - (2, 3)\cdot(4, 5) = 31 - 23 = 8$
17 $(12, 15)\cdot(-1, 2) = 18 \quad 3[(4, 5)\cdot(-1, 2)] = 3\cdot 6 = 18$
19 $\mathbf{v}\cdot\mathbf{w} = (v_1, v_2)\cdot(w_1, w_2) = v_1w_1 + v_2w_2 = w_1v_1 + w_2v_2 = (w_1, w_2)\cdot(v_1, v_2) = \mathbf{w}\cdot\mathbf{v}$
21 $(\mathbf{u} + \mathbf{v})\cdot\mathbf{w} = [(u_1, u_2) + (v_1, v_2)]\cdot(w_1, w_2) = (u_1 + v_1, u_2 + v_2)\cdot(w_1, w_2) = (u_1 + v_1)w_1 + (u_2 + v_2)w_2 = u_1w_1 + v_1w_1 + u_2w_2 + v_2w_2 = u_1w_1 + u_2w_2 + v_1w_1 + v_2w_2 = \mathbf{u}\cdot\mathbf{w} + \mathbf{v}\cdot\mathbf{w}$
23 $\|\mathbf{v} + \mathbf{w}\|^2 = (\mathbf{v} + \mathbf{w})\cdot(\mathbf{v} + \mathbf{w}) = \mathbf{v}\cdot(\mathbf{v} + \mathbf{w}) + \mathbf{w}\cdot(\mathbf{v} + \mathbf{w}) = \mathbf{v}\cdot\mathbf{v} + \mathbf{v}\cdot\mathbf{w} + \mathbf{w}\cdot\mathbf{v} + \mathbf{w}\cdot\mathbf{w} = \|\mathbf{v}\|^2 + 2\mathbf{v}\cdot\mathbf{w} + \|\mathbf{w}\|^2$
25 $\|\mathbf{v} + \mathbf{w}\|^2 - \|\mathbf{v} - \mathbf{w}\|^2 = (\|\mathbf{v}\|^2 + 2\mathbf{v}\cdot\mathbf{w} + \|\mathbf{w}\|^2) - (\|\mathbf{v}\|^2 - 2\mathbf{v}\cdot\mathbf{w} + \|\mathbf{w}\|^2) = 4\mathbf{v}\cdot\mathbf{w}$
27 $\frac{1}{2}\pi$ **29** arc cos $\frac{3}{5} \approx 53.13°$ **31** arc cos $\frac{3}{5} \approx 53.13°$
33 Take vertices $\mathbf{0}, \mathbf{v}, \mathbf{w}, \mathbf{v} + \mathbf{w}$, where $\|\mathbf{v}\| = \|\mathbf{w}\|$. Then one diagonal is $\mathbf{v} + \mathbf{w}$, the other is parallel to $\mathbf{w} - \mathbf{v}$. But

$$\begin{aligned}(\mathbf{v} + \mathbf{w})\cdot(\mathbf{w} - \mathbf{v}) &= \mathbf{v}\cdot\mathbf{w} - \mathbf{v}\cdot\mathbf{v} + \mathbf{w}\cdot\mathbf{w} - \mathbf{w}\cdot\mathbf{v} \\ &= -\|\mathbf{v}\|^2 + \|\mathbf{w}\|^2 = 0,\end{aligned}$$

so the diagonals are perpendicular.
35 19.47° **37** $\frac{1}{2}$W **39** 4.6 knots, 60° west of south

Section 4 page 450

1 $(\frac{1}{2}\sqrt{2}, -\frac{1}{2}\sqrt{2})\cdot\mathbf{x} = 0$ **3** $(\frac{3}{5}, -\frac{4}{5})\cdot\mathbf{x} = 0$ **5** $(\frac{3}{5}, \frac{4}{5})\cdot\mathbf{x} = \frac{2}{5}$ **7** $(\frac{12}{13}, \frac{5}{13})\cdot\mathbf{x} = \frac{1}{13}$
9 $(\frac{7}{25}, -\frac{24}{25})\cdot\mathbf{x} = 1$ **11** $(\frac{2}{13}\sqrt{13}, -\frac{3}{13}\sqrt{13})\cdot\mathbf{x} = \frac{4}{13}\sqrt{13}$
13 $(-\frac{2}{5}\sqrt{5}, \frac{1}{5}\sqrt{5})\cdot\mathbf{x} = \frac{1}{5}\sqrt{5}$ **15** $\dfrac{1}{\sqrt{a^2 + 1}}(-a, 1)\cdot\mathbf{x} = \dfrac{b}{\sqrt{a^2 + 1}}$
17 $(\frac{4}{5}, \frac{3}{5})\cdot\mathbf{x} = \frac{1}{5}$ **19** $(\frac{2}{5}\sqrt{5}, -\frac{1}{5}\sqrt{5})\cdot\mathbf{x} = \frac{1}{5}\sqrt{5}$ **21** $(\frac{3}{13}\sqrt{13}, \frac{2}{13}\sqrt{13})\cdot\mathbf{x} = \sqrt{13}$
23 2 **25** $\frac{11}{5}$ **27** $\frac{13}{5}$ **29** $\frac{119}{13}$ **31** $\frac{1}{2}\sqrt{2}$ **33** $\frac{1}{4}\pi$ **35** arc cos $\frac{4}{5} \approx 0.6435$
37 arc cos $\frac{63}{65} \approx 0.2487$ **39** arc cos $\frac{3}{10}\sqrt{10} \approx 0.3218$ **41** $(\frac{2}{5}\sqrt{5}, \frac{1}{5}\sqrt{5})\cdot\mathbf{x} = \frac{2}{5}\sqrt{5}$
43 $(\frac{5}{34}\sqrt{34}, -\frac{3}{24}\sqrt{34})\cdot\mathbf{x} = \frac{7}{17}\sqrt{34}$

Review Exercises page 450

1 $(2, 10)$ **3** $(3, -3)$ **5** $\mathbf{x} = (-2t - 3, -t + 1)$; $x - 2y = -5$
7 $x = 3t + 1, y = -2t - 1$ **9** $\frac{1}{3}\pi$ **11** $(\frac{40}{41}, -\frac{9}{41})\cdot\mathbf{x} = 3$ **13** $\frac{39}{5}$
15 $(\frac{3}{13}\sqrt{13}, \frac{2}{13}\sqrt{13})\cdot\mathbf{x} = \frac{11}{13}\sqrt{13}$

CHAPTER 12

Section 1 page 456

1 $(6, 4)$ **3** $(5, -2)$ **5** $(0, -8)$ **7** $(-4, 5)$ **9** $(-11, -5)$
11 $$\begin{aligned}(\bar{x}_1 - \bar{x}_2)^2 + (\bar{y}_1 - \bar{y}_2)^2 &= [(x_1 - h) - (x_2 - h)]^2 + [(y_1 - k) - (y_2 - k)]^2 \\ &= (x_1 - x_2)^2 + (y_1 - y_2)^2\end{aligned}$$
13 $x = \bar{x} + h, y = \bar{y} + k$, where $3h - 2k = 1$ **15** $\bar{x} = x + 1, \bar{y} = y$
17 $\bar{x} = x - \frac{1}{6}\pi, \bar{y} = y + 1$ **19** $\bar{\mathbf{x}}\cdot\mathbf{n} = p - \mathbf{h}\cdot\mathbf{n}$ **21** $(-4, -3)$ $(-3, 4)$
23 $(5, 12)$ $(12, 5)$ **25** $(1, 2)$; tangent **27** $(-3, 1)$ $(1, -3)$
29 $(\frac{1}{5} \pm \frac{2}{5}\sqrt{19}, \frac{2}{5} \mp \frac{1}{5}\sqrt{19})$ **31** $(-3, -5)$ $(-2, 2)$ **33** $(0, 0)$ $(7, -7)$
35 $(-2, -1)$; tangent **37** $(2, -4)$ $(6, -8)$ **39** none **41** $(1, 2)$; tangent
43 $(3, 1)$; tangent **45** $(\frac{3}{2} \pm \frac{1}{2}\sqrt{41}, \frac{3}{2} \mp \frac{1}{2}\sqrt{41})$ **47** $(-\frac{21}{8}, \pm\frac{3}{8}\sqrt{15})$ **49** none

Section 2 page 461

1 $x = -1$ **3** $y = x - \sqrt{2}$ **5** $y = -x\sqrt{3} + 2$ **7** $3x + 4y = 15$
9 $y - 2 = (\frac{4}{3} \pm \frac{1}{3}\sqrt{7})(x - 2)$ **11** $y - 1 = \frac{2}{3}(x + 5)$ and $y - 1 = -\frac{3}{2}(x + 5)$
13 $y - 2 = (\frac{4}{3} \pm \frac{1}{3}\sqrt{7})(x - 3)$ **15** $(x - x_0, y - y_0)\cdot(x_0, y_0) = 0$, so $x_0x + y_0y = x_0^2 + y_0^2 = r^2$. **17** Expand, then complete the square:

$$\mathbf{x}\cdot\mathbf{x} - (\mathbf{a} + \mathbf{b})\cdot\mathbf{x} = c^2 - \mathbf{a}\cdot\mathbf{b},$$
$$\|\mathbf{x} - \tfrac{1}{2}(\mathbf{a} + \mathbf{b})\|^2 = [\mathbf{x} - \tfrac{1}{2}(\mathbf{a} + \mathbf{b})]\cdot[\mathbf{x} - \tfrac{1}{2}(\mathbf{a} + \mathbf{b})]$$
$$= c^2 - \mathbf{a}\cdot\mathbf{b} + \tfrac{1}{4}(\mathbf{a} + \mathbf{b})\cdot(\mathbf{a} + \mathbf{b}) = c^2 - \tfrac{1}{4}\|\mathbf{a} - \mathbf{b}\|^2.$$

This is the circle with center $\frac{1}{2}(\mathbf{a} + \mathbf{b})$ and radius $(c^2 - \frac{1}{4}\|\mathbf{a} - \mathbf{b}\|^2)^{1/2}$ provided $c > \frac{1}{2}\|\mathbf{a} - \mathbf{b}\|$. **19** A quarter circle of radius 5. **21** The concentric circle of radius 10. **23** The circle with center $(1, 0)$ and radius $\frac{1}{2}$. **25** The circle with center $(2, 0)$ and radius 15.

Section 3 page 465

1, 3, 5

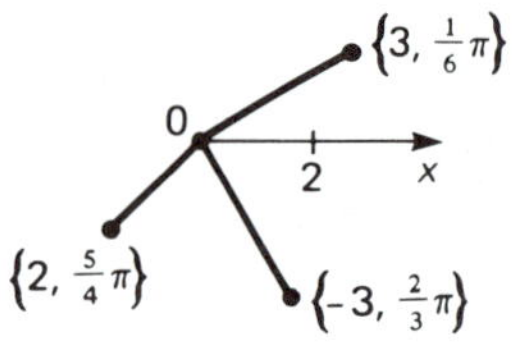

7

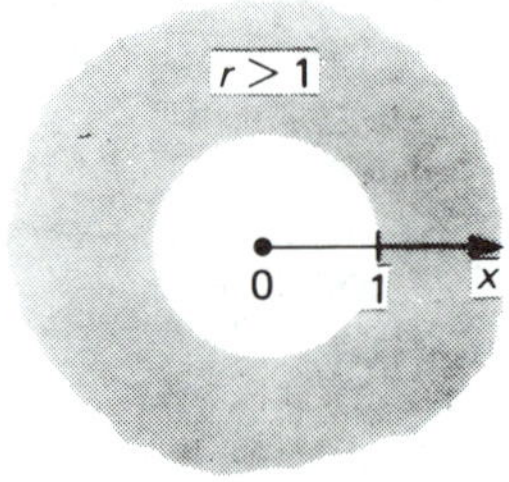

9

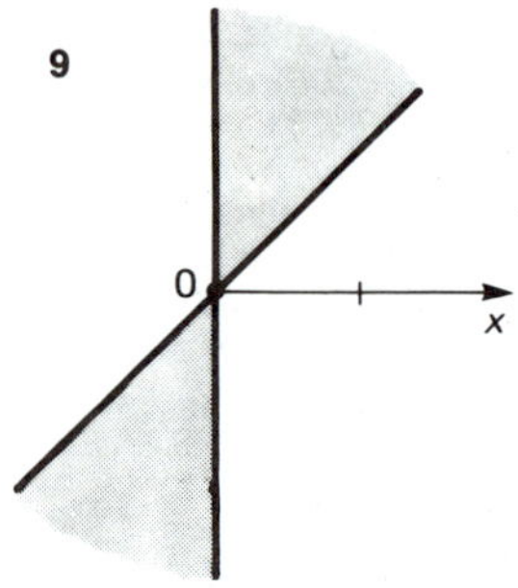

11 $(0, 1)$ **13** $(\frac{1}{2}\sqrt{3}, -\frac{1}{2})$ **15** $(-\sqrt{2}, -\sqrt{2})$ **17** $\{\sqrt{2}, \frac{1}{4}\pi\}$ **19** $\{\sqrt{2}, \frac{3}{4}\pi$
21 $\{2, -\frac{1}{6}\pi\}$ **23** $\theta = \frac{1}{4}\pi$ **25** $r\cos(\theta - \frac{1}{4}\pi) = \frac{1}{2}\sqrt{2}$

Section 4 page 469

1 $r = 4\cos\theta$ **3** $r = 2\sin\theta$

5

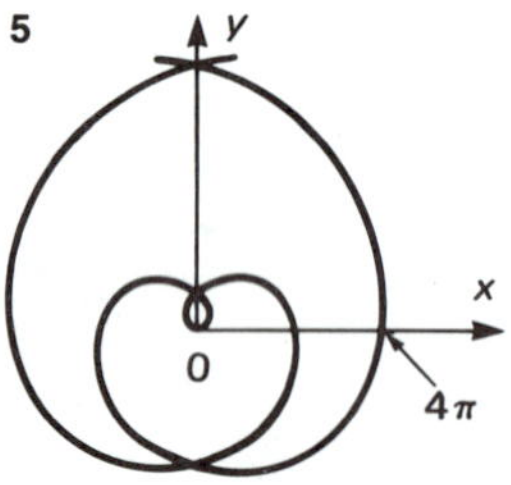

7

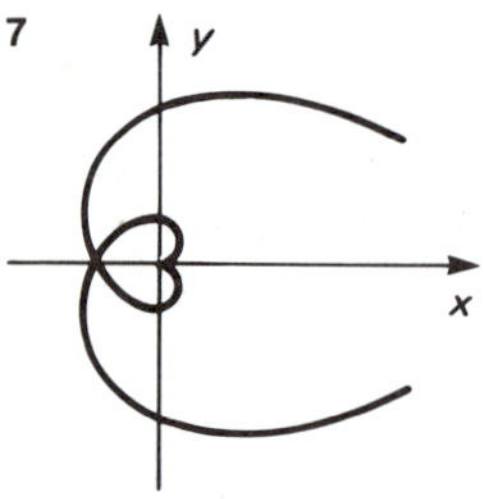

9

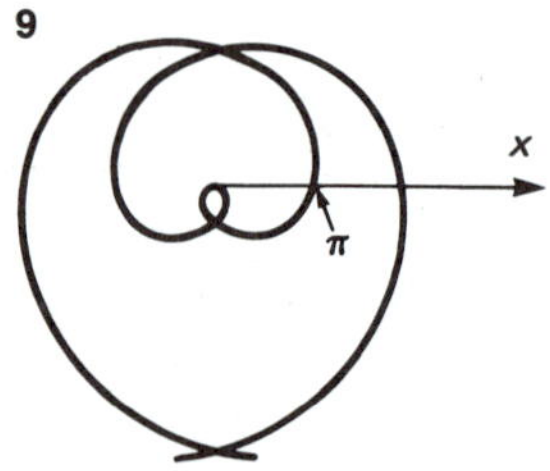

11

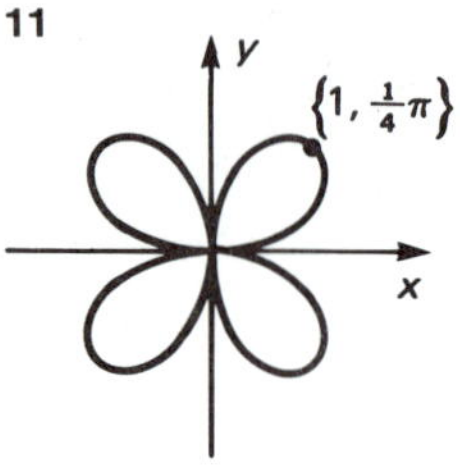

13

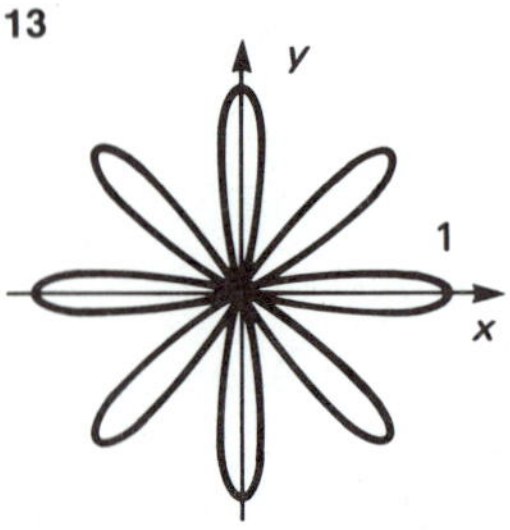

15

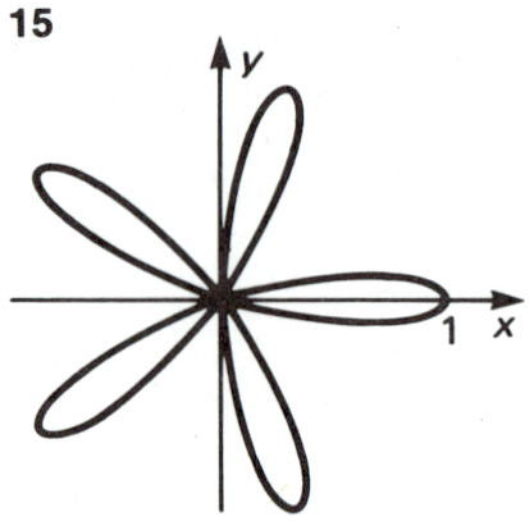

17

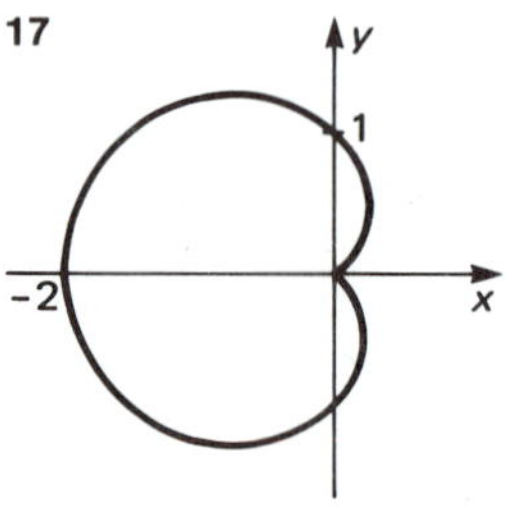

19

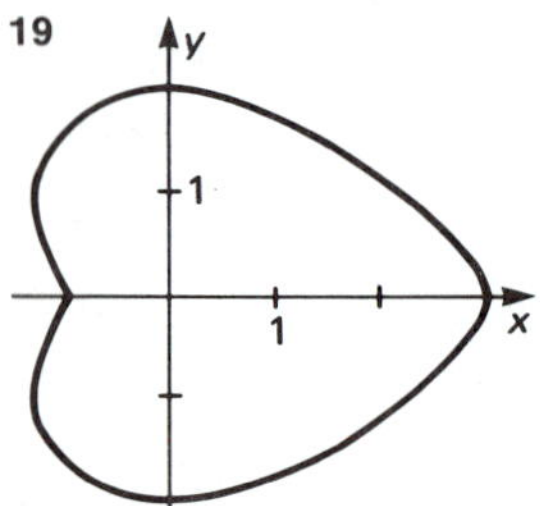

21

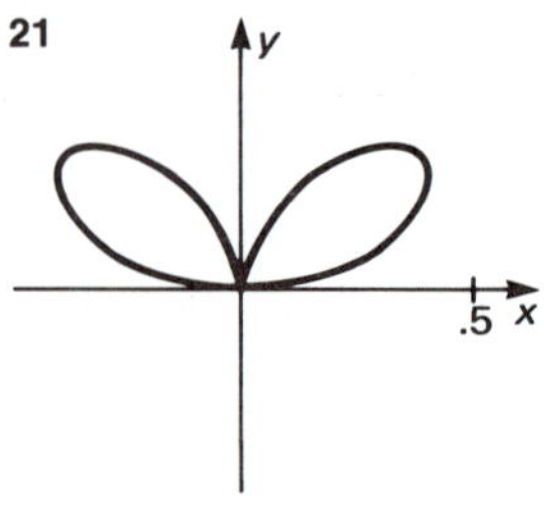

23
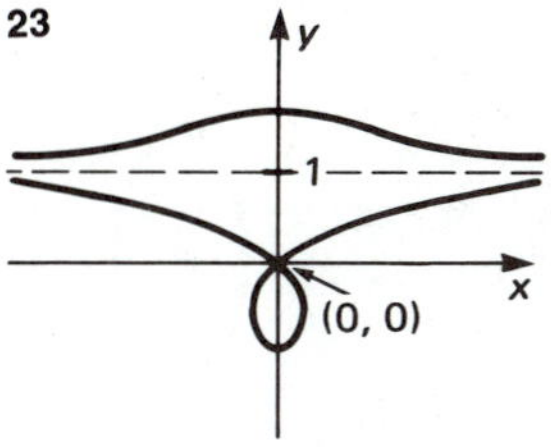

Section 5 page 472

1
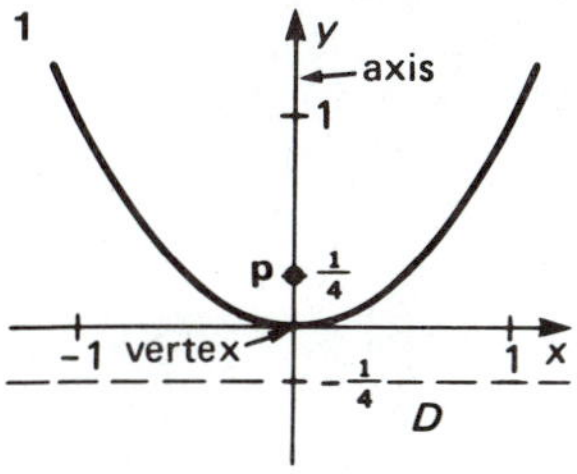

3
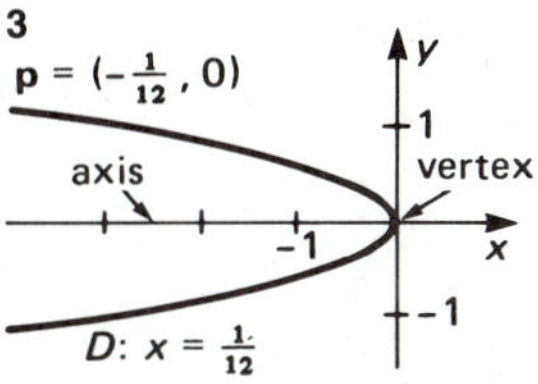

5
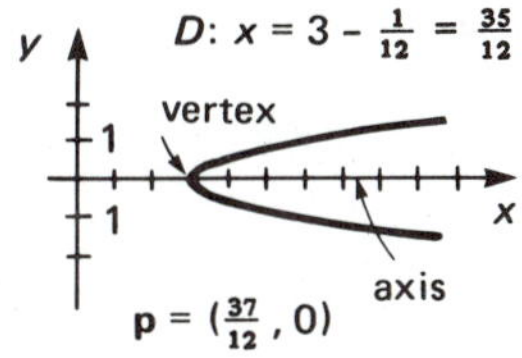

7
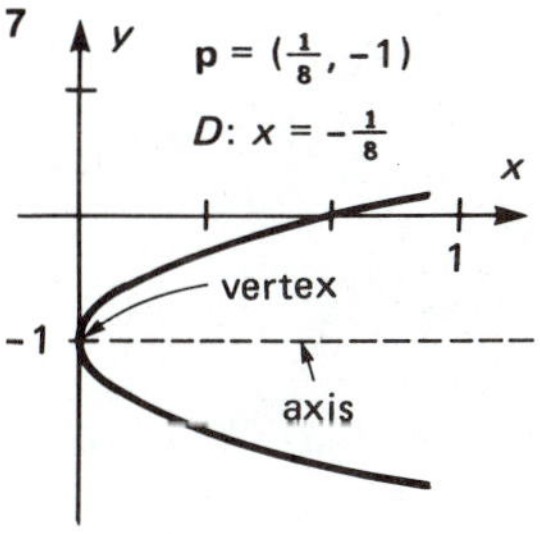

9
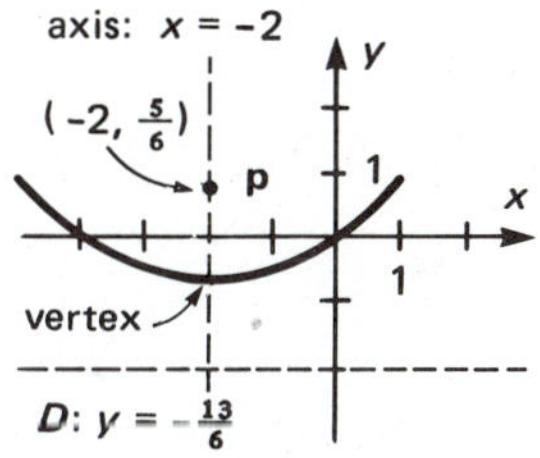

11 $y = 3x^2$ **13** $x - 1 = -(y - 2)^2$ **15** $y + 3 = \frac{1}{16}(x - 2)^2$ **17** $(-3, 9)$

19
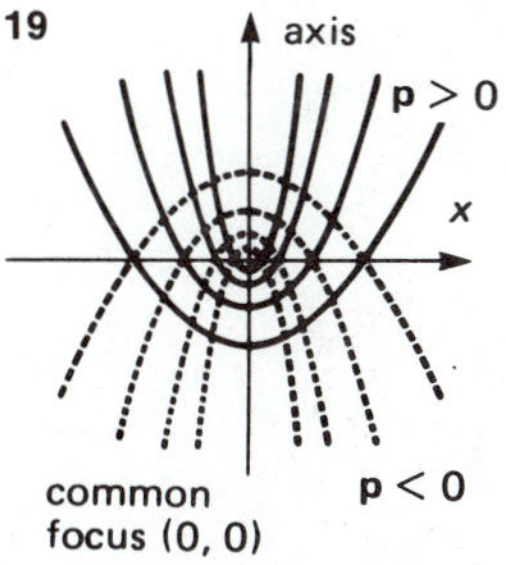

21 Eliminate y: $x^2 = 4ay = 2x_0x - 4ay_0 = 2x_0x - x_0^2$. Hence $(x - x_0)^2 = 0$, $x = x_0$, $y = x_0^2/4a$. The line is tangent to the parabola.

23 Eliminate t: $(v_0^2 \cos^2 \alpha)y = (v_0^2 \sin\alpha \cos\alpha)x - 16x^2$, the equation of a parabola.

25
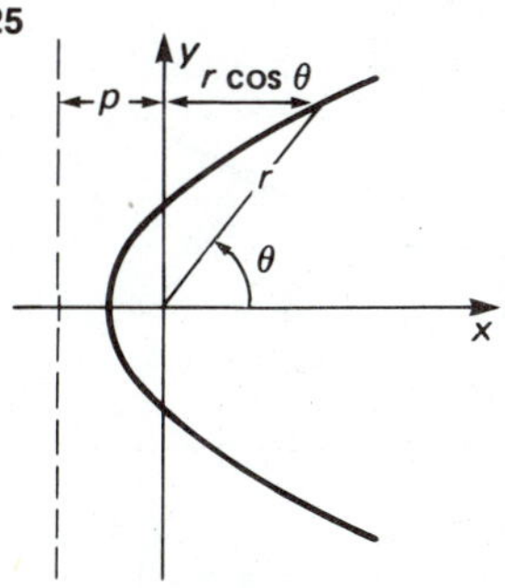

$r = r\cos\theta + p$ hence $r(1 - \cos\theta) = p$

27 $m = 2$; the line $y - 1 = 2(x - 1)$ is tangent to the parabola at $(1, 1)$.

Section 6 page 478

1

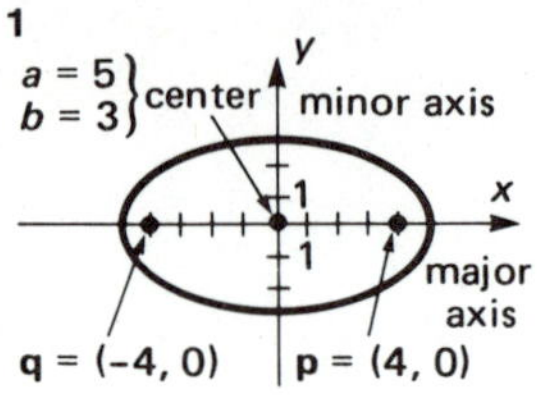

3

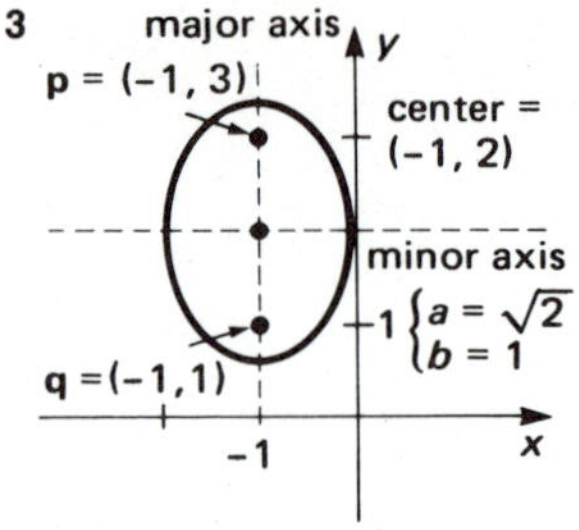

5

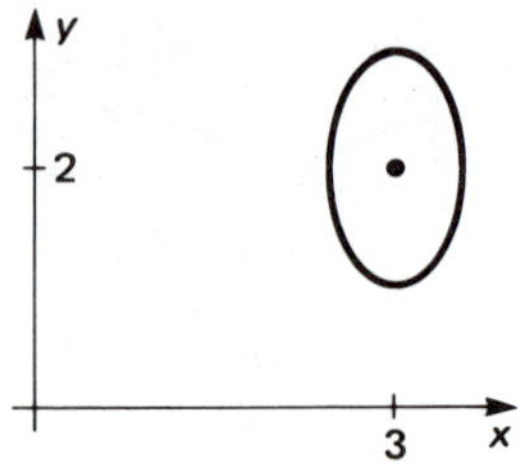

center: $(3, 2)$
foci: $(3, 2 \pm \frac{1}{2}\sqrt{2})$
vertices: $(3, 1), (3, 3)$,
$(3 \pm \frac{1}{2}\sqrt{2}, 2)$
$a = 1, b = \frac{1}{2}\sqrt{2}$

7 $\frac{1}{81}(x-1)^2 + \frac{1}{4}(y-4)^2 = 1$ **9** $\frac{1}{25}(x-5)^2 + \frac{1}{16}y^2 = 1$
11 $\frac{1}{16}(x-1)^2 + \frac{1}{12}y^2 = 1$ **13** Describe the ellipse by $\mathbf{x} = (a\cos\theta, b\sin\theta)$. Then $x^2 + y^2 = (a^2 - b^2)\cos^2\theta + b^2 \le (a^2 - b^2) + b^2 = a^2$, with "=" only if $\cos\theta = \pm 1$.
15 From the figure, $(\frac{1}{3}x)^2 + (\frac{1}{6}y)^2 = 1$, so the answer is the first quadrant portion of the ellipse $\frac{1}{9}x^2 + \frac{1}{36}y^2 = 1$.

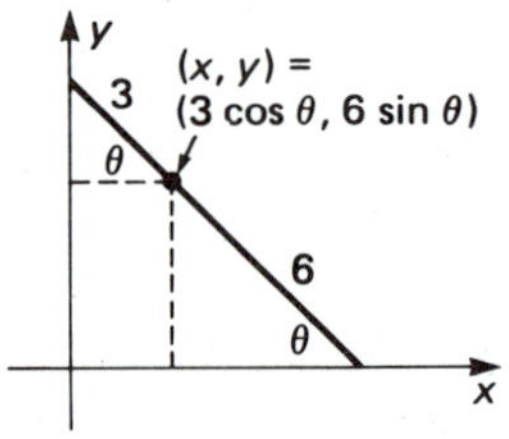

17 $\dfrac{(x + \frac{1}{2}a)^2}{(\frac{1}{2}a)^2} + \dfrac{y^2}{(\frac{1}{2}b)^2} = 1$; the ellipse with center $(-\frac{1}{2}a, 0)$, half the size of the original one, and parallel to it.

19

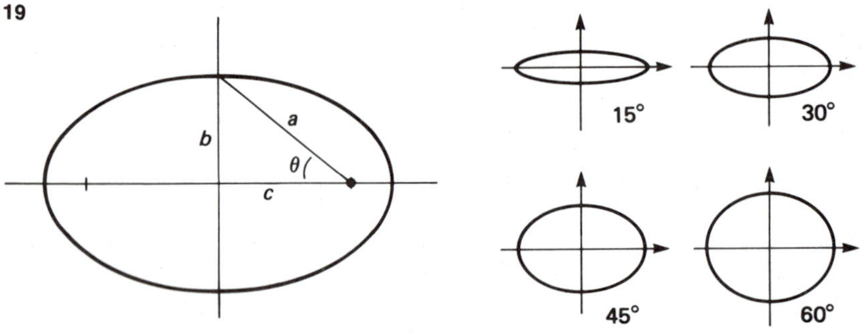

21 $e \approx 0.21$ **23** $x^2 + y^2 = \dfrac{(1 - t^2)^2 + (2t)^2}{(1 + t^2)^2} = \dfrac{(1 + t^2)^2}{(1 + t^2)^2} = 1$

Section 7 page 484

1

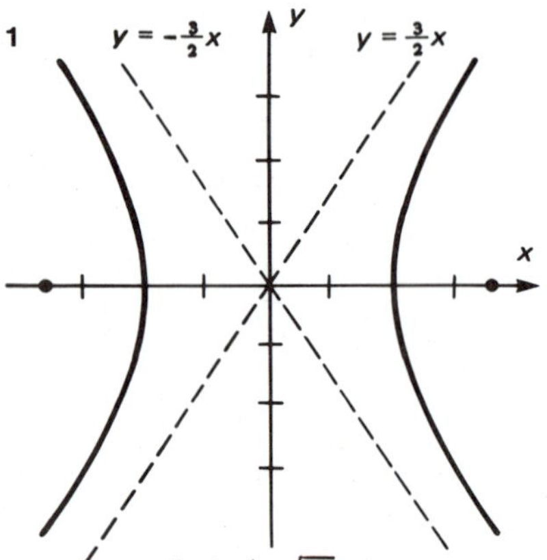

foci: $(\pm\sqrt{13}, 0)$
center: $(0, 0)$

3

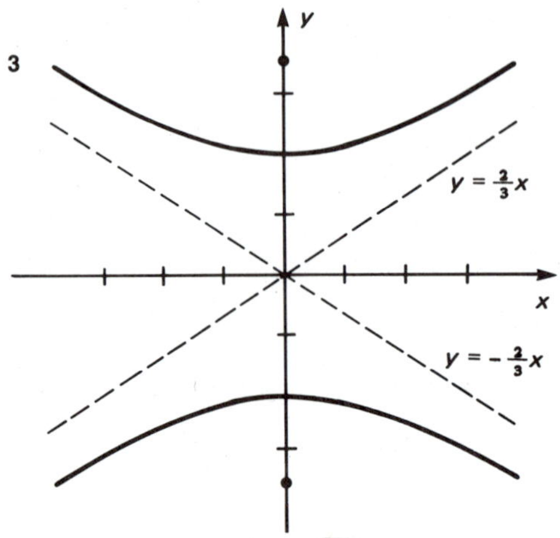

foci: $(0, \pm\sqrt{13})$
center: $(0, 0)$

5

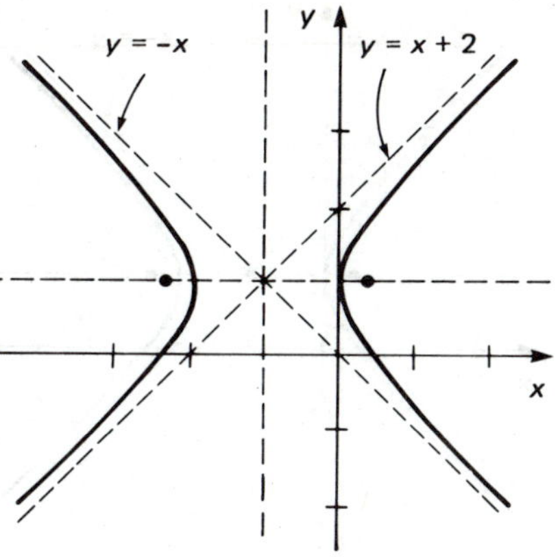

foci: $(-1 \pm \sqrt{2}, 1)$

center: $(-1, 1)$

7

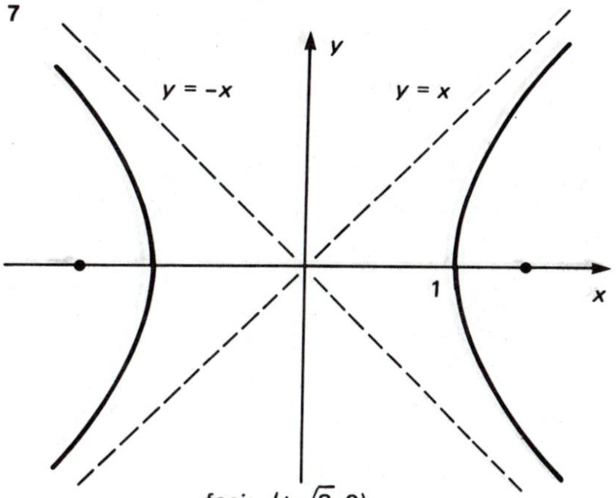

foci: $(\pm\sqrt{2}, 0)$

center: $(0, 0)$

9

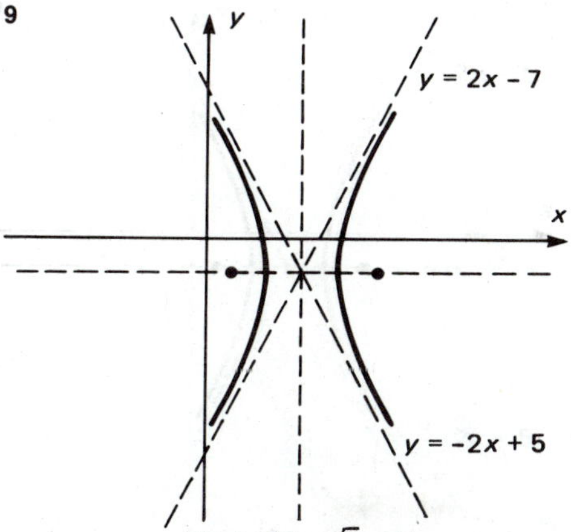

foci: $(3 \pm \sqrt{5}, -1)$

center: $(3, -1)$

11 $-\frac{1}{9}x^2 + \frac{1}{16}y^2 = 1$ **13** $\frac{1}{4}x^2 - \frac{1}{16}y^2 = 1$

15 $\frac{1}{3}(x - 1)^2 - \frac{1}{3}y^2 = 1$

17 Complete squares; the asymptotes are $y - \frac{1}{2}b = \pm(x + \frac{1}{2}a)$, two perpendicular lines.

19

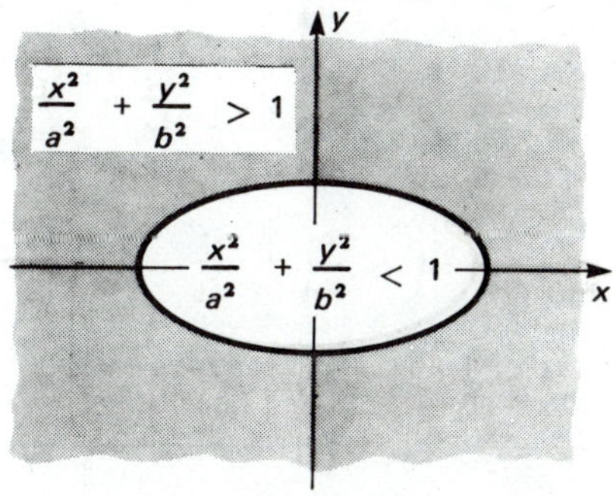

21 Let the circles be $\|\mathbf{x} - \mathbf{p}\| = r$ and $\|\mathbf{x} - \mathbf{q}\| = s$, with $\|\mathbf{p} - \mathbf{q}\| \geq r + s$. The desired locus is $\|\mathbf{x} - \mathbf{p}\| - r = \|\mathbf{x} - \mathbf{q}\| - s$, that is, $\|\mathbf{x} - \mathbf{p}\| - \|\mathbf{x} - \mathbf{q}\| = r - s$.

Section 8 page 491

1 $r = 2(\cos\theta + \sin\theta)$ **3** $r = 2\cos\theta + 2\sqrt{3}\sin\theta$ **5** $r = 10\cos\theta + 24\sin\theta$

7 $-\frac{1}{2}(\sqrt{2} - 1)\bar{x}^2 + \frac{1}{2}(\sqrt{2} + 1)\bar{y}^2 = 1$ **9** $\frac{1}{2}(\sqrt{2} + 1)\bar{x}^2 - \frac{1}{2}(\sqrt{2} - 1)\bar{y}^2 = 1$

11 ellipse $\alpha = \frac{1}{4}\pi$ $\frac{3}{2}\bar{x}^2 + \frac{1}{2}\bar{y}^2 = 1$ **13** hyperbola $\tan 2\alpha = \frac{1}{2}$ $\frac{1}{2}\bar{x}^2\sqrt{5} - \frac{1}{2}\bar{y}^2\sqrt{5} = 1$ **15** ellipse $\alpha = \frac{3}{8}\pi$ $(\frac{3}{2} + \frac{1}{2}\sqrt{2})\bar{x}^2 + (\frac{3}{2} - \frac{1}{2}\sqrt{2})\bar{y}^2 = 1$

17 parabola $\alpha = \frac{1}{4}\pi$ $2\bar{y}^2 - \bar{y}\sqrt{2} = \bar{x}\sqrt{2}$

19 hyperbola $\tan 2\alpha = -6$ $-(\frac{1}{2}\sqrt{37} - \frac{3}{2})\bar{x}^2 + (\frac{1}{2}\sqrt{37} + \frac{3}{2})\bar{y}^2 = 1$

21
$$\begin{aligned} x^2 + y^2 &= (\bar{x}^2\cos^2\alpha - 2\bar{x}\bar{y}\cos\alpha\sin\alpha + \bar{y}^2\sin^2\alpha) \\ &\quad + (\bar{x}^2\sin^2\alpha + 2\bar{x}\bar{y}\sin\alpha\cos\alpha + \bar{y}^2\cos^2\alpha) \\ &= \bar{x}^2(\cos^2\alpha + \sin^2\alpha) + \bar{y}^2(\sin^2\alpha + \cos^2\alpha) = \bar{x}^2 + \bar{y}^2. \end{aligned}$$

This means that rotation does not change the length of a vector.

23
$$\begin{aligned} \bar{a} + \bar{c} &= (a\cos^2\alpha + b\cos\alpha\sin\alpha + c\sin^2\alpha) \\ &\quad + (a\sin^2\alpha - b\sin\alpha\cos\alpha + c\cos^2\alpha) \\ &= a(\cos^2\alpha + \sin^2\alpha) + c(\sin^2\alpha + \cos^2\alpha) = a + c. \end{aligned}$$

Review Exercises page 492

1 $(x, y) = (\bar{x}, \bar{y}) + (2, 3)$ **3** $12(y + 2) = \pm 5(x - 8)$ **5** $\{\sqrt{2}, \frac{7}{4}\pi\}$

7 $(1, -\sqrt{3})$ **9** $r = 6\cos\theta + 8\sin\theta$

11
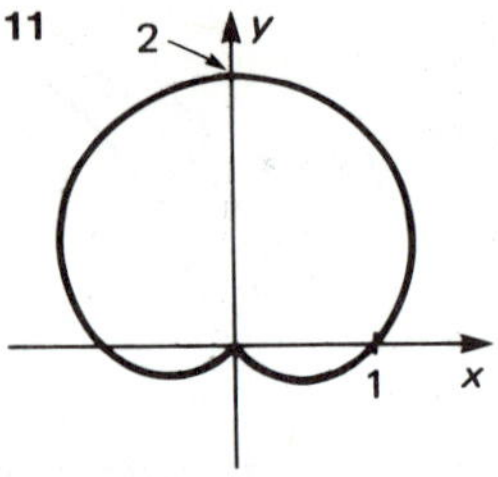

13
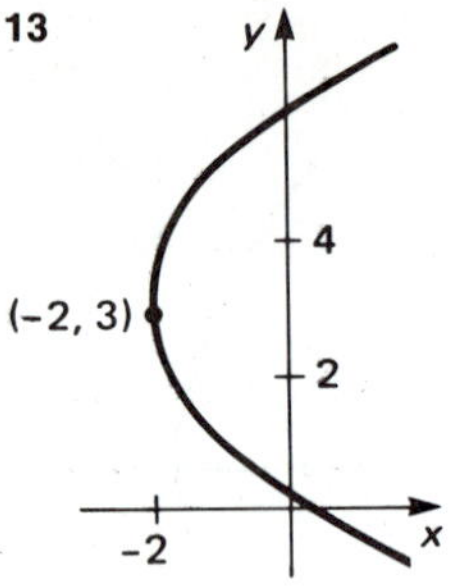

15
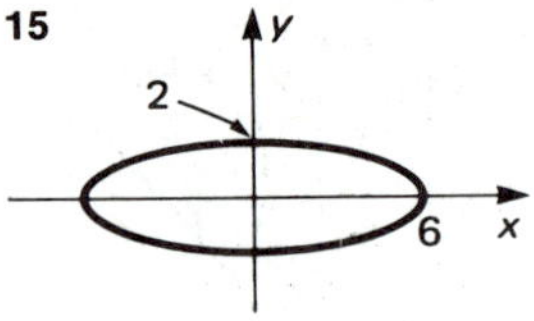

17
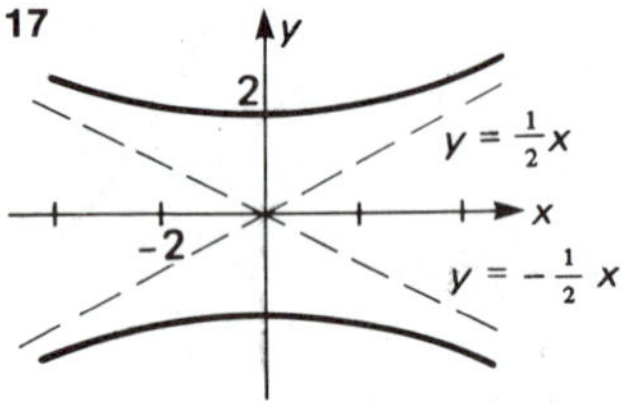

19
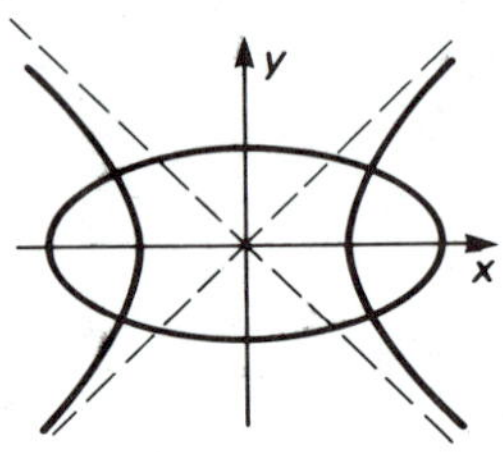

$(\frac{2}{5}\sqrt{10}, \pm\frac{1}{5}\sqrt{15})$
$(-\frac{2}{5}\sqrt{10}, \pm\frac{1}{5}\sqrt{15})$

CHAPTER 13

Section 1 page 499

1
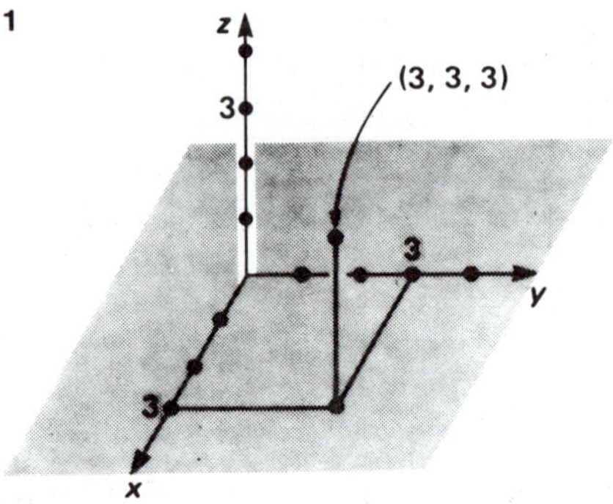

3
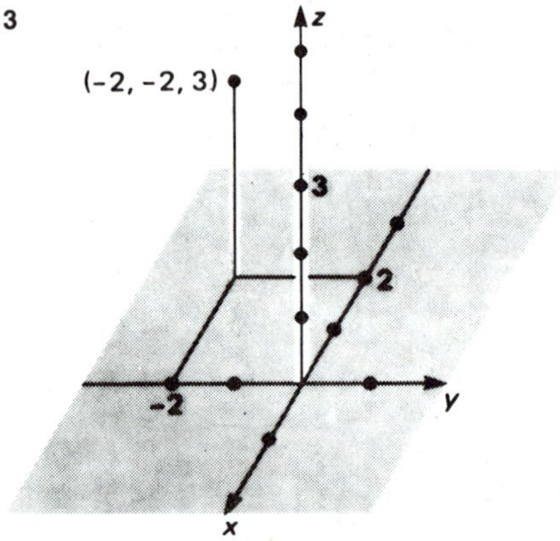

5 $(4, 2, 0)$ **7** $(12, 0, 4)$ **9** $(6, 6, -2)$ **11** $(1, 4, -1)$ **13** 2 **15** -8
17 $\sqrt{6}$ **19** $\sqrt{42}$ **21** $5\sqrt{2}$ **23** sphere, center $\mathbf{0}$, radius 1
25 $\arccos(-\frac{12}{25}) \approx 118.69°$ **27** $90°$ **29** $(1, 1, 0)$ and $(-2, 0, 1)$ for instance
31 $|\mathbf{x}_1 \cdot \mathbf{x}_2| = \|\mathbf{x}_1\| \, \|\mathbf{x}_2\| \, |\cos\theta| \leq \|\mathbf{x}_1\| \, \|\mathbf{x}_2\|$

Section 2 page 503

1 $\mathbf{x} = t\mathbf{a}$ is a line; $\mathbf{x} = \mathbf{0}$ for $t = 0$, and $\mathbf{x} = \mathbf{a}$ for $t = 1$. **3** $\mathbf{x} = (1 - t, t, 1)$
5 $\mathbf{x} = (0, 1 + t, 4 - 3t)$ **7** $\mathbf{x} \cdot (\frac{1}{3}, -\frac{2}{3}, \frac{2}{3}) = \frac{1}{3}$ **9** $\mathbf{x} \cdot (-\frac{8}{9}, \frac{1}{9}, -\frac{4}{9}) = 3$
11 $\mathbf{x} \cdot (\frac{1}{3}\sqrt{3}, \frac{1}{3}\sqrt{3}, \frac{1}{3}\sqrt{3}) = \sqrt{3}$ **13** $\mathbf{x} = (s + t, s + t, s - t)$
15 $\mathbf{x} = (2 + s + 2t, 1 - s + 2t, s + 3t)$ **17** $\mathbf{x} \cdot (\frac{1}{2}\sqrt{2}, -\frac{1}{2}\sqrt{2}, 0) = 0$
19 $\mathbf{x} \cdot (\frac{5}{42}\sqrt{42}, \frac{1}{42}\sqrt{42}, -\frac{4}{42}\sqrt{42}) = \frac{11}{42}\sqrt{42}$ **22** The parallel plane through $\mathbf{x}_0$ is $\mathbf{x} \cdot \mathbf{n} = \mathbf{x}_0 \cdot \mathbf{n}$, so the distance is $|\mathbf{x}_0 \cdot \mathbf{n} - p|$. **23** They are parallel if and only if the direction of the line is perpendicular to the normal to the plane, that is, $\mathbf{v} \cdot \mathbf{n} = 0$.
25 Let the point of intersection be $\mathbf{x} = \mathbf{x}_0 + t\mathbf{v}$. Then $(\mathbf{x}_0 + t\mathbf{v}) \cdot \mathbf{n} = p$. Solve for t:
$\mathbf{x}_0 \cdot \mathbf{n} + t\mathbf{v} \cdot \mathbf{n} = p \quad t = \dfrac{p - \mathbf{x}_0 \cdot \mathbf{n}}{\mathbf{v} \cdot \mathbf{n}}$, etc.
27 The plane through $\mathbf{y}_0$ with normal $\mathbf{u}$ is $\mathbf{x} \cdot \mathbf{u} = \mathbf{y}_0 \cdot \mathbf{u}$; it meets the line in the closest point $\mathbf{z}_0$. By Ex. 25, $\mathbf{z}_0 = \mathbf{x}_0 + [(\mathbf{y}_0 - \mathbf{x}_0) \cdot \mathbf{u}]\mathbf{u}$.

Section 3 page 509

1
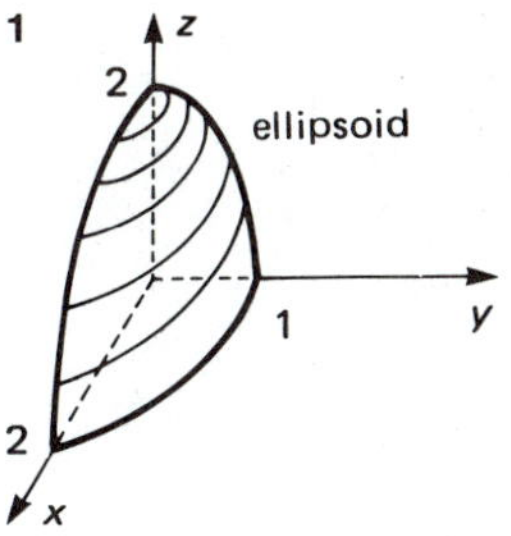

3
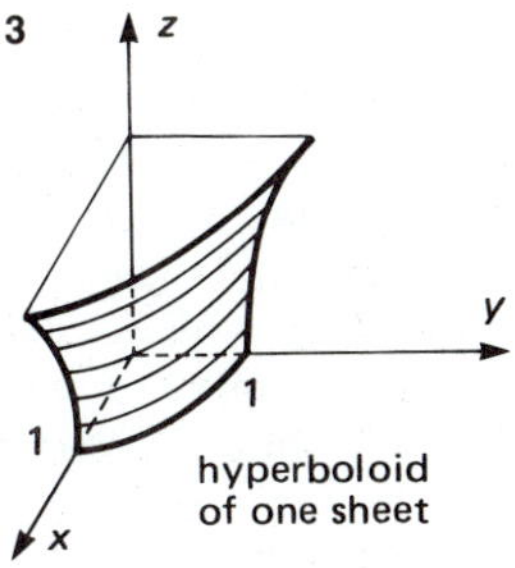

5
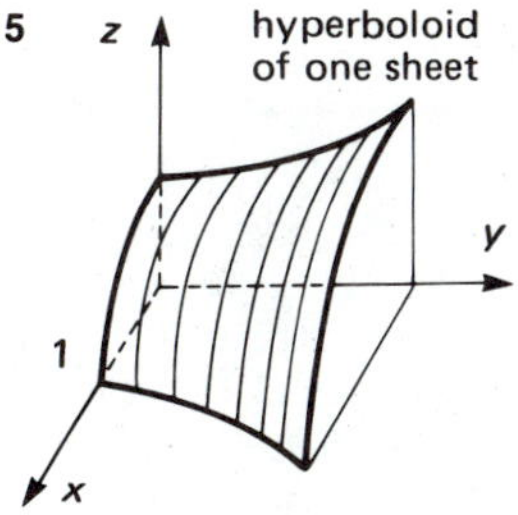

7
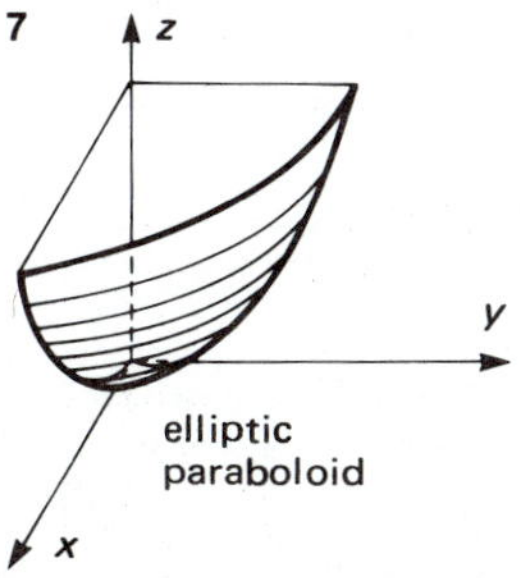

9
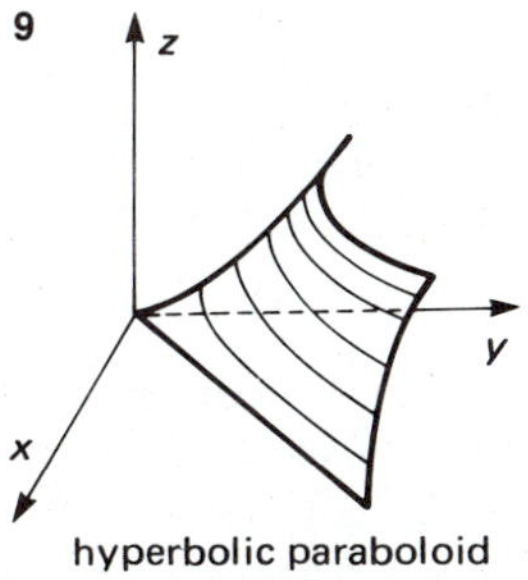

11 $z = (x + 1)^2 + y^2 - 1$; elliptic paraboloid with axis $x = -1$, $y = 0$.

13
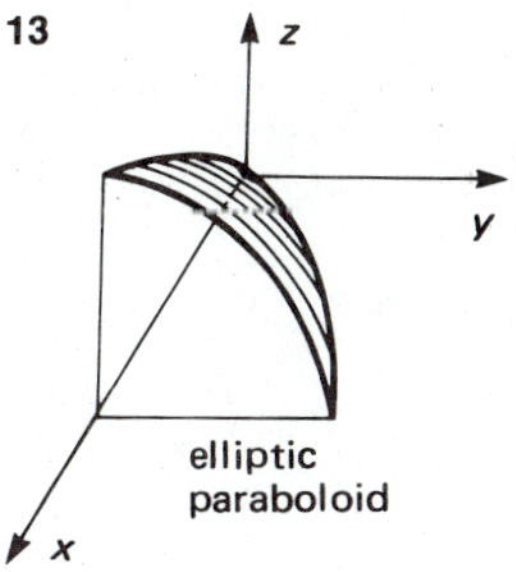

15
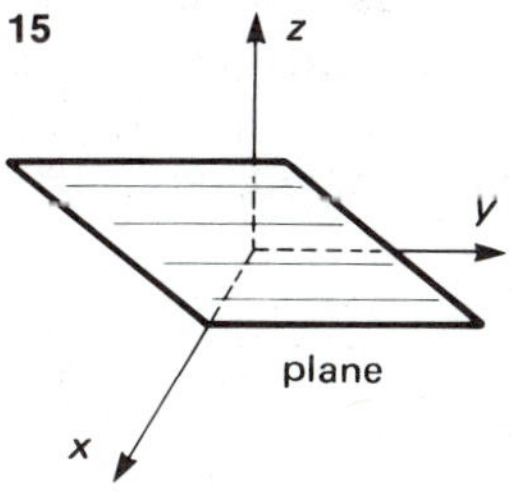

17
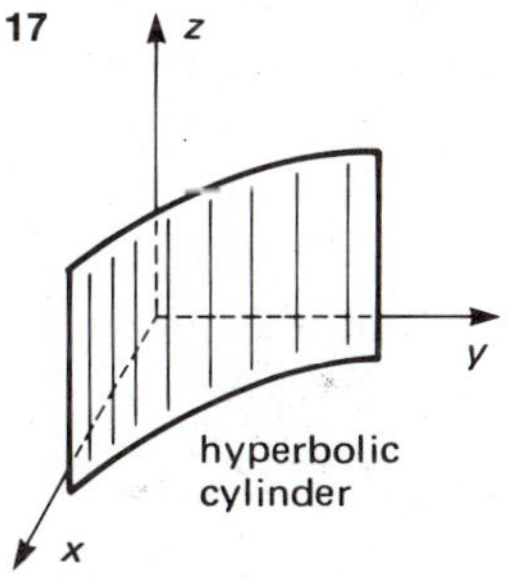

19
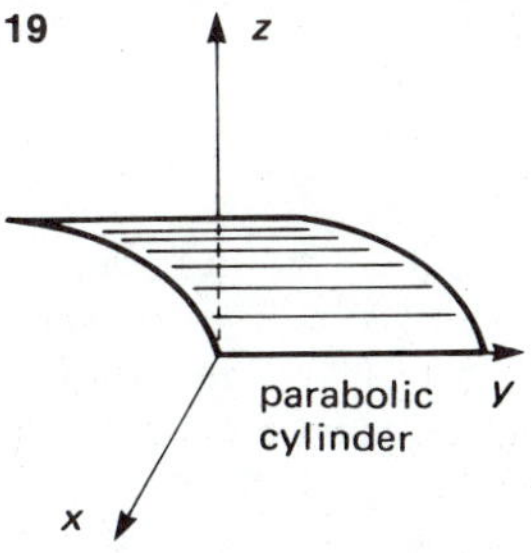

21
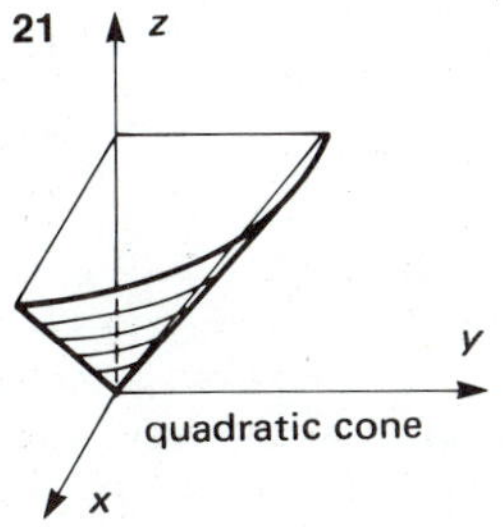

23
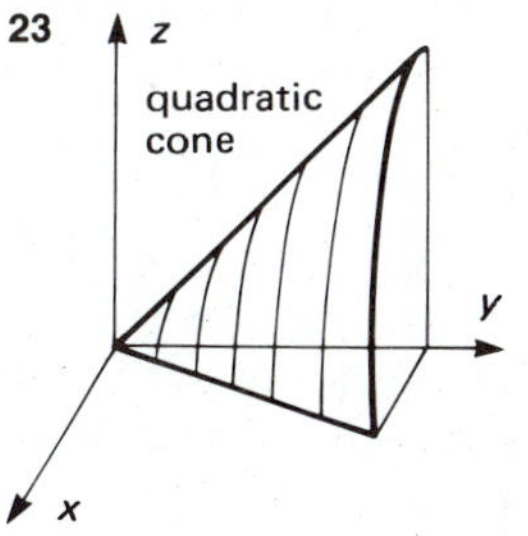

25 $f(x/z, y/z) = 0$ **27** $f(y, \sqrt{x^2 + z^2}) = 0$

Review Exercises page 509

1 $(\mathbf{c} - \mathbf{b})\cdot(\mathbf{a} - \mathbf{b}) = 0$; right angle at vertex $\mathbf{b}$ **3** $(\frac{4}{9}, -\frac{8}{9}, \frac{1}{9})\cdot\mathbf{x} = 3$; $(\frac{4}{3}, -\frac{8}{3}, \frac{1}{3})$
5 $(2, -1, 2)\cdot(3, -4, -5) = 0$, hence their normals are perpendicular.

7
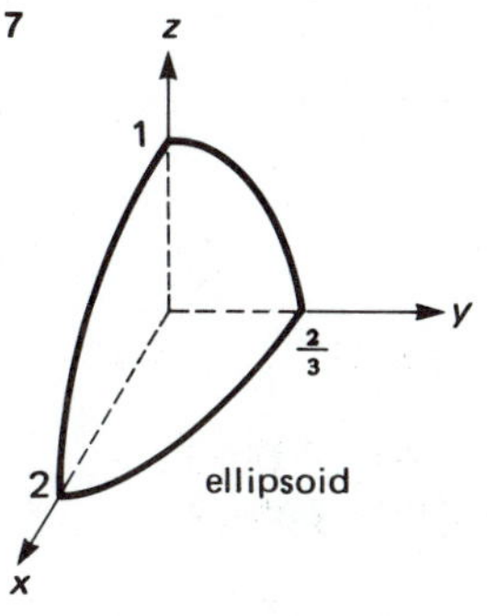

CHAPTER 14

Section 1 page 515

1 $9 + i$ **3** $-4 + 5i$ **5** $3 + 2i$ **7** $7 - i$ **9** $7 + 22i$ **11** $2i$
13 $\frac{3}{25} - \frac{4}{25}i$ **15** $1 + i$ **17** $\frac{22}{533} - \frac{7}{533}i$ **19** $-\frac{37}{145} - \frac{9}{145}i$ **21** $\frac{82}{85} + \frac{39}{85}i$
23 $3 - 3i$ **25** $8 - i$ **27** $-4 - 2i$ **29** 5 **31** $\sqrt{10}$ **33** $\sqrt{5} + \sqrt{17}$
35 $\frac{1}{5}\sqrt{5}$ **37** 1 **39** -1 **41** 1 **43** $(a + bi) + (c + di) =$
$(a + c) + (b + d)i = (c + a) + (d + b)i = (c + di) + (a + bi)$
45 $(a + bi)[(c + di) + (e + fi)] = (a + bi)[(c + e) + (d + f)i]$
$= [a(c + e) - b(d + f)] + [a(d + f) + b(c + e)]i$
$= [(ac - bd) + (ae - bf)] + [(ad + bc) + (af + be)]i$
$= [(ac - bd) + (ad + bc)i] + [(ae - bf) + (af + be)i]$
$= (a + bi)(c + di) + (a + bi)(e + fi)$.
47 $\overline{\overline{a + bi}} = \overline{a - bi} = a + bi$ **49** $|a + bi| = \sqrt{a^2 + b^2} \geq 0$
51 $|a - bi| = \sqrt{a^2 + (-b)^2} = \sqrt{a^2 + b^2} = |a + bi|$
53 $|\alpha\beta|^2 = (\alpha\beta)(\overline{\alpha\beta}) = \alpha\beta\overline{\alpha}\overline{\beta} = (\alpha\overline{\alpha})(\beta\overline{\beta}) = |\alpha|^2|\beta|^2$, hence $|\alpha\beta| = |\alpha|\,|\beta|$.

Section 2 page 523

1 $3(\cos\frac{3}{2}\pi + i\sin\frac{3}{2}\pi)$ **3** $\sqrt{2}(\cos\frac{7}{4}\pi + i\sin\frac{7}{4}\pi)$ **5** $2(\cos\frac{7}{6}\pi + i\sin\frac{7}{6}\pi)$
7 $6(\cos\frac{1}{6}\pi + i\sin\frac{1}{6}\pi)$ **9** $2(\cos 0 + i\sin 0)$ **11** $4(\cos\frac{2}{3}\pi + i\sin\frac{2}{3}\pi)$
13 $\sqrt{2}(\cos\frac{19}{12}\pi + i\sin\frac{19}{12}\pi)$ **15** $\frac{1}{4}\sqrt{2}(\cos\frac{17}{12}\pi + i\sin\frac{17}{12}\pi)$ **17** $\sqrt{3}, 3, 2\sqrt{3}, \frac{1}{3}\pi$
19 $\frac{1}{2}(\sqrt{3} - 1), -\frac{1}{2}(1 + \sqrt{3}), \sqrt{2}, \frac{19}{12}\pi$ **21** $3\sqrt{2}$ **23** $\sqrt{65}$ **25** $2\sqrt{5}$
27 $\frac{1}{2}\sqrt{2}$ **29** $\frac{2}{5}\sqrt{10}$ **31** $\frac{1}{29}\sqrt{290}$ **33** $\pm(\frac{1}{2}\sqrt{2} - \frac{1}{2}i\sqrt{2})$
35 $\pm\sqrt[4]{2}(\cos\frac{1}{8}\pi + i\sin\frac{1}{8}\pi)$ **37** $\pm\sqrt{2}(\cos\frac{11}{12}\pi + i\sin\frac{11}{12}\pi)$
39 $\pm\sqrt[4]{8}(\cos\frac{3}{8}\pi + i\sin\frac{3}{8}\pi)$ **41** Set $\alpha = x + yi$ and $\beta = u + vi$, so $\alpha\beta = (xu - yv) + (xv + yu)i$. Express $|\alpha|^2|\beta|^2 = |\alpha\beta|^2$ in terms of x, y, u, v. The relation follows.

43

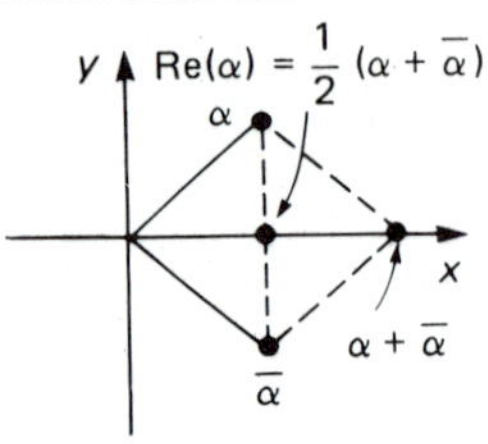

Section 3 page 528

1 $1 \pm 2i$ **3** $-\frac{1}{2} \pm \frac{1}{2}i\sqrt{23}$ **5** $\pm 2i, \pm i$ **7** $1, -\frac{1}{2} \pm \frac{1}{2}i\sqrt{3}$ **9** $\pm 1, \pm i$
11 $(x + 1)(x^2 - x + 1)$ **13** $z^5 + 1 = (z + 1)g(z)$ and $g(-1) = 5 \neq 0$, so $z + 1$ is not a factor of $g(z)$. Hence $(z + 1)^2$ is not a factor of $z^5 + 1$. **15** If $f(x)$ had only irreducible quadratic factors, then $f(x)$ would have even degree. Hence $f(x)$ has a real linear factor $x - r$, so $f(r) = 0$. **17** $a(x^2 - 4)(x^2 - 2x + 2)$ $a \neq 0$
19 $a(x^2 + b_1x + c_1)(x^2 + b_2x + c_2)$ $a \neq 0$, $b_1^2 - 4c_1 < 0$, $b_2^2 - 4c_2 < 0$
21 Set $\alpha = a + b\sqrt{2}$ $\beta = c + d\sqrt{2}$, a, b, c, d rational. Then
$\alpha \pm \beta = (a + b\sqrt{2}) \pm (c + d\sqrt{2}) = (a \pm c) + (b \pm d)\sqrt{2} \in \mathbf{T}$
$\alpha\beta = (a + b\sqrt{2})(c + d\sqrt{2}) = (ac + 2bd) + (ad + bc)\sqrt{2} \in \mathbf{T}$
23 $\alpha\alpha' = (a + b\sqrt{2})(a - b\sqrt{2}) = a^2 - 2b^2$. This is not zero, for if $a^2 - 2b^2 = 0$, then $a^2 = 2b^2$, $\sqrt{2} = (a/b)^2$. But $\sqrt{2}$ is irrational. Now

$$(a + b\sqrt{2})\frac{a - b\sqrt{2}}{a^2 - 2b^2} = \frac{a^2 - 2b^2}{a^2 - 2b^2} = 1,$$

so $$\alpha^{-1} = \frac{\alpha'}{\alpha\alpha'} = \frac{a}{a^2 - 2b^2} + \frac{-b}{a^2 - 2b^2}\sqrt{2} \in \mathbf{T}$$

Section 4 page 533

1 $-32 - 32i$ **3** -64 **5** $-\frac{1}{2} + \frac{1}{2}i\sqrt{3}$ **7** $\cos 15^\circ + i\sin 15^\circ$
9 $2^{-11}(1 - i\sqrt{3})$ **11** $\cos\frac{11}{9}\pi + i\sin\frac{11}{9}\pi$ **13** $\pm 1, \pm i$
15 $\zeta_0 = 1, \zeta_1 = \frac{1}{2} + \frac{1}{2}i\sqrt{3}, \zeta_2 = -\frac{1}{2} + \frac{1}{2}i\sqrt{3}, \zeta_3 = -1, \zeta_4 = -\frac{1}{2} - \frac{1}{2}i\sqrt{3}$,
$\zeta_5 = \frac{1}{2} - \frac{1}{2}i\sqrt{3}$ **17** $2, -1 \pm i\sqrt{3}$ **19** $\frac{1}{2}\sqrt{10}(1 + i), \frac{1}{2}\sqrt{10}(-1 \pm i)$
21 $-i, \pm\frac{1}{2}\sqrt{3} + \frac{1}{2}i$ **23** $2i, \pm\sqrt{3} - i$

25 $\sqrt[6]{2}[\cos(\frac{1}{12}\pi + \frac{2}{3}\pi k) + i\sin(\frac{1}{12}\pi + \frac{2}{3}\pi k)]$ $k = 0, 1, 2$
27 $\cos(\frac{1}{9}\pi + \frac{2}{3}\pi k) + i\sin(\frac{1}{9}\pi + \frac{2}{3}\pi k)$ $k = 0, 1, 2$
29 $\beta_0 = \sqrt[4]{8}(\cos\frac{11}{24}\pi + i\sin\frac{11}{24}\pi)$, $i\beta_0$, $-\beta_0$, $-i\beta_0$
31 $0 = \alpha^n - 1 = (\alpha - 1)(\alpha^{n-1} + \alpha^{n-2} + \cdots + \alpha + 1) = 0$ and $\alpha - 1 \neq 0$, so $\alpha^{n-1} + \cdots + \alpha + 1 = 0$. **33** $\alpha^5 = 1$ and $\alpha \neq 1$, hence $\alpha^4 + \alpha^3 + \alpha^2 + \alpha + 1 = 0$, $\alpha^2 + \alpha + 1 + \alpha^{-1} + \alpha^{-2} = 0$. But $\beta^2 = \alpha^2 + 2 + \alpha^{-2}$, so $(\beta^2 - 2) + \beta + 1 = 0$, $\beta^2 + \beta - 1 = 0$.
35 $\beta = \alpha + \alpha^{-1}$ $\alpha\beta = \alpha^2 + 1$ $\alpha^2 - \beta\alpha + 1 = 0$ **37** $(0, 0)$, $(1, 1)$, $(-\frac{1}{2} \pm \frac{1}{2}i\sqrt{3}, -\frac{1}{2} \mp \frac{1}{2}i\sqrt{3})$ **39** $\alpha^2 = 3 - 4i$, $\alpha^4 = -7 - 24i$. If z is any solution, then $(z/\alpha)^4 = z^4/\alpha^4 = 1$, so z/α is a 4-th root of unity, hence equal to ± 1 or $\pm i$. Therefore $z = \pm\alpha$ or $\pm i\alpha$. **41** The 4-th roots of unity are $1, i, -1, -i$. Since $1^1 = 1$ and $(-1)^2 = 1$, the roots 1 and -1 are not primitive. Since $i^1 \neq 1$, $i^2 \neq 1$, $i^3 \neq 1$, it follows that i is primitive, and similarly $-i$ is primitive. **43** The 6-th roots of unity are listed in the solution of Ex. 15 above. Of these, $\zeta_0, \zeta_2, \zeta_3, \zeta_4$ are not primitive because

$$\zeta_0{}^1 = 1 \qquad \zeta_2{}^3 = 1 \qquad \zeta_3{}^2 = 1 \qquad \zeta_4{}^3 = 1.$$

Both $\zeta_1 = \frac{1}{2} + \frac{1}{2}i\sqrt{3}$ and $\zeta_5 = \frac{1}{2} - \frac{1}{2}i\sqrt{3}$ are primitive. Since $\zeta_k = \zeta_1{}^k$ for $k = 0, \cdots, 5$, clearly ζ_1 is primitive. But $\zeta_5 = \zeta_1{}^{-1}$ is also primitive.
45 The 8-th roots of unity are

$$\zeta_k = \cos\frac{2\pi k}{8} + i\sin\frac{2\pi k}{8} \qquad k = 0, 1, \cdots, 7.$$

Of these, $\zeta_0, \zeta_2, \zeta_4, \zeta_6$ are the 4-th roots of unity, hence not primitive 8-th roots. The others, $\zeta_1, \zeta_3, \zeta_5, \zeta_7$ are primitive. For

$$\zeta_3{}^3 = \zeta_1, \qquad \zeta_5{}^5 = \zeta_1, \qquad \zeta_7{}^7 = \zeta_1.$$

47 The 10-th roots of unity are $1, \zeta, \zeta^2, \cdots, \zeta^9$, where $\zeta = \cos\frac{2}{10}\pi + i\sin\frac{2}{10}\pi$. The roots $1, \zeta^2, \zeta^4, \zeta^6, \zeta^8$ are 5-th roots of unity, hence not primitive, and $\zeta^5 = -1$ is also not primitive. The others, $\zeta, \zeta^3, \zeta^7, \zeta^9$ are primitive. Note that

$$(\zeta^3)^7 = \zeta \qquad (\zeta^7)^3 = \zeta \qquad (\zeta^9)^9 = \zeta$$

49 The primitive 6-th roots of unity are $\frac{1}{2} \pm \frac{1}{2}i\sqrt{3}$ by Ex. 43. By the quadratic formula, they are the roots of $z^2 - z + 1 = 0$. **51** Write $z^9 - 1 = (z^3 - 1)(z^6 + z^3 + 1)$. Each 9-th root of unity satisfies $z^9 - 1 = 0$. Those which are cube roots of unity satisfy $z^3 - 1 = 0$; each of the other 6 roots satisfies $z^6 + z^3 + 1 = 0$. (They are all primitive.)

Review Exercises page 534

1 $-\frac{13}{10} - \frac{11}{10}i$ **3** $5 - i$ **5** i **7** $5(\cos\frac{3}{2}\pi + i\sin\frac{3}{2}\pi)$
9 -1, $\frac{1}{2} \pm \frac{1}{2}i\sqrt{3}$ **11** $2^{14}(\frac{1}{2} + \frac{1}{2}i\sqrt{3})$ **13**

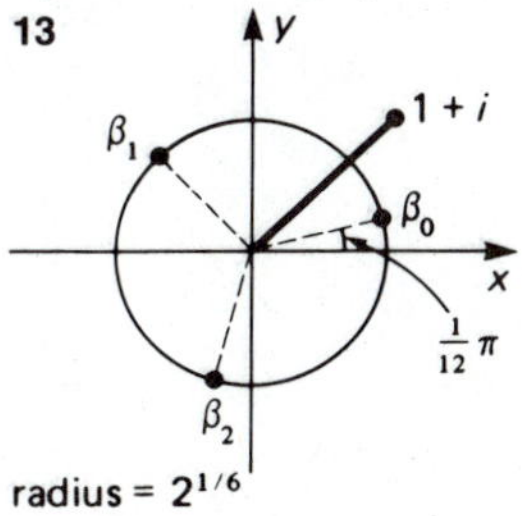

15 α^{-1} rotates clockwise on the circle $|z| = \frac{1}{2}$ **17** Set $\omega = \dfrac{\gamma - \alpha}{\gamma - \beta}$. Then $|\omega| = \left|\dfrac{\gamma - \alpha}{\gamma - \beta}\right| = \dfrac{|\gamma - \alpha|}{|\gamma - \beta|} = 1$ and $\arg\omega = \pm 60° = \pm\frac{1}{3}\pi$, hence

$$\omega = \cos\left(\pm\frac{1}{3}\pi\right) + i\sin\left(\pm\frac{1}{3}\pi\right) = \cos\left(\pm\frac{2\pi}{6}\right) + i\sin\left(\pm\frac{2\pi}{6}\right),$$

so ω is a 6-th root of unity (primitive, actually; see Ex. 41 of Section 4).

B.1 TABLES

Table 1 Three-Place Mantissas for Common Logarithms

N	0	1	2	3	4	5	6	7	8	9
1.0	000	004	009	013	017	021	025	029	033	037
1.1	041	045	049	053	057	061	064	068	072	076
1.2	079	083	086	090	093	097	100	104	107	111
1.3	114	117	121	124	127	130	134	137	140	143
1.4	146	149	152	155	158	161	164	167	170	173
1.5	176	179	182	185	188	190	193	196	199	201
1.6	204	207	210	212	215	217	220	223	225	228
1.7	230	233	236	238	241	243	246	248	250	253
1.8	255	258	260	262	265	267	270	272	274	276
1.9	279	281	283	286	288	290	292	294	297	299
2.0	301	303	305	307	310	312	314	316	318	320
2.1	322	324	326	328	330	332	334	336	338	340
2.2	342	344	346	348	350	352	354	356	358	360
2.3	362	364	365	367	369	371	373	375	377	378
2.4	380	382	384	386	387	389	391	393	394	396
2.5	398	400	401	403	405	407	408	410	412	413
2.6	415	417	418	420	422	423	425	427	428	430
2.7	431	433	435	436	438	439	441	442	444	446
2.8	447	449	450	452	453	455	456	458	459	461
2.9	462	464	465	467	468	470	471	473	474	476
3.0	477	479	480	481	483	484	486	487	489	490
3.1	491	493	494	496	497	498	500	501	502	504
3.2	505	507	508	509	511	512	513	515	516	517
3.3	519	520	521	522	524	525	526	528	529	530
3.4	531	533	534	535	537	538	539	540	542	543
3.5	544	545	547	548	549	550	551	553	554	555
3.6	556	558	559	560	561	562	563	565	566	567
3.7	568	569	571	572	573	574	575	576	577	579
3.8	580	581	582	583	584	585	587	588	589	590
3.9	591	592	593	594	595	597	598	599	600	601
4.0	602	603	604	605	606	607	609	610	611	612
4.1	613	614	615	616	617	618	619	620	621	622
4.2	623	624	625	626	627	628	629	630	631	632
4.3	633	634	635	636	637	638	639	640	641	642
4.4	643	644	645	646	647	648	649	650	651	652
4.5	653	654	655	656	657	658	659	660	661	662
4.6	663	664	665	666	667	667	668	669	670	671
4.7	672	673	674	675	676	677	678	679	679	680
4.8	681	682	683	684	685	686	687	688	688	689
4.9	690	691	692	693	694	695	695	696	697	698
5.0	699	700	701	702	702	703	704	705	706	707
5.1	708	708	709	710	711	712	713	713	714	715
5.2	716	717	718	719	719	720	721	722	723	723
5.3	724	725	726	727	728	728	729	730	731	732
5.4	732	733	734	735	736	736	737	738	739	740
N	0	1	2	3	4	5	6	7	8	9

Table 1 Three-Place Mantissas for Common Logarithms (*Continued*)

N	0	1	2	3	4	5	6	7	8	9
5.5	740	741	742	743	744	744	745	746	747	747
5.6	748	749	750	751	751	752	753	754	754	755
5.7	756	757	757	758	759	760	760	761	762	763
5.8	763	764	765	766	766	767	768	769	769	770
5.9	771	772	772	773	774	775	775	776	777	777
6.0	778	779	780	780	781	782	782	783	784	785
6.1	785	786	787	787	788	789	790	790	791	792
6.2	792	793	794	794	795	796	797	797	798	799
6.3	799	800	801	801	802	803	803	804	805	806
6.4	806	807	808	808	809	810	810	811	812	812
6.5	813	814	814	815	816	816	817	818	818	819
6.6	820	820	821	822	822	823	823	824	825	825
6.7	826	827	827	828	829	829	830	831	831	832
6.8	833	833	834	834	835	836	836	837	838	838
6.9	839	839	840	841	841	842	843	843	844	844
7.0	845	846	846	847	848	848	849	849	850	851
7.1	851	852	852	853	854	854	855	856	856	857
7.2	857	858	859	859	860	860	861	862	862	863
7.3	863	864	865	865	866	866	867	867	868	869
7.4	869	870	870	871	872	872	873	873	874	874
7.5	875	876	876	877	877	878	879	879	880	880
7.6	881	881	882	883	883	884	884	885	885	886
7.7	886	887	888	888	889	889	890	890	891	892
7.8	892	893	893	894	894	895	895	896	897	897
7.9	898	898	899	899	900	900	901	901	902	903
8.0	903	904	904	905	905	906	906	907	907	908
8.1	908	909	910	910	911	911	912	912	913	913
8.2	914	914	915	915	916	916	917	918	918	919
8.3	919	920	920	921	921	922	922	923	923	924
8.4	924	925	925	926	926	927	927	928	928	929
8.5	929	930	930	931	931	932	932	933	933	934
8.6	934	935	936	936	937	937	938	938	939	939
8.7	940	940	941	941	942	942	943	943	943	944
8.0	944	945	945	946	946	947	947	948	948	949
8.9	949	950	950	951	951	952	952	953	953	954
9.0	954	955	955	956	956	957	957	958	958	959
9.1	959	960	960	960	961	961	962	962	963	963
9.2	964	964	965	965	966	966	967	967	968	968
9.3	968	969	969	970	970	971	971	972	972	973
9.4	973	974	974	975	975	975	976	976	977	977
9.5	978	978	979	979	980	980	980	981	981	982
9.6	982	983	983	984	984	985	985	985	986	986
9.7	987	987	988	988	989	989	989	990	990	991
9.8	991	992	992	993	993	993	994	994	995	995
9.9	996	996	997	997	997	998	998	999	999	1000
N	0	1	2	3	4	5	6	7	8	9

Table 2 Trigonometric Functions of Degrees

deg	sin	cos	tan	cot	
0	.000	.000	.000		90
1	.017	.000	.017	57.290	89
2	.035	.999	.035	28.636	88
3	.052	.999	.052	19.081	87
4	.070	.998	.070	14.301	86
5	.087	.996	.087	11.430	85
6	.105	.995	.105	9.514	84
7	.122	.993	.123	8.144	83
8	.139	.990	.141	7.115	82
9	.156	.988	.158	6.314	81
10	.174	.985	.176	5.671	80
11	.191	.982	.194	5.145	79
12	.208	.978	.213	4.705	78
13	.225	.974	.231	4.331	77
14	.242	.970	.249	4.011	76
15	.259	.966	.268	3.732	75
16	.276	.961	.287	3.487	74
17	.292	.956	.306	3.271	73
18	.309	.951	.325	3.078	72
19	.326	946	.344	2.904	71
20	.342	.940	.364	2.747	70
21	.358	.934	.384	2.605	69
22	.375	.927	.404	2.475	68
23	.391	.921	.424	2.356	67
24	.407	.914	.445	2.246	66
25	.423	.906	.466	2.145	65
26	.438	.899	.488	2.050	64
27	.454	.891	.510	1.963	63
28	.469	.883	.532	1.881	62
29	.485	.875	.554	1.804	61
30	.500	.866	.577	1.732	60
31	.515	.857	.601	1.664	59
32	.530	.848	.625	1.600	58
33	.545	.839	.649	1.540	57
34	.559	.829	.675	1.483	56
35	.574	.819	.700	1.428	55
36	.588	.809	.727	1.376	54
37	.602	.799	.754	1.327	53
38	.616	.788	.781	1.280	52
39	.629	.777	.810	1.235	51
40	.643	.766	.839	1.192	50
41	.656	.755	.869	1.150	49
42	.669	.743	.900	1.111	48
43	.682	.731	.933	1.072	47
44	.695	.719	.966	1.036	46
45	.707	.707	1.000	1.000	45
	cos	sin	cot	tan	deg

Table 3 Inverse Trigonometric Functions to Degrees

x	arc sin	arc cos	arc tan	x	arc sin	arc cos	arc tan
.00	0.00	90.00	0.00	.50	30.00	60.00	26.57
.01	0.57	89.43	0.57	.51	30.66	59.34	27.02
.02	1.15	88.85	1.15	.52	31.33	58.67	27.47
.03	1.72	88.28	1.72	.53	32.01	57.99	27.92
.04	2.29	87.71	2.29	.54	32.68	57.32	28.37
.05	2.87	87.13	2.86	.55	33.37	56.63	28.81
.06	3.44	86.56	3.43	.56	34.06	55.94	29.25
.07	4.01	85.99	4.00	.57	34.75	55.25	29.68
.08	4.59	85.41	4.57	.58	35.45	54.55	30.11
.09	5.16	84.84	5.14	.59	36.16	53.84	30.54
.10	5.74	84.26	5.71	.60	36.87	53.13	30.96
.11	6.32	83.68	6.28	.61	37.59	52.41	31.38
.12	6.89	83.11	6.84	.62	38.32	51.68	31.80
.13	7.47	82.53	7.41	.63	39.05	50.95	32.21
.14	8.05	81.95	7.97	.64	39.79	50.21	32.62
.15	8.63	81.37	8.53	.65	40.54	49.46	33.02
.16	9.21	80.79	9.09	.66	41.30	48.70	33.42
.17	9.79	80.21	9.65	.67	42.07	47.93	33.82
.18	10.37	79.63	10.20	.68	42.84	47.16	34.22
.19	10.95	79.05	10.76	.69	43.63	46.37	34.61
.20	11.54	78.46	11.31	.70	44.43	45.57	34.99
.21	12.12	77.88	11.86	.71	45.23	44.77	35.37
.22	12.71	77.29	12.41	.72	46.05	43.95	35.75
.23	13.30	76.70	12.95	.73	46.89	43.11	36.13
.24	13.89	76.11	13.50	.74	47.73	42.27	36.50
.25	14.48	75.52	14.04	.75	48.59	41.41	36.87
.26	15.07	74.93	14.57	.76	49.46	40.54	37.23
.27	15.66	74.34	15.11	.77	50.35	39.65	37.60
.28	16.26	73.74	15.64	.78	51.26	38.74	37.95
.29	16.86	73.14	16.17	.79	52.19	37.81	38.31
.30	17.46	72.54	16.70	.80	53.13	36.87	38.66
.31	18.06	71.94	17.22	.81	54.10	35.90	39.01
.32	18.66	71.34	17.74	.82	55.08	34.92	39.35
.33	19.27	70.73	18.26	.83	56.10	33.90	39.69
.34	19.88	70.12	18.78	.84	57.14	32.86	40.03
.35	20.49	69.51	19.29	.85	58.21	31.79	40.36
.36	21.10	68.90	19.80	.86	59.32	30.68	40.70
.37	21.72	68.28	20.30	.87	60.46	29.54	41.02
.38	22.33	67.67	20.81	.88	61.64	28.36	41.35
.39	22.95	67.05	21.31	.89	62.87	27.13	41.67
.40	23.58	66.42	21.80	.90	64.16	25.84	41.99
.41	24.20	65.80	22.29	.91	65.51	24.49	42.30
.42	24.83	65.17	22.78	.92	66.93	23.07	42.61
.43	25.47	64.53	23.27	.93	68.43	21.57	42.92
.44	26.10	63.90	23.75	.94	70.05	19.95	43.23
.45	26.74	63.26	24.23	.95	71.81	18.19	43.53
.46	27.39	62.61	24.70	.96	73.74	16.26	43.83
.47	28.03	61.97	25.17	.97	75.93	14.07	44.13
.48	28.69	61.31	25.64	.98	78.52	11.48	44.42
.49	29.34	60.66	26.10	.99	81.89	8.11	44.71
				1.00	90.00	0.00	45.00
x	arc sin	arc cos	arc tan	x	arc sin	arc cos	arc tan

Note: For $x > 1$, use arc tan $x = 90° -$ arc tan $(1/x)$.

INDEX

A

B

C

E

F

G

H

I

L

M

N

O

P

Q

R

ALGEBRA

$(x+y)^2 = x^2 + 2xy + y^2$

$(x+y)(x-y) = x^2 - y^2$

$(1-x)(1+x+x^2+\cdots+x^{n-1}) = 1-x^n$

$\sqrt{2} \approx 1.41421\ 356$

$\sqrt{3} \approx 1.73205\ 081$

$\sqrt{5} \approx 2.23606\ 798$

Quadratic formula $ax^2 + bx + c = 0$

$$x = \frac{-b \pm \sqrt{b^2 - 4ac}}{2a}$$

Factorial $n! = 1 \cdot 2 \cdot 3 \cdots n$

Permutations $P_{n,k} = \dfrac{n!}{(n-k)!}$

Combinations $\dbinom{n}{k} = \dfrac{n!}{k!\,(n-k)!}$

Binomial Theorem $(x+y)^n = \displaystyle\sum_{k=0}^{n} \binom{n}{k} x^{n-k} y^k$

EXPONENTIALS AND LOGARITHMS

$a^{x+y} = a^x a^y$

$(ab)^x = a^x b^x$

$(a^x)^y = a^{xy}$

$e \approx 2.71828\ 183$

$\log_a(xy) = \log_a x + \log_a y$

$\log_a(x/y) = \log_a x - \log_a y$

$\log_a x^r = r \log_a x$

Natural growth $X = X_0 a^t \quad (a > 1)$

Natural decay $X = X_0 a^{-t} \quad (a > 1)$

Compound interest

$A = P(1+i)^n$

Annuity

$A = Ps_{\overline{n}|i}$

$$s_{\overline{n}|i} = \frac{(1+i)^n - 1}{i}$$

Payment of debts

$A = Pa_{\overline{n}|i}$

$$a_{\overline{n}|i} = \frac{1-(1+i)^{-n}}{i}$$

Arithmetic Progression

$a_1 = a, \quad a_2 = a+d, \cdots, a_n = a + (n-1)d$

$S = a_1 + a_2 + \cdots + a_n = \frac{1}{2} n(a_1 + a_n)$

Geometric Progression

$a_1 = a, \ a_2 = ar, \cdots, a_n = ar^{n-1}$

$S = a_1 + \cdots + a_n = a\,\dfrac{1-r^n}{1-r} \qquad (r \neq 1)$

GEOMETRY

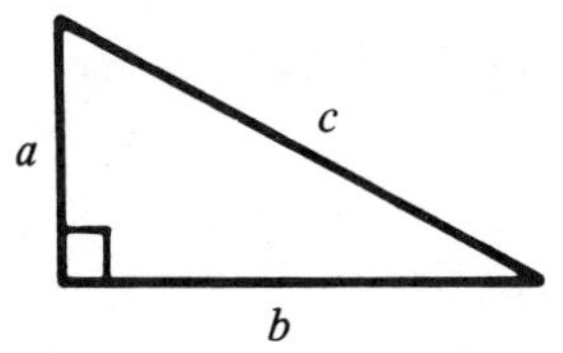

Special triangles

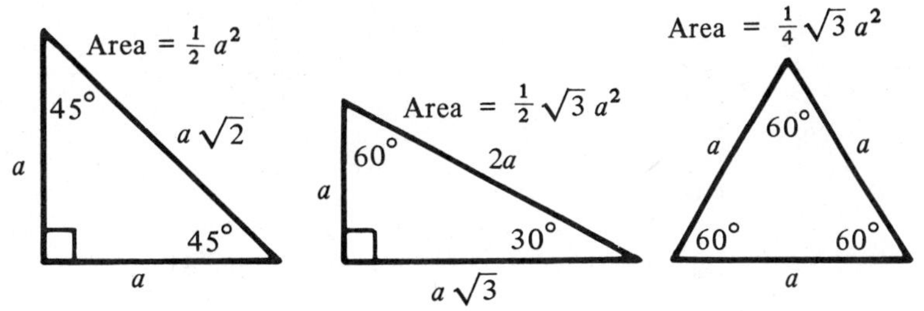

Pythagorean Theorem:

$c^2 = a^2 + b^2$

Some Pythagorean triangles (right triangles with integer sides)

a	3	5	7	8	9	11	12	13	15	16	20
b	4	12	24	15	40	60	35	84	112	63	21
c	5	13	25	17	41	61	37	85	113	65	29

Inscribed angles

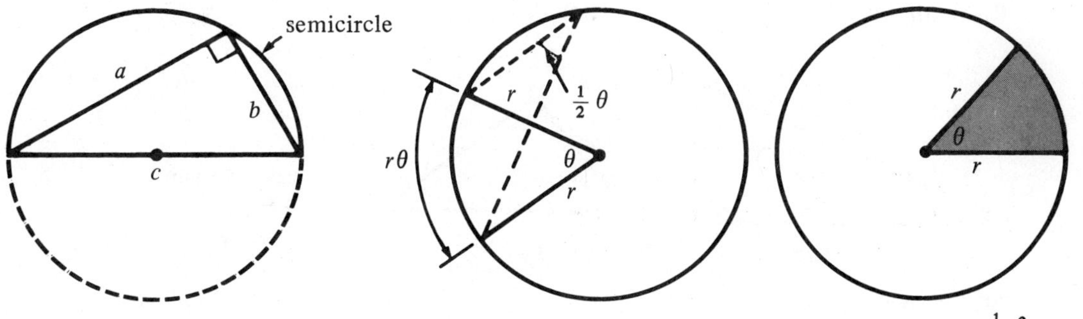

area of sector = $\frac{1}{2} r^2 \theta$

MATRIX MULTIPLICATION

$A = [a_{ij}]$

$i = 1, 2, \cdots, m \qquad j = 1, 2, \cdots, n$

$B = [b_{jk}]$

$j = 1, 2, \cdots, n \qquad k = 1, 2, \cdots, p$

$C = AB = [c_{ik}]$

$$c_{ik} = \sum_{j=1}^{n} a_{ij} b_{jk}$$

TRIGONOMETRY

$\tan\theta = \dfrac{\sin\theta}{\cos\theta}$ $\qquad$ $\cot\theta = \dfrac{1}{\tan\theta} = \dfrac{\cos\theta}{\sin\theta}$

$\sec\theta = \dfrac{1}{\cos\theta}$ $\qquad$ $\csc\theta = \dfrac{1}{\sin\theta}$

$$\begin{aligned} \pi \text{ rad} &= 180^\circ \\ 1 \text{ rad} &\approx 57.29578^\circ \\ 1^\circ &\approx 0.01745329 \text{ rad} \\ \pi &\approx 3.14159\ 2654 \end{aligned}$$

$\sin^2\theta + \cos^2\theta = 1$ $\qquad$ $1 + \tan^2\theta = \sec^2\theta$ $\qquad$ $1 + \cot^2\theta = \csc^2\theta$

$$\begin{aligned} \sin(\alpha+\beta) &= \sin\alpha\cos\beta + \cos\alpha\sin\beta \\ \cos(\alpha+\beta) &= \cos\alpha\cos\beta - \sin\alpha\sin\beta \end{aligned}$$

$$\tan(\alpha+\beta) = \frac{\tan\alpha + \tan\beta}{1 - \tan\alpha\tan\beta}$$

$$\begin{aligned} \sin 2\theta &= 2\sin\theta\cos\theta \\ \cos 2\theta &= \cos^2\theta - \sin^2\theta \end{aligned}$$

$$\tan 2\theta = \frac{2\tan\theta}{1 - \tan^2\theta}$$

θ°	0°	15°	30°	45°	60°	75°	90°
$\sin\theta^\circ$	0	$\frac{1}{4}(\sqrt{6}-\sqrt{2})$	$\frac{1}{2}$	$\frac{1}{2}\sqrt{2}$	$\frac{1}{2}\sqrt{3}$	$\frac{1}{4}(\sqrt{6}+\sqrt{2})$	1
$\cos\theta^\circ$	1	$\frac{1}{4}(\sqrt{6}+\sqrt{2})$	$\frac{1}{2}\sqrt{3}$	$\frac{1}{2}\sqrt{2}$	$\frac{1}{2}$	$\frac{1}{4}(\sqrt{6}-\sqrt{2})$	0
$\tan\theta^\circ$	0	$2-\sqrt{3}$	$\frac{1}{3}\sqrt{3}$	1	$\sqrt{3}$	$2+\sqrt{3}$	undef.

Right triangle

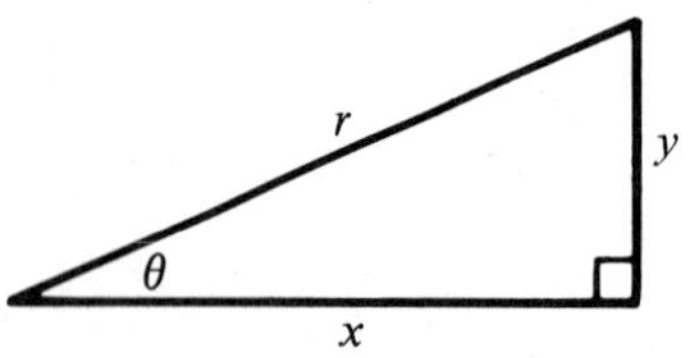

$\cos\theta = x/r$ $\qquad$ $\sec\theta = r/x$

$\sin\theta = y/r$ $\qquad$ $\csc\theta = r/y$

$\tan\theta = y/x$ $\qquad$ $\cot\theta = x/y$

Oblique triangle

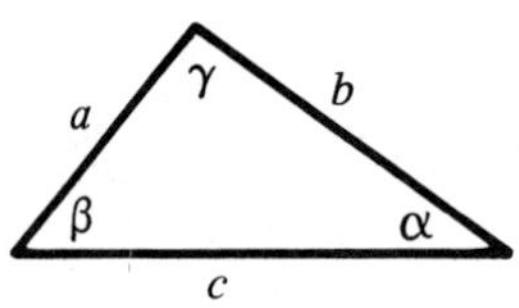

Law of sines

$$\frac{\sin\alpha}{a} = \frac{\sin\beta}{b} = \frac{\sin\gamma}{c}$$

Law of cosines

$$c^2 = a^2 + b^2 - 2ab\cos\gamma$$

Area formulas

$\text{Area} = \frac{1}{2}ab$

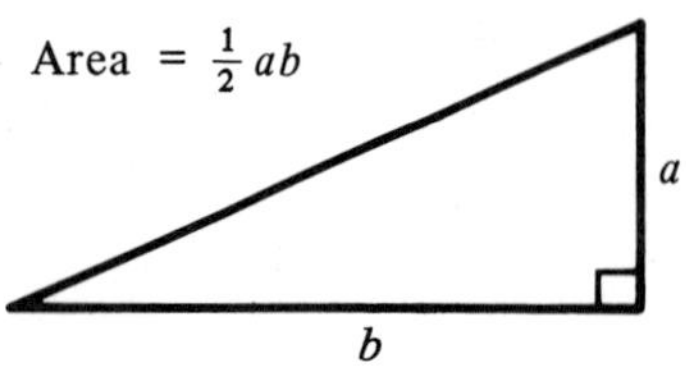

$\text{Area} = \frac{1}{2}bh = \frac{1}{2}ab\sin\gamma = \sqrt{s(s-a)(s-b)(s-c)}$

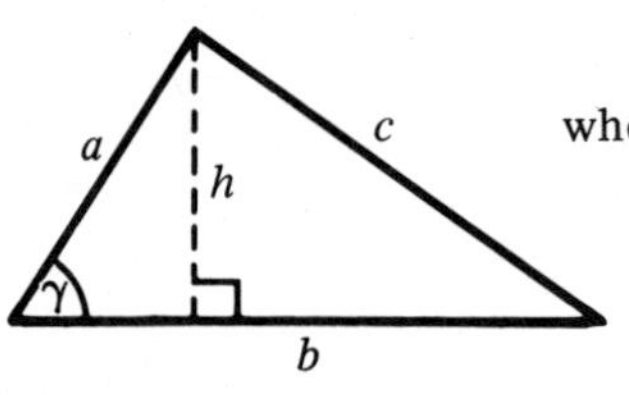

where $s = \frac{1}{2}(a+b+c)$

POLAR COORDINATES

$$\begin{cases} x = r\cos\theta \\ y = r\sin\theta \end{cases} \qquad \begin{cases} x^2 + y^2 = r^2 \\ \cos\theta = \dfrac{x}{r} \qquad \sin\theta = \dfrac{y}{r} \end{cases}$$

VECTORS

$\mathbf{u} = (a, b) \qquad \mathbf{v} = (c, d)$

$\mathbf{u} + \mathbf{v} = (a + c, b + d)$

$\mathbf{u}\cdot\mathbf{v} = ac + bd = \|\mathbf{u}\| \cdot \|\mathbf{v}\| \cos\theta$

$\|\mathbf{u}\|^2 = a^2 + b^2$

Triangle inequality

$\|\mathbf{u} + \mathbf{v}\| \leqslant \|\mathbf{u}\| + \|\mathbf{v}\|$

COMPLEX NUMBERS

$i^2 = -1$

$(a + bi) + (c + di) = (a + c) + (b + d)i$

$(a + bi)(c + di) = (ac - bd) + (ad + bc)i$

$\overline{a + bi} = a - bi$

$|a + bi|^2 = a^2 + b^2$

$\overline{\alpha\beta} = \bar{\alpha}\bar{\beta}$

$|\alpha\beta| = |\alpha|\cdot|\beta|$

Triangle inequality
$|\alpha + \beta| \leqslant |\alpha| + |\beta|$

Polar form $\qquad \alpha = a + bi \neq 0$

$$\alpha = r(\cos\theta + i\sin\theta) \qquad \begin{cases} a = r\cos\theta \\ b = r\sin\theta \end{cases} \qquad \begin{cases} r = \sqrt{a^2 + b^2} \\ \cos\theta = a/r \qquad \sin\theta = b/r \end{cases}$$

De Moivre's Theorem

$$[r(\cos\theta + i\sin\theta)]^n = r^n(\cos n\theta + i\sin n\theta)$$